ENCYCLOPEDIA OF STATISTICAL SCIENCES

VOLUME 1

A to Circular Probable Error

ENCYCLOPEDIA OF STATISTICAL SCIENCES

VOLUME 1

A to CIRCULAR PROBABLE ERROR

A WILEY-INTERSCIENCE PUBLICATION

John Wiley & Sons

NEW YORK · CHICHESTER · BRISBANE · TORONTO · SINGAPORE

Library of Congress Cataloging in Publication Data:
Main entry under title:
Encyclopedia of statistical sciences.

"A Wiley-Interscience publication."
Contents: v. 1. A–Circular probable error.
1. Mathematical statistics—Dictionaries.
2. Statistics—Dictionaries. I. Kotz, Samuel.
II. Johnson, Norman Lloyd. III. Read, Campbell B.
QA276.14.E5 519.5'03'21 81-10353
ISBN 0-471-05546-8 (v. 1) AACR2

Printed in the United States of America

10 9 8 7 6 5 4 3 2 1

CONTRIBUTORS

H. Ahrens, *Zentralinstitut für Mathematik und Mechanik der Adw, Berlin, East Germany*. Biometrical Journal

F. B. Alt, *University of Maryland, College Park, Maryland*. Bonferroni Inequalities and Intervals

O. D. Anderson, *University of Nottingham, Nottingham, England*. Box-Jenkins Model

D. F. Andrews, *University of Toronto, Toronto, Canada*. Andrews Function Plots

S. Asmussen, *University of Copenhagen, Copenhagen, Denmark*. Branching Processes

C. R. Baker, *University of North Carolina, Chapel Hill, North Carolina*. Absolute Continuity

V. P. Barabba, *Bureau of the Census, Washington, D.C.* Bureau of the Census

G. A. Barnard, *University of Waterloo, Ontario, Canada*. Causation

D. Basu, *Florida State University, Tallahassee, Florida*. Basu's Theorems

R. J. Beckman, *Los Alamos Scientific Laboratory, Los Alamos, New Mexico*. Chemistry, Statistical Methods in

B. Benjamin, *City University, London, England*. Census

V. P. Bhapkar, *University of Kentucky, Lexington, Kentucky*. Chi-Square Tests

G. K. Bhattacharyya, *University of Wisconsin, Madison, Wisconsin*. Bennett's Bivariate Sign Test

Y. M. M. Bishop, *Harvard University, Cambridge, Massachusetts*. Biometric Society

K. O. Bowman, *Union Carbide Corporation, Oak Ridge, Tennessee*. Approximations to Distributions

L. D. Broemeling, *Oklahoma State University, Stillwater, Oklahoma*. Box and Cox Transformation

F. B. Cady, *Cornell University, Ithaca, New York*. Biometrics

J. M. Cameron, *Wheaton, Maryland*. Calibration

G. J. Chaitin, *IBM, Yorktown Heights, New York*. Algorithmic Information Theory

J. Chambers, *Bell Laboratories, Murray Hill, New Jersey*. Algorithms, Statistical

S. K. Chatterjee, *Calcutta Statistical Association, Calcutta, India*. Calcutta Statistical Association Bulletin

H. Chernoff, *Massachusetts Institute of Technology, Cambridge, Massachusetts*. Chernoff Faces

E. Çinlar, *Princeton University, Princeton, New Jersey*. Chung Processes

W. G. Cochran, *Harvard University, Cambridge, Massachusetts*. Biometric Society

D. R. Cox, *Imperial College, London, England*. Binary Data; Biometrika

H. T. David, *Iowa State University, Ames, Iowa*. Chance (I)

N. R. Draper, *University of Wisconsin, Madison, Wisconsin*. Backward Elimination Selection Procedure.

B. S. Duran, *Texas Tech University, Lubbock, Texas.* Barton-David Test; Capon Test

M. R. Frankel, *City University of New York, New York.* Balanced Repeated Replications

G. H. Freeman, *National Vegetable Research Station, Warwick, England.* Agriculture, Statistics in

W. A. Fuller, *Iowa State University, Ames, Iowa.* Area Sampling

K. R. Gabriel, *University of Rochester, Rochester, New York.* Biplots

J. Galambos, *Temple University, Philadelphia, Pennsylvania.* Characterizations of Distributions

J. Gani, CSIRO, *University of Kentucky, Lexington, Kentucky.* Applied Probability; Applied Probability Journals

A. E. Gelfand, *University of Connecticut, Storrs.* Archaeology, Statistics in

J. D. Gibbons, *University of Alabama, University, Alabama.* Brown-Mood Median Test

R. E. Glaser, *Lawrence Livermore National Laboratory, Livermore, California.* Bartlett's Test of Homogeneity of Variances

A. Goldman, *University of Nevada, Las Vegas, Nevada.* Blackjack

W. A. Golomski, *Chicago, Illinois.* American Society for Quality Control (ASQC)

I. J. Good, *Virginia Polytechnic Institute and State University, Blacksburg, Virginia.* Axioms of Probability

B. G. Greenberg, *University of North Carolina, Chapel Hill, North Carolina.* Biostatistics

S. J. Haberman, *University of Chicago, Chicago, Illinois.* Association, Measures of

D. M. Hawkins, *National Research Institute for Mathematical Statistics, Pretoria, South Africa.* Branch-and-Bound Methods

T. P. Hettmansperger, *Pennsylvania State University, University Park, Pennsylvania.* Bowker's Test for Symmetry

C. C. Heyde, CSIRO, *Canberra, Australia.* Australian Journal of Statistics

M. Hills, *British Museum, London, England.* Anthropology, Statistics in; Allometry

D. V. Hinkley, *Stanford University, Stanford, California.* Annals of Statistics

W. Hoeffding, *University of North Carolina, Chapel Hill, North Carolina.* Asymptotic Normality

R. V. Hogg, *University of Iowa, Iowa City, Iowa.* Adaptive Methods

G. M. Jenkins, *Lancaster, England.* Autoregressive-Integrated Moving Average (ARIMA) Models; Autoregressive-Moving Average (ARMA) Models

V. M. Joshi, *University of Western Ontario, London, Ontario, Canada.* Admissibility

J. D. Kalbfleisch, *University of Waterloo, Waterloo, Ontario, Canada.* Ancillary Statistics

G. K. Kanji, *Sheffield City Polytechnic, Sheffield, England.* Bulletin in Applied Statistics (BIAS)

G. V. Kass, *University of the Witwatersrand, Johannesburg, South Africa.* Automatic Interaction Detection (AID) Techniques

J. R. Kettenring, *Bell Laboratories, Murray Hill, New Jersey.* Canonical Analysis

G. G. Koch, *University of North Carolina, Chapel Hill.* Chi-Square Tests; Chi-Square Tests, Numerical Examples

H. C. Kraemer, *Stanford University School of Medicine, Stanford, California.* Biserial Correlation

K. Krickeberg, *U.E.R. de Mathematiques, Logique Formelle et Informatique, Université Rene Descartes, Paris, France.* Bernoulli Society

R. G. Laha, *Bowling Green State University, Bowling Green, Ohio.* Characteristic Functions

H. O. Lancaster, *The University of Sydney, Sydney, New South Wales, Australia.* Chi-Square Distribution

F. C. Leone, *American Statistical Association, Washington, D.C.* American Statistical Association

E. A. Lew, *Punta Gorda, Florida.* Actuarial Statistics—Life

D. V. Lindley, *University College, London, England.* Bayesian Inference

J. I. McCool, *SFK Industries, Inc., King of Prussia, Pennsylvania.* Censored Data

B. H. Margolin, *National Institute for Environmental Health Sciences, Research Triangle Park, North Carolina.* Blocks, Balanced Incomplete; Blocks, Randomized Complete

D. H. Mellor, *Darwin College, Cambridge University, England.* Chance (II)

W. Morris, *Iowa State University, Ames, Iowa.* Chance (I)

J. Neter, *University of Georgia, Athens, Georgia.* American Statistican, The

H. Niederhausen, *Stanford University, Stanford, California.* Butler–Smirnov Test

R. M. Norton, *College of Charleston, Charleston, South Carolina.* Arc-Sine Distribution

J. L. Norwood, *Bureau of Labor Statistics, Washington, D.C.* Bureau of Labor Statistics

J. K. Ord, *Pennsylvania State University, University Park, Pennsylvania.* Aggregation

H. D. Patterson, *University of Edinburgh, Edinburgh, Scotland.* Changeover Designs

S. C. Pearce, *The University, Canterbury, England.* Analysis of Variance; Analysis of Covariance

S. K. Perng, *Kansas State University, Manhattan, Kansas.* Bahadur Efficiency

C. B. Read, *Southern Methodist University, Dallas, Texas.* Catastrophe Theory

G. K. Robinson, *Australian Road Research Board, Nunawading, Victoria, Australia.* Behrens-Fisher Problem

R. N. Rodriguez, *General Motors Laboratories, Warren, Michigan.* Burr Distributions

E. G. Schilling, *General Electric Company, Cleveland, Ohio.* Acceptance Sampling

G. A. F. Seber, *University of Auckland, Auckland, New Zealand.* Capture-Recapture Methods

E. Seneta, *University of Sydney, Sydney, New South Wales, Australia.* Abbe, Ernst; Bernstein, Sergei Natanovich; Bienaymé, Irenée-Jules; Boltzmann, Ludwig Edward; Boscovich, Ruggiero Giuseppe; Cauchy, Augustin-Louis; Chebyshev, Pafnuty Lvovich; Chuprov, Alexander Alexandrovich

R. J. Serfling, *Johns Hopkins University, Baltimore, Maryland.* Asymptotic Expansion

G. Shafer, *University of Kansas, Lawrence, Kansas.* Belief Functions; Bernoulli's, The

L. R. Shenton, *University of Georgia, Athens, Georgia.* Approximations to Distributions

W. L. Smith, *University of North Carolina, Chapel Hill, North Carolina.* Birth and Death Processes

B. Sowan, *Imperial College, London, England.* Biometrika

M. A. Stephens, *Simon Fraser University, British Columbia, Canada.* Anderson-Darling Test for Goodness of Fit

S. Stigler, *Center for Advanced Study in the Behavioral Sciences, University of Chicago, Chicago, Illinois.* Arithmetic Mean

M. E. Stokes, *University of North Carolina, Chapel Hill, North Carolina.* Chi-Square Tests, Numerical Examples

G. P. H. Styan, *McGill University, Montreal, Canada.* Canadian Journal of Statistics, The (La Revue Canadienne de Statistique)

L. Takács, *Case Western Reserve University, Cleveland, Ohio.* Ballot Problems

H. Taylor, *Cornell University, Ithaca, New York.* Brownian Motion

G. L. Tietjen, *Los Alamos Scientific Laboratory, Los Alamos, New Mexico.* Chemistry, Statistical Methods in

D. S. Tracy, *University of Windsor, Ontario, Canada.* Angle Brackets

R. K. Tsutakawa, *University of Missouri, Columbia, Missouri.* Bioassay, Statistical Methods in

G. J. G. Upton, *University of Essex, Colchester, England.* Categorical Data

G. L. Yang, *University of Maryland, College Park, Maryland.* Biometric Functions

S. L. Zabell, *Northwestern University, Evanston, Illinois.* Bortkiewicz, Ladislaus von

PREFACE

The purpose of this encyclopedia is to provide information about an extensive selection of topics concerned with statistical theory and the applications of statistical methods in various more or less scientific fields of activity. This information is intended primarily to be of value to readers who do not have detailed information about the topics but have encountered references (either by field or by use of specific terminology) that they wish to understand. The entries are not intended as condensed treatises containing all available knowledge on each topic. Indeed, we are on guard against placing too much emphasis on currently fashionable, but possibly ephemeral, matters. The selection of topics is also based on these principles. Nevertheless, the encyclopedia was planned on a broad basis—eight volumes, each of approximately 550 pages—so that it is possible to give attention to nearly all the many fields of inquiry in which statistical methods play a valuable (although not usually, or necessarily, a predominant) role.

Beyond the primary purpose of providing information, we endeavored to obtain articles that are pleasant and interesting to read and encourage browsing through the volumes. There are many contributors, for whose cooperation we are grateful, and a correspondingly wide range of styles of presentation, but we hope that each is attractive in its own way. There is also, naturally and inevitably, a good deal of variation among the (mathematical and technical-scientific) levels of the entries. For some topics, considerable mathematical sophistication is needed for adequate treatment; for others, it is possible to avoid heavy reliance on mathematical formulation.

We realize that even an eight-volume compendium cannot incorporate all of the terms, notions, and procedures that have appeared in statistical literature during the last century. There are also contributions by scientists who paved the way, as early as the seventeenth century, toward the statistical sciences as they are known today. We endeavored to include historical background and perspective when these seem important to the development of statistical methods and ideas.

It is to be expected that most readers will disagree with the relative amount of emphasis accorded to certain fields, and will find that some topics of considerable interest have been omitted. While this may reflect a lack of judgment or knowledge (or both) on our part, it is inevitable, because each person has a specific, idiosyncratic viewpoint on statistical matters (as on others). Our intention is to mirror the state of the art in the last quarter of the twentieth century, including terms (in particular mathematical) that found a place in the language of statistical methodology during its formative years.

We have two ways of cross-referencing: First, when a possibly unfamiliar term appears in an entry, reference to another entry is indicated by an asterisk, or by direct reference (e.g., *See* HISTOGRAMS). An asterisk sometimes refers to the preceding word but quite frequently to the preceding phrase. For example, " . . . random variable*" refers to the entry on random variables rather than on variables. We feel that this notation is the simplest possible. Second, most articles conclude with a list of related entries of potential interest for further reading. These two sets of cross-references may overlap but are usually not identical. The starred items are for utility, whereas the list is more for interest. Neither set is exhaustive and we encourage individual initiative in searching out further related entries.

Since our primary purpose is to provide information, we neither avoid controversial topics nor encourage purely polemic writing. We endeavor to give fair representation to different viewpoints but cannot even hope to approximate a just balance (if such a thing exists).

In accordance with this primary purpose, we believe that the imposition of specific rules of style and format, and levels of presentation, must be subordinate to the presentation of adequate and clear information. Also, in regard to notation, references, and similar minutiae, we did not insist on absolute uniformity although we tried to discourage very peculiar deviations that might confuse readers.

The encyclopedia is arranged lexicographically in order of entry titles. There are some inconsistencies; for example, we have "CHEMISTRY, STATISTICS IN" but "STATISTICS IN ASTRONOMY." This simply reflects the fact that the encyclopedia is being published serially, and the second of these entries was not available when the first volume was in production. (This volume does, however, contain the "dummy" entry "ASTRONOMY, STATISTICS IN *See* STATISTICS IN ASTRONOMY.")

We are indeed fortunate that Professor Campbell B. Read joined us as Associate Editor on October 1, 1980. Professor Read's active participation in the editorial process and the numerous improvements he contributed to this project have been invaluable. The Co-Editors-in-Chief express their sincerest appreciation of his expertise.

We also express our thanks to the members of the Advisory Board for their valuable advice and expert suggestions; to the Editorial Assistant, Ms. June Maxwell, for her devotion and for contributions to the project far beyond the call of duty; and last, but certainly not least, to all the contributors, who responded enthusiastically to our call for partnership in this undertaking.

Unsigned entries are contributed by the Editors—Samuel Kotz, Norman L. Johnson, and Campbell B. Read—either jointly or individually.

SAMUEL KOTZ
NORMAN L. JOHNSON

College Park, Maryland
Chapel Hill, North Carolina
January 1982

ENCYCLOPEDIA OF STATISTICAL SCIENCES

VOLUME 1

A to Circular Probable Error

ABAC

A graph from which numerical values may be read off, usually by means of a grid of lines corresponding to argument values.

(NOMOGRAM)

ABACUS

A simple instrument to facilitate numerical computation. There are several forms of abacus. The one in most common use at present is represented diagramatically in Fig. 1. It consists of a rectangular framework *ABCD* with a cross-piece *PQ* parallel to the longer sides, *AB* and *CD*, of the rectangle. There are a number (at least eight, often more) of thin rods or wire inserted in the framework and passing through *PQ*, parallel to the shorter sides, *AD* and *BC*. On each rod there are threaded four beads between *CD* and *PQ*, and one bead between *PQ* and *AB*.

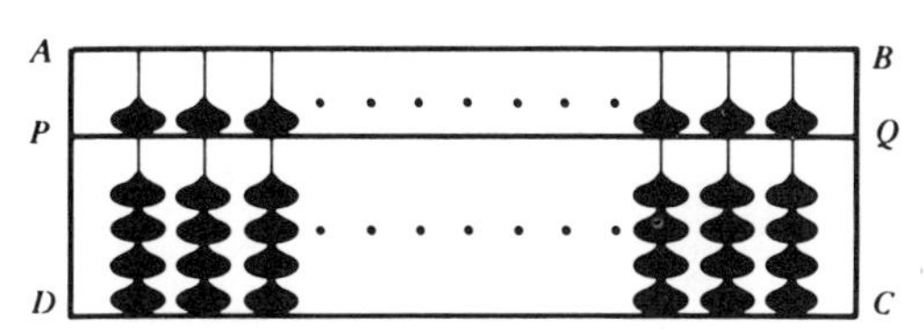

Figure 1. Diagrammatic representation of the form of abacus presently in common use.

Analogously to the meaning of position in our number system, the extreme right-hand rod corresponds to units; the next to the left, tens; the next to the left, hundreds; and so on. Each bead in the lower rectangle (*PQCD*) counts for 1, when moved up, and each bead in the upper rectangle (*ABQP* counts for 5. The number shown in Fig. 2 would be 852 if beads on all rods except the three extreme right-hand ones are as shown for the three extreme left-hand rods (corresponding to "zero").

The Roman abacus consisted of a metal plate with two sets of parallel grooves, the lower containing four pebbles and the upper one pebble (with a value five times that of

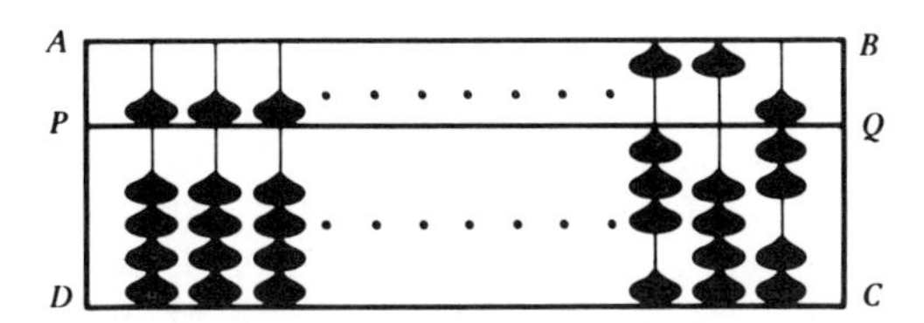

Figure 2. Abacus that would be showing the number 852 if beads on all rods except the three extreme right-hand ones are as shown for the three extreme left-hand rods (corresponding to "zero").

each pebble in the corresponding groove of the lower set). The Japanese and Chinese abacus (still in use) consists of a frame with beads on wires. The Russian abacus, which originated in the sixteenth century (the modern version in the eighteenth century), is also still in use.

Bibliography

Dilson, J. (1968). *The Abacus: A Pocket Computer*. St. Martin's Press, New York.

Gardner, M. (1979). *Mathematical Circus*. Alfred A. Knopf, New York, Chap. 18.

Pullan, J. (1969). *The History of the Abacus*. F. A. Praeger, New York.

ABBE, ERNST

Born: January 23, 1840, in Eisenach, Germany.

Died: January 14, 1905, in Jena, Germany.

Contributed to: theoretical and applied optics, astronomy, mathematical statistics.

The recognition of Abbe's academic talent by those in contact with him overcame a childhood of privation and a financially precarious situation very early in his academic career, when he completed "On the Law of Distribution of Errors in Observation Series," his inaugural dissertation for attaining a lectureship at Jena University at the age of 23 [1]. This dissertation, partly motivated by the work of C. F. Gauss*, seems to contain his only contributions to the probability analysis of observations subject to error.These contributions constitute a remarkable anticipation of later work in distribution theory and time-series* analysis, but they were overlooked until the late 1960s [5, 8], and almost none of the early bibliographies on probability and statistics (a notable exception being ref. 10) mention this work. In 1866, Abbe was approached by Carl Zeiss, who asked him to establish a scientific basis for the construction of microscopes; this was the beginning of a relationship that lasted throughout his life, and from this period on his main field of activity was optics [9] and astronomy.

Abbe shows, first, that the quantity $\Delta = \sum_{i=1}^{n} Z_i^2$, where Z_i, $i = 1, \dots, n$, are n independently and identically distributed $N(0, 1)$ random variables, is described by a chi-square* density with n degrees of freedom [5, 8], although this discovery should perhaps be attributed to I. J. Bienaymé* [4]. Second, again initially by means of a "discontinuity factor" and then by complex variable methods, Abbe obtains the distribution of $\Theta = \sum_{j=1}^{n} (Z_j - Z_{j+1})^2$, where $Z_{n+1} = Z_1$, and ultimately that of Θ/Δ, a ratio of quadratic forms* in $Z_1, \dots, Z_n$ very close in nature to the definition of what is now called the first circular serial correlation coefficient, * and whose distribution under the present conditions is essentially that used to test the null hypothesis of Gaussian white noise* against a first-order autoregression alternative, in time-series* analysis [3]. (The distribution under such a null hypothesis was obtained by R. L. Anderson in 1942.) Knopf [6] expresses Abbe's intention in his dissertation as being to seek a numerically expressible criterion to determine when differences between observed and sought values in a series of observations are due to chance alone.

References

[1] Abbe, E. (1863). *Über die Gesetzmässigkeit der Vertheilung der Fehler bei Beobachtungsreihen*. Hab. schrift., Jena (Reprinted as pp. 55–81 of ref. 2.)

[2] Abbe, E. (1906). *Gesammelte Abhandlungen*, Vol. 2. G. Fischer, Jena.

[3] Hannan, E. J. (1960). *Time Series Analysis*. Methuen, London, pp. 84–86.

[4] Heyde, C. C. and Seneta, E. (1977). *I. J. Bienaymé: Statistical Theory Anticipated*. Springer-Verlag, New York, p. 69.

[5] Kendall, M. G. (1971). *Biometrika*, **58**, 369–373. (Sketches Abbe's mathematical reasoning in relation to the contributions to mathematical statistics.)

[6] Knopf, O. (1905). *Jahresber. Dtsch. Math.-Ver.*, **14**, 217–230. [One of several obituaries by his associates; nonmathematical, with a photograph.

Another is by S. Czapski (1905), in *Verh. Dtsch. Phys. Ges.*, **7**, 89–121.]

[7] Rohr, L. O. M. von (1940). *Ernst Abbe*. G. Fischer, Jena. (Not seen.)

[8] Sheynin, O. B. (1966). *Nature (Lond.)*, **211**, 1003–1004. (Notes Abbe's derivation of the chi-square density.)

[9] Volkman, H. (1966). *Appl. Opt.*, **5**, 1720–1731. (An English-language account of Abbe's life and contributions to pure and applied optics; contains two photographs of Abbe, and further bibliography.)

[10] Wölffing, E. (1899). *Math. Naturwiss. Ver. Württemberg* [Stuttgart], *Mitt.*, (2) **1**, 76–84. [Supplements the comprehensive bibliography given by E. Czuber (1899), in *Jahresber. Dtsch. Math.-Ver.*, **7** (2nd part), 1–279.]

(CHI-SQUARE DISTRIBUTION
QUADRATIC FORMS
SERIAL CORRELATION COEFFICIENT
TIME-SERIES ANALYSIS)

E. SENETA

ABEL'S FORMULA

(Also known as the Abel identity.) If each term of a sequence of real numbers $\{a_i\}$ can be represented in the form $a_i = b_i c_i$, $i = 1, \ldots, n$, then $a_1 + a_2 + \cdots + a_n$ can be expressed as

$$s_1(b_1 - b_2) + s_2(b_2 - b_3) + \cdots + s_{n-1}(b_{n-1} - b_n),$$

where $s_i = c_1 + \cdots + c_i$. Equivalently,

$$\sum_{k=n}^{m} b_k c_k = B_m c_{m+1} - B_{n-1} c_n + \sum_{k=n}^{m} B_k (c_k - c_{k+1}),$$

where $B_k = \sum_{l=1}^{k} b_l$.

This representation is usually referred to as Abel's formula, due to Norwegian mathematician Niels Henrik Abel (1802–1829). (The continuous analog of this formula is the formula of integration by parts.) It is useful for manipulations with finite sums.

Bibliography

Knopp, K. (1951). *Theory and Application of Infinite Series*, 2nd ed. Blackie, London/Dover, New York.

ABSOLUTE ASYMPTOTIC EFFICIENCY (AAE) *See* ESTIMATION, POINT

ABSOLUTE CONTINUITY (of Measures on Infinite-Dimensional Linear Spaces)

Absolute continuity of measures, the Radon–Nikodym theorem*, and the Radon–Nikodym derivative* are subjects properly included in any basic text on measure and integration. However, both the mathematical theory and the range of applications can best be appreciated when the measures are defined on an infinite-dimensional linear topological space. For example, this setting is generally necessary if one wishes to discuss hypothesis testing* for stochastic processes with infinite parameter set. In this article we first define basic concepts in the area of absolute continuity, state general conditions for absolute continuity to hold, and then specialize to the case where the two measures are defined on either a separable Hilbert space or on an appropriate space of functions. Particular attention is paid to Gaussian measures.

The following basic material is discussed in many texts on measure theory; see, e.g., ref. 23. Suppose that (Ω, β) is a measurable space, and that μ_1 and μ_2 are two probability measures on (Ω, β). μ_1 is said to be absolutely continuous with respect to μ_2 ($\mu_1 \ll \mu_2$) if A in β and $\mu_2(A) = 0$ imply that $\mu_1(A) = 0$. This is equivalent to the following: $\mu_1 \ll \mu_2$ if and only if for every $\epsilon > 0$ there exists $\delta > 0$ such that $\mu_2(A) < \delta$ implies that $\mu_1(A) \leqslant \epsilon$. Similar definitions of absolute continuity can be given for nonfinite signed measures; this article, however, is restricted to probability measures. When $\mu_1 \ll \mu_2$, the Radon–Nikodym theorem states that there exists a real-valued β-measurable function f such that $\mu_1(A) = \int_A f d\mu_2$ for all A in β. The function f, which belongs to $L_1[\Omega, \beta, \mu_2]$ and is unique up to μ_2-equivalence, is called the Radon–Nikodym derivative of μ_1 with respect to μ_2, and is commonly denoted by $d\mu_1/d\mu_2$. In statistical and engineering applications $d\mu_1/d\mu_2$ is usually called the likelihood ratio*, a term that has its genesis in maximum likelihood estimation*.

Absolute continuity and the Radon–Nikodym derivative have important applications in statistics. For example, suppose that $X : \Omega \to \mathbb{R}^N$ is a random vector. Suppose also that under hypothesis H_1 the distribution function of X is given by $F_1 = \mu_1 \circ X^{-1}$ $[F_1(x) = \mu_1\{\omega : X(\omega) \leqslant x\}]$, whereas under H_2, X has the distribution function $F_2 = \mu_2 \circ X^{-1}$. F_i defines a Borel measure on $\mathbb{R}^1$; one says that F_i is induced from μ_i by X. A statistician observes one realization (sample path) of X, and wishes to design a statistical test to optimally decide in favor of H_1 or H_2. Then, under any of several classical decision criteria of mathematical statistics (e.g., Bayes risk, Neyman–Pearson*, minimum probability of error), an optimum decision procedure* when $\mu_1 \ll \mu_2$ is to form the test statistic* $\Lambda(X) = [dF_1/dF_2](X)$ and compare its value with some constant, C_0; the decision is then to accept H_2 if $\Lambda(X) \leqslant C_0$, accept H_1 if $\Lambda(X) > C_0$. The value of C_0 will depend on the properties of F_1 and F_2 and on the optimality criterion. For more details, see HYPOTHESIS TESTING*.[1]

Two probability measures μ_1 and μ_2 on (Ω, β) are said to be equivalent $(\mu_1 \sim \mu_2)$ if $\mu_1 \ll \mu_2$ and $\mu_2 \ll \mu_1$. They are orthogonal, or extreme singular $(\mu_1 \perp \mu_2)$ if there exists a set A in β such that $\mu_2(A) = 0$ and $\mu_1(A) = 1$. For the hypothesis-testing problem discussed above, orthogonal induced measures permit one to discriminate perfectly between H_1 and H_2. In many practical applications, physical considerations rule out perfect discrimination. The study of conditions for absolute continuity then becomes important from the aspect of verifying that the mathematical model is valid.

In the framework described, the random vector has range in $\mathbb{R}^N$. However, absolute continuity, the Radon–Nikodym derivative, and their application to hypothesis-testing problems are not limited to such finite-dimensional cases. In fact, the brief comments above on hypothesis testing apply equally well when X takes its value in an infinite-dimensional linear topological space, as when $X(\omega)$ represents a sample path* from a stochastic process* (X_t), $t \in [a, b]$. (The infinite-dimensional case does introduce interesting mathematical complexities that are not present in the finite-dimensional case.)

GENERAL CONDITIONS FOR ABSOLUTE CONTINUITY

We shall see later that special conditions for absolute continuity can be given when the two measures involved have certain specialized properties, e.g., when they are both Gaussian. However, necessary and sufficient conditions for absolute continuity can be given that apply to any pair of probability measures on any measurable space (Ω, β). Further, if (Ω, β) consists of a linear topological space Ω and the smallest σ-field β containing all the open sets (the Borel σ-field), then additional conditions for absolute continuity can be obtained that apply to any pair of probability measures on (Ω, β). Here we give one well-known set of general necessary and sufficient conditions. First, recall that if (Ω, β, P) is a probability space and F a collection of real random variables on (Ω, β), then F is said to be uniformly integrable with respect to P [23] if the integrals $\int_{\{\omega : |f(\omega)| \geqslant c\}} |f(\omega)|\, dP(\omega)$, $c > 0$, f in F, tend uniformly to zero as $c \to \infty$. An equivalent statement is the following: F is uniformly integrable (P) if and only if

(a) $$\sup_F \int_\Omega |f(\omega)|\, dP(\omega) < \infty$$

and

(b) For every $\epsilon > 0$ there exists $\delta > 0$ such that $P(A) < \delta$ implies that
$$\sup_F \int_A |f(\omega)|\, dP(\omega) \leqslant \epsilon.$$

Theorem 1. Suppose that μ_1 and μ_2 are two probability measures on a measurable space (Ω, β). Suppose that $\{\mathcal{F}_n, n \geqslant 1\}$ is an increasing family of sub-σ-fields of β such that β is the smallest σ-field containing $\cup_n \mathcal{F}_n$. Let μ_i^n be the restriction of μ_i to $\mathcal{F}_n$. Then $\mu_1 \ll \mu_2$ if and only if

(a) $\quad \mu_1^n \ll \mu_2^n \quad$ for all $n \geqslant 1$,

and

(b) $\quad \{d\mu_1^n / d\mu_2^n, n \geqslant 1\}$
is uniformly integrable (μ_2).

When $\mu_1 \ll \mu_2$, then $d\mu_1 / d\mu_2 = \lim_n d\mu_1^n / d\mu_2^n$ almost everywhere (a.e.) $d\mu_2$.

Condition (a) of Theorem 1 is obviously necessary. The necessity of (b) follows from the fact that $\{d\mu_1^n / d\mu_2^n, \mathcal{F}_n : n \geqslant 1\}$ is a martingale* with respect to μ_2. This property, and the martingale convergence theorem, yield the result that $d\mu_1 / d\mu_2 = \lim_n d\mu_1^n / d\mu_2^n$ a.e. $d\mu_2$. Sufficiency of (a) and (b) follows from the second definition of uniform integrability given above and the assumption that β is the smallest σ-field containing $\cup_n \mathcal{F}_n$.

Conditions (a) and (b) of Theorem 1 are also necessary and sufficient for $\mu_1 \ll \mu_2$ when the family of increasing σ-fields $(\mathcal{F}_t)$ has any directed index set.

A number of results frequently used to analyze absolute continuity can be obtained from Theorem 1. This includes, for example, Hájek's divergence criterion [20] and Kakutani's theorem on equivalence of infinite product measures [29] (a fundamental result in its own right).

The conditions of Theorem 1 are very general. However, in one respect they are somewhat unsatisfactory. They usually require that one specify an infinite sequence of Radon–Nikodym derivatives $\{d\mu_1^n / d\mu_2^n, n \geqslant 1\}$. It would be preferable to have a more direct method of determining if absolute continuity holds. One possible alternative when the measures are defined on a separable metric space involves the use of chacteristic functions*. The characteristic function of a probability measure defined on the Borel σ-field of a separable metric space completely and uniquely specifies the measure [38]. Thus in such a setting, two characteristic functions contain all the information required to determine whether absolute continuity exists between the associated pair of measures. The use of characteristic functions offers a method for attacking the following problem. For a given measure μ on (Ω, β) determine the set $\mathcal{P}_\mu$ of all probability measures on (Ω, β) such that $\nu \ll \mu$ for all ν in $\mathcal{P}_\mu$. Some results on this problem are contained in ref. 3; further progress, especially detailed results for the case of a Gaussian measure μ on Hilbert space, would be useful in several important applications areas (detection of signals in noise, stochastic filtering*, information theory*).

PROBABILITY MEASURES ON HILBERT SPACES

There has been much recent activity in the study of probability measures on Banach spaces [1, 4, 5, 31]. Here we restrict attention to the case of probabilities on Hilbert spaces; this is the most important class of Banach spaces for applications, and the theory is relatively well developed in this setting.

Let $\mathbf{H}$ be a real separable Hilbert space with inner product $\langle \cdot, \cdot \rangle$ and Borel σ-field Γ. Let μ be a probability measure on Γ. For any element y in $\mathbf{H}$, define the distribution function F_y by $F_y(a) = \mu\{x : \langle y, x \rangle \leqslant a\}$, a in $(-\infty, \infty)$. μ is said to be Gaussian if F_y is Gaussian for all y in $\mathbf{H}$. It can be shown that for every Gaussian μ there exists a self-adjoint trace-class nonnegative linear operator R_μ in $\mathbf{H}$ and an element m_μ in $\mathbf{H}$ such that

$$\langle y, m_\mu \rangle = \int_{\mathbf{H}} \langle y, x \rangle \, d\mu(x) \tag{1}$$

and

$$\langle R_\mu y, v \rangle = \int_{\mathbf{H}} \langle y - m_\mu, x \rangle \langle v - m_\mu, x \rangle \, d\mu(x) \tag{2}$$

for all y and v in $\mathbf{H}$. R_μ is called the covariance (operator) of μ, and m_μ is the mean (element). Conversely, to every self-adjoint nonnegative trace-class operator R_μ and element m in $\mathbf{H}$ there corresponds a unique Gaussian measure μ such that relations (1) and (2) are satisfied. Non-Gaussian measures μ may also have a covariance operator R_μ and mean element m_μ satisfying (1) and (2); however, the covariance R_μ need not be trace-class. For more details on probability measures on Hilbert space, see refs. 17, 38, and 53.

Elegant solutions to many problems of classical probability theory (and applications) have been obtained in the Hilbert space framework, with methods frequently making use of the rich structure of the theory of linear operators. Examples of such problems include Sazanov's solution to obtaining necessary and sufficient conditions for a complex-valued function on $\mathbf{H}$ to be a characteristic function* [49]; Prohorov's conditions for weak compactness of families of probability measures, with applications to convergence of stochastic processes [43]; the results of Mourier on laws of large numbers* [34]; the results of Fortét and Mourier on the central limit theorem* [15, 34]; and conditions for absolute continuity of Gaussian measures. The latter problem is examined in some detail in the following section. The study of probability theory in a Hilbert space framework received much of its impetus from the pioneering work of Fortét and Mourier (see refs. 15 and 34, and the references cited in those papers). Their work led not only to the solution of many interesting problems set in Hilbert space, but also to extensions to Banach spaces and more general linear topological spaces [1, 4, 5, 15, 31, 34].

The infinite-dimensional Hilbert spaces $\mathbf{H}$ most frequently encountered in applications are $L_2[0, T]$ $(T < \infty)$ and l_2. For a discussion of how Hilbert spaces frequently arise in engineering applications, *see* STATISTICAL COMMUNICATION THEORY. In particular, the interest in Gaussian measures on Hilbert space has much of its origin in hypothesis-testing and estimation problems involving stochastic processes: detection and filtering of signals embedded in Gaussian noise. For many engineering applications, the noise can be realistically modeled as a Gaussian stochastic process with sample paths almost surely (a.s.) in $L_2[0, T]$ or a.s. in l_2. When $\mathbf{H}$ is $L_2[0, T]$, a trace-class covariance operator can be represented as an integral operator whose kernel is a covariance function. Thus suppose that (X_t), $t \in [0, T]$, is a measurable zero-mean stochastic process on (Ω, β, P), inducing the measure μ on the Borel σ-field of $L_2[0, T]$; $\mu(A) = P\{\omega : X(\omega) \in A\}$. Then $E\int_0^T X_t^2(\omega)\,dt < \infty$ if and only if μ has a trace-class covariance operator R_μ defined by $[R_\mu f](t) = \int_0^T R(t, s) f(s)\,ds$, f in $L_2[0, T]$, where R is the covariance function of (X_t). If R_μ is trace-class, then $E\int_0^T X_t^2(\omega)\,dt = \text{trace } R_\mu$.

ABSOLUTE CONTINUITY OF PROBABILITY MEASURES ON HILBERT SPACE

If $\mathbf{H}$ is finite-dimensional and μ_1 and μ_2 are two zero-mean Gaussian measures on Γ, it is easy to see that μ_1 and μ_2 are equivalent if and only if their covariance matrices have the same range space. However, if $\mathbf{H}$ is infinite-dimensional, this condition (on the ranges of the covariance operators) is neither necessary nor sufficient for $\mu_1 \sim \mu_2$. The study of conditions for absolute continuity of two Gaussian measures on function space has a long and active history. Major early contributions were made by Cameron and Martin [6, 7] and by Grenander [18]. The work of Cameron and Martin was concerned with the case when one measure is Wiener measure (the measure induced on $C[0, 1]$ by the Wiener process*) and the second measure is obtained from Wiener measure by an affine transformation. Grenander obtained conditions for absolute continuity of a Gaussian measure (induced by a stochastic process with continuous covariance) with respect to a translation. Segal [50] extended the work of Cameron and Martin to a more general class of affine transformations of Wiener measure. Segal also obtained [50] conditions for absolute continuity of Gaussian "weak distributions." These necessary and sufficient conditions can be readily applied to obtain sufficient conditions for equivalence of any pair of Gaussian measures on **H**; they can also be used to show that these same conditions are necessary. Complete and general solutions to the absolute continuity problem for Gaussian measures were obtained by Feldman [12] and Hájek [21]. Their methods are quite different. The main result, in each paper, consists of two parts: a "dichotomy theorem," which states that any two Gaussian measures are either equivalent or orthogonal; and conditions that are necessary and sufficient for equivalence. The following theorem for Gaussian measures on Hilbert space is a modified version of Feldman's result [12]; several proofs have been independently obtained (Kallianpur and Oodaira [30], Rao and Varadarajan [44], Root [45]).

Theorem 2. Suppose that μ_1 and μ_2 are two Gaussian measures on Γ, and that μ_i has covariance operator R_i and mean m_i, $i = 1$, 2. Then:

1. either $\mu_1 \sim \mu_2$ or $\mu_1 \perp \mu_2$;
2. $\mu_1 \sim \mu_2$ if and only if all the following conditions are satisfied:
   (a) range $(R_1^{1/2}) =$ range $(R_2^{1/2})$;
   (b) $R_1 = R_2^{1/2}(I + T)R_2^{1/2}$, where I is the identity on **H** and T is a Hilbert–Schmidt operator in **H**:
   (c) $m_1 - m_2$ is in range $(R_1^{1/2})$.

Various specializations of Theorem 2 have been obtained; see the references in refs. 8 and 47. Two of the more interesting special cases, both extensively analyzed, are the following: (1) both measures induced by stationary Gaussian stochastic processes; (2) one of the measures is Wiener measure. In the former case, especially simple conditions can be given when the two processes have rational spectral densities; see the papers by Feldman [13], Hájek [22], and Pisarenko [40, 41]. In this case, when the two measures have the same mean function, $\mu_1 \sim \mu_2$ if and only if $\lim_{|\lambda|\to\infty} f_1(\lambda)/f_2(\lambda) = 1$, where f_i is the spectral density* of the Gaussian process inducing μ_i. Moreover, this occurs if and only if the operator T appearing in Theorem 2 is also trace-class [22]. For the case where one of the measures is Wiener measure, see the papers by Shepp [51], Varberg [54, 55], and Hitsuda [24].

The problem of determining the Radon–Nikodym derivative for two equivalent Gaussian measures on a Hilbert space has been studied, especially by Rao and Varadarajan [44]. For convenience, we use the notation of Theorem 2 and assume now that all covariance operators are strictly positive. In the case where the Hilbert space is finite-dimensional, the log of the Radon-Nikodym derivative $d\mu_1/d\mu_2$ (log-likelihood ratio*) is easily seen to be a quadratic-linear form; that is, $\log \Lambda(X) = \langle x, Wx \rangle + \langle x, b \rangle +$ constant, where the linear operator $W = \frac{1}{2}(R_2^{-1} - R_1^{-1})$, $b = R_1^{-1}m_1 - R_2^{-1}m_2$, and $\log \equiv \log_e$. However, when **H** is infinite-dimensional, the log-likelihood ratio need not be a quadratic-linear form defined by a bounded linear operator. This holds true even if the operator T of Theorem 2 is not only Hilbert–Schmidt, but is also trace class. However, when T is Hilbert-Schmidt, one can always express the log of the Radon–Nikodym derivative as an almost surely convergent series [44]. The essential difficulty in characterizing the likelihood ratio for infinite-dimensional Hilbert space is that the operators R_1 and R_2 cannot have *bounded* inverses and these two inverses need not have the same domain of definition. Even if range $(R_1) =$ range(R_2), so that $R_2^{-1} - R_1^{-1}$ is defined on range (R_1), it is not necessary that $R_2^{-1} - R_1^{-1}$ be bounded on range (R_1).

In the finite-dimensional case, if $R_1 = R_2$, then $\log \Lambda(X) = \langle x, b \rangle +$ constant, with b defined as above, so that the log-likelihood ratio is a bounded linear form. This need not be the case for infinite-dimensional Hilbert space; in general, $\log \Lambda(X)$ will be a bounded linear form (when $R_1 = R_2$) if and only if $m_1 - m_2$ is in the range of R_1. As can be seen from Theorem 1, this condition is strictly stronger than the necessary and sufficient condition for $\mu_1 \sim \mu_2$, which (with $R_1 = R_2$) is that $m_1 - m_2$ be in range $(R_1^{1/2})$.

If the two measures are induced by stationary Gaussian processes with rational spectral densities, expressions for the likelihood ratio can be given in terms of the spectral densitites; see the papers by Pisarenko [41] and Hájek [22].

In many applications, only one of the two measures can be considered to be Gaussian. For this case, a useful sufficient condition for absolute continuity is given in ref. 2. This condition can be applied when the two measures are induced by stochastic processes (X_t) and (Y_t), where (Y_t) is a function of (X_t) and a process (Z_t) that is independent of (X_t). In particular, if (X_t) is Gaussian and $(Y_t) = (X_t + Z_t)$, then conditions for absolute continuity can be stated in terms of sample path properties of the (Z_t) process (absolute continuity, differentiability, etc.). Such conditions can often be verified in physical models by knowledge of the mechanisms generating the observed data, when the distributional properties of the (Z_t) process are unknown. When (X_t) is the Wiener process on $[0, T]$, conditions for absolute continuity of the induced measures on $L_2[0, T]$ can be obtained from the results of refs. 10, 27, and 28. Some of these results do not require independence of (X_t) and (Z_t).

Other results on absolute continuity of measures on Hilbert space have been obtained for infinitely divisible measures [16], measures induced by stochastic processes with independent increments [16], admissible translations of measures [42, 52], and for a fixed measure and a second measure obtained from the first measure by a nonlinear transformation [16]. With respect to admissible translates, Rao and Varadarajan [44] have shown that if μ is a zero-mean measure having a trace-class covariance operator, R, then the translate of μ by an element y is orthogonal to μ if y is not in range $(R^{1/2})$. A number of these results are collected in the book by Gihman and Skorohod [17], which also contains much material on basic properties of probability measures on Hilbert space, and on weak convergence*. The book by Kuo [33] contains not only basic material on probability measures on Hilbert spaces (including absolute continuity), but also an introduction to some topics in probability on Banach spaces.

ABSOLUTE CONTINUITY OF MEASURES INDUCED BY STOCHASTIC PROCESSES

Many problems involving stochastic processes are adequately modeled in the framework of probability measures on Hilbert space, provided that the sample paths of each process of interest belong almost surely to some separable Hilbert space. However, this condition is not always satisfied; even when it is satisfied, one may prefer conditions for absolute continuity stated in terms of measures on $\mathbb{R}^T$ (the space of real-valued functions on T), where T is the parameter set of the process. For example, a class of stochastic processes frequently considered are those having almost all paths in $D[0, 1]$. $D[0, 1]$ is the set of all real-valued functions having limits from both left and right existing at all points of $(0, 1)$, with either left-continuity or right-continuity at each point of $(0, 1)$, and with a limit from the left (right) existing at $1(0)$. $D[0, 1]$ is a linear metric space* under the Skorohod metric [38], but this metric space is not a Hilbert space.

The general conditions for absolute continuity stated in Theorem 1 apply in any setting. Moreover, necessary and sufficient conditions for equivalence of measures (most frequently on $\mathbb{R}^T$) induced by two Gaussian stochastic processes can be stated in a number of ways: The reproducing kernel Hilbert space (r.k.H.s.) of the two covariance functions [30, 37, 39]; operators and elements in an L_2 space of real-valued random functions [12]; operators and elements in an L_2-space of random variables [46]; and tensor products [35]. Hájek's conditions for absolute continuity in terms of the divergence [21] apply to the general case. Sato [48] has stated conditions for absolute continuity in terms of a representation for all Gaussian processes whose induced measure on $\mathbb{R}^T$ is equivalent to the measure induced by a given Gaussian process. Several of these results are presented in [8]. Many other papers on absolute continuity for measures induced by two Gaussian processes have appeared; space does not permit an attempt at a complete bibliography.

Use of the r.k.H.s. approach to study linear statistical problems in stochastic processes was first explicitly and systematically employed by Parzen; the r.k.H.s. approach was also implicit in the work of Hájek (see the papers by Hájek [22] and Parzen [39] and their references).

For non-Gaussian processes, results on absolute continuity have been obtained for Markov processes* [16, 32], diffusion processes* [36], locally infinitely divisible processes [16], semimartingales* [25], point processes* [26], and non-Gaussian processes equivalent to the Wiener process [9, 10, 27, 28].

Dudley's result [9] is of particular interest to researchers interested in Gaussian measures. Suppose that (W_t) is the Wiener process on $[0, 1]$ with zero mean and unity variance parameter, and that $\beta(\cdot, \cdot)$ is a continuous real-valued function on $\mathbb{R} \times [0, 1]$. Let $Y_t = \beta(W_t, t)$. Dudley shows in ref. 9 that the measure on function space induced by (Y_t) is absolutely continuous with respect to Wiener measure if and only if $\beta(u, t) = u + \phi(t)$ or $\beta(u, t) = -u + \phi(t)$, where ϕ is in the r.k.H.s. of the Wiener covariance $\min(t, s)$. The methods used to prove this result rely heavily on some of the special properties of the Wiener process, such as the fact that (W_t) has the strong Markov property*, and laws of the iterated logarithm* for the Wiener process (obtained in ref. 9). A characterization of admissible β's for other Gaussian processes with continuous paths would be of much interest; such characterizations would necessarily require a different approach, and this problem is very much open at present.

The absolute continuity problem discussed in refs. 10, 27, and 28 has received much attention, partly because of its connection to signal detection* and nonlinear filtering*. One considers a measurable process (Y_t) defined by $Y_t = \int_0^t h_s\, ds + W_t$, $0 \leqslant t \leqslant T$, where (W_t) is a zero-mean Wiener process and (h_s) is a stochastic process with sample paths a.s. in $L_1[0, T]$. Let μ_Y and μ_W be the measures induced by (Y_t) and (W_t) on the space of continuous functions on $[0, 1]$. Conditions for $\mu_Y \ll \mu_W$, $\mu_Y \sim \mu_W$, and results on the Radon–Nikodym derivative have been obtained in refs. 10, 27, and 28. In the special case where (h_s) is independent of (W_t), a sufficient condition for $\mu_Y \sim \mu_W$ is that $\int_0^T h_s^2\, ds < \infty$ for almost all sample paths of (h_s). This condition is also sufficient for $\mu_Y \ll \mu_W$ if the process (h_s) is only assumed independent of future increments of (W_t).

Finally, we mention a result of Fortét [14], who has obtained a sufficient condition for orthogonality of two measures when one is Gaussian, expressed in terms of the r.k.H.s. of the two covariances. Suppose that μ_i is a probabliity measure on $\mathbb{R}^T$, $T = [0, 1]$, with r.k.H.s. H_i and mean function m_i. Then if μ_1 is Gaussian, μ_1 and μ_2 are orthogonal unless both the following conditions are satisfied: (a) $H_1 \subset H_2$; and (b) $m_1 - m_2 \in H_2$.

NOTE

1. The ubiquitous nature of the Radon–Nikodym derivative in various hypothesis-testing applications can be attributed to its being a necessary and sufficient statistic* [11].

References

[1] Aron, R. M. and Dineen, S., eds. (1978). Vector Space Measures and Applications, I. *Lect. Notes Math.*, **644**. (Contains most of the papers on probability theory presented at the Conference on Vector Space Measures and Applications, Dublin, 1978.)

[2] Baker, C. R. (1973). *Ann. Prob.*, **1**, 690–698.

[3] Baker, C. R. (1979). *Lect. Notes Math.*, **709**, 33–44.

[4] Beck, A., ed. (1976). Probability in Banach spaces. *Lect. Notes Math.*, **526**. (Collection of papers on various topics; Proceedings of First International Conference on Probability in Banach Spaces, Oberwolfach, 1975.)

[5] Beck, A., ed. (1979). Probability in Banach Spaces, II. *Lect. Notes Math.* **709**. (Proceedings of Second International Conference on Probability in Banach Spaces, Oberwolfach, 1978.)

[6] Cameron, R. H. and Martin, W. T. (1944). *Ann. Math.*, **45**, 386–396.

[7] Cameron, R. H. and Martin, W. T. (1945). *Trans. Amer. Math. Soc.*, **58**, 184–219.

[8] Chatterji, S. D. and Mandrekar, V. (1978). In *Probabilistic Analysis and Related Topics*, Vol. 1, A. T. Bharucha-Reid, ed. Academic Press, New York, pp. 169–197. (Extensive bibliography.)

[9] Dudley, R. M. (1971). *Zeit. Wahrscheinlichkeitsth.*, **20** 249–258; correction: *ibid.*, **30** (1974), 357–358.

[10] Duncan, T. E. (1970). *Inf. Control*, **16**, 303–310.

[11] Dynkin, E. B. (1951). *Uspehi Mat. Nauk* (N.S.), **6**, 68–90; translation: *Select. Transl. Math. Statist. Prob.* **1**, 17–40, (1961).

[12] Feldman, J. (1958). *Pacific J. Math.*, **8**, 699–708; correction: *ibid.*, **9**, 1295–1296 (1959).

[13] Feldman, J. (1960). *Pacific J. Math.*, **10**, 1211–1220.

[14] Fortét, R. (1973). *Ann. Inst. Henri Poincaré*, **9**, 41–58.

[15] Fortét, R. and Mourier, E. (1955). *Studia Math.*, **15**, 62–79.

[16] Gihman, I. I. and Skorohod, A. V. (1966). *Russ. Math. Surv.* (*Uspehi Mat. Nauk.*), **21**, 83–156.

[17] Gihman, I. I. and Skorohod, A. V. (1974). *The Theory of Stochastic Processes*, Vol. 1. Springer-Verlag, New York. (Basic material on probability on metric spaces; many results on absolute continuity; also has extensive treatment of weak convergence of probability measures, and applications to convergence of stochastic processes.)

[18] Grenander, U. (1950). *Ark. Mat.*, **1**, 195–277.

[19] Grigelonis, B. (1973). *Lect. Notes Math.*, **330**, 80–94.

[20] Hájek, J. (1958). *Czech. Math. J.*, **8**, 460–463.

[21] Hájek, J. (1958). *Czech. Math. J.*, **8**, 610–618.

[22] Hájek, J. (1962). *Czech. Math. J.*, **12**, 404–444.

[23] Hewitt, E. and Stromberg, K. (1965). *Real and Abstract Analysis*. Springer-Verlag, New York.

[24] Hitsuda, M. (1968). *Osaka J. Math.*, **5**, 299–312.

[25] Jacod, J. and Mémin, J. (1976). *Zeit Wahrscheinlichkeitsth.*, **35**, 1–37.

[26] Kabanov, Yu. M., Liptser, R. S., and Shiryayev, A. N. (1973). *Lect. Notes Math.*, **550**, 80–94.

[27] Kadota, T. T. and Shepp, L. A. (1970). *Zeit. Wahrscheinlichkeitsth.*, **16**, 250–260.

[28] Kailath, T. and Zakai, M. (1971). *Ann. Math. Statist.*, **42**, 130–140.

[29] Kakutani, S. (1948). *Ann. Math.*, **49**, 214–224.

[30] Kallianpur, G. and Oodaira, H. (1963). In *Time Series Analysis* (Proc. 1962 Symp.), M. Rosenblatt, ed. Wiley, New York, pp. 279–291.

[31] Kuelbs, J., ed. (1978). *Probability on Banach Spaces*. Marcel Dekker, New York. (Five papers in areas of central limit theorem, Gaussian measures, martingales.)

[32] Kunita, H. (1976). *Lect. Notes Math.* **511**, 44–77.

[33] Kuo, H.-H. (1975). Gaussian Measures in Banach Spaces. *Lect. Notes Math.*, **463**. (Basic material on Gaussian measures on Hilbert space, including absolute continuity; also contains results on abstract Wiener spaces for Gaussian measures on Banach space.)

[34] Mourier, E. (1953). *Ann. Inst. Henri Poincaré*, **13**, 161–244.

[35] Neveu, J. (1968). *Processus aléatoires gaussiens*. University of Montreal Press, Montreal. (Results on Gaussian processes, including absolute continuity. Extensive bibliography.)

[36] Orey, S. (1974). *Trans. Amer. Math. Soc.*, **193**, 413–426.

[37] Pan Yi-Min (1972). *Select. Transl. Math. Statist. Prob.*, **12**, 109–118; translation of *Shuxue Jinzhan*, **9**, 85–90 (1966).

[38] Parthasarathy, K. R. (1967). *Probability Measures on Metric Spaces*. Academic Press, New York. (Excellent collection of results available through the mid-1960s in various areas of probability on metric spaces (not including absolute continuity), plus foundations.)

[39] Parzen, E. (1963). In *Time Series Analysis* (Proc. Symp. 1962), M. Rosenblatt, ed. Wiley, New York, pp. 155–169.

[40] Pisarenko, V. (1961). *Radio Eng. Electron.*, **6**, 51–72.

[41] Pisarenko, V. (1965). *Theory Prob. Appl.*, **10**, 299–303.

[42] Pitcher, T. S. (1963). *Trans. Amer. Math. Soc.*, **108**, 538–546.

[43] Prohorov, Yu. V. (1956). *Theory Prob. Appl.*, **1**, 157–214.

[44] Rao, C. R. and Varadarajan, V. S. (1963). *Sankhya A*, **25**, 303–330.

[45] Root, W. L. (1963). In *Time Series Analysis* (Proc. 1962 Symp.), M. Rosenblatt, ed. Wiley, New York, 292–346.

[46] Rosanov, Yu. V. (1962). *Theory Prob. Appl.*, **7**, 82–87.

[47] Rosanov, Yu. V. (1971). Infinite-Dimensional Gaussian Distributions. *Proc. Steklov Inst. Math.*, No. 108. [Monograph containing results on absolute continuity, measurable linear functionals (including a zero–one law for linear manifolds). Extensive bibliography of papers on absolute continuity published prior to 1968.]

[48] Sato, H. (1967). *J. Math. Soc. Japan*, **19**, 159–172.

[49] Sazanov, V. V. (1958). *Theory Prob. Appl.*, **3**, 201–205.

[50] Segal, I. E. (1958). *Trans. Amer. Math. Soc.*, **88**, 12–41.

[51] Shepp, L. A. (1966). *Ann. Math. Statist.*, **37**, 321–354.

[52] Skorohod, A. V. (1970). *Theory Prob. Appl.*, **15**, 557–580.

[53] Skorohod, A. V. (1974). *Integration in Hilbert Space*. Springer-Verlag, New York. (Basic results concerning measures on Hilbert space; many results on absolute continuity.)

[54] Varberg, D. E. (1964). *Trans. Amer. Math. Soc.*, **113**, 262–273.

[55] Varberg, D. E. (1966). *Notices Amer. Math. Soc.*, **13**, 254.

(COMMUNICATION THEORY
GAUSSIAN PROCESSES
LIKELIHOOD RATIO
MARTINGALES
MEASURE THEORY
PROBABILITY
STOCHASTIC PROCESSES)

CHARLES R. BAKER

ABSOLUTE DEVIATION

The numerical value of the difference between two quantities regardless of its sign. If $\hat{\theta}$ is an estimate of θ, its absolute deviation from θ is $|\theta - \hat{\theta}|$.

(CHEBYSHEV'S INEQUALITY
MEAN DEVIATION
MOMENTS)

ABSOLUTE MOMENT

The expected value of the modulus (absolute value) of a random variable X, raised to power r is its rth absolute (crude) moment

$$\nu'_r = E[|X|^r].$$

The quantity

$$\nu_r = E[|X - E[X]|^r]$$

is the rth absolute central moment of X.

ν_1 is the mean deviation*.

ABSORBING MARKOV CHAINS

An absorbing Markov chain is that which contains one or more absorbing states, i.e., states for which the probability of leaving is zero. For any state of the chain there should be a nonzero probability of moving to one or more (although not necessarily all) of the absorbing states.

Bibliography

Bharucha-Reid, A. T. (1960). *Elements of the Theory of Markov Processes and Their Applications*. McGraw-Hill, New York.

Isaacson, D. L. and Madsen, R. W. (1976). *Markov Chains: Theory and Applications*. Wiley, New York.

Karlin, S. (1966). *A First Course in Stochastic Processes*. Academic Press, New York.

(MARKOV PROCESSES)

ACCELERATED LIFE TESTING *See* LIFE TESTING

ACCEPTABLE QUALITY LEVEL (AQL)

This is usually defined as the maximum percent defective (or the maximum number of defects per 100 units) that can be considered satisfactory for a process average.

Bibliography

Brownlee, K. A. (1965). *Statistical Theory and Methodology in Science and Engineering*, 2nd ed. Wiley, New York.

Johnson, N. L. and Leone, F. (1977). *Statistics and Experimental Design in Engineering and the Physical Sciences*, 2nd ed., Vol. 1. Wiley, New York, Chap. 10.

Juran, J. M., ed. (1964). *Quality-Control Handbook*, 2nd ed. McGraw-Hill, New York (first ed., 1953).

[ACCEPTANCE PROCESS LEVEL (APL)
LOT TOLERANCE PERCENT
DEFECTIVE (LTPD)
QUALITY CONTROL]

ACCEPTANCE ERROR

A term used in the theory of testing hypotheses* to denote a decision to accept a hypothesis H_0 when that hypothesis is not valid. It is also called a Type I error*.

(HYPOTHESIS TESTING
SIGNIFICANCE LEVEL)

ACCEPTANCE NUMBER

Given a sampling plan*, the acceptance number c denotes the maximum number of defective items that can be found in the sample without leading to rejection of the lot.

Bibliography

Duncan, A. J. (1974). *Quality Control and Industrial Statistics*, 4th ed. Richard D. Irwin, Homewood, Ill.

Standards Committee of ASQC* (1971). ASQC Standard A2 (rev. 1978).

[ACCEPTABLE QUALITY LEVEL (AQL)
CONSUMER'S RISK
QUALITY CONTROL
SAMPLING PLAN]

ACCEPTANCE PROCESS LEVEL (APL)

This is a fundamental notion in quality control. It is the process level* most remote from the standard that still yields product quality that we are willing to accept with high probability. Since most specifications are two-sided, (i.e., requiring characteristics to lie within a specified tolerance band, it is usually appropriate to specify both an upper and a lower APL.

Bibliography

Brownlee, K. A. (1965). *Statistical Theory and Methodology in Science and Engineering*, 2nd ed. Wiley, New York.

Duncan, A. J. (1974). *Quality Control and Industrial Statistics*, 4th ed. Richard D. Irwin, Homewood, Ill.

[QUALITY CONTROL
REJECTABLE PROCESS LEVEL (RPL)
SAMPLING PLAN]

ACCEPTANCE PROCESS ZONE

The band around the nominal value between the upper and lower acceptance process level (APL)* values.

[ACCEPTANCE PROCESS LEVEL (APL)]

ACCEPTANCE REGION (in Testing Hypotheses)

A hypothesis test* divides the space $\mathfrak{T}$ of a test statistic* T into two complementary regions, C (the critical region) and $\mathfrak{T} - C$. The region $\mathfrak{T} - C$ is called the acceptance region. This region is characterized by the property that if the *value* of the test statistic falls into this region the hypothesis under test is accepted.

(CRITICAL REGION
HYPOTHESIS TESTING)

ACCEPTANCE SAMPLING

The term "acceptance sampling" relates to the acceptance or rejection of a product or process on the basis of sampling inspection*. It has been pointed out that "sampling inspection is the process of evaluating the quality of material by inspecting some but not all of it" [4]. Its methods constitute decision rules for the disposition or sentencing of the product sampled. In this sense it may be contrasted with survey sampling*, the purpose of which is largely estimation*.

Sampling plans*, which specify sample size* and acceptance criteria, are fundamental to acceptance sampling. Such plans may be based on a simple dichotomous classification of conformance or nonconformance of a quality characteristic to specified criteria (attributes plans) or on a comparison of statistics computed from quantitative measurements to numerical criteria developed from the specifications and from assumptions about the shape and nature of the distribution of individual measurements (variables plans*). An example of the former is the attributes plan: sample 50 items and accept the lot of material from which the sample was taken if two or fewer items are found nonconforming; reject otherwise. An example of the latter is the variables plan: sample 12 items and accept the lot if the sample mean is more than 2 standard deviations above the lower specification limit; reject otherwise.

When a process parameter such as the mean* or standard deviation* is specified, the sampling plan resolves itself simply into a test of hypothesis. *See* HYPOTHESIS TESTING. That is, the sampling plan might be a *t*-test* if the plan is imposed to assure that the process mean conforms to a specified value. These tests are called variables plans for process parameter and are commonly used in the sampling of bulk materials. *See* BULK SAMPLING. More complicated situations arise when it is necessary to test the proportion of product beyond or between specification limits through the use of a measurement criterion. Such tests are referred to as variables plans for proportion nonconforming.

The operating characteristic (OC) curve* (complement of the power* curve) is the primary measure of performance of a sampling plan. It shows the probability of acceptance as a function of the value of the quality characteristic. As such, it provides a description of the protection afforded by the plan as well as a vehicle for comparison of various acceptance sampling plans and procedures. Two types of OC curves are distinguished. The type A operating characteristic curve relates to the inspection of individual lots of product and shows lot probability of acceptance as a function of lot quality*. In attributes inspection, it is computed from the hypergeometric distribution*. The type B operating characteristic curve relates to the process that produced the product to be inspected and shows the proportion of lots accepted in a continuing series as a function of the process average. In attributes inspection, it is computed either from the binomial distribution* when inspection is for proportion nonconforming, or from the Poisson distribution* when inspection is for nonconformities per unit. Details of the nature and construction of type A and type B operating characteristic curves are given in ref. 2.

Often the performance of a plan is characterized by two points on the OC curve: a producer's quality level with high probability of acceptance and a consumer's quality level with low probability of acceptance. The corresponding risks are called the producer's risk* (of rejection) and the consumer's risk* (of acceptance). (The producer's risk is conventionally taken to be 0.05 and the consumer's risk is 0.10.) The ratio of the consumer's quality level to the producer's quality level is called the operating (or discrimination) ratio. It describes the steepness of the OC curve and hence the capability of the plan to distinguish between acceptable and unacceptable quality. So-called two-point plans* can be derived from the producer's and consumer's quality levels through their operating ratios. A third point on the OC curve that is commonly referenced is the indifference quality level*, at which the probability of acceptance is 0.50. Sets of plans have been developed using the indifference quality level and the relative slope of the OC curve at that point.

Acceptance sampling procedures progress far beyond the simple single sampling plan to other, more complex procedures. Double sampling plans* allow the possibility of two samples, a second sample being taken if the results of the first sample are not sufficiently definitive. This concept can be extended to multiple sampling plans, involving more than two samples. Sequential procedures* are applied in acceptance sampling to achieve the excellent discrimination and economy of sample size associated with such methods. These plans may be used in both attributes and variables inspection.

Various measures and associated curves

have been developed to describe the properties of sampling plans. The average sample number (ASN)* curve describes the average sample size for various quality levels when using double*, multiple*, sequential*, or other procedures. The average outgoing quality (AOQ)* curve shows the average proportion of nonconforming product the consumer will receive if rejected lots are 100% inspected plotted against quality level. Its maximum is called the average outgoing quality limit (AOQL)*. Under such a procedure, the average total inspection curve shows the total number of units inspected in both sampling and 100% inspection and can be used to estimate and compare inspection loads.

For a continuing sequence of lots sampling plans may be combined into sampling schemes consisting of two or more plans used together with switching rules that establish the procedure for moving from one of the plans to another. Schemes can be constructed to give protection for both the producer and the consumer which is superior to that of the constituent plans with a reduction of sample size for the protection afforded. Such schemes are usually specified by an acceptable quality level (AQL)* which, when exceeded, will eventually lead to a switch to a tighter plan with consequent economic and psychological pressure on the supplier to improve the quality of the product. Sampling schemes have their own OC curves. Sometimes options for discontinuation of inspection are incorporated into the procedure. Sampling schemes may be combined into sampling systems that select specific schemes by prescribed rules. The most important sampling systems are military standards MIL-STD-105D for attributes and MIL-STD-414 for variables. Their civilian counterparts are ANSI Z1.4 and ANSI Z1.9 in the United States and international standards ISO 2859 and ISO 3951, respectively.

The variety of approaches in acceptance sampling is almost limitless. Continuous sampling plans are used on streams of output where lots are difficult to define. Chain sampling* plans link the criteria for the immediate sampling plan to past results. Grand lot sampling procedures combine samples from lots that have been shown to be homogeneous, to achieve larger sample size and greater discrimination. Skip-lot plans provide for the inspection of only a fraction of the lots submitted. Acceptance control charts* can be used to visually portray the results of inspection through the medium of the control chart. Bayesian plans introduce economic considerations and prior results and estimates into the sampling equation. Special plans have been developed for various areas of application, such as compliance testing, reliability and life testing, and safety inspection. *See* SAMPLING PLANS.

The application and administration of sampling inspection demands a broad range of knowledge of statistical methodology, because the determination of what, where, when, how and to what quality levels the inspection is to be carried out is largely empirical. Acceptance sampling procedures are an integral part of quality control practice and serve to distinguish between quality levels as a vehicle for quality improvement. In application, however, they should, where possible, be supplemented and eventually supplanted by statistical techniques for process quality control (such as control charts) in an effort to *prevent* the occurrence of nonconforming material rather than to *detect* it after it has been produced.

The history and development of acceptance sampling is described in detail by Dodge [1], who originated and developed many acceptance sampling procedures. The statistical methodology of acceptance sampling has been treated specifically by Schilling [5] and in the context of industrial statistics as a whole by Duncan [3].

References

[1] Dodge, H. F. (1969). *J. Quality Tech.*, **1**: Part I, Apr., 77–88; Part II, July, 155–162; Part III, Oct., 225–232; **2**: Part IV, Jan., 1970, 1–8.

[2] Dodge, H. F. and Romig, H. G. (1959). *Sampling Inspection Tables—Single and Double Sampling*, 2nd ed. Wiley, New York.

[3] Duncan, A. J. (1974). *Quality Control and Industrial Statistics*, 4th ed. Richard D. Irwin, Homewood, Ill.

[4] Freeman, H. A., Friedman, M., Mosteller, F., and Wallis, W. A., eds. (1948). *Sampling Inspection Principles: Procedures and Tables for Single, Double and Sequential Sampling in Acceptance Inspection and Quality Control*. McGraw-Hill, New York.

[5] Schilling, E. G. (1981). *Acceptance Sampling in Quality Control*. Marcel Dekker, New York.

[AVERAGE OUTGOING QUALITY LIMIT (AOQL)
CONTROL CHARTS
DODGE–ROMIG LOT TOLERANCE TABLES
OC CURVES
QUALITY CONTROL
SAMPLING PLANS]

EDWARD G. SCHILLING

ACCURACY

Accuracy of an observation or a statistic derived from a number of observations has to do with how close the value of the statistic is to a supposed "true value." It is to be distinguished from precision*, which indicates the resolving power of a measuring device and is usually given by the number of decimal places reported in the measurements made with this device. In estimation theory accuracy measures how close the estimate $\hat{\theta}$ of parameter θ is to the "true value" of θ.

Accuracy can be measured arbitrarily: for example, by the mean square error* of $\hat{\theta}$; precision, on the other hand, might be measured by the variance of $\hat{\theta}$.

(DEVIATION
ERROR
MEAN DEVIATION
MEAN SQUARE ERROR
PRECISION
REPEATABILITY
REPRODUCIBILITY
VARIABILITY
VARIANCE)

ACCURACY (in Neyman's sense)

Let T be a statistic* based on a sample from a population having an unknown parameter θ, and let $(L_1(T), L_2(T))$ be a confidence interval* for θ. If

$$Q(\theta_0) = \Pr\left[L_1(T) \leqslant \theta_0 \leqslant L_2(T) | \theta \right],$$

then $Q(\cdot)$ is defined to be the Neyman accuracy of the confidence interval and is used in comparing two confidence intervals with equal confidence coefficients. The interval with the smaller Neyman accuracy is said to be more selective; *see* CONFIDENCE INTERVALS AND REGIONS and the bibliographic references and comments.

Bibliography

Ghosh, B. K. (1975). *J. Amer. Statist. Ass.*, **70**, 457–462. [Uses $Q(\cdot)$ as a criterion in comparing sequential interval estimators.]

Kendall, M. G. and Stuart, A. (1967). *The Advanced Theory of Statistics*, 2nd ed., Vol. 2. Hafner, New York. (Discussion in Secs. 20.19 and 20.20.)

Neyman, J. (1937). *Philos. Trans. R. Soc. A*, **236**, 333–380. (Original paper.)

Zacks, S. (1971). *The Theory of Statistical Inference*. Wiley, New York. (An advanced mathematical discussion appears in Sec. 10.1.)

C. B. READ

ACCURACY (in Wolfowitz's sense)

Let T be a statistic* based on a sample from a population having an unknown parameter θ, and let $(L_1(T), L_2(T))$ be a confidence interval for θ. *See* CONFIDENCE INTERVALS AND REGIONS. If

$$W(a,b) = aE\left\{(L_1(T) - \theta)^2\right\} + bE\left\{(L_2(T) - \theta)^2\right\},$$

then $W(\cdot, \cdot)$ is defined to be the Wolfowitz accuracy of the confidence interval, and measures how close the confidence limits L_1 and L_2 are to the true value of θ; see the bibliographic references and comments.

Bibliography

Ghosh, B. K. (1975). *J. Amer. Statist. Ass.*, **70**, 457–462. [Uses $W(\cdot, \cdot)$ as a criterion in comparing sequential interval estimators.]

Harter, H. L. (1964). *J. Amer. Statist. Ass.*, **59**, 1133–1140. (The efficiency of competing confidence intervals is measured, inter alia, by the ratio of their Wolfowitz accuracies when $a = b = 1$.)

Wolfowitz, J. (1950). *Ann. Math. Statist.*, **21**, 218–230. (Original paper in which the concept is developed.)

C. B. Read

ACTUARIAL STATISTICS—LIFE

The statistics used by actuaries in evaluating risks, calculating liabilities, and planning the financial course of insurance, pensions, and similar arrangements are frequently referred to as actuarial statistics. Such statistics relate to fundamental notions of risk* and risk transfer*. Life actuarial statistics are those developed for life insurance and pensions; they generally pertain to long-term risks and are commonly used in projection of long-range results. Nonlife actuarial statistics are those developed for casualty insurance where risks are usually short term and often subject to wide fluctuations.

Life actuarial statistics center on the calculation of mortality*, morbidity and associated decrements, with the data base obtained for the most part from life insurance, health insurance, and pensions experience. Such statistics are used to describe changes in insured populations; determine the exposed to risk; and compile mortality, morbidity, and other decremental rates, such as withdrawal rates. They take into consideration the effects of selection exercised by insurers and by individuals seeking coverage. The life table* is the principal actuarial tool for evaluating the consequences of different mortality, morbidity, and other decremental rates as well as various forms of selection.

In addition to mortality and morbidity statistics, actuaries make use of statistics relating to the rates of return on investments and to the rates of expense involved in conducting life, health, or pension programs. The latter two basic elements in life insurance and pensions are less tractable to forecasting than are mortality rates, and past experience is less of a guide to their future course. Accordingly, in financial projections involving these elements actuaries rely less on statistics of past experience and more on judgmental considerations of the margins necessary for unanticipated contingencies in the future. It is noteworthy, however, that the rationale implicit in life table* analysis lends itself admirably to the study of past experience under different forms of investment and it has been so used to advantage.

Compilation of mortality rates has long been a primary concern of actuaries. The first person to bear the title of actuary, William Morgan (1775) [21], systematically assembled and examined the mortality experience of the Equitable Life Assurance Society of London and noted the lower death rates after examination for life insurance, which he correctly ascribed to the effects of selection by the insurer. The first government actuary in Great Britain, John Finlaison (1822) [21], analyzed the favorable mortality among annuitants in order to insure the solvency of the National Debt Office. Benjamin Gompertz, actuary of the Alliance of London, reasoning on physiological grounds, advanced in 1825 [8] perhaps the most discerning mathematical formulation of changes in mortality with age, which now bears his name; the Gompertz model* for the intensity of mortality at the precise moment of age X, known as the force of mortality*, is $\mu_x = Bc^X$, and this formula implies a geometric progression of death rates with age, which is intended to represent the deterioration in the ability of human beings to withstand death. Some 35 years later, William Makeham [18] proposed to add to the geometric component of the force of mortality postulated by Gompertz a constant component intended to reflect the causes of death due to chance*.

Both the Gompertz and the Makeham* formulas produce excessively high mortality rates at ages over 90, probably because the population surviving to these ages is intrinsically heterogeneous. A more closely fitting formula for the force of mortality

$$\mu_X = \frac{A + Bc^X}{1 + Dc^X}$$

was developed by Wilfred Perks in 1932 [23].

Heligman and Pollard [10] have fitted the curve

$$q_x p_x = A^{(x+B)^c} + De^{-E(\ln x - \ln F)^2} + GH^x$$

to Australian national mortality (1946–1948, 1960–1962, and 1970–1972) with some success.

These mathematical models of mortality are deterministic in form, having been conceived in the classical actuarial approach, which treats death rates as average or expected values.

Brillinger has demonstrated that these formulas can be derived from statistical theory, as has been done in life testing* [5]. According to Brillinger, if the human body be regarded as made up of many components whose lifetimes* are independently and identically distributed and if a human being dies at the first failure of a component, then the force of mortality* will be approximately one of the following:

1. Bc^X
2. $A/(B - X)^{c+1}$
3. $H/(X - B)^{c-1}$

Although these formulas assume independence and identical distribution of the many components comprising the human body, they hold up reasonably well for substantial deviations from these assumptions.

Brillinger suggests that the most general formula for the force of mortality under his assumptions would be in the form

$$\sum_i H_i(X - B_i)^{c_i - 1} + \sum_j A_j/(B_j - X)^{c_j+1} + \sum_k E_k d_k^X.$$

Stochastic theories of mortality which bring in not only assumptions about physiological processes but also their random variability have been proposed by Sacher and Trucco (1962) [26] and Beard (1964) [2]. A comprehensive stochastic theory of life contingencies, beginning with a random variable of time until death of a person at age X, is in the making [12]. The Committee on the Theory of Risk of the Society of Actuaries has pioneered in studies of the fluctuations in mortality (and other elements) as they affect the conduct of the life insurance business [31].

The classical actuarial approach is well illustrated by the construction of a life table*. For this purpose the average or expected value of the probability of a person at age x dying within a 1-year interval q_x is calculated for each successive year of age to be shown in the life table. The value of q_x at each age is commonly referred to as the annual mortality rate or simply as the mortality rate at age x. Fundamentally, however, the focus is on the annualized force of mortality $\mu_x = -(dl_x/dx)\log x$, where l_x is the population at exact age x. Because of the many changes in a population of insured lives or pensioners over a 1-year interval, it is necessary to make appropriate assumptions about the shape of the mortality curve over this interval. Among the assumptions put forward are (1) a uniform distribution* of deaths, (2) a constant force of mortality* [29], and (3) the assumption due to Balducci [9] that

$$1 - t q_{x+t} = (1 - t) q_x \qquad (0 \leqslant t \leqslant 1).$$

Greville has advanced a method for estimating q_x which does not require the concept of exposed to risk [9], but relies solely on the observed changes in a population over the 1-year interval. This is similar in concept to the Kaplan–Meier* product limit [16] widely used by biometricians but does not involve a limit.

Neither Greville's product method nor the Kaplan–Meier product limit are practical for mortality investigations of insured lives or pensioners, because they would be too laborious considering the great number of subintervals within each year interval and the large volume of data spread over many years. The concept of exposed to risk simplifies calculations of mortality rates and either the Balducci assumption or the constant force of mortality over the 1-year interval may be used to facilitate determination of the exposed to risk. There is extensive actuarial literature on methods of computing the exposed to risk [1].

The actuarial formulas for exposed to risk* used in deriving death rates are readily generalized to multiple decrements.

The assumption of a constant force of mortality over a 1-year interval, explored by F. A. Weck, has great advantages when two or more decrements are involved; the basic formula remains unchanged irrespective of the number of decrements, the interval of exposure, or the age range considered.

Multiple-decrement tables* serve many purposes but primarily as a model for demonstrating the impact on a population of several forces operating over finite intervals. The effects of selection can be regarded as a consequence of interacting decrements. For instance, in situations involving withdrawals associated with lower-than-average mortality, the selection on withdrawal produces increased mortality in the persisting population; in situations involving early retirements associated with higher-than-average mortality, the early retirements produce decreased mortality in the continuing population.

Multiple-decrement tables are commonly employed to chart the progress of life insurance and pension funds.

Virtually all of the corporate planning for life insurance requires financial projections that reflect the combined effects of mortality rates, withdrawal rates, expense rates, and rates of return on investment. Separate models may be necessary for agency and marketing operations and for special types of insurance such as those involving equity investments. The result is a network of interconnected multiple-decrement tables. Selected average values of the various rates involved are used to forecast the outcomes under low, medium, and high assumptions. However, to visualize the potential financial strains, it may be advisable to include random fluctuations, especially those in investment values. The Committee on the Theory of Risk of the Society of Actuaries has produced computerized multiple-decrement models of a stock life insurance company embodying a stochastic approach to the principal factors [24].

In pension programs with survivor benefits and under the Social Security System, it has been necessary to make long-range projections of the rates of retirement, rates of widowhood and remarriage, and rates of birth and orphanhood, pay scales, and variations in employment. Ever since the inception of the program, the responsibility for such projections within the Social Security System has rested in the Office of the Actuary (and its predecessor actuarial units) [22]. The actuarial studies of the Office of the Actuary of the Social Security Administration provide a series of increasingly sophisticated statistical compilations which in the past two decades have come to include disability, hospitalization, and cost of medical care.

After mortality or other decremental rates have been extracted from observations, it may be advisable to adjust (i.e., graduate) them so as to bring out their underlying character. Graduation* is desirable to reduce the effect of random fluctuations and also to provide a set of smooth values, showing a more regular progression age by age and duration by duration, to facilitate actuarial calculations [20]. A rate extracted from observations may be regarded as a sum of the true value of the rate and a random error due to sampling. Testing for nonrandom errors* is therefore in order before any smoothing is undertaken. Many types of graduation formulas have been proposed, which differ essentially in the weights given to smoothness and goodness of fit*. Smoothness is customarily expressed as the sum of the third or higher differences of the graduated rates; goodness of fit is measured by the sum of the differences between the ungraduated and graduated rates. Graduation procedures range from the relatively simple, such as graphical adjustments and curve fitting*, to formulas based on interpolation* [11] and to still-more-sophisticated formulas based on difference* equations [30] or a Bayesian approach [17]. Graduation differs from other statistical estimation problems in that it depends on assumptions as to the true values of the rates which are not contained in the observational data. These outside assumptions may be regarded as the prior distribution* in Bayesian inference*.

The principal end products of compiling mortality, morbidity, and other decremental rates have been (1) life and multiple-decrement tables; (2) models of life insurance, health insurance, and pension programs; and (3) mortality investigations of numerous factors that influence mortality.

In the United Kingdom and the United States most national life tables have for many years been constructed by actuaries [3]. Models of life insurance companies are a more recent development; as previously mentioned, some are based on stochastic conceptions of mortality and other elements.

Medico-actuarial studies have probably received the most widespread attention in the scientific community and in the news media. In the United States these studies go back to the 1890s, when systematic efforts were first made to place the evaluation of life risks on a scientific basis. Comprehensive investigations of the mortality among insured lives began with the Specialized Mortality Investigation of 1901 [27], which covered the experience for a number of occupations, physical impairments, and disabilities as well as family histories, over a 30-year period. The Medico-Actuarial Investigation, completed in 1918, dealt with the experience over the period 1885–1909 on a much larger scale [19]; it produced a mass of quantitative data bearing on the long-term risks among overweight men, men with histories of various common diseases, and men in hazardous occupations.

The Impairment Study of 1929 [13] (and its 1931 supplement [28]) covered a 20-year period and updated the extra mortality associated with a wide variety of physical impairments, including elevated blood pressure. Further mortality studies of physical impairments followed in 1936 [14], 1938 [4], and 1951 [15]. The Build and Blood Pressure Study of 1959 reported on the experience among $4\frac{1}{2}$ million insured lives traced for periods up to 22 years. Nearing completion at this time is a sequel to this study [7] that carries forward the experience among $4\frac{1}{2}$ million insured lives from 1954 to 1972. Another mortality investigation of insured lives according to physical impairments is underway. These investigations have all aimed to isolate and measure the effects of specific risk factors on mortality over a long period of time. The hypothesis that the total mortality risk can reasonably be represented by a linear compound of a number of independent risk factors (or combinations thereof) has been tested and found to be extremely useful for insurance purposes. The findings of the medico-actuarial studies have yielded unique information for public health and medicine, particularly as to the long-term hazards of overweight and hypertension.

The Society of Actuaries (and its predecessor bodies, notably the Actuarial Society of America) have played a leading role in the accumulation of a comprehensive body of knowledge on mortality* and morbidity. The Reports Issues of the Transactions of the Society of Actuaries* [25] provide a continuing flow of information on current mortality and morbidity among insured lives. These reports also include data on mortality of annuitants and pensioners, mortality among persons covered by group life insurance in many industries, and morbidity under various forms of health insurance.

References

[1] Batten, R. W. (1978). *Mortality Table Construction*. Prentice-Hall, Englewood Cliffs, N. J. (Basic textbook on mortality table construction.)

[2] Beard, R. E. (1964). *Trans. XVI Int. Congr. Actuaries*, London/Edinburgh, p. 463.

[3] Benjamin, B. and Haycocks, H. W. (1970). *The Analysis of Mortality and Other Actuarial Statistics*. Cambridge University Press, Cambridge. (Basic text on mortality analysis.)

[4] *Blood Pressure Study 1938* (1938). Society of Actuaries, Chicago.

[5] Brillinger, R. D. (1961). *Amer. Soc. Actuaries*, **13**, 116. (Excellent discussion of the link between statistical theory and actuarial models of mortality.)

[6] *Build and Blood Pressure Study 1959* (1960). Society of Actuaries, Chicago. (Standard for mortality investigations of build and blood pressure.)

[7] *Build and Blood Pressure Study 1979* (1980). Society of Actuaries, Chicago.

[8] Gompertz, B. (1825). *Philos. Trans. R. Soc. Lond.*, **115**, 513. (Nineteenth-century approach to mortality laws.)

[9] Greville, T. N. E. (1978). *Actuarial Res. Clearing House (ARCH), No. 2*, p. 4. (Elaboration of fundamental examination of concepts involved in mortality rates.)

[10] Heligman, L. and Pollard, J. H. (1980). *J. Inst. Actuaries*, **107**, 49.

[11] Henderson, R. (1938). *Mathematical Theory of Graduation*. Actuarial Society of America, Chicago. (Advanced text on graduation by interpolation and summation.)

[12] Hickman, J. C. (1964). *Trans. Soc. Actuaries*, **14**, 1. (Derivation of basic actuarial formulas for premiums and reserves in terms of statistical theory.)

[13] *Impairment Study 1929* (1929). Society of Actuaries, Chicago.

[14] *Impairment Study 1936* (1936). Society of Actuaries, Chicago.

[15] *Impairment Study 1951* (1954). Society of Actuaries, Chicago. (Standard for medico-actuarial studies.)

[16] Kaplan, E. L. and Meier, P. (1958). *J. Amer. Statist. Ass.*, **53**, 457. (Approach to estimates of survival from statistical theory.)

[17] Kimeldorf, G. S. and Jones, D. A. (1967). *Trans. Amer. Soc. Actuaries*, **17**, 33. (Bayesian approach to graduation.)

[18] Makeham, W. M. (1860). *J. Inst. Actuaries*, **8**, 301.

[19] *Medico-Actuarial Mortality Investigation* (1918). Society of Actuaries, Chicago.

[20] Miller, M. D. (1946). *Elements of Graduation*. Society of Actuaries, Chicago. (Introductory text on graduation.)

[21] Mitchell, R. B. (1974). *From Actuarius to Actuary*. Society of Actuaries, Chicago. (Essentially, a history of actuarial science.)

[22] Myers, R. J. (1975). *Trans. Amer. Soc. Actuaries*, **37**, 155.

[23] Perks. W. (1953). *J. Inst. Actuaries*, **63**, 12.

[24] Report of Committee on Theory of Risk of the Society of Actuaries on Study Note on Provision of the Risks of Adverse Deviation, June 1974. (Basic study note on risk theory for actuaries.)

[25] Reports issues of the *Transactions of the Society of Actuaries*.

[26] Sacher, G. A. and Trucco, E. (1962). *Ann. N. Y. Acad. Sci.*, **96**, 985.

[27] *Specialized Mortality Investigation* (1901). Society of Actuaries, Chicago.

[28] *Supplement to Medical Impairment Study 1929* (1931). Society of Actuaries, Chicago.

[29] Weck, F. A. (1947). *Rec. Amer. Inst. Actuaries*, **36**, 23. (Fundamental inquiry into assumptions in calculations of mortality rates.)

[30] Whittaker, E. T. and Robinson, G. (1944). *The Calculus of Observations*, 4th ed. Macmillan, New York. (Mathematics underlying graduation by difference equations.)

[31] Wooddy, J. C. (1968). *Report on Mortality Fluctuations Study Based on Data Supplied by the Mortality Committee of the Society of Actuaries*. Society of Actuaries, Chicago.

(GRADUATION
LIFE INSURANCE
LIFE TABLES
MORTALITY
MULTIPLE-DECREMENT TABLES
RISK ANALYSIS
RISK TRANSFER
SOCIAL INSURANCE
STOCHASTIC THEORY OF RISK)

EDWARD A. LEW

ADAPTIVE METHODS

In adaptive statistical inference, we use the sample to help us select the appropriate type of statistical procedure needed for the situation under consideration. For a simple illustration of this, say that we use the sample kurtosis* K as a selector statistic [3]. One adaptive point estimator, T, for the center of a distribution would be

$$T = \begin{cases} \text{midrange*}, & K \leqslant 2, \\ \text{arithmetic mean*}, & 2 < K < 5, \\ \text{median*}, & 5 \leqslant K. \end{cases}$$

That is, if the sample looks as if it arises from a short-tailed distribution, the average of the largest and smallest items of the sample is used as our estimator. If it looks like a long-tailed situation, the median is used. Otherwise, our estimate is the arithmetic mean (average) $\bar{x}$.

To generalize this illustration somewhat, suppose that we have a whole family (not necessarily finite) of possible distributions. Within this family of distributions, take a few representative ones, say $F_1, F_2, \ldots, F_k$. Now, for each of these k distributions, suppose that we can find a good statistic to make the inference under consideration. Let us say that these respective statistics are $T_1, T_2, \ldots, T_k$. We observe the sample from a distribution; and with a selector statistic, say Q, we determine which one of $F_1, F_2, \ldots, F_k$ seems closest to the underlying distribution from which the sample arises. If Q suggests that we have been sampling from F_i, then we would use the statistic T_i; or if Q suggests that we might be someplace between F_i and F_j, then we could use a combination of T_i and T_j; or more generally, Q could dictate a statistic that is a linear combination of all the statistics, $T_1, T_2, \ldots, T_k$: let us say

$$T = \sum_{i=1}^{k} W_i T_i, \qquad \sum_{i=1}^{k} W_i = 1,$$

where the weights $W_1, W_2, \ldots, W_k$ are functions of the statistic Q. If it looks more like the sample arises from F_i, then, of course, the weight W_i would be large.

Consider a very simple example in which we are trying to choose the best of three types of concrete [10]. The compression strengths were tested after bars of each type had been exposed to severe environmental conditions for a period of 1 year. Seven ($n = 7$) observations were taken for each type of cement, where the observations are the breaking strengths of the bars measured in pounds per square inch. Let us denote the order statistics* of a sample by $y_1 \leqslant y_2 \leqslant \cdots \leqslant y_7$. However, since we do not know from what underlying distribution these arose, we choose three representative distributions: the short-tailed uniform* using the midrange $(y_1 + y_7)/2$ as an estimate of center, the normal* using the average $\bar{x}$ as the estimate, and the long-tailed double exponential* with a modified median $(3y_3 + 8y_4 + 3y_5)/14$ as the statistic. These statistics were computed for each of the three samples and are given in Table 1 together with the original data.

It is interesting to note that using the midrange or median statistics, concrete B looks to be the best, whereas $\bar{x}$ suggests concrete C. Accordingly, a selector statistic

Table 1

Concrete	A	B	C
Ordered	5060	5625	4880
Observations	5398	6020	6030
	5820	6270	6290
	6131	6636	6372
	6400	6880	6920
	7527	7337	8320
	7560	8170	8581
Midrange	6310.0	6897.5	6730.5
Mean	6270.86	6705.43	6770.43
Modified median	6122.00	6609.86	6471.86

is needed, and we use

$$Q = \frac{(y_7 - y_1)/2}{\sum |y_i - M|/7},$$

the ratio of one-half of the range divided by the mean deviation* from the sample median M. (Q is defined somewhat differently when $n > 20$.) The average of the three Q values is computed to obtain $\bar{Q} = 1.876$. The midrange, average, or median is selected respectively, according to whether $\bar{Q}$ falls below, in between, or above

$$2.08 - (2/n) \quad \text{and} \quad 2.96 - (5.5/n);$$

that is, with $n = 7$, 1.794 and 2.174. (The formulas for these cutoffs have been determined empirically.) Since $\bar{Q} = 1.876$, it seems as if the distribution has fairly normal tails; thus the statistic $\bar{x}$ chooses concrete C as the best.

We must understand, however, that the inference under consideration is not necessarily a point estimate in the general situation. We could be considering a confidence interval* or a test of hypothesis*. Moreover, making an inference in this manner, that is, selecting the underlying distribution and then making the inference from the same data, can certainly destroy certain probabilities that are of interest in statistics. For example, if we are constructing a nominal 95% confidence interval, we can actually spoil the confidence coefficient* by such a procedure, so that it might actually be 0.80 or even 0.70. Or if we are making a test of a statistical hypothesis, the significant level might not be $\alpha = 0.05$, but 0.15 or 0.25. Despite this fact, however, the adaptive idea is useful in good data analysis; therefore, it is necessary for us to adjust our theories to the applications. That is, we want our theories to support the applications, not oppose them.

This forces us to look at some of the difficulties associated with the corresponding sampling distribution theory. Let us say that θ is the location parameter* and we are interested in testing the hypothesis $H_0 : \theta = \theta_0$ against the hypothesis $H_1 : \theta > \theta_0$. Again, suppose that we have a family of distributions for which θ is the location parameter of each member of the family. If we are sampling from F_i, say, we would reject the hypothesis H_0 and accept the alternative hypothesis H_1 if some statistic, say Z_i, was greater than or equal to c_i; i.e., $Z_i \geqslant c_i$, $i = 1, 2, \ldots, k$. Therefore, our adaptive test might be something like this: reject H_0 and accept H_1 if

$$Z = \sum_{i=1}^{k} W_i Z_i \geqslant c,$$

where $W_1, W_2, \ldots, W_k$ are functions of some selector statistic, say Q. The significance level of the test is then

$$\Pr\left| \sum_{i=1}^{k} W_i Z_i \geqslant c \,|\, H_0 \right|.$$

This probability is difficult to compute, so let us first consider a special and easier situation in which each of the W's is equal to 0 or 1. Of course, only one W_i can equal 1, and the rest must equal 0. Thus if Q suggests that F_i is the underlying distribution, then we will use Z_i. That is, if $Q \in R_i$, where $R_1, R_2, \ldots, R_k$ are appropriate mutually exclusive* and exhaustive* sets, we will select Z_i for the test statistic. Under these conditions, the significance level would be

$$\sum_{i=1}^{k} \Pr[\, Q \in R_i \quad \text{and} \quad Z_i \geqslant c_i \,|\, H_0].$$

If each of the individual tests is made at the 0.05 significance level, it has been observed in practice that this significance level is frequently somewhat larger than that nominal significance level of 0.05.

There is a certain desirable element of model building in this entire procedure; that is, we observe the data and select the model that seems appropriate, and then we make the statistical inference* for the situation under consideration. However, there can be some cheating in doing this; that is, if we construct the model from given data and then make a test of hypothesis using those data, our nominal significance level is not necessarily the correct one. Moreover, even some researchers carry this to an extreme by selecting the test procedure that favors what they want (usually rejection of the null hypothesis*). They might then quote a significance level of 0.05, while the real α, for the overall selection and testing procedure might be higher than 0.25.

There is a method, however, of "legalizing" this cheating. Suppose that the selector statistic Q and each Z_i are independent under the null hypothesis H_0. Then the significance level is

$$\sum_{i=1}^{k} \Pr[\, Q \in R_i \quad \text{and} \quad Z_i \geqslant c_i \,|\, H_0]$$

$$= \sum_{i=1}^{k} \Pr[\, Q \in R_i \,|\, H_0] \Pr[\, Z_i \geqslant c_i \,|\, H_0]$$

$$= \alpha \sum_{i=1}^{k} \Pr[\, Q \in R_i \,|\, H_0] = \alpha,$$

provided that each individual test is made at the nominal significance level α. That is, this common significance level α is exactly the same as the overall significance level. The important feature of this is to make certain that the selector statistic Q is independent of the test statistic. One elegant way of achieving this is through distribution-free (nonparametric) methods* [5].

To illustrate the beauty of the nonparametric methods in these situations, let us consider the two-sample problem. Suppose that we have two independent continuous-type distributions, F and G. The null hypothesis H_0 is the equality of the two corresponding functions. Say that the sample X_1,

$X_2, \ldots, X_m$ arises from F, and the sample $Y_1, Y_2, \ldots, Y_n$ arises from G. We suggest three nonparametric statistics that can be used to test this null hypothesis [2]. The first, Tukey's quick test*, is used when the underlying distributions have short tails, like those of the uniform distribution*. Tukey's statistic is

$$T_1 = (\# Y\text{'s} > \text{largest } X) + (\# X\text{'s} < \text{smallest } Y).$$

A large T_1 would suggest the alternative hypothesis H_1 that the Y's tend to be larger than the X's. Thus we reject H_0 and accept H_1 if T_1 is greater than or equal to c_1, where

$$\Pr[T_1 \geqslant c_1 \mid H_0] = \alpha.$$

The second statistic T_2 is that of Mann, Whitney, and Wilcoxon. This statistic is a good one in case the underlying distributions have middle tails, like those of the normal* or logistic* distributions. After combining the two samples, we determine the ranks of the Y's in the combined sample; say those ranks are $R_1, R_2, \ldots, R_n$. One form of the Mann–Whitney–Wilcoxon statistic* is

$$T_2 = \sum_{i=1}^{n} R_i.$$

Now we reject H_0 and accept H_1 if T_2 is greater than or equal to c_2, where

$$\Pr[T_2 \geqslant c_2 \mid H_0] = \alpha.$$

The third statistic is that associated with the median test. It is

$T_3 = \# Y$'s greater than the combined sample median. We reject H_0 if that statistic, T_3, is greater than or equal to c_3, where

$$\Pr[T_3 \geqslant c_3 \mid H_0] = \alpha.$$

Each of the probabilities denoted by α in these three tests does not depend on the form of the underlying continuous distribution, and sometimes these tests are called distribution-free tests.

For an example of each of these statistics, refer to data on the three types of concrete, and let the samples from A and B represent, respectively, the X and Y values with $m = n = 7$. The computed statistics are $T_1 = 3$, $T_2 = 59$, with $T_3 = 4$. Let us now consider an adaptive procedure that selects one of these three statistics. Considering the combined sample (i.e., the X's and Y's together) use a selector statistic, say Q, and decide whether we have short-tailed distributions, in which case we use the T_1 test; middle-tailed distributions, in which case we use the T_2 test; or long-tailed distributions, in which case we use the T_3 test. It turns out that the overall (selecting and testing) significance level will also equal α because each of T_1, T_2, and T_3 is independent of Q. The reason we have this independence under H_0 is that the order statistics of the combined sample are complete, sufficient statistics for the underlying "parameter," the common distribution $F = G$. Moreover, it is well known that the complete, sufficient statistics for $F = G$ are then independent of statistics whose distributions do not depend upon $F = G$, such as T_1, T_2, and T_3. However, the selector statistic Q is a function of the complete, sufficient statistics, and thus it is also independent of each of the statistics T_1, T_2, and T_3, under H_0. Incidentally, in our example, using the Q and $\bar{Q}$ associated with the illustration about concrete, the statistics T_2 for middle-tailed distributions would be selected and $T_2 = 59$ has a p-value of 0.288. Thus the null hypothesis would not be rejected at the significance level of $\alpha = 0.05$.

Although these nonparametric methods can be generalized to multivariate situations such as regression*, many statisticians do not find them extremely satisfactory in data analysis. Possibly the newer robust statistics show more promise in adaptation; some of them are "almost distribution-free" and lend themselves better to data analysis. Although it is impossible to give many details on robustness in this short article, the idea is illustrated with the trimmed mean.

Suppose that we are attempting to make an inference about the center θ of a symmetric distribution. Let $X_1 \leqslant X_2 \leqslant \cdots \leqslant X_n$ represent the items of a random sample, ordered according to magnitude. The

β-trimmed mean is

$$\bar{X}_\beta = \frac{1}{h} \sum_{i=g+1}^{n-g} X_i,$$

where β is usually selected so that $g = \eta\beta$ is an integer (otherwise, $g = [\eta\beta]$, the greatest integer in $\eta\beta$) and where $h = n - 2g$. Of course, $\bar{X}_{\beta=0} = \bar{X}$.

It is well known that

$$Z = \frac{\bar{X} - 0}{S/\sqrt{n-1}},$$

where $S^2 = \sum_{i=1}^{n}(X_i - \bar{X})^2/n$, has a t-distribution* with $n - 1$ degrees of freedom provided that the sample arises from a normal distribution. However, even though the underlying distribution is nonnormal (without really long tails), Z still has a distribution fairly close to this t-distribution. This is what we mean by "almost distribution-free." Now it is not so well known, but true [11], that

$$Z_\beta = \frac{\bar{X}_\beta - 0}{\sqrt{SS(\beta)/h(h-1)}},$$

where

$$\begin{aligned} SS(\beta) = (g+1)\left(X_{g+1} - \bar{X}_\beta\right)^2 \\ + \left(X_{g+2} - \bar{X}\right)^2 + \cdots \\ + \left(X_{n-g-1} - \bar{X}\right)^2 \\ + (g+1)\left(X_{n-g} - \bar{X}_\beta\right)^2, \end{aligned}$$

has an approximate t-distribution with $h - 1$ degrees of freedom for many underlying distributions, so that Z_β is almost distribution-free. Of course, $Z_{\beta=0} = Z$.

In an adaptive procedure using some Z_β to make an inference about θ, a selector statistic, such as the kurtosis K or Q, can be used to choose an appropriate β. This β will be larger for larger values of K and Q. In making inferences about θ based upon a selected Z_β, the overall confidence coefficient or the overall significance level will deviate somewhat from the nominal one. However, these deviations are not great; in many instances we have found that α equals something like 0.06 rather than the nominal $\alpha = 0.05$. Thus we can place great reliability on the level of the resulting inferences.

These adaptive and robust methods have been extended to multivariate situations and the interested reader is referred to some of the following articles and their references for further study. The future seems bright for adaptive methods, and these will bring applications and theory closer together.

References

[1] Andrews, D. F. et al. (1972). *Robust Estimates of Location*. Princeton University Press, Princeton, N. J.

[2] Conover, W. F. (1971). *Practical Nonparametric Statistics*. Wiley, New York.

[3] Hogg, R. V. (1967). *J. Amer. Statist. Ass.*, **62**, 1179–1186.

[4] Hogg, R. V. (1974). *J. Amer. Statist. Ass.*, **69**, 909–927.

[5] Hogg, R. V., Fisher, D. M., and Randles, R. H. (1975). *J. Amer. Statist. Ass.*, **70**, 656–661.

[6] Hogg, R. V. (1979). *Amer. Statist.*, **33**, 108–115.

[7] Huber, P. J. (1973). *Ann. Math. Statist.*, **43**, 1041–1067.

[8] Huber, P. J. (1973). *Ann. Statist.*, **1**, 799–821.

[9] Jaeckel, L. A. (1971). *Ann. Math. Statist.*, **42**, 1540–1552.

[10] Randles, R. H., Ramberg, J. S., and Hogg, R. V. (1973). *Technometrics*, **15**, 769–778.

[11] Tukey, J. W. and McLaughlin, D. H. (1963). *Sankhyā A*, **25**, 331–352.

(DISTRIBUTION-FREE METHODS
EXPLORATORY DATA ANALYSIS
ROBUST ESTIMATION)

ROBERT V. HOGG

ADDITION THEOREM

Let A_i and A_j be two events defined on a sample space. Then

$$\Pr\left[A_i \cup A_j\right] = \Pr\left[A_i\right] + \Pr\left[A_j\right] - \Pr\left[A_i \cap A_j\right],$$

where $\Pr[A_i \cup A_j]$ denotes the probability of A_i or A_j or both occurring, $\Pr[A_i]$ and $\Pr[A_j]$ denote respectively the probability of A_i and the probability of A_j, and $\Pr[A_i \cap A_j]$ denotes the probability of both A_i and A_j occurring.

The theorem is extended for the general case of n events as follows:

$$\Pr[A_1 \cup \cdots \cup A_n] = \sum_{i=1}^{n} \Pr[A_i]$$

$$- \sum_{i_1}^{n-1} \sum_{<i_2}^{n} \Pr[A_{i_1} \cap A_{i_2}]$$

$$+ \sum_{i_1}^{n-2} \sum_{<i_2}^{n-1} \sum_{<i_3}^{n} \Pr[A_{i_1} \cap A_{i_2} \cap A_{i_3}]$$

$$- \cdots + (-1)^{n+1} \Pr\left[\bigcap_{i=1}^{n} A_i\right].$$

It is also called *Waring's theorem*.

(BONFERRONI INEQUALITIES
BOOLE'S INEQUALITY
INCLUSION-EXCLUSION
METHOD)

ADMISSIBILITY

DEFINITION

Admissibility is a very general concept that is applicable to any procedure of statistical inference*. The statistical literature contains discussions of admissibility of estimators*, confidence intervals*, confidence sets, tests of hypotheses, sampling designs in survey sampling*, and so on. For each class of procedures, there is formulated a definition of admissibility which is appropriate for that class only. But all such definitions are based on a common underlying notion—that a procedure is admissible if and only if there does not exist within that class of procedures another one which performs uniformly at least as well as the procedure in question and performs better than it in at least one case. Here "uniformly" always means for all values of the parameter* (or parameters) that determines the (joint) probability distribution* of the random variables under investigation. It thus remains only to define how the condition of "performing as well as" is interpreted in each case. All such definitions are based closely on that of the admissibility of a decision rule* formulated in Abraham Wald's theory of statistical decision functions or decision theory*, as it is briefly called. In fact, the importance of the notion of admissibility in statistical theory rests on the adoption of the decision-theoretic approach to statistical problems formulated in Wald's theory.

DECISION THEORY

Wald's theory formulates the following general model for statistical decision making. Let $\mathcal{S}$ denote the sample space of the random variables under investigation, of which the true (joint) probability distribution is unknown, it being known only to belong to a family $\mathcal{P} = (P_\theta, \theta \in \Omega)$. Depending upon the object of the investigation (e.g., point or interval estimation or hypothesis testing*, etc.), there is a specific set $\mathcal{A}$ of all possible decisions a which the statistician may make. A decision rule δ is a function which prescribes for each sample point z how the decision would be made, i.e., some specific a chosen from $\mathcal{A}$. (δ may either assign a unique a to each z—such a rule is called a nonrandomized rule—or it may assign to each z a probability distribution δ_z on $\mathcal{A}$, the choice of a specific a being made according to that probability distribution by an independent random experiment.) Consequences of wrong decisions are allowed for by assuming a suitable nonnegative loss function* $\mathcal{L}(a, \theta)$. When the experiment, i.e., the taking of observations on the random variables, is repeated a large number of times under identical conditions, the average long-run loss converges to its "expectation" or "mean value" $\mathcal{R}(\theta, \delta)$ which is called the risk function*. Admissibility of decision rules is defined in terms of the risk function as follows:

A decision rule δ_2 is better than a rule δ_1 if

$$\mathcal{R}(\theta, \delta_2) \leqslant \mathcal{R}(\theta, \delta_1) \qquad \text{for all } \theta \in \Omega \quad (1)$$

and the strict inequality in (1) holds for at least one $\theta \in \Omega$.

A decision rule δ is admissible if there exists no rule δ_1 which is better than δ. *See* DECISION THEORY for further details.

A LIMITATION

A limitation of the admissibility principle may be noted here. The central problem of decision theory is: Under the conditions of a given decision problem, how should the choice of a decision rule be effected from the class of all decision rules? The admissibility criterion requires that inadmissible rules be left out of consideration. This leaves the class of admissible rules, which, however, is generally very large, and admissibility says nothing as to how a choice should be made from this large class. The choice is therefore made in practice by applying other statistical principles, such as unbiasedness*, minimaxity*, and invariance*, or by taking into consideration the statistician's prior beliefs* regarding the weight to be attached to the different possible values of the parameter. In the last-mentioned case, if the prior beliefs are expressed as a probability distribution τ on Ω, the risk function $\mathfrak{R}(\theta,\delta)$ integrated with respect to τ gives the Bayes risk $r(\tau,\delta)$. The appropriate decision rule, called the Bayes rule, is then the one that minimizes $r(\tau,\delta)$ for given τ. This is the Bayesian mode of inference. In Bayesian inference*, there is thus an optimum decision rule which often is unique, determined by the prior beliefs, and hence the concept of admissibility has less importance. In Wald's theory, however, the approach is non-Bayesian and based on the long-run frequency*. Thus the risk function $\mathfrak{R}(\theta,\delta)$ represents the average loss in the long run when the experiment is repeated a large number of times under identical conditions. There was the same approach in the earlier Neyman–Pearson* theories of interval estimation and hypothesis testing, which are included in Wald's theory as particular cases. This approach is often referred to as the N-P-W approach. The importance of the criterion of admissibility is thus related to the N-P-W approach.

Another point to be noted is that if a decision rule is inadmissible, there exists another that should be used in preference to it. But this does not mean that every admissible rule is to be preferred over any inadmissible rule. It is easy to construct examples of rules that are admissible but which it would be absurd to use. An example in point estimation is an estimator that is equal to some constant k whatever be the observations. The risk function then vanishes for $\theta = k$ and the estimator is an admissible one. (See ref. 11 for a more sophisticated example.)

PARTICULAR PROCEDURES

Any statistical inference procedure such as point or interval estimation* corresponds simply to a class of decision rules. But the definition of admissibility considered appropriate for a particular procedure may not always be equivalent to the decision theory definition in terms of a risk function. For example, consider interval estimation of the parameter θ in a probability density function $f(x,\theta)$ on the basis of independent observations. Let x denote collectively the observations and T_1, T_2 the statistics (functions of x) defining the confidence interval. [The pair (T_1, T_2) constitutes in this case the decision rule.] Admissibility for confidence intervals is defined as follows. "A set of confidence intervals $\{T_1, T_2\}$ is admissible if there exists no other set $\{T_1^*, T_2^*\}$ such that

(a) $$T_2^*(\mathbf{x}) - T_1^*(\mathbf{x}) \leqslant T_2(\mathbf{x}) - T_1(\mathbf{x}) \quad \text{for all } \mathbf{x},$$

(b) $$P_\theta\{T_1^*(\mathbf{x}) \leqslant \theta \leqslant T_2^*(\mathbf{x})\} \geqslant P_\theta\{T_1(\mathbf{x}) \leqslant \theta \leqslant T_2(\mathbf{x})\} \quad \text{for all } \theta,$$

the strict inequality in (b) holding for at least one θ. [The probabilities in (b) are the inclusion probabilities.] This definition is obviously not reducible to one based on the risk function, as there are two inequalities; moreover, the first of these is required to hold at each sample point **x**, there being no averaging over the sample space, which is an essential ingredient in the definition of a risk function.

In the case of point estimation*, on the other hand, the definition of admissibility is identical with that for decision rules. A loss function that is found reasonable and mathematically convenient, particularly in point estimation problems, is the squared error, i.e., $L(t,\theta) = c(t-\theta)^2$, where t is the estimate and c is any positive constant. Further if, as often is the case, the estimators are restricted to the class of unbiased* estimators, then the admissibility criterion reduces to one based on the variance or equivalently the efficiency* of the estimator.

In the case of hypothesis testing* of the null hypothesis $\theta = \theta_0$, the decision-theoretic definition of admissibility reduces to one based on the power function* in the Neyman–Pearson theory by a suitable choice of the loss function, namely, by putting $\mathcal{L}(\theta, a_1) = 0$ and $\mathcal{L}(\theta, a_2) = 1$, where for the value θ, a_1 is the correct decision and a_2 the incorrect one. The set $\mathcal{Q}$ consists in this case of only two points, corresponding to the rejection or nonrejection of the null hypothesis. (See ref. 2 for a paper dealing with admissibility of tests.)

SPECIAL TYPES OF ADMISSIBILITY

These are extensions of the basic notion of admissibility.

Strong and Weak Admissibility

In some cases it is found necessary to introduce weak and strong versions of admissibility, strong admissibility being based on a more stringent criterion. For example, in the problem of interval estimation (see the section "Particular Procedures"), if condition (a) is replaced by

$$(a^*) \qquad E_\theta\{T_2^* - T_1^*\} \leqslant E_\theta\{T_2 - T_1\} \quad \text{for all } \theta,$$

and the sign of strict inequality required to hold for at least one θ in either (a*) or in (b), we obtain a more stringent criterion as the set of alternatives is enlarged. (See ref. 8 for weak and strong admissibility of confidence sets.)

ε-Admissibility

A decision rule δ_0 is said to be ϵ-admissible if there exists no other decision rule δ_1 such that

$$\mathcal{R}(\theta, \delta_1) < \mathcal{R}(\theta, \delta_0) - \epsilon \qquad \text{for all } \theta \in \Omega.$$

(See "Decision Theory" section of this entry for definitions of these terms.)

ϵ-admissibility provides a measure of the extent by which an inadmissible rule falls short of being admissible. (See ref. 10 for an application.)

Uniform Admissibility

This term is special to survey sampling* theory. Earlier investigations had related mostly to the admissibility of a particular estimator e_1 under a given sampling design p_1. But in survey sampling, the choice of the sampling design is generally, subject to certain limitations of cost and time, within the statistician's control. This leads to the notion of the joint admissibility of the pair (e_1, p_1) within a class of pairs (e, p). (Such a pair is now called a sampling strategy or, more simply, a strategy.) It would be pointless to consider the joint admissibility within the class of all possible pairs (e, p), as then the only admissible sampling design would be that in which the whole population is observed with probability 1. It is therefore necessary to place a restriction on the class $\mathcal{C}$ of designs. The restrictions usually assumed are that the expected sample size or the expected sampling cost under p should not exceed certain limits, as these are the restraints that generally apply in practice. Of course, the particular sampling design p_1 must also satisfy the restriction. The term "uniform admissibility" denotes just this joint admissibility of an estimator e_1 and a sampling design

p_1, defined as usual, within a class of pairs (e, p) such that p belongs to a specified class $\mathcal{C}$ of designs. Note that the uniform admissibility of (e_1, p_1) is a stronger property than the admissibility of e_1 under the given design p_1, as the former implies the latter. (See refs. 14 and 15 for some recent related results.)

Hyperadmissibility

This notion is special to survey sampling theory* and denotes broadly that an estimator is admissible for the population as a whole and also for every possible subpopulation of that population (see ref. 6 for more details).

Admissibility within a Restricted Class

It is often necessary to consider the admissibility of a procedure within a restricted class of procedures. For example, in the case of point estimation, an unbiased estimator T_0 of a parameter θ is said to be admissible within the unbiased class if there exists no other unbiased estimator T_1 of θ that is "better" than T_0.

RELATIONS WITH COMPLETENESS* AND EFFICIENCY*

In Wald's theory, the notion of admissibility is intimately related to that of completeness. The theory requires that the statistician should restrict the choice of a decision rule to the class of all admissible rules. But in general, there is no simple characteristic that distinguishes the class of all admissible rules from that of all inadmissible ones. This leads to the notion of a complete class that contains all the admissible rules: "A class $\mathcal{C}$ of decision rules is said to be complete if given any rule δ_1 not in $\mathcal{C}$, there exists at least one rule δ_0 in $\mathcal{C}$ that is better than δ_1." (See "Decision Theory" section for definition of betterness.) Hence if a class of rules is known to be complete, the statistician may validly restrict the choice of decision rule to such class as all the excluded rules are necessarily inadmissible. Of course, after choosing a particular rule from a complete class, it would have to be tested for its admissibility. It is further shown in the theory that there exists a simply characterized class of decision rules (the class of generalized Bayes rules) which under very general conditions forms a complete class.

Essential completeness is a sharpening of the notion of completeness "A class $\mathcal{C}$ of decision rules is said to be essentially complete if given any rule δ_1 not in $\mathcal{C}$, there exists at least one rule δ in $\mathcal{C}$ which is as good as δ_1, i.e., such that $\mathcal{R}(\theta, \delta_0) \leqslant \mathcal{R}(\theta, \delta_1)$ for all $\theta \in \Omega$." Clearly, a statistician may validly restrict the choice of a decision rule to an essentially complete class if it exists. It is shown in the theory that "the class of decision rules based on a sufficient statistic is always essentially complete." This proposition provides the decision-theoretic justification for the sufficiency principle* in statistics. For some related propositions such as the Rao–Blackwell theorem*, see ref. 4.

STEIN'S RESULT

A notable result relating to admissibility is that of Stein [16]: For k independent normal variables, the sample means are jointly inadmissible for the population means with the squared errors as loss function if $k \geqslant 3$. The theoretical and practical implications of Stein's results are still a matter of debate. (See ref. 3. For a recent theoretical paper, see ref. 1.)

SURVEY SAMPLING*

Survey sampling essentially involves no new point of principle. The commonly considered estimation problem is to estimate the population total. If the squared error is taken as the loss function, as is often the case, the admissibility of an estimator is defined as follows: An estimator $(e(s, \mathbf{x})$ is admissible if there does not exist any other estimator $e'(s, \mathbf{x})$ such that

$$\sum_s p(s)\left[e'(s,\mathbf{x}) - T(\mathbf{x})\right]^2 \leqslant \sum_s p(s)\left[e(s,\mathbf{x}) - T(\mathbf{x})\right]^2 \quad \text{for all } \mathbf{x}$$

and the strict inequality holds for at least one $\mathbf{x}$. Here $\mathbf{x} = (x_1, x_2, \ldots, x_N)$ denotes the population vector and is the parameter, $T(\mathbf{x}) = \sum_{i=1}^{N} x_i$, $p(s)$ is the probability of the sample s under the chosen sampling design, and N denotes the number of units in the population. The estimator $e(s, \mathbf{x})$ must, of course, depend only on the x_i observed in the sample. See SURVEY SAMPLING for further details.

The following important general results have been proved recently. (The lateness of the results is a consequence of the fact the correct model for survey sampling was developed only after 1950; see SURVEY SAMPLING.)

1. The sample mean is admissible as the estimator of the population mean in the entire class of all estimators whatever the sampling design, and for a very wide class of loss functions [7].
2. The Horwitz–Thompson estimator* is always admissible in the restricted class of unbiased estimators [5].

Suppose that samples are taken independently from k different finite populations. Are the sample means together jointly admissible for the population means with squared error as loss function? It is found that they are. Thus in the case of finite populations, an effect corresponding to Stein's result for the multivariate normal* population does not occur. This is a very recent result [9].

References

[1] Alam, K. (1975). *J. Multivariate Anal.*, **5**, 83–95.

[2] Brown, L. D., Cohen, A., and Strawderman, W. E. (1979). *Ann. Statist.*, **3**, 569–578.

[3] Efron, B. and Morris, C. (1973). *J. R. Statist. Soc. B*, **35**, 379–421.

[4] Ferguson, T. S. (1967). *Mathematical Statistics: A Decision-Theoretical Approach.* Academic Press, New York.

[5] Godambe, V. P. and Joshi, V. M. (1965). *Ann. Math. Statist.*, **36**, 1707–1722.

[6] Hanurav, T. V. (1968). *Ann. Math. Statist.*, **39**, 621–641.

[7] Joshi, V. M. (1968). *Ann. Math. Statist.*, **39**, 606–620.

[8] Joshi, V. M. (1969). *Ann. Math. Statist.*, **40**, 1042–1067.

[9] Joshi, V. M. (1979). *Ann. Statist.*, **7**, 995–1002.

[10] Kagan, A. M. (1970). *Sankhyā* A, **32**, 37–40.

[11] Makani, S. M. (1977). *Ann. Statist.*, **5**, 544–546.

[12] Neyman, J. (1937). *Philos. Trans. R. Soc. Lond. A*, **236**, 333–380.

[13] Neyman, J. and Pearson, E. S. (1933). *Philos. Trans. R. Soc. Lond. A*, **231**, 289–337.

[14] Scott, A. J. (1975). *Ann. Statist.*, **3**, 489–491.

[15] Sekkappan, R. M. and Thompson, M. E. (1975). *Ann. Statist.*, **3**, 492–499.

[16] Stein, C. (1956). *Proc. 3rd Berkeley Symp. Math. Stat. Prob.*, Vol. 1. University of California Press, Berkeley, Calif., pp. 197–206.

[17] Wald, A. (1950). *Statistical Decision Functions.* Wiley, New York.

(BAYESIAN INFERENCE
DECISION THEORY
ESTIMATION
HYPOTHESIS TESTING
STATISTICAL INFERENCE
SURVEY SAMPLING)

V. M. JOSHI

ADVANCES IN APPLIED PROBABILITY
See APPLIED PROBABILITY JOURNALS

AGGREGATE

The word "aggregate" has several meanings. As a verb, it means putting together, or combining, elements that usually differ in some notable respect. As a noun, it is used to describe the result of this process. The word is also sometimes used as a synonym for "total," as in "aggregate production" and "aggregate debt."

In geology, and especially in mining engineering, the word is specifically applied to collections of samples of ore.

(ARITHMETIC MEAN
GEOLOGY, STATISTICS IN
MIXTURES)

AGGREGATE INDEX NUMBERS *See* INDEX NUMBERS

AGGREGATION

Aggregation may be a phenomenon of direct interest, as in the study of biological populations, or it may reflect the necessary reduction of primary data to produce a usable statistical summary, as in the construction of index numbers*. Since the two topics are quite distinct, we consider them separately.

AGGREGATION AS AN OBSERVABLE PHENOMENON

It is often true that events (individuals) cluster in time or space or both (e.g., larvae hatching from eggs laid in a mass, aftershocks of an earthquake). Thus if the random variable of interest is the number of events occurring in an interval of time (or in a selected area), the clustering is manifested in a greater probability of extreme events (large groups) than would be expected otherwise. Alternatively, individual members of a population may be in close proximity because of environmental conditions. In either case, the population is said to be aggregated.

A standard initial assumption (corresponding to the absence of aggregation) is that the random variable follows a Poisson distribution*, and various indices have been proposed to detect departures from the Poisson process*. These methods are based upon data collected either as quadrat counts or as measurements of distance (or time) from randomly selected individuals (or points) to the nearest individual, known as nearest-neighbor distances. For example, the index of dispersion is defined as $I = s^2/m$, where m and s^2 denote the sample mean and variance. For the Poisson process, $E(I \mid m > 0) \doteq 1$; values of I significantly greater than 1 suggest aggregation; $I < 1$ is indicative of regular spacing of the individuals [5, Chap. 4]. Other measures, based upon both quadrat counts and distances, are summarized in Pielou [15, Chaps. 8 and 10] and Cormack [6]. When different kinds of individual (e.g., species) have different aggregation patterns, this make inferences about population characteristics such as diversity* much more difficult.

If the Poisson process is used to describe parents (or centers), each parent may give rise to offspring (or satellites). If these clusters are independent but have identical size distributions, the resulting distribution for the total count is a (Poisson) randomly stopped sum distribution*. If environmental heterogeneity is postulated, a compound distribution*, usually based on the Poisson, is appropriate. For both classes of Poisson-based distributions, $I > 1$. These standard distributions lack an explicit spatial or temporal dimension, for which a dispersal mechanism must be incorporated. The resulting model, known as a center-satellite process, has three components: a Poisson process for locating the cluster center, a distribution to generate the number of satellites, and a dispersal distribution to describe displacements from the center. This class of processes was introduced by Neyman and Scott [14] and is mathematically equivalent to the class of doubly stochastic Poisson processes defined for heterogeneity (see Bartlett [3, Chap. 1]).

A more empirical approach to aggregation is that of Taylor [18], who suggests that the population mean, μ, and variance, σ^2, are related by the power law:

$$\sigma^2 = A\mu^b, \qquad A > 0, \quad b > 0.$$

It is argued that values of b greater than 1 (the Poisson value) reflect density dependence in the spatial pattern of individuals. Although this view has been contested (see the discussion in Taylor [18]), a substantial body of empirical evidence has been presented in its support [19].

Knox [12] developed a test to detect the clustering of individuals in space and time, which may be formulated as follows. Suppose that n individuals (e.g., cases of a disease) are observed in an area during a time period. If cases i and j are less than a specified critical distance from one another, set the indicator variable $w_{ij} = 1$; otherwise, set $w_{ij} = 0$. Similarly, if i and j occur within a specified time of one another, set $y_{ij} = 1$; otherwise, set $y_{ij} = 0$. Then the space–time interaction coefficient is

$$\text{STI} = \sum_{i \neq j} \sum w_{ij} y_{ij}$$

For example, for a disease such as measles, we might consider cases within 1 mile of each other occurring 10 days or less apart (the length of the latent period). If n_S and n_T denote the number of adjacent pairs in space and time, respectively, and both are small relative to n, then the conditional distribution of STI given n_S, n_T, and n is approximately Poisson with expected value $n_S n_T / n$. The test has been extended to several spatial and temporal scales by Mantel [13]. For further details, see Cliff and Ord [5, Chaps. 1 and 2].

AGGREGATION AS A STATISTICAL METHOD

Aggregation in this sense involves the compounding of primary data in order to express them in summary form. Also, such an exercise is necessary when a model is specified at the micro (or individual) level but the usable data refer to aggregates. Then the question that arises is whether the equations of the micro model can be combined in such a way as to be consistent with the macro (or aggregate) model to which the data refer.

We may wish to compound individual data records, such as consumers' expenditures, or to combine results over time and/or space. The different cases are described in turn.

Combining Individual Records

Consider a population of N individuals in which the ith individual ($i = 1, \ldots, N$) has response Y_i to input x_i of the form

$$Y_i = f(x_i, \beta_i) + \epsilon_i$$

where β_i denotes a (vector of) parameter(s) specific to the ith individual and ϵ_i denotes a random-error term. For example, the equation may represent the consumer's level of expenditure on a commodity given its price. Then the total expenditure is $Y = \sum Y_i$ (summed over $i = 1, \ldots, N$), and the average input is $x = \sum x_i / N$.

In general, it is not possible to infer an exact relationship between Y and x from the micro relations. The few results available refer to the linear aggregation of linear equations. Theil [20] showed that when f denotes a linear function so that

$$Y_i = \alpha_i + \beta_i x_i + \epsilon_i,$$

perfect aggregation is possible, in that we may consider a macro relation of the form

$$Y = \alpha + \beta x^* + \epsilon,$$

where $\alpha = \sum \alpha_i$, $\beta = \sum \beta_i$, $x^* = \sum \beta_i x_i / \beta$, and $\epsilon = \sum \epsilon_i$. That is, we must use the weighted average x^* rather than the natural average x. Further, a different aggregation procedure is required for each regressor variable and for the same regressor variable with different response variables [2, Chap. 20; 20, Chap. 2]. If we use the natural average, the macro relation is

$$Y = \alpha + \beta x + N \operatorname{cov}(x_i, \beta_i) + \epsilon,$$

where the covariance is evaluated over the N members of the population and represents the aggregation bias. This bias is small, for example, when x is the price variable in a consumer demand equation, but may be much more substantial when x denotes consumers' income in such a relationship. When the micro relationship is a nonlinear function of x_i, the nonlinearity will generate a further source of aggregation bias. It must be concluded that exact aggregation is rarely possible, although the bias may be small in many cases. For further discussion and recent developments of the theory, see Ijiri [11].

Aggregation of Groups

Instead of forming a macro relation from a known group of individuals, we may wish to identify suitable groups from a finer classification of individuals. This is a necessary step in the construction of broad industrial classifications for use in input–output systems. See ECONOMETRICS. Blin and Cohen [4] propose a method of cluster analysis* for solving this problem.

Temporal Aggregation

Variates may be continuous or summed over a unit time interval, although the variate is recorded only as an aggregate over periods of r units duration. For a model that is linear in the regressor variables and has time-invariant parameters, aggregation is straightforward provided that there are no lagged variables. However, if

$$Y_t = \alpha + \beta x_{t-k} + \epsilon_t$$

for some $k > 0$ and k not a multiple of r, exact aggregation is not possible; any aggregated model will involve x-values for two or more time periods [20, Chap. 4]. Such models are often formulated using distributed lags*. Also, Granger and Morris [9] show that the autoregressive–moving average (ARMA) models* are often appropriate in this case.

The aggregation of time series* exhibiting positive autocorrelation tends to increase the values of the various test statistics and thereby give rise to overoptimistic assessments of the model [17]. However, Tiao and Wei [21] have shown that whereas aggregation can considerably reduce the efficiency of parameter estimators*, it has much less effect upon prediction efficiency. Indeed, it has been shown that there are circumstances where forecasts from aggregate relations may be more accurate than an aggregate of forecasts from micro relations [1, 10].

The discussion so far has assumed that β does not vary over time. For a discussion of aggregation when there is a change of regime (time-varying parameters), see Goldfeld and Quandt [8, Chap. 4].

Spatial Aggregation

Many problems in spatial aggregation are similar to those of time series, but they are further compounded by the (sometimes necessary) use of areas of irregular shape and different size. Yule and Kendall [22] first showed how different aggregations of spatial units affect estimates of correlation*. Cliff and Ord [5, Chap. 5] give a general review of methods for estimating the autocovariance and cross-correlation functions using a nested hierarchy of areal sampling units. The estimation of these functions for irregular areas depends upon making rather restrictive assumptions about the nature of interaction between areas.

Cliff and Ord considered data from the *London Atlas*; a 24×24 lattice of squares of side 500 meters was laid over the Greater London area and the percentage of land used for commercial (X), industrial (Y), office (Z), and other purposes was recorded for each square. The correlation between each X and Y for different combinations of grid squares were as follows:

	Size of Spatial Unit			
	1×1	2×2	4×4	8×8
corr(X, Z)	0.19	0.36	0.67	0.71
corr(Y, Z)	0.09	0.16	0.33	0.34

The correlation functions exhibit an element of mutual exclusion at the smallest levels, and the positive correlation for larger spatial units indicates the general effects of areas zoned for housing and nonhousing purposes. Here, as in time series, we must think in terms of a distance or time-dependent correlation function and not a unique "correlation" between variables.

A PERSPECTIVE

Aggregation appears both as a phenomenon of interest in its own right and as a necessary evil in modeling complex processes. The center-satellite models have proved useful in astronomy [14], in ecology [3, 15], in geography [5], and several other disciplines. At the present time, such processes offer a flexible tool for simulation work, although further work on the theory and analysis of such processes is desirable; the data analytic distance methods of Ripley [16] represent a useful step in the right direction. In epidemiology [12, 13] and hydrology (models of storms, etc.) the development of clusters in both space and time is important, although relatively little work exists to date. The work of Taylor [18] represents a challenge to the theoretician, as useful models generating the empirical regularities observed by Taylor are still lacking.

Econometricians seem to be turning away from the view that an aggregate model is the sum of its parts and placing more emphasis upon aggregated models per se. The nonlinearities of the aggregation procedure, combined with the complexities of the underlying processes [8], suggest that aggregated models with time-dependent parameters are likely to play an increasing role in economics and other social sciences.

Where the level of aggregation is open to choice [4], further work is needed to identify suitable procedures for combining finer units into coarser ones. Similar problems arise in quadrat sampling* [15, p. 222].

The use of a sample to estimate the mean over an area or volume is of interest in the geosciences (e.g., drillings in an oil field). Estimators for such aggregates are based upon a variant of generalized least squares* known as "kriging"*; see Delfiner and Delhomme [7] and the papers of Matheron cited therein for further details.

In all these areas, much remains to be discussed about the statistical properties of the estimators currently used, and there is still plenty of scope for the development of improved methods.

References

[1] Aigner, D. J., and Goldfeld, S. M. (1974). *Econometrica*, **42**, 113–134.

[2] Allen, R. G. D. (1959). *Mathematical Economics*. Macmillan, London. (Outlines the aggregation problem in econometric modeling and sets it in the context of other aspects of mathematical economics; written at an intermediate mathematical level.)

[3] Bartlett, M. S. (1975). *The Statistical Analysis of Spatial Pattern*. Chapman & Hall, London. (A concise introduction to the theory of spatial point processes and lattice processes with a variety of applications in ecology.)

[4] Blin, J. M. and Cohen, C. (1977). *Rev. Econ. Statist.* **52**, 82–91. (Provides a review and references on earlier attempts to form viable aggregates as well as new suggestions.)

[5] Cliff, A. D. and Ord, J. K. (1981). *Spatial Processes: Models, Inference and Applications*. Pion, London. (Discusses aggregation problems in the context of spatial patterns with examples drawn from ecology and geography; written at an intermediate mathematical level.)

[6] Cormack, R. M. (1979). In *Spatial and Temporal Processes in Ecology*, R. M. Cormack and J. K. Ord, eds. International Co-operative Publishing House, Fairland, Md., pp. 151–211. (An up-to-date review of spatial interaction models with an extensive bibliography. Other papers in the volume cover related aspects.)

[7] Delfiner, P. and Delhomme, J. P. (1975). In *Display and Analysis of Spatial Data*, J. C. Davis and J. C. McCullagh, eds. Wiley, New York; pp. 96–114. (This volume contains several other papers of general interest on spatial processes.)

[8] Goldfeld, M. and Quandt, R. E. (1976). *Studies in Nonlinear Estimation*. Cambridge, Mass.: Ballinger.

[9] Granger, C. W. J. and Morris, J. J. (1976). *J. R. Statist. Soc. A*, **139**, 246–257.

[10] Grunfeld, Y. and Griliches, Z. (1960). *Rev. Econ. Statist.* **42**, 1–13.

[11] Ijiri, Y. (1971). *J. Amer. Statist. Ass.*, **66**, 766–782. (A broad review of aggregation for economic models and an extensive bibliography.)

[12] Knox, E. G. (1964). *Appl. Statist.*, **13**, 25–29.

[13] Mantel, N. (1967). *Cancer Res.*, **27**, 209–220.

[14] Neyman, J. and Scott, E. L. (1958). *J. R. Statist. Soc. B*, **20**, 1–43. (The seminal paper on clustering processes.)

[15] Pielou, E. C. (1977). *Mathematical Ecology*. Wiley, New York. (Describes methods for the measurement of aggregation among individuals; extensive bibliography.)

[16] Ripley (1977).
[17] Rowe, R. D. (1976). *Int. Econ. Rev.*, **17**, 751–757.
[18] Taylor, L. R. (1971). In *Statistical Ecology*, Vol. 1, G. P. Patil et al., eds. Pennsylvania State University Press, University Park, Pa., pp. 357–377.
[19] Taylor, L. R. and Taylor, R. A. J. (1977). *Nature*, (*Lond.*), **265**, 415–421.
[20] Theil, H. (1954). *Linear Aggregates of Economic Relations*. North-Holland, Amsterdam. (The definitive work on aggregation for economic models.)
[21] Tiao, G. C. and Wei, W. S. (1976). *Biometrika*, **63**, 513–524.
[22] Yule, G. U. and Kendall, M. G. (1965). *An Introduction to the Theory of Statistics*. Charles Griffin, London.

(DIVERSITY INDICES
ECONOMETRICS
STATISTICAL ECOLOGY)

J. K. ORD

AGRESTI COEFFICIENT (TAU) *See* MULTIPLE ASSOCIATION.

AGRICULTURE, STATISTICS IN

The area covered by this topic is so vast that whole volumes have been written on it, e.g., ref. 8—and that is only an introduction. The use of statistical techniques in agricultural research goes back many years, and indeed agriculture was one of the areas in which modern analytical techniques were first devised. The interchange of ideas between statistical and agricultural science has been of mutual benefit to both subjects. This continues to the present day, and all that is done here is to point out particular topics of joint importance to the two sciences.

HISTORY

The earliest paper describing what may be thought of as a statistically designed agricultural experiment appears to be that of Cretté de Palluel [4]. This concerned an experiment on the fattening of sheep in which 16 animals, four each of four different breeds, were fed on four diets, one of each breed per diet. The animals were killed at four monthly intervals so that the experiment could be regarded, in modern terms, either as a $\frac{1}{4}$ replicate* of a 4^3 factorial* or a 4×4 Latin square*. This experiment, which antedates the founding of modern agricultural research stations by more than half a century, shows in a simple form the principles of good experimental design* and analysis.

Agricultural journals have been in existence in their present form since the early years of this century, and many now have statisticians on their editorial boards. Thus the *Journal of Agricultural Science* has been published in Cambridge since 1905 and deals with many branches of agriculture and animal husbandry. The *Journal of Agricultural Research* was founded in Washington in 1913 and changed to *Agronomy Journal* in 1949, reflecting its prime concern with crops. *Tropical Agriculture* has been published since 1924 in Trinidad, and the *Indian Journal of Agricultural Science* since 1931 in New Delhi. These two deal primarily with tropical agriculture, as does *Experimental Agriculture*, which started in 1930 in Oxford as the *Empire Journal of Experimental Agriculture* and dropped its imperial connections in 1965. All these journals have a long and honorable history of statistical writing, from early papers on the methodology of the analysis of field data in the *Journal of Agricultural Science* in the 1920s to several papers on the techniques of intercropping trials in *Experimental Agriculture* in the late 1970s.

Courses on statistical methods applied to agriculture have been taught for many years, one of the first being those by G. W. Snedecor* at Iowa State College as early as 1915. However, the first statistician appointed to work at an agricultural research station was R. A. Fisher*, who went to Rothamsted Experimental Station in 1919. Within a few years Fisher had developed the technique of analysis of variance* for use in analyzing the results of agricultural experiments; he was also quick to emphasize the importance of replication and randomization in field trials and introduced the randomized block design*. A good summary of Fisher's early work is given by Yates [12].

PRESENT POSITION

From the 1930s onward, statistical methods for agricultural use have been greatly extended, both by the introduction of new techniques and by their use in agricultural research throughout the world. Thus, at Rothamsted, Fisher's colleague and successor F. Yates introduced more complex experimental designs. Among others, Yates recognized the importance of extensive experimentation: Crowther and Yates [5] gave a comprehensive summary of fertilizer trials in northern Europe from 1900 to that time. Yates also used statistical methods in surveys of agricultural practice, from 1944 onward [14]. Again, these statistical techniques were initially employed in agronomy and crop husbandry, but similar principles were soon applied to experiments with animals, despite their often greater expense and difficulty. A comprehensive statement of the part statistics, and statisticians, can play in planning field experiments was given by Finney [7], and the position since then has changed only in detail, not in broad outline.

METHODS OF EXPERIMENTAL DESIGN AND ANALYSIS

The main techniques used in practice for design and analysis of agricultural experiments continue to be based largely on Fisherian principles. Thus, since all agricultural work is subject to biological variability, treatments in comparative experiments are replicated in space, and sometimes in time also. Further, the application of any treatment to a particular set of plants or animals, or piece of ground, is usually randomized, possibly with some restrictions, although systematic designs* are sometimes used for particular purposes. These same principles are used, to a lesser degree, in the design of surveys, the random element occurring in the selection of units to be sampled.

The most commonly used experimental design for field trials is the randomized block* design, in which the area of land available for experimentation is divided into *blocks*, within which it is hoped that soil conditions are reasonably uniform; the blocks are subdivided into *plots* to which treatments* are applied. (The names "block" and "plot," now widely used in experimental design, reflect the agricultural context in which they were first applied.) There are three main lines of development of practical designs, in the directions of factorial experimentation*, incomplete block designs*, and row and column designs*. Full details are given in the relevant articles elsewhere, but there are whole books devoted to the topic of experimental design, e.g., Cochran and Cox [2] for ways of allocating treatments to plots and Cox [3] for other aspects of the planning of practical experiments.

The standard technique for analyzing the results of agricultural experiments is the analysis of variance*. Although this has its critics and is certainly not universally applicable, it remains the usual method for assessing whether the variation among a group of treatments is greater than would occur if all the observed effects were due to chance. However, this technique occupies only the middle range of the examination of experimental results: it is first necessary to summarize observed data to see whether they have any meaning at all, and it is frequently desirable to synthesize the observed results into more formal models, which may advance agricultural theory as well as practice.

Since it is common to take many records on an agricultural crop or animal, the first task is to sort out those on which to conduct a formal statistical analysis. For example, if a crop is harvested over a long period of time (e.g., tomatoes or coffee), does one wish to analyze total yield, or early yield, or indeed the proportion of the total yield in a specified time? Again, there may be derived variables of interest: in experiments with animals, it could be the digestibility of the feed or the butterfat content of the milk. In pest and disease control trials it is often far more important to determine the damage on a crop than to assess the total yield, damaged and undamaged together. All these preliminaries are a vital part of the statistical assessment of a trial*; also, noting apparently anomalous values may help to pinpoint errors in recording, or alternatively, lead to the discovery of quite unsuspected effects.

Formal statistical analysis is not always necessary when the main purpose of a trial is just to obtain preliminary information for use in a further trial, for example at an early stage in a plant breeding project. However, it is common to conduct analyses of variance on trial results, if only to provide an assessment of residual variation* after allowing for treatment effects*. Some trials have treatments that are quantitative in nature, and the technique of regression* as well as analysis of variance will be useful at this formal stage. With two variables, judicious use of analysis of covariance* permits the effect of one variable on another to be assessed and allowed for. When, as is common, many variables have been recorded, multivariate methods of analysis (*see* MANOVA) may be used as an alternative to the separate analysis of each record.

Although analysis of variance and its derivatives are undoubtedly the methods most commonly used for data analysis, they are not the only ones; many other techniques may be used to supplement or replace them. Thus an important area for studying experimental techniques is the investigation of best plot sizes. An early study here was that by Fairfield Smith [6] of the relation between plot size and variability. Subsequent work has shown that it is also often necessary to take account of possible variation due to individual plants as well as the environment, while there are many nonstatistical factors that have to be considered in practice. Studies of animal breeding trials and components of variance (*see* VARIANCE COMPONENTS) have proceeded together ever since the work of Henderson [9] dealing with a nonorthogonal set of data on dairy cows. Many agricultural experiments are now conducted to provide data for testing a mathematical model, and there are biologically important models that do not fall conveniently into the linear form suitable for analysis-of-variance techniques. One example among many is the set of models describing the relations between crop yield and plant density, work on which is conveniently summarized by Willey and Heath [11]. There is now much interest in plant disease epidemiology, and although the earlier theoretical work, both biological and mathematical, was not relevant to practical agriculture, some of the more recent studies are, e.g., ref. 1. Finally, the design and analysis of series of trials often present problems different in kind from those for a single trial: for trials of crop varieties, references range in time from 1938 [13] to 1980 [10].

APPLICATION AREAS

There is now scarcely an agricultural experimental station anywhere in the world that does not use statistical techniques of the types outlined here; indeed, many have their own statisticians. This is true not only of the United States and the United Kingdom, where these methods started, but of other countries in the English-speaking world. The language barrier has proved no impediment, and striking advances have been made in many European countries, including the Netherlands, East Germany, and Poland. Further, the methods, although originating largely in the more developed countries with a temperate climate, have been used in tropical developing countries, such as India, Israel, and others in Asia, together with those in Africa and Latin America.

Experiments on many crops now use statistical methods; these include a wide range of temperate cereals, fruit, vegetables, and forage crops, and an even wider range of tropical cereals and plantation crops. Experiments in the area of animal husbandry and disease control also use statistical techniques (although the methods used on large and expensive long-lived animals cannot be identical with those on short-term annual crops). Surveys using statistical methods have been conducted on an equally wide range of temperate and tropical practices in agriculture and animal husbandry. Indeed, the use of statistical methods now permeates the whole of research and development in agriculture and related disciplines throughout the world.

References

[1] Butt, D. J. and Royle, D. J. (1974). In *Epidemics of Plant Diseases*, J. Kranz, ed., pp. 78–114.

[2] Cochran, W. G. and Cox, G. M. (1957). *Experimental Designs*, 2nd ed. Wiley, New York.

[3] Cox, D. R. (1958). *Planning of Experiments*. Wiley, New York.

[4] Cretté de Palluel (1788; English version by A. Young, 1790). *Ann. Agric.*, **14**, 133–139.

[5] Crowther, E. M. and Yates, F. (1941). *Emp. J. Exper. Agric.*, **9**, 77–97.

[6] Fairfield Smith, H. (1938). *J. Agric. Sci. Camb.*, **28**, 1–23.

[7] Finney, D. J. (1956). *J. R. Statist. Soc. A*, **119**, 1–27.

[8] Finney, D. J. (1972). *An Introduction to Statistical Science in Agriculture*, 4th ed. Blackwell, Oxford.

[9] Henderson, C. R. (1953). *Biometrics*, **9**, 226–252.

[10] Patterson, H. D. and Silvey, V. (1980). *J. R. Statist. Soc. A*, **143**, 219–240.

[11] Willey, R. W. and Heath, S. B. (1969). *Adv. Agron.*, **21**, 281–321.

[12] Yates, F. (1964). *Biometrics*, **20**, 307–321.

[13] Yates, F. and Cochran, W. G. (1938). *J. Agric. Sci. Camb.*, **28**, 556–580.

[14] Yates, F., Boyd, D. A., and Mathison, I. (1944). *Emp. J. Exp. Agric.*, **12**, 164–176.

(ANALYSIS OF COVARIANCE
ANALYSIS OF VARIANCE
EXPERIMENTAL DESIGN
FISHER, R. A.)

G. H. FREEMAN

AKAIKE'S CRITERION

A criterion used to decide on the order of a regression, where there is a natural sequence for introduction of successive predictor values (e.g., ARIMA*, polynomial regression*). It consists of choosing m to minimize the "final prediction error":

$$\frac{n+m+1}{n-m-1} \times (\text{residual mean square with } m \text{ predictors}),$$

where n is the sample size.

The same criterion may also be used in stepwise regression*.

Bibliography

Akaike, H. (1969). *Ann. Inst. Statist. Math. Tokyo*, **21**, 243.

Akaike, H. (1970). *Ann. Inst. Statist. Math. Tokyo*, **22**, 203.

ALGEBRA OF EVENTS

Let Ω be a space whose points correspond to the possible outcomes of a random experiment. Certain subsets of Ω are called *events*, and *probability* is assigned to these subsets. A collection $\mathfrak{F}$ of subsets of Ω is called an *algebra* (the term *field* is also used) if the following conditions are satisfied:

(a) The space Ω belongs to $\mathfrak{F}$ ($\Omega \in \mathfrak{F}$).

(b) The collection $\mathfrak{F}$ is closed under complementation and finite union. Formally:

b_1: If $A \in \mathfrak{F}$, then the complement $\bar{A}$ (also denoted as A^c) belongs to $\mathfrak{F}$.

b_2: If $A_1, \ldots, A_n \in \mathfrak{F}$, then union $\cup_{i=1}^n A_i$ (also denoted as $A_1 \cup \cdots \cup A_n) \in \mathfrak{F}$.

[Since $\overline{(\cup_{i=1}^n \bar{A}_i)} = \cap_{i=1}^n A_i$, b_1 and b_2 imply that an algebra is also closed under finite intersection.]

If in place of b_2 we require $\mathfrak{F}$ to be closed under *countable* union, namely, if $A_1, A_2, \ldots \in \mathfrak{F}$, then $\cup_{i=1}^\infty A_i \in \mathfrak{F}$, the collection $\mathfrak{F}$ is called a *σ-algebra* (or *σ-field*). The notion of the σ-algebra of events is a basic concept for theoretical probability theory.

(AXIOMS OF PROBABILITY)

ALGORITHM

A rule for performing a calculation—usually, although not necessarily, numerical. For example, one might have algorithms for classificatory purposes, as well as for evaluation of roots of determinantal equations. Algorithms do not provide any background for the calculations to which they refer, either in terms of motivation or justification.

Algorithms for specific purposes are described in separate entries, in particular in the article ALGORITHMS, STATISTICAL.

ALGORITHMIC INDEPENDENCE *See* ALGORITHMIC INFORMATION THEORY

ALGORITHMIC INFORMATION THEORY

The Shannon entropy* concept of classical information theory* [9] is an ensemble notion; it is a measure of the degree of ignorance concerning which possibility holds in an ensemble with a given a priori probability distribution*

$$H(p_1, \ldots, p_n) \equiv - \sum_{k=1}^{n} p_k \log_2 p_k.$$

In algorithmic information theory the primary concept is that of the *information content* of an individual object, which is a measure of how difficult it is to specify or describe how to construct or calculate that object. This notion is also known as *information-theoretic complexity*. For introductory expositions, see refs. 1, 4, and 6. For the necessary background on computability theory and mathematical logic, see refs. 3, 7, and 8. For a more technical survey of algorithmic information theory and a more complete bibliography, see ref. 2. See also ref. 5.

The original formulation of the concept of algorithmic information is independently due to R. J. Solomonoff [22], A. N. Kolmogorov* [19], and G. J. Chaitin [10]. The information content $I(x)$ of a binary string x is defined to be the size in bits (binary digits) of the smallest program for a canonical universal computer U to calculate x. (That the computer U is universal means that for any other computer M there is a prefix μ such that the program μp makes U do exactly the same computation that the program p makes M do.) The *joint information* $I(x, y)$ of two strings is defined to be the size of the smallest program that makes U calculate both of them. And the *conditional* or *relative information* $I(x \mid y)$ of x given y is defined to be the size of the smallest program for U to calculate x from y. The choice of the standard computer U introduces at most an $O(1)$ uncertainty in the numerical value of these concepts. [$O(f)$ is read "order of f" and denotes a function whose absolute value is bounded by a constant times f.]

With the original formulation of these definitions, for most x one has

$$I(x) = |x| + O(1) \tag{1}$$

(here $|x|$ denotes the length or size of the string x, in bits), but unfortunately

$$I(x, y) \leqslant I(x) + I(y) + O(1) \tag{2}$$

holds only if one replaces the $O(1)$ error estimate by $O(\log I(x)I(y))$.

Chaitin [12] and L. A. Levin [20] independently discovered how to reformulate these definitions so that the subadditivity property (2) holds. The change is to require that the set of meaningful computer programs be an instantaneous code, i.e., that no program be a prefix of another. With this modification, (2) now holds, but instead of (1) most x satisfy

$$\begin{aligned} I(x) &= |x| + I(|x|) + O(1) \\ &= |x| + O(\log |x|). \end{aligned}$$

Moreover, in this theory the decomposition of the joint information of two objects into the sum of the information content of the first object added to the relative information of the second one given the first has a different form than in classical information theory. In fact, instead of

$$I(x, y) = I(x) + I(y \mid x) + O(1), \tag{3}$$

one has

$$I(x, y) = I(x) + I(y \mid x, I(x)) + O(1). \tag{4}$$

That (3) is false follows from the fact that $I(x, I(x)) = I(x) + O(1)$ and $I(I(x) \mid x)$ is unbounded. This was noted by Chaitin [12] and studied more precisely by Solovay [12, p. 339] and Gač [17].

Two other concepts of algorithmic information theory are *mutual* or *common information* and *algorithmic independence*. Their importance has been emphasized by Fine [5, p. 141]. The mutual information content of two strings is defined as follows:

$$I(x : y) \equiv I(x) + I(y) - I(x, y).$$

In other words, the mutual information* of two strings is the extent to which it is more economical to calculate them together than to calculate them separately. And x and y are said to be algorithmically independent if their mutual information $I(x : y)$ is essentially zero, i.e., if $I(x, y)$ is approximately equal to $I(x) + I(y)$. Mutual information is symmetrical, i.e., $I(x : y) = I(y : x) + O(1)$. More important, from the decomposition (4) one obtains the following two alternative expressions for mutual information:

$$\begin{aligned} I(x : y) &= I(x) - I(x \mid y, I(y)) + O(1) \\ &= I(y) - I(y \mid x, I(x)) + O(1). \end{aligned}$$

Thus this notion of mutual information, although it applies to individual objects rather than to ensembles, shares many of the formal properties of the classical version of this concept.

Up until now there have been two principal applications of algorithmic information theory: (a) to provide a new conceptual foundation for probability theory and statistics by making it possible to rigorously define the notion of a *random sequence**, and (b) to provide an information-theoretic approach to metamathematics and the limitative theorems of mathematical logic. A possible application to theoretical mathematical biology is also mentioned below.

A random or patternless binary sequence x_n of length n may be defined to be one of maximal or near-maximal complexity, i.e., one whose complexity $I(x_n)$ is not much less than n. Similarly, an infinite binary sequence x may be defined to be random if its initial segments x_n are all random finite binary sequences. More precisely, x is random if and only if

$$\exists c \forall n [I(x_n) > n - c]. \tag{5}$$

In other words, the infinite sequence x is random if and only if there exists a c such that for all positive integers n, the algorithmic information content of the string consisting of the first n bits of the sequence x, is bounded from below by $n - c$. Similarly, a *random real number* may be defined to be one having the property that the base 2 expansion of its fractional part is a random infinite binary sequence.

These definitions are intended to capture the intuitive notion of a lawless, chaotic, unstructured sequence. Sequences certified as random in this sense would be ideal for use in Monte Carlo* calculations [14], and they would also be ideal as one-time pads for Vernam ciphers or as encription keys [16]. Unfortunately, as we shall see below, it is a variant of Gödel's famous incompleteness theorem that such certification is impossible. It is a corollary that no pseudorandom number* generator can satisfy these definitions. Indeed, consider a real number x, such as $\sqrt{2}$, π, or e, which has the property that it is possible to compute the successive binary digits of its base 2 expansion. Such x satisfy

$$I(x_n) = I(n) + O(1) = O(\log n)$$

and are therefore maximally nonrandom. Nevertheless, most real numbers are random. In fact, if each bit of an infinite binary sequence is produced by an independent toss of an unbiased coin, then the probability that it will satisfy (5) is 1. We consider next a particularly interesting random real number, Ω, discovered by Chaitin [12, p. 336].

A. M. Turing's theorem that the halting problem is unsolvable is a fundamental result of the theory of algorithms [4]. Turing's theorem states that there is no mechanical procedure for deciding whether or not an arbitrary program p eventually comes to a halt when run on the universal computer U. Let Ω be the probability that the standard computer U eventually halts if each bit of its program p is produced by an independent toss of an unbiased coin. The unsolvability of the halting problem is intimately connected to the fact that the halting probability Ω is a random real number, i.e., its base 2 expansion is a random infinite binary sequence in the very strong sense (5) defined above. From (5) it follows that Ω is normal (a notion due to E. Borel [18]), that Ω is a Kollectiv* with respect to all computable place selection rules (a concept due to R. von Mises and A. Church [15]), and it also follows that Ω satisfies all computable statistical tests of randomness* (this notion being due to P. Martin-Löf [21]). An essay by C. H. Bennett on other remarkable properties of Ω, including its immunity to computable gambling schemes, is contained in ref. 6.

K. Gödel established his famous incompleteness theorem by modifying the paradox of the liar; instead of "This statement is false" he considers "This statement is unprovable." The latter statement is true if and only if it is unprovable; it follows that not all true statements are theorems and thus that any formalization of mathematical logic is incomplete [3, 7, 8]. More relevant to algorithmic information theory is the paradox of "the smallest positive integer that cannot be specified in less than a billion words." The contradiction is that the phrase in quotes only has 14 words, even though at least 1 billion should be necessary. This is a version of the Berry paradox, first published by Russell [7, p. 153]. To obtain a theorem rather than a contradiction, one considers instead "the binary string s which has the shortest proof that its complexity $I(s)$ is greater than 1 billion." The point is that this string s cannot exist. This leads one to the metatheorem that although most bit strings are random and have information content approximately equal to their lengths, it is impossible to prove that a specific string has information content greater than n unless one is using at least n bits of axioms. See ref. 4 for a more complete exposition of this information-theoretic version of Gödel's incompleteness theorem, which was first presented in ref. 11. It can also be shown that n bits of assumptions or postulates are needed to be able to determine the first n bits of the base 2 expansion of the real number Ω.

Finally, it should be pointed out that these concepts are potentially relevant to biology. The algorithmic approach is closer to the intuitive notion of the information content of a biological organism than is the classical ensemble viewpoint, for the role of a computer program and of deoxyribonucleic acid (DNA) are roughly analogous. Reference 13 discusses possible applications of the concept of mutual algorithmic information to theoretical biology; it is suggested that a living organism might be defined as a highly correlated region, one whose parts have high mutual information.

General References

[1] Chaitin, G. J. (1975). *Sci. Amer.*, **232** (5), 47–52. (An introduction to algorithmic information theory emphasizing the meaning of the basic concepts.)

[2] Chaitin, G. J. (1977). *IBM J. Res. Dev.*, **21**, 350–359, 496. (A survey of algorithmic information theory.)

[3] Davis, M., ed. (1965). *The Undecidable—Basic Papers on Undecidable Propositions, Unsolvable Problems and Computable Functions*. Raven Press, New York.

[4] Davis, M. (1978). In *Mathematics Today: Twelve Informal Essays*, L. A. Steen, ed. Springer-Verlag, New York, pp. 241–267. (An introduction to algorithmic information theory largely devoted to a detailed presentation of the relevant background in computability theory and mathematical logic.)

[5] Fine, T. L. (1973). *Theories of Probability: An Examination of Foundations*. Academic Press, New York. (A survey of the remarkably diverse proposals that have been made for formulating probability mathematically. Caution: The material on algorithmic information theory contains some inaccuracies, and it is also somewhat dated as a result of recent rapid progress in this field.)

[6] Gardner, M. (1979). *Sci. Amer.*, **241** (5), 20–34. (An introduction to algorithmic information the-

ory emphasizing the fundamental role played by Ω.)

[7] Heijenoort, J. van, ed. (1977). *From Frege to Gödel: A Source Book in Mathematical Logic, 1879–1931.* Harvard University Press, Cambridge, Mass. (This book and ref. 3 comprise a stimulating collection of all the classic papers on computability theory and mathematical logic.)

[8] Hofstadter, D. R. (1979). *Gödel, Escher, Bach: An Eternal Golden Braid.* Basic Books, New York. (The longest and most lucid introduction to computability theory and mathematical logic.)

[9] Shannon, C. E. and Weaver, W. (1949). *The Mathematical Theory of Communication.* University of Illinois Press, Urbana, Ill. (The first and still one of the very best books on classical information theory.)

Additional References

[10] Chaitin, G. J. (1966). *J. ACM*, **13**, 547–569; **16**, 145–159 (1969).

[11] Chaitin, G. J. (1974). *IEEE Trans. Inf. Theory*, **IT-20**, 10–15.

[12] Chaitin, G. J. (1975). *J. ACM*, **22**, 329–340.

[13] Chaitin, G. J. (1979). In *The Maximum Entropy Formalism*, R. D. Levine and M. Tribus, eds. MIT Press, Cambridge, Mass., pp. 477–498.

[14] Chaitin, G. J. and Schwartz, J. T. (1978). *Commun. Pure Appl. Math.*, **31**, 521–527.

[15] Church, A. (1940). *Bull. AMS*, **46**, 130–135.

[16] Feistel, H. (1973). *Sci. Amer.*, **228** (5), 15–23.

[17] Gač, P. (1974). *Sov. Math. Dokl.*, **15**, 1477–1480.

[18] Kac, M. (1959). *Statistical Independence in Probability, Analysis and Number Theory.* Mathematical Association of America, Washington, D.C.

[19] Kolmogorov, A. N. (1965). *Problems of Inf. Transmission*, **1**, 1–7.

[20] Levin, L. A. (1974). *Problems of Inf. Transmission*, **10**, 206–210.

[21] Martin-Löf, P. (1966). *Inf. Control*, **9**, 602–619.

[22] Solomonoff, R. J. (1964). *Inf. Control*, **7**, 1–22, 224–254.

(ENTROPY
INFORMATION THEORY
MARTINGALES
MONTE CARLO METHODS
PSEUDO-RANDOM NUMBER GENERATORS
STATISTICAL INDEPENDENCE
TESTS OF RANDOMNESS)

G. J. CHAITIN

ALGORITHMS, STATISTICAL

Traditionally, in mathematics, the term "algorithm"* means "some special process for solving a certain type of problem" [3].[1] With the advent of automatic computing, the term was adopted to refer to the description of a process in a form suitable for implementation on a computer. Intuitively, an algorithm is useful in mathematics or in computing if the "type of problem" is well defined and if the "special process" can be used effectively for these problems. A reasonable definition of the term for our purposes is:

> An algorithm is a process for the solution of a type of problem, such that the process can be implemented computationally without significant difficulty and that the class of problems treated is computationally specific and well understood.

Statistical algorithms are those algorithms having useful application to problems encountered in statistics. They are not, it should be emphasized, restricted to algorithms written by or specifically for statisticians. Such a restriction would exclude a wide range of useful work and, unfortunately, would still include a number of inferior approaches to some problems.

In the general process of using computers to assist in statistical analysis of data, three aspects are frequently important: the recognition of the need for an algorithm (more generally, the role of algorithms in the overall approach); the attempt to find or implement a suitable algorithm; and judgments about the quality of an algorithm. Let us consider each of these questions in turn.

ALGORITHMS AND STATISTICAL COMPUTING

The importance of good algorithms derives from their role as building blocks supporting reliable, flexible computing. Statisticians (and equally, physicists, chemists, engineers, and other users of computers) have tended to plunge in with ad hoc attacks on specific computing problems, with relatively little use of existing algorithms or research in computing. Many arguments, some of them quite sound, support this approach. The end user is interested in the "answer" (in our case, the statistical analysis), not in the process that produces it. Particularly at early stages of statistical computing, the statistician was often not familiar with computing, either in the sense of a user or in the more important sense of understanding some of the basic principles of computation. The problem to be solved often appeared straightforward, with a solution that was qualitatively obvious to the statistician. In this case, finding or developing an algorithm seems a waste of valuable time. Furthermore, it may not be at all obvious how a statistical problem can be formulated in appropriate terms for algorithms which frequently were devised for other problem areas.

Paradoxically, some of the statistical systems and packages developed to assist statisticians aggravate the tendency to take ad hoc rather than algorithmic approaches. The many conveniences of using high-level systems make it tempting to rig an intuitively plausible solution within the system rather than reach outside to find a high-quality algorithm for the problem. In many systems, the process of integrating such an algorithm into the system may require a high degree of programming skill and knowledge of the system's inner workings.

Although arguments for casual solutions to statistical computing problems have some force, there are stronger arguments that statisticians should try to integrate high-quality algorithms into their computing. Two arguments are particularly important. First, the use of good computational algorithms generally improves the use of our own time, in spite of the widely held intuition to the contrary. Second, the quality and the defensibility of the statistical analysis of data is eventually inseparable from the quality of the underlying computations.

Support for the first argument is that well-chosen algorithms will not only increase the chance that a particular statistical computation succeeds relatively quickly, but will usually greatly simplify the (inevitable) process of adapting the computation to new data or to a change in the analysis. As for the second argument, this is asserting both that the statistician should understand what an analysis has produced, in clear and precise terms, and also that the operational steps should be communicable and independently reproducible by others. Well-defined and correct computations are needed if statistical analysis is to satisfy fundamental criteria of scientific validity. This requires, in turn, that computations in statistical systems and specially programmed data analysis be based on algorithms that are accepted as correct implementations of valid computational methods.

As computing evolves, statisticians should be able to combine the convenience of statistical systems with the use of high-quality algorithms. Statistical systems increasingly incorporate good algorithms for the common operations. Advances in techniques of language design and implementation can simplify the process of integrating new algorithms into such systems, gradually merging the process of user programming and system extension [2].

SPECIFYING ALGORITHMS

(This section is directed largely to persons writing algorithms.) Our definition of an algorithm requires that it be suitable for implementation on a computer but deliberately does not restrict the form in which the algorithm is specified. The specification may be translatable mechanically into computer steps, i.e., may be given in a programming language. Alternatively, the specification may be instructions that someone familiar with a programming language can understand and implement. An important goal of computer science is to merge the two forms by improving programming languages and the art of algorithm design to the point that algorithms can be presented in a computer-readable form which is at the same time comprehensible to reasonably well informed human beings. Steps toward this goal, such as techniques of structured programming, are of value in that they increase the chance that one can understand what an algorithm does and hence the degree of confidence in its correctness.

The most convenient specification of an algorithm would intuitively seem to be in a programming language that is locally available, so that a running program could in principle be generated directly. This convenience has to be tempered by the need to understand the algorithm and occasionally by the unsuitability of the common programming languages (e.g., FORTRAN) to handle certain problems (e.g., random number generation*). Clear verbal descriptions are still important as supplements to program code. In some cases, semiverbal presentations can be used either as supplements or as replacement for actual code. Two styles of semiverbal description are used: natural language statements organized into numbered steps, usually with iteration among the steps; and "pidgin" programming languages, with most of the description identical to some language, but with natural language inserted where the actual code would be harder to understand and with details omitted.

Given that an algorithm is to be presented in a programming language, which one will be most helpful? The overwhelming majority of published and otherwise generally circulated algorithms are written in FORTRAN, at least for scientific computing. Presenting an algorithm in this language is then likely to make it implementable widely (at least on the larger computers) and allow it to be used with many existing programs. Other, older languages are less frequently used. ALGOL60 was designed specifically to bridge the previously mentioned gap between readability and implementability; however, at the time of its design, it could take only a partial step in this direction. Although many early published algorithms were written in ALGOL60 (for a time the only accepted language for algorithm sections), FORTRAN has largely taken over, in spite of its deficiencies in generality and readability. Many of the ALGOL60 algorithms were subsequently translated into FORTRAN (some examples will be mentioned in the section "Finding Algorithms"). Other languages, such as PL-1 and COBOL, have at most marginal relevance to algorithms for statistics.

Three other languages do, however, need to be considered: APL, BASIC, and PASCAL. These are all important for interactive computing, particularly on the smaller machines. APL is widely used in statistical analysis; its advantages are its interactive nature and a general, convenient approach to arrays*. Some existing algorithms written in APL have been ad hoc and poorly designed. Nevertheless, there is a large community of users. Also, improvements in APL have made the description of some calculations more attractive (in particular, the inclusion in APL of some key operators to support the kind of calculations done in regression* and multivariate analysis*).

BASIC is also interactive, but in appearance is a (simplified) language of the FORTRAN family. It shares the advantages of availability on small computers, relative ease of initial learning, and a sizable user community. As with APL, the language has suffered at times from algorithms written without enough understanding of the problem. Both languages have been somewhat neglected by the computer-science community involved in developing high-quality algorithms. The neglect is a combination of professional isolation and some intrinsic flaws in the languages themselves. For example, both languages are rather clumsy for expressing the iterative calculations that most algorithms involve. BASIC, in addition, may make the process of separate definition of algorithms difficult.

PASCAL is again oriented to the use of small, interactive computers and can be learned fairly easily. It derives, however, from the ALGOL family of languages. PAS-

CAL is a simple, structured language, well adapted to writing many types of algorithms in a clear and readable form. One of its attractions, in fact, is to the portion of the computer-science community interested in writing programs whose correctness can be formally verified. For these reasons, PASCAL is perhaps the most attractive new language for the specification of algorithms. At the time of writing, however, its applications to statistical computing are miminal. Applications of PASCAL are mostly to nonnumerical problems. Its importance as a vehicle for statistical algorithms is largely in the future.

FINDING ALGORITHMS

There are several sources for statistical algorithms, with no simple process for searching them all. Algorithms from the various sources will tend to differ in reliability and in the convenience of implementation. Roughly in descending order of overall reliability, the major sources are:

Published Algorithm Sections

Several computing journals have published algorithms in one of a set of accepted programming languages (typically FORTRAN and ALGOL60). These algorithms have been independently refereed and (in principle) tested. They should conform to specified requirements for quality (see the next section) established by journal policy. The journal *Applied Statistics** publishes such an algorithm section specifically for statistical computing. Some statistical algorithms have also appeared in *Communications in Statistics** (*B*). Major general algorithm sections appear in *Transactions on Mathematical Software* and *The Computer Journal*. The publication *Collected Algorithms of the Association of Computing Machinery* reprints the former set of general algorithms and contains an important cumulative index, covering most published algorithm sections as well as many algorithms published separately in scientific journals.

General Algorithm Libraries

These are collections of algorithms, usually distributed in machine-readable form, for a wide range of problems. Although the algorithms are often the work of many people, the libraries usually exert some central editorial control over the code. As a result, from the user's viewpoint, greater uniformity and simplicity can be achieved. However, the distributors may not be as disinterested judges of the library contents as are editors of algorithm sections. Confidence in the quality of the library rests to a large extent on evaluation of the organization distributing it. The International Mathematical and Statistical Library (IMSL), specifically oriented to statistical algorithms, is distributed by an independent organization in suitable FORTRAN source for many computer systems. The National Algorithm Group (NAG) is a publicly sponsored British organization designed to coordinate the distribution and development of algorithms. In this work it has had the cooperation of a number of professional groups, including the Royal Statistical Society*. A number of scientific laboratories also maintain and distribute general algorithm libraries, e.g., the PORT library (Bell Laboratories), Harwell Laboratory (U.K. Atomic Energy Research Establishment), and the National Physical Laboratory.

Specialized Algorithm Packages

These are less general collections of algorithms than the previous. They provide a range of solutions to a set of related problems, frequently in greater detail than that provided by general libraries. In addition, they attack some problem areas that tend to be ignored by published algorithms, such as graphics*. Questions of reliability will be similar to the general algorithm libraries. A series of specialized packages has been developed with the cooperation of Argonne National Laboratories, covering topics such as eigenvalue problems, linear equations, and function approximation. Graphics packages include the GR-Z package (Bell Laboratories) for data analysis and the DISSPLA package (a general-purpose system distributed commercially).

Scientific Journals

In addition to published algorithm sections, many published papers contain algorithm descriptions, either in one of the semiverbal forms or in an actual programming language. A qualified referee should have examined the paper, but unless given an explicit statement, it is probably unwise to assume that the algorithm has been independently implemented and tested. Nevertheless, there are a number of problems for which the only satisfactory published algorithms are of this form (e.g., some random number generation* techniques).

Unpublished Papers; Program Sharing

These categories are perhaps last resorts—least in average quality but certainly not least in quantity. It may be that more algorithms exist in these forms than in all other categories combined. They are usually not refereed, except unintentionally by users, and one should expect to spend time testing them before putting them into regular use. Simply finding out about the algorithms requires considerable effort. Of most help are library search techniques and centralized clearing houses for technical reports (such as the National Technical Information Service in the United States).

With increased familiarity, the process of searching the various sources will become more straightforward. Services provided by technical libraries, such as literature searches (now often computerized and relatively inexpensive) and centralized listings of papers, books, and memoranda, are extremely valuable. Modern library personnel are often very knowledgeable and helpful in searching through the jungle of technical literature. Of course, once one or more algorithms have been found, there remains the question of whether they are adequate and, if not, what steps can be taken to improve or replace them.

THE QUALITY OF ALGORITHMS

The problem of evaluating algorithms has no simple solution. For most statistical applications, a sensible judgment about algorithms requires some understanding of the computational methods being used. The discussion in Chambers [1] and in the further references cited there provides background to some of the computational methods important for statistics. Although it is tempting to hope that some mechanical evaluation of algorithms could resolve their quality thorougly, this is rarely the case. Most problems are too complex for an evaluation that treats the algorithm as a black box*; i.e., as a phenomenon to be judged only by its empirical performance, without regard for the techniques used. A tendency to use only this approach to evaluate statistical software is regrettable, particularly since it reinforces the overall ad hoc approach which has been detrimental to statistical computing in the past.

In the process of evaluating algorithms, both empirically and in terms of the method used, one may apply some general guidelines. Four helpful classes of questions are the following.

IS THE ALGORITHM USEFUL? Does it solve the problem at hand? Is it general enough for all the cases likely to be encountered? Will it adapt to similar problems to be encountered later, or will a new algorithm have to be found essentially from scratch?

IS THE ALGORITHM CORRECT? Will it run successfully on all the cases? If not, will it detect and clearly indicate any failures? For numerical calculations, what guarantees of accuracy are available? (If not theoretical estimates beforehand, are there at least reliable measures of accuracy after the fact?) We emphasize again that such judgments require understanding of what numerical methods can do to solve the problem.

HOW HARD IS IT TO IMPLEMENT AND USE? Does the form of the algorithm require considerable local effort (e.g., because the algorithm is written in English or in a programming language not locally available)? Does the algorithm as implemented make inconvenient assumptions (such as limits on the size of problem that can be handled)? Is it written portably, or are there features that will need to be changed locally? Most important, is the algorithm comprehensible, so that there is some hope of fixing problems or making modifications after one is committed to its use?

IS THE ALGORITHM EFFICIENT? Will its requirements for storage space, running time, or other computer resources be modest enough to make it practical for the problems at hand? Are there convenient, general estimates of these requirements? Issues of efficiency are often overemphasized, in the sense that the human costs involved in the previous questions are far more important in most applications. Nevertheless, we can still encounter problems that exceed the capacity of current computing, and it is good to be careful of such situations. As with accuracy, it is important to understand what computing science can do for the problem. Both theoretical estimates (of the order of difficulty, frequently) and empirical estimates are helpful.

NOTE

1. For the general reader, the complete Oxford English Dictionary gives algorithm (preferably algorism) as meaning the arabic numerals, with the chastening added meaning of a cypher or nonentity.

References

[1] Chambers, J. M. (1977). *Computational Methods for Data Analysis*. Wiley, New York.

[2] Chamber, J. M. (1980). *Amer. Statist.*, **34**, 238–243.

[3] James, G. and James, R. C. (1959). *Mathematics Dictionary*. D. Van Nostrand, Princeton, N. J.

Bibliography

The following is a selection of references, by subject, to some of the more useful algorithms for statistical applications. In most cases, the algorithms are presented in the form of subprograms or procedures in some programming language. A few are (reasonably precise) verbal descriptions to be followed by the reader in implementing the procedure. There is, of course, no assertion that these are "best" algorithms. Most of them do present a good combination of reliability, generality, and simplicity.

In addition to these references, several organizations provide algorithm libraries. Two sources that should be mentioned specifically for statistical and scientific computing are:

International Mathematical and Statistical Libraries, Inc. (IMSL), 7500 Bellaire Boulevard, Houston, Tex. 77036, USA.

The Numerical Algorithms Group (NAG), 7 Banbury Road, Oxford OX2 6NN, England.

Fourier Transforms

Singleton, R. C. (1968). *Commun. ACM*, **11**, 773–779.

Singleton, R. C. (1969). *IEEE Trans. Audio Electroacoust.*, **17**, 93–103.

Graphics

Akima, H. (1978). *ACM Trans. Math. Software*, **4**, 148–159.

Becker, R. A. and Chambers, J. M. (1977). *The GR-Z System of Graphical Subroutines for Data Analysis*. Write to Computer Information Library, Bell Laboratories, Murray Hill, N. J. 07974.

Crane, C. M. (1972). *Computer J.*, **15**, 382–384.

Doane, D. P. (1976). *Amer. Statist.*, **30**, 181–183.

Lewart, C. R. (1973). *Commun. ACM*, **16**, 639–640.

Newman, W. M. and Sproull, R. F. (1979). *Principles of Interactive Computer Graphics*, 2nd ed. McGraw-Hill, New York.

Scott, W. (1979). *Biometrika*, **66**, 605–610.

Nonlinear Models

Brent, R. P. (1973). *Algorithms for Minimization without Derivatives*. Prentice-Hall, Englewood Cliffs, N. J.

Gill, P. E. and Murray, W. (1970). A Numerically Stable Form of the Simplex Algorithm. *Rep. Math. No. 87*, National Physical Laboratory, Teddington, England.

Lill, A. S. (1970). *Computer J.*, **13**, 111–113; also, *ibid.*, **14**, 106, 214 (1971).

O'Neill, R. (1971). *Appl. Statist.*, **20**, 338–345.

Shanno, D. F. and Phua, K. H. (1976). *ACM Trans. Math. Software*, **2**, 87–94.

Numerical Approximation

Cody, W. J., Fraser, W., and Hart, J. F. (1968). *Numer. Math.*, **12**, 242–251.

Hart, J. F., Cheney, E. W., Lawson, C. L., Maehly, H. J., Mesztenyi, C. K., Rice, J. R., Thacher, H. C., and Witzgall, C. (1968). *Computer Approximations*. Wiley, New York.

Numerical Integration

Blue, J. L. (1975). Automatic Numerical Quadrature: DQUAD. *Comp. Sci. Tech. Rep. No.* 25, Bell Laboratories, Murray Hill, N. J.

Gentleman, W. M. (1972). *Commun. ACM*, **15**, 353–355.

Numerical Linear Algebra

Businger, P. A. and Golub, G. H. (1969). *Commun. ACM*, **12**, 564–565.

Wilkinson, J. H. and Reinsch, C. eds. (1971). *Handbook for Automatic Computation Vol.* 2: *Linear Algebra*. Springer-Verlag, Berlin. (Contains a wide selection of algorithms, many subsequently used in the EISPAK package developed at Argonne Laboratories.)

Programming

Ryder, B. G. (1974). *Software—Pract. Exper.*, **4**, 359–377.

Sande, G. (1975). *Proc. 8th Comp. Sci. Statist. Interface Symp.* Health Sciences Computing Facility, UCLA, Los Angeles, Calif., pp. 325–326.

Random Numbers

Ahrens, J. H. and Dieter, U. (1972). *Commun. ACM*, **15**, 873–882.

Ahrens, J. H. and Dieter, U. (1974). *Computing*, **12**, 223–246.

Ahrens, J. H. and Dieter, U. (1974). Acceptance–Rejection Techniques for Sampling from the Gamma and Beta Distributions. *Tech. Rep. No. AD*-782478, Stanford University, Stanford, Calif. (Available from National Technical Information Service.)

Chambers, J. M., Mallows, C. L., and Stuck, B. W. (1976). *J. Amer. Statist. Ass.*, **71**, 340–344.

Kinderman, A. J. and Monahan, J. F. (1977). *ACM Trans. Math. Software*, **3**, 257–560.

Regression

Barrodale, I. and Roberts, F. D. K. (1974). *Commun. ACM*, **17**, 319–320.

Chambers, J. M. (1971). *J. Amer. Statist. Ass.*, **66**, 744–748.

Daniel, J. W., Gragg, W. B., Kaufman, L., and Stewart, G. W. (1976). *Math. Comp.*, **30**, 772–795.

Gentleman, W. M. (1974). *Appl. Statist.*, **23**, 448–454.

Wampler, R. H. (1979). *ACM Trans. Math. Software*, **5**, 457–465.

Sorting

Brent, R. P. (1973). *Commun. ACM*, **16**, 105–109.

Chambers, J. M. (1971). *Commun. ACM*, **14**, 357–358.

Loesser, R. (1976). *ACM Trans. Math. Software*, **2**, 290–299.

Singleton, R. C. (1969). *Commun. ACM*, **12**, 185–186.

Utilities

Fox, P., Hall, A. D., and Schryer, N. L. (1978). The PORT Mathematical Subroutine Library. *ACM. Trans. Math. Software*, **4**, 104–126. (Also Bell laboratories, Murray Hill, N. J., *Computing Sci. Tech. Rep. No.* 47.)

Kernighan, B. W. and Plauger, P. J. (1976). *Software Tools*. Addison-Wesley, Reading, Mass.

(COMPUTERS AND STATISTICS
STATISTICAL PACKAGES)

JOHN M. CHAMBERS

ALIAS

When two (or more) parameters affect the distribution of a test statistic* in similar ways, each is said to be an alias of the other(s). The term is especially associated with fractional factorial designs*, in the analysis of which certain sums of squares have distributions that can reflect the existence of any one, or some, of a number of different effects.

(CONFOUNDING
FRACTIONAL FACTORIAL DESIGNS)

ALIAS GROUP *See* FRACTIONAL FACTORIAL DESIGNS

ALIAS MATRIX *See* FRACTIONAL FACTORIAL DESIGNS

ALIASING *See* CONFOUNDING

ALLOKURTIC CURVE

A curve with "unequal" curvature, or a skewed*, as distinguished from an isokurtic* curve (which has equal curvature and is symmetrical). This term is seldom used in modern statistical literature.

(KURTOSIS
SKEWNESS)

ALLOMETRY

It is rare, in nature, to observe variation in size without a corresponding variation in shape. This is true during the growth of an organism when radical shape changes are commonplace; when comparing different species from the same family; and even when comparing mature individuals from the same species. The quantitative study of this relationship between size and shape is known loosely as allometry and the main tool is the log-log plot*. If X and Y are two dimensions that change with size, then the way each changes relative to the other is best studied by plotting $\log X$ vs. $\log Y$. In the past this was usually done on special log-log graph paper, but calculators have rendered such devices obsolete. Natural algorithms will be used in this article (and are recommended).

Some examples are shown in Figs. 1 to 3. In Fig. 1 the points represent different individuals, each measured at one point during growth. In Fig. 2 the points refer to mature individuals. In Fig. 3 the points refer to different species, and X and Y now refer to mean values (or some other typical values) for the species. The value of the log-log plot

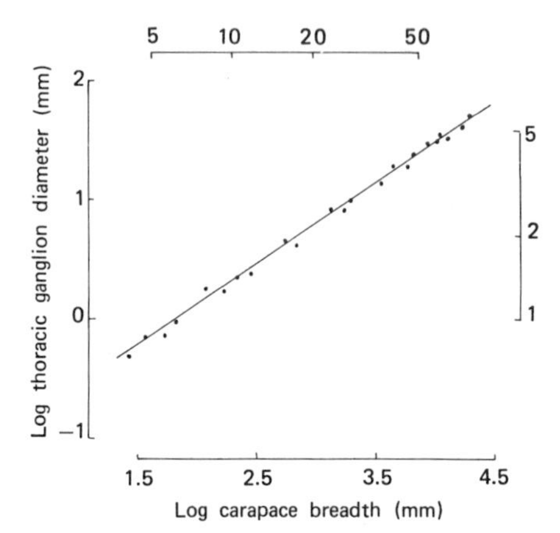

Figure 1. Growth of crabs (redrawn from Huxley [10]). The slope of the line is approximately 0.6.

is that it provides a simple summary of departures from *isometric** size variation, i.e., variation in which geometric similarity is maintained. If X and Y are both linear dimensions, then isometric variation corresponds to a constant ratio Y/X, which in turn corresponds to a line of slope 1 on the log-log plot. If Y is linear but X is a volume (or weight), then isometric variation corresponds to a constant value for $Y/X^{1/3}$, i.e., a line of slope 0.33 in the log-log plot. In a similar fashion the slope is 0.50 when Y is linear and X an area and 0.67 when Y is an

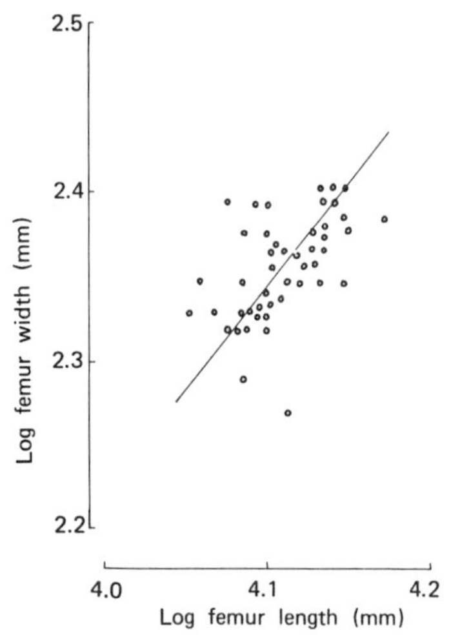

Figure 2. Variation between mature pine martens (redrawn from Jolicoeur [11]). The slope of the line is approximately 1.3.

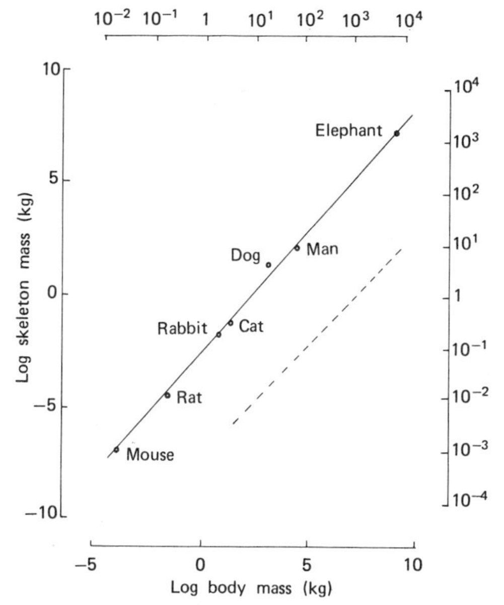

Figure 3. "Mouse-to-elephant" line (redrawn from Schmidt-Nielsen, p. 5, in Pedley [21]). The dashed line has slope 1.

area and X a volume. Slopes of X on Y are the reciprocals of the slopes of Y on X.

The log-log plot was first used systematically by Huxley [9, 10] and Teissier [23]. They found that for growth studies, using a wide variety of dimensions and organisms, the plot could often be adequately summarized by a straight line, i.e., a power law of the type $Y = bX^{\alpha}$ in original units. The coefficient α is the slope in the log-log plot and $\log b$ is the intercept: $\log Y = \alpha \log X + \log b$. For linear dimensions $\alpha = 1$ corresponds to isometry* and $\alpha \neq 1$ is referred to as allometric growth: positive if $\alpha > 1$ and negative if $\alpha < 1$. Departures from a simple straight-line relationship are not uncommon, the best known being where there is a sudden change in relative growth rate during development.

Some importance was initially attached to providing a theoretical basis for the power law, but today the relationship is recognized as being purely empirical. It is the "roughly linear relationship" so widely used in biology, but it happens to be in log units because this is the natural scale on which to study departures from isometric size variation. Linear plots using original units have also been used, but these have the disadvantage that isometry now corresponds to a line passing through the origin rather than a line of slope 1 (in log units), and there are usually no measurements in the region of the origin.

The log-log plot has also been widely used when comparing species in the general study of the effect of scale on form and function. Once again a linear relationship often proves adequate, but it is not only departures from isometry that are important but also deviations of individual species from the allometric line.

Gould [6] should be consulted for further details about the history of allometry and for a review of applications.

STATISTICAL METHODS FOR ESTIMATING A STRAIGHT LINE

The central problem is to estimate the straight-line relationship displayed in the log-log plot. To avoid too many "logs," the values of $\log X$ and $\log Y$ will be referred to as x and y. The equation of a straight line passing through (x_0, y_0) with slope α(y on x) is $y - y_0 = \alpha(x - x_0)$. This corresponds to the power law $Y = bx^{\alpha}$, where $\log b = y_0 - \alpha x_0$. Figure 4 shows four different approaches to fitting a line. In the first two the lines are chosen to minimize the sum of squares of deviations* $\sum d^2$ and correspond to regression* of y on x, and x on y. In the third the line minimizing $\sum d^2$ is called the major axis and in the fourth the value of

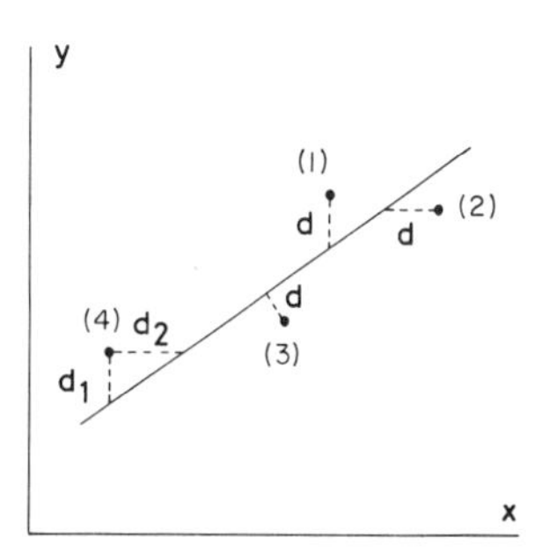

Figure 4. Four methods of fitting a straight line.

$\sum d_1 d_2$ is minimized to produce the reduced major axis [14] or D-line [25]. All four lines pass through the centroid of points $(\bar{x}, \bar{y})$ and differ only in their values for α the slope of y on x. All the estimates may be expressed in terms of r, S_x, and S_y, the sample correlation* and standard deviations* of x and y. They are:

1. $\hat{\alpha} = rS_y/S_x$ (regression of y on x)
2. $\hat{\alpha} = r^{-1}S_y/S_x$ (regression of x on y)
3. $\hat{\alpha} =$ slope of the major axis (given below)
4. $\hat{\alpha} = S_y/S_x$ with sign the same as that of r.

The two regression estimates differ from one another and there is no natural way of resolving the question of which to choose. For this reason regression theory is not as helpful in allometry as in other branches of applied statistics. The slope of the major axis is found by first obtaining the positive root of the equation in t,

$$t/(1 - t^2) = r\lambda/(1 - \lambda^2), \tag{1}$$

where $\lambda = S_y/S_x$ or S_x/S_y, whichever is less than 1. If $\lambda = S_y/S_x < 1$, then t is the slope in units of y on x. If $\lambda = S_x/S_y < 1$, then t is the slope in units of x on y and t^{-1} is the slope in units of y on x. When r is negative, the same numerical value is used for t but the sign is now negative.

Of the last two estimates the most popular has been the reduced major axis. The major axis has been criticized because it depends on the ratio $\lambda = S_y/S_x$ in a nonlinear way so that a change of units for x and y does not affect t in the same way as it affects λ. This has no validity when x and y refer to log measurements because a change of units for the measurements X and Y leaves r, S_x, and S_y unchanged. The major axis has also been criticized as being more difficult to compute than the others, but in fact (1) is easy to solve either as a quadrate or graphically. For values of λ in the range 0.2 to 0.8 a good approximation is $t \simeq \lambda + 0.3 \log r$. As λ approaches 1, the right-hand side of (1) tends to ∞, so t tends to 1. There is no general agreement as to which of these two estimates is to be preferred. Fortunately, when r is high, they give very similar answers.

Statistical sampling properties of the estimates are based on the bivariate normal distribution*. Approximate large sample standard errors* are given by Kermack and Haldane [14]:

$$SE(\hat{\alpha}) \simeq \sqrt{\left(\frac{1 - r^2}{r^2 n}\right)}\, t$$

for the major axis

$$SE(\hat{\alpha}) \simeq \sqrt{\left(\frac{1 - r^2}{n}\right)}\, \frac{S_y}{S_x}$$

for the reduced major axis.

Since $t > S_y/S_x$, the standard error for the major axis is always greater than that for the reduced major axis (considerably greater if r is low). This is an argument in favor of the reduced major axis, but such an estimate would make little sense if r were too low. When $S_x = S_y$ and $r = 0$, it amounts to drawing a line of slope 1 through a circular cloud of points. At least the large sampling error of the major axis serves as a warning that the line is difficult to determine when r is low. As a numerical example, consider the case $r = 0.8$, $n = 20$, $\lambda = S_y/S_x = 0.6$. The slope of the major axis is found from $t/(1 - t^2) = 0.75$, a quadratic equation yielding $t = 0.5352$ ($t = \tan\{\frac{1}{2}\tan^{-1}(2 \times 0.75)\}$ on a calculator). The slope of the reduced major axis is $\lambda = 0.6$ and the approximation for t is $\lambda + 0.3 \log 0.8 = 0.5331$. The two estimates for α are 0.54 ± 0.09 (major axis) and 0.60 ± 0.08 (reduced major axis). When $r = 0.4$ they become 0.33 ± 0.17 and 0.60 ± 0.12, respectively.

An alternative to basing estimates directly on the bivariate normal distribution* is to assume that x and y would lie exactly on a straight line if it were not for "errors." The errors are due to biological variation rather than to measurement errors* in this context. The model is called the linear functional relationship model. It is possible to estimate

α provided that the ratio of error variances is assumed known. In fact, if they are assumed equal, the estimate turns out to be the major axis. However, no sampling theory based on the model is available and the approach has not proved very fruitful. Further details are given in Sprent [22] and Kendall and Stuart [13]. Bartlett's estimate of slope also belongs to this class [2]. It is an example of the use of an instrumental variate* (see Kendall and Stuart [13]). The estimate has enjoyed some popularity in allometry, but unfortunately the assumptions underlying its use are not usually fulfilled unless the scatter* is low. A detailed numerical example of the use of Bartlett's estimate has been given by Simpson et al. [21].

STATISTICAL TESTS

The most commonly required test is for departure from isometry. For both the major axis and reduced major axis, $\alpha = 1$ implies that $\sigma_y = \sigma_x$ (in terms of population parameters), and this is most conveniently tested for by computing $z = y - x$ and $w = y + x$ for each individual. Since $\text{cov}(z, w) = \sigma_y^2 - \sigma_x^2$, the test for $\sigma_y = \sigma_x$ is equivalent to testing for zero correlation* between z and w. If the isometric value of α is 0.33 (for example) rather than 1, then $z = y - 0.33x$ and $w = y + 0.33x$ are used to test whether $\sigma_y = 0.33\sigma_x$ in the same way.

Confidence intervals* for α may be obtained using the approximate large sample standard errors for $\log \hat{\alpha}$. These are easily derived from those given earlier and are

$$SE(\log \hat{\alpha}) \simeq \sqrt{\left(\frac{1 - r^2}{r^2 n}\right)}$$

for the major axis

$$SE(\log \hat{\alpha}) \simeq \sqrt{\left(\frac{1 - r^2}{n}\right)}$$

for the reduced major axis.

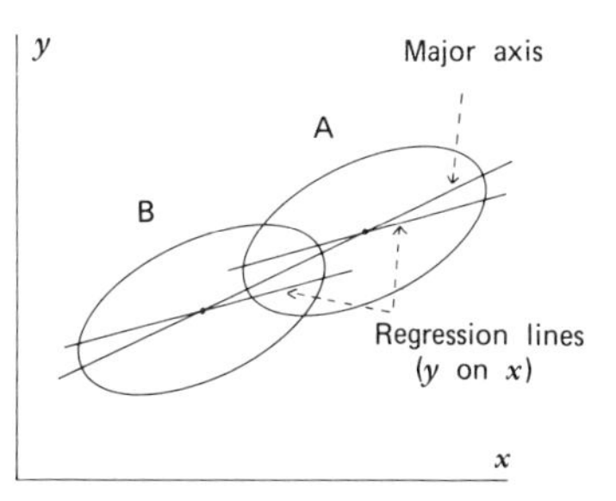

Figure 5. Allometric extension.

There is some advantage to using the SE of $\log \hat{\alpha}$ rather than $\hat{\alpha}$ since the formulas do not involve the population value α. A better approximation for the reduced major axis has been given by Clarke [4]. See also Jolicoeur [12].

An important question in allometry is whether or not one group of individuals or species is an allometric extension of another. Figure 5 illustrates a case where this is so and also shows why the use of different lines can give different answers to this question. Group A is a linear extension of group B only along the major axis, not along the regression line of y on x. There seems to be no generally recommended statistical test for this situation. Analysis of covariance* would be used with regression lines,* and this suggests a rough test for use with the major axis. First obtain a pooled estimate* of the common slope of the axis in the two groups as $\hat{\alpha} = \frac{1}{2}(\hat{\alpha}_1 + \hat{\alpha}_2)$ and then test for differences between groups in a direction perpendicular to this slope, ignoring any sampling error in $\hat{\alpha}$. This is equivalent to calculating $z = y - \hat{\alpha}x$ for each individual (or species) in each group and then testing whether the mean of z differs between groups.

Statistical tests are often omitted in allometry and there are a number of reasons for this. One is that in many growth studies the random error* is small compared to the systematic change and the conclusions are apparent. Another is that where points represent species, which cannot be thought of as randomly sampled from a population, the relevance of statistical sampling theory is slight. Even in growth studies the individuals are rarely a proper random sample from a defined population, so that significance tests* play a less important role than they do in randomized experiments*. Finally, the inadequacy of statistical methods based on the bivariate normal distribution or the linear functional relationship model is an important factor.

MULTIVARIATE ALLOMETRY

When more than two measurements of an organism are studied, an overall view of the joint variation is desirable. For p measurements $X_1, \ldots, X_p$ the log-log plot generalizes to a plot of $x_1 = \log X_1, \ldots, x_p = \log X_p$ in p dimensions, a useful concept even though it is not possible to actually plot the points. Isometry corresponds to the direction vector $\boldsymbol{\alpha}_0 = (1/\sqrt{p}, \ldots, 1/\sqrt{p})$ with the usual provision that if X_i represents an area or volume it is replaced by $X_i^{1/2}$ or $X_i^{1/3}$, respectively. Allometry again corresponds to departures from isometry, but clearly the possibilities are considerably wider in p dimensions than in two dimensions. A further problem is that there is no longer a large body of empirical evidence to suggest that departures from isometry are usually adequately summarized by a straight line. However, such a departure is a sensible starting point, and it may be summarized by a direction vector $\boldsymbol{\alpha} = (\alpha_1, \ldots, \alpha_p)$, where $|\boldsymbol{\alpha}| = 1$ (i.e., $\sum \alpha_i^2 = 1$). The angle between $\boldsymbol{\alpha}$ and $\boldsymbol{\alpha}_0$ is given by $\cos\theta = \sum \alpha_i/\sqrt{p}$.

The direction $\boldsymbol{\alpha}$ may be estimated using either the major or the reduced major axis. In the latter $\hat{\alpha}$ is taken to be the direction vector proportional to $(S_1, \ldots, S_p)$ where S_i is the sample standard deviation of x_i. Using the major axis, $\hat{\alpha}$ is taken equal to the direction of the first eigenvector* of the covariance matrix* of $x_1, \ldots, x_p$. This eigenvector is often referred to as the first principal component*. Corruccini and Henderson [5] give a good example of the use of the major axis. The linear functional relationship model has also been generalized to many variables [8, 22], but there are few examples of its actual use.

The statistical sampling theory for many variables encounters a major difficulty that does not exist in the bivariate case. This is the need to specify the error structure about the allometric line. In the bivariate case variation perpendicular to the line is in one dimension only and may be summarized by a standard deviation. In p dimensions the space perpendicular to the allometric line has $p - 1$ dimensions, so that covariation as well as variation must be specified. The simplest thing is to assume zero covariance perpendicular to the line, which implies that all eigenvalues* of the covariance matrix apart from the first are equal. A fairly straightforward test for this hypothesis has been described by Morrison [16, p. 250]. The eigenvector corresponding to the first eigenvalue is the allometric direction, and for this to be a sensible summary of the data the first eigenvalue must be considerably greater than the $p - 1$ (equal) eigenvalues. This is the same qualitative judgment that has to be made in the bivariate case. The first eigenvalue is the variance in the allometric direction and the others are the variances in directions perpendicular to the line. Variation along the line should be much greater than that about the line. Isometry corresponds to a further specialization of this situation to the case where the first eigenvector is $(1/\sqrt{p}, \ldots, 1/\sqrt{p})$. Kshirsagar [15] gives an overall χ^2 test* for both the isometric direction and independent variation about the line and also shows how to partition the χ^2 into its two components. Morrison [16] describes a test for the isometric direction which does not rest on the assumption that all other eigenvalues are equal. Anderson [1] should be consulted for the detailed derivation of the tests described in Morrison's book.

Mosimann [17, 18] has proposed a more general approach which avoids the very restrictive assumption of independent variation orthogonal to $\boldsymbol{\alpha}_0$ or $\boldsymbol{\alpha}$. He defines size as a function G of the original measurements $\mathbf{X}$, with the property that $G(\mathbf{X}) > 0$ and $G(a\mathbf{X}) = aG(\mathbf{X})$ for $a > 0$. A shape vector $\mathbf{Z}(\mathbf{X})$ is defined to be any dimensionless vector with $p - 1$ components. Isometry is then defined to be statistical independence* of shape and size, and it is shown that for a given choice of size, either all shape vectors are independent of size or none are. Mosimann also shows that if shape is independent of one measure of size, then it cannot be independent of any other measure of size.

A test for isometry of shape with respect to a given size can be carried out on a log scale using multiple regression* techniques to test independence*. For example, if $(x_1, \ldots, x_p)$ are the measurements on a log scale, then a common measure of size is $\bar{x} = \sum x_i/p$ and a possible shape vector is $(x_1 - \bar{x}, \ldots, x_{p-1} - \bar{x})$. If R^2 is the multiple correlation* between $\bar{x}$ and $(x_1 - \bar{x}, \ldots, x_{p-1} - \bar{x})$ based on n individuals, then $\{R^2(n-p)\}/\{(1-R^2)(p-1)\}$ may be used to test the independence of shape and size. Provided that $(x_1, \ldots, x_p)$ have a multivariate normal distribution*, the statistic has the F-distribution* with $p-1$, $n-p$ degrees of freedom when the null hypothesis is true. Mosimann and James [19] give an example of this test, but no attempt is made to quantify departures from isometry, and this seems to be a weakness of the approach so far.

The problem of whether one group of points is an allometric extension of another in p dimensions does not seem to have been dealt with explicitly. There have been some advances in the wider problem of making comparisons between groups, ignoring variation in certain given directions (e.g., the isometric direction). Burnaby [3] has extended discriminant analysis* to deal with this situation and Gower [7] has extended Burnaby's work to cover the case where the direction to be ignored is an estimated allometric direction.

References

[1] Anderson, T. W. (1963). *Ann. Math. Statist.*, **34**, 122–148.

[2] Bartlett, M. S. (1949). *Biometrics*, **5**, 207–212.

[3] Burnaby, T. P. (1966). *Biometrics*, **22**, 96–110.

[4] Clarke, M. R. B. (1980). *Biometrika*, **67**, 441–446.

[5] Corruccini, R. S. and Henderson, A. M. (1978). *Amer. J. Phys. Anthropol.*, **48**, 203–208.

[6] Gould, S. J. (1966). *Biol. Rev.*, **41**, 587–640.

[7] Gower, J. C. (1976). *Bull. Geol. Inst. Univ. Upps.* (N.S.), **7**, 1–10.

[8] Hopkins, J. W. (1966). *Biometrics*, **22**, 747–760.

[9] Huxley, J. S. (1924). *Nature (Lond.)*, **114**, 895–896.

[10] Huxley, J. S. (1932). *Problems of Relative Growth*. Methuen, London.

[11] Jolicoeur, P. (1963). *Growth*, **27**, 1–27.

[12] Jolicoeur, P. (1968). *Biometrics*, **24**, 679–682.

[13] Kendall, M. G. and Stuart, A. (1967). *The Advanced Theory of Statistics*, Vol. 2., 2nd ed., Charles Griffin, London.

[14] Kermack, K. A. and Haldane, J. B. S. (1950). *Biometrika*, **37**, 30–41.

[15] Kshirsagar, A. M. (1961). *Biometrika*, **48**, 397–407.

[16] Morrison, D. F. (1976). *Multivariate Statistical Methods* (2nd ed.), McGraw-Hill, New York.

[17] Mosimann, J. E. (1970). *J. Amer. Statist. Ass.*, **65**, 930–945.

[18] Mosimann, J. E. (1975). Statistical problems of size and shape, I, pp. 187–217. Statistical problems of size and shape, II, pp. 219–239. In *Statistical Distributions in Scientific Work*, Vol. 2, G. P. Patil, S. Kotz, and J. K. Ord, eds. D. Reidel, Dordrecht.

[19] Mosimann, J. E. and James, F. C. (1979). *Evolution*, **33**, 444–459.

[20] Pedley, T. J. (1977). *Scale Effects in Animal Locomotion*. Academic Press, New York.

[21] Simpson, G. G., Roe, A., and Lewontin, R. C. (1960). *Quantitative Zoology* (rev. ed.). Harcourt Brace, New York.

[22] Sprent, P. (1969). *Models in Regression and Related Topics*. Methuen, London.

[23] Teissier, G. (1931). *Trav. Stn. Biol. Roscoff*, **9**, 227–238.

[24] Teissier, G. (1948). *Biometrics*, **4**, 14–53.

Further Reading

D'Arcy Thompson's book *Growth and Form* (Cambridge University Press, Cambridge, 1917, 1942) provides the standard introduction to problems of scale in form and function. An account of some recent work is given in *Scale Effects in Animal Locomotion* by T. J. Pedley (Academic Press, New York, 1977). S. J. Gould's article "Allometry and size in ontogeny and phylogeny" (*Biol. Rev.*,

41, 587–640) gives a comprehensive survey of the uses of allometry, and Sprent's 1972 article "The mathematics of size and shape" (*Biometrics*, **28**, 23–37) reviews the mathematical and statistical side. D. L. Pilbeam and S. J. Gould, in their 1974 contribution "Size and scaling in human evolution" (*Science*, **186**, 892–901), give an interesting account of the use of allometry to investigate evolutionary relationships. A variety of different allometric studies are reported in *Problems of Relative Growth* by J. S. Huxley (Methuen, London, 1932), and a good discussion of some of the problems with allometry is given in *Essays on Growth and Form Presented to D'Arcy Wentworth Thompson*, edited by W. E. Le Gros Clark and P. B. Medewar (Clarendon Press, Oxford, 1945).

(ANTHROPOLOGY, STATISTICS IN
PRINCIPAL COMPONENT ANALYSIS
REGRESSION ANALYSIS
SIZE AND SHAPE DISTRIBUTIONS)

M. HILLS

ALMOST CERTAIN CONVERGENCE

See CONVERGENCE OF SEQUENCES OF DISTRIBUTIONS

ALTERNATIVE HYPOTHESIS

A hypothesis that differs from the hypothesis being tested is an alternative hypothesis, usually one to which it is hoped that the test used will be sensitive. Alternative hypotheses should be chosen having regard to (1) what situation(s) are likely to arise if the hypothesis tested is not valid and (2) which ones, among these situations, it is of importance to detect.

Usually, a whole class of alternative hypotheses, rather than a single one, is used.

(CRITICAL REGION
HYPOTHESIS TESTING
LEVEL OF SIGNIFICANCE
NULL HYPOTHESIS
POWER)

AMERICAN SOCIETY FOR QUALITY CONTROL (ASQC)

The United States is generally considered the country that founded modern quality control. The work of Walter Shewhart of the American Telephone and Telegraph Company and his associates George Edwards, Harold Dodge, and Harry Romig forms the nucleus around which the movement grew. However, it was the crisis of World War II that gave impetus to the field. Business managers realized that government-mandated quality control programs for defense products had an equally important application in civilian products.

During World War II a series of short courses were conducted throughout the country on statistical quality control*. Those who attended were encouraged to get together to exchange ideas and to reinforce their new-found knowledge. A series of local societies and several regional ones were founded throughout the country. In 1946, they formed a confederation called the American Society for Quality Control, which acts as the primary professional society for the United States, Mexico, and Canada. The headquarters office, initially in New York, was later transferred to Milwaukee, Wisconsin, where it is now located.

The ASQC has encouraged community colleges and universities to offer courses in the field. Despite the demand, most institutions of higher education choose not to have separate curricula; rather, courses are woven into other curricula.

The society is organized in modified matrix fashion. There are 174 local geographic sections throughout major industrial areas of the United States. They hold about eight meetings per year. Some members also choose to join industrial or subdiscipline-oriented divisions. They usually hold one or more national conferences per year. The divisions are Chemical; Textile and Needle Trades; Aerospace and Defense; Automotive; Nuclear; Biomedical; Food, Drug, and Cosmetic; Administrative Applications; Electronics; Energy; Inspection; Reliability; and Statistics.

The 30,000 members (1979) are classified as Members, Senior Members, or Fellows.

Honors are presented by the society at the national, divisional, and sectional levels. Best known are the Shewhart Medal, the Edwards Medal, the Grant Award, and the Brumbaugh Award.

The initial journal of ASQC was *Industrial Quality Control**, later renamed *Quality Progress**. The *Journal of Quality Technology** has been published since 1969. *Technometrics** is a journal published jointly with the American Statistical Association. In addition, the *Transactions* of the Annual Technical Conference are found to be a useful source of practical information. Many other publications are published centrally and by the divisions, conference boards, and local sections.

In the early days of the ASQC the emphasis was primarily on the statistical basis of sampling schemes* for raw materials coming into a plant or warehouse. There was also interest in reducing the cost of final inspection. The early emphasis was on detection of nonconforming products. A bit later the emphasis changed to the prevention of defects. Shewhart's work in the economic control of quality* in manufacturing provided the basis for this work. Today, managerial, motivational, and engineering aspects get a more balanced hearing. The ASQC is the primary vehicle for teaching these concepts. It has an important educational arm which conducts courses in quality and reliability throughout the United States. In addition, educational materials are available for persons wishing to conduct courses under local auspices. Examinations for certification as Certified Quality Engineer (CQE), Certified Reliability Engineer (CRE), Certified Quality Technician (CQT), and Certified Reliability Technician (CRT) are available throughout the world for both members and nonmembers.

The savings due to reliable quality control programs initiated since World War II are approaching $1 trillion worldwide.

In recent years ASQC has become involved in coordinating national quality standards on behalf of the American National Standards Institute. These have worldwide impact.

The society maintains close relations with societies associated with quality in various parts of the world. These include the European Organization for Quality, the Japanese Union of Scientists and Engineers, the New Zealand Organization for Quality Assurance, the Australian Organization for Quality Control, and a variety of others.

The Milwaukee staff is at present (1980) under the direction of Executive Director Wayne Kost.

(*JOURNAL OF QUALITY TECHNOLOGY*
QUALITY CONTROL
TECHNOMETRICS)

W. A. Golomski

AMERICAN STATISTICAL ASSOCIATION

The American Statistical Association, founded in 1839 as a nonprofit corporation, has as its purpose "to foster, in the broadest manner, statistics and its applications, to promote unity and effectiveness of effort among all concerned with statistical problems, and to increase the contribution of statistics to human welfare." It is a professional association whose membership is open to individuals with interest and background in the development and use of statistics in both methodology and application.

The association is governed by an elected board of directors, which represents geographical areas, subject-matter areas, and the offices of president, past president, president-elect, secretary-treasurer, and three vice-presidents. There is also a council made up of representation from each of the chapters of the association as well as the sections, as noted below. There are presently 63 chapters, three in Canada and the remainder in the United States. Membership at the end of 1979 was approximately 14,000 with about 600 in Canada and 1000 in other countries.

There are approximately 1500 student members.

A central office is maintained in Washington, D.C. (806 15th Street NW, Washington, D.C. 20005). The staff consists of approximately 20 individuals, including an executive director, managing director, and the usual support staff. Technical editing for some of its journals is maintained at this office.

The chapters of the ASA have individual programs throughout the year. These vary from one or two to as many as 20 meetings in a single year. Each chapter is autonomous in its program of activities. Chapters vary in size from 25 to 1800 members. A chapter may sponsor a seminar of one or more days or a short course in a subject-matter area or a set of methodological techniques. Chapters often collaborate with local or regional units of other professional associations.

Within ASA there are eight sections and two subsections. These represent different areas of activity and include Business and Economics, Social Statistics, Statistical Education, Physical and Engineering Sciences, Biometrics, Statistical Computing, Survey Research Methods and Biopharmaceutics. The two subsections are Teaching of Statistics in the Health Sciences, and Statistics in Marketing. The sections and subsections develop various activities which are useful to the statistical practitioner and the researcher. These include (1) cosponsorship of regional meetings with other professional associations, (2) review of statistical computer packages*, (3) development of visual-aids materials, (4) appraisal of surveys, and (5) recommendations of statistical curricula in health sciences, computer sciences, industrial statistics, etc. Sections cosponsor symposia, workshops, and special topic meetings. A number of the sections participate in advisory capacities in such areas as national and international standards, educational programs, and federal statistical activities.

One of the strengths of the association is its committees, which vary from advisory boards to government agencies to joint committees with other professional associations. There are approximately 35 committees within the association. Although some are the usual "in-house" committees and some are short-term ad hoc committees, many relate to activities and programs outside the statistical profession, as well as key issues and concerns on national and international levels. The Committee on National and International Statistical Standards maintains direct participation in reviewing international standards and attending international meetings on standards. The Committee on Statistics and the Environment collaborates with the Natural Resources Law Section of the American Bar Association in developing workshops in statistics and increasing the level of understanding between these two groups. The Committee on Law and Criminal Justice Statistics reviews the programs and structure of statistical data collection and analysis in the U.S. Department of Justice. Advisory committees serve the U.S. Bureau of the Census* and the Energy Information Administration of the Department of Energy. Possibilities for similar committees are continually being studied. The areas of privacy and confidentiality, as well as the statistical components of legislative action, are serviced by active committees. For the latter, a consortium of 10 professional associations has been formed. A major effort is being exerted by a joint committee of the American Statistical Association and the National Council of Teachers of Mathematics on developing the teaching of Statistics at the secondary school level. Some of its activities are in close communication with similar groups in other countries.

The association holds an annual meeting. Often, this is in cooperation with other statistical and related societies, e.g., the Biometric Society*, Eastern and Western North American Regions, and the Institute of Mathematical Statistics*. This is usually preceded by a two-day short course in statistical methodology. There is also joint sponsorship in national and regional activities with such professional organizations as the American Society for Quality Control (ASQC)*, the Society of Actuaries, etc.

The association engages in programs of research and development in various areas of statistics in the types of activities that would not ordinarily be considered appropriate for a particular university or research institution. At times, the association engages in national and international projects in an attempt to make statistics even more useful to the profession and to the public at large. An example of this is the international seminar on the Transfer of Methodology between Academic and Government Statisticians.

Some of the programs in which the ASA has been engaged have far-reaching consequences in the areas of social science, survey research methods, physical and engineering sciences, and health sciences, to state a few. For example, beginning in 1977, the association has engaged in a joint venture with the U.S. Bureau of the Census, "Research to Improve the Social Science Data Base." This includes research fellows and trainees in time series*, demography*, computer software, editing* of large data sets, and a number of related areas. Another research program concerns the quality of surveys.

Educational programs, often in conjunction with other professional associations, include the development of resource material and visual aids for the teaching of statistics*, a visiting lecturer program, development and testing of a statistical curriculum for secondary schools, and short courses. The association also sponsors research fellowships and traineeships in which an individual spends a major portion of a year at a federal statistical agency. It is planned to extend this program to industrial establishments. In the area of continuing education, videotapes of short courses and special lectures are made available to a wide segment of professional statisticians, not only in ASA chapters, but at universities and industrial organizations.

In 1976, in collaboration with ASA members in several Latin American countries, the association initiated a program of symposia and site visits to stimulate the development of statistics in these countries. The result has been increased professional activity and the development of new statistical associations in some of the host countries. Additional such programs are contemplated.

The American Statistical Association is affiliated with a number of national and international organizations. It is one of approximately 50 national statistical associations affiliated with the International Statistical Institute*. Furthermore, it maintains representation in five of the sections of the American Association for the Advancement of Science (AAAS). Through its representatives the ASA is active in some AAAS programs, such as scientific freedom and human rights and the international consortium of professional associations. The ASA maintains representation on the councils of the American Federation of Information Processing Societies, the Conference Board of the Mathematical Sciences, the Social Science Research Council, and the National Bureau of Economic Research.

The association issues six publications: the *Journal of the American Statistical Association**, *The American Statistician**, *Amstat News*, *Technometrics** (jointly with the American Society for Quality Control), the *Journal of Educational Statistics** (jointly with the American Educational Research Association), and *Current Index to Statistics** (with the Institute of Mathematical Statistics*). Beyond this, there are five proceedings of its annual meetings by sections and various reports on conferences, symposia, and research programs.

The American Statistical Association is dedicated to be of service to its members abroad as well as its members within the United States and Canada, and has from time to time assisted other associations in the development of programs. Its services extend to the entire profession and to society in general. New programs are developed as needs become evident. Occasionally, the Association is called upon to provide an independent review or appraisal of statistical programs or the statistical content of a critical program in science or technology.

(*AMERICAN STATISTICIAN*
THE JOURNAL OF THE AMERICAN STATISTICAL ASSOCIATION)

F. C. LEONE

AMERICAN STATISTICIAN, THE

The American Statistician is one of three principal publications of the American Statistical Association* (ASA), the other two being the *Journal of the American Statistical Association** (*JASA*) and *Amstat News*. *JASA* is devoted largely to new developments in statistical theory and its extensions to a wide variety of fields of application; *Amstat News* is concerned primarily with current news items. *The American Statistician*, in turn, emphasizes the professional development of ASA members by publishing articles that will keep persons apprised of new developments in statistical methodology through expository and tutorial papers dealing with subjects of widespread interest to statistical practitioners and teachers.

The origins of *The American Statistician* can be traced to the *American Statistical Association Bulletin*, which, from 1935 until 1947, served as the organ of information relating to ASA chapter activities, members, annual meetings, and employment opportunities for statisticians. At the end of World War II, the need was recognized for a more ambitious publication, providing expanded news coverage as well as serving the needs for professional development, and publication of *The American Statistician* was authorized. The first issue appeared in August 1947.

The contents of the first issue provide a preview of the major emphases that prevailed for many years. A major portion of the issue (11 of a total of 25 pages) was devoted to news items, including information about ASA programs, government statistics, other statistical societies, chapter activities, and news about members. The articles dealt with statistical applications in engineering and process control ("Statistical Engineering"* by Tumbleson), electronic computing ("New High-Speed Computing Devices" by Alt), and the teaching of statistics* ("A Well-Rounded Curriculum in Statistics" by Neiswanger and Allen). In addition, a note by Haemer on graphic presentation, the first in a series, appeared, and a "Questions and Answers" department to serve as a consulting forum, edited by Mosteller, was begun. The early emphasis on teaching of statistics continued throughout the period 1947–1979. Similarly, the early recognition of electronic computation was matched by continuing attention to statistical computing, with particular emphasis on this area found since the mid-1970s. A department "Statistical Computing" was begun in 1974, and a subdepartment "New Developments in Statistical Computing" was initiated in 1977.

Teaching of statistics has always been a major focus of *The American Statistician*. The department "The Teacher's Corner" was begun in 1962. In 1977, a subdepartment "Accent on Teaching Materials" was initiated.

The professional interests of statisticians were the focus of several early papers. In the 1970s, renewed interest in statistical consulting has led to more frequent articles on this subject. It is interesting to note that an article "The Outlook for Women in Statistics," by Zapoleon, appeared in 1948.

Occasionally, special symposia papers have been published, such as a series of papers on "Reliability and Usability of Soviet Statistics" (1953) and the papers of a symposium on unemployment statistics (1955). During 1981, the "Proceedings of the Sixth Symposium on Statistics and the Environment" is to be published as a special issue of *The American Statistician*.

Table 1 presents the number of pages in each volume from 1947 to 1980. The size of a volume, after some initial fluctuations, grew slowly but steadily from 1949 to 1954, and then remained fairly stable until 1959, when *The American Statistician* again became substantially larger. Beginning in 1965, further growth occurred until 1974, when there was a reduction in size. The year 1974 represented a major turning point for *The American Statistician*. Beginning in that year, all news items were moved to the new publication *Amstat News*. In 1973, these items accounted for about one-third of the issues' pages, and the drop in the number of

Table 1 Number of Pages and Entries in Volumes 1–34 of *The American Statistician*, 1947–1980

Year	Volume	Number of Pages	Number of Entries
1947[a]	1	64	13
1948	2	156	35
1949	3	90	14
1950	4	114	28
1951	5	111	19
1952	6	134	28
1953	7	138	19
1954	8	154	19
1955	9	152	20
1956	10	129	17
1957	11	153	17
1958	12	158	18
1959	13	182	24
1960	14	186	29
1961	15	178	20
1962	16	185	16
1963	17	202	29
1964	18	194	32
1965	19	253	31
1966	20	238	35
1967	21	242	50
1968	22	230	59
1969	23	254	64
1970	24	254	42
1971	25	304	59
1972	26	306	60
1973	27	248	53
1974	28	177	45
1975	29	185	48
1976	30	216	50
1977	31	198	42
1978	32	160	35
1979	33	248	75
1980	34	257	82

[a]Only three issues in 1947.

pages published from 1973 to 1974 corresponds largely to this shift. Since 1974, *The American Statistician* has devoted its pages exclusively to papers and notes on statistics. Also, the number of yearly issues was reduced to four from five, which had been the number of issues published each year beginning in 1949. Since 1974, the number of pages published annually has again tended to grow.

The substantial growth in the number of pages published in the 1960s was accompanied by increases in the number of articles and notes published in each volume. Table 1 presents the number of entries in each volume. An entry is essentially an article or a note. Since 1964, the mean number of entries per volume has been 49, compared to 21 entries per volume for the period 1947–1963. Indeed, 73 percent of all entries in the period 1947–1979 were published since 1964.

The special departments of *The American Statistician* have varied during the period 1947–1980. For example, the two initial departments "Questions and Answers" (a statistical consulting* forum) and "Hold That Line" (concerned with graphic presentation*) were terminated in 1953 and 1951, respectively. The department "Questions and Answers" was revived in 1954, but was modified from a consulting forum to a section containing essays on a variety of statistical subjects. The new "Questions and Answers" department was edited by Ernest Rubin from 1954 until 1973, when the department was terminated. Rubin wrote many of the essays published during this period. "Letters to the Editor" began as a regular department in 1961 and has remained an important department since then.

The major departments in existence in 1980 are "The Teacher's Corner" (with the subdepartment "Accent on Teaching Materials"), "Statistical Computing*" (with the subdepartment "New Developments in Statistical Computing"), and "Letters to the Editor."

Only seven persons have served as editor of *The American Statistician* since 1947. They are Sylvia C. Weyl (1947–1953), Almarin Phillips (1953–1956, 1960), Morris Hamburg (1956–1959, 1960–1967, 1969–1972), Ernest Rubin (1968), Donald F. Morrison (1972–1975), John Neter (1976–1980), and Gary G. Koch (1981–1983). By April 1948, an editorial committee of six persons was organized to assist the editor. In addition, department editors continued to be used. In 1954, the editorial committee gave way to associate editors. Seven associate edi-

Table 2 Contents of February 1980 Issue of *The American Statistician*

Title of Paper	Author(s)
R. A. Fisher and the Design of Experiments, 1922–1926	Box
Business Directories: Findings and Recommendations of the ASA Committee on Privacy and Confidentiality	ASA Committee on Privacy and Confidentiality
Classroom and Platform Performance	Mosteller
The Teaching of Statistics: Content versus Form	Kempthorne
The Role of Nonparametrics in Introductory Statistics Courses	Noether
We Need Both Exploratory and Confirmatory	Tukey
A Program of Teaching and Consultation in Research Methods and Statistics for Graduate Students in Nursing	Brogan
When Speaking to Washington, Tell the Truth, the Whole Truth, and Nothing But the Truth and Do So Intelligibly	Bross
Getting People to Use Statistics Properly	Hooke
On the History of the Use of Geometry in the General Linear Model	Herr
In Memoriam: Gertrude Mary Cox	Sukhatme
Three Simple Inductive Proofs in Probability and Statistics	Feldman and Fox
Computer Generation of Data Sets for Homework Exercises in Simple Regression	Searle and Firey
The Equivalence of Two Rules of Classification for Two Populations	Poole

tors were utilized at the beginning, and this number grew to 22 by the end of 1979.

The editorial review process for manuscripts in effect in 1980 is similar to that used by many other professional journals. The editor reviews a newly received manuscript for its suitability for the journal. All manuscripts that are potentially suitable for the journal are assigned to an associate editor for review. The associate editor reviews the manuscript and sends it to two appropriate referees for a careful review. The referees send their evaluations to the associate editor, who then makes a recommendation to the editor. The editor makes the final publication decision. Suitable manuscripts are usually returned to the author for revision at this stage. When a revised manuscript has been resubmitted to the editor, the same review process is followed again.

Submissions in 1980 ran close to 300 manuscripts. The acceptance rate for manuscripts submitted is about 23 percent. The median time for conducting the first review of a manuscript is about 3 months, and over 90 percent of these first reviews are completed within 6 months.

The publication policy for *The American Statistician* is developed by the ASA Committee for Publications. The policy in effect during the late 1970s calls for articles of general interest on (1) important current national and international statistical problems and programs, (2) public policy matters of interest to the statistical profession, (3) training of statisticians, (4) statistical practice, (5) the history of statistics*, and (6) the teaching of statistics. In addition, expository and tutorial papers on subjects of widespread interest to statistical practitioners and teachers are strongly encouraged. Table 2 contains the contents of a typical issue.

The American Statistician has been indexed three times. In 1964, an author and chronological index was published for entries during 1947–1963. In 1974, an author index for the period 1964–1973 was published. A comprehensive author and title index, the latter with entries for all important words in the title, was published in 1979 covering entries for the period 1947–1978. Since 1977, an annual author index is published in the November issue.

In 1980, the circulation of *The American Statistician* was about 14,000 copies. Until the end of 1980 the editor was John Neter, Department of Quantitative Business Analysis, College of Business Administration, University of Georgia, Athens, Georgia 30602. The editor for the period 1981–1983 is Gary G. Koch, Department of Biostatistics, University of North Carolina, Chapel Hill, North Carolina 27514.

JOHN NETER

AMSTAT NEWS *See AMERICAN STATISTICIAN, THE*

ANALYSIS OF COVARIANCE

The analysis of covariance is a special form of the analysis of variance* and mathematically need not be distinguished from it, although there are differences in utilization. (Any reader who is unfamiliar with the analysis of variance is advised to read the article on that topic before proceeding.) Using the analysis of covariance, an experiment or other investigation is planned and the analysis of variance is sketched out, based upon the model

$$\mathbf{y} = \mathbf{M}\boldsymbol{\theta} + \boldsymbol{\eta},$$

where $\mathbf{y}$ is the vector of n data and $\boldsymbol{\eta}$ of n independent residuals*, $\boldsymbol{\theta}$ is a vector of p parameters, and $\mathbf{M}$ is an $n \times p$ matrix relating to the data to the parameters. It is then realized that there are q variates that could be measured which might explain some of the variation in $\mathbf{y}$, so the model is extended to read

$$\mathbf{y} = \mathbf{M}\boldsymbol{\theta} + \mathbf{D}\boldsymbol{\beta} + \boldsymbol{\eta} = (\mathbf{M} \vdots \mathbf{D}) \begin{bmatrix} \boldsymbol{\theta} \\ \cdots \\ \boldsymbol{\beta} \end{bmatrix} + \boldsymbol{\eta},$$

where $\mathbf{D}$ is an $n \times q$ matrix of supplementary data and $\boldsymbol{\beta}$ is a vector of q regression coefficients*, one appropriate to each variate. By this extension it is hoped to improve the estimate of $\boldsymbol{\theta}$. Following the nomenclature usual in correlation* and regression*, the values of y make up the dependent variate and those in the columns of $\mathbf{D}$ the independent variates.

As has been said, that is not different in its essentials from an ordinary analysis of variance. Thus there is nothing novel in introducing parameters that are not themselves under study but might serve to explain irrelevant variation. The blocks of an experimental design* will serve as an example. Further, the effect of blocks can be removed in either of two ways. If there were three of them, three block parameters could be introduced in θ, probably with some implied constraint to reduce them effectively to two. Alternatively, two independent variates, x_1 and x_2, could be introduced, such that x_1 was equal to $+1$ in block I, to -1 in block II, and to 0 in block III, while x_2 took the values $+1$, $+1$, and -2 in the three blocks, respectively. The outcome would be the same. Where a variate is thus derived from characteristics of the design rather than from measurement it is called a "pseudo-variate"*. The device is one that links the analyses of variance and covariance as a single technique.

Nevertheless, the user will continue to see them as different, usually thinking of the analysis of variance as the form visualized at the inception of the investigation and of the analysis of covariance as a means of coping with accidents and afterthoughts.

HISTORY AND DEVELOPMENT

The idea of allowing for an independent variate originated with Fisher* [4], who unfortunately did not appreciate that a covariance adjustment necessarily introduces nonorthogonality. The derivation of standard errors* of means is due to Wishart [6]. Bartlett [1] considerably extended the usefulness by introducing pseudo-variates for incomplete data. Later developments have tended to assimilate the method to the analysis of variance.

FORM OF THE CALCULATIONS

Nowadays, there are numerous computer packages able to carry out the necessary calculations. Nevertheless, some more detailed understanding of them can be helpful.

To take first the case of only one independent variate, the analysis of variance for its data, x, will give a sum of squared deviations* for error that has a quadratic form, i.e., it can be written as $\mathbf{x}'\mathbf{H}\mathbf{x}$, where $\mathbf{H}$ is some positive semidefinite matrix derived from the design. A corresponding quantity, $\mathbf{y}'\mathbf{H}\mathbf{y}$, exists for the dependent variate, $\mathbf{y}$. It will also be necessary to know $\mathbf{y}'\mathbf{H}\mathbf{x} = \mathbf{x}'\mathbf{H}\mathbf{y}$. Then, for a single independent variate, β, the regression coefficient of y on x, equals $\mathbf{y}'\mathbf{H}\mathbf{x}/\mathbf{x}'\mathbf{H}\mathbf{x}$. Also, the sum of squared deviations is reduced from $\mathbf{y}'\mathbf{H}\mathbf{y}$ with f degrees of freedom to $\mathbf{y}'\mathbf{H}\mathbf{y} - (\mathbf{y}'\mathbf{H}\mathbf{x})^2/(\mathbf{x}'\mathbf{H}\mathbf{x})$ with $(f-1)$ when all values of y are adjusted to a standard value of x. This new mean-squared deviation* will be written as σ^2.

In the analysis of covariance the variation in x is regarded as a nuisance because it disturbs the values of y, the variate actually under study. Accordingly, any mean of y, e.g., a treatment mean*, is adjusted to a standard value of x. If the corresponding mean of x differs from this standard by d, the mean of y needs to be adjusted by βd. Similarly, if a difference of means of y is under study and the corresponding means of x differ by d, the same adjustment needs to be applied to make the y-means comparable.

An adjustment of βd will have a variance of $\sigma^2 d^2/(\mathbf{x}'\mathbf{H}\mathbf{x})$. If no adjustment had taken place, the variance of the y-mean (or difference of y-means) would have been, say, $A(\mathbf{y}'\mathbf{H}\mathbf{y})/f$, where A is a constant derived from the design. After adjustment the corresponding figure is $[A + d^2/(\mathbf{x}'\mathbf{H}\mathbf{x})]\sigma^2$, which is not necessarily a reduction, although sometimes the advantage will be considerable.

These results are readily generalized to cover p independent variates. Let $\mathbf{C}$ be a $(p+1)\times(p+1)$ matrix; the first row and the first column relate to the dependent variate and the others to the independent variates taken in some standard order. The element in the row for variate u and the column for variate v is $\mathbf{u'Hv}$. Then writing $\mathbf{C}$ in partitioned form,

$$\mathbf{C} = \begin{pmatrix} Y & \mathbf{P'} \\ \mathbf{P} & \mathbf{X} \end{pmatrix},$$

the new error sum of squared deviations is $Y - \mathbf{P'X}^{-1}\mathbf{P}$ with $(f - p)$ degrees of freedom, thus giving σ^2, and the vector of regression coefficients, β, is $\mathbf{X}^{-1}\mathbf{P}$. If an adjustment of $\boldsymbol{\beta}'\mathbf{d}$ is applied to a mean, it will have a variance of $\mathbf{d'X}^{-1}\mathbf{d}\sigma^2$.

Some special points need attention. For example, in the analysis of variance some lines, e.g., the treatment line in the analysis of data from an experiment in randomized blocks*, can be obtained directly without a second minimization. This is not so when covariance adjustments are introduced; it is necessary first to find $E = \mathbf{y'Hy} - (\mathbf{x'Hy})^2/\mathbf{x'Hx}$ and then to ignore treatments in order to find $\mathbf{yH_0'y}$, $\mathbf{x'H_0y}$, and $\mathbf{x'H_0x}$, where $\mathbf{H}_0$ is some other matrix, and to attribute $(E_0 - E)$ to treatments, where $E_0 = \mathbf{y'H_0y} - (\mathbf{x'H_0y})^2/\mathbf{x'H_0x}$; i.e., it is necessary to allow for the possibility that β_0, the regression coefficient when treatments are included with error, will be different from β, the regression coefficient* based on error alone. Some have argued against this complication on the grounds that the two cannot really be different, but reflection shows that they could be. In an agricultural field experiment, for example, the error may derive chiefly from differences in availabile nutrient from one part of the field to another. If treatments are the exaggeration of such differences by fertilizer applications, all may be well, but if they are something entirely different, such as pruning or a change of variety, it is not reasonable to expect that β_0 will be the same as β. Now that computation is so easy, the complication should be accepted at all times. It is, in any case, required by the mathematics.

Similar complications arise in split-plot* situations, because the two regression coefficients derived from the two error lines, one for main plots and the other from subplots, are often different. Since the two errors can be made up from quite different sources, i.e., the balance of different kinds of uncontrolled variation depends on the plot size, a comparison of the two regression coefficients can be illuminating. The difficulties are chiefly those of presentation and are much relieved if an intelligible explanation can be given for why adjustments, which may necessarily depend upon different regressions, behave as they do.

NUMERICAL EXAMPLE

The experiment described in the article on the analysis of variance has an available independent variate, namely x, the number of boxes of fruit, measured to the nearest tenth of a box, for the four seasons previous to the application of treatments. Full data are set out in Table 1. There need be no difficulty about the sums of squared deviations. The sums of products of deviations are here found simply by multiplying corresponding deviations from each plot and adding. In general, wherever in the calculation of sums of squared deviations a function of x or y is squared, the sum of products of deviations is found by multiplying the function of x by the corresponding function of y. To cope with the plot feigned to be missing, it will be convenient to use a pseudo-variate, w, that has the value 1 for treatment A in block 1, and 0 elsewhere. Once that is done it does not matter what values for x and y are assigned to the missing plot*. Here, where it is intended to calculate the analysis with the plot first included and later excluded, it will be convenient to use the actual values throughout—8.2 and 287, respectively.

Allowing only for blocks, sums of squares and products are:

	y	w	x
y	24,182	51.50	710.1
w	51.50	0.8333	1.100
x	710.1	1.100	31.63

Table 1 Yields from a Soil Management Trial on Apple Trees

	Block[a]							
	I		II		III		IV	
Treatment	x	y	x	y	x	y	x	y
A	8.2	287	9.4	290	7.7	254	8.5	307
B	8.2	271	6.0	209	9.1	243	10.1	348
C	6.8	234	7.0	210	9.7	286	9.9	371
D	5.7	189	5.5	205	10.2	312	10.3	375
E	6.1	210	7.0	276	8.7	279	8.1	344
S	7.6	222	10.1	301	9.0	238	10.5	357

Source. These data were first presented by S. C. Pearce [5] and have been considered in some detail by D. R. Cox [3] and C. I. Bliss [2].

[a] x represents boxes of fruit per plot in the 4 years preceding the application of treatments; y, the crop weight in pounds during the 4 years following.

Allowing for treatments as well, they are:

	y	w	x
y	23432	42.75	688.3
w	42.75	0.6250	0.958
x	688.3	0.958	24.23

Ignoring w and the missing plot for the moment, the sum of squared deviations allowing only for blocks is

$$24{,}182 - (710.1)^2/31.63 = 8240$$

with 19 degrees of freedom. Allowing for treatments also, it is

$$23{,}432 - (688.3)^2/24.23 = 3880$$

with 14 degrees of freedom. The analysis of variance now reads

Source	d.f.	Sum of Squares	Mean Square	F
Treatments	5	4360	872	3.15
Error	14	3880		
Treatments + error	19	8240		

The new picture is very different. The F-value* of 3.15 is significant ($P < 0.05$). Clearly, the independent variate has effected an improvement. If there is any doubt, an analysis of variance establishes it, namely,

Source	d.f.	Sum of Squares	Mean Square	F
Regression	1	19,552	19,552	70.58
Error	14	3,880	277	
Regression + error	15	23,432		

The regression coefficient is 688.3/24.23 = 28.41. Adjusting treatment means for y to a standard value of $x = 8.308$ (its mean), that for A is

$$274.5 - 28.41(8.450 - 8.308) = 280.5$$

and for S

$$279.5 - 28.41(9.300 - 8.308) = 251.3.$$

The last figure shows a large effect of the adjustment. It appears that the randomization* had assigned treatment S to some heavily cropping trees, as the values of x show, and it had therefore shown up better than it should. To take the difference between the adjusted means of treatments A and S, i.e., $29.2 = 280.5 - 251.3$, the standard error is

$$\sqrt{277\left\{\frac{1}{4} + \frac{1}{4} + \frac{(9.300 - 8.450)^2}{24.23}\right\}}$$
$$= 12.1,$$

which suggests that A is, in fact, a more fruitful treatment than S. The other treatments can be investigated in the same way. The covariance adjustment has had two beneficial effects. It has markedly reduced the error variance and it has given better-based treatment means.

The case of the "missing plot" could have been dealt with in the same way. Adjusting y by w instead of x, the error sum of squared deviations would have been $23{,}432 - (42.75)^2/0.6250 = 20{,}508$ with 14 degrees of freedom, as before. For treatments and error together, the corresponding figure would have been $24{,}182 - (51.50)^2/0.8333 = 20{,}999$ with 19 degrees of freedom, which leaves 491 with 5 degrees of freedom for treatments. The error variance is now 1465 = (20,508/14). The method has the advantage of giving easily a figure for the standard error of the difference between treatment means. Thus, that for treatment A and any other treatment is

$$\sqrt{1465\left\{\frac{1}{4}+\frac{1}{4}+\frac{(0.2500-0.0000)^2}{0.6250}\right\}}$$
$$= 29.6$$

There is, however, no objection to using two independent variates, e.g., both w and x. In that case the new sum of squared deviations ignoring treatments is

$$24{,}182 - (51.50 \quad 710.1)\begin{pmatrix} 0.8333 & 1.100 \\ 1.100 & 31.63 \end{pmatrix}^{-1}\begin{pmatrix} 51.50 \\ 710.1 \end{pmatrix}$$
$$= 7295 \quad \text{with} \quad 18 \text{ degrees of freedom.}$$

For the error alone the corresponding figure is 3470 with 13 degrees of freedom. That leads to the following analysis of variance:

Source	d.f.	Sum of Squares	Mean Square	F
Treatments	5	3825	765	2.87
Error	13	3470	267	
Treatments + error	18	7295		

CHOICE OF INDEPENDENT VARIATES

Modern computer packages permit the simultaneous use of several independent variates, and thus they extend the usefulness of the techniques. Nevertheless, the temptation to introduce every available independent variate is to be resisted.

The original use of covariance was to effect adjustments where some disturbing factor had not been controlled. That remains the most common application. For example, a suspicion may arise that a thermostat is not functioning properly. Pending its replacement, the investigator may introduce periodic measurements of temperature, which are then used as an independent variate in the analysis of data. Assuming that there is a straight-line relationship between the data, y, and the temperature, the outcome might be much the same apart from the loss of a degree of freedom from the error and an arbitrary loss of regularity in the standard errors*. Again, in an agricultural context it may be found that an experiment has been laid down on variable soil, a crisis that could be resolved by a covariance adjustment on measurements of soil texture or soil acidity. Sometimes, too, there are quantities that cannot be controlled precisely. It is not possible, for example, to grow test animals of exactly the same body weight or to find patients with identical blood counts. In these instances covariance adjustments may be used as a matter of course.

At one time the technique had a bad reputation, arising, it is said, from one much-publicized example. The story goes that a field experiment was conducted on the yield of barley, a covariance adjustment being made on germination rates. As a result the error sum of squared deviations* was gratifyingly reduced, but the treatment effects also disappeared. An obvious explanation is that the treatments had affected yield by way of the germination rates and in no other manner. If so, the conclusion could have been of some importance. The story is not a warning against using covariance adjustments at all; indeed, it shows their value in revealing mechanisms, but it does warn against facile interpretations, especially when the independent variate does not justify its name but is dependent upon the treatments.

If the covariate is measured before the application of treatments, which are then allocated at random, no question need arise. Nor does any arise when the aim is to find what the treatments do to the dependent variate apart from the obvious indirect effect through the independent. Sometimes, however, the user cannot be sure whether the treatments do or do not affect the independent variate*. It is wise in such cases to be very cautious. It is true that subjecting the figures to an analysis of variance may decide the matter. If, however, the proposed independent variate is an array of ones and zeros, indicating the presence or absence of some feature, the user is unlikely to obtain any useful guidance. Even a more amenable variate may give an inconclusive response.

CONSTANCY OF THE REGRESSION COEFFICIENT

It is by no means obvious that the regression coefficients will be independent of the treatments, and in some instances the assumption may verge on the absurd. Occasionally, the difficulty can be met by transformation of the variates. Where that is not feasible, an algebraic solution by least squares is usually not possible. Thus, with data from a designed experiment it is simple to fit separate regression coefficients if the design is completely randomized but not if there are blocks, as usually there will be. Although with the aid of a computer some kind of minimization can often be achieved, it is open to question whether a constant block effect regardless of treatments in conjunction with variable regression coefficients makes a convincing and realistic model.

ACCIDENTS AND MISHAPS

It has been recognized for a long time that the analysis of covariance provides a theoretically sound way of dealing with data from damaged experiments. For example, suppose that an experiment has been designed in such a way that a program is available for calculating an analysis of variance, but m plots (i.e., units) have been lost. (It must be reasonable to assume that the loss is not a result of the treatments that have been applied.) Each gap in the data is filled with a convenient value such as zero, the treatment mean, or even an arbitrary number. It is now required to estimate the deviation between the value for the missing plot given by the method of least squares and the value that has been assigned. That is done by writing down a pseudo-variate for each missing value. It equals zero for all plots except the one to which it refers, when it equals 1. A covariance adjustment on the pseudo-variates will give a correct analysis, the regression coefficient of the dependent variate on any pseudo-variate being minus the required deviation for the plot. The method has several advantages. For one thing, unlike many methods for dealing with incomplete data, it gives a correct F-value for any effect. For another, it gives correct standard errors for treatment contrasts*. Also, it obtains degrees of freedom without special adjustment.

A similar problem can arise when the data for any plot (or unit) is the sum of an unspecified number of observations, such as weighings. If someone makes a mistake, there can be doubt whether a certain observation belongs to this plot or that. The difficulty can sometimes be resolved by attributing it first to one and then to the other. If residuals look extraordinary in one case but are unremarkable in the other, the difficulty is over, but sometimes doubt will remain. A similar problem arises when samples are taken, say for chemical analyses, and some labels are lost. The samples can still be analyzed and a total found. In both examples it is possible to state the total for the two plots without knowing how to apportion it between them. It suffices to attribute the total to one plot and zero to the other and to adjust by a pseudo-variate equal to $+1$ and -1 for the two plots and to zero for all others. If three plots are involved in the muddle, two pseudo-variates are required. The total is attributed to one plot and zero to the others. The first pseudo-variate equals $+2$ for the plot with the total, -1 for the other two involved, and zero for the others. It therefore serves to apportion a correct amount to the plot credited with the total. A second pseudo-variate apportions the rest between the other two affected plots, being equal to $+1$ and -1 for those plots and to zero for all others. The method can be extended easily to more complicated cases.

Adjustments of this sort can be made in conjunction. Thus provided that the program can manage so many, it is permissible to have some independent variates for adjustment by related quantities, others to allow for missing values, and still others to apportion mixed-up values. All these adjustments can be made correctly and simultaneously.

TRANSFORMATION OF VARIATES

It should be noted that the independent variates, despite their name, play a role in the model analogous to that of **M** rather than that of y, i.e., no assumptions are involved about their distributions, which are perhaps known, but nothing depends upon them. Accordingly, there is no need to seek variance-stabilizing transformations* for them. It is, however, still necessary to consider if y needs one, since not only must the elements of $\boldsymbol{\eta}$ be distributed independently but also they should have equal variances. In the case of the independent variates, the need is for them to be linearly related to y (or to the transformation of y) and that may call for a transformation of a different kind. Alternatively, it may be desirable to introduce the same variate twice but in different forms. Thus, it has already been mentioned that a field experiment on variable land might be improved by an adjustment on soil acidity. However, unless the species is one that favors extremely acid or extremely alkaline soils, there will almost certainly be an optimal value somewhere in the middle of the range of acidity and it would be sensible to introduce both soil pH and its square to allow the fitting of a parabola.

The correct choice of transformation for the independent variate is especially important if the intention is to enquire how far treatments affect the dependent variate other than through the independent. It is then essential to fit the right relationship; an inept choice of transformations can do harm.

An additional variate does little harm but it is not wise to load an analysis with adjustments in the hope of something emerging. A further independent variate should be included only for good reason, but it should not be omitted if there are good reasons for regarding it as relevant. If it does nothing, there should be reserve about taking it out again. Clearly, if variates are included when they reduce the error variance and excluded if they do not, bias must result. Also, in presenting results it is more convincing to report that some quantity or characteristic was allowed for but had in fact made little difference than to ignore it.

LEVEL OF ADJUSTMENT OF AN INDEPENDENT VARIATE

Basically, the method consists of estimating the partial regression coefficients* of the dependent variate upon each of the independent variates and then adjusting the dependent variate to correspond to standard values of the independent variates. What these standard values should be requires some thought.

First, as long as only linear relations are in question, it does not much matter, because differences in adjusted means of the dependent variate will be unaffected, although not the means themselves. If, however, the first independent variate is some measured quantity, x, and the second is x^2 introduced to allow for curvature, a computer package, which does not know that the two are related, will adjust the first to $\bar{x}$, the mean of

x, and the second to $(\bar{x}^2)$, which will not be the same as $(\bar{x})^2$. Probably little harm will have been done, but the point needs to be noted.

Pseudo-variates do not usually cause much trouble with their standard values as long as only differences are in question. When they are used for missing plots, the adjustment should be to zero, corresponding to the presence of the plot and not to a mean value, if actual means of the dependent variate are to relate to those given by other methods.

INTERPRETATION OF AN ANALYSIS OF COVARIANCE

An analysis of covariance having been calculated, the first step usually is to look at the error-mean-squared deviation to see if it is as small as was hoped. If it is not, there are two possibilities. One is that, as with the analysis of variance, some important source of variation has been left in error; the other comes from the independent variates having had little effect. The latter case needs to be noted because a research team can go on using adjustments believing them to be a sovereign remedy, even though in fact they do no good. To test the matter formally, it is sufficient to carry out an F-test* using the two minimizations provided by the analysis of variance with and without the independent variates. Sometimes only one of the independent variates is in question; it may be helpful to repeat the calculations omitting that variate to see if it has really had a useful effect.

The testing of effects is the same as for the analysis of variance. They should be examined in logical order and tables prepared to show all effects of importance. Where there is any possibility of an independent variate having been affected by the treatments, it may be advisable to examine the position using the analysis of variance.

The standard error of a treatment mean depends party upon the design and the error variance, as in the analysis of variance, and partly on the magnitude of the adjustments that have been required. In one sense these modifications of standard errors are a help. Cases commonly dealt with using pseudo-variates, missing plots, for example, require modifications that are not easy to make except by covariance and then they are made automatically. On the other hand, a measured independent variate will give arbitrary variation in the standard errors, which can be very awkward, for instance, in a multiple comparison test*. Incidentally, some computer packages disguise the situation by giving a common mean standard error for a set of quantities which, but for the adjustments, would all have been determined with the same precision. Although often convenient, the practice can also be misleading.

References

[1] Bartlett, M. S. (1937). *J. R. Statist. Soc. Suppl. 4*, 137–183.

[2] Bliss, C. I. (1967). *Statistics in Biology*, Vol. 2. McGraw-Hill, New York, Chap. 20.

[3] Cox, D. R. (1958). *Planning of Experiments*. Wiley, New York, Chap. 4.

[4] Fisher, R. A. (1935). *The Design of Experiments*. Oliver & Boyd, Edinburgh. (A passage was added to later editions of *Statistical Methods for Research Workers* prior to the appearance of *The Design of Experiments*.)

[5] Pearce, S. C. (1953). *Field Experimentation with Fruit Trees and Other Perennial Plants*. Commonwealth Bureau of Horticulture and Plantation Crops, Farnham Royal, Slough, England, App. IV.

[6] Wishart, J. (1936). *J. R. Statist. Soc. Suppl. 3*, 79–82.

Further reading

The general literature on the subject is rather scanty. *Biometrics** devoted Part 2 of Volume 13 (1957) to a series of papers on the analysis of covariance. Later *Communications in Statistics** similarly devoted Volume A8, Part 8 (1979) to the topic. There is a valuable account of the method in C. I. Bliss's *Statistics in Biology*, Volume II, Chapter 20 (McGraw-Hill, New York). Also, there are short but illuminating descriptions by D. J. Finney in *An Introduction to Statis-*

tical Science in Agriculture (Munksgaard, Copenhagen, 1962) and by D. R. Cox in Chapter 4 of *Planning of Experiments* (Wiley, New York, 1958).

(AGRICULTURE, STATISTICS IN
ANALYSIS OF VARIANCE
DESIGN OF EXPERIMENTS
FACTOR ANALYSIS
GENERAL LINEAR MODEL
REGRESSION ANALYSIS)

S. C. PEARCE

ANALYSIS OF VARIANCE

The analysis of variance is best seen as a way of writing down calculations rather than as a technique in its own right. For example, a significance of a correlation coefficient*, r, based on n pairs of observations, can be tested in several ways. Using the analysis of variance the calculations are set out thus:

Source	Degrees of Freedom	Sum of Squared Deviations	Mean Squared Deviation
Correlation	1	Sr^2	Sr^2
Error	$n-2$	$S(1-r^2)$	$S(1-r^2)/(n-2)$
Total	$n-1$	S	

The significance is then judged by an F-test*, i.e., from the ratio of the two mean squared deviations, which is $(n-2)r^2/(1-r^2)$ with one and $(n-2)$ degrees of freedom. By using a standard format, standard tests are available, thus obviating the need for numerous formulas and a range of tables, but in general the same conclusions will be reached however the calculations are set out.

HISTORY AND DEVELOPMENT

In its origins the method was devised by Fisher [1] for the study of data from agricultural experiments. At first the only designs, i.e., randomized blocks* and Latin squares*, were orthogonal, but later F. Yates [4] showed how to deal with those that were nonorthogonal, like balanced incomplete blocks*, and those with a factorial structure of treatments [5]. From this point development was rapid, so that the analysis of variance was being used for a wide range of problems that involved the studying of a linear hypothesis*. The original problem of analyzing data from block designs acquired a new dimension from the use of matrices, due to Tocher [3], and the solution of the normal equations in terms of generalized inverses*. *See* GENERAL LINEAR MODEL.

THE METHOD DESCRIBED

In essentials the method depends upon the partition of both degrees of freedom and the sums of squared deviations between a component called "error" and another, which may be termed the "effect," although generally it will have a more specific name. Thus, in the example above the effect was the correlation. The nomenclature should not mislead. The sum of squared deviations for the effect is influenced by error also, which is thought of as an all-pervading uncertainty or noise* distributed so that, in the absence of the effect i.e., on the null hypothesis*, the expectation of the two sums of squared deviations will be in the ratio of their respective degrees of freedom. Hence the mean squared deviations, i.e., the sums of squares divided by their degrees of freedom, should have similar expectations. If, however, the effect does exist, it will inflate its own mean squared deviation but not that of the error, and if large enough, will lead to significance being shown by the F-test. In this context, F equals the ratio of the mean squared deviations for the effect and for error.

In practice, analyses of variance are usually more complicated. In the example, the total line was obtained by minimizing the sum of squared residuals*, $\sum_i \eta_i^2$, in

$$y_i = \alpha + \eta_i,$$

whereas that for error came from minimizing a similar quantity in

$$y_i = \alpha + \beta x_i + \eta_i.$$

In short, the test really investigated the existence or non-existence of β. In the example α is common to both minimizations, representing as it does a quantity needed to complete the model but not itself under test, whereas β was in question. That is the general pattern. Thus in a randomized block design the blocks form a convenient "garbage can" where an ingenious experimenter can dispose of unwanted effects such as spatial position, different observers, sources of material, and much else that would disturb the experiment if not controlled. Consequently, they must be allowed for, although no one is studying the contents of garbage cans. There will also be parameters for treatments*, which are under study. Minimization with respect to the block parameters alone gives a measure of the remaining variation, i.e., that due to treatments and uncontrollable error. A further minimization on block and treatments parameters together gives the error line, that for treatments being found by difference. (The fact that it can be found more easily by direct calculation obscures its real origins.) The block line, relating as it does to variation that has been eliminated, is not really relevant but is ordinarily included.

Such an analysis is called "intrablock"*, studying as it does variation within blocks and discarding any between them. In some instances it is possible to derive an "interblock"* analysis, in which the block parameters are regarded as random variables. The procedure then is to minimize the sum of their squares, both when the treatment parameters are included and when they are excluded. The additional information can be worthwhile, but not necessarily so. For example, if each block is made up in the same way with respect to treatments, a study of the differences between blocks can provide no information about the effects of treatments. Also, unless the number of blocks appreciably exceeds that of treatments, there will not be enough degrees of freedom to determine the interblock error properly. Not least, if good use has been made of blocks as garbage cans, the distribution of their parameters must be regarded as arbitrary.

Complications arise when there are several effects. Here it is advisable to form the error allowing for them all, although that will be considered in more detail below. The problems arise rather in deciding the order of testing, but that is often a matter of logic rather than statistics. For example, with a factorial design* of treatments, if it appears that there is an interaction* of factors A and B, the conclusion should be that the response to the various levels of A depends on the level of B, and *vice versa*. If that is so, there is no point in examining the main effects of factors A and B, since each relates to the response to the levels of one factor when it has been averaged over levels of the other. The only true interpretation must rest upon a two-way table* of means. Again, if the example is extended to cover parabolic effects, i.e.,

$$y_i = \alpha + \beta x_i + \gamma x_i^2 + \eta_i,$$

and if it appears that γ should be included in the model, i.e., the relationship of x_i and y_i is not a straight line, there need be no detailed study of β, since it has no meaning except as the slope of such a line. However, it is always necessary to consider what really is under test. For example, it could be that

$$y_i = \alpha + \gamma x_i^2 + \eta_i$$

was the expected relationship and doubts had arisen whether βx_i was not needed as well. The analysis of variance is an approach of wonderful subtlety, capable of adaptation to a wide range of problems. It is used to best advantage when it reflects the thinking and questioning that led to the inception of the investigation in the first place. Consequently, each analysis should be individual and should study each question in logical order. *There is no place for automated procedures, as if all research programs raised the same questions*. Also, although the analysis of variance had its origins in the testing of hypotheses*, there is no reason for leaving it there. It can shed light on the sources of experimental error*; it can suggest confidence limits* for means and differences of means and much else. In the hands of a thoughtful user it has unimagined potentiality; as an unthinking process it leads to few rewards.

NUMERICAL EXAMPLE

The data in Table 1 [2] represent yields per plot from an apple experiment in four randomized blocks, I through IV. There were six treatments. One of them, S, was the standard practice in English apple orchards of keeping the land clean during the summer, letting the weeds grow up in the fall, and turning them in for green manure in the spring. The rest, A through E, represented alternative methods in which the ground was kept covered with a permanent crop. The interest then lay in finding out if any of the other methods showed any improvement over S.

It is first necessary to find the sum of squared deviations ignoring treatments and considering only blocks. The estimated value for each plot, i, is

$$y_i = \beta_j + \eta_i,$$

where β_j is the parameter for the block, j, in which it finds itself. It will quickly appear that the sum of $\eta_i^2 = (y_j - \beta_j)^2$ is minimized where β_i is taken to be the appropriate block mean, i.e.,

$$\beta_1 = 235.50, \quad \beta_2 = 248.50,$$
$$\beta_3 = 268.67, \quad \beta_4 = 350.33.$$

With these values known it is possible to write down η_i for each plot; e.g., those for blocks I and II and treatments A and B are

$$\begin{matrix} 51.50 & 41.50 & \cdots \\ 35.50 & -39.50 & \cdots \\ \vdots & \vdots & \end{matrix}$$

The sum of these quantities squared comes to 24, 182; it has 20 (= 24 − 4) degrees of freedom, since there are 24 data to which four independent parameters have been fitted. It is now required to fit treatment parameters as well, i.e., to write

$$y_i = \beta_j + \gamma_k + \eta_i.$$

In a randomized block design* in which all treatment totals are made up in the same way with respect to block parameters, i.e., the design is orthogonal*, it is sufficient to estimate a treatment parameter, γ_k, as the difference of the treatment mean from the general mean. The table of deviations, η_i, now starts

$$\begin{matrix} 42.75 & 32.75 & \cdots \\ 43.50 & -31.50 & \cdots \\ \vdots & \vdots & \end{matrix}$$

The sum of these squares is now 23, 432 with 15 degrees of freedom, because an additional five degrees of freedom have been used to estimate how the six treatment means diverge from the general mean. (Note that only five such quantitites are independent. When five have been found, the sixth is known.) The analysis of variance is therefore

Source	d.f.	Sum of Squares	Mean Square
Treatment	5	750	150
Error	15	23, 432	1562
Treatments + error	20	24, 182	

Table 1 Yields in Pounds per Plot over Four Seasons from an Experiment on Soil Management with Apple Trees

Treatment	Blocks I	II	III	IV	Totals
A	287	290	254	307	1138
B	271	209	243	348	1071
C	234	210	286	371	1101
D	189	205	312	375	1081
E	210	276	279	344	1109
S	222	301	238	357	1118
Totals	1413	1491	1612	2102	6618

There is no suggestion that the treatment mean square has been inflated relative to the error, and therefore no evidence that the treatments in general have had any effect. However, the study really relates to comparisons between A through E and S. In view of the orthogonality of the design and the fact that each treatment mean is based on four data, the variance of a difference of two means is $(\frac{1}{4} + \frac{1}{4})1562 = 781$, the standard error being the square root of that quantity, i.e., 27.9. There is no question of any other treatment being an improvement on the standard because all except A give smaller means. (However, the situation can be changed; *see* ANALYSIS OF COVARIANCE.)

The analysis above was for an orthogonal design. If, however, the datum for treatment A in block 1 had been missing, a more complicated situation would have arisen. (It is true that in practice a missing plot value would be fitted, but it is possible to carry out a valid analysis of variance without doing that.) The deviations allowing only for blocks start

$$\begin{array}{ccc} — & 32.75 & \cdots \\ 45.80 & -47.50 & \cdots \\ \vdots & \vdots & \end{array}$$

the mean for block I being now 225.2. The sum of squared deviations is 20,999 with 19 degrees of freedom. The so-called normal equations*, derived from the block and treatment totals, are

$$\begin{array}{ll} 1126 = 5\beta_1 + \sum\gamma - \gamma_1 & 1138 = \sum\beta - \beta_1 + 3\gamma_1 \\ 1491 = 6\beta_2 + \sum\gamma & 1071 = \sum\beta \qquad + 4\gamma_2 \\ 1612 = 6\beta_3 + \sum\gamma & 1101 = \sum\beta \qquad + 4\gamma_3 \\ 2102 = 6\beta_4 + \sum\gamma & 1081 = \sum\beta \qquad + 4\gamma_4 \\ & 1109 = \sum\beta \qquad + 4\gamma_5 \\ & 1118 = \sum\beta \qquad + 4\gamma_6, \end{array}$$

where $\sum\beta = \beta_1 + \beta_2 + \beta_3 + \beta_4$ and $\sum\gamma = \gamma_1 + \gamma_2 + \gamma_3 + \gamma_4 + \gamma_5 + \gamma_6$. At this point it is necessary to say that no one in practice solves the normal equations as they stand because better methods are available, e.g., the Kuiper–Corsten iteration* or the use of generalized inverses*. However, there is no objection in principle to a direct solution, the difficulty being that there are not enough equations for the parameters. An equation of constraint, e.g.,

$$3\gamma_1 + 4\gamma_2 + 4\gamma_3 + 4\gamma_4 + 4\gamma_5 + 4\gamma_6 = 0,$$

can always be used. Further, the one just suggested, which makes the treatment parameters sum to zero over the whole experiment, is very convenient. It follows that

$$\begin{array}{ll} \beta_1 = 224.34 & \gamma_1 = -5.74 \\ \beta_2 = 248.74 & \gamma_2 = -5.39 \\ \beta_3 = 268.91 & \gamma_3 = 2.11 \\ \beta_4 = 350.57 & \gamma_4 = -2.89 \\ & \gamma_5 = 4.11 \\ & \gamma_6 = 6.36. \end{array}$$

Hence subtracting the appropriate parameters from each datum, the values of η_i are

$$\begin{array}{ccc} — & 47.00 & \cdots \\ 52.05 & -34.35 & \cdots \\ \vdots & \vdots & \end{array}$$

The sum of their squares is now 20,508 with 14 degrees of freedom, yielding the following analysis of variance:

Source	d.f.	Sum of Squares	Mean Square
Treatments	5	491	98
Error	14	20,508	1465
Treatments + error	19	20,999	

MULTIPLE COMPARISONS*

Over the years there has grown up an alternative approach to testing in the analysis of variance. As long as there are only two treatments or two levels of a factor, the *F*-test has a clear meaning but if there are three or more, questions arise as to where the differences are. Some care is needed here, because background knowledge is called for. If, for example, it had appeared that different varieties of wheat gave different percentages of a vegetable protein in their grain, the result would surprise no one and would merely indicate a need to assess each variety separately. At the other extreme, if the treatments formed a highly structured set, it might be quite obvious what should be investigated next. Thus if it had appeared that the outcome of a chemical reaction depended upon the particular salt used to introduce a metallic element, the science of chemistry is so developed that a number of lines of advance could probably be suggested immediately, some of which might receive preliminary study from further partition of the treatment line. (Of course, no body of data can confirm a hypothesis suggested by itself.) Sometimes, however, the experimenter is in the position of suspecting that there could be a structure but he does not know what it is. It is then that multiple comparison tests have their place. Like much else in the analysis of variance, their unthinking use presents a danger, but that is not to deny them a useful role. Also, an experimenter confronted with a jumble of treatment means that make no sense could well adopt, at least provisionally, the treatment that gave the highest mean, but would still want to know how it stood in relation to its nearest rivals. It might be so much better that the others could be discarded, or it might be so little better that further study could show that it was not really to be preferred. Such a test can be useful. Like others, it can be abused.

COMPOSITION OF ERROR

In its original form the analysis of variance was applied to data from agricultural field experiments designed in randomized blocks. The error was then clearly identified as the interaction of treatments and blocks, i.e., the F-test investigated the extent to which treatment differences were consistent from block to block. If a difference of means was much the same in each block, it could clearly be relied upon; if it had varying values, the significance was less well established. In other instances the error may be the interaction of treatments and some other factor such as occasion or operator. In all such cases the purport of the test is clear.

Sometimes, however, the error is less open to interpretation, being little more than a measure of deviations from a hypothesis that is itself arbitrary. Thus, if the agricultural field experiment had been designed in a Latin square*, the error would have been made up of deviations from parameters for treatments added to those for an underlying fertility pattern, assumed itself to derive from additive effects of rows and columns. If, as is usually the case, there is no prior reason why the fertility pattern should have that form, there is a danger of the error sum of squared deviations being inflated by sources that will not affect the treatment line, thus reversing the usual position. In fact, it is not unknown for the error mean square deviation to be the larger, and sometimes there is good reason for it.

The subject of error is sometimes approached by distinguishing fixed effects* from random*. In the first, the levels are determinate and reproducible, like pressure or temperature, whereas in the latter they are to be regarded as a random selection of possible values, like the weather on the successive days of an investigation. For many purposes the distinction is helpful. The ideal, as has been suggested, is an error that represents the interaction between a fixed effect and a random effect of the conditions over which generalization is required, but other possibilities can be recognized. For example, an error made up of interactions between fixed effects is nearly useless unless it can be assumed that there will be no real interaction.

Further questions arise when confounding* is introduced. The experimenter may have decided that there can be no interaction of factors A, B, and C and would then be ready to have it confounded. The experimenter may get rid of it in that way if possible, but if it remains in the analysis it should be as part of error. However, if the experimenter believes that $A \times B \times C$ never exists, he or she cannot believe that its value depends upon the level of D, so $A \times B \times C \times D$ also should either be confounded or included in error. Such decisions can be made readily enough before analysis begins; they are more difficult afterward. Plainly, it would be wrong to start with an error that was acceptable and add to it those high-order interactions* that were small and to exclude from it those that were large. The error that finally resulted would obviously be biased*. On the other hand, there are occasions when sight of the data convinces the experimenter that his or her preconceptions were wrong. In that case the experimenter usually does well to confine the analysis to an indubitable error, e.g., the interaction of blocks and treatments, and to regard all else as subject to testing. As with the Latin square*, mistaken assumptions about the composition of error can lead to the inflation of its sum of squared deviations.

The question of what is and what is not error depends to some extent upon the randomization*. Let there be b blocks of xy treatments, made up factorially by x levels of factor X and y levels of factor Y. The obvious partition of the $(bxy - 1)$ degrees of freedom between the data is

Blocks	$(b-1)$
X	$(x-1)$
Y	$(y-1)$
$X \times Y$	$(x-1)(y-1)$
Blocks $\times$ $X(I)$	$(b-1)(x-1)$
Blocks $\times$ $Y(II)$	$(b-1)(y-1)$
Blocks $\times$ $X \times Y(III)$	$(b-1)(x-1)(y-1)$.

No one need object if each of the effects X, Y and X × Y is compared with its own interaction with blocks, i.e., with I, II, and III, respectively. If, however, all treatment combinations have been subject to the same randomization procedure, the components I, II, and III can be merged to give a common error with $(b-1)(xy-1)$ degrees of freedom. In a split-plot* design, where X is allocated at random to the main plots, II and III can together form a subplot error, leaving I to form that for main plots. Given a strip-plot* (or crisscross) design, on the other hand, each component of error must be kept separate. The example illustrates the old adage: "As the randomization is, so is the analysis." Sometimes there are practical difficulties about randomization and they can raise problems in the analysis of data, but they are not necessarily insuperable ones. For example, in the split-plot case it may not be feasible to randomize the main plot factor, X, in which case it is vitiated and so is its interaction with blocks (component I), but it might still be permissible to use the subplot analysis*. Again, if the subplot factor, Y, has to be applied systematically (it might be different occasions on which the main plot was measured), component II may be vitiated, but X × Y can still be compared with component III.

SUITABILITY OF DATA

With so many computer packages available it is easy to calculate an analysis of variance that is little better than nonsense. Some thought therefore needs to be given to the data before entering the package.

Strictly, the data should be continuous. In fact, it is usually good enough that they spread themselves over 10 or more points of a scale. When they do not, e.g., as with a body of data that consists mostly of ones and zeros, the sum of squared deviations for error is inflated relative to those for effects with consequent loss of sensitivity. Discrete data often come from Poisson* or binomial* distributions and may call for one of the variance-stabilizing transformations* to be considered below.

It is also required that the residuals* (the η_i in the examples) should be distributed independently and with equal variance. Independence is most important. Where the data come from units with a spatial or temporal relationship, independence is commonly achieved by a randomization of the treatments. If that is impracticable, the analysis of variance in its usual forms is better avoided. Equality of variance is more problematical. Where treatments are equally replicated*, the F-test is fairly robust* against some treatments giving a higher variance than others. If, however, attention is directed to a subset of treatments, as happens with multiple comparisons* and can happen when the treatment line is partitioned, it is necessary to have a variance estimated specifically for that subset or serious bias could result. Here two main cases need to be distinguished. Some treatments may involve more operations than others and may therefore give rise to larger errors. For example, the injection of a plant or animal can itself give rise to variation that is absent when other methods of administration are adopted. In the other case, the variance for any treatment bears a functional relationship to the mean value for that treatment.

To take the first case, if the error is the interaction of treatment and blocks, it is an easy matter to partition the treatment line into three parts: (1) between the groups, one with higher and one with lower variance; (2) within one group; and (3) within the other. The error can then be partitioned accordingly. Depending upon circumstances, it may be easier to regard each component of treatments as having its own error or to concoct a variance for each contrast of interest between the treatment means. Alternatively, if the treatments of high and low variance are associated with two levels of a factor, e.g., administration by injection as compared with some other method, a better way and one that places fewer constraints on the nature of error may be to group the data into pairs according to the two levels, a pair being the same with respect to other factors, and to analyze separately the sum and difference of data from the pair. In effect, that is virtually the same as regarding the design as if it were in split plots, the factor associated with the divergent variances being the one in subplots.

In the other case, where the variance of any observation depends upon the mean, the usual solution is to find a variance-stabilizing transformation*. Thus, if the variance is proportionate to the mean, as in a Poisson distribution*, it may be better to analyze the square roots of the data rather than the data themselves. Such transformations can be useful, especially when they direct attention to some quantity more fundamental than that measured. Thus given the end product of a growth process, it is often more profitable to study the mean growth rate than the final size, because that is what the treatments have been affecting, and similarly the error has arisen because growth rates have not been completely determined. The approach can, however, cause problems when the transformation is no more than a statistical expedient, especially when it comes to the interpretation of interactions. Thus suppose that an untreated control has led to 90% of insects surviving, applications of insecticide A had given a survival rate of 60% while B had given 45%. A declaration that the two insecticides had not interacted would lead most people to suppose that A and B in combination had given a survival rate of 30% ($30 = 60 \times 45/90$). If the data are analyzed without transformation, a zero interaction would imply 15% for the combination ($15 = 60 + 45 - 90$). Using the appropriate angular transformation, the figures become (0) 71.6, (A) 50.8, (B) 42.1, leading to an expectation for (AB) of 21.3 for zero interaction. This last figure corresponds to 13.2%, which is even further from the ordinary meaning of independence*.

INTERPRETATION OF AN ANALYSIS OF VARIANCE

As has been said, the analysis of variance is a subtle approach that can be molded to many ways of thought. It is at its best when the questions implicit in the investigation are answered systematically and objectively. It is not a method of "data snooping," the treatment line being partitioned and re-partitioned until something can be declared "significant." Nor is it rigid, as if there were some royal way to be followed that would ensure success. Its function is to assist a line of thought, so any unthinking procedure will be unavailing.

The first step is usually to look at the mean squared deviation for error, which sums up the uncontrolled sources of variation. An experienced person may note that its value is lower than usual, which could suggest that improved methods have paid off, or it could be so large as to show that something had gone wrong. If that is the case, an examination of residuals may show which observations are suspect and provide a clue for another time. It may even lead to a positive identification of the fault and the exclusion of some data. The fault, however, must be beyond doubt; *little credence attaches to conclusions based on data selected by the investigator to suit his own purposes*.

Next, it is wise to look at the line for blocks or whatever else corresponds to the control of extraneous variation. It is possible for a research team to fill their garbage cans again and again with things that need not have been discarded. If the block line rarely shows any sign of having been inflated, it could well be that a lot of trouble is being taken to control sources of variation that are of little importance anyway. Also, if the sources of variation are indeed so little understood, it could be that important components are being left in error.

All this is a preliminary to the comparison of treatment effects* and error. Here some thought is needed. First, the partition of the treatment line into individual effects may be conventional but irrelevant to immediate needs. For example, in a $2 \times p$ factorial set of treatments, there may be no interest in the main effects* and interactions*, the intention being to study the response of the factor with two levels in p different conditions. Even if the partition is suitable, the order in which the effects should be studied needs consideration. As has been explained, a positive response to one test may render others unnecessary.

However, the need may not be for testing at all. The data may give a set of means, and it is only necessary to know how well they have been estimated. Even if the need is for testing, there are occasions when an approach by multiple comparisons* is called for rather than by F-test. Also, there are occasions when a significance test* has proved negative, but interest centers on its power*; that is, the enquiry concerns the probability of the data having missed a difference of specified size, supposing that it does exist.

A difficult situation arises when a high-order interaction appears to be significant but without any support from the lower-order interactions contained in it. Thus if $A \times B \times C$ gives a large value of F while $B \times C$, $A \times C$, $A \times B$, A, B, and C all seem unimportant, it is always possible that the 1 : 20 chance, or whatever it may be, has come off. Before dismissing awkward effects, however, it is as well to look more closely at the data. The whole interaction could depend upon one observation that is obviously wrong. If, however, all data for a particular treatment combination go the same way, whether up or down, there is the possibility of some complicated and unsuspected phenomenon that requires further study.

Anyone who interprets an analysis of variance should watch out for the inflated error. If there is only one treatment effect and that is small, little can be inferred, but if several F-values for treatment effects are well below expectation, it is at least possible that the error has been badly conceived. For example, blocks may have been chosen so ineptly that, instead of bringing together similar plots or units, each is composed of dissimilar ones. Again, the distribution of the residuals may be far from normal. Randomization* may have been inadequate, or the error may represent deviations from an unlikely model. The matter should not be left. A valuable pointer might be obtained to the design of better investigations in the future.

References

[1] Fisher, R. A. (1925). *Statistical Methods for Research Workers*. Oliver & Boyd, Edinburgh.

[2] Pearce, S. C. (1953). *Field Experimentation with Fruit Trees and Other Perennial Plants. Tech. Commun. Bur. Hort. Plantation Crops*, Farnham Royal, Slough, England, **23**. App IV.

[3] Tocher, K. D. (1952). *J. R. Statist. Soc. B*, **14**, 45–100.

[4] Yates, F. (1933). *J. Agric. Sci.*, **23**, 108–45.

[5] Yates, F. (1937). The Design and Analysis of Factorial Experiments. *Tech. Commun. Bur. Soil Sci. Rothamsted*, **35**.

Further Reading

A standard work is H. Scheffé's book *The Analysis of Variance* (Wiley, New York, 1959). Other useful books are *Statistical and Experimental Design in Engineering and the Physical Sciences* by N. L. Johnson and F. C. Leone (Wiley, New York, 1966), especially Volume 2, and *The Linear Hypothesis: A General Theory* by G. A. F. Seber (Charles Griffin, London, 1966). In *Experiments: Design and Analysis* (Charles Griffin, London, 1977), J. A. John and M. H. Quenouille discuss the analysis of variance for many standard designs, and G. B. Wetherill's *Intermediate Statistical Methods* (Chapman & Hall, London, 1980) looks with some care at various models. In Chapter 9 of their book, *Applied Regression Analysis* (Wiley, New York, 1966), N. R. Draper and H. Smith show the relationship of the analysis of variance to regression methods. Other useful references are Chapter 3 of C. R. Rao's *Advanced Statistical Methods in Biometric Research*, and Chapter 4 of G. A. F. Seber's *Linear Regression Analysis*, both published by Wiley, New York, the first in 1952 and the second in 1976.

(AGRICULTURE, STATISTICS IN
ANALYSIS OF COVARIANCE
CONFOUNDING
DESIGN AND ANALYSIS OF EXPERIMENTS
F-TESTS
GENERAL LINEAR MODEL
MULTIPLE COMPARISONS
REGRESSION ANALYSIS)

S. C. PEARCE

ANCILLARY STATISTICS

Let X be an observable vector of random variables with probability density function* (PDF) $f_X(x;\theta)$, where θ is an unknown parameter taking values over a space Λ. If the distribution of the statistic, or vector of statistics, $A = a(X)$ is not dependent upon θ, then A is said to be ancillary for (the estimation of) θ. Suppose now that X is transformed in a 1 : 1 manner to the pair (S, A). The joint PDF of S, A can be written

$$f_{S\,|\,A}(s;\theta\,|\,a)f_A(a), \qquad (1)$$

where the second term is free of θ. As the examples below will make apparent, an ancillary statistic often indexes the precision of an experiment; certain values of A indicate that the experiment has been relatively informative about θ; others indicate a less informative result. For this reason, it is often argued that procedures of inference should be based on the conditional distribution* of the data given the observed value of the ancillary. For example, estimation* and hypothesis-testing* procedures are based on the first term of (1), and their frequency properties are evaluated over the reference set in which $A = a$ is held fixed at its observed value.

In the simplest context, ancillary statistics arise in experiments with random sample size.

Example 1. Contingent upon the outcome of a toss of an unbiased coin, either 1 or 10^4 observations are taken on a random variable Y which has a $N(\theta, 1)$ distribution. The sample size N is ancillary and, in estimating or testing hypotheses about θ, it seems imperative that N be regarded as fixed at its observed value. For example, a size α test of $\theta = 0$ versus the one-sided alternative $\theta > 0$ has critical region* $\overline{Y} > z_\alpha$ if $N = 1$ or $\overline{Y} > 10^{-2}z_\alpha$ if $N = 10^4$, where $\overline{Y}$ is the sample mean and z_α is the upper α point of a $N(0, 1)$ distribution. Conditional on the observed value of N, this test is uniformly most powerful* (UMPT), and since this test is conditionally of size α for each N, it is also unconditionally of size α. It is discomforting, however, that power considerations when applied to the unconditional experiment do not lead to this test [5].

A second example illustrates further the role of ancillary statistics.

Example 2. Let $X_{1:n}, \ldots, X_{n:n}$ be the order statistics* of a random sample from the density $f(x - \theta)$, where f is of specified form and $-\infty < \theta < \infty$. It is easily seen from considerations of the group of location transformations that the statistics $A_i = X_{i:n} - X_{1:n}$, $i = 2, \ldots, n$, are jointly ancillary for θ. R. A. Fisher* [7] describes these statistics as giving the "configuration" of the sample and their observed values can be viewed as being descriptive of the observed likelihood* function. For example, if $n = 2$ and $f(z) = \pi^{-1}(1 + z^2)^{-1}$, $-\infty < z < \infty$, the likelihood is unimodal if $A_2 \leqslant 1$ and bimodal if $A_2 > 1$, while $\overline{X} = (X_1 + X_2)/2$ is the maximum likelihood estimator* in the first case and a local minimum in the second. Here again, a conditional approach to the inference problem is suggested.

By an easy computation, the conditional density of $M = X_{1:n}$ given $A_2 = a_2, \ldots, A_n = a_n$ is

$$c \prod_{1}^{n} f(m + a_i - \theta), \qquad (2)$$

where $a_1 = 0$ and c is a constant of integration. The choice of the minimum M is arbitrary; the maximum likelihood estimator T or any other statistic that measures location may be used in its place. All such choices will lead to equivalent conditional inferences*. The essential point is that inferences can be based on the conditional distribution*(2).

Both of the foregoing examples suggest that, at least in certain problems, a conditionality principle is needed to supplement other statistical principles. The approach then seems clear: In evaluating the statistical evidence, repeated sampling criteria should be applied to the conditional experiment defined by setting ancillary statistics at their observed values. As will be discussed in the section "Nonuniqueness and Other Problems," however, there are some difficulties in applying this directive.

It should be noted that some authors require that an ancillary A be a function of the minimal sufficient statistic* for θ. This is discussed further in the section just mentioned and in "Conditionality and the Likelihood Principle."

RECOVERY OF INFORMATION

Ancillary statistics were first defined and discussed by Fisher [7], who viewed their recognition and use as a step toward the completion of his theory of exhaustive estimation [8, pp. 158 ff.]. For simplicity, we assume that θ is a scalar parameter and that the usual regularity conditions apply so that the Fisher information* may be written

$$I_X(\theta) = E\left[\partial \log f_X(x;\theta)/\partial\theta\right]^2$$
$$= -E\left[\partial^2 \log f_X(x;\theta)/\partial\theta^2\right].$$

If the maximum likelihood estimator $T = \hat{\theta}(X)$ is sufficient* for θ, then $I_T(\theta) = I_X(\theta)$ for all $\theta \in \Lambda$. Fisher calls T exhaustive because all information is retained in reducing the data to this scalar summary.

It often happens, however, that T is not sufficient and that its sole use for the estimation of θ entails an information loss measured by $I_X(\theta) - I_T(\theta)$, which is nonnegative for all θ and positive for some θ. Suppose, however, that T can be supplemented with a set of ancillary statistics A such that (T, A) are jointly sufficient* for θ. The conditional information in T given $A = a$ is defined as

$$I_{T|A=a}(\theta)$$
$$= -E\left[\partial^2 \log f_{T|A}(t;\theta|a)/\partial\theta^2 \,|\, A = a\right]$$
$$= -E\left[\partial^2 \log f_{T,A}(t,a;\theta)/\partial\theta^2 \,|\, A = a\right],$$

since $f_A(a)$ is free of θ. Thus since T, A are jointly sufficient,

$$E\left[I_{T|A}(\theta)\right] = I_{T,A}(\theta) = I_X(\theta).$$

The average information in the conditional distribution of T given A is the whole of the information in the sample. The use of the ancillary A has allowed for the total recovery of the information on θ. Depending on the particular observed outcome $A = a$, however, the conditional information $I_{T|A=a}(\theta)$ may be greater or smaller than the expected information $I_X(\theta)$.

Viewed in this way, an ancillary statistic A quite generally specifies the informativeness of the particular outcome actually observed.

To some extent, the usefulness of an ancillary is measured by the variation in $I_{T|A}(\theta)$.

Although only a scalar parameter θ has been considered above, the same general results hold also for vector parameters. In this case, $I_X(\theta)$ is the Fisher information matrix* and $I_X(\theta) - I_T(\theta)$ is nonnegative definite. If T is the vector of maximum likelihood estimators, A is ancillary and T, A are jointly sufficient, the conditional information matrix, $I_{T|A}(\theta)$, has expectation $I_X(\theta)$ as above.

NONUNIQUENESS AND OTHER PROBLEMS

Several difficulties arise in attempting to apply the directive to condition on the observed values of ancillary statistics.

1. There are no general constructive techniques for determining ancillary statistics.
2. Ancillaries sometimes exist which are not functions of the minimal sufficient statistic*, and conditioning upon their observed values can lead to procedures that are incompatible with the sufficiency principle.
3. There is, in general, no maximal ancillary.

In this section we look at problems 2 and 3. It should be noted that in certain problems (e.g., Example 2), group invariance* arguments provide a partial solution to problem 1. Even in such problems, however, there can be ancillaries present which are not invariant. An interesting example is given by Padmanabhan [10]. In the context of Example 2 with $f(\cdot)$ a standard normal density* and $n = 2$, he defines the statistic

$$B = \begin{cases} X_1 - X_2 & \text{if } X_1 + X_2 \geqslant 1, \\ X_2 - X_1 & \text{if } X_1 + X_2 < 1, \end{cases}$$

and shows that B is ancillary but not invariant.

Basu [2] has given several examples of nonunique ancillary statistics. The first of these concerns independent bivariate normal* variates $(X_i, Y_i), i = 1, \ldots, n$, with means 0, variances 1, and correlation ρ. In this example, $(X_1, \ldots, X_n)$ and $(Y_1, \ldots, Y_n)$ are each ancillary and conditional inference would clearly lead to different inferences on ρ. This is an example of problem 2 above, and to avoid this difficulty many authors require that the ancillary be a function of the minimal sufficient statistic (e.g., ref. 5). If an initial reduction to the minimal sufficient set, $\sum X_i^2 + \sum Y_i^2$ and $\sum X_i Y_i$, is made, there appears to be no ancillary present.

Not all examples of nonuniqueness are resolved by this requirement. Cox [4] gives the following example, which derives from another example of Basu.

Example 3. Consider a multinomial distribution* on four cells with respective probabilities $(1-\theta)/6$, $(1+\theta)/6$, $(2-\theta)/6$, and $(2+\theta)/6$, where $|\theta| < 1$. Let $X_1, \ldots, X_4$ represent the frequencies in a sample of size n. Each of the statistics

$$A_1 = X_1 + X_2 \qquad A_2 = X_1 + X_4$$

is ancillary for θ, but they are not jointly ancillary.

If conditional inference* is to be useful in such problems, methods for selecting from among competing ancillaries are needed. Cox [4] notes that the usefulness of an ancillary is related to the variation in $I_{T|A}(\theta)$ (see the preceding section) and suggests (again with scalar θ) that the ancillary be chosen to maximize

$$\text{var}\{I_{T|A}(\theta)\}.$$

In general, this choice may depend on θ. In the example above, however, Cox shows that A_1 is preferable to A_2 for all θ. The choice of variance as a measure of variation is, of course, arbitrary.

Barnard and Sprott [1] argue that the ancillary's role is to define the shape of the likelihood function, and that, in some problems, invariance* considerations lead to a straightforward selection between competing ancillaries. In the example above, the estimation problem is invariant under reflections with $\theta \leftrightarrow -\theta$ and $X_1 \leftrightarrow X_2, X_3 \leftrightarrow X_4$. Under this transformation, the ancillary A_1 is invariant while $A_2 \leftrightarrow n - A_2$. Thus under a natural group of transformations, the ancillary A_1 and not A_2 is indicated. This type of argument suggests that *invariance**, and not ancillarity, is the key concept.

CONDITIONALITY AND THE LIKELIHOOD PRINCIPLE*

In a fundamental paper, Birnbaum [3] formulates principles of sufficiency, conditionality, and likelihood and shows that the sufficiency* and conditionality principles are jointly equivalent to the likelihood principle. In this section we outline Birnbaum's arguments and some of the subsequent work in this area.

Birnbaum introduces the concept of the "total evidential meaning" (about θ) of an experiment E with outcome x and writes $\mathcal{E}v(E,x)$. Total evidential meaning is left undefined but the principles are formulated with reference to it, as follows.

The Sufficiency Principle (S). Let E be an experiment with outcomes x, y and t be a sufficient statistic. If $t(x) = t(y)$, then

$$\mathcal{E}v(E,x) = \mathcal{E}v(E,y).$$

This principle (S) is almost universally accepted by statisticians, although some would limit the types of experiments E to which the principle applies.

Let $L(\theta; x, E)$ denote the likelihood* function of θ on the data x from experiment E.

The Likelihood Principle* (L). Let E_1 and E_2 be experiments with outcomes x_1 and x_2, respectively. Suppose further that

$$L(\theta; x_1, E_1) \propto L(\theta; x_2, E_2).$$

Then $\mathcal{E}v(E_1, x_1) = \mathcal{E}v(E_2, x_2)$.

This principle (L) (sometimes called the strong likelihood principle) asserts that only the observed likelihood is relevant in assessing the evidence. It is in conflict with methods of significance testing*, confidence interval procedures, or indeed any methods that are based on repeated sampling*. The sufficiency principle is sometimes called the weak likelihood principle since, by the sufficiency of the likelihood function, it is equivalent to the application of the likelihood principle to a single experiment.

An experiment E is said to be a mixture experiment* with components E_h if, after relabeling of the sample points, E may be thought of as arising in two stages. First, an observation h is made on a random variable H with known distribution, and then x_h is observed from the component experiment E_h. The statistic H is an ancillary statistic.

The Conditionality Principle (C). Let E be a mixture experiment with components E_h. Then

$$\mathcal{E}v(E,(h,x_h)) = \mathcal{E}v(E_h, x_h).$$

This principle asserts that the inference we should draw in the mixture experiment with outcome (h, x_h) should be the same as that drawn from the simpler component experiment E_h when x_h is observed.

Birnbaum's Theorem. (S) + (C)$\Leftrightarrow$(L).

Proof. It follows immediately that (L) implies (C). Also (L)$\Rightarrow$(S) since, by the factorization theorem*, two outcomes giving the same value of a sufficient statistic yield proportional likelihood functions. To show that (C) and (S) together imply (L), let E_1 and E_2 be two experiments with outcomes x_1 and x_2, respectively, such that $L(\theta; x_1, E_1) \propto L(\theta; x_2, E_2)$. Let $\Pr[H = 1] = 1 - \Pr[H = 2] = p$ be a specified nonzero probability. In the mixture experiment E with components E_h, $h = 1, 2$, the outcomes $(H = 1, x_1)$ and $(H = 2, x_2)$ give rise to proportional likelihoods. Since the likelihood function is itself minimally sufficient, (S) implies that

$$\mathcal{E}v(E,(H=1,x_1)) = \mathcal{E}v(E,(H=2,x_2)).$$

On the other hand, (C) implies that

$$\mathcal{E}v(E,(H=h,x_h)) = \mathcal{E}v(E_h,x_h), \qquad h=1,2,$$

and hence $\mathcal{E}v(E_1,x_1) = \mathcal{E}v(E_2,x_2)$. Thus (S) + (C) ⇒ (L).

Since almost all frequency-based methods of inference contradict the likelihood principle, this result would seem to suggest that sufficiency and conditionality procedures are jointly incompatible with a frequency theory. It should be noted, however, that the foregoing argument pertains to the symmetric use of sufficiency and conditionality arguments. The likelihood principle would appear to follow only under such conditions.

Durbin [6] restricted the conditionality principle to apply only after an initial reduction to the minimal sufficient statistic. He defined a reduced experiment E' in which only T is observed and considered a revised conditionality principle (C′) which applied to this reduced experiment. In essence, this restricts attention to ancillary statistics that are functions of the minimal sufficient statistic. This change apparently obviates the possibility of deducing (L).

A second approach [9] classifies ancillaries as being experimental or mathematical. The former are ancillary by virtue of the experimental design and the purpose of the investigation. They are ancillary statistics regardless of the parametric model chosen for the chance setup being investigated. Mathematical ancillaries, on the other hand, are ancillary because of the particular parametric model assumed. In the examples given above, N is an experimental ancillary in Example 1 while the ancillaries A_1 and A_2 in Example 3 are mathematical. Example 2 may be interpreted in two ways. If this is a measurement model* whereby the response X arises as a sum, $X = \theta + e$, e being a real physical entity with known distribution, then the ancillaries $A_2, \ldots, A_n$ are experimental. (The purpose of the experiment is to determine the physical constant θ.) More usually, an experiment with data of this type is designed to determine the distribution of X and the model $f(x-\theta)$ is a preliminary specification. In this case, the ancillaries are mathematical. The primary principle is taken to be the experimental conditionality principle and other principles (e.g., sufficiency) are applied only after conditioning on any experimental ancillaries present.

References

[1] Barnard, G. A. and Sprott, D. A. (1971). In *Waterloo Symposium on Foundations of Statistical Inference*. Holt, Rinehart and Winston, Toronto, pp. 176–196. (Investigates relationships between ancillaries and the likelihood function.)

[2] Basu, D. (1964). *Sankhyā A*, **26**, 3–16. (Discusses the problem of nonuniqueness and gives several interesting examples.)

[3] Birnbaum, A. (1962). *J. Amer. Statist. Ass.* **57**, 269–306. (A fundamental paper dealing with the formalization of inference procedures and principles.)

[4] Cox, D. R. (1971). *J. R. Statist. Soc. B*, **33**, 251–255.

[5] Cox, D. R. and Hinkley, D. V. (1974). *Theoretical Statistics*, Chapman & Hall, London. (Probably the most complete text reference on ancillary statistics.)

[6] Durbin, J. (1970). *J. Amer. Statist. Ass.* **65**, 395–398.

[7] Fisher, R. A. (1936). *Proc. Amer. Acad. Arts Sci.* **71**, 245–258. (First discussion of the existence and use of ancillary statistics.)

[8] Fisher, R. A. (1973). *Statistical Methods and Scientific Inference*, 3rd ed. Oliver & Boyd, Edinburgh. (A discussion of the recovery of information and the use of ancillaries.)

[9] Kalbfleisch, J. D. (1975). *Biometrika*, **62**, 251–268.

[10] Padmanabhan, A. R. (1977). *Amer. Statist.* **31**, 124.

(BASU THEOREMS
CONDITIONAL INFERENCE
FIDUCIAL INFERENCE
LIKELIHOOD
STATISTICAL INFERENCE
SUFFICIENCY)

JOHN D. KALBFLEISCH

ANDERSON–DARLING TEST FOR GOODNESS OF FIT

The Anderson–Darling statistic is a member of the group of goodness-of-fit* statistics which has come to be known as empirical distribution function (EDF) statistics [6] because they measure the discrepancy between the empirical distribution function* of a given sample and the theoretical distribution to be tested. It is designed to test that random variable X has a continuous cumulative distribution function $F_X(x;\theta)$; θ is a vector of one or more parameters entering into the distribution function. Thus for the normal distribution*, the vector $\theta = (\mu, \sigma^2)$. Where there is no ambiguity, $F(x;\theta)$ or $F(x)$ will be written for $F_X(x;\theta)$. The empirical distribution function* (EDF) is defined as $F_n(x)$ = (proportion of sample $\leqslant x$), and a family of statistics measuring the discrepancy between $F_n(x)$ and $F(x;\theta)$ is the Cramér–von Mises family*

$$W^* = \int_{-\infty}^{\infty} [F_n(x) - F(x;\theta)]^2 \psi(x)\,dx; \tag{1}$$

the function $\psi(x)$ gives a weighting to the squared difference between $F_n(x)$ and $F(x;\theta)$.

The Anderson–Darling statistic is W^* with $\psi(x) = [\{F(x;\theta)\}\{1 - F(x;\theta)\}]^{-1}$. This weight function has the effect of giving greater importance to observations in the tail than do other EDF statistics, counteracting the fact that $F_n(x) - F(x;\theta)$ approaches zero in each tail. Since tests of fit are often needed implicitly or explicitly to guard against wayward observations in the tails, the statistic is a recommended one, with generally good power* properties over a wide range of alternative distributions when $F(x;\theta)$ is not the true distribution [6, 9].

The numerical calculation of the Anderson–Darling statistic A^2 is done by the following two steps: Calculate

$$z_i = F(x_{(i)};\theta) \qquad (i = 1, \ldots, n). \tag{2}$$

Then

$$A^2 = -\left\{ \sum_{i=1}^{n} (2i-1)[\ln z_i + \ln(1 - z_{n+1-i})] \right\} / n - n. \tag{3}$$

Note that since the $x_{(i)}$ are in ascending order, the z_i will also be in ascending order, although the usual notation of order statistics has now been omitted.

GOODNESS-OF-FIT TEST* FOR A COMPLETELY SPECIFIED CONTINUOUS DISTRIBUTION

For the calculations above, the tested distribution $F(x;\theta)$ must be completely specified, i.e., the parameters in θ must be known. When this is the case we describe the situation as case 0. The statistic A^2 was introduced by Anderson and Darling [1, 2] and for case 0 they gave the asymptotic distribution and a table of percentage points*. Large values of A^2 will indicate a bad fit. The distribution of A^2 for a finite sample rapidly approaches the asymptotic distribution and for practical purposes this distribution can be used for sample sizes greater than 5. A table of percentage points is given in Table 1*a*. To make the goodness-of-fit test, A^2 is calculated as in (2) and (3) and compared with these percentage points; the null hypothesis* that random variable X has the distribution $F(x;\theta)$ is rejected at level α if A^2 exceeds the appropriate percentage point at this level.

ASYMPTOTIC THEORY OF THE ANDERSON–DARLING STATISTIC

The distribution of A^2 for case 0 is the same for all distributions tested. This is because the probability integral transformation* is made at (2), and the values of z_i are ordered values from a uniform distribution* with limits 0 and 1. A^2 is therefore a function of ordered uniform random variables. When θ contains unknown components, the z_i given by transformation (2) using an estimate $\hat{\theta}$ instead of θ will not be ordered uniform random variables, and the distribution theory of A^2 (as for all other EDF statistics) becomes substantially more difficult. In general, the distribution of A^2 will depend on n and also on the values of the unknown parameters.

Table 1 Modifications A^* of A^2, and Upper Tail Percentage Points, for the Following Test Situations: (*a*) Case 0 (Completely Specified Distribution); (*b*) Normal Distribution; (*c*) Exponential Distribution

Modification A^*	Upper Tail Percentage Level α			
	0.10	0.05	0.025	0.01
(*a*) None needed	1.933	2.492	3.070	3.857
(*b*) Case 1[a]	0.894	1.087	1.285	1.551
Case 2[a]	1.743	2.308	2.898	3.702
Case 3: $A^* = A^2(1.0 + 0.75/n + 2.25/n^2)$	0.631	0.752	0.873	1.035
(*c*) $A^* = A^2(1.0 + 0.3/n)$	1.062	1.321	1.591	1.959

[a]For cases 1 and 2, normal distribution tested, no modifications have been calculated. The percentage points given can be used with unmodified A^2 for $n > 20$.

Fortunately, an important simplification occurs when unknown components of θ are location and scale parameters* only; then the distribution of an EDF statistic, with an appropriate estimate for θ, will depend on the distribution tested, but not on the specific values of the unknown parameters. This simplification makes it worthwhile to calculate the asymptotic theory and percentage points for A^2, for special distributions with location* and scale* parameters, and this has been done for the normal* and exponential* distributions by Stephens [6, 7] and Durbin et al. [3]; for the extreme value distribution* (these can be used for the Weibull distribution* also) and for the logistic distribution* by Stephens [8, 10]; and for the gamma distribution* with unknown scale parameter but known shape parameter, by Pettitt and Stephens [4]. Many authors (for references, see Stephens [6]) have provided Monte Carlo* points for finite n for EDF statistics. Stephens, in the papers quoted, has used these points to provide modifications to the statistics. These are functions of the statistic and of n which can then be used with the asymptotic percentage points only. Thus each test situation needs only one line of percentage points.

The first step in testing goodness of fit for any of these distributions is to estimate the unknown parameters. This should be done by maximum likelihood*, for the modifications and asymptotic theory to hold. Suppose that $\hat{\theta}$ is the vector of parameters, with any unknown parameters estimated as above. Vector $\hat{\theta}$ replaces θ in (2), to give the z_i, and A^2 is always calculated from (3). A^2 is then modified as described, to give A^*, and A^* is compared with the line of percentage points given. These are the percentage points of the asymptotic distribution of A_2 for the test situation considered.

TESTS FOR THE NORMAL DISTRIBUTION*

The procedure will be illustrated for tests for the normal distribution. Following Stephens [6], we distinguish three cases:

Case 1. The mean μ is unknown and is estimated by $\bar{x}$, the sample mean, but σ^2 is known.

Case 2. The mean μ is known, and σ^2 is estimated by $\sum_i (x_i - \mu)^2/n$ ($= s_1^2$, say).

Case 3. Both parameters are unknown and are estimated by $\bar{x}$ and $s^2 = \sum_i (x_i - \bar{x})^2 /(n-1)$.

For these cases the calculation of z_i is done

in two stages. First, w_i is calculated from

$$w_i = \frac{x_{(i)} - \bar{x}}{\sigma} \quad \text{(case 1)},$$

$$w_i = \frac{x_{(i)} - \mu}{s_1} \quad \text{(case 2)},$$

$$w_i = \frac{x_{(i)} - \bar{x}}{s} \quad \text{(case 3)},$$

then z_i is the cumulative probability of a standard normal distribution, to the value w_i, found from tables or computer routines. The value of A^2 is then calculated from (3). To make the test, use the modification and percentage points given in Table 1*b*, for the appropriate case.

Illustration

The following value of men's weights in pounds, first given by Snedecor, were used by Shapiro and Wilk [5] as an illustration of another test for normality: 148, 154, 158, 160, 161, 162, 166, 170, 182, 195, 236. The mean is 172 and the standard deviation 24.95. For a test for normality (case 3), the values of w_i begin $w_1 = (148\text{–}172)/24.95 = -0.962$, and the corresponding z_1 is, from tables, 0.168. When all the z_i have been found, the formula (3) gives $A^2 = 1.007$. Now to make the test, the modification in Table 2 first gives

$$A^* = A^2(1.0 + 0.75/11.0 + 0.25/121.0)$$
$$= A^2(1.086) = 1.095;$$

when this value is compared with the percentage points in Table 1*b*, for case 3, the sample is seen to be significant at the 1 percent level.

TESTS FOR THE EXPONENTIAL DISTRIBUTION*

The distribution tested has CDF $F(x) = 1 - \exp(-x/\beta), x > 0$, described as Exp($x$, β), with β an unknown positive constant. Maximum likelihood gives $\hat{\beta} = \bar{x}$, so that z_i are found from $z_i = 1 - \exp(-x_{(i)}/\bar{x})$, $i = 1, \dots, n$. A^2 is calculated from (3), modified to give A^* by the formula in Table 1*c*, and A^* is compared with the percentage points in Table 1*c*.

For the more general exponential distribution given by $F(x) = 1 - \exp\{-(x - \alpha)/\beta\}, x > \alpha$, when both α and β are unknown, a convenient property of the distribution may be used to return the test situation to the case just described above. The transformation $y_{(i)} = x_{(i+1)} - x_{(1)}$ is made, for $i = 1, \dots, n-1$; the $n-1$ values of $y_{(i)}$ are then used to test that they come from $\exp(y; \beta)$ as just described. The substitution to $y_{(i)}$ reduces the sample size by one, but eliminates α very straightforwardly.

TESTS FOR THE EXTREME-VALUE, WEIBULL, AND LOGISTIC DISTRIBUTIONS

The extreme value distribution is $F(x;\theta) = \exp[-\exp\{-(x - \alpha)/\beta\}]$ $(-\infty < x < \infty)$, with $\theta = (\alpha, \beta)$ and α and β constants, β positive. Tests for the three cases when α, β or both are unknown have been given by Stephens [8]. They can be used also for the Weibull distribution

$$F(x;\theta) = 1 - \exp\left[-\{(x - \alpha)/\beta\}^{\gamma}\right] \quad (x > \alpha), \qquad (4)$$

with $\theta = (\alpha, \beta, \gamma)$; β and γ must be positive. When α is known, the substitution $Y = -\ln(X - \alpha)$ gives, for the distribution function of Y, $F(y) = \exp[-\exp\{-(y - \alpha')/\beta'\}]$ $(y > \alpha')$, where $\beta' = 1/\gamma$ and $\alpha' = -\ln \beta$, so that Y has the extreme-value distribution considered above. If, in addition to α, γ is known in (4), the substitution $Y = -\ln(X - \alpha)$ gives an extreme-value distribution for Y with scale parameter β' known; if α and β are known in (4), the substitution gives an extreme-value distribution for Y with location parameter α' known.

The logistic distribution is $F(x;\theta) = [1 + \exp\{-(x - \alpha)/\beta\}]^{-1} (x > \alpha)$, with $\theta = (\alpha, \beta)$; α and β are constants, with β positive. Stephens [10] has provided tests for this distribution.

TESTS FOR THE GAMMA DISTRIBUTION WITH KNOWN SHAPE PARAMETER

The density under test is $f(x;\theta) = \{\Gamma(m) \times \beta^m\}^{-1} x^{m-1} e^{-x/\beta}, x > 0$, and the distribution is $F(x;\theta) = \int_0^x f(t;\theta)\,dt$. The parameter vector $\theta = (m, \beta)$ contains m as shape parameter and β as scale parameter. Pettitt and Stephens [4] have given tests for this distribution with known m.

CONCLUSION

It was stated above that A^2 has generally good power properties in many test situations. Stephens [6, 9] verifies this for tests of normality and for tests for exponentiality. Other test statistics are sometimes more powerful, although not overwhelmingly so, and they are often more difficult to calculate. The combination of ease of calculation and of making the test, and good power, make A^2 an attractive statistic for goodness of fit.

References

[1] Anderson, T. W. and Darling, D. A. (1952). *Ann. Math. Statist.*, **23**, 193–212.

[2] Anderson, T. W. and Darling, D. A. (1954). *J. Amer. Statist. Ass.*, **49**, 765–769.

[3] Durbin, J., Knott, M., and Taylor, C. C. (1975). *J. R. Statist. Soc. B*, **37**, 216–237.

[4] Pettitt, A. N. and Stephens, M. A. (1976). EDF Statistics for Testing for the Gamma Distribution with Application to Testing for Equal Variances. Unpublished.

[5] Shapiro, S. S. and Wilk, M. B. (1965). *Biometrika*, **52**, 591–611.

[6] Stephens, M. A. (1974). *J. Amer. Statist. Ass.* **69**, 730–737.

[7] Stephens, M. A. (1976). *Ann. Statist.*, **4**, 357–369.

[8] Stephens, M. A. (1977). *Biometrika*, **64**, 583–588.

[9] Stephens, M. A. (1978). Goodness of Fit with Special Reference to Tests for Exponentiality. *Tech. Rep. No. 262*, Dept. of Statistics, Stanford University, Stanford, Calif.

[10] Stephens, M. A. (1979). *Biometrika*, **66**, 591–595.

(CRAMÉR–VON MISES TEST
DEPARTURE FROM NORMALITY, TESTS FOR
GOODNESS-OF-FIT)

MICHAEL A. STEPHENS

ANDREWS FUNCTION PLOTS

Function plots are displays of multivariate data in which all dimensions of the data are displayed. Each observation is displayed as a line or function running across the display. The plots are useful in detecting and assessing clusters* and outliers*. Statistical properties of the plots permit tests of significance* to be made directly from the plot.

The display of data of more than two dimensions requires special techniques. Symbols may be designed to represent simultaneously several dimensions of the data. These may be small symbols used in a scatter plot with two dimensions of the data giving the location of the symbol of the page. Anderson [1] gives examples of such glyphs*. Patterns involving one or more of the plotting dimensions are most easily detected.

Alternatively, these may be larger symbols displayed separately. Chernoff faces* are an example of this type. Although no two dimensions have special status as plotting coordinates, the detection of patterns is more awkward. Function plots are a method of displaying large (page size) symbols simultaneously.

CONSTRUCTION OF PLOTS

Although only few statisticians have experience in displaying items of more than three dimensions, all statisticians are familiar with displays of functions $f(t)$. These may be considered as infinite dimensional. This suggests a mapping of multivariate data, observation by observation, into functions and then displaying the functions. Many such mappings are possible, but the mapping proposed here has many convenient statistical properties.

For each observation involving k dimensions $\mathbf{x}' = (x_1, \ldots, x_k)$, consider the function

$$f_{\mathbf{x}}(t) = x_1\sqrt{2} + x_2 \sin t + x_3 \cos t + x_4 \sin 2t + x_5 \cos 2t + \ldots$$

plotted for values $-\pi < t < \pi$. Each observation contributes one line running across the display. The completed display consists of several such lines.

STATISTICAL PROPERTIES

The mapping $x \to f_x(t)$ preserves distances. For two points x, y the equation

$$\sum_{i=1}^{k} (x_i - y_i)^2 = \pi^{-1} \int_{-\pi}^{\pi} [f_x(t) - f_y(t)]^2 \, dt$$

implies that two functions will appear close *if and only if* the corresponding points are close.

The mapping is linear. This implies that

$$\overline{f_x}(t) = f_{\bar{x}}(t).$$

If the data have been scaled so that the variates are approximately independent with the same variance σ^2, then the variance of $f_x(t)$ is constant, independent of t, or almost constant. Since

$$\begin{aligned} \operatorname{var}(f_x(t)) &= \sigma^2\left(\tfrac{1}{2}\pi + \sin^2 t + \cos^2 t + \sin^2 2t + \cos^2 2t + \ldots\right) \\ &= \sigma^2\left(\tfrac{1}{2} + k/2 + R\right), \end{aligned}$$

where $R = 0$ if k is odd and $\sin^2[(k+1)/2]$ if k is even. This relation may be used to produce and display confidence bands* and tests for outliers*. These tests may be made for preselected values of t or marginally for all values of t. Scheffé's method of multiple comparison* may be used here.

EXAMPLE

Figure 1 is a plot of the Fisher iris data. These data consist of observations of four variables (log units) on 150 iris flowers. The example is commonly used to demonstrate multivariate techniques. Figure 1 clearly demonstrates the separation of one group.

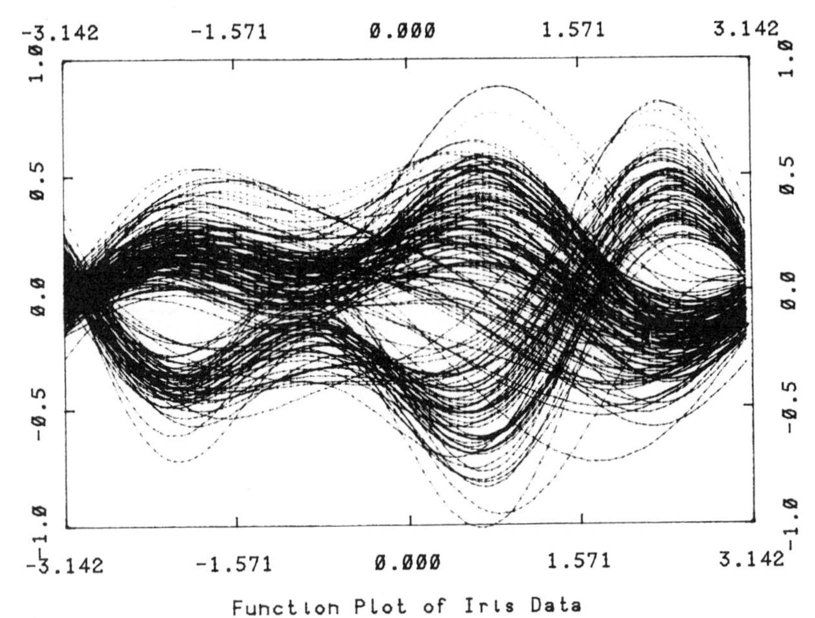

Figure 1 All species—150 observations.

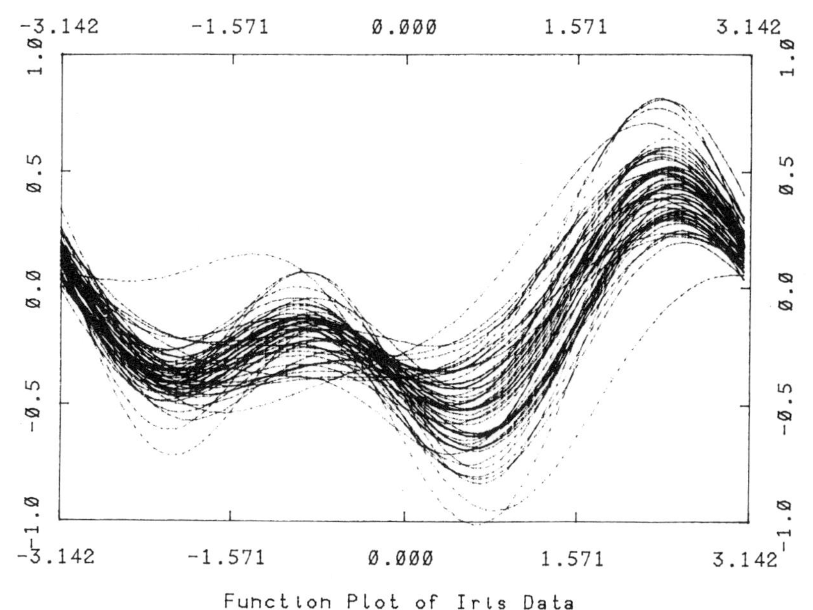

Figure 2 One species—50 observations.

This group consists of one species. This is verified in Fig. 2, which is the plot of this species alone. Note the presence of two "outliers" represented by two straggling lines.

FURTHER NOTES

In some applications, with large data sets, the data may be summarized for each value of t by selected order statistics* of the values $f_x(t)$. Thus a complex plot may be reduced to a plot of the median, the quartiles, the 10% points, and the outlying observations. The order statistics were chosen so that the lines will be almost equally spaced for Gaussian (normal) data.

The order of the variables included in the specification of the function has no effect on the mathematical or statistical properties, although it does affect the visual appearance of the display. Some experience suggests that dominant variables should be associated with the lower frequencies.

General Reference

[1] Anderson, E. (1960). *Technometrics*, **2**, 387–392.

Bibliography

Andrews, D. F. (1972). *Biometrics*, **28**, 125–136.

Chernoff, H. (1973). *J. Amer. Statist. Ass.* **68**, 361–368.

Fisher, R. A. (1936). *Ann. Eugen.* (*Lond.*), **7** (Pt. II), 179–188.

Gnanadesikan, R. (1977). *Methods for Statistical Data Analysis of Multivariate Observations*. Wiley, New York.

(CHERNOFF FACES
GRAPHICAL REPRESENTATION OF DATA
MULTIVARIATE ANALYSIS)

D. F. Andrews

ANGLE BRACKETS

Tukey [4] used angle brackets for symmetric means. These are power product sums* (augmented monomial symmetric functions) divided by the number of terms forming the sum, thus giving the mean power product [5, p. 38].

If observations in a sample are $x_1, \ldots, x_n$, the power product sum $[P] = [p_1 \cdots p_\pi] = \sum_{\neq}^{n} x_i^{p_1} x_j^{p_2} \cdots x_l^{p_\pi}$, where the sum is over all permutations of the subscripts, and no two subscripts are equal.

The number of terms forming the sum is $n(n-1)\ldots(n-\pi+1)=n^{(\pi)}$. The symmetric mean or angle bracket is then $\langle P\rangle = [P]/n^{(\pi)}$. Thus

$$\langle 1\rangle = \frac{1}{n}[1] = \frac{1}{n}\sum_{1}^{n} x_i = \bar{x},$$

$$\langle r\rangle = \frac{1}{n}[r] = \frac{1}{n}\sum_{1}^{n} x_i^r = m'_r,$$

$$\langle rs\rangle = \frac{1}{n^{(2)}}[rs] = \frac{1}{n(n-1)}\sum_{i\neq j}^{n} x_i^r x_j^s,$$

$$\langle 11\rangle = \frac{1}{n^{(2)}}[11] = \frac{1}{n(n-1)}\sum_{i\neq j}^{n} x_i x_j.$$

But

$$[1]^2 = [2] + [11]$$

as

$$\left(\sum_{1}^{n} x_i\right)^2 = \sum_{1}^{n} x_i^2 + \sum_{i\neq j}^{n} x_i x_j.$$

Hence

$$\langle 11\rangle = \frac{1}{n-1}\{n\langle 1\rangle^2 - \langle 2\rangle\}.$$

Tukey [4] and Schaeffer and Dwyer [3] give such recursion formulas for the computation of angle brackets. Elementary examples of computing angle brackets are given in Keeping [2]. Two angle brackets may be multiplied by the rule

$$\langle r\rangle\langle s\rangle = \frac{n-1}{n}\langle rs\rangle + \frac{1}{n}\langle r+s\rangle.$$

A similar symmetric mean may be defined for the population $x_1, \ldots, x_N$, and denoted by $\langle P\rangle_N = [P]_N/N^{(\pi)}$, where $[P]_N$ denotes the power product sum $\sum_{\neq}^{N}$ for the population. Then, if E_N denotes the expected value for the finite population, it follows from an argument of symmetry [3,4] that $E_N\langle P\rangle = \langle P\rangle_N$. Tukey [4] calls this property "inheritance on the average." It makes angle brackets attractive in the theory of sampling from finite populations*, as sample brackets are unbiased estimates* of corresponding population brackets.

Every expression that is (1) a polynomial, (2) symmetric, and (3) inherited on the average can be written as a linear combination of angle brackets with coefficients that do not depend on the size of the set of numbers concerned [4].

Since Fisher's k-statistic* k_p is defined as $k_p = \sum_P (-1)^{\pi-1}(\pi-1)!C(P)\langle P\rangle$ and the finite population K-parameter is $K_p = \sum_P (-1)^{\pi-1}(\pi-1)!C(P)\langle P\rangle_N$, it directly follows that $E_N(k_p) = K_p$. For infinite populations, $E\langle P\rangle = \mu'_{p_1}\cdots\mu'_{p_\pi}$, and hence $E(k_p) = \kappa_p$. *See* FISHER'S k-STATISTICS for more details.

Tukey [5] defines polykays* by a symbolic multiplication of the k-statistics written as linear combinations of angle brackets. The symbolic product of the brackets is a bracket containing the elements in the brackets multiplied, i.e., $\langle p_1\cdots p_\pi\rangle \circ \langle q_1\cdots q_\chi\rangle = \langle p_1\cdots p_\pi q_1\cdots q_\chi\rangle$. Thus $k_{21} = k_2 \circ k_1 = \{\langle 2\rangle - \langle 11\rangle\}\circ\langle 1\rangle = \langle 21\rangle - \langle 111\rangle$. Tukey [4] uses angle brackets in the consideration of randomized (or random) sums. Hooke [1] extends them to generalized symmetric means for a matrix.

References

[1] Hooke, R. (1956). *Ann. Math. Statist.*, **27**, 55–79.

[2] Keeping, E. S. (1962). *Introduction to Statistical Inference*. D. Van Nostrand, Princeton, N. J.

[3] Schaeffer, E., and Dwyer, P. S. (1963). *J. Amer. Statist. Ass.*, **58**, 120–151.

[4] Tukey, J. W. (1950). *J. Amer. Statist. Ass.*, **45**, 501–519.

[5] Tukey, J. W. (1956). *Ann. Math. Statist.*, **27**, 37–54.

(FISHER'S k-STATISTICS
POLYKAYS
POWER PRODUCT SUMS)

D. S. TRACY

ANGULAR TRANSFORMATION

The variance of a binomial proportion (H) is $p(1-p)/n$, which is a function of the expected value p.

Using the variance-equalizing transformation*, the arc-sine transformation

$$\phi(H) = 2\sqrt{n}\arcsin\sqrt{H}$$

is obtained. It is called the angular transformation because "arc sine" is an angle.

The variance of $\phi(H)$ is approximately 1 and it is approximately normally distributed. These transformations were analyzed by R. A. Fisher [3] and are particularly useful in analysis of variance* and regression analysis* involving frequencies. Tables of $\phi(H)$ (in radians) are given in Hald [5]. (Example: $p = 0.600$ gives $\phi(H) = 1.7722$ radians.) An empirical study of the accuracy of this transformation was carried out by Freeman and Tukey [4]. Mosteller and Tukey [6] facilitated the use of this transformation by developing a binomial probability paper* for problems related to testing significance in proportions and in other applications. The modified transformation arcsin $2\sqrt{n+\frac{1}{2}}\,/\sqrt{(x+\frac{3}{8})/(n+\frac{3}{4})}$ is somewhat more effective in equalizing variances. See Anscombe [1].

References

[1] Anscombe, F. J. (1948). *Biometrika*, **35**, 246–254.

[2] Bartlett, M. S. (1947). *Biometrics*, **3**, 39–51.

[3] Fisher, R. A. (1954). *Biometrics*, **10**, 130–139.

[4] Freeman, M. F. and Tukey, J. W. (1950). *Ann. Math. Statist.*, **21**, 607–611.

[5] Hald, A. (1952). *Statistical Tables and Formulas*. Wiley, New York.

[6] Mosteller, F. and Tukey, J. W. (1949). *J. Amer. Statist. Ass.*, **44**, 174, 212.

[7] Mosteller, F. and Youtz, C. (1961). *Biometrika*, **48**, 433–440.

(BINOMIAL PROBABILITY PAPER
FREEMAN–TUKEY TRANSFORMATION
SQUARE-ROOT TRANSFORMATION
VARIANCE STABILIZING TRANSFORMATIONS)

ANIMAL SCIENCE *See* STATISTICS IN ANIMAL SCIENCE

ANNALS OF EUGENICS *See* HUMAN GENETICS, ANNALS OF

ANNALS OF HUMAN GENETICS *See* HUMAN GENETICS, ANNALS OF

ANNALS OF MATHEMATICAL STATISTICS *See* ANNALS OF STATISTICS

ANNALS OF PROBABILITY

One of the two current official journals of the Institute of Mathematical Statistics* (IMS). It started publication in 1973 and at present (1980) is published bimonthly in February, April, June, August, October, and December. The first editor was R. Pyke (1973–1975). The present editor is R. M. Dudley (M.I.T.). Volume 7 (1979) contained 1100 pages.

The journal concentrates on theoretical papers dealing with modern probability theory. Papers on properties of statistical distributions and limit theorems that appear in this journal may often be found to be useful to theoretical statisticians.

ANNALS OF STATISTICS

The *Annals of Statistics (AS)*, first published in 1973, is one of the two journals resulting from a split of the old *Annals of Mathematical Statistics*, the other journal being the *Annals of Probability**. The two current *Annals* are official publications of the Institute of Mathematical Statistics* (IMS).

The original *Annals of Mathematical Statistics (AMS)* started before the Institute existed, and in fact was originally published by the American Statistical Association* in 1930. At that time it had become apparent that the *Journal of the American Statistical Association* (JASA)* could not adequately represent the interests of mathematically inclined statisticians, who were beginning to do important research. Willford King, writing in a prefatory statement to the first issue of *AMS*, said:

> The mathematicians are, of course, interested in articles of a type which are not intelligible to the non-mathematical readers of our Journal. The Editor of our Journal [JASA] has, then, found it a puzzling problem to satisfy both classes of readers.

> Now a happy solution has appeared. The Association at this time has the pleasure of presenting to its mathematically inclined members the first issue of the ANNALS OF MATHEMATICAL STATISTICS, edited by Prof. Harry C. Carver of the University of Michigan. This Journal will deal not only with the mathematical technique of statistics, but also with the applications of such technique to the fields of astronomy, physics, phychology, biology, medicine, education, business, and economics. At present, mathematical articles along these lines are scattered through a great variety of publications. It is hoped that in the future they will be gathered together in the *Annals*.

The seven articles that followed in that first issue covered a very wide range indeed, as their titles suggest:

Remarks on Regression

Synopsis of Elementary Mathematical Statistics

Bayes Theorem

A Mathematical Theory of Seasonal Indices

Stieltjes Integrals in Mathematical Statistics

Simultaneous Treatment of Discrete and Continuous Probability by Use of Stieltjes Integrals

Fundamentals of Sampling Theory

Harry Carver, the founding editor, took on sole responsibility for the young journal when in 1934 the *ASA* stopped its financial support. He continued to publish privately until 1938 when the IMS took over the financial responsibility. Actually, the IMS had come into existence in 1935, and the *Annals* had been its official publication from the start. But Carver, a prime mover in starting the IMS, insisted that it not be tied down by support of the journal.

After the crucial period of Carver's editorship, S. S. Wilks was appointed editor, a post he held from 1938 until 1949. Since that time the editorship has been held for 3-year periods by Wilks's appointed successors (Table 1).

Table 1

Editors of *Annals of Mathematical Statistics*

H. C. Carver (1930–1938)
S. S. Wilks (1938–1949)
T. W. Anderson (1950–1952)
E. L. Lehmann (1953–1955)
T. E. Harris (1955–1958)
W. H. Kruskal (1958–1961)
J. L. Hodges, Jr. (1961–1964)
D. L. Burkholder (1964–1967)
Z. W. Birnbaum (1967–1970)
I. Olkin (1970–1972)

Editors of *Annals of Statistics*

I. Olkin (1972–1973)
I. R. Savage (1974–1976)
R. G. Miller, Jr. (1977–1979)
D. V. Hinkley (1980–1982); current editor

Managing Editors of *AMS* and *AS*

P. L. Meyer (1964–1969)
K. J. C. Smith (1969–1974)
D. R. Truax (1974–1980)

Each editor has headed a distinguished editorial board, but none compares to the illustrious board established in 1938: Wilks, Craig, Neyman (co-editors), Carver, Cramér, Deming, Darmois, R. A. Fisher*, Fry, Hotelling*, von Mises, Pearson, Rietz, and Shewhart.

With this auspicious start, the *AMS* became the focal point for developments in theoretical statistics, particularly those developments associated with the general mathematical theories of estimation, testing, distribution theory, and design of experiments. The full impact of the *Annals*, past and present, is clearly seen in the bibliographies of most books on theoretical statistics. Many of the newer statistical methods can be traced back to pioneering research papers in *AMS*.

After some 40 years of growing strength and size, the *AMS* was split in 1972 during the editorship of I. Olkin, who continued as first editor of *AS*. Each volume of *AS* consists of 1300-plus pages devoted entirely to research articles. (The news items and notices that used to appear in *AMS* until 1972 are now published in the *IMS Bulletin*, issued bimonthly. The journal currently does not publish book reviews.) A volume currently consists of six bimonthly issues. All papers are refereed under the general guidelines of editorial policy, a normal review involving an associate editor and two solicited referees. In recent years the number of new articles submitted to the journal in any one year has been close to 400, of which about 150 are accepted after an average review time of about 6 months plus revision time. The period from acceptance to publication, once well in excess of a year, is now down to 8 months as a result of concerned action by both the IMS council and the editorial board.

The general editorial policy of *AMS* has recently shifted slightly from policies in effect during the last decade. In fact, the present policy is very much in tune with the following statement, made in the 1938 volume of *JASA*:

> The *Annals* will continue to be devoted largely to original research papers dealing with topics in the mathematical theory of statistics, together with such examples as may be useful in illustrating or experimentally verifying the theory. However, in view of the purpose of the Institute of Mathematical Statistics which, interpreted broadly, is to stimulate research in the mathematical theory of statistics and to promote cooperation between the field of pure research and fields of application, plans are being made to extend the scope of the *Annals* to include expository articles from time to time on various fundamental notions, principles, and techniques in statistics. Recognizing that many theoretical statistical problems have their origin in various fields of pure and applied science and technology, papers and shorter notes dealing with theoretical aspects of statistical problems arising in such fields will be welcomed by the editors.

AS continues to recognize its singular and historic role as publisher of general theory, while reflecting the impact that theory does and should have on practical problems of current interest. For that reason, relevance and novelty are at least as important as mathematical correctness.

As has always been the case, *AS* has a concern to represent the whole membership of the IMS, including particularly those overseas and the younger generation of recent graduates. The current membership of IMS, some 2800 in number, makes up a truly international audience to which are added about 100 new student members every year. The interests of this audience, while perhaps similar in spirit to those in 1930, are very different in detail. This is reflected in the contents of current and forthcoming issues of *AS*, which contain papers on the following topics:

Identification of an Element of a Large Population in the Presence of Noise

Application of Dirichlet Distributions to Contingency Tables

A Sequential Clinical Trial for Testing $p_1 = p_2$

The Role of Exchangeability in Inference

A Bayesian Nonparametric Approach to Reliability Estimation

Estimation of the Order of Linear Time Series Models

Estimation in Multivariate Errors in Variables Regression Models

Theory of Non-linear Smoothers

as well as papers on robustness, optimal design, nonparametric density estimation, and many other diverse topics of current interest.

(ANNALS OF PROBABILITY
INSTITUTE OF MATHEMATICAL STATISTICS)

D. V. HINKLEY

ANNALS OF THE INSTITUTE OF STATISTICAL MATHEMATICS

Published in English by the Institute of Statistical Mathematics (4-6-7 Minami Azabu, Minato-ku, Tokyo, Japan 106) and is devoted mainly to theoretically oriented papers in statistics. Each issue consists of Part A (theory and methods) and Part B (applications, which is usually much smaller). Three issues are published per year. The journal originated in 1949; its first volume contained 160 pages and 10 contributions exclusively by Japanese statisticians. Volume 30 (1978) was comprised of 504 pages and some 40 contributions, of which about 15 were by "foreign" statisticians. The founding editor-in-chief is K. Matusita, who is at present the editor of Part A. (H. Akaike and S. Nishira serve as the editors for Part B.)

ANOCOVA TABLE *See* ANALYSIS OF COVARIANCE; ANOVA TABLE

ANOVA TABLE

An ANOVA table is a conventional way of presenting the results of an analysis of variance*. There are usually four columns, headed

1. Source (of variation)
2. Degrees of freedom*
3. Sum of squares
4. Mean square*

Columns 1 and 2 reflect the size and pattern of the data being analyzed and the model being used. Column 4 is obtained by dividing the entry (in the same row) in column 3 by that in column 2.

Sometimes there is a fifth column, giving the ratios of mean squares to a residual mean square* (or mean squares). These statistics are used in applying the standard F-test* used in the analysis of variance.

The value of the ANOVA table is not only in its convenient and tidy presentation of the quantities used in applying analysis-of-variance tests. The juxtaposition of all the quantities used for a number of different tests (or the mean squares for many different sources of variation) can provide valuable insight into the overall structure of variation. For example, in the analysis of a factorial experiment*, the groups of interactions* of specified order can provide evidence of relatively great variation arising when a particular factor (or group of factors) is involved.

The term "ANOCOVA table" is also used (although rather infrequently) to describe similar tables relevant to the analysis of covariance*.

(ANALYSIS OF VARIANCE)

ANSARI—BRADLEY *W*-STATISTICS

This statistic is used in a distribution-free rank-sum test of the hypothesis of equality of scale parameters (dispersions)*, where the underlying continuous populations have a common median*. (The case when the assumption of common medians is not applicable can be handled by the Moses rank-like test*; see [6].)

The data consist of $N = m + n$ observations, $X_1, \ldots, X_m$ and $Y_1, \ldots, Y_n$. We assume a model of the form $X_i = \sigma_1 e_i + \mu$, $i = 1, \ldots, m$, and $Y_j = \sigma_2 e_{m+j} + \mu$, $j = 1, \ldots, n$, where e_i $(i = 1, \ldots, N)$ are mutually independent unobservable random variables from the same continuous population with zero median, μ is the *common* median of the X and Y population, and σ_i, $i = 1, 2$, are scale parameters. In other words, the only difference between X and Y population is a difference in scale parameters. (If the assumptions are not satisfied, the test is *conservative* in that it rejects the null hypothesis* less frequently than expected, provided that $m \approx n$.)

To obtain the W-statistic the results from two samples are arranged in ascending rank order while retaining the identity of the separate samples. The W-statistic is the sum of the rank scores assigned to members of the smaller sample according to the following scheme. Assign rank 1 to each of the smallest and the largest observation in the combined sample, rank 2 to the second smallest and the second largest, and so on. For even N the array of ranks is $1, 2, \ldots,$

$N/2, N/2, \ldots, 2, 1$; for odd N the array of ranks is $1, 2, 3, \ldots (N-1)/2, (N+1)/2, (N-1)/2, \ldots 3, 2, 1$. If R_i denotes the rank of X_i (the "test" sample), then the test statistic $W = \sum_{i=1}^{m} R_i$.

The sampling distribution $f(W; n; m)$ of the Ansari–Bradley [1] statistic under the null hypothesis H_0: $\sigma_1 = \sigma_2$ satisfies

$$f(W;n;m) = f(W;n,m-1) + f(W-N/2;n-1,m) \quad \text{for even } N$$

and

$$f(W;n;m) = f(W;n,m-1) + f\left(W - \frac{N+1}{2}; n-1, m\right) \quad \text{for odd } N.$$

A computer program in the ISO FORTRAN language for generating the null distribution of the W-statistic was published by Dinneen and Blakesley [3]. For very large samples the normal approximation based on statistics W^* is adequate. The statistic W^* is defined by

$$W^* = \frac{W - E_0(W)}{\{\text{var}_0(W)\}^{1/2}}$$

$$= \begin{cases} \dfrac{W - [m(m+n+2)/4]}{\{mn(m+n+2)(m+n-2)/[48(m+n-1)]\}^{1/2}} & \text{if } m+n \text{ is even,} \\[2ex] \dfrac{W - \{m(m+n+1)^2/[4(m+n)]\}}{\{mn(m+n+1)[3+(m+n)^2]/[48(m+n)^2]\}^{1/2}} & \text{if } m+n \text{ is odd.} \end{cases}$$

When H_0 is true, the statistic W^* has an asymptotic $(\min(n,m))$ standard normal distribution.

The Siegel–Tukey test* is of similar nature, but uses a different system for assigning ranks; see [4].

References

[1] Ansari, A. R. and Bradley, R. A. (1960). *Ann. Math. Statist.*, **31**, 1174–1189.

[2] Bradley, J. V. (1968). *Distribution-Free Statistical Tests*. Prentice-Hall, Englewood Cliffs, N.J.

[3] Dinneen, L. C. and Blakesley, B. C. (1976). *Appl. Statist.*, **25**, 75–77.

[4] Eeden, C. van (1964). *J. Amer. Statist. Ass.*, **59**, 105–119.

[5] Hollander, M. and Wolfe, D. A. (1973). *Nonparametric Statistical Methods*. Wiley, New York. Chap. 5.

[6] Moses, L. E. (1963). *Ann. Math. Statist.*, **34**, 973–983.

(MANN–WHITNEY–WILCOXON TEST
RANK-SUM TESTS
SIEGEL–TUKEY TEST)

ANTHROPOLOGY, STATISTICS IN

Statistical methods were introduced into physical anthropology by Quetelet and Galton during the nineteenth century. From their work grew the "Biometrics School," headed by K. Pearson* and Weldon, whose members studied the variation between local races in an attempt to clarify the processes of inheritance and evolution. Human skulls, in particular, were intensively studied because of the availability of historical material. The statistical treatment of data from skulls raised many new problems, and Pearson [23] took one of the first steps in multivariate analysis* by introducing the coefficient of racial likeness*, a statistic based on all measurements and used to assess the significance of differences between groups. There is less emphasis on craniometry today than there was in Pearson's time, but the basic statistical problem is still with us. How should variation in the shape of complex objects such as bones be described?

Pearson believed passionately in measurement* as the basis of all science, although he recognized that most advances in the study of shape had actually relied on visual comparisons. Unfortunately, the measurement of shape is very difficult. In the majority of published examples the procedure has been to identify common landmarks on the objects and then to measure angles and linear distances. The hope is that a statistical summary of these data will embody a summary of shape. The early biometricians were restricted to simple statistical summaries by primitive computing equipment, but techniques of multivariate analysis are now commonly used to attempt a more explicit description of shape variation within a group and to make comparisons between groups.

Landmarks are hard to find on some objects, and an alternative approach is to record the coordinates of a large number of points on the object. Measures of shape must then be based on geometric properties of the surfaces containing the points, and should be independent of which points are chosen. The technique has so far been limited to outlines from sections of the original objects for which curves rather than surfaces are relevant. This has been for practical rather than theoretical reasons. A plot of radial distance versus angle has been used to summarize outlines, but this has the disadvantage of depending on the choice of origin. An alternative is a plot of tangent direction versus distance round the outline [27]. In both cases the plots may be described quantitatively using Fourier series*. The geometric approach is well reviewed in Bookstein [5].

SHAPE AND SIZE

The special problem of determining the extent to which shape is related to size is referred to as allometry*. Apart from this concern, size variation is usually of little interest in studies of shape. However, linear distances inevitably reflect the size of an object so that size variation can be a nuisance. Since only the relative magnitude of distances is important for shape, the distances are often replaced by ratios of one distance to another. If shape is related to size, such ratios still might well be related to size, but the degree of relationship will be much less than for the original distances. It is usual to choose one distance that is strongly influenced by size and to use this as the denominator when expressing other distances as ratios. Mosimann [18] and Corruccini [7] have suggested the use of a symmetric function of all the size-dependent variables as the denominator.

VARIATION WITHIN A GROUP

Measurements made on an object are regarded as a vector of observations x on a vector of variables X. The individual variables in X are referred to as $X_1, \ldots, X_v$. Data from a group of n objects consists of n vectors, $x_1, \ldots, x_n$, which together form a $n \times v$ data matrix. The rows of this matrix, which are the vectors x_i, may be represented as n points in v-space and the columns as v points in n-space. The two representations are sometimes referred to as Q and R, respectively. The Euclidean metric is used in both spaces so that in row space (Q) the distance between two objects is, in matrix notation, $(\mathbf{x} - \mathbf{y})^T(\mathbf{x} - \mathbf{y})$. Thus two objects that are close in row space are similar in respect of the v measurements.

The usual statistical summary is based on the mean* and standard deviation* of each variable, together with correlations* between pairs of variables. This depends on the distribution being roughly multivariate normal*, an assumption that may be partially tested by inspecting the distribution of each variable separately (which should be normal) and each possible bivariate plot (which should be linear). If there is no correlation, then the variation in shape is uninteresting: objects vary, but not in any consistent way. Suppose now that all objects are roughly the same size. Then a high correlation (positive or negative) is regarded as evidence that the two variables are constrained by the necessity for the object to stay in the same class of shapes and that jointly they are measuring a single aspect of shape. A negative correlation can be converted to a positive one by using the reciprocal of a measurement or ratio, or the complement of an angle, and this is usually done to ensure positive correlations as far as possible. If a group of variables has high positive intercorrelations, then the group is taken to be measuring a single aspect of shape. Different aspects of shape will have relatively low correlation, by definition. Statistically, this amounts to grouping the variables on the basis of their correlations. It may be done by eye for a small number of variables or by extracting principal components for a larger number ($v > 10$). If the objects do not have the same size, then the interpretation of correlations depends on the nature of the variables. Correlation among linear distances will almost certainly be partly, perhaps largely, due to size variation. Correlation among angles and ratios will generally indicate constraints of shape.

Principal component analysis* extracts the components $(Z_1, \dots, Z_v)$ from the covariances* between the original variables. Each component is a linear combination of the variables $(X_1, \dots, X_v)$. If $Z_1 = a_1X_1 + \cdots + a_vX_v$, then a_i is called the loading* of X_i on the first component and is proportional to the covariance between Z_1 and X_i. When the data have been standardized by reducing each variable by its mean and scaling each to have unit standard deviation, then covariance equals correlation. In this case the loadings are used to group variables according to their correlations with the first few components and hence with each other. Sometimes the procedure is reversed and the first few components are "named" according to the variables they are associated with. This is logically equivalent to grouping the variables*. Howells [14] gives a good example.

If the observed values of X for an object are substituted in the expression for Z_1, the result is a *score* for that object on Z_1. The scores on Z_1 and Z_2 may be plotted, using rectangular axes, to give an approximation to the representation of objects in row space. The quality of the approximation depends on how much of the overall variability between objects has been reproduced by Z_1 and Z_2. The plot is useful for spotting any lack of homogeneity* in the group of objects. (Rao [25] gives a very detailed account of the different uses of principal components.)

VARIATION BETWEEN GROUPS

If the comparison of several groups is to be meaningful, each must have a representative shape. In other words, the groups must be homogeneous, displaying some variation, but not too much. We shall assume that there are k groups in all, distinguishing between them by using different letters $(x, y, z, \dots)$ to refer to a typical vector of measurements in each group. The vector of means for each group is regarded as representing a mean shape. For example, McLearn et al. [16] reconstruct a typical profile from the mean of measurements taken from a large number of individual profiles of the human face. In this example the result still looked like a human face, i.e., $\bar{x}$ satisfied the same geometric constraints as each x_i in $\bar{x} = \sum x_i/n$, but since these constraints are in general nonlinear, this will not always be the case.

When comparing groups there are two kinds of questions: comparison between pairs of groups and an overall comparison. The latter requires some metric* enabling one to decide whether group x is closer to y than to z, and a wide variety have been proposed. A good review is given in Weiner [26]. The most commonly used metric is now $(\mathbf{x} - \mathbf{y})^T\mathbf{\Sigma}^{-1}(\mathbf{x} - \mathbf{y})$, where $\mathbf{\Sigma}$ is a positive definite $p \times p$ matrix. This may be interpreted as follows. If a multivariate normal density with covariance $\mathbf{\Sigma}$ is centered at x, then all points y that are equiprobable in this density are equidistant from x in this metric. If $\mathbf{\Sigma} = \mathbf{I}$, the contours of equal probability are spheres; otherwise, they are ellipsoids. The metric is satisfactory only for groups with similar patterns of covariance, in which case $\mathbf{\Sigma}$ is taken equal to $\mathbf{S}$, the pooled covariance matrix within groups. This leads to D^2, equal to $(\bar{\mathbf{x}} - \bar{\mathbf{y}})^T\mathbf{S}^{-1}(\bar{\mathbf{x}} - \bar{\mathbf{y}})$, as the measure of distance between groups. It is clear from the derivation that the comparison of two large values of D^2 is unlikely to be satisfactory. In fact, although D^2 takes account of correlations within groups, it is, like all metrics, rather a blunt instrument when used for assessing affinity.

With a lot of groups the pairs of D^2 values can be confusing, so a visual overall picture of the interrelationships between groups is produced using principal components*. A $k \times v$ data matrix in which the rows are now the group means is used and principal component analysis is carried out in the D^2 metric [12]. The resulting components are called canonical variates or sometimes discriminant functions*. Scores for each group on the first two or three canonical variates are plotted using rectangular axes. Good examples of the use of canonical variates are those of Ashton, et al. [2], Day and Wood [9], and Oxnard [19].

If something is known about the function of the objects, then it may be possible to order the groups according to this function, at least roughly, e.g., low, medium, and high. We assume that function is not quantifiable, so that only a rough ordering is possible. The plot will show whether or not this ordering is associated with the major dimensions of shape variation. Of course, there may well be other measures of shape more highly associated with function than the first few canonical variates. Aspects of shape that are highly variable are not necessarily those most highly associated with function.

The D^2-metric itself is not without disadvantages. First the assumption that the pattern of variation is constant for all groups is inherently unlikely with shape studies. Second, when a large number of variables is used to ensure that shape is adequately described, they often contain redundancies which lead to nearly singular matrices $\mathbf{S}$ and hence to very large and meaningless D^2-values. Some progress has been made in overcoming these and other difficulties. To avoid distortion from studying scores on just the first two or three canonical variates, Andrews [1] has suggested representing each group by a weighted combination of trigonometric functions of θ with weights equal to the scores on all v canonical variates. As θ varies, each group is represented by a smooth curve. Burnaby [6] has shown how the D^2-metric may be adjusted to measure change only in certain directions in row space. This can be useful when it is required to avoid a direction corresponding to growth or size change. Penrose [24] showed how the adjusted D^2-metric becomes equal to the Euclidean metric when all correlations are equal. To avoid D^2 altogether, some authors have pooled the groups and studied the extent to which the individual objects can be clustered, either visually using principal components and the Euclidean metric, or automatically using various clustering* algorithms [20]. The techniques of multidimensional scaling* [15] and principal coordinates* [11] are also relevant here.

FOSSILS

The stimulus to study variation in modern groups of man often comes from a particularly important fossil find. If canonical variates for the modern groups have been evaluated, then scores on these variates can be obtained for the fossil, and it can be placed on the plot of modern group means relative to the first two or three canonical variates. This plot indicates its position relative to the modern groups, but the result must be treated with caution. Both the modern groups and the fossil should fit well into the two- or three-dimensional plot, for otherwise their relative positions will be distorted. If the fit is poor, then the actual D^2 distances of the fossil from the modern groups should be compared, but if these are all large, then the assessment of affinity is bound to be unsatisfactory. Day and Wood [9], after associating a canonical variate with function, used it to predict function for a fossil that was very different from the modern groups used to derive the canonical variate. This situation is in some ways analogous to that encountered when using a regression line* to predict values outside the range on which the line was based.

PREVALENCE OF ATTRIBUTES*

Consider v attributes of an object measured by $X_1, \ldots, X_v$, where these now take only two possible values (presence/absence). The comparison between groups rests on a comparison of the prevalence for each attribute. If $p(X_i)$ is the prevalence for X_i, then the difference between two groups is measured on the transformed scale $\theta = \sin^{-1}\sqrt{p}$. On this scale the standard deviation of θ

is approximately $1/(4n)$. *See* ANGULAR TRANSFORMATION. If the v attributes are independent, then v differences, $\theta_i - \theta_i'$, may be combined to provide an overall distance $d^2 = \sum(\theta_i - \theta_i')^2$. Berry and Berry [4] give an example based on attributes of the skull. Edwards [10] has generalized this to cover attributes with more than two states, such as blood groups.

STATISTICAL TESTS AND PREDICTION

Since there is rarely any element of randomization* in data collection for physical anthropology, the role of significance tests* is less important than that of description. Three tests are commonly performed: equality of covariance matrices between groups, difference between two groups based on D^2, and zero intercorrelation within a set of variables. All three are based on the multivariate normal distribution*. Full details are given in Morrison [17].

A vector **x** which is incomplete cannot be used in multivariate analysis* without the missing values being replaced by some estimates. The group means for the relevant variables are sometimes used, but a better method is to use the set of complete vectors to predict the missing values using multiple regression* [3].

Predicting the sex of bones can be dealt with statistically if reference groups of known sex are available. Each unknown bone is allocated to the closer of the male and female reference groups using the D^2-metric. Day and Pitcher–Wilmott [8] give an example. The maturity of bones is assessed from characteristics that can be ordered with respect to maturity. Scaling techniques* are used to make the assessment quantitative [13].

References

[1] Andrews, D. F. (1972). *Biometrics*, **28**, 125–136.

[2] Ashton, E. H., Healy, M. J. R., and Lipton, S. (1957). *Proc. R. Soc. Lond. B*, **146**, 552–572.

[3] Beale, E. M. L. and Little, R. J. A. (1975). *J. R. Statist. Soc. B*, **37**, 129–145.

[4] Berry, A. C. and Berry, R. J. (1967). *J. Anat.*, **101**, 361–379.

[5] Bookstein, F. L. (1978). The Measurement of Biological Shape and Shape Change. *Lect. Notes Biomath.*, **24**. Springer-Verlag, Berlin.

[6] Burnaby, T. P. (1966). *Biometrics*, **22**, 96–110.

[7] Corruccini, R. S. (1973). *Amer. J. Phys. Anthropol.*, **38**, 743–754.

[8] Day, M. H. and Pitcher-Wilmott, R. W. (1975). *Ann. Hum. Biol.*, **2**, 143–151.

[9] Day, M. H. and Wood, B. A. (1968). *Man*, **3**, 440–455.

[10] Edwards, A. W. F. (1971). *Biometrics*, **27**, 873–881.

[11] Gower, J. C. (1966). *Biometrika*, **53**, 325–338.

[12] Gower, J. C. (1966). *Biometrika*, **53**, 588–590.

[13] Healy, M. J. R. and Goldstein, H. (1976). *Biometrika*, **63**, 219–229.

[14] Howells, W. W. (1972). In *The Functional and Evolutionary Biology of Primates: Methods of Study and Recent Advances*, R. H. Tuttle, ed. Aldine-Atherton, Chicago, pp. 123–151.

[15] Kruskal, J. B. (1964). *Psychometrika*, **29**, 1–27.

[16] McLearn, I., Morant, G. M., and Pearson, K. (1928). *Biometrika*, **20B**, 389–400.

[17] Morrison, D. F. (1967). *Multivariate Statistical Methods*. McGraw-Hill, New York.

[18] Mosimann, J. E. (1970). *J. Amer. Statist. Ass.*, **65**, 930–945.

[19] Oxnard, C. E. (1973). *Form and Pattern in Human Evolution*. University of Chicago Press, Chicago.

[20] Oxnard, C. E. and Neely, P. M. (1969). *J. Morphol.*, **129**, 1–22.

[21] Pearson, E. S. (1936). *Biometrika*, **28**, 193–257.

[22] Pearson, E. S. (1938). *Biometrika*, **29**, 161–248.

[23] Pearson, K. (1926). *Biometrika*, **18**, 105–117.

[24] Penrose, L. S. (1954). *Ann. Eugen.* (*Lond.*), **18**, 337–343.

[25] Rao, C. R. (1964). *Sankhyā A*, **26**, 329–358.

[26] Weiner, J. S. (1972). *The Assessment of Population Affinities*, J. S. Weiner, ed. Clarendon Press, Oxford.

[27] Zahn, C. T. and Roskies, R. Z. (1972). *IEEE Trans. Computers*, **C-21**, 269–281.

Further Reading

The book by Oxnard [19] provides the best available general introduction to the subject, and contains a good bibliography. The series of papers by Howells (1951 onward; see the bibliography in ref. [19]) contain good examples of most of the important methodological contributions made during this period. The use of canonical variates is well explained in Ashton et al. [2]. Bookstein's monograph [5] provides a useful antidote to the uncritical use of interlandmark distances; although more mathematical than the other references cited it is well worth the effort. Technical details about principal components and canonical variates are best obtained from Rao [25], Gower [11, 12], and Morrison [17]. For historical details the early volumes of *Biometrika* should be consulted, particularly Volume 1 and the account of Karl Pearson's life and work given by E. S. Pearson [21, 22].

(ALLOMETRY
CLUSTER ANALYSIS
CORRELATION
DISCRIMINANT ANALYSIS
MULTIDIMENSIONAL SCALING
MULTIVARIATE ANALYSIS
PATTERN RECOGNITION
PRINCIPAL COMPONENT ANALYSIS
REGRESSION)

M. HILLS

ANTIMODE

The opposite of mode* in the sense that it corresponds to a (local) minimum frequency. As with the mode, it is sometimes desired that the name should be applied only to global, and not to local minima. The more common use, however, includes local minima.

Note that whereas $x = 1$ is a mode of the PDF*,

$$f_X(x) = \begin{cases} 2x & 0 \leqslant x \leqslant 1, \\ 0 & \text{elsewhere,} \end{cases}$$

$x = 0$ is not an antimode of this PDF. On the other hand,

$$f_X(x) = \begin{cases} |x| & -1 \leqslant x \leqslant 1, \\ 0 & \text{elsewhere,} \end{cases}$$

has an antimode at $x = 0$ (and modes at $x = -1$ and $x = 1$).

The antimode itself refers to the frequency (or PDF) at the antimodal value of the argument.

(MEASURES OF LOCATION
MODE)

ANTITHETIC VARIATES

If T_1 and T_2 are unbiased estimators of a parameter θ, then $T = \frac{1}{2}(T_1 + T_2)$ is also unbiased and has variance $\frac{1}{4}\{\text{var}(T_1) + \text{var}(T_2) + 2\,\text{cov}(T_1, T_2)\}$. This variance is reduced (for fixed $\text{var}(T_1)$ and $\text{var}(T_2)$) by reducing $\text{cov}(T_1, T_2)$ and making the correlation between T_1 and T_2 negative and as large, numerically, as possible. Pairs of variates constructed with this aim in view are called *antithetic variates*.

The concept arose in connection with estimation of integrals by simulation* (Monte Carlo) experiments [2, 3]. The following example [1, p. 61] may help to clarify ideas. If X is uniformly distributed* between 0 and 1, then for any function $g(x)$,

$$E[g(X)] = \int_0^1 g(x)\,dx,$$

so $g(X)$ is an unbiased estimator* of $\int_0^1 g(x)\,dx$. So is $g(1 - X)$, since $(1 - X)$ is also uniformly distributed between 0 and 1. If $g(x)$ is a monotonic function of x, $g(X)$ and $g(1 - X)$ are negatively correlated and are "antithetic variates." In particular,

$$\text{var}(g(X) + g(1 - X)) \leqslant \tfrac{1}{2}\,\text{var}(g(X)).$$

The construction of a variate antithetic to $g(X)$ can be extended in a simple way to cases when the function $g(x)$ is not monotonic, but the interval 0 to 1 can be split into a finite number of intervals in each of which $g(x)$ is monotonic.

The method can also be applied to estimation of multivariate integrals by a straightforward extension. Use of antithetic variables can be a powerful method of increasing accuracy of estimation from simulation in appropriate situations.

General References

[1] Hammersley, J. M. and Handscomb, D. C. (1964). *Monte Carlo Methods.* Methuen, London.

[2] Hammersley, J. M. and Mauldon, J. G. (1956). *Proc. Camb. Philos. Soc.*, **52**, 476–481.

[3] Hammersley, J. M. and Morton, K. W. (1956). *Proc. Camb. Philos. Soc.*, **52**, 449–475.

[4] Tukey, J. W. (1957). *Proc. Camb. Philos. Soc.*, **53**, 923–924.

(QUADRATURE
SIMULATION)

APL *See* STATISTICAL SOFTWARE

APPLIED PROBABILITY

Applied probability is that field of mathematical research and scholarship in which the theory and calculus of probability are applied to real-life phenomena with a random component. Such applications encompass a broad range of problems originating in the biological, physical, and social sciences, as well as engineering and technology.

The term "applied probability" first appeared as the title of the proceedings of a symposium on the subject, held by the American Mathematical Society in 1955 [3]. It became popular through its use by the *Methuen Monographs in Applied Probability and Statistics*, edited from 1959 by M. S. Bartlett. The two fields are closely related: applied probability is concerned primarily with modeling random phenomena* (*see* STOCHASTIC PROCESSES), while statistics serves to estimate parameters* and test the goodness of fit* of models to observed data. Bartlett has expressed the opinion that neither field could exist without the other; this is a viewpoint shared by many applied probabilists.

There are currently several periodicals publishing material in applied probability; the principal ones in order of their dates of first publication are *Teoriya Veroyatnostei i ee Primeneniya* (1956), (English translation: *Theory of Probability and Its Applications**) *Zeitschrift für Wahrscheinlichkeitstheorie* (1962), the *Journal of Applied Probability** (1964), *Advances in Applied Probability** (1969), *Stochastic Processes and Their Applications* (1973), and *Annals of Probability** (1973). Other mathematical or statistical journals publish the occasional paper in applied probability, and there is considerable discussion of applied probability models in journals of biology, physics, psychology, operations research, and engineering. Among these are *Theoretical Population Biology*, the *Journal of Statistical Physics**, *Psychometrika, Operations Research**, the *Journal of Hydrology*, and the *Journal of the Institute of Electrical Engineers.*

It is impossible to give a comprehensive description of current work in applied probability; perhaps a few illustrative examples selected at random from the recent literature in each of the biological, physical, social, and technological areas will serve to indicate the breadth of the field.

BIOLOGICAL SCIENCES: BIRD NAVIGATION; OPTIMAL HUNTING

David Kendall [2] recently investigated some interesting models of bird navigation. Ornithologists have surmised that birds navigate instinctively by reference to the sun and stars; Kendall constructed two models to simulate such navigation realistically.

The first is referred to as the Manx model, after the Manx shearwater, which flies across the Atlantic to its European breeding grounds. In this model the bird flies laps of approximately 20 miles each, at a speed of roughly 40 mph. At the end of each lap it redirects itself toward its goal but commits an angular error, having the von Mises or wrapped normal distribution*. When it arrives within a radius of approximately 10 miles of its home, it recognizes its destination and heads directly for it. The second Bessel model allows both the lap length and the deflection from the correct direction to be random. Under appropriate conditions, the two models converge to Brownian motion*.

In his paper, Kendall analyzes bird data on which to base his models and tests these by repeated simulations*, leading to graphical representations of Manx and Bessel flights. He compares these models theoretically and numerically, carries out some diffusion* approximations, and concludes with a lengthy study of the hitting times to the circumference of the homing target. Kendall's work is a model of what is most illuminating in the applied probabilist's approach to a scientific problem. Practical data are carefully considered, and a suitable model is found to fit it. The apprentice applied probabilist could well base his methods and style on Kendall's work.

Another of the many interesting problems which arise in the biological context is that of determining optimal hunting or harvesting policies for animal populations. Abakuks [1] has discussed such a policy for a population growing according to a stochastic logistic* scheme, subject to natural mortality*. The object is to maximize the long-term average number of animals hunted or harvested per unit time. It is shown that there is a critical population size x_c such that hunting or harvesting is optimal if and only if the population is greater or equal to this number.

PHYSICAL SCIENCES: ISING LATTICES

In statistical mechanics*, one may need to determine the partition function for large lattices of points representing crystals, particles, or atoms. The Ising model, which helps to characterize qualitative changes at critical parameter values, is therefore important in theoretical physics.

Consider a rectangular lattice of $N = m \times n$ points in a plane; these points are labeled 1 to N. At each site i, the random variable X_i may take the values ± 1, where $+1$ may, for example, correspond to the formation of a crystal. The joint distribution of the site variables is given by

$$P\{\mathbf{X} = \mathbf{x}\} = k^{-1}(a)\exp(a \sum x_i x_j),$$

where $k(a) = \sum_{\mathbf{x}} \exp(a \sum x_i x_j)$ is the partition function, and the $\sum x_i x_j$ is taken over all nearest-neighbor pairs; $\mathbf{X}$ is a simple Markov random field.

Pickard [4] has recently obtained some limit theorems for the sample correlation* between nearest neighbors in an Ising lattice for the noncritical case. This provides a model for asymptotic testing and estimation of the correlation between nearest neighbors, based on experimental data.

SOCIAL SCIENCES: MANPOWER SYSTEMS; ECONOMIC OPTIMIZATION

A firm or company is a hierarchically graded manpower system, usually modeled by a Markov chain*. The probabilities p_{ij} of this chain denote annual promotion rates from grade i to grade j of the firm; the state of the graded manpower system is described from year to year by the numbers of individuals in each grade.

The simpler manpower models are mainly linear, but Vassiliou [7] has considered a high-order nonlinear Markovian model for promotion based on three principles. The first is the ecological principle that promotions should be proportional to suitable staff available for them, as well as to vacancies for promotion. The second is that resignations from different grades are different, and the third is an inertia principle which prescribes that when there is a reduced number of vacancies, promotions may still be made faster than the ecological principle suggests.

Difference equations* for the mean number of individuals in each grade are obtained, and the model is used to provide detailed numerical forecasts for probabilities of promotion in organizations with five grades. A comparison is made between predictions based on an earlier linear model*, and the present nonlinear model*; actual data from a large British firm are found to be adequately described by the latter.

Optimization may be important in a variety of sociological contexts, some purely economic, others more technological. Vered and

Yechiali [8] have recently studied the optimization of a power system for a private automatic telephone exchange (PABX). Several maintenance policies are considered, which depend on the set of parameters (m, u, μ, v) of the PABX system, where n is the number of independent rectifiers in parallel, m the minimal number of units that will keep the system operative, μ their mean lifetime before failure (*see* FAILURE TIME DISTRIBUTIONS), and v the period of time between regular maintenance visits. Repairs are carried out every v units of time, or when the system fails. The authors determine the optimal (m, n, μ, v) to minimize costs for a required level of reliability* of the PABX system, and provide tables of numerical results for the optimal parameters in both the cases of periodic and emergency maintenance.

TECHNOLOGICAL SCIENCES: RELIABILITY

All engineering components have a failure point; it is therefore of importance to study the reliability of mechanical or electronic parts to determine the probability of their failure times. In this context, two recent problems, the first studied by Szász [5] concerning two lifts, and the second by Taylor [6] on the failure of cables subjected to random loads, will be of interest.

Szász [5] examines a building that has two lifts working independently of each other. The functioning of each lift forms an alternating renewal process* with working-time distribution F and repair-time distribution G. Suppose that this latter distribution $G = G(x, \epsilon)$ depends on $\epsilon > 0$ in such a way that its mean $\int x\, dG(x, \epsilon)$ tends to zero as $\epsilon \to 0$. The author sets out to find the asymptotic distribution of the first instant τ^{ϵ} at which both lifts are simultaneously out of order, as $\epsilon \to 0$.

It is shown that under certain conditions, as $\epsilon \to 0$, the normalized point process W^{ϵ}: $w_1^{\epsilon} < w_2^{\epsilon} < \cdots$, where $w_k^{\epsilon} = \epsilon \tau_k^{\epsilon}$, tends to a Poisson process* with parameter $2\lambda^{-2}$, where $\lambda = \int x\, dF(x)$. Thus subject to certain very general conditions, one must expect breakdowns of both lifts to occur according to a Poisson process. For the favorable case in which λ is very large, this process will have a very small mean.

Taylor [6] is concerned with the reliability of deep-sea cables made up of fiber bundles, and the effect on them of random loads generated by waves that rock the ocean vessels deploying them. A simple model with random loads is studied, subject to power-law breakdown, such that the failure time T under constant load L follows the negative exponential distribution*

$$\Pr[T > x] = \exp(-KL^{\rho}x) \qquad (x \geqslant 0),$$

where $K > 0$ and $\rho \geqslant 1$.

The asymptotic distribution of T under random loads is derived and Taylor shows that random loads have a significant effect on the lifetime of a cable. The loss in mean lifetime cannot be predicted from the first few moments of the load process; it depends on the entire marginal probability distribution of the load, as well as the power-law exponent ρ. It is shown that the asymptotic variance of the lifetime has two components, the first due to the variation of individual fibers, and the second to the variation of the load.

EXTENT AND FUTURE OF THE FIELD

It is of interest to know the methods of probability theory that are most used in attacking problems of applied probability. The most commonly applied areas are found to be Markov chains and processes (including diffusion processes*), branching processes* and other stochastic processes* (mostly stationary), limit theorems*, distribution theory and characteristic functions*, methods of geometrical probability*, stopping times, and other miscellaneous methods.

To list the subsections of the main categories given in the first four sections exhaustively would be impossible, but in the biological sciences one finds that population processes, mathematical genetics, epidemic theory, and virology are the major subfields. In the technological sciences, operations research*, queueing theory*, storage (*see* DAM THEORY), and traffic theory are possibly the most active areas.

Applied probability is a very broad subject; as we have seen, it encompasses real-life problems in a variety of scientific and other fields. Although the subject feeds on practical problems, it requires a very high level of theoretical competence in probability. In solving these problems, every approach that proves successful is a useful one. Classical mathematical analysis, numerical analysis, statistical calculations, limit theorems*, simulation*, and every other branch of mathematics are legitimate weapons in the search for a solution. Applied probability, although a relatively small branch of mathematics, relies on the resources of the entire subject. It maintains itself successfully without specific affiliation to any particular school or tradition, whether it be British, French, Russian, or North American, while drawing on the best aspects of them all. Its strength lies in the universality of its traditions and the versatility of its mathematical methods.

A further point of importance is the delicate interrelation of theory and practice in applied probability. Without practice (which involves computation and statistics), applied probability is trivial; without theory, it becomes shallow. Close contact is required with experiment and reality for the healthy development of the subject. The collection and analysis of data cannot be avoided, and a certain amount of numerical work will always prove necessary. In attacking problems of applied probability there is a complete cycle from the examination of data to the development of a theoretical model, followed by the statistical verification of the model and its subsequent refinement in the light of its goodness of fit*.

It seems possible that too much effort has been diverted into model building for its own sake, as well as in following through the mathematical refinements of new models. The further development of applied probability requires consideration of real-life problems and the validation of models for these based on observed data. Research workers in the field are aware that only by paying close attention to data and considering genuine problems can their contributions to the subject achieve full scientific stature. For interested readers, a selected bibliography is appended.

References

Note: The following references are all highly technical.

[1] Abakuks, A. (1979). *J. Appl. Prob.*, **16**, 319–331.

[2] Kendall, D. G. (1974). *J. R. Statist. Soc. B*, **36**, 365–402.

[3] McColl, L. A., ed. (1957). *Applied Probability* [Proc. Symp. Appl. Math., **7** (1955)]. McGraw-Hill, New York (for the American Mathematical Society.)

[4] Pickard, D. (1976). *J. Appl. Prob.*, **13**, 486–497.

[5] Szász, D. (1977). *Ann. Prob.*, **5**, 550–559.

[6] Taylor, H. M. (1979). *Adv. Appl. Prob.*, **11**, 527–541.

[7] Vassiliou, P.-C. G. (1978). *J. R. Statist. Soc. A*, **141**, 86–94.

[8] Vered, G. and Yechiali, U. (1979). *Operat. Res.*, **27**, 37–47.

Bibliography

Bailey, N. T. J. (1975). *The Mathematical Theory of Infectious Diseases and Its Applications*. Charles Griffin, London. (Detailed survey of modeling of infectious diseases.)

Barlow, R. E. and Proschan, F. (1965). *Mathematical Theory of Reliability*. Wiley, New York. (Basic principles of reliability.)

Cohen, J. W. (1969). *The Single Server Queue*. North-Holland, Amsterdam. (Comprehensive treatise on queueing.)

Ewens, W. J. (1969). *Population Genetics*. Methuen, London. (Short monograph on mathematical genetics.)

Iosifescu, M. and Tautu, P. (1973). *Stochastic Processes and Applications in Biology and Medicine*, 2 vols. Springer-Verlag, Berlin. (Compendium of stochastic methods in biology and medicine.)

Keyfitz, N. (1968). *Introduction to Mathematics of Population*. Addison-Wesley, Reading, Mass. (Basic principles of demography.)

Kimura, M. and Ohta, T. (1971). *Theoretical Aspects of Population Genetics*. Monographs in Population Biology 4. Princeton University Press, Princeton, N.J. (Monograph on population genetics with emphasis on diffusion.)

Newell, G. F. (1971). *Applications of Queueing Theory*. Chapman & Hall, London. (Short practical account of queueing applications.)

Pielou, E. C. (1969). *An Introduction to Mathematical Ecology*. Wiley-Interscience, New York. (Basic principles of mathematical ecology.)

Pollard, J. H. (1973). *Mathematical Models for the Growth of Human Populations*. Cambridge University Press, Cambridge. (Good introduction to human population models.)

Ross, S. M. (1970). *Applied Probability Models with Optimization Applications*. Holden-Day, San Francisco. (Elementary text with broad range of examples.)

Syski, R. (1960). *Introduction to Congestion Theory in Telephone Systems*. Oliver & Boyd, Edinburgh. (Stochastic processes in telephone traffic.)

Takács, L. (1962). *Introduction to the Theory of Queues*. Oxford University Press, New York. (Introductory treatment of queueing theory.)

Thompson, C. J. (1972). *Mathematical Statistical Mechanics*. Macmillan, New York. (Introduction to statistical mechanics.)

(DAMAGE MODELS
DAM THEORY
ECOLOGICAL STATISTICS
GENETICS, STATISTICS IN
QUEUEING THEORY
RELIABILITY THEORY
RENEWAL PROCESSES
STATISTICS IN CRYSTALLOGRAPHY)
STOCHASTIC PROCESSES

J. GANI

APPLIED PROBABILITY JOURNALS

[*Journal of Applied Probability (JAP); Advances in Applied Probability (AAP)*]

JAP is an international journal which first appeared in June 1964; it is published by the Applied Probability Trust, Sheffield, in association with the London Mathematical Society. The journal contains research papers and notes on applications of probability theory to the biological, physical, social, and technological sciences. A volume of approximately 800 to 900 pages is published each year, consisting of four issues, which appear in March, June, September, and December. *JAP* considers research papers not exceeding 20 printed pages in length, and short communications of 7 to 8 printed pages in the nature of notes or brief accounts of work in progress.

AAP is a companion publication of the Applied Probability Trust launched in 1969. It publishes review and expository papers in applied probability, as well as mathematical and scientific papers of interest to probabilists. A volume of approximately 800 to 900 pages is published each year; it also consists of four issues appearing in March, June, September, and December. *AAP* considers review papers; longer papers in applied probability which may include expository material, expository papers on branches of mathematics of interest to probabilists; papers outlining areas in the biological, physical, social, and technological sciences in which probability models can be usefully developed; and papers in applied probability presented at conferences that do not publish their proceedings. Occasionally, a special *AAP* supplement is published to record papers presented at a conference of particular interest.

HISTORY

In 1962, the editor-in-chief of the two journals, J. Gani, made his first attempts to launch the *Journal of Applied Probability*. At that time, a large number of papers on applications of probability theory were being published in diverse journals dealing with general science, statistics, physics, applied mathematics, economics, and electrical engineering, among other topics. There were then only two probability journals, the Russian *Teoriya Veroyatnostei* (English translation: *Theory of Probability and Its Applications**) and the German *Zeitschrift für Wahrscheinlichkeitstheorie**. Neither of these specialized in applications of probability theory, although papers in applied probability occasionally appeared in them.

Having assembled an editorial board in 1962, Gani attempted to launch the *Journal of Applied Probability* with assistance from the Australian National University and the

Australian Academy of Science. He was not successful in this, and it was only after raising private contributions from himself, Norma McArthur, and E. J. Hannan, both of the Australian National University, that he was able to provide half the finance necessary for the publication of *JAP*. In May 1963, with the support of D. G. Kendall of the University of Cambridge, he was able to persuade the London Mathematical Society to donate the remaining half of the funds required, and thus collaborate in the publication of the journal.

In February 1964, agreement was reached that the ownership of *JAP* should be vested in the Applied Probability Trust, a nonprofit-making organization for the advancement of research and publication in probability, and more generally mathematics. The four trustees were to include the three Australian sponsors of the journal and one trustee nominated by the London Mathematical Society. At the moment (1980) these four trustees are G. E. H. Reuter for the London Mathematical Society (represented in Australia by P. A. P. Moran), Norma McArthur, E. J. Hannan, and J. Gani. The agreement was ratified legally on June 1, 1964, although de facto collaboration had already begun several months before.

By the end of 1963 the Trustees had agreed that the printing of *JAP* should be entrusted to the Jerusalem Academic Press, Israel. The first issue appeared in June 1964. Every effort was made to prevent the time lag between submission and publication of a paper exceeding 15 months; this has since become an established policy for both *JAP* and *AAP*.

Initially, papers were solicited for the journal, but it was not long before the flow of submissions became fairly heavy. Volume 1 (1964) of *JAP* totaled 399 pages in two issues; by 1967 these had grown to three, and by 1971 to four issues per annum. *AAP* was initially published in 1969, also as an annual volume of two issues; it rapidly grew to three issues in 1972 and four in 1974. The two journals have since maintained their approximate sizes of 800 to 900 pages per annum, each in four quarterly issues.

ORGANIZATION

The office of the Applied Probability Trust is located at the University of Sheffield, England. The total staff of the journal consists of an executive editor, Mavis Hitchcock, appointed in 1965, a technical editor (assisted by a free-lance proofreader), and two part-time clerical assistants. This staff is responsible for the entire business of the Trust and for the processing of all papers submitted to the journals. Currently, *JAP* is printed in Israel by the Jerusalem Academic Press, and *AAP* is printed in Northern Ireland by the Universities Press (Belfast) Ltd.

Papers may be submitted to the Editor-in-Chief, the Applied Probability Trust office in Sheffield, or any one of the journal editors. A copy of the author's correspondence should always be sent to the Trust office for the record. When a paper is submitted, it is sent to one or more referees for their comments; the paper is accepted, revised, or rejected primarily on the basis of these. The final responsibility for acceptance of a paper lies with the editor-in-chief.

After a paper has been accepted, it is marked up for the printers by the technical editor; where necessary, diagrams and tables prepared by the authors are also annotated for the printers. The marked-up copy is then sent to the appropriate printer for setting. First proofs are sent to the authors; their corrections are incorporated by the technical editor together with those made by the Trust's proofreader in a proof copy that is returned to the printer. Second proofs are corrected by the technical editor. (The journals have been able to keep the time lag between submission of the accepted version of a paper and printing to not more than 15 months.)

JOURNAL STATISTICS

The Applied Probability Trust office has been in the habit of publishing a journal index at regular intervals. The first appeared for the *JAP* in 1971 (Vols. 1–7, 1964–1970). The second was produced in 1974 for *JAP*

Table 1 Distribution of Papers in the Applied Probability Journals by Subject, 1974–1976 and 1964–1976

Main Subsections	Number of Papers (%)			
	1974–1976		1964–1976	
Probability theory	268 (59.7)		602 (52.6)	
Applied probability	181 (40.3)		542 (47.4)	
Biological sciences		56 (12.5)		154 (13.5)
Physical sciences		14 (3.1)		42 (3.7)
Social sciences		7 (1.6)		17 (1.5)
Technological sciences		104 (23.2)		329 (28.8)
Total	449		1144	

Vols. 8–10 (1971–1973) and *AAP* Vols. 1–5 (1969–1973), and the third index was published in 1977 for *JAP* Vols. 11–13 (1974–1976) and *AAP* Vols. 6–8 (1974–1976). It is proposed to continue publishing an index for both journals at 3-year intervals.

In Table 1, the general contents of the journals are listed for all the years of publication 1964–1976 and for the years 1974–1976. It will be seen from the percentages of *JAP* and *AAP* papers in various categories that there has been an increasing probability theory content in the journals over recent years and a decrease in applied probability. This trend has continued steadily since the inception of the journals, despite constant encouragement to authors to submit material more closely related to applied probability topics. Table 2 analyzes the number of papers in each of 19 detailed categories and

Table 2 Detailed Analysis of Papers by Subject 1974–1976 and 1964–1976

Subject	Number of Papers					
	1974–1976			1964–1976		
Probability theory	268			602		
Branching processes		57			101	
Markov chains		34			75	
Markov processes		22			62	
Other stochastic processes		76			153	
Limit theorems		15			54	
Distribution theory		16			38	
Geometrical probability		18			40	
Miscellaneous		30			79	
Applied probability	181			542		
Biological sciences		56			154	
Population processes			11			36
Mathematical genetics			19			54
Epidemiology			16			42
Virology			1			11
Miscellaneous			9			11
Physical sciences		14			42	
Social sciences		7			17	
Technological sciences		104			329	
General and operations research			22			60
Queueing theory			61			192
Storage			16			52
Traffic theory			5			25
Total	449			1144		

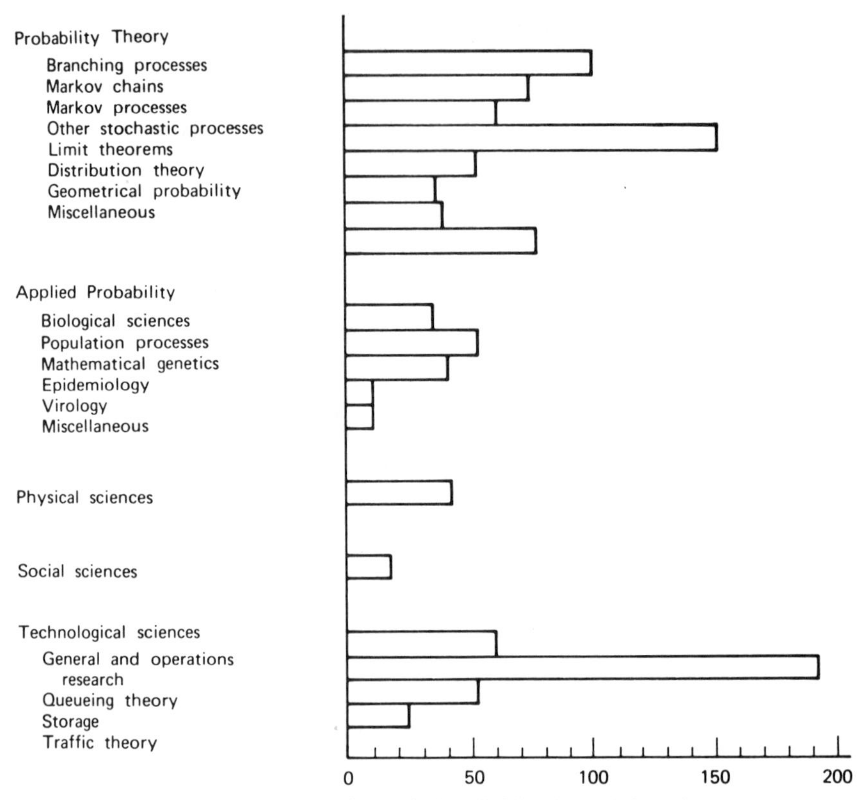

Figure 1 Numbers of papers in the applied probability journals by subject, 1964–1976.

indicates the changes that occurred in 1974–1976 compared with the content of the journals from 1964 to 1976. Figure 1 displays graphically the number of papers in the applied probability journals by subject for 1964–1976.

The journals are endeavoring to maintain a balance of approximately 50% probability theory arising from applications, and 50% applied probability subdivided into categories for the biological, physical, social, and technological sciences. The number of papers submitted to the journals continues to grow, but there has clearly been a shift in research interests toward theory since the *JAP* was first published in 1964.

It is clearly impossible for the *JAP* and *AAP* to collect together every diverse strand of the subject, but it is nevertheless worth making the effort, since it helps probabilists to find most of the applied material they require in a few periodicals. Among these are the two mentioned earlier, as well as the more recent *Stochastic Processes and Their Applications**, and the IMS's *Annals of Probability**, both first published in 1973. It is the policy of *JAP* and *AAP* to contribute to future development. See the final section of the article APPLIED PROBABILITY.

The interested reader may wish to refer to articles in the following bibliography.

Bibliography

Gani, J. (1975). *Proc. Conf. Directions Math. Statist.*, S. G. Ghurye, ed. Suppl. *Adv. Appl. Prob.*, **7**(3), 38–49.

Gani, J. and Spier, A. (1965). *Amer. Statist.*, **19**(4), 18–21.

Journal of Applied Probability Index No. 1 (1971). Applied Probability Trust, Sheffield, England.

Journal of Applied Probability Index No. 2 (1974). Applied Probability Trust, Sheffield, England.

Journal of Applied Probability Index No. 3 (1977). Applied Probability Trust, Sheffield, England.

Kendall, D. G. (1964). Journal of Applied Probability 1, No. 1 (June 1964). *Biometrika*, **51**, 529. (A review of the first issue.)

Pyke, R. (1975). *Proc. Conf. Directions Math. Statist.*, S. G. Ghurye, ed. Suppl. *Adv. Appl. Prob.*, 7(3), 17–37.

J. GANI

APPLIED STATISTICS *See* JOURNAL OF THE ROYAL STATISTICAL SOCIETY, SER. C

APPROXIMATIONS TO DISTRIBUTIONS

HISTORICAL DEVELOPMENTS

The main currents of thought in the development of the subject can be traced to the works of Laplace* (1749–1827), Gauss* (1777–1855), and other scientists of a century or so ago. The normal* density played an important role because of its alleged relation to the distribution of errors of measurement, and it is said that Gauss was a strong "normalist" to the extent that departures were due to lack of data. Natural and social phenomena called for study, including the search for stability, causality, and the semblence of order. Quetelet* (1796–1874) exploited the binomial* and normal* distributions, paying particular attention to the former in his investigations of data reflecting numbers of events. Being well versed in mathematics and the natural sciences (he was a contemporary of Poisson*, Laplace, and Fourier), he searched for order in social problems, such as comparative crime rates and human characteristics, posing a dilemma for the current notions of free will. His contributions to social physics, including the concepts of stability and stereotypes, paved the way for modern sociology. Lexis (1837–1914) opened up fresh avenues of research with his sequences of dependent trials, his aim being to explain departures from the binomial exhibited by many demographic* studies.

Urn problems*, involving drawing of balls of different colors, with and without replacement, have played an important part in distributional models. Karl Pearson* [58] interested himself in a typical discrete distribution (the hypergeometric*) originating from urn sampling, and was led to the formulation of his system of distributions by a consideration of the ratio of the slope to ordinate of the frequency polygon*. Pearson's system (see the section "The Pearson System"), including a dozen distinct types, was to play a dominant role in subsequent developments.

The effect of transformation, with natural examples such as the links between distribution of lengths, areas, and volumes, also received attention; for example, Kapteyn [40] discussed the behavior of a nonlinear mapping of the normal.

Which models have proved useful and enlightening? It is impossible to answer this question purely objectively, for utility often depends on available facilities. Statistical development over this century has reacted specifically, first to all applied mathematics, and second, to the influence of the great digital computer invasion. So models exploiting mathematical asymptotics have slowly given way to insights provided by massive numerical soundings.

For example, the Pearson system probability integrals, involving awkward quadratures* for implementation, have recently [56] been tabulated and computerized. Its usefulness has become obvious in recent times. Similarly, the Johnson [35] translation system*, at one time prohibitive in numerical demands, now has extensive tabulated solutions, which, however, present no problem on a small computer.

Questions of validation of solutions are still mostly undecided. As in numerical quadrature, different appropriate formulas on differential grids form the basis for error analysis, and so for distributional approximation, we must use several approaches for comparison; a last resort, depending on the circumstances, would be simulation* studies.

In fitting models by moments it should be kept in mind that even an infinite set of moments may not determine a density uniquely; it turns out that there are strange nonnull functions, mathematical skeletons, for which all moments are zero. This aspect of the problem, usually referred to as the problem of moments, has been discussed by Shohat and Tamarkin [66].

On the other hand, from Chebyshev-type inequalities*, distributions having the same first r moments, cannot be too discrepant in their probability levels. For the four-moment case, the subject has been studied by Simpson and Welch [67]. They consider $\Pr[x < y] = \alpha$ and the problem of bounds for y given α. Also, it should be kept in mind that moments themselves are subject to constraints; thus for central moments, $\mu_4 \mu_2 - \mu_3^2 \geqslant \mu_2^3$.

TEST STATISTICS IN DISTRIBUTION

The distribution of test statistics (Student's t*, the standard deviation*, the coefficient of variation*, sample skewness*, and kurtosis*) has proved a problem area and excited interest over many decades. Here we are brought face to face with the fact that normality rarely occurs, and in the interpretation of empirical data conservatism loosened its grip slowly and reluctantly.

Outstanding problems are many. For example, consider statistics which give information over and above that supplied by measures of scale and location. Skewness and kurtosis are simple illustrations, the former being assessed by the third central moment, and the latter by the fourth. The exact distributions of these measures under the null hypothesis, normal universe sampled, are still unknown, although approximations are available (see "Illustrations"). Similarly, Student's t and Fisher's F*, although known distributionally under normality, are in general beyond reach under alternatives. These problems direct attention to approximating distributions by mathematical models.

In passing, we note that whereas large sample assumptions usually slant the desired distribution toward a near neighborhood of normality, small-sample assumptions bring out the amazing intricacies and richness of distributional forms. McKay [44] showed the density of the skewness in samples of four from a normal universe to be a complete elliptic integral; Geary [30] shows the density for samples of five and six to be a spline function*, and remarkably, smoothness of density becomes evident for larger samples. Recently, Hoq et al. [33] have derived the exact density of Student's ratio, a noncentral version, for samples of three and four in sampling from an exponential distribution*. Again, spline functions* appear, and it becomes evident that the problem presents insuperable difficulties for larger samples.

In this article we discuss the following systems:

1. K. Pearson's unimodal curves based on a differential equation.
2. Translation systems* based on a mapping of a normal, chi-squared*, or other basic density.
3. Perturbation models based on the normal density, arising out of studies of the central limit theorem*.
4. Multivariate models.
5. Discrete distributions*.

Literature

Johnson and Kotz [37] give an up-to-date comprehensive treatment, the first of its kind, with many references. Discrete, continuous univariate, and multivariate distributions are discussed. A comprehensive account of distributions, with many illuminating examples and exercises, is provided by Kendall and Stuart [42]. Patil and Joshi [53] give a comprehensive summary with most important properties. Bhattacharya and Ranga Rao [4] give a theoretical and mathematical account of approximations related to the central limit theorem*. Some specialized approaches are given in Crain [16] and Dupuy [22].

THE PEARSON SYSTEM

The Model

The model is defined by solutions of the differential equation

$$y' = -(x+a)y/(Ax^2+Bx+C), \quad (1)$$

$y(\cdot)$ being the density. We may take here (but not necessarily in the following discussion) $E(x)=0$, $\operatorname{var}(x)=1$, so that $E(x^3) = \sqrt{\beta_1}$, $E(x^4)=\beta_2$, where the skewness $\beta_1 = \mu_3^2/\mu_2^3$ and the kurtosis $\beta_2 = \mu_4/\mu_2^2$. Arranging (1) as

$$x^s(Ax^2+Bx+C)y' + x^s(x+a)y = 0 \quad (2)$$

and integrating, using $s=0,1,2,3$, we find

$$A = (2\beta_2 - 3\beta_1 - 6)/\Delta,$$
$$B = \sqrt{\beta_1}\,(\beta_2+3)/\Delta,$$
$$C = (4\beta_2 - 3\beta_1)/\Delta, \qquad a = B,$$
$$\Delta = 10\beta_2 - 12\beta_1 - 18.$$

[If $\Delta = 0$, then define $A\Delta=\alpha$, $B\Delta=\beta$, $C\Delta=\gamma$, so that (1) becomes

$$y' = 2y\sqrt{\beta_1}\Big/\left(x^2 - 2x\sqrt{\beta_1} - 3\right),$$

leading to a special case of type 1.]

Note that $\sqrt{\beta_1}$, β_2 uniquely determine a Pearson density. Moreover, it is evident that solutions of (1) depend on the zeros of the quadratic denominator. Although K. Pearson identified some dozen types, it is sufficient here, in view of the subsequent usage of the system, to describe the main densities. We quote noncentral (μ_s') or central (μ_s) moments for convenience, and expressions for $\sqrt{\beta_1}$ and β_2.

Normal:

$$y(x) = (2\pi)^{-1/2}\exp\left(-\tfrac{1}{2}x^2\right) \qquad (x^2<\infty). \quad (3a)$$

$$\mu_1' = 0, \qquad \mu_2 = 1,$$
$$\sqrt{\beta_1} = 0, \qquad \beta_2 = 3.$$

Type 1 (or Beta*):

$$y(x) = \Gamma(a+b)x^{a-1}(1-x)^{b-1}/(\Gamma(a)\Gamma(b)) \quad (0<x<1;\, a,b>0). \quad (3b)$$

$$\mu_1' = a/\alpha_0, \qquad \mu_2 = ab/\left(\alpha_0^2\alpha_1\right),$$
$$\mu_3 = 2ab(b-a)/\left(\alpha_0^3\alpha_1\alpha_2\right),$$
$$\mu_4 = 3ab\left(ab(\alpha_0-6)+2\alpha_0^2\right)/\left(\alpha_0^4\alpha_1\alpha_2\alpha_3\right);$$
$$\alpha_s = a+b+s.$$

Gradient at $x=0$ is finite if $a>1$.
Gradient at $x=1$ is finite if $b>1$.

$$\sqrt{\beta} \geqslant 0 \qquad \text{if } b \geqslant a.$$

Type I may be U-, Ω-, or J-shaped according to the values of a and b.

Type III (Gamma*, Chi-Squared*):

$$y(x) = (x/a)^{\rho-1}\exp(-x/a)/(a\Gamma(\rho)) \quad (0<x<\infty;\, a,\rho>0). \quad (3c)$$

$$\mu_1' = a\rho, \qquad \mu_2 = a^2\rho,$$
$$\sqrt{\beta_1} = 2/\sqrt{\rho}, \quad \beta_2 = 3 + 6/\rho.$$

If the density is zero for $x<-s$, then Type III becomes

$$y(x) = ((x+s)/a)^{\rho-1} \cdot \exp(-(x+s)/a)/(a\Gamma(\rho)) \quad (-s<x<\infty),$$

and the only modification in the moments is that $\mu_1' = s + a\rho$. For other types, see Elderton and Johnson [26].

Recurrence for Moments

For the standardized central moments $\nu_s = \mu_s/\sigma^s$ ($\mu_2=\sigma^2$), we have from (1),

$$\nu_{s+1} = \frac{s}{D_s}\left\{(\beta_2+3)\nu_s\sqrt{\beta_1} + (4\beta_2 - 3\beta_1)\nu_{s-1}\right\}$$
$$(s=1,2,\ldots;\, \nu_0=1, \nu_1=0), \quad (4)$$

where $D_s = 6(\beta_2 - \beta_1 - 1) - s(2\beta_2 - 3\beta_1 - 6)$. Note that if $2\beta_2 - 3\beta_1 - 6 < 0$, then $D_s > 0$ since $\beta_2 - \beta_1 - 1 > 0$. Thus in the Type I region of the (β_1, β_2) plane, all moments exist, whereas below the Type III line, $2\beta_2 - 3\beta_1 - 6 > 0$, only a finite number of moments exists; in this case the highest moment ν_s occurs when $s = [x] + 1$, where $x = 6(\beta_2 - \beta_1 - 1)/(2\beta_2 - 3\beta_1 - 6)$ and $[x]$ refers to the integer part of x.

Evaluation of Percentage Points

First, for a given probability level α, we may seek t_α, where

$$\int_{t_\alpha}^{\infty} y(x)\,dx = \alpha \qquad (5)$$

and $y(\cdot)$ is a solution of the Pearson differential equation, with $\alpha = 1$ when $t_\alpha = -\infty$. It is possible to solve (5) by fragmentation, employing tailor-made procedures for each subregion of the $(\sqrt{\beta_1}, \beta_2)$ space. An alternative, encompassing the whole system, is to use (5) and (1) as simultaneous equations for the determination of t_α. Computer programs have been constructed by Amos and Daniel [1] and Bouver and Bargmann [5].

The converse problem of finding α given t_α has also been solved by the two approaches.

This problem of the relation between t_α and α, as one might expect, transparently reflects the fashions and facilities available over its history of a century or so. Computerized numerical analysis*, in modern times, has dwarfed the quadrature* and inverse interpolation* problems involved, and directed attention away from applied mathematical expertise. Nonetheless, there is a rich literature on probability integral problems, much of it relating to various aspects of closest approximation.

The Johnson et al. [39] tables have been available for two decades and give lower and upper points at the percent levels 0.1, 0.25, 0.5, 1.0, 2.5, 5.0, 10.0, 25.0, and 50.0 in terms of (β_1, β_2); the tables are given with additions in the Pearson and Hartley tables [56]. Interpolation is frequently necessary, however. Approximation formulas for standard percent levels 1.0, 5.0, etc., have been given by Bowman and Shenton [8, 9].

Illustrations

Pearson curves have been fitted to distributions of the following statistics.

SKEWNESS STATISTIC. $\sqrt{b_1}$, defined as $m_3/m_2^{3/2}$, where for the random sample $X_1, X_2, \ldots, X_n$, $m_i = \sum(X_j - \bar{X})^i/n$, and $\bar{X}$ is the sample mean. For sampling from the normal, Fisher [28] demonstrated the independence of $\sqrt{b_1}$ and m_2, from which exact moments can be derived. E. S. Pearson [55] gives comparisons of approximations, including Pearson Type VII, for $n = 25(5)40(10)60$ and probability levels $\alpha = 0.05$ and 0.01.

D'Agostino and Tietjen [20] also give comparisons for $n = 7, 8, 15, 25, 35$ at $\alpha = 0.1, 0.05, 0.025, 0.01, 0.005$, and 0.001.

Mulholland [47] in a remarkable study of $\sqrt{b_1}$ has developed approximations to its density for samples $4 \leqslant n \leqslant 25$. The work uses an iterative integral process based on examination of the density discontinuities and to a certain extent follows earlier work of this kind by Geary [30].

When nonnormal sampling is involved, independence property of $\bar{X}$ and S^2 breaks down, and a Taylor series* for $\sqrt{b_1}$ and its powers can be set up leading to an asymptotic series in n^{-1}. Examples of Pearson approximations have been given by Bowman and Shenton [9] and Shenton and Bowman [65].

KURTOSIS STATISTIC b_2. This is defined as m_4/m_2^2, and the independence of $\bar{X}$ and S^2 in normal sampling enables exact moments to be evaluated. Pearson approximations are given in Pearson [54].

NONCENTRAL χ^2. For independent $X_i \in N(0, 1)$ define

$$\chi'^2 = \sum(a_i + X_i)^2.$$

Pearson approximations are given in Pearson and Hartley [56, pp. 10–11, 53–56]; see also Solomon and Stephens [68].

WATSON'S U_N^{2*}. For approximations to this goodness-of-fit statistic, see Pearson and Hartley [56, 77].

MISCELLANEOUS. Bowman et al. [11] have studied Pearson approximations in the case of Student's t under nonnormality.

Four moments of Geary's ratio* of the mean deviation to the standard deviation in normal samples show the density to be near the normal for $n \geqslant 5$; Pearson curves give an excellent fit [7].

Discussion

In approximating a theoretical density by a four-moment Pearson density, we must remember that the general characteristics of the one must be reflected in the other. Bimodal and multimodal densities, for example, should in general be excluded. Also, for $\sqrt{b_1}$ and $n = 4$ under normality, the true distribution consists of back-to-back J-shapes. However, the Pearson curve is $\cap$-shaped (see McKay [44], and Karl Pearson's remarks). A comparison is given in Table 1.

Since the four-moment fit ignores end points of a statistic (at least they are not fitted in the model), approximations to extreme percentage points will deteriorate sooner or later. Thus, the maximum value of $\sqrt{b_1}$ is known [61] to be $(n-2)/\sqrt{n-1}$, so a Type 1 model will either over or under estimate this end point (see Table 1). Similarly, an associated matter concerns tail abruptness; Pearson models usually deteriorate at and near the blunt tail.

Again, exact moments of a statistic may be intractable. Approximations therefore induce further errors, but usually percentage points are not finely tuned to the measures of skewness and kurtosis.

Finally, in general a unimodal density near the normal ($\sqrt{\beta_1}$ small, $\beta_2 = 3$ approximately) should be well approximated by a Pearson density.

Literature

The classical treatise is Elderton's *Frequency Curves and Correlation* [25]. This has now been revised and appears as *Systems of Frequency Curves* [26] by W. P. Elderton and N. L. Johnson (Cambridge University Press). It contains a complete guide to fitting Pearson curves to empirical data, with comments on other approximation systems.

The basic papers by Karl Pearson are those of 1894 [57], 1895 [58], and 1901 [59]. They are of considerable interest in showing the part played by the normal law of errors and modifications of it; for example, Pearson argues that data might be affected by a second normal component to produce the appearance of nonnormality, and solves the two-component normal mixture problem completely. It has since been shown [10] that if exact moments are available (five are needed) then there may be one or two solutions to the problem, or quite possibly none at all. Pearson's examples note these cases.

Again, Pearson from time to time expresses his displeasure at the sloppy tabulation and abbreviation of data. One example he considers concerns the distribution of 8689 cases of enteric fever received into the Metropolitan Asylums Board Fever Hospitals, 1871–1893. There are 266 cases reported for "under age 5" and 13 for "over age 60." Fitting types I and III to the data by moments, he notes that both models suggest an unlikely age for first onset of the disease (-1.353 and -2.838 years, respectively) and similarly an unlikely upper limit to the duration of life (385 years). (It is quite possible that he was well aware of the small chances involved, and we should keep in mind the point that to arrive at bounds of any kind for the data was something not supplied by conventional normal theory modeling.)

It should be emphasized that a model for a distribution of experimental data encounters problems that are different from those

Table 1 Pearson and Johnson Approximations

Population	Statistic	Sample Size	A[a]	$\alpha = 0.01$	0.05	0.10	0.90	0.95	0.99
Exponential	$\sqrt{m_2}$	$n = 3^b$	E	0.045	0.109	0.163	1.293	1.610	2.340
			P	0.057	0.113	0.164	1.296	1.612	2.337
			P*	0.368	0.368	0.369	1.290	1.632	2.374
			S_B	0.019	0.102	0.164	1.289	1.609	2.358
		$n = 4$	E	0.095	0.180	0.245	1.342	1.632	2.303
			P	0.108	0.184	0.245	1.343	1.633	2.293
			P*	0.397	0.397	0.399	1.347	1.654	2.322
			S_B	0.083	0.178	0.246	1.340	1.632	2.308
	$v = (s_x/m_1')$	$n = 10$	M	0.481	0.587	0.648	1.231	1.351	1.617
			P	0.499	0.591	0.646	1.237	1.353	1.598
			P*	0.531	0.600	0.650	1.238	1.356	1.601
			S_B^*	0.509	0.596	0.650	1.237	1.355	1.604
				$\alpha = 0.90$	0.950	0.975	0.990	0.995	0.999
Normal	$\sqrt{b_1}$	$n = 4^c$	E	0.831	0.987	1.070	1.120	1.137	1.151
			P	0.793	0.958	1.074	1.178	1.231	1.306
			S_B	0.791	0.955	1.071	1.179	1.237	1.327
		$n = 8$	E	0.765	0.998	1.208	1.452	1.606	1.866
			P	0.767	0.990	1.187	1.421	1.583	1.929
			S_U	0.767	0.990	1.187	1.421	1.583	1.929
				$\alpha = 0.01$	0.05	0.10	0.90	0.95	0.99
	b_2	$n = 10$	M	1.424	1.564	1.681	3.455	3.938	4.996
			P	1.495	1.589	1.675	3.471	3.931	4.921
			S_B	1.442	1.577	1.677	3.456	3.921	4.933
			D	1.39	1.56	1.68	3.53	3.95	5.00
			A_G	1.287	1.508	1.649	3.424	3.855	4.898

[a]E, exact; P, four-moment Pearson on statistic; P*, four-moment Pearson on square of statistic; M, Monte Carlo of 100,000 runs; A_G, Anscombe and Glynn [2]; D, D'Agostino and Tietjen [19]; E for $\sqrt{m_2}$ derived by Lam [43]; S_U, S_B, Johnson; S_B^*, Johnson on square of statistic.

[b]For $n = 3$, P* has density $c_0(m_2 - 0.136)^\alpha/(m_2 + 13.168)^\beta$, $\alpha = -0.740$, $\beta = 7.770$; P has density $c_1(\sqrt{m_2} - 0.024)^\alpha/(\sqrt{m_2} + 13.076)^\beta$, $\alpha = 0.738$, $\beta = 39.078$.

[c]For $n = 4$, P is the density $c_2(1.896^2 - b_1)^{3.26}$; from theory $\sqrt{b_1}$ max. is 1.155.

of the corresponding situation for a theoretical statistic. In the latter, errors are due solely to choice of model, although different solution procedures are possible; for example, parameters may be determined by moments, or by percentiles, and also in part from a knowledge of end points. However, in the case of experimental data, parameters are estimated by different procedures (least squares*, maximum likelihood*, moments*, etc.) and there is the question of biases*, variances, etc., and more detailed knowledge of the sampling distribution of the estimates.

Since higher moments of data are subject to considerable sampling errors, a recent [48] method of fitting is recommended when one (or both) end point(s) is known. The procedure calculates a modified kurtosis using the end point and four moments.

TRANSLATION SYSTEMS

Early Ideas

Suppose that the berries of a fruit have radii (measured in some decidable fashion) normally distributed $N(R, \sigma^2)$. Then surface area $S = \pi r^2$ will no longer be normal and indeed will be skewed distributionally. This idea was developed a little later than the introduction of the Pearson system by Kapteyn [40] in his treatment of skew frequency curves in biology. From a different point of view and using a hypothesis relating to elementary errors, Wicksell [71] traced the relation between certain transformation of a variate and a genetic theory of frequency.

Kapteyn considered the transformations

$$y = (X + h)^q - m \qquad [y \in N(0,1)]$$

with $-\infty < q < \infty$, the special case $q \to 0$ leading to a logarithmic transformation.

The development of the subject was perhaps retarded because of overanxiety to trace a transformation's relation to natural phenomena and mathematical difficulties. Moreover, the Kapteyn mapping proved intractable mathematically in most cases.

Johnson's Systems

Johnson [35] introduced two transformations of the normal density. His S_U system* relates to a hyperbolic sine function, has doubly infinite range, and

$$y = \sinh\left(\frac{X-\gamma}{\delta}\right) \qquad (6)$$
$$(-\infty < y < \infty, \delta > 0),$$

where $X \in N(0,1)$.

Similarly, the S_B system* relates to densities with bounded range, and

$$y = 1/(1 + e^{(\gamma - X)/\delta}) \qquad (0 < y < 1, \delta > 0), \qquad (7)$$

where again $X \in N(0,1)$.

These transformations are both one-to-one, and the S_U case readily yields to moment evaluation, whereas S_B does not.

For S_U, the first four moments are

$$\mu_1'(y) = -\sqrt{\omega}\sinh\Omega,$$
$$\mu_2(y) = (\omega - 1)(\omega\cosh(2\Omega) + 1)/2,$$
$$\mu_3(y) = -(\omega - 1)^2\sqrt{\omega}\{(\omega^2 + 2\omega)\sinh(3\Omega) + 3\sinh\Omega\}/4, \qquad (8)$$
$$\mu_4(y) = (\omega - 1)^2\{d_4\cosh(4\Omega) + d_2\cosh(2\Omega) + d_0\},$$

where

$$d_4 = \omega^2(\omega^4 + 2\omega^3 + 3\omega^2 - 3)/8,$$
$$d_2 = \tfrac{1}{2}\omega^2(\omega + 2),$$
$$d_0 = 3(2\omega + 1)/8, \qquad \text{and}$$
$$\ln\omega = 1/\delta^2, \quad \Omega = \gamma/\delta.$$

Note that since $\omega > 1$, $\mu_3(y)$ has the sign of $(-\Omega)$, and unlike the structure of the Pearson system, here all moments are functions of ω and Ω. (For the Pearson system, mean and variance are not affected by $\sqrt{\beta_1}$ and β_2, whereas they are for S_U.)

If Y is a variate, then set

$$Y = (y + p)/q \qquad (9a)$$

so that

$$\nu_1' = E(Y) = (\mu_1'(y) + p)/q$$
$$\nu_2 = \text{var}(Y) = (\mu_2(y))/q^2 \qquad (9b)$$

Determine p and q. The values of $\mu_1'(y)$ and $\mu_2(y)$ are set by equating the skewness ($\sqrt{\beta_1}$), and kurtosis (β_2) of Y to those of y; assistance here is given in Tables 34 and 35 in Pearson and Hartley [56].

Johnson's S_U system [35] immediately provides percentiles from those of the normal, and also an equivalent normal variate,

$$X = \gamma + \delta\sinh^{-1} y.$$

However, one must keep in mind that the S_U density is still only a four-moment approximation.

To fix the domain of validity of S_U, let $\Omega = -k$ and $k \to \infty$, so that the mean of y tends to ∞ and $\sigma^2 \sim \omega(\omega - 1)e^{2k}/4$. Then if $t = y/\sigma$ from (6), $X = c + \ln t$, which corresponds to a log-normal transformation, and from (8)

$$\sqrt{\beta_1} = (\omega + 2)\sqrt{\omega - 1},$$
$$\beta_2 = \omega^4 + 2\omega^3 + 3\omega^2 - 3, \qquad (10)$$

the parametric form of the boundary in the $(\sqrt{\beta_1}, \beta_2)$ plane.

Examples of S_U and S_B are given in Table 1.

Literature

An adequate description of S_U and S_B is given by Pearson and Hartley [56, pp. 80–

87], including tables to facilitate fitting. Moreover, the iterative scheme for the evaluation of the parameters of S_U given by Johnson [36] is readily programmed for a small computer. A rational fraction solution for ω, leading also to a value of Ω, has been developed by Bowman and Shenton [10]; a similar scheme is also available for the S_B-system [13].

SERIES DEVELOPMENTS

These originated in the nineteenth century, and are related to procedures to sharpen the central limit theorem*. For large n, for $z_1, z_2, \ldots, z_n$ mutually independent variates with common standard deviation σ, the distribution of $s = (\sum z)/(\sigma\sqrt{n})$ is approximately normal. A better approximation appears from Charlier's A-series [14],

$$\phi(x) + (a_3/3!)\phi^{(3)}(x) + (a_4/4!)\phi^{(4)}(x) + \cdots \quad (11)$$

where $\phi(x) = (2\pi)^{-1/2}\exp(-\frac{1}{2}x^2)$ is the standard normal density. [This approximation involves derivatives of $\phi(x)$.] Cramér [17, 18] has proved that certain asymptotic properties of (11) hold. Another version of the series development (11) takes the form $\Phi(d/dx)\phi(x)$, where

$$\Phi(t) \equiv \exp\sum \epsilon_j(-t)^j/j!, \quad (12)$$

the operator changing the cumulants* (κ_r) of $\phi(\cdot)$ to ($\kappa_r + \epsilon_r$). This, from Cramér's work (see his note and references in ref. 18), has similar asymptotic properties with respect to the central limit theorem*.

Since the derivatives of the normal density are related to Hermite polynomials* (*see* CHEBYSHEV–HERMITE POLYNOMIALS), (11) may be written

$$\left[1 - (a_3/3!)H_3(x) + (a_4/4!)H_4(x) - \cdots\right]\phi(x), \quad (13)$$

and if this approximates a density $f(x)$, then using orthogonality,

$$a_s = (-1)^s E\left[H_s(x)\right],$$

where the expectation* operator refers to $f(x)$. Thus in terms of cumulants,

$$a_3 = -\kappa_3, \quad a_4 = \kappa_4,$$

$$a_5 = -\kappa_5, \quad a_6 = \kappa_6 + 10\kappa_3^2,$$

etc., the coefficients a_1, a_2 being zero because of the use of the standard variate x.

If empirical data are being considered, only the first five or six terms of (13) can be contemplated because of large sampling errors of the moments involved. However, this problem does not arise in the approximation of theoretical structures, and cases are on record (for example, ref. 47) in which the twentieth polynomial has been included; quite frequently the density terms turn out to have irregular sign and magnitude patterns.

The Edgeworth* form [23, 24],

$$\phi(x)\left\{1 + (\kappa_3/3!)H_3(x) + (\kappa_4/4!)H_4(x) + (\kappa_5/5!)H_5(x) + \left(\kappa_6 + 10\kappa_3^2\right)H_6(x)/6! + \cdots\right\}$$

has one advantage over (13), in that when applied to certain statistics, the standardized cumulants κ_r may be shown to be of order $1/n^{(1/2)r-1}$, where n is the sample size, so that if carried far enough the coefficients will tend to exhibit a regular magnitude pattern.

Nonnormal Kernels

The basic function $\phi(\cdot)$ need not be normal, and Romanowsky [62] introduced generalizations such as gamma- and beta*-type densities, with associated Laguerre* and Jacobi* orthogonal polynomials. The normal density is, however, generally favored because of the simplicity of the Chebyshev–Hermite system of polynomials.

Cornish–Fisher Expansions

Using a differential series, such as (12) with normal kernel, Cornish and Fisher [15] derived a series for the probability integral of a variate whose cumulants are known. By inversion of series, they derived a series for the deviate at a given probability level in terms of polynomials in the corresponding normal deviate. Fisher and Cornish [29] extended the earlier study to include terms of order n^{-3}, these involving the eighth cumulant and polynomials of degree seven. Further generalizations are due to Finney [27], who treated the case of several variates; Hill and Davis [32], who gave a rigorous treatment indicating the procedure for the derivation of a general term; and Draper and Tierney [21], who tabulated the basic polynomials involved in the terms to order n^{-4} and cumulants up to κ_{10}; unpublished results of Hill and Davis for the higher-order polynomials agreed with those found by Draper and Tierney.

Some Applications

In his paper on testing for normality* (*see* DEPARTURES FROM NORMALITY, TESTS FOR), Geary [31] considered the distribution of t under nonnormality, basing it on what he called a differential series [expression (12)] with kernel the density of t^* under normality, i.e., $c_n(1 + t^2/(n-1))^{-n/2}$. He derived a series to order n^{-2} for the probability integral, using it cautiously for a sample of 10 under only moderate nonnormality. In an earlier study [30] searching for precise forms for the density of the sample skewness under normality, he used the Cornish–Fisher series to assess probability levels using cumulants up to the eighth. Mulholland [47] developed the subject further, providing a recursive scheme for evaluating at least theoretically any moment of $\sqrt{b_1}$, and a Charlier series for the probability integral up to polynomials of degree 20.

Series involving Laguerre polynomials and a gamma density have been exploited by Tiku. For example, his study [69] of the variance ratio* and Student's t [70], both under nonnormality, involve terms up to order n^{-2}.

Discussion

In approximating distributions, convergence* questions, and even to a less extent, asymptotic properties are irrelevant, for we are concerned with a finite number of terms, since very rarely can general terms be found. Thus questions of nonnegativity of the density arise [3, 64], and internal checks for closeness of approximation via convergence or otherwise are not available. At best the Charlier and Edgeworth series can only serve as backup models.

MULTIVARIATE DENSITIES

The Pearson system defined in (1) becomes, in bivariate form,

$$\frac{1}{y}\frac{\partial y}{\partial x_i} = \frac{P_i(x_1, x_2)}{Q_i(x_1, x_2)} \qquad (i = 1, 2),$$

where P_i and Q_i are functions of x_1 and x_2 of degrees 1 and 2, respectively. There are again several types, including the bivariate normal* with density

$$\phi(x_1, x_2) = (1-\rho^2)^{-1/2}(2\pi)^{-1} \times \exp\left\{\left(-\tfrac{1}{2}x^2 - \rho xy - \tfrac{1}{2}y^2\right)/(1-\rho^2)\right\}, \quad (\rho^2 < 1)$$

in standard form, which reduces to a product form when $\rho = 0$ (i.e., when the variates are uncorrelated).

Another well-known type is the Dirichlet* density,

$$y(x_1, x_2) = c x_1^{a-1} x_2^{b-1}(1 - x_1 - x_2)^{c-1} \qquad (a, b, c > 0)$$

with domain $x_1, x_2 > 0$, $x_1 + x_2 \leqslant 1$.

Series development based on the Charlier model take the form

$$\left\{1 + a_{10}\partial_1 + a_{01}\partial_2 + (a_{20}/2!)\partial_1^2 + a_{11}\partial_1\partial_2 + (a_{02}/2!)\partial_2^2 + \cdots\right\}\phi(x_1, x_2),$$

where $\partial_i \equiv \partial/\partial x_i$.

Similarly, there are multivariate translation systems, following the Johnson S_U and S_B models.

The Pearson system, for which marginal distributions are of Pearson form, can be fitted by moments, a new feature being the necessity to use product moments*, the simplest being related to the correlation between the variates. A similar situation holds for Charlier multivariate developments, and again, as with the single-variate case, negative frequencies can be a problem.

Literature

A sustained study of empirical bivariate data, at least of historical interest, is given in K. Pearson's 1925 paper [60].

A brief discussion of frequency surfaces will be found in Elderton and Johnson [26]. A more comprehensive treatment, including modern developments, is that of N. L. Johnson and S. Kotz [37]. Mardia's work [45] is a handy reference.

DISCRETE DISTRIBUTIONS

There is now a richness of variety of discrete distributions, and the time is long passed when there were possibly only three or so choices, the binomial*, the Poisson*, and the hypergeometric*; the geometric appearing infrequently in applications.

Just as a differential equation is used to define the Pearson system of curves, so analogs in finite differences* may be used to generate discrete density functions [41, 49, 58]. Again, if $G_i(t)$, $i = 1, 2, \dots,$ are the probability generating functions* of discrete variates, then new distributions arise from $G_1(G_2(t))$ and similar structures; for example, Neyman's contagious distributions* are related to choosing G_1 and G_2 as Poisson generating functions.

There are many applications where distribution approximation models are required for random count data* [51, 52]. We also mention occupancy problems* [38], meteorological phenomena relating to drought frequency, storm frequency, degree-days, etc. However, the need for approximating discrete distributions is not common, especially with the advent of computer facilities. A classical exception is the normal approximation to the binomial distribution when the index n is large; in this case, if the random variate is x, then with probability parameter p, we consider the approximation $(x - np)/\sqrt{npq}$ to be nearly normal (for refinements, see, e.g., Molenaar [46]).

Again the Poisson or binomial density functions may be used as the basis for Charlier-type approximation expansions. Thus for the Poisson function $\psi(x) = e^{-m}m^x/x!$, we consider

$$f(x) = \{1 + \alpha_2\nabla_x^2/2! + \alpha_3\nabla_x^3/3!t + \cdots\}\psi(x),$$

where $\nabla g(x) \equiv g(x) - g(x-1)$ is a backward difference*. This type of approximation may be considered for random variates taking the values $x = 0, 1, \dots,$ when moments exist and the density to be approximated is complicated.

Literature

Historical information has been given by Särndal [63], and comprehensive accounts are those of Johnson and Kotz [37] and Patil and Joshi [53]. A short account, with new material, is given by J. K. Ord [50].

References

[1] Amos, D. E. and Daniel, S. L. (1971). Tables of Percentage Points of Standardized Pearson Distributions. *Rep. No. SC-RR-71 0348*, Sandia Laboratories, Albuquerque, N.M.

[2] Anscombe, F. and Glynn, W. J. (1975). Distribution of the Kurtosis Statistics b_2 for Normal Samples. *Tech. Rep. 37*, Dept. of Statistics, Yale University, New Haven, Conn.

[3] Barton, D. E. and Dennis, K. E. (1952). *Biometrika*, **39**, 425–427.

[4] Bhattacharya, R. N. and Ranga Rao, R. (1976). *Normal Approximations and Asymptotic Expansions*. Wiley, New York.

[5] Bouver, H. and Bargmann, R. (1976). *Amer. Stat. Ass.: Proc. Statist. Computing Sect.*, pp. 116–120.

[6] Bowman, K. O., Beauchamp, J. J., and Shenton, L. R. (1977). *Int. Statist. Rev.*, **45**, 233–242.

[7] Bowman, K. O., Lam, H. K., and Shenton, L. R. (1980). *Reports of Statistical Application Res., Un. Japanese Scientists and Engineers*, **27**, 1–15.

[8] Bowman, K. O., Serbin, C. A., and Shenton, L. R. (1981). *Commun. Statist. B*, **10**(1), 1–15.

[9] Bowman, K. O. and Shenton, L. R. (1973). *Biometrika*, **60**, 155–167.

[10] Bowman, K. O. and Shenton, L. R. (1973). *Biometrika*, **60**, 629–636.

[11] Bowman, K. O. and Shenton, L. R. (1979). *Biometrika*, **66**, 147–151.

[12] Bowman, K. O. and Shenton, L. R. (1979). *Commun. Statist. B*, **8**(3), 231–244.

[13] Bowman, K. O. and Shenton, L. R. (1980). *Commun. Statist. B*, **9**(2), 127–132.

[14] Charlier, C. V. L. (1905). *Ark. Mat. Astron. Fys.*, **2**(15).

[15] Cornish, E. A. and Fisher, R. A. (1937). *Rev. Inst. Int. Statist.*, **4**, 1–14.

[16] Crain, B. R. (1977). *Siam J. Appl. Math.*, **32**, 339–346.

[17] Cramér, H. (1928). *Skand. Aktuarietidskr*, **11**, 13–74, 141–180.

[18] Cramér, H. (1972). *Biometrika*, **59**, 205–207.

[19] D'Agostino, R. B. and Tietjen, G. L. (1971). *Biometrika*, **58**, 669–672.

[20] D'Agostino, R. B. and Tietjen, G. L. (1973). *Biometrika*, **60**, 169–173.

[21] Draper, N. R. and Tierney, D. E. (1973). *Commun. Statist.*, **1**(6), 495–524.

[22] Dupuy, M. (1974). *Int. J. Computer Math. B*, **4**, 121–142.

[23] Edgeworth, F. Y. (1905). *Camb. Philos. Trans.*, **20**, 36–66, 113–141.

[24] Edgeworth, F. Y. (1907). *J. R. Statist. Soc.*, **70**, 102–106.

[25] Elderton, W. P. (1960). *Frequency Curves and Correlation*. Cambridge University Press, Cambridge.

[26] Elderton, W. P. and Johnson, N. L. (1969). *Systems of Frequency Curves*. Cambridge University Press, Cambridge.

[27] Finney, D. J. (1963). *Technometrics*, **5**, 63–69.

[28] Fisher, R. A. (1928). *Proc. Lond. Math. Soc.*, **2**, 30, 199–238.

[29] Fisher, R. A. and Cornish, E. A. (1960). *Technometrics*, 2, 209–225.

[30] Geary, R. C. (1947). *Biometrika*, **34**, 68–97.

[31] Geary, R. C. (1947). *Biometrika*, **34**, 209–242.

[32] Hill, G. W. and Davis, A. W. (1968). *Ann. Math. Statist.*, **39**, 1264–1273.

[33] Hoq, A. K. M. S., Ali, M. M., and Templeton, J. G. (1977). Distribution of Student's Ratio Based on the Exponential Distribution. *Working Paper No. 77-001*, Dept. of Industrial Engineering, University of Toronto, Toronto.

[34] Hotelling, H. (1961). *Proc. 4th Berkeley Symp. Math. Stat. Prob.*, Vol. 1. University of California Press, Berkeley, Calif., pp. 319–359.

[35] Johnson, N. L. (1949). *Biometrika*, **36**, 149–176.

[36] Johnson, N. L. (1965). *Biometrika*, **52**, 547–558.

[37] Johnson, N. L. and Kotz, S. (1972). *Distributions in Statistics: Continuous Multivariate Distributions*. Wiley, New York.

[38] Johnson, N. L. and Kotz, S. (1977). *Urn Models and Their Application*. Wiley, New York.

[39] Johnson, N. L., Nixon, E., Amos, D. E., and Pearson, E. S. (1963). *Biometrika*, **50**, 459–498.

[40] Kapteyn, J. C. (1903). *Skew Frequency Curves in Biology and Statistics*. Noordhoff, Groningen.

[41] Katz, L. (1963). *Proc. Int. Symp. Discrete Distrib.*, Montreal, pp. 175–182.

[42] Kendall, M. G. and Stuart, A. (1969). *The Advanced Theory of Statistics*, 3rd ed., Vol. 1: *Distribution Theory*. Charles Griffin, London/Hafner Press, New York.

[43] Lam, H. K. (1978). The Distribution of the Standard Deviation and Student's t from Non-normal Universes. Ph.D. dissertation, University of Georgia.

[44] McKay, A. T. (1933). *Biometrika*, **25**, 204–210.

[45] Mardia, K. V. (1970). Families of Bivariate Distributions. *Griffin's Statist. Monogr. No. 20.*

[46] Molenaar, W. (1970). Approximation to the Poisson, Binomial, and Hypergeometric Distribution Functions. *Math. Centre Tracts No. 31.* Mathematisch Centrum, Amsterdam.

[47] Mulholland, H. P. (1977). *Biometrika*, **64**, 401–409.

[48] Müller, P.-H. and Vahl, H. (1976). *Biometrika*, **63**, 191–194.

[49] Ord, J. K. (1967). *J. R. Statist. Soc. A*, **130**, 232–238.

[50] Ord, J. K. (1975). Families of Frequency Distributions. *Griffin's Statist. Monogr. No. 30.*

[51] Patil, G. P. (1965). *Classical and Contagious Discrete Distributions* (Proc. Int. Symp., Montreal). Pergamon Press, New York.

[52] Patil, G. P. ed. (1970). *Random Counts in Physical Science, Geological Science, and Business*, Vols. 1–3. The Penn State Statistics Series. Pennsylvania State University Press, University Park, Pa.

[53] Patil, G. P. and Joshi, S. W. (1968). *A Dictionary and Bibliography of Discrete Distributions.* Oliver & Boyd, Edinburgh.

[54] Pearson, E. S. (1963). *Biometrika*, **50**, 95–111.

[55] Pearson, E. S. (1965). *Biometrika*, **52**, 282–285.

[56] Pearson, E. S. and Hartley, H. O. (1972). *Biometrika Tables for Statisticians*, Vol. 2. Cambridge University Press, Cambridge.

[57] Pearson, K. (1894). *Philos. Trans. R. Soc. Lond. A*, **185**, 71–110.

[58] Pearson, K. (1895). *Philos. Trans. R. Soc. Lond. A*, **186**, 343–414.

[59] Pearson, K. (1901). *Philos. Trans. R. Soc. Lond. A*, **197**, 443–459.

[60] Pearson, K. (1925). *Biometrika*, **17**, 268–313.

[61] Pearson, K. (1933). *Biometrika*, **25**, 210–213.

[62] Romanowsky, V. (1925). *Biometrika*, **16**, 106–116.

[63] Särndal, C.-E. (1971). *Biometrika*, **58**, 375–391.

[64] Shenton, L. R. (1951). *Biometrika*, **38**, 58–73.

[65] Shenton, L. R. and Bowman, K. O. (1975). *J. Amer. Statist. Ass.*, **70**, 349, 220–229.

[66] Shohat, J. A. and Tamarkin, J. E. (1943). *The Problem of Moments*. American Mathematical Society, New York.

[67] Simpson, J. A. and Welch, B. L. (1960). *Biometrika*, **47**, 399–410.

[68] Solomon, H. and Stephens, M. A. (1978). *J. Amer. Statist. Ass.*, **73**, 153–160.

[69] Tiku, M. L. (1964). *Biometrika*, **51**, 83–95.

[70] Tiku, M. L. (1971). *Aust. J. Statist.*, **13**(3), 142–148.

[71] Wicksell, S. D. (1917). *Ark. Mat. Astron. Fys.*, **12**(20).

Acknowledgment

This research was operated by Union Carbide Corporation under Contract W-7405-eng-26 with the U.S. Department of Energy.

(ASYMPTOTIC EXPANSION
CORNISH–FISHER EXPANSION
FUNCTIONS, APPROXIMATIONS TO
GRAM–CHARLIER EXPANSION
JOHNSON SYSTEM OF DISTRIBUTIONS
PEARSON SYSTEM OF DISTRIBUTIONS)

K. O. Bowman
L. R. Shenton

APPROXIMATIONS TO FUNCTIONS *See* FUNCTIONS, APPROXIMATIONS TO

A PRIORI DISTRIBUTION

The term *a priori distribution* is used to describe a distribution ascribed to a parameter in a model. It occurs most commonly in the application of Bayesian methods. The a priori distribution is usually supposed to be known exactly—and not to depend on unknown parameters of its own.

(BAYESIAN INFERENCE)

ARBITRARY ORIGIN *See* CODED DATA; CODING THEOREM

ARBITRARY SCALE *See* CODED DATA; CODING THEOREM

ARCHAEOLOGY, STATISTICS IN

The application of statistical thought to archaeology has been a slow process. This reluctance arises because (1) archaeological data rarely can be gathered in a well-designed statistical experiment; (2) describing empirical findings requires an expertise that is not easily modeled.

In recent years, however, the central problem of archaeology, generally labeled "typology," and the important related problem of "seriation" have received considerable mathematical and statistical attention, which we discuss herewith. The advent of the high-speed computer has made feasible analyses of large sets of archaeological data which were previously impracticable. The application of routine statistical methodology has been infrequent and nonsystematic (see ref. 5, Chap. 13). A recent article by Mueller [23] is noteworthy, illustrating the use of sampling schemes in archaeological survey. The important question of how to locate artifact sites in a region that cannot be totally surveyed is examined. Simple random sampling and stratified sampling are compared in conjunction with an empirical study.

The artifact provides the class of entities with which archaeology is concerned. Typology is concerned with the definition of artifact types. Since mode of usage is unobservable, definition arises from an assortment of qualitative and quantitative variables (e.g., shape, color, weight, length) yielding a list of attributes for each artifact. Artifact types are then defined in terms of "tight clusters of attributes" [19]. The definition of types is usually called taxonomy* and the methods of numerical taxonomy have come to be employed in archaeological typology. (One does find in the literature references to taxonomy for archaeological sites, an isomorphic problem to that of typology for artifacts.)

"Typological debate" [11] has run several decades, resulting in a voluminous literature (see ref. 5). The issues of contention include such matters as whether types are "real" or "invented" to suit the researcher's purposes, whether there is a "best" classification of a body of materials, whether types can be standardized, whether types represent "basic" data, and whether there is a need for more or fewer types. Statistical issues arise in the construction of a typology.

Krieger's effort is a benchmark in unifying the typological concept. An earlier article in this spirit by Gozodrov [10] is deservedly characterized by Krieger as "tentative and fumbling" [19, p. 271]. Krieger reveals the variance in published thought on the classification* issue through examination of the work on pottery description and on projectile style. He articulates the "typological method" and cites Rouse [28] as a good illustration. The earliest quantitative work is by Spaulding [30]. A previous paper by Kroeber [20] concerned itself solely with relating pairs of attributes. The usual χ^2 statistic* (*see* CHI-SQUARE TESTS) as well as other measures of association* were studied for 2×2 presence–absence attribute tables. From this lead Spaulding suggests, given attribute lists for each artifact in the collection, the preparation of cross-tabulations of all attributes (attribute categories are no longer restricted to presence–absence). Two-way tables are χ^2-tested, leading to the clustering of nonrandomly associated attributes. Artifact types are then defined by identifying classes of artifacts that exhibit sets of associated attributes.

Attempting to formalize Spaulding's technique led archaeologists into the realm of cluster analysis*. A basic decision in cluster analysis is whether the items or the components of the item data vectors are to be clustered. Specifically, are we clustering artifacts (Q-mode* of analysis, as it has been called) or attributes (R-mode* of analysis)? In defining a typology a Q-mode of analysis is appropriate, but the unfortunate use of the phrase "cluster of attributes" by both Krieger and Spaulding has resulted in a persistent confusion in the literature. Factor* and principal components* analyses have occasionally been employed as R-mode analyses to group attributes by significant dimensions interpretable as underlying features of the data.

Typology, then, involves discerning clusters of artifacts on the basis of similarity of their attributes. Similarity between artifacts is measured by similarity functions that arise in other (e.g., psychological and sociological) settings as well. Nonmathematically, a similarity function between two vectors reflects the closeness between components of the vectors as an inverse "distance." For a set of vectors the similarities between all pairs are arranged in a "similarity" matrix*. Beginning with such a matrix, clustering procedures are usually effected with the assistance of a computer.

The earliest computer typologies were done by single-link (nearest-neighbor) clustering [29, p. 180]. Links are assigned from largest similarities and clusters are derived from linked units. Unwelcome "chaining" often occurs, whence average linkage (weighted and unweighted) and complete linkage procedures have been suggested [29, p. 181]. Their sensitivity to spurious large similarities led Jardine and Sibson [15] to formulate double-link cluster analysis, but now "chaining" returns. Thus Hodson [13] proposes a K-means cluster analysis approach. The total collection is partitioned into a predetermined number of clusters. Rules are defined for transferring artifacts from one cluster to another until a "best" clustering is obtained. The procedure is repeated for differing initial numbers, with expertise determining the final number of types defined. The approach can accommodate very large collections.

Similarities implicitly treat attributes in a heterarchical manner in defining types. Whallon [34] suggests that often a hierarchy of importance among attributes exists and that this is how archaeologists feel their way to defining types. Attribute trees are defined where presence or absence of an attribute creates a branch. Employing χ^2 values computed over appropriate 2×2 tables, the sequence of attributes forming the tree is achieved and also the definition of types from such sequences. Cell frequency problems plague many of the χ^2 values. Read [25] formalizes hierarchical classification in terms of partitions of a set of items and allows both discrete and continuous attributes. Clark [3] extends these ideas, assuming discrete attributes and setting them in appropriate higher-order contingency tables* to which log-linear models* are fitted.

In summary, then, a typology is usually obtained by clustering* (through an appropriate procedure) artifacts having similar attributes and defining types through these attributes. Recent hierarchical classification approaches show promise for data sets exhibiting weak clustering.

The next natural step in archaeological enterprise is the comparison of artifact collections, the process called seriation. In broadest terms, seriation consists of arranging a set of collections in a series with respect to similarity of the component artifacts to infer ordering in some nonobservable (usually time) dimension. The collections will typically be grave lots or assemblages. The chronological inference is drawn by the assumption that the degree of similarity between two collections varies inversely with separation in time. Such an assumption implicitly requires a "good" typology for the collections. Of course, other dimensions (e.g., geographic, cultural) may also affect the degree of similarity between collections and confound a "time" ordering. Techniques such as stratigraphy*, dated inscriptions, cross-ties with established sequences, or radiocarbon dating, if available, would thus preempt seriation. If, in addition to order, one wishes relative distance (in units of time) between the collections, we have a scaling problem as well. The seriation literature is quite extensive. Sterud [33] and Cowgill [6] provide good bibliographies.

The general principles of sequence dating originate with Flinders Petrie [24]. Brainerd [2] and Robinson [27] in companion papers set forth the first formalized mathematical seriation procedure. Robinson offers the methodology with examples and is credited as first to have linked similarities with sequencing. Brainerd provides the archaeological support for the method as well as the interpretation of its results. Some earlier formal attempts in the literature include Spier [31], Driver and Kroeber [8], and Rouse [28]. An assortment of earlier ad hoc seriations are noted by Brainerd [2, p. 304], who comments that they were often qualified as provisional pending stratigraphic support. Kendall [17], making more rigorous the ideas of Petrie, sets the problem as one of estimation. The observed collections Y_i are assumed independent with independent components Y_{ij} indicating the number of occurrences of the jth artifact type in collection i. Each Y_{ij} is assumed Poisson* distributed with mean μ_{ij}, a function of parameters reflecting abundance, centrality, and dispersion. P, the permutation of the Y_i's yielding the true temporal order, is also a parameter. A maximum likelihood* approach enables maximization over the μ_{ij}'s independently of P and yields a scoring function $S(P)$ to be maximized over all permutations. But for as few as 15 collections, exhaustive search for the maximum is not feasible.

Similarities help again the similarity matrix now being between collections, described by vectors, with components noting incidence or abundance of artifact types. Using similarities, an order is specified up to reversibility, with expertise then directing it. Labeling the similarity matrix by F, the objective of a seriation is to find a permutation of the rows and columns to achieve a matrix A with elements a_{ij} such that a a_{ij} increases in j for $j < i$; a_{ij} decreases in j for $j > i$. A similarity matrix having this form has been called a Robinson matrix; the process of manipulating F to this form has been called petrifying. A permutation achieving this form must be taken as ideal under our assumptions but one need not exist. Practically, the goal is to get "close" (in some sense) to a Robinson form.

Taking Robinson's lead, archaeologists

such as Hole and Shaw [14], Kuzara et al. [21], Ascher and Ascher [1], and Craytor and Johnson [7], studying large numbers of collections, develop orderings with elaborate computer search procedures (e.g., rules to restrict the search, sampling from all permutations, trial-and-error manipulations). Kendall [18], making these approaches more sophisticated, develops the "horseshoe method," based upon a multidimensional scaling* program in two dimensions. Both theory and examples suggest that with repeated iterations of such a program, a two-dimensional figure in the shape of a horseshoe may be expected if the data are amenable to a seriation. The horseshoe is then unfolded to give a one-dimensional order. Kadane [16] also suggests a computer-based approach by relating the problem of finding a "best" permutation to the traveling salesman problem*. In both cases one seeks a minimum-path-length permutation for a set of points, a problem for which effective computer solutions exist.

Sternin [32] takes a more mathematical tack. He sets the model $F = PAP^T + E$ where P is an unknown permutation matrix and E an error matrix accounting for the possible inability to restore F to exactly a Robinson form. Sternin argues that for certain types of Robinson matrices (e.g., exponential, Green's, and Toeplitz matrices*), the components of eigenvectors corresponding to the two largest eigenvalues will exhibit recognizable patterns. With $E = 0$, F and A have the same eigenvalues*, so Sternin suggests rearranging the components of the corresponding eigenvectors* of F to these patterns.

Gelfand [9] presents two "quick and dirty" techniques. Both methods guarantee the obtaining of the ideal P, if one exists. The better method takes each collection in turn as a reference unit, sequencing all other collections about it. After orienting each of these sequences in the same direction, a final order is obtained by "averaging" these sequences. The averaging should reduce the effect of E and yield a sequence invariant to the original order. An index of fit for a permutation similar to the stress measure used in multidimensional scaling is given, enabling comparison of orders. If an ideal permutation exists, it will minimize this index. Renfrew and Sterud [26] describe a "double-link" method analogous to double-link clustering.

In summary, if an ideal seriation exists, it can be found. If not, but if the data are sufficiently "one-dimensional," the foregoing techniques yield orders from which, with minor modifications suggested by expertise or index of fit, a "best" sequence can be produced.

We now turn to brief discussion of an example. The La Tène Cemetery at Munsingen-Rain near Berne, Switzerland, has proved a rich source of archaeological evidence and has been discussed in numerous articles over the past 15 years. (Hodson [12] provides the definitive work.) The excavation consists of 59 "closed-find" graves. Within these graves were found considerable numbers of fibulae, anklets, bracelets, etc. These ornamental items are typical of the more complex kinds of archaeological material in providing a wide range of detail that allows almost infinite variation within the basic range. A typology for these items was developed employing single-link cluster analysis, average-link cluster analysis, and a principal components analysis*. As a result, some 70 varieties or "types" were defined. A 59×70 incidence matrix of types within graves was created and converted to a 59×59 similarity matrix between graves. This matrix has been seriated using both the Kendall horseshoe method and Gelfand's technique. The unusual, almost linear form of the cemetery implies a geographical sequencing, which enabled Hodson to establish a very satisfactory seriation. The serial orders obtained by Kendall and by Gelfand in the absence of this information are both in good agreement with Hodson's.

In conclusion, the two books by Clarke [4, 5] provide the best current picture of quantitative work in archaeology. Specifically, the articles by Hill and Evans [11] and by Cowgill [6] in the earlier book present excellent synopses on typology and seriation, respectively. The article by Hodson [13] is delightful in bringing some very sophisticated statistical thought to these problems. Finally, the volume from the conference in Mamaia [22] documents a very significant dialogue between archaeologists and statisticians and mathematicians. It bodes well for future analytic work in archaeology.

References

[1] Ascher, M. and Ascher, R. (1963). *Amer. Anthropol.*, **65**, 1045–1052.

[2] Brainerd, G. W. (1951). *Amer. Antiq.*, **16**, 301–313.

[3] Clark, G. A. (1976). *Amer. Antiq.*, **41**, 259–273.

[4] Clarke, D. L. (1972). *Models in Archaeology*. Methuen, London.

[5] Clarke, D. L. (1978). *Analytical Archaeology*, 2nd ed. Methuen, London.

[6] Cowgill, G. L. (1972). In *Models for Archaeology*, D. L. Clarke, ed. Methuen, London, pp. 381–424.

[7] Craytor, W. B. and Johnson, L. (1968). Refinements in Computerized Item Seriation. *Mus. Nat. History, Univ. Oreg. Bull.* 10.

[8] Driver, H. E. and Kroeber, A. L. (1932). *Univ. Calif. Publ. Amer. Archaeol. Ethnol.*, **31**, 211–256.

[9] Gelfand, A. E. (1971). *Amer. Antiq.*, **36**, 263–274.

[10] Gozodrov, V. A. (1933). *Amer. Anthropol.*, **35**, 95–103.

[11] Hill, J. N. and Evans, R. K. (1972). In *Models in Archaeology*, D. L. Clarke, ed. Methuen, London, pp. 231–274.

[12] Hodson, F. R. (1968). *The LaTène Cemetery at Munsingen-Rain*. Stämpfi, Berne.

[13] Hodson, F. R. (1970). *World Archaeol.*, **1**, 299–320.

[14] Hole, F. and Shaw, M. (1967). Computer Analysis of Chronological Seriation. *Rice Univ. Stud. No. 53(3)*, Houston.

[15] Jardine, N. and Sibson, R. (1968). *Computer J.*, **11**, 177.

[16] Kadane, J. B. (1972). Chronological Ordering of Archaeological Deposits by the Minimum Path Length Method. *Carnegie-Mellon Univ. Rep. No. 58* Carnegie-Mellon University, Dept. of Statistics, Pittsburgh, Pa.

[17] Kendall, D. G. (1963). *Bull. I.S.I.*, **40**, 657–680.

[18] Kendall, D. G. (1971). In *Mathematics in the Archaeological and Historical Sciences*. Edinburgh Press, Edinburgh, pp. 215–252.

[19] Krieger, A. D. (1944). *Amer. Antiq.*, **9**, 271–288.

[20] Kroeber, A. L. (1940). *Amer. Antiq.*, **6**, 29–44.

[21] Kuzara, R. S., Mead, G. R., and Dixon, K. A. (1966). *Amer. Anthropol.*, **68**, 1442–1455.

[22] *Mathematics in the Archaeological and Historical Sciences* (1971). Edinburgh Press, Edinburgh.

[23] Mueller, J. W. (1974). The Use of Sampling in Archaeology Survey. *Amer. Antiq. Mem. No. 28*.

[24] Petrie, W. M. Flinders (1899). *J. Anthropol. Inst.*, **29**, 295–301.

[25] Read, D. W. (1974). *Amer. Antiq.*, **39**, 216–242.

[26] Renfrew, C. and Sterud, G. (1969). *Amer. Antiq.*, **34**, 265–277.

[27] Robinson, W. S. (1951). *Amer. Antiq.*, **16**, 293–301.

[28] Rouse, I. (1939). Prehistory in Haiti, A Study in Method. *Yale Univ. Publ. Anthropol. No. 21*.

[29] Sokal, R. R. and Sneath, H. A. (1963). *Principles of Numerical Taxonomy*. W. H. Freeman, San Francisco.

[30] Spaulding, A. C. (1953). *Amer. Antiq.*, **18**, 305–313.

[31] Spier, L. (1917). An Outline for a Chronology of Zuni Ruins. *Anthropol. Papers Amer. Mus. Nat. History*, **18**, Pt. 3.

[32] Sternin, H. (1965). Statistical Methods of Time Sequencing. *Stanford Univ. Rep. No. 112*, Dept. of Statistics, Stanford University, Stanford, Calif.

[33] Sterud, G. (1967). Seriation Techniques in Archaeology. Unpublished M.S. thesis, University of California at Los Angeles.

[34] Whallon, R. (1972). *Amer. Antiq.*, **37**, 13–33.

(CLUSTER ANALYSIS
MULTIDIMENSIONAL SCALING
SIMILARITY MEASURES
TRAVELING SALESMAN PROBLEM)

ALAN E. GELFAND

ARC-SINE DISTRIBUTION

The arc-sine distribution is a name attributed to a discrete and several continuous probability distributions. The discrete and one of the continuous distributions are principally noted for their applications to fluctuations in random walks*. In particular, the discrete distribution describes the percentage of time spent "ahead of the game" in a fair coin tossing contest, while one of the continuous distributions has applications in the study of waiting times*. The distribution most appropriately termed "arc-sine" describes the location, velocity, and related attributes at random time of a particle in simple harmonic motion. Here "random time" means that the time of observation is independent of the initial phase angle, $0 \leqslant \theta_0 < 2\pi$.

The arc-sine distribution with parameter $b > 0$ has support $[-b, b]$ and PDF $\pi^{-1}(b^2 - x^2)^{-1/2}$ for $-b < x < b$. The position at random time of a particle engaged in simple harmonic motion with amplitude $b > 0$ has the arc-sine (b) distribution.

If X is an arc-sine (1) random variable (RV) and $b \neq 0$, then the RV $Y = bX$ has arc-sine ($|b|$) distribution. Salient features of this distribution are:

$$\text{moments:} \quad \begin{cases} EX^{2k} = 2^{-2k}\binom{2k}{k} \\ EX^{2k+1} = 0 \end{cases}$$

$$(k = 0, 1, 2, \dots)$$

$$\text{CDF:} \quad (\sin^{-1}x + \pi/2)/\pi \qquad (-1 < x < 1)$$

$$\text{characteristic function:} \quad Ee^{itX} = J_0(t),$$

where $J_0(t)$ is the Bessel function* of the first kind, of order 0, $\sum_{k=0}^{\infty}(-1)^k(t/2)^{2k}/(k!)^2$.

Let $\sim$ denote "is distributed as." In ref. 6, Norton showed that if X_1 and X_2 are independent arc-sine(b) RVs, then $b(X_1 + X_2)/2 \sim X_1X_2$, and in ref. 7 made the following conjecture. Let X_1 and X_2 be independent identically distributed RV's having all moments, and let F denote the common CDF. Then the only nondiscrete F for which $b(X_1 + X_2)/2 \sim X_1X_2$ is the arc-sine(b) distribution. This conjecture was proved by Shantaram [8].

Arnold and Groeneveld [1] proved several results. Let X be a symmetric RV. Then $X^2 \sim (1 + X)/2$ if and only if $X \sim$arc-sine(1). If X is symmetric and $X^2 \sim 1 - X^2$, then $X \sim 2X\sqrt{1 - X^2}$ if and only if $X \sim$arc-sine(1). If X_1 and X_2 are symmetric independent identically distributed RVs with $X_i^2 \sim 1 - X_i^2$, then $X_1^2 - X_2^2 \sim X_1X_2$ if and only if $X_i \sim$arc-sine(1).

Feller [3] discusses distributions that have acquired the arc-sine name. Set $u_{2k} = \binom{2k}{k}2^{-2k}$, $k = 0, 1, 2, \dots$ ($u_0 = 1$). Let X_k equal ± 1 according to the kth outcome in a fair coin tossing game, and let $S_n = \sum_{k=1}^{n} X_k$ denote the net winnings of a player through epoch $n (S_0 = 0)$. From epochs 0 through $2n$, let Z_{2n} denote that epoch at which the last visit to the origin occurs. Then Z_{2n} necessarily assumes only even values and $\Pr[Z_{2n} = 2k] = u_{2k}u_{2n-2k} = \binom{2k}{k}\binom{2n-2k}{n-k}2^{-2n}$, $k = 0, 1, \dots, n$. The probability distribution of the RV Z_{2n} is called the *discrete arc-sine distribution of order n*. Set $\Pr[Z_{2n} = 2k] = p_{2k,2n}$. The probability that in the time interval from 0 to $2n$ the S_j's are positive (the player is ahead) during exactly $2k$ epochs is $p_{2k,2n}$. This result is readily rephrased in terms of $x = k/n$, the proportion of the time the player is ahead. If $0 < x < 1$, the probability that at most $x(100)\%$ of the S_j's are positive tends to $2\pi^{-1}\sin^{-1}\sqrt{x}$ as $n \to \infty$. The corresponding PDF is $\pi^{-1}[x(1 - x)]^{-1/2}$, $0 < x < 1$, which has acquired the name "arc-sine density." Consideration of $p_{2k,2n}$ or the PDF shows that in a fair coin tossing game, being ahead one-half the time is the least likely possibility, and being ahead 0% or 100% of the time are the most likely possibilities. The probability that the first visit to the terminal value S_{2n} ocurs at epoch $2k$ (or $2n - 2k$) is $p_{2k,2n}$. In a game of $2n$ tosses the probability that a player's maximum net gain occurs for the first time at epoch k, where $k = 2r$ or $k = 2r + 1$, is $\frac{1}{2}p_{2r,2n}$ for $0 < k < 2n$, u_{2n} for $k = 0$, and $\frac{1}{2}u_{2n}$ for $k = 2n$.

Feller also notes related results in other settings. Let $X_1, X_2, \dots$ be independent symmetric RVs with common continuous CDF F. Let K_n denote the epoch (index) at which the maximum of $S_0, S_1, \dots, S_n$ is first attained. Then $\Pr[K_n = k] = p_{2k,2n}$ and, for fixed $0 < \alpha < 1$, as $n \to \infty \Pr[K_n < n\alpha] \to 2\pi^{-1}\sin^{-1}\sqrt{\alpha}$. The number of strictly positive terms among $S_1, \dots, S_n$ has the same distribution as K_n.

Standard beta* densities with support $[0, 1]$ and having form $f_\alpha(x) = [B(1 - \alpha, \alpha)]^{-1}x^{-\alpha}(1 - x)^{\alpha - 1}$, $0 < \alpha < 1$, are called *generalized arc-sine densities*. When $\alpha = \frac{1}{2}$, f_α is the "arc-sine density" $\pi^{-1}[x(1 - x)]^{-1/2}$, $0 < x < 1$, mentioned earlier. Such PDFs play a role in the study of waiting times*. For example, let $X_1, X_2, \dots$ be positive independent RVs with common CDF F and

$S_n = \sum_{k=1}^{n} X_k$. Let N_t denote the random index for which $S_{N_t} \leqslant t < S_{N_t+1}$. Define $Y_t = t - S_{N_t}$. A result of Dynkin [2] is that if $0 < \alpha < 1$ and $1 - F(x) = x^{-\alpha}L(x)$, where $L(tx)/L(t) \to 1$ as $t \to \infty$, then the variable Y_t/t has limiting distribution with PDF f_α. Horowitz [4] extended Dynkin's result to semilinear Markov processes*. Imhof [5] considers the case in which t denotes time and $\{X(t) : 0 \leqslant t \leqslant T\}$ is a stochastic process* satisfying certain conditions. If V denotes the elapsed time until the process reaches a maximum, then

$$\Pr[V < \alpha T] = 2\pi^{-1}\sin^{-1}\sqrt{\alpha}\,.$$

References

[1] Arnold, B. and Groeneveld, R. (1980). *J. Amer. Statist. Ass.*, **75**, 173–175. (Treats some characterizations of the distribution.)

[2] Dynkin, E. B. (1961). *Select. Transl. Math. Statist. Probl.*, **1**, 171–189. (Requires probability theory.)

[3] Feller, W. (1966). *An Introduction to Probability Theory and Its Applications*. Wiley, New York. (Provides a good survey of the arc-sine distributions.)

[4] Horowitz, J. (1971). *Ann. Math. Statist.*, **42**, 1068–1074. (Requires elementary stochastic processes.)

[5] Imhof, J. P. (1968). *Ann. Math. Statist.*, **39**, 258–260. (Requires probability theory.)

[6] Norton, R. M. (1975). *Sankhyā*, *A*, **37**, 306–308. (Gives a characterization.)

[7] Norton, R. M. (1978). *Sankhyā*, *A*, **40**, 192–198. (Treats primarily moment properties of discrete RVs.)

[8] Shantaram, R. (1978). *Sankhyā*, *A*, **40**, 199–207. (Uses combinatorial identities.)

(CHARACTERIZATIONS OF DISTRIBUTIONS
RANDOM WALKS
WAITING TIMES)

R. M. NORTON

ARC-SINE TRANSFORMATION *See* ANGULAR TRANSFORMATION

AREA SAMPLING

An area sample is a sample with primary sampling units* that are well-defined fractions of the earth's surface. The sampling frame* can be visualized as a map that has been subdivided into N nonoverlapping subareas that exhaust the total area of interest. The N distinct subareas are the primary sampling units. *See* SAMPLING for a discussion of the meanings of primary sampling unit and sampling frame. The sampling units in area sampling are often called *segments* or *area segments*. The list of all area segments is the *area frame*.

Area samples are used to study characteristics of the land, such as the number of acres in specific crops, the number of acres under urban development, the number of acres covered with forest, or the fraction of cultivated acres subject to severe water erosion. Area sampling is an integral part of the U.S. Department of Agriculture's method of estimating acreages and yields of farm crops. *See* AGRICULTURE, STATISTICS IN.

An example of a recent large-scale area sample is the study of the potential for increasing the cropland of the United States conducted by the U.S. Soil Conservation Service. See Dideriksen et al. [1] and Goebel [2]. Area sampling is used heavily in forestry*. See Husch et al. [6] and Labau [13].

Area samples are also used when the observation units are persons or institutions for which a list is not available. For example, area frames are used in studies of the general population in the United States and in other countries where current lists of residents are not maintained. The Current Population Survey of the U.S. Bureau of the Census*, from which statistics on unemployment are obtained, is an area sample of the population of the United States.

Area sampling developed apace with probability sampling*. Mahalanobis* [14] in a discussion of area sampling described the contribution of Hubbock [5], who was responsible for a 1923 study that specified methods of locating a random sample* of areas used in estimating the yield of rice. King [12] cites a number of European studies of the 1920s and 1930s that used a type of area sampling.

In 1943, a large project was undertaken by the Statistical Laboratory of Iowa State

College in cooperation with the Bureau of Agricultural Economics, U.S. Department of Agriculture, to design a national area sample of farms in the United States. The name *Master Sample* was applied to the project. The Bureau of the Census* also cooperated in the project and developed an area sample of cities, which, together with the Master Sample of rural areas, was used as a sample of the entire population. The materials developed in the Iowa State project, updated for changes in culture, are still used in the creation of area samples of the rural part of the United States. Stephan [18] provides an excellent review of sampling history.

The basic idea of area sampling is relatively simple, but the efficient implementation of the method requires some sophistication. It must be possible for the field worker (enumerator) to identify the boundaries of each area segment. Thus roads, streets, streams, fences, and other "natural boundaries" are used to define the segments, whenever possible. Aerial photographs, county highway maps, and street maps are materials commonly used in area sampling. Aerial photographs are particularly useful for studies outside heavily urbanized areas. *See* CENSUS.

A precise set of rules associating the elements of the population with the area segments must be developed when the population of interest is not a characteristic of the land itself. For a study of households, the households whose residence is on a given area segment are associated with that segment. Even for this relatively simple rule, problems arise. One problem is the definition of a household. There is also the problem of defining the primary residence in the case of multiple residences. For samples of farms or other businesses, it is common practice to associate the business with the segment on which the "headquarters" is located. See, for example, Jabine [7] for a set of rules of association.

In studies of the characteristics of the land itself, the definition of boundaries of the area segments is very important. A phenomenon called "edge bias" has been identified in empirical studies of crop yields. It has been observed that field workers tend to include plants near the boundary of a plot. Therefore, yields based on small areas are often biased upward. See Sukhatme [19] and Masuyama [16].

Any subdivision of the study area into segments will, theoretically, provide an unbiased estimator* of the population total. But the variances of the estimator obtained for two different partitions of the study area may be very different. Therefore, the design of efficient area samples requires that the area segments be as nearly equal in "size" as is possible. In this context, size is a measure assigned to the area segments that is correlated with the characteristic of interest. An example of a measure of size is the number of households reported in the most recent census. The measure of size used in the Master Sample of Agriculture was the number of dwelling units indicated on county highway maps. This number was correlated with the number of farm headquarters.

In area samples used to study populations over time the size of the area segment will change. Gray and Platek [3] discuss methods of modifying the design in such situations.

Part of the cost of designing area samples is the cost of obtaining information on the estimated size of the area units from recent photos, maps, censuses, city directories, field visits, etc. In designing an area sample, the practitioner must balance the cost of obtaining additional information, the smaller variance of estimates obtained from units of nearly equal estimated size, and the practical requirement for boundaries that can be identified by the field worker.

References

[1] Dideriksen, R. I., Hidlebaugh, A. R., and Schmude, K. O. (1977). Potential Cropland Study. *Statist. Bull. No. 578*, U.S. Dept. of Agriculture.

[2] Goebel, J. J. (1967). *Proc. Social Statist. Sect. Amer. Statist. Ass.*, 1976, pp. 350–354.

[3] Gray, C. B. and Platek, R. (1968). *J. Amer. Statist. Ass.*, **63**, 1280–1297.

[4] Houseman, E. E. (1975). Area Frame Sampling in Agriculture. *U.S. Dept. Agric. Bull. SRS No. 20.*

[5] Hubbock, J. A. (1927). Sampling for Rice Yield in Bihar and Orissa. *Bull. No. 166*, Agricultural Research Institute, Pusa, Government of India. Reprinted in *Sankhyā*, **7**, 281–294 (1947).

[6] Husch, B., Miller, C. I., and Beers, T. W. (1972). *Forest Mensuration*. Ronald Press, New York.

[7] Jabine, T. B. (1965). In *Estimation of Areas in Agricultural Statistics*. Food and Agriculture Organization of the United Nations, Rome, pp. 136–175.

[8] Jessen, R. J. (1942). Statistical Investigation of a Sample Survey for Obtaining Farm Facts. *Res. Bull. 304*, Iowa Agric. Exper. Stat., Iowa State College, Ames, Iowa.

[9] Jessen, R. J. (1945). *J. Amer. Statist. Ass.*, **40**, 45–56.

[10] Jessen, R. J. (1947). *J. Farm Econ.*, **29**, 531–540.

[11] Jessen, R. J. (1978). *Statistical Survey Techniques*. Wiley, New York.

[12] King, A. J. (1945). *J. Amer. Statist. Ass.*, **40**, 38–45.

[13] Labau, V. J. (1967). Literature on the Bitterlich Method of Forest Cruising. *U.S. Forest Serv. Res. Paper PNW-47*, U.S. Dept. of Agriculture.

[14] Mahalanobis, P. C. (1944). *Philos. Trans. Ser. B*, **231**, 329–451.

[15] Mahalanobis, P. C. (1947). *Sankhyā*, **7**, 269–280.

[16] Masuyama, M. (1954). *Sankhyā*, **14**, 181–186.

[17] Monroe, J. and Finkner, A. L. (1959). *Handbook of Area Sampling*. Chilton, New York.

[18] Stephan, F. F. (1948). *J. Amer. Statist. Ass.*, **43**, 12–39.

[19] Sukhatme, P. V. (1947). *J. Amer. Statist. Ass.*, **42**, 297–310.

(AGRICULTURE, STATISTICS IN
BUREAU OF THE CENSUS
FORESTRY, STATISTICS IN
SURVEY SAMPLING)

WAYNE A. FULLER

ARFWEDSON DISTRIBUTION

This is the distribution of the number (M_0, say) of zero values among k random variables $N_1, N_2, \ldots, N_k$ having a joint equiprobable multinomial distribution*. If the sum of the N's is n, then

$$\Pr[M_0 = m] = \binom{k}{m} \sum_{i=0}^{m} (-1)^i \binom{m}{i} \left(\frac{m-i}{k}\right)^n$$
$$= k^{-n}\binom{k}{m}\Delta^m \mathbf{0}^n$$
$$(m = 0, 1, \ldots, k-1).$$

It is a special occupancy distribution*, being the number of cells remaining empty after n balls have been distributed randomly among k equiprobable cells.

Bibliography

Arfwedson, G. (1951). *Skand. Aktuarietidskr*, **34**, 121–132.

Johnson, N. L. and Kotz, S. (1969). *Discrete Distributions*. Wiley, New York, p. 251.

(DIFFERENCES OF ZERO
MULTINOMIAL DISTRIBUTION
OCCUPANCY DISTRIBUTIONS)

ARIMA (p, d, q) *See* AUTOREGRESSIVE–INTEGRATED MOVING AVERAGE (ARIMA) MODELS

ARITHMETIC MEAN

The arithmetic mean of n quantities X_1, $X_2, \ldots, X_n$ is the sum of the quantities divided by their number n. It is commonly denoted by $\bar{X}$, or when it is desirable to make the dependence upon the sample size explicit, by $\bar{X}_n$. Symbolically,

$$\bar{X} = (1/n)\sum_{i=1}^{n} X_i.$$

An alternative form of the definition, useful for iterative calculations, is

$$\bar{X}_{n+1} = \bar{X}_n + (X_{n+1} - \bar{X}_n)/(n+1);$$
$$\bar{X}_0 = 0.$$

Historically, the arithmetic mean is one of the oldest algorithmic methods for combining discordant measurements in order to produce a single value, although even so, few well-documented uses date back to before the seventeenth century [3, 6, 8]. Today it is the most widely used and best understood data summary in all of statistics. It is included as a standard function on all but the simplest hand-held calculators, and it enjoyes the dual distinction of being the optimal method of combining measurements from several points of view, and being the least robust* such method according to others. Our discussion begins with the consideration of its distributional properties.

DISTRIBUTION

Suppose that the X_i are independent, identically distributed (i.i.d.) with CDF $F_X(x)$, mean μ, and variance σ^2. The distribution of $\bar{X}$ may be quite complicated, depending upon F_X, but it will always be true that

$$E[\bar{X}] = \mu, \tag{1}$$

$$\mathrm{var}(\bar{X}) = \sigma^2/n, \tag{2}$$

whenever these moments exist. For some distributions F_X, $\bar{X}$ possesses a distribution of simple form. If F_X is $N(\mu, \sigma^2)$, then $\bar{X}$ has a normal $N(\mu, \sigma^2/n)$ distribution*. If F_X is a Bernoulli (p) distribution*, then $n\bar{X}$ has a binomial bin(n, p) distribution*. If F_X is a Poisson (λ) distribution*, $n\bar{X}$ has a Poisson $(n\lambda)$ distribution. If F_X is a Cauchy distribution*, then $\bar{X}$ has the same Cauchy distribution. If F_X is a gamma distribution* or a chi-squared distribution*, then $\bar{X}$ has a gamma distribution*. The exact density of $\bar{X}$ when F_X is a uniform distribution* was derived as long ago as 1757 by T. Simpson [6, 8]; it is a complicated case of a B-spline*. For further information about the distribution of $\bar{X}$ for different parametric families F_X, see the entries under those distributions.

Since $\bar{X}$ is a sum of independent random variables, many aspects of its distribution are amenable to study by using generating functions*. In particular, the characteristic function* of $\bar{X}$ can be given in terms of the characteristic function $\phi(t)$ of F_X as $\phi_{\bar{X}}(t) = [\phi(n^{-1}t)]^n$.

Much is known about the asymptotic behavior of $\bar{X}_n$ for large n. *See* LAWS OF LARGE NUMBERS; LIMIT THEOREMS, CENTRAL. For example, the Kolmogorov strong law of large numbers* states that $\bar{X}_n \overset{\text{a.s.}}{\to} \mu$ as $n \to \infty$ if and only if $E|X_i| < \infty$ and $EX_i = \mu$. The classical central limit theorem states that if $\sigma^2 < \infty$, $n^{1/2}(\bar{X}_n - \mu)$ is asymptotically distributed as $N(0, \sigma^2)$. The degree of approximation that can be expected from this asymptotic result has received considerable study, although even the strongest available results are usually too pessimistic to be practically useful. *See* e.g., ASYMPTOTIC NORMALITY.

Various refinements to the normal approximation are also available. *See* EDGEWORTH EXPANSIONS.

If the measurements X_i are not independent, then of course the distribution of $\bar{X}$ may be more complicated. If the X_i form a stationary* sequence with $E[X_i] = \mu$ and $\mathrm{var}(X_i) = \sigma^2$, then $E[\bar{X}] = \mu$, but $\mathrm{var}(\bar{X})$ may be either larger or smaller than in the independent case. For example, if the X_i follow a first-order moving average process [with $X_i = \mu + a_i + \theta a_{i-1}$, where the a_i are i.i.d. with $E[a_i] = 0$, $\mathrm{var}(a_i) = \sigma_a^2$, and $\sigma^2 = (1 + \theta^2)\sigma_a^2$], then $\rho = \mathrm{corr}(X_i, X_{i+1}) = \theta/(1 + \theta^2)$ varies from -0.5 to 0.5, and

$$\mathrm{var}(\bar{X}) = (\sigma^2/n)[1 + 2(1 - (1/n))\rho], \tag{3}$$

which varies from σ^2/n^2 to $2\sigma^2/n - \sigma^2/n^2$. *See* TIME SERIES. If the measurements X_i are determined by sampling at random* from a finite population* of size N without replacement, then

$$\mathrm{var}(\bar{X}) = \frac{\sigma^2}{n}\left(\frac{N-n}{N-1}\right), \tag{4}$$

where σ^2 is the population variance. *See* FINITE POPULATIONS, SAMPLING FROM.

STATISTICAL PROPERTIES

The arithmetic mean $\bar{X}$ is usually considered as an estimator of a population mean μ: if the X_i are i.i.d. with CDF $F_X(x)$ and finite mean μ, then $\bar{X}$ is an unbiased estimator* of μ regardless of F_X (in this sense it is a nonparametric estimator of μ). The same is true if the X_i are sampled at random without replacement from a finite population with mean μ. Chebyshev's inequality* tells us $\Pr[|\bar{X} - \mu| > \epsilon] \leqslant \mathrm{var}(\bar{X})/\epsilon^2$, so $\bar{X}$ will in addition be a consistent estimator* of μ as long as $\mathrm{var}(\bar{X}) \to 0$ as $n \to \infty$, which will hold, for example, if $\sigma^2 < \infty$ in (2), (3), and (4). Laws of large numbers* provide several stronger consistency results. For the case of i.i.d. X_i with finite variances, (2) can be interpreted as stating that the precision of $\bar{X}$ as an estimator of μ increases as the square root of the sample size. In the i.i.d. case, the nonparametric unbiased estimator of var($\bar{X}$) is s^2/n, where s^2 is the sample variance* $(n-1)^{-1}\sum(X_i - \bar{X})^2$.

The arithmetic mean enjoys several optimality properties beyond unbiasedness and consistency. It is a special case of a least-squares estimator*; $\sum(X_i - c)^2$ is minimized by $c = \bar{X}$. As such, $\bar{X}$ has all the properties of least-squares estimators: The Gauss–Markov theorem* ensures that $\delta = \bar{X}$ minimizes $E(\delta - \mu)^2$ within the class of all linear unbiased estimators; when F_X is normal $N(\mu, \sigma^2)$, $\bar{X}$ is the maximum likelihood estimator of μ; and from a Bayesian* perspective, $\bar{X}$ is at the maximum of the posterior distribution* of μ for a uniform prior distribution*. (In fact, C. F. Gauss* proved in 1809 that this later property of $\bar{X}$ characterized the normal within the class of all location parameter families.)

The optimality of $\bar{X}$ as an estimator of a parameter of course depends upon the parametric family in question, but in the i.i.d. case there are several examples of F_X [including $N(\mu, \sigma_0^2)$, σ_0^2 known; Poisson (μ); Bernoulli (μ); the one-parameter exponential (μ)*], where $\bar{X}$ is the maximum likelihood estimator and a minimal sufficient* statistic. For a simple example of how the optimality of $\bar{X}$ depends upon the distribution and the criterion, however, see ref. 10.

Much attention to the arithmetic mean in recent years has focused upon its lack of robustness*, in particular, its sensitivity to aberrant measurements such as are likely to occur when sampling from heavy-tailed distributions. The most commonly noted example of this was noted as early as 1824 by Poisson: if F_X is a Cauchy distribution*, then $\bar{X}$ has the same Cauchy distribution (and thus no mean or variance no matter how large n is). Estimators such as the sample median* perform much more efficiently in this case. Indeed, even a small amount of heavy-tailed contamination* can in principle drastically effect the efficiency of $\bar{X}$ as an estimator of μ. Opinion is divided on the question of whether such contamination occurs in practice with a severity or frequency to dictate drastic remedy; see refs. 1, 5, and 8 for an airing of these and related issues. Meanwhile, a vast array of estimators have been devised that are less sensitive than $\bar{X}$ to extreme measurements; the simplest of these (the Winsorized mean* and the trimmed mean*) are, in fact, equivalent to the calculation of the arithmetic mean of a modified sample.

The arithmetic mean is frequently used as a test statistic for testing hypotheses about the mean μ, often in the form of Student's t statistic*, $t = \sqrt{n}(\bar{X} - \mu_0)/s$, and as the basis of confidence intervals* for μ.

RELATIONSHIP TO OTHER MEANS

Two classical means, the geometric* and the harmonic*, are related simply to the arithmetic mean. If $Y_i = \ln X_i$ and $Z_i = X_i^{-1}$, the geometric mean of the X_i is given by $(\prod X_i)^{1/n} = \exp(\bar{Y})$ and the harmonic mean is $(\bar{Z})^{-1}$. Hardy et al. [5] discuss inequality relationships between these and more general mean functions, the simplest being that if all X_i are positive, then $(\bar{Z})^{-1} \leqslant \exp(\bar{Y}) \leqslant \bar{X}$. *See* GEOMETRIC MEAN; HARMONIC MEAN. Many other means have been related to the arithmetic mean in less mathematically precise ways. The best known such relationship is that between the arithmetic mean $\bar{X}$, the median m, and the mode* M for empirical distributions that are unimodal and skewed to the right; it is frequently true that $M \leqslant m \leqslant \bar{X}$. Furthermore, a rough rule of thumb that goes back at least to 1917 (see refs. 2, 4, 7 and 12) observes that these means often satisfy, approximately, the relationship $(\bar{X} - M) = 3(\bar{X} - m)$. The arithmetic mean may also be viewed as the expected valued or mean of the empirical distribution* which places mass $1/n$ at each X_i, a fact that points to several other characterizations of $\bar{X}$: it is the center of gravity of the X_i; it is value such that the sum of the residuals about that value is zero $[\sum(X_i - \bar{X}) = 0]$; it is a functional of the empirical* cumulative distribution function F_n,

$$\overline{X} = \int x\, dF_n(x).$$

For other definitions for which the arithmetic mean is a special case, *see* INDEX NUMBERS, ORDER STATISTICS, ROBUST ESTIMATORS, AND WEIGHTED MEANS.

References

[1] Andrews, D. F., Bickel, P. J., Hampel, F. R., Huber, P. J., Rogers, W. H. and Tukey, J. W. (1972). *Robust Estimates of Location: Survey and Advances*. Princeton University Press, Princeton, N. J. (This book presents the results of a Monte Carlo study of a large number of estimates, including the arithmetic mean.)

[2] Doodson, A. T. (1917). *Biometrika*, **11**, 425–429.

[3] Eisenhart, C. The Development of the Concept of the Best Mean of a Set of Measurements from Antiquity to the Present Day. ASA Presidential Address, 1971, to appear. (The early history of the arithmetic mean.)

[4] Groeneveld, R. A. and Meeden, G. (1977). *Amer. Statist.*, **31**, 120–121.

[5] Hardy, G. H., Littlewood, J. E. and Polya, G. (1964). *Inequalities*. Cambridge University Press, Cambridge.

[6] Plackett, R. L. (1958). *Biometrika*, **45**, 130–135. Reprinted in *Studies in the History of Statistics and Probability*, E. S. Pearson and M. G. Kendall, eds. Charles Griffin, London, 1970. (The early history of the arithmetic mean.)

[7] Runnenburg, J. T. (1978). *Statist. Neerlandica*, **32**, 73–79.

[8] Seal, H. L. (1977). *Studies in the History of Statistics and Probability*, Vol. 2, M. G. Kendall and R. L. Plackett, eds. Charles Griffin, London, Chap. 10. (Originally published in 1949.)

[9] Stigler, S. M. (1977). *Ann. Statist.*, **5**, 1055–1098. (A Study of the performance of several estimates, including the arithmetic mean, on real data sets. Much of the discussion focuses on the characteristics of real data.)

[10] Stigler, S. M. (1980). *Ann. Statist.*, **8**, 931–934. (An early example is discussed which shows that the arithmetic mean need not be optimal, even for samples of size 2, and even if all moments exist.)

[11] Tukey, J. W. (1960). In *Contributions to Probability and Statistics*, I. Olkin et al., eds. Stanford University Press, Stanford, Calif., pp. 448–485.

[12] Zwet W. R. van (1979). *Statist. Neerlandica*, **33**, 1–5.

S. STIGLER

ARITHMETIC PROGRESSION

A sequence of numbers with constant difference between successive numbers. The mth member of the sequence a_m can be expressed as

$$a_m = a_1 + (m - 1)d,$$

where d is the constant difference.

(GEOMETRIC PROGRESSION)

ARITHMETIC TRIANGLE *See* COMBINATORICS

ARMA (p,q) MODELS *See* AUTOREGRESSIVE–MOVING AVERAGE (ARMA) MODELS

ARRAY

This term is applied to the distribution of sample values of a variable Y, for a fixed value of another variable X. It refers especially to the frequency distribution* formed by such values when set out in the form of a contingency table*. Such an array is formed only when the data are discrete or grouped.

(CONDITIONAL DISTRIBUTION
ORTHOGONAL ARRAY)

ARRAY MEAN

The arithmetic mean* of the values of a variable Y in a group defined by limits on the values of variables $X_1, \ldots, X_k$ (an array*). It is an estimate of the regression function* of Y on $X_1, \ldots, X_k$.

(ARRAY
CONTINGENCY TABLE
REGRESSION)

ARS CONJECTANDI *See* BERNOULLIS, THE

ARUMA MODELS *See* BOX–JENKINS MODEL

ASSIGNABLE CAUSE

In model building, effects of certain factors ("causes") are allowed for in construction of the model. Ideally, all causes likely to have noticeable effects should be so represented. Such causes are often called "assignable causes." A better term might be "recognized causes."

Usually, there are, in fact, effects arising from causes that are not allowed for ("assigned") in the model. It is hoped that these will not be seriously large; they are supposed to be represented by random variation* included in the model.

Note that not all assignable causes may be actually used ("assigned") in the model. In the interests of simplicity, causes with recognized potential for effect may be omitted if the magnitudes of the effects are judged likely to be small.

(MODEL BUILDING)

ASSOCIATION, MEASURES OF

Measures of association are numerical assessments of the strength of the statistical dependence of two or more qualitative variables. The common measures can be divided into measures for nominal polytomous variables* and measures for ordinal polytomous variables*.

MEASURES FOR NOMINAL VARIABLES

The most common measures of association for nominal variables are measures of prediction analogous in concept to the multiple correlation coefficient* of regression analysis*.

Consider two polytomous random variables X and Y with respective finite ranges I and J. A measure of prediction $\phi_{Y \cdot X}$ of Y given X depends on a measure Δ of the dispersion of a polytomous random variable. Such a measure is always nonnegative, with $\Delta_Y = 0$ if and only if Y is essentially constant; i.e., $\Delta_Y = 0$ if and only if for some $j \in J, p_{\cdot j} = \Pr[Y = j] = 1$. The measure does not depend on the labeling* of elements. Formally, one may require that $\Delta_Y = \Delta_{\sigma(Y)}$ if σ is a one-to-one transformation from J into a finite set J'. The added requirement is imposed that the conditional dispersion* $\Delta_{Y \cdot X}$ of Y given X not to exceed the unconditional dispersion Δ_Y of Y. Here $\Delta_{Y \cdot X}$ is the expected value of $\Delta_{Y \cdot X}(X)$, and $\Delta_{Y \cdot X}(i)$ is the dispersion of Y given that $X = i \in I$. [The definition of $\Delta_{Y \cdot X}(i)$ when $p_{i \cdot} = \Pr[X = i] = 0$ does not matter.] The measure of prediction

$$\phi_{Y \cdot X} = 1 - \Delta_{Y \cdot X}/\Delta_Y$$

compares the conditional dispersion of Y given X to the unconditional dispersion of Y, just as the multiple correlation coefficient* compares the expected conditional variance of the dependent variable to its unconditional variance. The measure $\phi_{Y \cdot X}$ is well defined if Y is not essentially constant. When $\phi_{Y \cdot X}$ is defined, $0 \leqslant \phi_{Y \cdot X} \leqslant 1$, with $\phi_{Y \cdot X} = 0$ if X and Y are independently distributed and $\phi_{Y \cdot X} = 1$ if X is an essentially perfect predictor* of Y, i.e., if for some function k from I to J, $\Pr[Y = k(X)] = 1$.

Two common examples of such measures of prediction are the λ coefficient of Guttman* [6] and of Goodman and Kruskal [1] and the τ-coefficient of Goodman and Kruskal* [1]. Let $p_{ij} = \Pr[X = i, \ Y = j]$, $i \in I, j \in J$. Then

$$\lambda_{Y \cdot X} = \frac{\left(\sum_{i \in I} \max_{j \in J} p_{ij} - \max_{j \in J} p_{\cdot j}\right)}{\left(1 - \max_{j \in J} p_{\cdot j}\right)},$$

$$\tau_{Y \cdot X} = \frac{\left(\sum_{i \in I} \sum_{j \in J} p_{ij}^2 / p_{i \cdot} - \sum_{j \in J} p_{\cdot j}^2\right)}{\left(1 - \sum_{j \in J} p_{\cdot j}^2\right)}.$$

In the last formula, 0/0 is defined as 0 to ensure that $p_{ij}^2/p_{i \cdot}$ is always defined.

In the case of $\lambda_{Y \cdot X}$, the measure $\Delta_Y = 1 - \max_{j \in J} p_{\cdot j}$ is the minimum probability of error from a prediction that Y is a constant k, while $\Delta_{Y \cdot X} = 1 - \sum_{i \in I} \max_{j \in J} p_{ij}$ is the minimum probability of error from a prediction that Y is a function $k(X)$ of X. In the case of $\tau_{Y \cdot X}$, $\Delta_Y = 1 - \sum_{j \in J} p_{\cdot j}^2$ is the probability that the random variable Y' does not equal Y, where Y' and Y are independent and identically distributed (i.i.d.). Similarly, $\Delta_{Y \cdot X} = 1 - \sum_{i \in I} \sum_{j \in J} p_{ij}^2 / p_{i\cdot}$ is the probability that $Y' \neq Y$, where given X, Y and Y' are conditionally i.i.d.

Other measures of this type are also available. For example, Theil [18] has considered the measure

$$\eta_{Y \cdot X}$$
$$= - \sum_{i \in I} \sum_{j \in J} p_{ij} \log\left(\frac{p_{ij}}{p_{i\cdot}\, p_{\cdot j}} \right) \Big/ \sum_{j \in J} p_{\cdot j} \log p_{\cdot j},$$

based on the entropy* measure $\Delta_Y = -\sum_{j \in J} p_{\cdot j} \log p_{\cdot j}$ and the conditional entropy measure

$$\Delta_{Y \cdot X} = - \sum_{i \in I} \sum_{j \in J} p_{ij} \log(p_{ij} / p_{i\cdot}).$$

In these measures, $0/0 = 0$ and $0 \log 0 = 0$.

The coefficient $\lambda_{Y \cdot X}$ has an attractively simple interpretation in terms of prediction*; however, $\lambda_{Y \cdot X}$ has the possible disadvantage that $\lambda_{Y \cdot X}$ may be 0 even if X and Y are dependent. In contrast, $\eta_{Y \cdot X}$ and $\tau_{Y \cdot X}$ are only 0 if X and Y are independent.

Partial and Multiple Association

As in Goodman and Kruskal [1, 3], generalizations to cases involving three or more polytomous variables are straightforward. Consider a new polytomous variable W with finite range H. If $\Delta_{Y \cdot WX}$ denotes the conditional dispersion of Y given the polytomous vector (W, X), then the multiple association coefficient $\phi_{Y \cdot WX}$ may be defined as $1 - \Delta_{Y \cdot WX} / \Delta_Y$. The partial association of Y and W given X may then be defined as

$$\phi_{Y \cdot W | X} = 1 - \Delta_{Y \cdot WX} / \Delta_{Y \cdot X}.$$

Thus $\phi_{Y \cdot W|X}$ measures the additional predictive power of W given that X has already been used as a predictor of Y. If W and Y are conditionally independent given X, then $\phi_{Y \cdot W|X} = 0$. If X is not an essentially perfect predictor of Y but (W, X) is an essentially perfect predictor of Y, then $\phi_{Y \cdot W|X} = 1$. In general, if X is not an essentially perfect predictor of Y, one has $0 \leqslant \phi_{Y \cdot W|X} \leqslant 1$ and

$$1 - \phi_{Y \cdot WX} = (1 - \phi_{Y \cdot X})(1 - \phi_{Y \cdot Z|X}).$$

Symmetric Measures

A measure of prediction $\phi_{Y \cdot X}$ of Y given X is not generally equal to the corresponding measure $\phi_{X \cdot Y}$ for prediction of X by Y. This behavior can be contrasted with the square ρ^2 of the correlation coefficient* of two continuous random variables U and V. In the continuous case, ρ^2 measures the power of U as a predictor of V and the power of V as a predictor of U. In cases in which a symmetric measure is desired, Goodman and Kruskal [1] propose measures of the form

$$\phi_{XY} = 1 - (\Delta_{Y \cdot X} + \Delta_{X \cdot Y}) / (\Delta_Y + \Delta_X).$$

For example,

$$\begin{aligned} \lambda_{XY} = \Big(&\textstyle\sum_{i \in I} \max_{j \in J} p_{ij} + \sum_{j \in J} \max_{i \in I} p_{ij} \\ &- \max_{j \in J} p_{\cdot j} - \max_{i \in I} p_{i\cdot} \Big) \\ &\cdot \left(2 - \max_{j \in J} p_{\cdot j} - \max_{i \in I} p_{i\cdot}\right)^{-1}. \end{aligned}$$

The measure ϕ_{XY} is defined if either X or Y is not essentially constant. The coefficient ϕ_{XY} ranges between 0 and 1, with $\phi_{XY} = 0$ if X and Y are independent and $\phi_{XY} = 1$ if and only if for some functions k from I to J and m from J to I, $\Pr[Y = k(X)] = \Pr[X = m(Y)] = 1$.

Standardization

In some cases, it is desirable to standardize the marginal distributions* of X and Y before computation of a measure of assocation. For example, one may wish to find $\phi_{Y' \cdot X'}$, where X' has some standard reference distribution such as the uniform distribution* of I and the conditional distribution of Y' given that X' is identical to the conditional distribution of Y given X. If $p'_{i\cdot} = \Pr[X' = i]$, $i \in I$, and $p'_{\cdot j} = \sum_{i \in I} (p'_{i\cdot} / p_{i\cdot}) p_{ij}$, $j \in J$, where $p_{i\cdot} = 0$ only when $p'_{i\cdot} = 0$, then

$$\lambda_{Y' \cdot X'} = \frac{\left(\sum_{i \in I} (\max_{j \in J} p_{ij}) p'_{i\cdot} / p_{i\cdot} - \max_{j \in J} p'_{\cdot j}\right)}{\left(1 - \max_{j \in J} p'_{\cdot j}\right)}.$$

Similarly, one may consider a measure $\phi_{Y^* \cdot X^*}$, where Y^* has the same standard marginal distributions and the conditional distribution of X^* given Y^* is the same as the conditional distribution of X given Y. More thorough standardization is also possible, as in Mosteller [13]. One may consider $\phi_{U \cdot V}$, where U and V have standard marginal distributions and $\Pr[U = i, \ V = j] = s_i t_j p_{ij}$, $i \in I, j \in J$, for some s_i, $i \in I$, and t_j, $j \in J$.

Optimal Prediction

Measures such as $\phi_{Y \cdot X}$ always have interpretations in terms of optimal prediction in the following sense. Some nonnegative and possibly infinite function $A_j(j, \mathbf{q})$ is defined for $j \in J$ and $\mathbf{q}$ in the simplex* S_J of vectors $\mathbf{q} = \langle q_j : j \in J \rangle$ with nonnegative coordinates with sum $\sum_{j \in J} q_j = 1$. This function represents a loss incurred if a probabilistic prediction $\mathbf{q}$ is made for Y and $Y = j$. The function is such that

$$d_J(\mathbf{q}) = \sum_{j \in J} q_j A_J(j, \mathbf{q}) \leqslant \sum_{j \in J} q_j A_J(j, \mathbf{q}') \qquad (\mathbf{q}, \mathbf{q}' \in S). \quad (1)$$

The dispersion Δ_Y of Y is $d_J(\mathbf{p}_Y)$, where $\mathbf{p}_Y = \langle p_{\cdot j} : j \in J \rangle$. Thus Δ_Y is the minimum expected loss achievable by prediction of Y without knowledge of X. Similarly, $\Delta_{Y \cdot X}(i)$ is $d_J(\mathbf{p}_{Y \cdot X}(i))$, where $\mathbf{p}_{Y \cdot X}(i) = \langle p_{ij}/p_{i \cdot} : j \in J \rangle$. Thus $\Delta_{Y \cdot X}(i)$ is the minimum expected loss achievable in prediction of Y, given that it is known that $X = i$, and $\Delta_{Y \cdot X}$ is the minimum expected loss achievable in prediction of Y given that X is known.

In the case of $\lambda_{Y \cdot X}$, one may define

$$A_J(j, \mathbf{q}) = \begin{cases} 1, & j \notin B(\mathbf{q}), \\ 1 - 1/m(\mathbf{q}), & j \in B(\mathbf{q}), \end{cases}$$

where $j \in B(\mathbf{q})$ if $q_j = \max_{k \in J} q_k$ and $m(\mathbf{q})$ is the number of elements in $B(\mathbf{q})$. In the typical case in which q_j has a unique maximum at a coordinate $j = \rho(\mathbf{q})$, the penalty $A_J(j, \mathbf{q})$ is 1 for $j \neq \rho(\mathbf{q})$ and 0 otherwise. In the case of $\tau_{Y \cdot X}$, $A_J(j, \mathbf{q})$ is the squared distance $\sum_{k \in K} (\delta_{kj} - q_k)^2$, where δ_{kj} is 1 for $k = j$ and δ_{kj} is 0 for $k \neq j$, while for $\eta_{Y \cdot X}$, one has

$$A_J(j, \mathbf{q}) = -\log q_j.$$

The loss function $A_J(j, \mathbf{q})$ is almost uniquely determined by the dispersion measure d_J. If $d_J(\alpha \mathbf{q})$ is defined to be $\alpha d_J(\mathbf{q})$ for $\alpha \geqslant 0$ and $\mathbf{q} \in S_J$, then d_J is a concave function on the set O_J of vectors $\mathbf{q} = \langle q_j : j \in J \rangle$ with all coordinates nonnegative. As in Savage [16], it follows that (1) is satisfied by $A_J(j, \mathbf{q})$, $j \in J$, $\mathbf{q} \in S_J$, if and only if for all $\mathbf{q} \in S_J$ and $\mathbf{q}' \in O_J$,

$$d_J(\mathbf{q}') \leqslant d_J(\mathbf{q}) + \sum_{j \in J} A_J(j, \mathbf{q})(q'_j - q_j).$$

As in Rockafellar [15], the vector $A_J(\mathbf{q}) = \langle A_J(j, \mathbf{q}) : j \in J \rangle$ is called a supergradient of d_J at $\mathbf{q}$. Some $A_J(\mathbf{q})$ exists at each $\mathbf{q} \in S_J$. If $\mathbf{q}$ is an element in the simplex S_J with all coordinates positive and if d_J is differentiable at $\mathbf{q}$, then $A_J(j, \mathbf{q})$ must be the partial derivative at $\mathbf{q}$ of d_J with respect to q_j. Thus the $A_J(j, \mathbf{q})$, $j \in J$, are uniquely determined and continuous at almost every point $\mathbf{q}$ on the simplex S_J. For example, in the case of $\eta_{Y \cdot X}$,

$$d_j(\mathbf{q}) = -\sum_{j \in J} q_j \log q_j + \left(\sum_{j \in J} q_j \right) \log \left(\sum_{j \in J} q_j \right)$$

for $\mathbf{q} \in O_J$. If all q_j are positive and $\sum_{j \in J} q_j = 1$, then $A(j, \mathbf{q}) = -\log q_j$ is the partial derivative at $\mathbf{q}$ of d_J with respect to q_j.

Estimation of Measures of Prediction

Typically bivariate measures of prediction in which standardization is not involved are estimated on the basis of a contingency table* $\mathbf{n} = \langle n_{ij} : i \in I, j \in J \rangle$ with a multinomial distribution of sample size N and probabilities $\mathbf{p} = \langle p_{ij} : i \in I, j \in J \rangle$. The estimate $\hat{\mathbf{p}} = N^{-1}\mathbf{n}$ of $\mathbf{p}$ is substituted for $\mathbf{p}$ in the formulas for $\phi_{Y \cdot X}$ and ϕ_{XY}. If $n_{\cdot j} = \sum_{i \in I} n_{ij}$, $j \in J$, $n_{i \cdot} = \sum_{j \in J} n_{ij}$, $i \in I$, $\hat{\mathbf{p}}_X = \langle N^{-1} n_{i \cdot} : i \in I \rangle$, $\hat{\mathbf{p}}_Y = \langle N^{-1} n_{\cdot j} : j \in J \rangle$, $\hat{\mathbf{p}}_{Y \cdot X}(i)$

$= \langle n_{ij}/n_{i\cdot} : j \in J \rangle$, $i \in I$, and $\hat{\mathbf{p}}_{X\cdot Y}(j) = \langle n_{ij}/n_{\cdot j} : i \in I \rangle, j \in J$, then

$$\hat{\phi}_{Y\cdot X} = 1 - N^{-1} \sum_{i \in I} n_{i\cdot} d_J(\hat{\mathbf{p}}_{Y\cdot X}(i))/d_J(\hat{\mathbf{p}}_Y),$$

$$\hat{\phi}_{XY} = 1 - N^{-1} \times \frac{[\sum_{i\in I} n_{i\cdot} d_J(\hat{\mathbf{p}}_{Y\cdot X}(i)) + \sum_{j \in J} n_{\cdot j} d_I(\hat{\mathbf{p}}_{X\cdot Y}(j))]}{d_J(\hat{\mathbf{p}}_Y) + d_I(\hat{\mathbf{p}}_X)}.$$

For example,

$$\hat{\lambda}_{Y\cdot X} = \Big(\sum_{i \in I} \max_{j \in J} n_{ij} - \max_{j\in J} n_{\cdot j} \Big) \Big/ \Big(N - \max_{j \in J} n_{\cdot j} \Big).$$

Extensions to multivariate measures are straightforward.

Normal Approximation

Normal approximations* for distributions of measures such as $\hat{\phi}_{Y\cdot X}$ and $\hat{\phi}_{XY}$ are readily obtained as in Goodman and Kruskal [3, 4]. Assume that d_J is differentiable at $\mathbf{p}_Y$ and at $\mathbf{p}_{Y\cdot X}(i)$ for $i \in I' = \{i \in I \colon p_{i\cdot} > 0\}$. [Alternatively, it suffices if $A(j, \cdot)$ is continuous at $\mathbf{p}_Y$ whenever $p_{\cdot j} > 0$ and $A(j, \cdot)$ is continuous at $p_{Y\cdot X}(i)$, $i \in I$, whenever $p_{ij} > 0$.] Assume that Y is not essentially constant. Then $N^{1/2}(\hat{\phi}_{Y\cdot X} - \phi_{Y\cdot X})$ has the large-sample distribution $N(0, \sigma^2(\phi_{Y\cdot X}))$, where $\sigma^2(\phi_{Y\cdot X})$ is the variance of the random variable

$$H(Y|X) = \big[(1 - \phi_{Y\cdot X}) d_J(Y, \mathbf{p}_Y) - d_J(Y, \mathbf{p}_{Y\cdot X}(X)) \big] / d_J(\mathbf{p}_Y).$$

Since $E[H(Y \mid X)] = 0$, $\sigma^2(\phi_{Y\cdot X}) = \sum_{i\in I} \sum_{j \in J} p_{ij} [H(j \mid i)]^2$. For examples of formulas, see Goodman and Kruskal [3, 4]. Assume, in addition, that d_I is differentiable at $\mathbf{p}_X$ and at $\mathbf{p}_{X\cdot Y}(j)$ for j such that $p_{\cdot j} > 0$. Thus $N^{1/2}(\hat{\phi}_{XY} - \phi_{XY})$ has large-sample distribution $N(0, \sigma^2(\phi_{XY}))$, where $\sigma^2(\phi_{XY})$ is the variance of

$$H(X, Y) = \{ (1 - \phi_{XY}) [d_I(X, \mathbf{p}_X) + d_J(Y, \mathbf{p}_Y)] - d_I(X, \mathbf{p}_{X\cdot Y}(J)) - d_J(Y, \mathbf{p}_{Y\cdot X}(I)) \} \cdot [d_I(\mathbf{p}_X) + d_J(\mathbf{p}_Y)]^{-1}.$$

Again $E[H(X, Y)] = 0$ and $\sigma^2(\phi_{XY}) = \sum_{i\in I} \sum_{j\in J} p_{ij} [H(i, j)]^2$. Since differentiability implies continuous differentiability in concave functions, $\sigma^2(\phi_{Y\cdot X})$ and $\sigma^2(\phi_{XY})$ possess consistent* estimates $\hat{\sigma}^2_{Y\cdot X}$ and $\hat{\sigma}^2_{XY}$ obtained by replacing p_{ij} by $\hat{p}_{ij} = N^{-1} n_{ij}$ in the relevant formulas. If $\sigma^2(\phi_{Y\cdot X}) > 0$, $0 < \alpha < 1$, and z_α is the upper $(\alpha/2)$-point of the $N(0, 1)$ distribution, then an approximate confidence interval* for $\phi_{Y\cdot X}$ of level α is

$$\big[\hat{\phi}_{Y\cdot X} - z_\alpha \hat{\sigma}(\phi_{Y\cdot X})/N^{1/2}, \hat{\phi}_{Y\cdot X} + z_\alpha \hat{\sigma}(\phi_{Y\cdot X})/N^{1/2} \big].$$

A similar argument applies to ϕ_{XY}.

Since $0 \leqslant \hat{\phi}_{Y\cdot X} \leqslant 1$, $\sigma(\phi_{Y\cdot X})$ must be 0 if a normal approximation applies and $\phi_{Y\cdot X}$ is 0 or 1. If all p_{ij} are positive and $\sigma(\phi_{Y\cdot X})$ is 0, then $\phi_{Y\cdot X} = 0$, for $\sigma_{Y\cdot X} = 0$ implies that $H(j \mid i)$ is always 0, so that

$$\begin{aligned} \Delta_Y &= \sum_{j\in J} p_{\cdot j} A(j, \mathbf{p}_Y) \\ &\leqslant \sum_{j \in J} p_{\cdot j} A(j, \mathbf{p}_{Y\cdot X}(i)) \\ &= (1 - \phi_{Y\cdot X}) \sum_{j\in J} p_{\cdot j} A(j, p_{\cdot j}) \\ &= (1 - \phi_{Y\cdot X}) \Delta_Y. \end{aligned}$$

In the special case of $\hat{\lambda}_{Y\cdot X}$, Goodman and Kruskal [3, 4] note that $\sigma^2(\lambda_{Y\cdot X})$ is defined whenever $m(\mathbf{p}_Y)$ and $m(\mathbf{p}_{Y\cdot X}(i))$, $i \in I$, are all 1, and $\sigma^2(\lambda_{Y\cdot X}) = 0$ only if $\lambda_{Y\cdot X}$ is 0 or 1. The normal approximation always applies to the estimate $\hat{\tau}_{Y\cdot X}$ and $\hat{\eta}_{Y\cdot X}$; however, a simple necessary and sufficient condition for $\sigma^2(\tau_{Y\cdot X})$ or $\sigma^2(\eta_{Y\cdot X})$ to be 0 appears difficult to find.

Sampling by Rows

An alternative sampling problem has also been considered in Goodman and Kruskal [3, 4], which is particularly appropriate for a standardized measure $\phi_{Y'\cdot X'}$ in which the conditional distribution of Y' given $X' = i \in I$ is the same as the conditional distribution of Y given X and X' has a known marginal distribution with $p_i' = \Pr[X' = i]$. Let each row $\langle n_{ij}: j \in J \rangle$, $i \in I$, have an independent multinomial* distribution with sample size $N_i > 0$ and probabilities $\langle p_{ij}/p_{i\cdot} : j \in J \rangle$. For simplicity, assume that each $p_{i\cdot}$ is positive. Let $N = \sum N_i$, let N_i/N approach $p_{i\cdot}$, and let N approach infinity. Consider the standardized estimate

$$\hat{\phi}_{Y'\cdot X'} = 1 - \sum_{i \in I} p_{i\cdot}' d_J(\hat{\mathbf{p}}_{Y\cdot X}(i)) \Big/ d_J(\hat{\mathbf{p}}_Y'),$$

where $\hat{\mathbf{p}}'_Y = \langle \hat{p}'_{\cdot j}: j \in J \rangle$, $\hat{p}'_{\cdot j} = \sum_{i \in I} \cdot p_{i\cdot}' n_{ij}/N_i$, $j \in J$, and $\hat{\mathbf{p}}_{Y\cdot X}(i) = \langle n_{ij}/N_i: j \in J \rangle$ for $i \in I$. Assume that d_J is differentiable at $\mathbf{p}_Y' = \langle p'_{\cdot j}: j \in J \rangle$ and at $\mathbf{p}_{Y\cdot X}(i)$, $i \in I$. Then $N^{1/2}(\hat{\phi}_{Y'\cdot X'} - \phi_{Y'\cdot X'})$ has an approximate $N(0, \sigma^2(\phi_{Y'\cdot X'}))$ distribution, where $\sigma^2(\phi_{Y'\cdot X'})$ is the expected conditional variance $\sum_{i \in I} \sum_{j \in J} p_{ij}[H'(j \mid i)]^2 - \sum_{i \in I}[\sum_{j \in J} H'(j \mid i) p_{ij}]^2 / p_{i\cdot}$ of $H'(Y \mid X)$ given X. Here for $i \in I, j \in J$,

$$H'(j \mid i) = (p_{i\cdot}'/p_{i\cdot})\big[(1 - \phi_{Y'\cdot X'})A_J(Y, \mathbf{p}_Y') - A_J(Y, \mathbf{p}_{Y\cdot X}(i))\big].$$

In the special case $p_{i\cdot}' = p_{i\cdot}$, one has $\phi_{Y'\cdot X'}$ equal to the unstandardized coefficient $\phi_{Y\cdot X}$, $H'(j \mid i) = H(j, i)$, and $\sigma^2(\phi_{Y'\cdot X'}) \leqslant \sigma^2(\phi_{Y\cdot X})$.

Clearly, $\sigma^2(\phi_{Y'\cdot X'}) = 0$ if $\phi_{Y'\cdot X'}$ is 0 or 1. In the case of $\lambda_{Y'\cdot X'}$, $\sigma^2(\lambda_{Y'\cdot X'}) = 0$ if and only if $\phi_{Y'\cdot X'}$ is 0 or 1, as noted in Goodman and Kruskal [4]. More generally, $\sigma^2(\phi_{Y'\cdot X'}) = 0$ implies that $\phi_{Y'\cdot X'} = 0$ if all probabilities p_{ij} are positive. The proof is only slightly changed from the corresponding proof for $\sigma_{Y\cdot X}$.

Older Measures

Numerous older measures of association between nominal variables are reviewed by Goodman and Kruskal [1, 2] and by Kendall and Stuart [10, pp. 556–561]. The most common are based on the chi-square* statistic. They include the mean square contingency

$$\phi^2 = \sum_{i \in I} \sum_{j \in J} (p_{ij} - p_{i\cdot} p_{\cdot j})^2 / (p_{i\cdot} p_{\cdot j})$$

and the coefficient of contingency $C = [\phi^2/(1 + \phi^2)]^{1/2}$ of Pearson [14] and Tschuprow's [19] coefficient $T = [\phi^2/\nu^{1/2}]^{1/2}$. In the last expression $\nu = (r-1)(s-1)$, I has r elements, and J has s elements. These measures lack the functional interpretations available in the case of $\phi_{Y\cdot X}$ or ϕ_{XY}.

MEASURES OF ASSOCIATION FOR ORDINAL POLYTOMOUS VARIABLES

The most commonly used measure of association for ordinal polytomous variables is the γ coefficient of Goodman and Kruskal [1–5]. Assume that the ranges I of X and J of Y are well ordered, so that if i and i' are in I, then $i < i'$, $i = i'$, or $i > i'$ and if j and j' are in J, then $j < j'$, $j = j'$, or $j > j'$. Let (X_1, Y_1) and (X_2, Y_2) be independently distributed pairs with the same distribution as (X, Y). Let $C = 2\Pr[X_1 > X_2$ and $Y_1 > Y_2]$ be the probability that either $X_1 > X_2$ and $Y_1 > Y_2$ or $X_1 < X_2$ and $Y_1 < Y_2$, so that (X_1, Y_1) and (X_2, Y_2) are concordant. Let $2D = 2\Pr[X_1 > X_2$ and $Y_2 > Y_1]$ be the probability that either $X_1 > X_2$ and $Y_1 < Y_2$ or $X_1 < X_2$ and $Y_1 > Y_2$, so that (X_1, Y_1) and (X_2, Y_2) are discordant. Then

$$\gamma_{XY} = \frac{C - D}{C + D}.$$

The coefficient is defined if $p_{ij} > 0$ and $p_{i'j'} > 0$ for some $i, i' \in I$ and $j, j' \in J$ with $i \neq i'$ and $j \neq j'$. One has $-1 \leqslant \gamma_{XY} \leqslant 1$, with $\gamma_{XY} = 0$ under independence of X and Y. For γ_{XY} to be 1, the nonzero p_{ij} must have an ascending staircase pattern, so that if $i < i'$, $p_{ij} > 0$, and $p_{i'j'} > 0$, then $j \leqslant j'$, while if $j < j'$, $p_{ij} > 0$, and $p_{i'j'} > 0$, then $i \leqslant i'$. Similarly, γ_{XY} can only be -1 if the nonzero p_{ij} have a descending staircase pattern. In the special case in which I and J have the two elements 1 and 2, γ_{XY} is the *coefficient of association* $(p_{11}p_{22} - p_{12}p_{21})/(p_{11}p_{22} + p_{12}p_{21})$ of *Yule** [20]. In this special case, $\gamma_{XY} = 0$ if and only if X and Y are independent, $\gamma_{XY} = 1$ only if $\Pr[X = Y] = 1$, and $\gamma_{XY} = -1$ only if $\Pr[X = Y] = 0$.

The measure γ_{XY} only considers pairs $(X_1,$

$Y_1)$ and (X_2, Y_2) in which $X_1 \neq X_2$ and $Y_1 \neq Y_2$. An alternative approach by Somers [17] considers all pairs with just $X_1 \neq X_2$. One obtains the asymmetric coefficient

$$\gamma_{Y \cdot X} = (C - D)/\Pr[X_1 \neq X_2].$$

Again $-1 \leqslant \gamma_{Y \cdot X} \leqslant 1$, with $\gamma_{Y \cdot X} = 0$ if X and Y are independent. The coefficient $\gamma_{Y \cdot X}$ can only be -1 (or 1) if γ_{XY} is -1 (or 1) and if for each $j \in J$, p_{ij} is positive for no more than one $i \in I$. For further variants on γ_{XY}, see Kendall and Stuart [10, pp. 561–565].

Estimation of γ_{XY} and $\gamma_{Y \cdot X}$ is straightforward given a table $\mathbf{n} = \langle n_{ij} : i \in I, j \in J \rangle$ with a multinomial distribution with sample size N and probabilities $\mathbf{p} = \langle p_{ij} : i \in I, j \in J \rangle$. Let $\hat{C}$ be the sum of all products $2n_{ij}n_{i'j'}/N^2$ such that $i < i'$ and $j < j'$, and let $\hat{D}$ be the sum of all products $2n_{ij}n_{i'j'}/N^2$ such that $i < i'$ and $j < j'$. Then $\hat{\gamma}_{XY} = (\hat{C} - \hat{D})/(\hat{C} + \hat{D})$ and $\hat{\gamma}_{Y \cdot X} = (\hat{C} - \hat{D})/[1 - \sum_{i \in I}(n_{i \cdot}/N)^2]$. As noted in Goodman and Kruskal [3, 4] $N^{1/2}(\hat{\gamma}_{XY} - \gamma_{XY})$ has an approximate $N(0, \sigma^2(\phi_{XY}))$ distribution, with

$$\sigma^2(\phi_{XY}) = \frac{16}{(C + D)^4} \sum_{i \in I} \sum_{j \in J} p_{ij}(CS_{ij} - DR_{ij})^2$$

$$\leqslant 2(1 - \gamma_{XY}^2)/(C + D),$$

$$S_{ij} = \Pr[X > i \text{ and } Y < j] + \Pr[X < i \text{ and } Y > j],$$

$$R_{ij} = \Pr[X > i \text{ and } Y > j] + \Pr[X < i \text{ and } Y < j].$$

Similarly, $N^{1/2}(\hat{\gamma}_{Y \cdot X} - \gamma_{Y \cdot X})$ has an approximate $N(0, \sigma^2(\gamma_{Y \cdot X}))$ distribution with $E = \Pr[X_1 \neq X_2, Y_1 \neq Y_2]$ and

$$\sigma^2(\gamma_{Y \cdot X}) = \frac{4}{(C + D + E)^4} \sum_{i \in I} \sum_{j \in J} p_{ij} \cdot \left[(C - D)(1 - p_{i \cdot}) - (C + D + E)(S_{ij} - R_{ij})\right]^2.$$

One has $\sigma^2(\gamma_{XY}) = 0$ if $|\gamma_{XY}| = 1$ and $\sigma^2(\gamma_{Y \cdot X}) = 0$ if $|\gamma_{Y \cdot X}| = 1$. If all p_{ij} are positive, then $\sigma^2(\gamma_{XY}) > 0$. In the special case of $\gamma_{XY} = 0$, one has $\gamma_{Y \cdot X} = 0$,

$$\sigma^2(\gamma_{XY}) = \frac{4}{(C + D)^2} \sum_{i \in I} \sum_{j \in J} p_{ij}(S_{ij} - R_{ij})^2,$$

$$\sigma^2(\gamma_{Y \cdot X}) = \frac{4}{(C + D + E)^2} \sum_{i \in I} \sum_{j \in J} p_{ij} \times (S_{ij} - R_{ij})^2.$$

Kendall's τ^*, Spearman's ρ_s^*, and Goodman and Kruskal's γ^*

In contrast to the nominal measures of association, Goodman and Kruskal's γ coefficient remains well defined if the respective ranges I and J of X and Y are infinite and $\Pr[X = i] = \Pr[Y = j] = 0$ for any $i \in I$ and $j \in J$. The coefficient γ_{XY} is then Kendall's [8] τ measure $\tau_k = C - D$. It remains true that $-1 \leqslant \tau_k \leqslant 1$, with $\tau_k = 0$ when X and Y are independent. Estimation of τ_k is, however, best described in terms of independent pairs (X_t, Y_t), $1 \leqslant t \leqslant N$, with common distribution (X, Y). Then Kendall's τ statistic $\hat{\tau}_k$ is $2(N_c - N_d)/[N(N - 1)]$, where there are N_c s and t with $1 \leqslant s < t \leqslant N$ such that (X_s, Y_s) and (X_t, Y_t) are concordant ($X_s < X_t$ and $Y_s < Y_t$ or $X_t < X_s$ and $Y_t < Y_s$) and there are N_d s and t with $1 \leqslant s < t \leqslant N$ such that (X_s, Y_s) and (X_t, Y_t) are discordant ($X_s < X_t$ and $Y_s > Y_t$ or $X_t < X_s$ and $Y_t > Y_s$). As N becomes large $N^{1/2}(\hat{\tau}_k - \tau_k)$ has an approximate $N(0, \sigma^2(\tau_k))$ distribution. Here $\sigma^2(\tau_k) = 16(F - C^2)$ and

$$F = \Pr[X_1 > X_2, X_1 > X_3, Y_1 > Y_2, Y_1 > Y_3] + \Pr[X_1 < X_2, X_1 < X_3, Y_1 < Y_2, Y_1 < Y_3]$$

is the probability that both (X_2, Y_2) and (X_3, Y_3) are concordant with (X_1, Y_1). If X and Y are independent, then $\sigma^2(\tau_k) = 4/9$. See Hoeffding [7] and Kruskal [12] for details.

Closely related to Kendall's τ is the Spearman rank correlation coefficient r_s. Assume that all X_t are distinct and all Y_t are distinct. Let R_t be the number of s with $X_s \leqslant X_t$, and let S_t be the number of t with $Y_s \leqslant Y_t$. Then

r_s is the sample correlation of the pairs (R_t, S_t), $1 \leqslant t \leqslant N$. The statistic r_s provides a measure

$$\rho_s = 6\{\Pr[X_1 > X_2, Y_1 > Y_3] - 1\}$$

of the probability that (X_1, Y_1) and (X_2, Y_3) are concordant. An alternative unbiased estimate of ρ_s is $\hat{\rho}_s = [(n+1)/(n-2)]r_s - [3/(n-2)]\hat{\tau}_k$, which has been termed the unbiased grade coefficient* by Konijn [11]. One has $-1 \leqslant \rho_s \leqslant 1$, with $\rho_s = 0$ under independence of X and Y. Both $N^{1/2}(\hat{\rho}_s - \rho_s)$ and $N^{1/2}(r_s - \rho_s)$ have limiting distribution $N(0, \sigma^2(\rho_s))$ as N becomes large. The formula for $\sigma^2(\rho_s)$ has been given by Hoeffding [7]. Since it is somewhat complicated, it will be omitted here in the general case. Under independence of X and Y, $\sigma^2(\rho_s) = 1$.

For further details concerning these and related measures, see Kendall [9] and Kruskal [12]. (*See* also KENDALL'S τ, SPEARMAN'S τ_s, AND GOODMAN AND KRUSKAL'S τ AND γ.)

NUMERICAL EXAMPLE

To illustrate results, consider Table 1, which can be found in Goodman and Kruskal [1], among many other references. Let X refer to eye color and let Y refer to hair color.

Some resulting estimated measures of association are listed in Table 2. In the case of the measures for ordered variables, eye color is ordered from blue to brown and hair color is ordered from fair to black.

Asymptotic standard deviations are based on the assumption that the counts in Table 1 have a multinomial distribution. The asymmetry in X and Y and the large variations in the sizes of measures are to be expected. As noted as early as Goodman and Kruskal [1], instincts developed from regression analysis* are not necessarily appropriate for assessment of the size of measures of association for ordinal or nominal polytomous variables.

Table 2 Estimated Measures of Association for Table 1

Measure	Estimate	Estimated Asymptotic Standard Deviation of Estimate
$\lambda_{Y \cdot X}$	0.192	0.012
$\lambda_{X \cdot Y}$	0.224	0.013
λ_{XY}	0.208	0.010
$\tau_{Y \cdot X}$	0.081	0.005
$\tau_{X \cdot Y}$	0.089	0.005
τ_{XY}	0.085	0.005
$\eta_{Y \cdot X}$	0.075	0.004
$\eta_{X \cdot Y}$	0.085	0.005
η_{XY}	0.080	0.004
γ_{XY}	0.547	0.013
$\gamma_{X \cdot Y}$	0.346	0.009
$\gamma_{Y \cdot X}$	0.371	0.010

Table 1

	Hair Color Group				
Eye Color Group	Fair	Red	Brown	Black	Total
Blue	1768	47	807	189	2811
Gray or green	946	53	1387	746	3132
Brown	115	16	438	288	857
Total	2829	116	2632	1223	6800

References

[1] Goodman, L. A. and Kruskal, W. H. (1954). *J. Amer. Statist. Ass.*, **49**, 732–764. (Goodman and Kruskal's λ, τ, and γ are defined and the basic criteria for measures of association are presented.)

[2] Goodman, L. A. and Kruskal, W. H. (1959). *J. Amer. Statist. Ass.*, **54**, 123–163. (A valuable historical review and bibliography are provided.)

[3] Goodman, L. A. and Kruskal, W. H. (1963). *J. Amer. Statist. Ass.*, **58**, 310–364. (Asymptotic distributions are obtained for estimates of Goodman and Kruskal's λ, τ, and γ. Further results appear in their 1972 paper.)

[4] Goodman, L. A. and Kruskal, W. H. (1972). *J. Amer. Statist. Ass.*, **67**, 415–421.

[5] Goodman, L. A. and Kruskal, W. H. (1979). *Measures of Association for Cross Classifications*. Springer-Verlag, New York. (A volume consisting of Goodman and Kruskal's four papers on measures of association, indicated above.)

[6] Guttman, L. (1941). In *The Prediction of Personal Adjustment* (Bull. 48), P. Horst et al., eds. Social Science Research Council, New York, pp. 253–318.

[7] Hoeffding, W. (1948). *Ann. Math. Statist.*, **19**, 293–325. (The methods presented here apply to numerous problems involving ordinal data.)

[8] Kendall, M. G. (1938). *Biometrika*, **30**, 277–283.

[9] Kendall, M. G. (1962). *Rank Correlation Methods*, 3rd ed. Charles Griffin, London. (A standard text on rank correlation statistics.)

[10] Kendall, M. G. and Stuart, A. (1967). *The Advanced Theory of Statistics*, 2nd ed. Vol. 2. Charles Griffin, London.

[11] Konijn, H. S. (1956). *Ann. Math. Statist.*, **27**, 300–323.

[12] Kruskal, W. H. (1958). *J. Amer. Statist. Ass.*, **53**, 814–861. (A very helpful review.)

[13] Mosteller, F. (1968). *J. Amer. Statist. Ass.*, **63**, 1–28.

[14] Pearson, K. (1904). Mathematical Contributions to the Theory of Evolution. XIII. On the Theory of Contingency and Its Relation to Association and Normal Correlation. *Drapers' Company Res. Mem., Biometric Ser. 1*. (Pearson introduces classical association measures related to the chi-square statistic* and relates them to the bivariate normal distribution.*)

[15] Rockafellar, R. T. (1970). *Convex Analysis*. Princeton University Press, Princeton, N.J.

[16] Savage, L. J. (1971). *J. Amer. Statist. Ass.*, **66**, 783–801.

[17] Somers, R. H. (1962). *Amer. Soc. Rev.*, **27**, 799–811.

[18] Theil, H. (1970). *Amer. J. Sociol.*, **76**, 103–154.

[19] Tschuprow, A. A. (1925). *Grundbegriffe und Grundprobleme der Korrelationstheorie*. Teubner, Leipzig.

[20] Yule, G. U. (1900). *Philos. Trans. R. Soc. Lond. A*, **194**, 257–319. (A valuable early paper on measurement of association of dichotomous variables.)

(CATEGORICAL DATA
CONTINGENCY TABLES
CORRELATION
DEPENDENCE, MEASURES OF
GOODMAN AND KRUSKAL'S τ AND γ
STANDARDIZATION)

S. J. Haberman

ASTRONOMY, STATISTICS IN *See* STATISTICS IN ASTRONOMY

ASYMMETRIC POPULATION

A population that is not symmetric. The property of asymmetry is also called "skewness." It should be noted, however, that measures of skewness* usually correspond to some particular feature of skewness. The moment ratio $\sqrt{\beta_1} = \alpha_3 = \mu_3/\mu_2^{3/2}$, for example, is indeed zero for symmetric populations, but it can also be zero for populations that are asymmetric.

It is better to limit the use of the adjective "asymmetric" to distributions, and not apply it to populations.

(SKEWNESS, MEASURES OF
SYMMETRIC POPULATIONS)

ASYMMETRY

The property of not being symmetrical.

(ASYMMETRIC POPULATION
SKEWNESS)

ASYMPTOTIC EXPANSION

Many test statistics* and estimators* have probability distributions that may be approximated quite well by a normal distribution* in the case of a sufficiently large sample size. This is of practical use, for example, in such problems as determining critical regions* for tests of specified sizes and determining confidence regions* with specified confidence coefficients. Let T_n denote a test statistic based on a sample $X_1, \ldots, X_n$ from a distribution F, let a_n and b_n be suitable normalizing constants, and let G_n denote the distribution of the normed statistic $(T_n - a_n)/b_n$. The "normal approximation" is expressed by

$$\lim_{n\to\infty} G_n(t) = \Phi(t) \qquad (-\infty < t < \infty), \quad (1)$$

where $\Phi(t) = (2\pi)^{-1/2} \int_{-\infty}^{t} \exp(-x^2/2)\,dx$, the standard normal distribution* function. Often (1) can be established under moderate regularity assumptions on the distribution function F and the functional form of the statistic T_n. *See* ASYMPTOTIC NORMALITY.

Of fundamental importance is the question of the error of approximation in (1) for a particular value of n. One useful type of answer is supplied by a "Berry–Esséen" rate*, namely an assertion of the form

$$\sup_{-\infty<t<\infty} |G_n(t) - \Phi(t)| = O(n^{-1/2}), \quad (2)$$

available under additional restrictions on F and the form of T_n. A more refined answer is given by an *expansion* of the error $G_n(t) - \Phi(t)$ in powers of $n^{-1/2}$. This requires additional restrictions on F and T_n. However, not only does it provide detailed information in (2), but it also supplies a way to replace $\Phi(t)$ by an improved approximation. Below we shall survey such expansions for the key special case that T_n is a *sum*, and then we shall comment briefly on other cases. For more details see the entries devoted to specific expansions.

Let $X_1, X_2, \ldots$ be independent and identically distributed random variables with distribution F, mean μ, variance σ^2, and characteristic function* ψ. Let G_n denote the distribution of the normed sum $(\sum_1^n X_i - n\mu)/n^{1/2}\sigma$. If Cramér's condition

$$\text{(C)} \qquad \lim_{|z|\to\infty} \sup |\psi(z)| < 1$$

is satisfied and the kth moment of F is finite, then

$$\left| G_n(t) - \Phi(t) - \sum_{j=1}^{k-3} P_{3j-1}(t) e^{-t^2/2} n^{-j/2} \right| < M n^{-(k-2)/2}, \quad (3)$$

where M is a constant depending on k and F but not on n or t, and $P_m(t)$ is a polynomial (essentially the *Chebyshev–Hermite* polynomial*) of degree m in t. Indeed, we have the expressions

$$P_2(t)e^{-t^2/2} = -(\lambda_3/3!)\Phi^{(3)}(t),$$
$$P_3(t)e^{-t^2/2} = (\lambda_4/4!)\Phi^{(4)}(t) + (10\lambda_3^2/6!)\Phi^{(6)}(t), \ldots,$$

where λ_j denotes the jth cumulant* of F [the coefficient of $(iz)^j/j!$ in the MacLaurin series expansion of $\log \psi(z)$]. We may express λ_j (essentially) as a polynomial in the moments of F, obtaining in particular

$$\lambda_3 = \frac{E[(X-\mu)^3]}{\sigma^3} = \gamma_1$$

and

$$\lambda_4 = \frac{E[(X-\mu)^4]}{\sigma^4} - 3 = \gamma_2,$$

known as the coefficients of skewness* and kurtosis*, respectively. Therefore, up to terms of order n^{-1}, which usually suffices in practical applications, the approximation given by (3) may be written conveniently in the form

$$G_n(t) \doteq \Phi(t) - \frac{\gamma_1}{6}(t^2-1)\phi(t)n^{-1/2} - \left[\frac{\gamma_2}{24}(t^3-3t) + \frac{\gamma_1^2}{72} \times (t^5 - 10t^3 + 15t)\right]\phi(t)n^{-1}, \quad (4)$$

with error $O(n^{-3/2})$ uniformly in t.

The expansion (3) is called the *Edgeworth expansion* for G_n. Corresponding expansions for the density g_n follow by replacing all functions of t by their derivatives. The assumption (C) is always satisfied if the distribution F has an absolutely continuous component. Analogs of (3) hold under alternative conditions on F, e.g., the case of a lattice distribution*. Versions also have been developed allowing the X_i's to have differing distributions or to be stationary dependent. Furthermore, other metrics besides $\sup_t |G_n(t) - \Phi(t)|$ have been treated. For extensive treatments of (3) and these various ramifications, see Wallace [12], Cramér [7], Ibragimov and Linnik [9], Petrov [10], Bhattacharya and Ranga Rao [3], and Bhattacharya [2].

An inverse problem related to (3) concerns the equation

$$G_n(t_p) = 1 - p,$$

where $0 < p < 1$. The solution t_p may be

expressed asymptotically as

$$t_p \sim \mu + \sigma w, \tag{5}$$

where w is given by an expansion (the *Cornish–Fisher**) which like (3) involves the quantities $\{\lambda_i\}$ and the Chebyshev–Hermite polynomials.

For detailed numerical illustration of the effectiveness of the expansions (3) and (5), see Abramowitz and Stegun [1], pp. 935–936, 955, 958–959. As noted in connection with (4), the improvement of the Edgeworth approximation over simply the normal approximation can be attributed to use of the coefficients of skewness and kurtosis; this provides a convenient intuitive basis for assessing the potential degree of improvement. Numerical illustration related to (4) is provided by Bickel and Doksum [5].

Finally, let us consider statistics other than sums. For such cases the question of asymptotic normality*, (1), has received extensive treatment. Secondarily, rather recently, the associated Berry–Esséen rates, (2), have received attention. Consequently, results of types (1) and (2) are now available for several important wide classes of statistics: U-statistics*; von Mises differentiable statistical functions*; linear functions of order statistics; M-estimates*; and rank statistics*. Detailed exposition may be found in Serfling [11]. *See* also ASYMPTOTIC NORMALITY. However, except for isolated results, the question of asymptotic expansions analogous to (3) has only very recently gained intensive interest and development. For comments on Edgeworth expansions for rank statistics, such as the two-sample Wilcoxon statistic*, see Hájek and Šidák [8], Sec. IV.4.2. Multivariate Edgeworth-type expansions are discussed by Chambers [6]. For a review of recent activity, see Bickel [4].

References

[1] Abramowitz, M. and Stegun, I. A., eds. (1965). *Handbook of Mathematical Functions*. U.S. Government Printing Office, Washington, D.C.

[2] Bhattacharya, R. N. (1977). *Ann. Prob.*, **5**, 1–27.

[3] Bhattacharya, R. N. and Ranga Rao, R. (1976). *Normal Approximation and Asymptotic Expansions*. Wiley, New York.

[4] Bickel, P. J. (1974). *Ann. Statist.*, **2**, 1–20.

[5] Bickel, P. J. and Doksum, K. A. (1977). *Mathematical Statistics*. Holden-Day, San Francisco.

[6] Chambers, J. M. (1967). *Biometrika*, **54**, 367–383.

[7] Cramér, H. (1970). *Random Variables and Probability Distributions*, 3rd ed. Cambridge University Press, Cambridge.

[8] Hájek, J. and Šidák, Z. (1967). *Theory of Rank Tests*. Academic Press, New York.

[9] Ibragimov, I. A. and Linnik, Y. V. (1971). *Independent and Stationary Sequences of Random Variables*. Wolters-Noordhoff, Groningen.

[10] Petrov, V. V. (1975). *Sums of Independent Random Variables*. Springer-Verlag, New York.

[11] Serfling, R. J. (1980). *Approximation Theorems of Mathematical Statistics*. Wiley, New York.

[12] Wallace, D. (1958). *Ann. Math. Statist.*, **29**, 635–654.

(APPROXIMATIONS TO DISTRIBUTIONS
ASYMPTOTIC NORMALITY
CORNISH–FISHER AND EDGEWORTH EXPANSIONS)

R. J. SERFLING

ASYMPTOTIC NORMALITY

The exact distribution of a statistic is usually highly complicated and difficult to work with. Hence the need to approximate the exact distribution by a distribution of a simpler form whose properties are more transparent. The limit theorems* of probability theory provide an important tool for such approximations. In particular, the classical central limit theorems* state that the sum of a large number of independent random variables is approximately normally distributed under general conditions (see the section "Central Limit Theorems for Sums of Independent Random Variables"). In fact, the normal distribution* plays a dominating role among the possible limit distributions. To quote from Gnedenko and Kolmogorov [18, Chap. 5]: "Whereas for the convergence of distribution functions of sums of independent variables to the normal law only restrictions of a very general kind, apart from that of being infinitesimal (or asymptotically constant), have to be imposed on the summands, for the convergence to another limit law some very special properties are required of the summands." Moreover, many statistics behave asymptotically like sums of independent random variables (see the fifth, sixth, and seventh sections). All of this helps to explain the importance of the normal distribution* as an asymptotic distribution.

Suppose that the statistics T_n, $n = 1, 2, \ldots$, when suitably normed, have the standard normal limit distribution; i.e., for some constants $b_n > 0$ and a_n and for every real x we have

$$\Pr[(T_n - a_n)/b_n \leqslant x] \to \Phi(x) \qquad \text{as } n \to \infty \tag{1}$$

where

$$\Phi(x) = (2\pi)^{-1/2} \int_{-\infty}^{x} e^{-y^2/2}\, dy.$$

Then we say that T_n is asymptotically normal with mean a_n and variance b_n^2, or asymptotically normal (a_n, b_n^2). [Note that a_n and b_n^2 need not be the mean and the variance of T_n; indeed, (1) may hold even when T_n has no finite moments.]

It can be shown that if (1) holds for every x, the convergence is uniform in x, so that

$$\sup_{-\infty < x < \infty} \left| \Pr[(T_n - a_n)/b_n \leqslant x] - \Phi(x) \right| \to 0 \qquad \text{as } n \to \infty. \tag{2}$$

[This is due to the continuity of $\Phi(x)$.]

The knowledge that (1) or (2) holds is not enough for most statistical applications. For one thing, the statistician wants to know how large n has to be in order that the limit distribution may serve as a satisfactory approximation. Also, if the distribution of T_n depends on unknown parameters, the statistician wants to know how the values of the parameters affect the speed of convergence to the limit. Both goals are met, to some extent, by the Berry–Esseen theorem* (see below) and related results discussed in "Remainder Term in the Central Limit Theorem."

When the approximation provided by the limit distribution* is unsatisfactory, asymptotic expansions*, treated in the third section, may prove more helpful.

Conditions for the convergence of the moments of a statistic to the corresponding moments of its limit distribution are briefly discussed in the fourth section. The fifth section deals with the distributions of functions of asymptotically normal random variables. The asymptotic normality of functions of independent random variables and of sums of dependent random variables is considered in the sixth and seventh sections, respectively. The final section deals with functional central limit theorems, which are concerned with asymptotic distributions of random functions.

CENTRAL LIMIT THEOREMS* FOR SUMS OF INDEPENDENT RANDOM VARIABLES

The following classical central limit theorem for the partial sums of an infinite sequence of independent, identically distributed (i.i.d.) random variables is due to Lindeberg.

Theorem 1. Let $X_1, X_2, \ldots$ be an infinite sequence of i.i.d. random variables with finite mean a and positive and finite variance σ^2. Then, as $n \to \infty$, $X_1 + \cdots + X_n$ is asymptotically normal $(an, \sigma^2 n)$.

In the following theorem only finite sequences of independent (not necessarily identically distributed) random variables are involved, which makes it better adapted to most applications.

Theorem 2. For each $N = 1, 2, \ldots$ let $X_{N1}, X_{N2}, \ldots, X_{Nn}$ be $n = n(N)$ independent random variables with finite p-th moments, for some $p > 2$. Let $B_N = \sum_j \operatorname{var}(X_{Nj})$ (the index j runs from 1 to n). If

$$B_N^{-p/2} \sum_j E|X_{Nj} - EX_{Nj}|^p \to 0 \qquad \text{as } N \to \infty, \tag{3}$$

then $\sum_j X_{Nj}$ is asymptotically normal $(\sum_j EX_{Nj}, B_N)$.

This theorem is due to Liapunov. Condition (3) may be replaced by a weaker condition, due to Lindeberg, which does not assume finite moments of order > 2 (see ref. 32).

For a general central limit theorem for sums of independent random variables which assumes no finite moments and for other central limit theorems, see ref. 32, Chap. IV, Sec. 4.

Multidimensional central limit theorems give conditions for the convergence of the distribution of a sum of independent random vectors to a multivariate normal distribution*; see Cramér [10] and Uspensky [40].

If the sums of independent random variables have probability densities, the latter will converge, under certain conditions, to a normal probability density. For results of this type, known as local* central limit theorems, see ref. 32.

REMAINDER TERM IN THE CENTRAL LIMIT THEOREM

The following result, due to Esseen [17], gives an explicit upper bound for the difference between the distribution function of a sum of independent random variables and the normal distribution function.

Theorem 3. Let $X_1, \dots, X_n$ be independent random variables,

$$EX_j = 0, \qquad E|X_j|^3 < \infty, \qquad (j = 1, \dots, n),$$

and let

$$B_n = \sum_{j=1}^{n} EX_j^2 > 0, \qquad L_n = \sum_{j=1}^{n} E|X_j|^3 / B_n^{3/2}.$$

Then

$$\left| \Pr\left[B_n^{-1/2} \sum_{j=1}^{n} X_j \leqslant x \right] - \Phi(x) \right| \leqslant CL_n \qquad \text{for all } x, \qquad (4)$$

where C is a numerical constant.

The assumption $EX_j = 0$ is made merely to simplify the notation. If $EX_j = a_j$, replace X_j by $X_j - a_j$ in the statement of the theorem.

The least value of C for which (4) holds is not known. It is known [2] that (4) is true with $C = 0.7975$ and is not true with $C < 0.4097$.

Note that Theorem 3 involves only one finite sequence of independent random variables and is not a limit theorem. It easily implies Theorem 2 with $p = 3$.

Under the further assumption that $X_1, \dots, X_n$ are identically distributed, with $EX_1 = 0$, $EX_1^2 = \sigma^2$, inequality (4) simplifies to

$$\left| \Pr\left[n^{-1/2}\sigma^{-1} \sum_{j=1}^{n} X_j \leqslant x \right] - \Phi(x) \right| \leqslant Cn^{-1/2}\sigma^{-3}E|X_1|^3. \qquad (5)$$

This inequality was also derived by Berry [4] and is known as the Berry–Esseen inequality.

The upper bounds in (4) and (5) do not depend on x. S. V. Nagaev has shown that inequality (5) is still true (except perhaps for the value of the constant C) if the right side is multiplied with $(1 + |x|)^{-3}$. For this and related results, see ref. 32.

For extensions of these results to sums of independent random vectors, see ref. 5.

ASYMPTOTIC EXPANSIONS*

Let $X_1, X_2, \dots$ be i.i.d. random variables, $EX_1 = 0$, $0 < \sigma^2 = EX_1^2 < \infty$,

$$F_n(x) = \Pr\left[\sum_{j=1}^{n} X_j \leqslant x\sigma n^{1/2} \right].$$

By Theorem 1, $F_n(x) \to \Phi(x)$ as $n \to \infty$. However, the approximation of $F_n(x)$ by $\Phi(x)$ is often too crude to be useful. There are expansions of the difference $F_n(x) - \Phi(x)$ in powers of $n^{-1/2}$ that may provide more accurate approximations.

The form of the expansion depends on whether the random variable X_1 is lattice* or nonlattice. [A random variable X is called a lattice random variable if, for some numbers $h > 0$ and a, the values of $(X - a)/h$ are integers; the largest h with this property is called the maximum span. Otherwise, X is nonlattice.]

Theorem 4. If the random variables $X_1, X_2, \dots$ are i.i.d., nonlattice, and have a finite third moment, then

$$F_n(x) = \Phi(x) + \Phi'(x)\, Q_1(x) n^{-1/2} + o(n^{-1/2}) \qquad (6)$$

uniformly in x. Here $\Phi'(x) = (2\pi)^{1/2} \exp \cdot (-x^2/2)$ is the standard normal density function and

$$Q_1(x) = \frac{1}{6} \frac{EX_1^3}{\sigma^3} (1 - x^2).$$

For a proof and for extensions of (6) involving higher powers of $n^{-1/2}$, see refs. 5 and 32. Expansions of this type have been studied by Chebyshev*, Edgeworth, Cramér, Esseen, and others.

Theorem 5. If $X_1, X_2, \ldots$ are i.i.d. lattice random variables taking the values $a + kh$ $(k = 0, \pm 1, \pm 2, \ldots)$, where h is the maximum span, and have a finite third moment, then

$$F_n(x) = \Phi(x) + \Phi'(x)(Q_1(x) + S_1(x)) n^{-1/2} + o(n^{-1/2}) \qquad (7)$$

uniformly in x. Here

$$S_1(x) = \frac{h}{\sigma} S\left(\frac{x\sigma n^{1/2} - an}{h} \right),$$

$$S(x) = [x] - x + \tfrac{1}{2},$$

and $[x]$ is the largest integer $\leqslant x$.

This theorem is due to Esseen [17]; see also ref. 18. For an extension of (7) that involves higher powers of $n^{-1/2}$, see refs. 5 and 32.

Asymptotic expansions of the distribution function and the probability density function of a sum of independent random variables that need not be identically distributed are also treated in ref. 32.

CONVERGENCE OF MOMENTS

If a statistic T_n has a normal limit distribution, its moments need not converge to the corresponding moments of the latter; in fact, T_n need not have any finite moments.

If the conditions of Theorem 2 with a fixed $p > 2$ are satisfied then for all positive integers $q \leqslant p$, the qth absolute moment of $\sum_j (X_{Nj} - EX_{Nj})/B_N^{1/2}$ converges to the corresponding moment of the standard normal distribution; see S. N. Bernstein* [3] and Hall [23]. A similar result is due to Zaremba [41]. Bounds for the remainder terms in such limit theorems for moments have been obtained by von Bahr [1] and Hall [23], among others. An interesting discussion of the convergence of moments of certain statistics can be found in Cramér [9, Chap. 27].

FUNCTIONS OF ASYMPTOTICALLY NORMAL RANDOM VARIABLES

We often encounter statistics that are functions of sample moments or of generalized sample moments of the form $M_n = n^{-1} \sum_{j=1}^n g(X_j)$. If the X_j are i.i.d., $Eg(X_1) = a$, var $g(X_1) = \sigma^2$ $(0 < \sigma^2 < \infty)$, then M_n is asymptotically normal $(a, \sigma^2/n)$.

Theorem 6. Let the random variables M_n, $n \geqslant 1$, be asymptotically normal $(a, \sigma^2/n)$. If $H(x)$ is a function of the real variable x whose derivative $H'(x)$ exists and is $\neq 0$ and continuous at $x = a$, then $H(M_n)$ is asymptotically normal $(H(a), H'(a)^2\sigma^2/n)$.

This result can be extended to functions of k moment-like statistics which are asymptotically k-variate normal. We state the extension for $k = 2$.

Theorem 7. Let the random vectors (M_{1n}, M_{2n}), $n \geqslant 1$, be asymptotically bivariate normal* with mean (a_1, a_2) and covariances σ_{ij}/n, $i, j = 1, 2$. If $H(x, y)$ is a function of the real variables x and y whose partial derivatives at (a_1, a_2),

$$H_1 = \partial H(x, y)/\partial x|_{(a_1, a_2)},$$

$$H_2 = \partial H(x, y)/\partial y|_{(a_1, a_2)},$$

exist and are not both zero, and which has a total differential at (a_1, a_2), so that

$$H(x, y) = H(a_1, a_2) + H_1 x + H_2 y + x\epsilon_1(x, y) + y\epsilon_2(x, y),$$

where $\epsilon_i(x, y) \to 0$ as $(x, y) \to (a_1, a_2)$ $(i = 1,$

2), then $H(M_{1n}, M_{2n})$ is asymptotically normal with mean $H(a_1, a_2)$ and variance $(H_1^2\sigma_{11} + 2H_1H_2\sigma_{12} + H_2^2\sigma_{22})/n$.

Proofs of these or closely related results can be found in refs. 9 and 26.

Note that the conditions of Theorems 6 and 7 are such that $H(M_n)$ and $H(M_{1n}, M_{2n})$ can be approximated by the linear terms of their Taylor expansions*. If the linear terms vanish and they can be approximated by the quadratic terms, the asymptotic distribution will be that of a quadratic form* in normal random variables.

ASYMPTOTIC NORMALITY OF FUNCTIONS OF INDEPENDENT RANDOM VARIABLES

Let $T_n = T_n(X_1, \ldots, X_n)$ be a function of the independent random variables $X_1, \ldots, X_n$. Suppose that $ET_n^2 < \infty$.

*Hájek's projection lemma** approximates T_n by the statistic

$$\hat{T}_n = \sum_{j=1}^{n} E[T_n \mid X_j] - (n-1)ET_n,$$

which is a sum of independent random variables. By the corollary of that entry we have

Theorem 8. Let the stated assumptions be satisfied for all n. Suppose that $\hat{T}_n$ is asymptotically normal $(E\hat{T}_n, \operatorname{var}\hat{T}_n)$ and that

$$(\operatorname{var}\hat{T}_n)/\operatorname{var}(T_n) \to 1 \qquad \text{as } n \to \infty.$$

Then T_n is asymptotically normal $(ET_n, \operatorname{var} T_n)$.

Hájek [21] and Dupač and Hájek [14] used the projection lemma to prove the asymptotic normality of a simple linear rank statistic*,

$$\sum_{j=1}^{n} a_n(j)b_n(R_{nj}),$$

where $a_n(j)$, $b_n(j)$ are constants, $R_{n1}, \ldots, R_{nn}$ are the respective ranks* of $X_{n1}, \ldots, X_{nn}$, and, for each n, $X_{n1}, \ldots, X_{nn}$ are mutually independent, continuously distributed random variables. (For details, see the papers cited.) On the asymptotic normality of linear ranks statistics*, see also refs. 20 and 22. Compare also the end of the following section. Related results on multivariate linear rank statistics have been obtained in Ruymgaart and van Zuijlen [37] and the papers there cited.

Another class of statistics whose asymptotic normality can be proved with the help of Hájek's lemma are the U-statistics*,

$$U_n = \frac{1}{\binom{n}{m}} \sum_{1 \leqslant j_1 < \cdots < j_m \leqslant n} f(X_{j_1}, \ldots, X_{j_m}),$$

where m is a fixed integer, $n \geqslant m$, the X_j are mutually independent random variables, and f is a real-valued function, symmetric in its m arguments; see ref. 24. A Berry–Esseen type bound for U-statistics* is derived in ref. 7.

Linear combinations of functions of order statistics* are asymptotically normal under general conditions; see Ruymgaart and van Zuijlen [36] and other work there cited.

There are statistics T_n which satisfy the conditions of Hájek's lemma and are asymptotically normally distributed, but whose asymptotic normality cannot be established by means of Theorem 8. A simple example is

$$T_n = X_1X_2 + X_2X_3 + \cdots + X_{n-1}X_n,$$

where the X_j are i.i.d. with a finite second moment. This is a special case of a sum of 1-dependent random variables; see the following section.

On the asymptotic normality of the sum of a random number of independent random variables (which is of interest in sequential analysis*), see ref. 8.

ASYMPTOTIC NORMALITY OF SUMS OF DEPENDENT RANDOM VARIABLES

Sums of independent random variables are asymptotically normal under general conditions. We may expect that the asymptotic normality will be preserved if the summands are allowed to be weakly dependent* in a suitable sense.

One way of expressing weak dependence

is in terms of conditional expectations. For example, let $X_1, X_2, \ldots$ be a sequence of (possibly dependent) random variables and let $S_n = X_1 + \cdots + X_n$. Suppose that the Liapunov condition (3) with $N = n$, $X_{Nj} = X_j$ is satisfied and that $EX_j = 0$. In ref. 29, Sec. 31, it is shown that if, in addition, the conditional moments $E[X_j \mid S_{j-1}]$ and $E[X_j^2 \mid S_{j-1}]$ differ sufficiently little (in a specified sense) from the corresponding unconditional moments, then S_n is asymptotically normal. For other, related results, see ref. 29.

Dvoretzky [15] has shown that sums of dependent random variables are asymptotically normal under conditions such as those in the central limit theorems for sums of independent random variables (e.g., Theorem 2), except that quantities such as means, and the like, are replaced by conditional means, and the like, the conditioning being relative to the preceding sum.

Another notion of weak dependence that has proved fruitful is the following. For simplicity we restrict ourselves to stationary sequences (X_n), so that, for all n, the joint distribution of $X_{h+1}, \ldots, X_{h+n}$ does not depend on h. A stationary sequence (X_n) is said to satisfy the strong mixing condition if there are numbers $d(r)$ converging to 0 as $r \to \infty$ such that

$$|\Pr[A \cap B] - \Pr[A]\Pr[B]| \leqslant d(n-m)$$

for any events A and B determined by conditions on the random variables X_k, $k \leqslant m$ and X_k, $k \geqslant n$, respectively, and for all m, n $(m < n)$.

Rosenblatt [34] has shown that the partial sums $X_1 + \cdots + X_n$ of a stationary sequence satisfying the strong mixing condition are asymptotically normal under conditions on some of their moments. For other sufficient conditions, see ref. 27.

A simple example of a sequence satisfying the strong mixing condition is an m-dependent sequence. The sequence $(X_n, n \geqslant 1)$ is said to be m-dependent if for all integers $1 \leqslant r \leqslant s < t \leqslant u$ the random vectors $(X_r, \ldots, X_s)$ and $(X_t, \ldots, X_u)$ are independent whenever $t - s > m$. A central limit theorem for sums of m-dependent random variables was proved in ref. 26. An improved version is due to Orey [30]. On Berry–Esseen type bounds for sums of m-dependent random variables, see Shergin [38].

Sums of m-dependent random variables and U-statistics have the feature in common that some subsets of the summands are mutually independent. A central limit theorem for more general sums of this type is due to Godwin and Zaremba [19].

For a central limit theorem for Markov chains* under conditions related to strong mixing, see Rosenblatt [35]. Sums of martingale differences are asymptotically normal under appropriate conditions; see, e.g., ref. 8.

Finally, we mention some of the so-called combinatorial central limit theorems, which have uses in sampling from a finite population* and in rank statistics. Let the random vector $(R_{n1}, \ldots, R_{nn})$ be uniformly distributed on the $n!$ permutations of the integers $1, \ldots, n$, and let $a_n(j)$, $b_n(j)$, $j = 1, \ldots, n$, be real numbers. Then the sums

$$\sum_{j=1}^{n} a_n(j) b_n(R_{nj})$$

are asymptotically normal under certain conditions on the $a_n(j)$, $b_n(j)$; see ref. 20. A similar result [25] holds for sums of the form

$$\sum_{j=1}^{n} a_n(j, R_{nj}).$$

FUNCTIONAL CENTRAL LIMIT THEOREMS

Functional central limit theorems form a far-reaching extension of the classical central limit theorems. We confine ourselves to a brief description of some typical results in this area.

Let $X_1, X_2, \ldots$ be i.i.d. random variables with mean 0 and variance 1. Let $S_0 = 0$, $S_n = X_1 + \cdots + X_n$, $n \geqslant 1$, and define for $0 \leqslant t \leqslant 1$

$$Y_n(t) = n^{-1/2} S_{[nt]} + n^{-1/2}(nt - [nt]) X_{[nt]+1},$$

where $[nt]$ is the largest integer $\leqslant nt$. Thus for given values of $n, X_1, \ldots, X_n$, $Y_n(t)$ is a continuous, piecewise linear function of t such that $Y_n(j/n) = n^{-1/2}S_j$ for $j = 0, 1, \ldots, n$.

Now let $W(t)$, $0 \leqslant t \leqslant 1$, be the standard Brownian motion* process (Wiener process*) on $[0, 1]$. Thus for each fixed $t \in (0, 1]$ the random variable $W(t)$ is normally distributed with mean 0 and variance t ($W(0) = 0$), and for any finitely many points $t_1 < t_2 < \cdots < t_k$ in $[0, 1]$ the increments $W(t_2) - W(t_1), W(t_3) - W(t_2), \ldots, W(t_k) - W(t_{k-1})$ are mutually independent. Each increment $W(t_j) - W(t_{j-1})$ is normally distributed with mean 0 and variance $t_j - t_{j-1}$. These facts determine the joint (normal) distribution of $W(t_1), \ldots, W(t_k)$. It is known that the random function $W(t)$, $0 \leqslant t \leqslant 1$, is continuous with probability 1.

By a theorem of Donsker [11] the random functions Y_n converge in distribution*, as $n \to \infty$, to the random function W. The exact meaning of this statement is explained, e.g., in Billingsley [6a]. *See also* CONVERGENCE OF DISTRIBUTIONS. It has the important implication that for a large class of functionals $h(f)$ of a continuous function $f(t)$, $0 \leqslant t \leqslant 1$, the distributions of the random variables $h(Y_n)$ converge to that of $h(W)$. A trivial example is $h(f) = f(1)$. The implication that the distributions of $Y_n(1) = n^{-1/2}S_n$ converge to that of $W(1)$ is essentially equivalent to the central limit theorem, Theorem 1. A more interesting functional to which Donsker's theorem applies is $h(f) = \max_{0 \leqslant t \leqslant 1} f(t)$. Since $\max_{0 \leqslant t \leqslant 1} Y_n(t) = n^{-1/2}\max(0, S_1, \ldots, S_n)$, Donsker's theorem implies that

$$\lim_{n\to\infty} \Pr\left[n^{-1/2}\max(0, S_1, \ldots, S_n) \leqslant x\right] = \Pr\left[\max_{0 \leqslant t \leqslant 1} W(t) \leqslant x\right]. \tag{8}$$

A proof of Donsker's theorem can be found in Billingsley [6a], where also other applications of the theorem are discussed and other similar theorems are proved.

Donsker's theorem and theorems of a similar type are called functional central limit theorems.

For the limit in (8) we have

$$\Pr\left[\max_{0 \leqslant t \leqslant 1} W(t) \leqslant x\right] = \max(2\Phi(x) - 1, 0). \tag{9}$$

This can be proved from the properties of the Wiener process, or by applying the so-called invariance principle (not to be confused with the invariance principle* in statistical inference*). Donsker's theorem, just as the central limit theorem with $a = 0$ and $\sigma^2 = 1$, assumes only that the X_n are i.i.d. with mean 0 and variance 1. Thus in either theorem the limit is invariant in this class of distributions of X_n. Once it is known that the limit (8) exists, it can be evaluated directly by choosing the distribution of X_n in a convenient way; for details, see Billingsley [6a]. The idea of the invariance principle was first conceived by Erdös and Kac [16].

We conclude with another functional central limit theorem. Let $X_1, X_2, \ldots$ be i.i.d. random variables with common distribution function $F(t)$. Let $F_n(t)$ be the empirical distribution function* corresponding to the sample $X_1, \ldots, X_n$. Define the random function $Z_n(t)$, t real, by

$$Z_n(t) = n^{1/2}(F_n(t) - F(t)).$$

First suppose that the X_n are uniformly distributed* with $F(t) = t$, $0 \leqslant t \leqslant 1$. In this case $F_n(t) - F(t) = 0$ outside of $[0, 1]$, and we may restrict t to the interval $[0, 1]$.

Let $W^0(t)$, $0 \leqslant t \leqslant 1$, be the Brownian* bridge process. This is the Gaussian process* on $[0, 1]$ whose distribution is specified by the requirements

$$EW^0(t) = 0,$$

$$EW^0(s)W^0(t) = \min(s, t) - st.$$

In the present case (F uniform) the random functions Z_n converge in distribution to the random function W^0, in a similar sense as the convergence of Y_n to W; see Donsker [12] or Billingsley [6a]. (An important difference is that the functions Z_n are not continuous as the Y_n are.) One implication is that

$$\lim_{n\to\infty} \Pr\left[\sup_t n^{1/2}|F_n(t) - F(t)| \leqslant x\right] = \Pr\left[\sup_t |W^0(t)| \leqslant x\right]. \tag{10}$$

Earlier, Kolmogorov [28] proved, by a different method, that the limit in (10) equals

$$1 - 2\sum_{k=1}^{\infty} (-1)^{k+1} e^{-2k^2x^2} \qquad (11)$$

for $x > 0$. Thus the probability on the right of (10) is equal to (11). The present approach to deriving results such as (10) was heuristically described by Doob [13] and made rigorous by Donsker [12].

The case where $F(t)$ is an arbitrary continuous distribution function can be reduced to the uniform case (by noting that $X_n' = F(X_n)$ is uniformly distributed on [0, 1]), and (10) remains valid. For the case where $F(t)$ is any distribution function on [0, 1], see Billingsley [6a].

The general foundations underlying the functional central limit theorems were laid by Prohorov [33] and Skorohod [39] and are expounded in the books of Billingsley [6a, 6b]. See also Parthasarathy [31] and Loève [29, Chap. 13].

References

[1] Bahr, B. von (1965). *Ann. Math. Statist.*, **36**, 808–818.

[2] Beek, P. van (1972). *Zeit. Wahrscheinlichkeitsth.*, **23**, 187–196.

[3] Bernstein, S. N. (1939). *Dokl. Akad. Nauk SSSR* (Compt. rend.), **24**, 3–8.

[4] Berry, A. C. (1941). *Trans. Amer. Math. Soc.*, **49**, 122–136.

[5] Bhattacharya, R. N. and Ranga Rao, R. (1976). *Normal Approximation and Asymptotic Expansions*. Wiley, New York. (Thorough treatment of normal approximations and asymptotic expansions of distributions of sums of independent random variables and random vectors, with emphasis on error bounds.)

[6a] Billingsley, P. (1968). *Convergence of Probability Measures*. Wiley, New York.

[6b] Billingsley, P. (1971). *Weak Convergence of Measures*. SIAM, Philadelphia.

[7] Callaert, H. and Janssen, P. (1978). *Ann. Statist.*, **6**, 417–421.

[8] Chow, Y.-S. and Teicher, H. (1978). *Probability Theory: Independence, Interchangeability, Martingales*. Springer-Verlag, New York. (Includes careful proofs of representative results on asymptotic normality.)

[9] Cramér, H. (1946). *Mathematical Methods of Statistics*. Princeton University Press, Princeton, N.J. (A classic text, still unsurpassed.)

[10] Cramér, H. (1970). *Random Variables and Probability Distributions*, 3rd ed. Cambridge Tracts in Mathematics and Mathematical Physics No. 36. Cambridge University Press, Cambridge (first ed., 1937). (Concise monograph in the classical vein.)

[11] Donsker, M. (1951). An Invariance Principle for Certain Probability Limit Theorems. *Mem. Amer. Math. Soc. No. 6*.

[12] Donsker, M. (1952). *Ann. Math. Statist.*, **23**, 277–281.

[13] Doob, J. L. (1949). *Ann. Math. Statist.*, **20**, 393–403.

[14] Dupač, V. and Hájek, J. (1969). *Ann. Math. Statist.*, **40**, 1992–2017.

[15] Dvoretzky, A. (1972). *Proc. 6th Berkeley Symp. Math. Statist. Prob.*, Vol. 2. University of California Press, Berkeley, Calif., pp. 513–535.

[16] Erdös, P. and Kac, M. (1946). *Bull. Amer. Math. Soc.*, **52**, 292–302.

[17] Esseen, C.-G. (1945). *Acta Math.*, **77**, 1–125. (A fundamental paper on asymptotic normality.)

[18] Gnedenko, B. V. and Kolmogorov, A. N. (1968). *Limit Distributions for Sums of Independent Random Variables*, rev. ed. (Translated from the Russian by K. L. Chung.) Addison-Wesley, Reading, Mass. (A classic monograph on the subject of the title.)

[19] Godwin, H. J. and Zaremba, S. K. (1961). *Ann. Math. Statist.*, **32**, 677–686.

[20] Hájek, J. (1961). *Ann. Math. Statist.*, **32**, 501–523.

[21] Hájek, J. (1968). *Ann. Math. Statist.*, **39**, 325–346.

[22] Hájek, J. and Šidák, Z. (1967). *Theory of Rank Tests*. Academic Press, New York. (Includes results on asymptotic normality of rank statistics.)

[23] Hall, P. (1978). *J. Aust. Math. Soc. A*, **25**, 250–256.

[24] Hoeffding, W. (1948). *Ann. Math. Statist.*, **19**, 293–325. (On *U*-statistics.)

[25] Hoeffding, W. (1951). *Ann. Math. Statist.*, **22**, 558–566.

[26] Hoeffding, W. and Robbins, R. (1948). *Duke Math. J.*, **15**, 773–780. (On sums of *m*-dependent random variables.)

[27] Ibragimov, I. A. and Linnik, Yu. V. (1971). *Independent and Stationary Sequences of Random Variables*. (Translated from the Russian and edited by J. F. C. Kingman.) Wolters-Noordhoff, Groningen. (Includes central limit theorems for sums of stationary random variables.)

[28] Kolmogorov, A. N. (1933). *G. Ist. Att.*, **4**, 83–91.

[29] Loève, M. (1977). *Probability Theory*, 4th ed., Vols. 1 and 2. Springer-Verlag, New York. (Includes treatment of sums of independent and dependent random variables.)

[30] Orey, S. (1958). *Duke Math. J.*, **25**, 543–546.

[31] Parthasarathy, K. R. (1967). *Probability Measures on Metric Spaces*. Academic Press, New York.

[32] Petrov, V. V. (1975). *Sums of Independent Random Variables*. (Translated from the Russian by A. A. Brown.) Springer-Verlag, New York. (Contains a wealth of information on the subject of the title, with proofs of the more important results.)

[33] Prohorov, Yu. V. (1956). *Theory Prob. Appl.*, **1**, 157–214.

[34] Rosenblatt, M. (1956). *Proc. Natl. Acad. Sci. USA*, **42**, 43–47.

[35] Rosenblatt, M. (1971). *Markov Processes: Structure and Asymptotic Behavior*. Springer-Verlag, New York. (Contains a chapter on the central limit theorem for Markov processes.)

[36] Ruymgaart, F. H. and van Zuijlen, M. C. A. (1977). *Ned. Akad. Wetensch. Proc. A*, **80**, 432–447.

[37] Ruymgaart, F. H. and van Zuijlen, M. C. A. (1978). *Ann. Statist.*, **6**, 588–602.

[38] Shergin, V. V. (1979). *Teor. Veroyatn. Ee Primen.*, **24**, 781–794. (In Russian. English translation to appear in *Theory Prob. Appl.*, **24**, 782–796.)

[39] Skorohod, A. V. (1956). *Theory Prob. Appl.*, **1**, 261–290.

[40] Uspensky, J. V. (1937). *Introduction to Mathematical Probability*. McGraw-Hill, New York. (This early text includes much material not found in other books.)

[41] Zaremba, S. K. (1958). *Math. Zeit.*, **69**, 295–298.

(CONVERGENCE OF DISTRIBUTIONS
ERGODIC THEORY
LIMIT THEOREMS
LIMIT THEOREMS, CENTRAL)

W. HOEFFDING

ATTRIBUTE

In statistical usage this term often has the connotation "nonquantitative." It is applied to characteristics that are not easily expressed in numerical terms: e.g., temperament, taste, and species. The term *qualitative character* is used synonymously.

The *theory of attributes* (see Chapters 4 to 6 of Yule and Kendall [1]) is mainly concerned with analysis of contingency tables* and categorical data*.

Reference

[1] Yule, G. U. and Kendall, M. G. (1950). *Introduction to the Theory of Statistics*, 14th ed. Hafner, New York, Charles Griffin, London. (The early editions of this work, by G. U. Yule alone, contain a considerably longer discussion of attributes.)

AUSTRALIAN JOURNAL OF STATISTICS

The *Australian Journal of Statistics* was founded by the (then) Statistical Society of New South Wales in 1959. From the outset it has appeared three times a year, the three issues constituting a volume. The size of the *Journal* has grown from 102 pages in Vol. 1 (1959) and 132 pages in Vol. 2 (1960) to 178 pages in Vol. 11 (1969), 231 pages in Vol. 15 (1973), and 287 pages in Vol. 20 (1978). The founding editor of the *Journal* was H. O. Lancaster, who served from 1959–1971. He was followed by C. R. Heathcote (1971–1973), C. C. Heyde (1973–1978), and E. J. Williams (1978–). The current editorial address is: The Editor, *Australian Journal of Statistics*, Department of Statistics, University of Melbourne, Parkville, Vic. 3052, Australia.

The editorial policy of the journal has changed little since its foundation. It is aimed to achieve a balance between theoretical and applied articles in the following areas: (1) mathematical statistics, econometrics, and probability theory; (2) new applications of established statistical methods; (3) applications of newly developed methods; (4) case histories of interesting practical applications; (5) studies of concepts (particularly in economic and social fields) defined in terms suitable for statistical measurement; (6) sources and applications of Australian statistical data; and (7) matters of general interest, such as surveys of the applications of statistics in broad fields. No ranking is implied in this list. Contributions may be in the form of articles, short notes, or letters to the editor. The journal also publishes book reviews, in the form of both extensive critical reviews and short book notices. A news and notes section regularly appeared in the journal until 1977, but this function has been taken over by a separate publication, the *Statistical Society of Australia Newsletter*, which first appeared in May 1977.

An international perspective and coverage is intended for the journal and contributions from nonmembers of the Statistical Society of Australia have always been welcomed. All submissions are treated equally on their scientific merits within their broad area. All papers are refereed, normally by one referee, and in many cases referees outside Australia are used. The majority of submissions come from outside Australia. Since 1975 the submissions from Australia have fluctuated between 25 and 40% of the total. Correspondingly, of the papers finally published, the proportion from Australia has fluctuated between 35 and 60% of the total. The contents of a recent issue (Vol. 20, No. 2, August 1978) included: "A Component of Variance due to Competition" by W. T. Federer and O. O. Ladipo; "A Randomization Problem in Forming Designs with Superimposed Treatments" by D. A. Preece, R. A. Bailey, and H. D. Patterson; "On the Relationship Between the Optimal Extrapolative and the Expectations Generating Methods" by J. McDonald; "Notes on the Australian Occupational Classification as a Social Science Research Tool" by P. McDonnell, F. Lancaster Jones, and P. Duncan-Jones; "From Social Statistics to Social Indicators" by R. V. Horn; "Asymptotically Equivalent Forms of Hodges–Lehmann Type Estimates in Regression" By B. M. Brown and D. G. Kildea: "A Linear Model with Errors Lacking a Variance II" by R. L. Chambers and C. R. Heathcote; "On the Expected Values of Record Values" by H. N. Nagaraja; and "On the Bivariate Triangular Distribution" by R. C. Griffiths.

At the time of establishment of the journal, the Statistical Society of New South Wales, based in Sydney and founded in 1947, was the only society of its kind in Australia. It assumed the responsibility for starting the journal, which, as its name implies, was intended to serve the statistical profession in Australia. The then president of the Statistical Society of New South Wales, P. C. Wickens, wrote in a foreword to the first issue of the journal: "It is hoped . . . that it will not be long before statistical societies will be firmly established in other States, and when this occurs it will undoubtedly be necessary to reconsider the management of the Journal."

The hopes expressed in this statement were not long in coming to fruition. The Statistical Society of Canberra was formed in 1961. In October 1962 the Statistical Society of Australia was formed by the amalgamation of the New South Wales and Canberra Societies, which then became branches of the main society. At this stage the journal became the responsibility of the new society, but its editor and editorial policy remained unchanged. Further branches of the society were later formed in Victoria (1964), Western Australia (1964), and South Australia (1967). In 1976, responsibility for the journal was assumed by the Australian Statistical Publishing Association Inc., whose membership is coterminous with membership of the Central Council of the Statistical Society of Australia.

C. C. HEYDE

AUTOMATIC INTERACTION DETECTION (AID) TECHNIQUES

Automatic interaction detection (AID) is a technique for analyzing a large quantity of data (whose number of cases is typically hundreds or thousands) by subdividing it into disjoint exhaustive subsets so as best to "explain" a dependent variable on the basis of a given set of categorized predictors. Although first noted under a different name in 1959 [1], the current computer versions were introduced by a series of papers dating from 1963, starting with one by J. N. Morgan and J. A. Sonquist [8]. Since its inception AID has grown in popularity and found application (and misapplication) in many applied fields. It can be used as an end analysis in itself, or as a prior screening device to sift the variables and draw attention to certain interactions for specific inclusion in subsequent analysis by other methods.

All AID techniques (normally implemented as computer programs) operate in a stepwise manner, first subdividing the total data base according to one of the predictors* (chosen by the algorithm to maximize some given criterion), then reexamining separately and subdividing each of the groups formed by the initial subdivision, and continuing in a like manner on each new subgroup formed until some stopping criterion is reached. Some versions have a "look-ahead" feature which allows possible subdivisions of subdivisions to be examined before the primary subdivision is effected—a procedure that may be theoretically desirable but is currently not cost-effective or usually worthwhile for the majority of data bases met in practice. This results in a tree-like structure called a dendrogram* in AID literature.

The "automatic" of AID refers to the computer program making all the decisions as to which predictor to use when, and how. This attribute can be countermanded by the user in some versions of AID. The "interaction" in AID refers to its design property that each subgroup formed in the analysis is treated *individually*. Thus the technique does not constrain itself to producing "additive" or "symmetric" models, although these may result anyway.

DEPENDENT VARIABLE

The various types of AID are distinguished by the nature of the dependent variable. Standard AID [10] operates on one that is an ordinal scalar, while the multivariate extension to an ordinal vector is handled by MAID [3]. A nominal scalar-dependent variable can be analyzed by "chi-squared* AID" CHAID [6] or "theta AID" THAID [7]. The theta statistic of the latter technique is the total proportion of observations that belong to a modal category in an AID subdivision. Thus, theta is bounded below by d^{-1} for a d-category dependent variable, and bounded above by unity.

Table 1 Classical-Type Criteria for Various Dependent Variables

Type of Dependent Variable	Criterion
Ordinal scalar	t-test*, ANOVA*
Ordinal scalar and covariate	F-test*, ANOCOVA*
Nominal	Chi-square*
Ordinal vector	Hotelling's T^2*, MANOVA*
Ordinal vector and covariates	MANOCOVAR*

The criterion for a "good" subdivision that "explains" the data depends on the nature of the dependent variable. Table 1 lists the possible types of dependent variable together with the appropriate classical-type criterion for each one. The criteria are all interrelated in that the suitable special case of each type (e.g., ordinal vector of dimension 1, a dichotomous nominal variable) reduces to one of the others in the list. Other criteria have been proposed (e.g., THAID above), but as they are dissimilar to the classical criteria (even asymptotically), they are in need of theoretical backing.

Superficially, AID bears resemblence to stepwise regression* both in intent and procedure. A comparison of AID with this and other statistical techniques is given in ref. 10.

PREDICTORS

The categorized predictors are used to determine the possible subdivisions of the data at any one stage, and hence result in the subdivisions being meaningful in that a particular subgroup can be labeled according to which categories of which predictors define the subgroup. Anything from 2 to 100 predictors are typically used in a single analysis (depending also on the size of the data base), each predictor having from 2 up to about 10 categories (depending on type), although predictors with more categories have appeared in the literature.

The categories of each predictor are grouped by AID to define the subdivisions (in many versions this reduction is *always* to two subdivisions—i.e., a binary split), the allowable grouping determined by the type of predictor. These types, which carry names peculiar to AID, include monotonic (ordered categories in which only adjacent categories may be grouped), floating (monotonic plus an additional category that may be grouped with any of the others—usually used for missing information or "don't-knows"), free (nominal categories with no restriction on grouping), interleaved (a number of monotonic categories used for combining different ordered scales; special cases include the aforementioned types), and cyclic (monotonic with "wraparound," useful for *U**- or inverted *U*-distributions*). Modern theory has concentrated on the behavior of AID on monotonic and free predictors, where not unsurprisingly the Bonferroni inequality* applied to a dichotomous predictor produces reasonable conservative significance tests of subdivision differences for the multicategory case. The theory of piecewise-regression* and clustering* are applicable to monotonic and free predictors, respectively, but unfortunately their asymptotic theories, which dwell on the increase in the number of categories (in our terminology), are not relevant, as the practical application of AID involves a small number of categories.

CRITICISM

Many criticisms have been leveled at AID-type techniques, the two major ones being that (1) AID produces too many subdivisions or idiosyncratic results (based on experiments with random data in which *no* subdivisions should be produced), and (2) the interpretation of AID results are often fallacious. The first criticism is based on earlier versions of AID that contained no valid hypothesis test as to whether the subdivisions produced were statistically significant. Further, the older versions of AID made no distinction between the types of the predictors involved when choosing one on which to base the subdivisions—thus introducing a bias in favor of using multicategory free predictors. Appropriate significance testing for standard AID has been developed in ref. 5. The latest versions of AID introduce testing as part and parcel of the technique, preferring to subdivide on the most "significant" rather than most "explanatory" predictor and thus remove both the aforementioned bias and the generation of nonsignificant splits. Ideally, a stringent significance level should be used to take into account the number of predictors present in the analysis, using e.g., Boole's (Waring) inequality*.

The second criticism is not so much an indictment of AID but rather of the ignorance displayed by some of its users. Added to this is the inadequate manner in which many current users present their AID results. Examples are completely omitting mention of sample sizes in the various subdivisions, the statistical significance* of these subdivisions, or appropriate auxiliary information that would allow a reader to repair this lack, at least in his or her own mind. The dendrogram resulting from an AID analysis is so appealing that a tendency has arisen for users to ignore the possible existence of competing predictors that would be revealed by an examination of the AID statistics produced for each predictor both before and after a subdivision. Certain apparent "interactions" quoted in the literature (see, e.g., the criticism [2] of the study [4] could well be spurious for this reason, and in certain circumstances, small changes in the data could cause different predictors to be selected by AID, with consequent different conclusions reached by a naive user. This fault is probably magnified by the "automatic" nature of AID mentioned above.

Finally, among other possible limitations,

the appropriateness of an AID analysis can be no better than the appropriateness of the splitting criterion as given in Table 1, and all the assumptions inherent in the theory behind these criteria and their properties naturally carry through to AID.

EXAMPLE

Figure 1 depicts the dendrogram* resulting from a CHAID analysis of 798 first-year commerce students enrolled at the University of the Witwatersrand. The dependent variable is the student's midyear mark (June 1979) in the course Mathematics and Statistics IB (M & S1B), classified into the three groups: passed ($> 49\%$), borderline (40–49%), and failed ($< 40\%$).

Table 2 gives details of the first stage of the analysis and indicates the predictive power and optimal grouping of the categories of each predictor. Predictors 2 and 5 (whether a local matriculant, and sex) are immediately discarded since there is no significant difference between the groups, as indicated by a significance level of $p = 1$. Predictors 3, 6, 7, and 8, while possessing an optimal grouping as shown on the right-hand side of the table, are nevertheless not considered significant. Note that predictor 6, which has nine categories optimally grouped into two groups, is ostensibly "significant" since the 2×3 contingency table so formed has $p = 0.0014$; however, taking into account the optimization that went into forming this table, a conservative estimate of the "significance" using a Bonferroni inequality* is $p = 0.35$—clearly *not* significant. Predictor 11 (midterm test mark) is clearly the best with conservative $p \leqslant 3.0 \times 10^{-49}$, but was included in the analysis for information only, and precluded from forming the basis of a subdivision of the data since the purpose of the analysis is to predict on the basis of information available before the commencement of the academic year. Predictor 1 is the best usable predictor ($p \leqslant 6.6 \times 10^{-16}$) and divides the data into four groups.

Figure 1 displays in detail the four-way subdivision of the total group from which it is clear that the passrate declines from 80% for the students with high (A and B) matriculation mathematics marks, through 59 and 46% for those with intermediate marks (C and D) down to 29% for the lower marks (E). Those students for whom no mark is available (14 such students coded "?", mainly foreign or older students) are interestingly enough grouped with the poorest students.

The analysis then continued with each of the four subgroups. Information on each of the predictors similar to that in Table 2 was produced, from which further details are available. The students with mathematics symbol C are further divided in Fig. 1 according to their pre-test, where it is seen that those who did not attend the test (it was not compulsory, and implies that the students were skipping classes even at this early stage) perform worse than those who did—no matter what their mark! The groups with mathematics marks D and E were each further subdivided according to their mark in the same course last year (clearly only available for repeat students).

Finally, in the lowest level in Fig. 1, there are two further subdivisions. Although they are technically significant ($p \leqslant 0.016$ and $p \leqslant 0.0065$) they should be considered marginal, since these levels make no allowance for the number of predictors (effectively 10) examined. (They do, however, take into account the type and number of categories in the predictor used to define the subdivision.) Nevertheless, considering the pretest, it is comforting to note the similar poor performance of students who did not attend it, on the two occasions where this predictor was used.

This is merely a brief summary of some of the information and conclusions available from an AID-type analysis. The secondary details concerning the predictive power and optimum grouping of the categories within each predictor for each of the subdivisions provide valuable insight to the structure of the data and interrelationships of the predictors.

```
                              *********
                              * N=798 *
                              *F: 30% *
                              *B: 24% *
                              *P: 47% *
                              *********
                                  |
                                  |P<6.6E-16
                     MATRICULATION | MATHEMATICS
   ---------------------------------------------------------------------
   |                  |                          |                     |
   |A,B               |C                         |D                    |?,E
*********          *********                 *********             *********
* N= 97 *          * N=162 *                 * N=264 *             * N=275 *
*F: 12% *          *F: 21% *                 *F: 28% *             *F: 42% *
*B:  8% *          *B: 20% *                 *B: 26% *             *B: 29% *
*P: 80% *          *P: 59% *                 *P: 46% *             *P: 29% *
*********          *********                 *********             *********
                       |                         |                     |
                       |P<.0003                  |P<9.8E-5             |P<2.1E-7
                   PRE | TEST               PREV | MARK           PREV | MARK
              ----------------             ------------         ---------------
              |              |             |          |         |             |
              |NO MARK       |F,B,P        |30+       |<30,?    |10+          |<10,?
          *********     *********     *********  ********* *********     *********
          * N= 22 *     * N=140 *     * N= 41 *  * N=223 * * N= 63 *     * N=212 *
          *F: 54% *     *F: 16% *     *F:  7% *  *F: 32% * *F: 14% *     *F: 50% *
          *B: 23% *     *B: 19% *     *B:  8% *  *B: 29% * *B: 30% *     *B: 29% *
          *P: 23% *     *P: 65% *     *P: 85% *  *P: 39% * *P: 56% *     *P: 21% *
          *********     *********     *********  ********* *********     *********
                            |                                                |
                            |P<.016                                          |P<.0065
                    TYPE OF | MATHEMATICS                                PRE | TEST
                      --------------                                   -------------
                      |            |                                   |           |
                      |HIGHER      |STANDARD                           |?,F        |B,P
                  *********    *********                         *********   *********
                  * N=111 *    * N= 29 *                         * N=181 *   * N= 31 *
                  *F: 11% *    *F: 34% *                         *F: 54% *   *F: 32% *
                  *B: 19% *    *B: 21% *                         *B: 30% *   *B: 23% *
                  *P: 70% *    *P: 45% *                         *P: 17% *   *P: 45% *
                  *********    *********                         *********   *********
F = FAIL
B = BORDERLINE
P = PASS

THE SIGNIFICANCE LEVELS QUOTED ARE CONSERVATIVE IN THAT THEY TAKE INTO ACCOUNT THE TYPE OF THE PREDICTOR
  USED AND THE NUMBER OF ITS CATEGORIES, BUT DOES NOT REFLECT THE POSSIBLE NUMBER OF PREDICTORS AVAILABLE
  AT THAT STAGE.
```

Figure 1 Dendrogram of CHAID analysis of first-year commerce students.

Table 2 First Stage of CHAID Analysis

Predictor	Number of Categories	Type	Ostensible Significance Level	Conservative Significance Level	Optimum Number of Groups	Optimum Grouping[a]
1. Matriculation mathematics (A, high mark; E, low)	6	Floating	3.0×10^{-17}	6.6×10^{-16}	4	AB C D ?E
2. Local matriculant? (Y, yes; N, no)	2	Monotonic	1.0	1.0	1	YN
3. Matriculation English (A, high mark; E, low)	6	Floating	0.018	0.40	3	?A B CDE
4. Year of matriculation (4, up to 1974; 5, 1975; 8, 1978)	6	Floating	4.1×10^{-5}	9.1×10^{-4}	3	8 7 ?654
5. Sex (M, male: F, female)	3	Floating	1.0	1.0	1	?MF
6. Course and rules registered under (G, general; BCom. full time; etc.	9	Free	0.0014	0.35	2	G/7C9 LOPT
7. Year of study (1, first; . . . ; 5, fifth or more)	6	Floating	0.048	0.43	2	1 23?45
8. Type of matriculation mathematics (H, higher; S, standard)	3	Floating	0.036	0.11	2	?H S
9. Previous M & S1B mark (?, not a repeat student; 1, first; etc.)	9	Floating	1.1×10^{-9}	1.4×10^{-8}	2	23456 ?78
10. Pre-test (F, failed; B, borderline; P, passed)	4	Floating	8.7×10^{-11}	4.3×10^{-10}	2	?F BP
11. First midterm (April) test (F, failed; B, borderline; P, passed)	4	Floating	6.1×10^{-50}	3.0×10^{-49}	3	?F B P

[a]The symbol "?" refers to missing information and is the "floating" category.

STATE OF THE ART AND FUTURE DEVELOPMENTS

The underlying theory behind valid hypothesis testing in AID is still embryonic. At present only standard AID has provision for a covariate; the other versions have yet to be so extended. Computer installations with powerful interactive terminals or personalized computer systems do not need the "automatic" decision making of AID. Instead, they could offer the researcher the opportunity to introduce additional background information and take various decisions dynamically along the lines of ref. 9. Such a feature is still to be implemented in AID.

References

[1] Belson, W. (1959). *Appl. Statist.*, **8**, 65–75.

[2] Doyle, P. (1973). *Operat. Res. Quart.*, **24**, 465–467.

[3] Gillo, M. W. and Shelly, M. W. (1974). *J. Amer. Statist. Ass.*, **69**, 646–653. (MAID.)

[4] Heald, G. I. (1972). *Operat. Res. Quart.*, **23**, 445–457.

[5] Kass, G. V. (1975). *Appl. Statist.*, **24**, 178–189. (Theory and additional references to AID.)

[6] Kass, G. V. (1980). *Appl. Statist.*, **29**, 119–127. (CHAID.)

[7] Morgan, J. N. and Messenger, R. C. (1973). THAID—a sequential analysis program for the analysis of nominal scale dependent variables. ISR, University of Michigan, Ann Arbor, Mich. (THAID.)

[8] Morgan, J. N. and Sonquist, J. A. (1963). *J. Amer. Statist. Ass.*, **58**, 415–434.

[9] Press, L. I., Rogers, M. S. and Shure, G. H. (1969). *Behav. Sci.*, **14**, 364–370.

[10] Sonquist, J. A., Baker, E. L. and Morgan, J. N. (1971). Search for Structure (Alias-AID-III). ISR, University of Michigan, Ann Arbor, Mich. (Mainly a user manual, but contains background material, examples, and further references.)

(CLASSIFICATION TECHNIQUES
COMPUTERS AND STATISTICS
PREDICTION
STEPWISE REGRESSION)

G. V. KASS

AUTOREGRESSIVE–INTEGRATED MOVING AVERAGE (ARIMA) MODELS

An important class of models for describing a single time series* z_t is the class of autoregressive–moving average models* referred to as ARMA(p,q)* models.

$$(z_t - \mu) = \phi_1(z_{t-1} - \mu) + \dots + \phi_p(z_t - \mu) + a_t - \theta_1 a_{t-1} - \dots - \theta_q a_{t-q}, \tag{1}$$

where the notation in (1) implies that (a) z_t is the original time series, or some suitable nonlinear transformation of it (such as a logarithm or a square root); (b) z_t is a stationary* time series with a fixed mean μ; (c) a_t is a random residual series, which can also be interpreted as the series of one-step-ahead forecast errors; and (d) $\phi_1, \dots, \phi_p$, $\theta_1, \dots, \theta_q$, μ are parameters to be estimated from the data. Alternatively, autoregressive-moving average (ARMA) models* may be written in terms of the backward-shift operator* B, such that $B^j z_t = z_{t-j}$, $B^j a_t = a_{t-j}$, as follows:

$$(z_t - \mu) = \frac{1 - \theta_1 B - \dots - \theta_q B^q}{1 - \phi_1 B - \dots - \phi_p B^p} a_t \tag{2}$$

$$= \frac{\theta(B)}{\phi(B)} a_t \tag{3}$$

Thus the ARMA(p,q) model represents the time series, or a suitable nonlinear transformation, as the output from a linear filter whose input is a random series and whose *transfer function** is a rational function of the backward-shift operator B.

The model (1) is not of immediate practical use because very few real-world time series are stationary time series in statistical equilibrium about a fixed mean μ. Instead, they are characterized by random changes in their level, slope, etc., and by the presence of seasonal patterns which also evolve with time. Traditional methods of handling such diverse behavior involve the decomposition of the time series into a "trend"*, a "seasonal component"*, and a "residual component." After removal of the trend and seasonal component, it is customary to describe the residual component by means of a stationary ARMA(p, q) model of the form (1). Such an approach suffers from the following disadvantages: (a) it is arbitrary as to what is called a trend and a seasonal component; (b) removal of the trend and seasonal component introduces additional autocorrelation into the residual component; and (c) the assumptions normally made about the behavior of the trend and seasonal component are unrealistic.

To overcome these difficulties a new class of models, called autoregressive–integrated moving average models, referred to as ARIMA models, has been developed to describe, *under the umbrella of one model*, trends, seasonality, and residual random behavior [1]. Moreoever, such models for describing nonstationary* time series contain flexible structures which allow the trend and seasonal component to be nondeterministic, i.e., their statistical properties evolve in time. In addition, iterative methods have been developed [1] for identifying (or specifying), estimating* (or fitting*) and checking (or criticizing) such models given the data.

NONSEASONAL ARIMA MODELS

Consider the first-order autoregressive model

$$(z_t - \mu) = \phi(z_{t-1} - \mu) + a_t \qquad (4)$$

or, in backward-shift-operator notation ($B^j z_t = z_{t-j}$,

$$(1 - \phi B)(z_t - \mu) = a_t,$$

which is stationary if $|\phi| < 1$. The solution of the difference equation* (4) may be written as the sum of the complementary function, i.e., the solution of $(1 - \phi B)(z_t - \mu) = 0$, and a particular integral, i.e., any function that satisfies (4). Relative to a time origin $t = 0$, the solution of (4) thus becomes

$$(z_t - \mu) = \phi^t(z_0 - \mu) + \sum_{j=1}^{t} \phi^{j-1} a_j. \qquad (5)$$

If $|\phi| > 1$, the first term in (5) dominates and the growth of the series is explosive. Although such explosive nonstationarity occurs in some situations (such as bacterial growth), for most practical situations it is convenient to work with a less severe form of nonstationarity. This can be achieved by setting $\phi = 1$ in (4) and (5), which then become

$$(1 - B)z_t = z_t - z_{t-1} = a_t, \qquad (6)$$

$$z_t = z_0 + \sum_{j=1}^{t} a_j, \qquad (7)$$

i.e., z_t is a random walk* model. More generally, we consider a nonstationary ARMA(p, q) model

$$\phi'(B)(z_t - \mu) = \theta(B)a_t, \qquad (8)$$

where $\phi'(B)$ is a nonstationary autoregressive operator. To prevent explosive nonstationarity, we impose the restriction that d of the factors of $\phi'(B)$ are unity, i.e., $\phi'(B) = \phi(B)(1 - B)^d$. Model (8) then becomes

$$\phi(B)(1 - B)^d z_t = \theta(B)a_t \qquad (9)$$

where $\phi(B)$ is a stationary autoregressive* operator and $\theta(B)$ is an invertible moving average* operator, as in a stationary autoregressive–moving average model. Since $(1 - B)z_t = z_t - z_{t-1} = \nabla z_t$, where ∇ is the backward difference operator, the model (9) can also be written

$$\phi(B)\nabla^d z_t = \theta(B)a_t \qquad (\nabla^0 = 1). \qquad (9)$$

Model (9) implies that whereas z_t is a nonstationary series, its dth difference $w_t = \nabla^d z_t$ is stationary and can be described by an autoregressive–moving average model. The model (9) is called an *autoregressive–integrated moving average model* or ARIMA(p, d, q) model, where p is the number of parameters in the autoregressive operator, d is the number of times that the series has to be differenced to induce stationarity, and q is the number of parameters in the moving average operator. Provided that the series does not contain seasonality, the ARIMA model (9) with small values of p, d, and q is capable of describing a wide range of practically occurring time series. When $d > 0$, the stationary series $w_t = \nabla^d z_t$ will usually have a zero mean. However, a useful generalization of (9) can be obtained by allowing w_t to have a nonzero mean, i.e.,

$$\phi(B)(\nabla z_t - \mu) = \theta(B)a_t. \quad (10)$$

With $d > 0$, model (10) is capable of describing a *deterministic* polynomial trend of degree d as well as a stochastic nonstationary component. For example, when $d = 1$, the model is capable of describing nonstationary stochastic behavior over and above an underlying linear growth rate.

SPECIAL CASES OF NONSEASONAL ARIMA(P, D, Q) MODELS

With $p = 0$, $d = 1$, $q = 1$, $\mu = 0$, model (10) becomes

$$\nabla z_t = (1 - \theta B)a_t \quad (11)$$

i.e., a nonstationary series whose first difference is stationary and can be represented as a first-order moving average model. Model (11) can be inverted to give

$$z_t = (1 - \theta)(z_{t-1} + \theta z_{t-2} + \theta^2 z_{t-3} + \cdots) + a_t. \quad (12)$$

Thus the one-step-ahead forecast of z_t from origin $(t - 1)$ is an exponentially weighted moving average of past values of the series. By solving the difference equation, model (11) may also be written

$$\begin{aligned} z_t &= a_t + (1 - \theta)(a_{t-1} + a_{t-2} + \cdots) \\ &= a_t + l_{t-1} \end{aligned} \quad (13)$$

Although the series has no fixed mean, at a given time it has a local level l_{t-1} which is updated from time $(t - 1)$ to time t according to

$$l_t = l_{t-1} + (1 - \theta)a_t.$$

Thus when the random shock a_t occurs, a proportion $(1 - \theta)a_t$ of it is absorbed into the "level" of the series and the remaining proportion θa_t is "lost" from the system.

With $p = 0$, $d = 1$, $q = 1$, $\mu \neq 0$, model (10) becomes

$$\nabla z_t - \mu = (1 - \theta B)a_t. \quad (14)$$

The solution of the difference equation (14) may be written as a complementary function (the solution of $\nabla z_t - \mu = 0$) and the particular integral (13), i.e.,

$$z_t = c + \mu t + a_t + (1 - \theta)(a_{t-1} + a_{t-2} + \cdots) \quad (15)$$

and thus contains a "deterministic drift" term.

With $p = 0$, $d = 2$, $q = 2$, $\mu = 0$, model (10) becomes

$$\nabla^2 z_t = (1 - \theta_1 B - \theta_2 B^2)a_t. \quad (16)$$

Thus z_t and ∇z_t are nonstationary series, and second-order differencing $\nabla^2 z_t$ is necessary to induce stationarity. It may be shown [1, p. 111] that model (16) implies that the series has a local "level" l_t and a local "slope" s_t which are updated from time $t - 1$ to time t by the new random shock a_t according to

$$\begin{aligned} l_t &= l_{t-1} + s_{t-1} + (1 + \theta_2)a_t, \\ s_t &= s_{t-1} + (1 - \theta_1 - \theta_2)a_t. \end{aligned}$$

SEASONAL ARIMA MODELS

One of the deficiencies in handling seasonal time series in the past has been the absence of parametric models to describe seasonal behavior. A new class of models [3] for describing seasonality as well as nonstationary trends can be obtained by modification of the nonseasonal ARIMA model (10). Suppose that data become available at monthly intervals and that they are set out in the form of a two-way table in which the columns denote months and the rows denote years. The series for a particular column, say March, may contain a trend but is not seasonal in its behavior. Hence it is reasonable to link the observation for March in this year to observations in previous Marches by an ARIMA(P, D, Q) model of the form (9):

$$\Phi(B^s)\nabla_s^D z_t = \Theta(B^s)\alpha_t, \quad (17)$$

where the autoregressive and moving average operators are now polynomials in B^s of degrees P and Q, respectively, and s is the seasonal period and equals 12 in this case. Also, the nonseasonal difference operator ∇ in (9) is replaced by the seasonal difference operator $\nabla_s z_t = z_t - z_{t-s}$ in (17) and ∇_s^D denotes the Dth seasonal difference. In general, the error terms α_t in models of the form (17) fitted to each month separately would not be random since the behavior of the series in March of this year will usually depend not only on what happened in previous Marches but also on the behavior of the series in February, January, etc., of this year. To describe this monthly dependence, we can use the nonseasonal ARIMA model

$$\phi(B)\nabla^d \alpha_t = \theta(B)a_t, \qquad (18)$$

where a_t is now a random series and $\phi(B)$ and $\theta(B)$ are polynomials in B of degrees p and q, respectively. Substituting (18) in (17), and allowing for the possibility that the differenced series may have a nonzero mean μ, we obtain the ARIMA$(p,d,q) \times (P, D, Q)$ multiplicative model

$$\begin{aligned} \phi_p(B)\Phi_P(B^s)(\nabla^d \nabla_s^D z_t - \mu) \\ = \theta_q(B)\Theta_Q(B^s)a_t, \end{aligned} \qquad (19)$$

where the subscripts on the operators denote the degrees of the polynomials involved.

In some cases, it may be better to work with a nonmultiplicative model in which the autoregressive operator, or the moving average operator, or both operators, cannot be factorized into a product of nonseasonal and seasonal operators, as in (19); for example, the right-hand side of (19) might take the form

$$(1 - \theta_1 B - \theta_{12} B^{12} - \theta_{13} B^{13})a_t.$$

Seasonal models of the form (19) may be fitted to data with a range of seasonal periods: e.g., daily data ($s = 7$), weekly data ($s = 12$), and quarterly data ($s = 4$). Moreover, several seasonal periods may occur simultaneously; e.g., hourly traffic data may display a cycle over a day ($s = 24$) and a further cycle over a week ($s = 168$). In such examples it may be necessary to add further seasonal autoregressive, moving average, and differencing operators to the model.

BUILDING ARIMA MODELS

Figure 1a shows part of a series consisting of the logarithms of the electricity consumption in one country. The series contains an upward trend and an annual cycle. ARIMA models of the form (19) may be fitted to data, using an iterative cycle of identification, estimation, and checking, as described below.

Initial analysis [4] suggested that to achieve a homoscedastic* distribution of the residuals a_t, it is necessary to apply a logarithmic transformation $\ln z_t$ to the data before fitting a model of the form (19). Alongside the plot of $\ln z_t$ shown in Fig. 1a is a plot of the autocorrelation function r_k of $\ln z_t$ as a function of the lag k. The autocorrelation function fails to damp out with the lag k and is indicative of nonstationarity [1]. Figure 1b shows the autocorrelation function of the nonseasonal difference $\nabla \ln z_t$. This autocorrelation function has peaks at 12, 24, 36, . . . , indicating nonstationarity with respect to the seasonal behavior and suggesting that further seasonal differencing is needed. Figure 1c shows the autocorrelation function of $\nabla\nabla_{12}\ln z_t$. This function contains no obvious trends, implying that the differenced and transformed series $w_t = \nabla\nabla_{12}\ln z_t$ is stationary. The next step is to arrive at an initial guess of the seasonal and nonseasonal autoregressive and moving average structure needed to explain the autocorrelation function of w_t. The autocorrelation functions of autoregressive–moving average models are characterized by a discrete number of spikes corresponding to the moving average part of the model and damped exponentials and/or damped sine waves corresponding to the autoregressive part of the model. The largest autocorrelations r_k in Figure 1c occurs at lags 1 and 12, suggesting

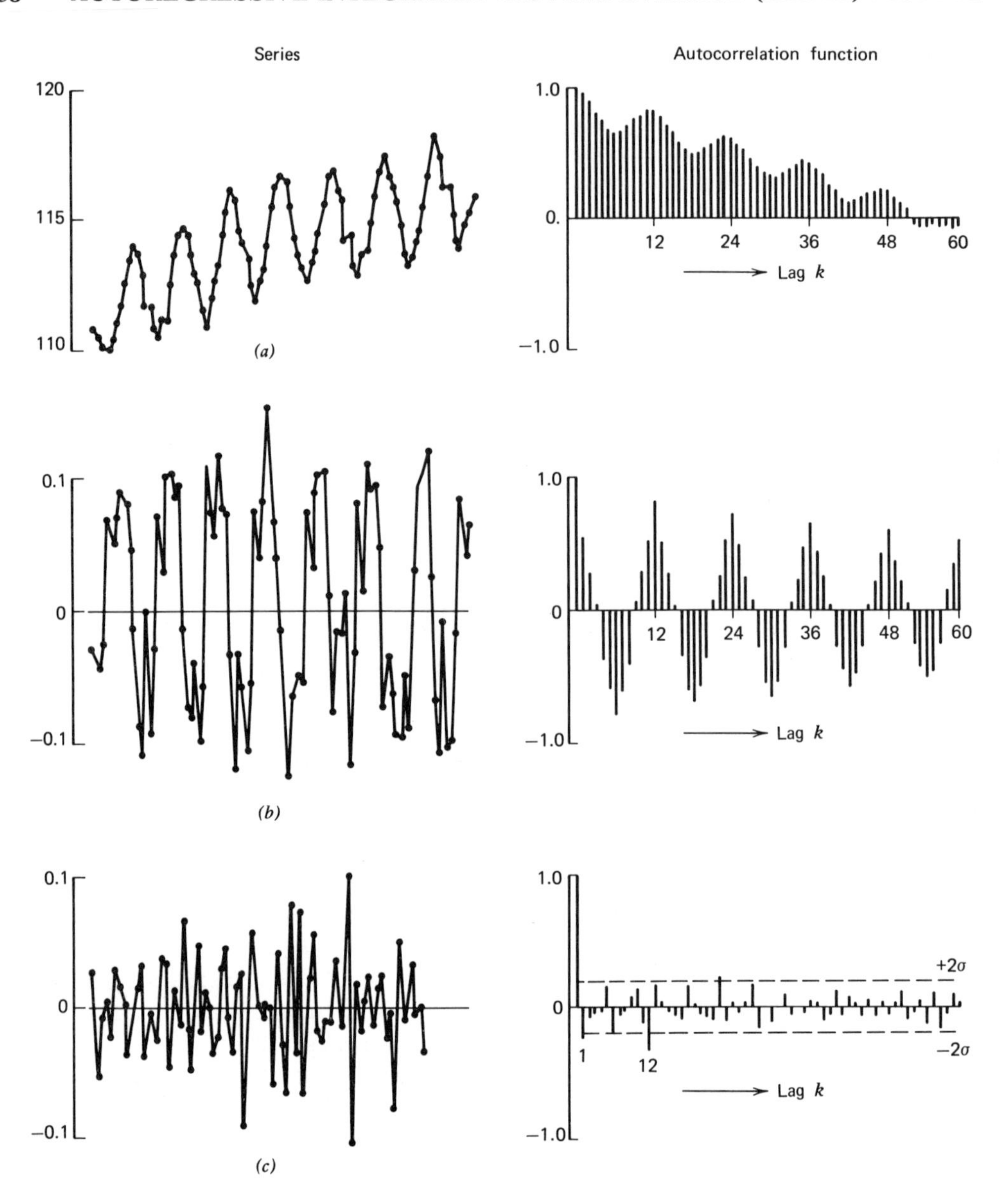

Figure 1 Various differences of the logarithms of national electricity consumption series, together with their corresponding autocorrelation functions: (*a*) $\ln Y_t$; (*b*) $\nabla \ln Y_t$; (*c*) $\nabla\nabla_{12} \ln Y_t$. Reproduced with the permission of GJP Publications from *Practical Experiences with Modeling and Forecasting Time Series* by Gwilym M. Jenkins.

an initial model of the form

$$\nabla\nabla_{12}\ln z_t = (1 - \theta B)(1 - \Theta B^{12})a_t, \quad (20)$$

where we may take as initial estimates of the parameters, $\theta = 0.30$ (based on $r_1 = -0.27$) and $\hat{\Theta} = 0.35$ (based on $r_{12} = -0.33$), using a procedure described in Box and Jenkins [1].

The initial model structure (20) may be fitted to the data by iterative calculation of the maximum likelihood* estimates, starting from the initial values given above (see ref. 1, pp. 269–284, for the likelihood* function of autoregressive–moving average models). The fitted model, based on $N = 96$ observations, was

$$\nabla\nabla_{12}\ln Y_t = \underset{\pm 0.08}{(1-0.73B)}\underset{\pm 0.05}{(1-0.83B^{12})}\,a_t \tag{21}$$

with estimated residual variance $\sigma_a^2 = 0.0006481 (\sigma_a = 0.0255)$. The $\pm$ values underneath the estimated parameters denote the 1-standard-error* limits.

Examination of the residuals a_t in (21) showed that none of the residuals was large compared with their standard deviation $\sigma_a = 0.0255$ and that 5 residuals out of 83 fell outside $\pm 2\sigma_a$, in reasonable accord with expectation. The largest autocorrelations $r_a(k)$ of the residuals a_t occurred at lags 6 and 24, suggesting some evidence of model inadequacy. However, further elaboration of the model revealed little improvement on model (21). Further details of how this model was built, and how it was elaborated to a transfer function* model relating electricity consumption to temperature, have been given by Jenkins [4].

Model (21) may be used to forecast future values z_{t+l} for each lead time $l = 1, 2, 3, \ldots$ from the current origin t by writing it at time $t + l$ in the form

$$\begin{aligned}\ln z_{t+l} - \ln z_{t+l-1} - \ln z_{t+l-12} + \ln z_{t+l-13} \\ = a_{t+l} - 0.73a_{t+l-1} - 0.83a_{t+l-12} \\ +0.61a_{t+l-13}\end{aligned} \tag{22}$$

and then taking conditional expectations at time t, bearing in mind that the conditional expectation of future values of the random series a_{t+l} for $l > 0$ are zero. For example, when $l = 1$, the one-step-ahead forecast $\hat{z}_t(1)$ can be calculated from

$$\begin{aligned}\ln \hat{z}_t(1) = \ln z_t + \ln z_{t-11} - \ln z_{t-12} \\ - 0.73a_t - 0.83a_{t-11} + 0.61a_{t-12},\end{aligned} \tag{23}$$

where $a_{t-j} = \ln z_{t-j} - \ln \hat{z}_{t-j-1}(1)$ for $j \geqslant 0$. Thus the forecasts for each lead time l can be generated recursively, together with the standard deviations of the forecast errors $e_t(l) = \ln \hat{z}_{t+l} - \ln z(l)$ (see ref. 1). In the example above, further improvements in forecasting accuracy could be expected by introducing into the model other variables which are related with electricity consumption: e.g., temperature, industrial production, price. Such *transfer function* models are discussed in Box and Jenkins [1] and in Jenkins [4].

MULTIVARIATE ARIMA MODELS

Univariate ARIMA models may be generalized to deal with mutual interaction between several nonstationary time series. To illustrate the possibilities, consider two time series z_{1t} and z_{2t}. First, nonlinear transformation and nonseasonal differencing may be needed to produce stationary time series

$$w_{1t} = \nabla^{d_1} z_{1t}, \qquad w_{2t} = \nabla^{d_2} z_{2t}. \tag{24}$$

Then it might be possible to describe the resulting stationary vector by a multivariate autoregressive–moving average model*

$$\begin{bmatrix}\phi_{11}(B) & \phi_{12}(B)\\ \phi_{21}(B) & \phi_{22}(B)\end{bmatrix}\begin{bmatrix}w_{1t}-\mu_1\\ w_{2t}-\mu_2\end{bmatrix} = \begin{bmatrix}\theta_{11}(B) & \theta_{12}(B)\\ \theta_{21}(B) & \theta_{22}(B)\end{bmatrix}\begin{bmatrix}a_{1t}\\ a_{2t}\end{bmatrix}, \tag{25}$$

where a_{1t} and a_{2t} are the one-step-ahead forecast errors or residuals for z_{1t} and z_{2t}, respectively. If the forecasts are to be optimal, a_{1t} must be a random series, a_{2t} a random series, and a_{1t} and a_{2t} mutually uncorrelated series except possibly at simultaneous times. The model defined by (24) and (25) is an example of an ARIMA($\mathbf{P}, \mathbf{d}, \mathbf{Q}$) model, where $\mathbf{P}$ is a matrix whose elements (p_{ij}) define the degrees of the polynomials $\phi_{ij}(B)$ in the autoregressive matrix, the vector $\mathbf{d}$ has elements d_i which define the degrees of differencing needed to induce stationarity of the time series*, and $\mathbf{Q}$ is a matrix whose elements (q_{ij}) define the degrees of the polynomials $\theta_{ij}(B)$ in the moving average matrix. The foregoing models may also be generalized to deal with seasonality [2, 4], and may be generalized by introducing explanatory variables* to explain the simultaneous behavior of the vector $\mathbf{z}_t$ of time series, leading to multiple output–multiple input transfer function models [4].

References

[1] Box, G. E. P. and Jenkins, G. M. (1970). *Time Series Analysis: Forecasting and Control.* Holden-Day, San Francisco (2nd ed., 1976).

[2] Box, G. E. P., Hillmer, S. C., and Tiao, G. C. (1976). *NBER Census Conf. Seasonal Time Ser.*, Washington, D. C.

[3] Box, G. E. P., Jenkins, G. M., and Bacon, D. W. (1967). In *Advanced Seminar on Spectral Analysis of Time Series, B. Harris, ed.*, Wiley, New York, pp. 271–311.

[4] Jenkins, G. M. (1979). *Practical Experiences with Modelling and Forecasting Time Series*. GJP Publications, St. Helier, N.J.

Further Reading

The basic theoretical properties of ARIMA models are given in *Time Series Analysis: Forecasting and Control* by G. E. P. Box and G. M. Jenkins (Holden-Day, San Francisco, 1970), together with practical procedures for identifying, fitting, and checking such models. Further accounts are given in *Applied Time Series Analysis for Managerial Forecasting* by C. R. Nelson (Holden-Day, San Francisco, 1973), *Time Series Analysis and Forecasting: The Box–Jenkins Approach* by O. D. Anderson (Butterworth, London, 1975), and *Forecasting Economic Time Series* by C. W. J. Granger and P. Newbold (Academic Press, New York, 1977). Theoretical and practical accounts of multivariate ARIMA models are given in *Practical Experiences with Modelling and Forecasting Time Series* by G. M. Jenkins (GJP Publications, St. Helier, N.J., 1979). Numerous papers in the field are published in *Statistical Literature* and are listed in *Current Index to Statistics** (CIS).

(AUTOREGRESSIVE–MOVING AVERAGE (ARMA) MODELS
FORECASTING
MODEL BUILDING
SEASONALITY
TIME SERIES
TRANSFER FUNCTIONS)

G. M. JENKINS

AUTOREGRESSIVE MODELS *See* AUTOREGRESSIVE–MOVING AVERAGE (ARMA) MODELS

AUTOREGRESSIVE–MOVING AVERAGE (ARMA) MODELS

Data occurring in the form of time series* occur in many branches of the physical sciences, social sciences, and engineering. Figure 1 shows an example of a single time series* in which the value z_t of a certain variable (the average number of spots on the sun's surface) is plotted against time t. Earlier approaches to analyzing time series were based on decomposing the variance of the series into components associated with different frequencies, based on Fourier analysis* and leading more recently to methods based on spectral analysis*.

Alternative historical approaches were based on building a model for the time series in the time domain. The main motivation for building such a model was to forecast future values of the time series given its past history. However, such a model could be used (a) to gain a better understanding of the mechanisms generating the series, (b) to smooth the random variation in the series, and (c) to allow for dependence in the series when the data were used for other statistical purposes, such as testing the differences between the means of two sets of data or relating one time series to another as in some form of regression analysis*. The first practically useful models for describing time series were the autoregressive models introduced by G. U. Yule and G. Walker (see below) and the moving average* models introduced by Slutsky, Wold, and others. Later, it was recognized that more general

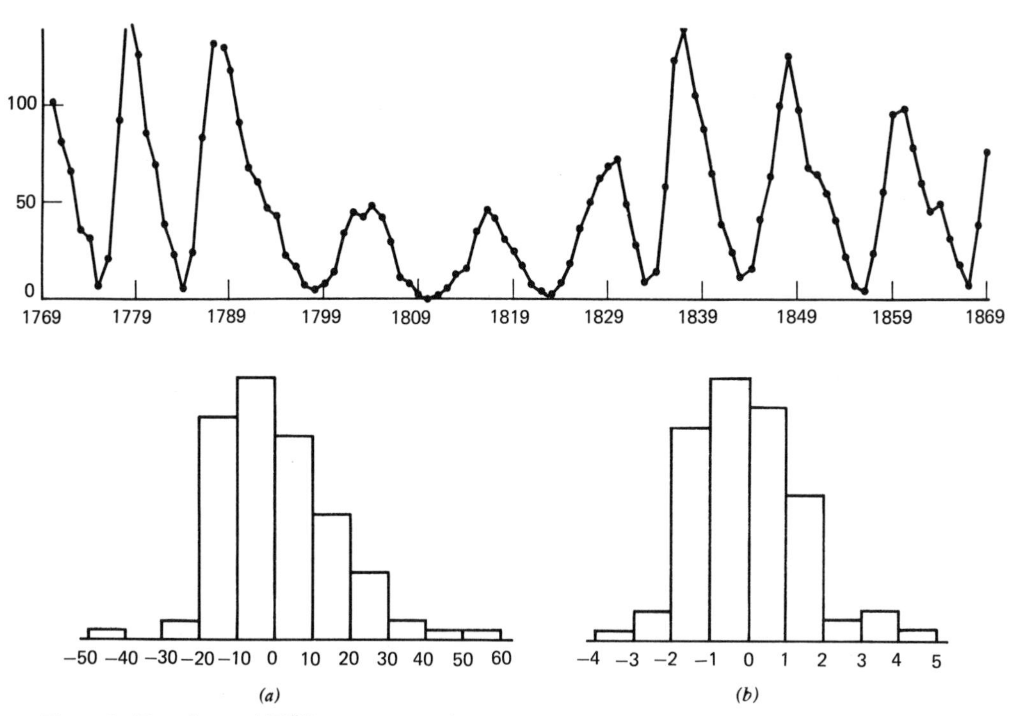

Figure 1 Plot of annual Wölfer sunspot numbers (1770–1869), together with histograms of residuals a_t from: (*a*) model $(z_t - 46.9) = 1.42(z_{t-1} - 46.9) - 0.73(z_{t-2} - 46.9) + a_t$; (*b*) model $(\sqrt{z_t} - 7.4) + 1.41(\sqrt{z_{t-1}} - 7.4) - 0.70(\sqrt{z_{t-2}} - 7.4) + a_t$.

structures could be obtained by combining autoregressive and moving average models, leading to autoregressive–moving average (ARMA) models.

GENERAL LINEAR MODEL*

When forecasting* an observed time series z_t from a knowledge of its past history, it is natural to think of the forecast as being obtained by applying a set of weights to past values of the time series. Thus, the one-step-ahead forecast of z_t made from *origin* $(t - 1)$ may be written

$$\text{forecast} = \pi_1 z_{t-1} + \pi_2 z_{t-2} + \pi_3 z_{t-3} + \cdots, \tag{1}$$

where π_j is the weight applied to the previous observation z_{t-j} in order to forecast z_t. When the future observation z_t comes to hand, it follows that

$$z_t = \text{forecast} + \text{forecast error} \tag{2}$$

Substituting for the forecast from (1) and denoting the forecast error by a_t, (2) becomes

$$z_t = \pi_1 z_{t-1} + \pi_2 z_{t-2} + \pi_3 z_{t-3} + \cdots + a_t. \tag{3}$$

If the forecast one step ahead of z_t is the best possible, then the forecast errors a_t, a_{t-1}, a_{t-2} ... should be a *random series**, or *white noise**. If not, it should be possible to forecast the forecast errors and add this forecast to forecast (1) to obtain a better forecast. Model (3), with a_t a random series, is called a linear model*. In practice, the forecast errors a_t may depend on the level of the series, in which case a better representation is obtained by using a nonlinear transformation of z_t in (3), such as a log or square-root transformation*. From now on it will be assumed that the notation used in (3) denotes a representation for z_t or a suitable nonlinear transformation of z_t chosen so as to make the forecast errors a_t homoscedastic*. Although the representation (3) provides a useful general way of modeling a time series, it suffers from the disadvantage that it contains a large (potentially infinite) number of weights or parameters π_i. Since it would be impossible to estimate very accurately such a large number of weights, a practical solution to time-series problems requires a more parsimonious representation, containing as few parameters as possible. Such economy in parameterization can be achieved using autoregressive and moving average models.

PURE AUTOREGRESSIVE MODELS

From now on it is assumed that the time series is *stationary**, i.e., that it is in statistical equilibrium about a fixed mean μ and that it possesses, among other properties, a constant variance and a covariance structure which depends only on the difference k (or lag) between two time points. Suppose also that the weights π_i applied to past observations in the representation (3) are zero beyond a certain point p. Then, writing the series as a deviation about its mean μ, (3) becomes

$$(z_t - \mu) = \phi_1(z_{t-1} - \mu) + \phi_2(z_{t-2} - \mu) + \cdots + \phi_p(z_{t-p} - \mu) + a_t, \quad (4)$$

where the finite set of weights or parameters ϕ_i may be estimated from the data. In words, (4) implies that the current deviation of the time series from its mean is a linear combination of the p previous deviations plus a random residual* a_t. The analogy between (4) and a multiple regression model* should be noted. Because the regressor variables in (4) are lagged values of the series itself and not distinct variables, (4) is called an *autoregressive* model of order p, or an AR(p). Model (4) also implies that the best forecast of z_t made from origin $(t - 1)$ is a linear combination of the past p-values of the series.

Introducing the backward-shift operator* B, such that $Bz_t = z_{t-1}$, $B^j z_t = z_{t-j}$, (4) may be written in the alternative form

$$(1 - \phi_1 B - \phi_2 B^2 - \cdots - \phi_p B^p)(z_t - \mu) = a_t. \quad (5)$$

Thus an AR(p) model is characterized by an operator

$$\phi(B) = (1 - \phi_1 B - \phi_2 B^2 - \cdots - \phi_p B^p), \quad (6)$$

which is a polynomial of degree p in the backward-shift operator B. The polynomial (6) may have real factors of the form $(1 - G_i B)$ or complex factors corresponding to complex roots of $\phi(B) = 0$. Complex factors indicate the presence of a quasi-cyclic* component in the data. Such cycles* do not have fixed periods, as in a sine wave, but are subject to random changes in amplitude, phase, and period. The fact that complex roots in (6) produce quasi-cyclical behavior in z_t may be seen by noting that if $p = 2$ and $a_t = 0$, the solution of the difference equation* (6) is a damped sine wave, as in the motion of a damped simple pendulum. When the zero on the right-hand side is replaced by a random series a_t, the sine wave is prevented from damping out by a series of random shocks, producing randomly disturbed sinosoidal behavior.

Autoregressive models were first suggested by G. U. Yule [9] who used a second-order model to describe the annual series of Wölfer sunspot numbers. Figure 1 shows a plot of this series, based on the annual average of daily readings, for the period 1770–1869. The fitted model* is

$$(z_t - 46.9) = \underset{\pm 0.07}{1.42}(z_{t-1} - 46.9) - \underset{\pm 0.07}{0.73}(z_{t-2} - 46.9) + a_t, \quad (7)$$

where the $\pm$ values underneath the estimated parameters are their estimated standard error limits. The variance of the residuals a_t can be estimated together with the parameters μ, ϕ_1, and ϕ_2 [2] and was σ_a^2

= 228.0 in this example. The operator $(1 - 1.42B + 0.73B^2)$ corresponding to (7) has complex factors with a period p that can be calculated from

$$\cos \frac{2\pi}{p} = \frac{\phi_1}{2\sqrt{-\phi_2}} = \frac{1.42}{2\sqrt{0.73}}$$

and is 10.65 years. Figure 1 also shows the histogram* of the residuals a_t corresponding to model (7). The distribution is skew, suggesting that a transformation of the data is needed before fitting a model. Using an approach due to Box and Cox [1](*see* BOX AND COX TRANSFORMATION), it may be shown that a better representation is obtained using the following model based on a square-root transformation*:

$$\left(\sqrt{z_t} - 7.4\right) = \underset{\pm 0.07}{1.41}\left(\sqrt{z_{t-1}} - 7.4\right) - \underset{\pm 0.07}{0.70}\left(\sqrt{z_{t-2}} - 7.4\right) + a_t \tag{8}$$

with $\sigma_a^2 = 1.994$. Note that the parameter estimates are changed only very slightly by transformation. Its main affect is to shrink the peaks and stretch the troughs in the series, resulting in a more symmetric distribution* of the residuals, as shown in Fig. 1. The estimate of the average period corresponding to model (8) is 11.05 years, much closer than the previous value of 10.65 years to the period quoted by meteorologists for this series.

PURE MOVING AVERAGE MODELS

For autoregressive models the π-weights in the representation (8) have a cut-off after p, where p is the order of the model. In some situations it may be more appropriate to apply steadily declining weights to generate the forecasts rather than weights which have an abrupt cut-off. Such a pattern of weights may be obtained, e.g., by using a moving average model

$$z_t - \mu = a_t - \theta a_{t-1} = (1 - \theta B)a_t. \tag{9}$$

Inverting (9), we obtain

$$a_t = \frac{1}{1 - \theta B}(z_t - \mu)$$

and provided that $|\theta| < 1$ and $|B| < 1$, this expression may be expanded to give

$$a_t = (1 + \theta B + \theta^2 B^2 + \cdots)(z_t - \mu), \tag{10}$$

so that the π-weights decay exponentially. More generally, a moving average model of order q, or MA(q) is defined by

$$(z_t - \mu) = a_t - \theta_1 a_{t-1} - \theta_2 a_{t-2} - \cdots - \theta_q a_{t-q} \tag{11}$$

where a_t is a random series. The model (11) contains q parameters $\theta_1, \theta_2, \ldots, \theta_q$, which can be estimated from the data. It implies that the current deviation of the series z_t from its mean μ is a linear combination of the current and q previous random shocks a_t (or one-step-ahead forecast errors) which have entered the system. In backward shift notation, (11) may be written as $(z_t - \mu) = \theta(B)a_t$, where the moving average operator $\theta(B)$ is given by

$$\theta(B) = 1 - \theta_1 B - \theta_2 B^2 - \cdots - \theta_q B^q. \tag{12}$$

The model (11) has π-weights consisting of a mixture of real exponentials, corresponding to real factors of $\theta(B)$, and of damped sine waves, corresponding to complex factors of $\theta(B)$.

MIXED AUTOREGRESSIVE–MOVING AVERAGE MODELS

Result (10) shows that an MA(1) can be written as an autoregressive model of infinite order. If θ is small, say $\theta = 0.3$, then from a practical point of view the infinite series in (10) can be truncated after the term in B since it would require a long length of series to detect the parameter 0.09 in the next term $0.09B^2$ in the expansion. However, if θ is moderate or large, several terms would be needed in (10) to provide an adequate approximation to the single-parameter model

(9). Thus if the moving average model were incorrectly specified as an autoregressive model, involving several parameters, the estimates of the parameters in the autoregressive representation would tend to have high standard errors and be highly correlated.

Conversely, the AR(1) model

$$(1 - \phi B)(z_t - \mu) = a_t \qquad (13)$$

can be written as an infinite-order moving average model

$$(z_t - \mu) = a_t + \phi a_{t-1} + \phi^2 a_{t-2} + \cdots, \qquad (14)$$

and hence estimation problems will be encountered if an autoregressive model is incorrectly specified as a moving average model. To achieve parsimony in parameterization in a given practical situation, it may be necessary to include both autoregressive and moving average terms in the model. Thus the mixed autoregressive–moving average model, or ARMA(p,q), is defined by [8]

$$\begin{aligned}(z_t - \mu) &= \phi_1(z_{t-1} - \mu) + \cdots \\ &\quad + \phi_p(z_{t-p} - \mu) + a_t - \theta_1 a_{t-1} \\ &\quad - \cdots - \theta_q a_{t-q}. \end{aligned} \qquad (15)$$

Written in terms of the backward shift operator, (15) becomes

$$(z_t - \mu) = \frac{1 - \theta_1 B - \cdots - \theta_q B^q}{1 - \phi_1 B - \cdots - \phi_p B^p} a_t. \qquad (16)$$

The form (15) represents the time series z_t (or an appropriate nonlinear transformation of z_t) as the output from a linear filter whose input is a random series and whose *transfer function** is a rational function of the backward shift operator B. In words, (15) implies that the current deviation of the time series from its mean is a linear combination of the p previous devations and of the current and q previous residuals a_t (or one-step-ahead forecast errors). The ARMA(p,q) model (15) is capable of generating π-weights in (1), the first p of which follow no fixed pattern and the remainder of which lie on a curve that is a mixture of damped exponentials and sine waves. Table 1 shows special cases of the general ARMA(p,q) model of the kind that frequently arise in practice.

STATIONARITY AND INVERTIBILITY CONDITIONS

The parameters in the ARMA(p,q) model (16) must satisfy the following two conditions. (a) For z_t to be written in the π-weight form (1), i.e.,

$$\left(1 - \sum_{j=1}^{\infty} \pi_j B^j\right)(z_t - \mu) = \theta^{-1}(B)\phi(B)z_t = a_t \qquad \text{for} |B| < 1,$$

the factors of $\theta(B)$ must be less than unity in modulus (the invertibility condition). This condition implies that the forecast weights π_j die out; i.e., the forecast depends less on what has happened in the distant past than in the recent past. (b) For z_t to be written in the ψ-weight form

$$(z_t - \mu) = \left(1 + \sum_{j=1}^{\infty} \psi_j B^j\right) a_t = \phi^{-1}(B)\theta(B)a_t \qquad \text{for} |B| < 1,$$

the factors of $\theta(B)$ must be less than unity in modulus (the stationarity condition). This condition implies that the series is stationary with finite variance. Table 2 shows the characteristic shapes of the π-weights and ψ-weights for the AR(p), MA(q), and ARMA(p,q) models.

AUTOCORRELATION FUNCTIONS

The autocovariance functions $\gamma_k = E[(z_t - \mu)(z_{t+k} - \mu)]$ shows how the dependence between neighboring values of the series varies with the lag* k. It may be calculated from the autocovariance generating function

$$\begin{aligned}\gamma(B) &= \sum_{k=-\infty}^{\infty} \gamma_k B^k \\ &= \sigma_a^2 \phi^{-1}(B)\phi^{-1}(B^{-1})\theta(B)\theta(B^{-1}). \end{aligned} \qquad (17)$$

Table 1 Some Simple Special Cases of the Autoregressive–Moving Average Model

(p, q)	Nature of Model	Mathematical Form of Model	Backward-Shift-Operator Form of Model
(1, 0)	First-order autoregressive	$z_t - \mu = \phi_1(z_{t-1} - \mu) + a_t$	$z_t - \mu = \frac{1}{1 - \phi_1 B} a_t$
(2, 0)	Second-order autoregressive	$z_t - \mu = \phi_1(z_{t-1} - \mu) + \phi_2(z_{t-2} - \mu) + a_t$	$z_t - \mu = \frac{1}{1 - \phi_1 B - \phi_2 B^2} a_t$
(0, 1)	First-order moving average	$z_t - \mu = a_t - \theta_1 a_{t-1}$	$z_t - \mu = (1 - \theta_1 B) a_t$
(0, 2)	Second-order moving average	$z_t - \mu = a_t - \theta_1 a_{t-1} - \theta_2 a_{t-2}$	$z_t - \mu = (1 - \theta_1 B - \theta_2 B^2) a_t$
(1, 1)	First-order autoregressive, first-order moving average	$z_t - \mu = \phi_1(z_{t-1} - \mu) + a_t - \theta_1 a_{t-1}$	$z_t - \mu = \frac{1 - \theta_1 B}{1 - \phi_1 B} a_t$

Table 2 shows the characteristic patterns of the autocorrelation functions $\rho_k = \gamma_k/\gamma_0$ of AR(p), MA(q), and ARMA (p, q) models. Such patterns may be used to provide an initial guess of the structure of an observed time series [2].

PARTIAL AUTOCORRELATION FUNCTIONS

A complementary tool to the autocorrelation function for identifying the structure of an ARMA(p, q) model is the partial autocorre-

Table 2 Summary of Properties of AR, MA, and ARMA Models

	AR(p) Models	MA(q) Models	ARMA(p, q) Models
π-weights and partial autocorrelations function	Cutoff after p; follow no fixed pattern	Infinite; mixture of damped exponentials and sine waves	First p values follow no fixed pattern; there after, mixture of damped exponentials and sine waves
ψ-weights and autocorrelation function	Infinite; mixture of damped exponentials and sine waves	Cutoff after q; follow no fixed pattern	First q values follow no fixed pattern; there after, mixture of damped exponentials and sine waves

lation function s_k [2]. The partial autocorrelation function may be estimated by fitting autoregressive models of orders 1, 2, 3, . . . , k to a time series and picking out the estimates $s_1, s_2, \ldots, s_k$ of the *last* parameter in the model. Table 2 shows the partial autocorrelation function shapes corresponding to AR(p), MA(q), and ARMA(p, q) models. The duality in the properties of autoregressive and moving average models should be noted.

MULTIVARIATE AUTOREGRESSIVE–MOVING AVERAGE MODELS

If $\mathbf{z}_t$ denotes an m-vector of mutually interacting time series, the univariate ARMA (p, q) model (16) may be generalized to

$$\boldsymbol{\phi}(B)(\mathbf{z}_t - \boldsymbol{\mu}) = \boldsymbol{\theta}(B)\mathbf{a}_t, \qquad (18)$$

where $\boldsymbol{\phi}(B)$ is an autoregressive matrix whose elements $\phi_{ij}(B)$ are autoregressive operators, $\boldsymbol{\mu}$ a vector of mean values, $\boldsymbol{\theta}(B)$ a moving average matrix with elements $\theta_{ij}(B)$, and $\mathbf{a}_t$ a vector of random series that are mutually uncorrelated. For further discussion of the properties of multivariate ARMA models, the reader is referred to Quenouille [7], Hannan [4], Box and Tiao [3], and Jenkins [5, 6].

Since time series occurring in practice are rarely stationary with fixed means, ARMA models are of limited use in describing practical situations. The modifications necessary to make them practically useful are discussed under autoregressive–integrated moving average (ARIMA) models*.

References

[1] Box, G. E. P. and Cox, D. R. (1964). *J. Roy. Statist. Soc. Ser. B*, **26**, 211–252.

[2] Box, G. E. P. and Jenkins, G. M., (1970). *Time Series Analysis; Forecasting and Control*. Holden-Day, San Francisco (2nd ed., 1976).

[3] Box, G. E. P. and Tiao, G. C., (1977). *Biometrika*, **64**, 355–366.

[4] Hannan, E. J., (1970). *Multiple Time Series*, Wiley, New York.

[5] Jenkins, G. M., (1975). *Proc. 8th Int. Biometric Conf.*, Constanta, Romania, 1974. Editura Academici Republicii Socialists Romania, 53.

[6] Jenkins, G. M., (1979). *Practical Experiences with Modelling and Forecasting Time Series*, GJP Publicationa, St. Helier, N. J.

[7] Quenouille, M. H., (1957). *The Analysis of Multiple Time Series*, Charles Griffin, London.

[8] Wold, H., (1938). *A Study in the Analysis of Stationary Time Series*, Almqvist & Wiksell, Stockholm (2nd ed., 1953).

[9] Yule, G. U., (1927). *Philos. Trans. Roy. Soc. A.*, **226**, 267.

Further Reading

The first balanced account of theoretical and practical aspects of autoregressive–moving average models is given in the book by Wold [8]. Box and Jenkins [2] summarize the properties of these models and also give practical guidelines for identifying, fitting, and checking such models given the data. Pioneering work on multivariate models is to be found in Quenouille [7], and Hannan [4] discusses the theoretical background for both univariate and multivariate models. Practical guidelines for building multivariate autogressive–moving average models have been given by Jenkins [6].

(AUTOREGRESSIVE–INTEGRATED MOVING AVERAGE (ARIMA) MODELS
FORECASTING
MOVING AVERAGES MODELS
TIME SERIES
TRANSFER FUNCTION)

G. M. JENKINS

AVAILABILITY

A property of a system, defined as the proportion of time the system is functioning (properly). If failure and repair times are each distributed exponentially* with expected values θ and ϕ, then the availability is $\theta/(\theta + \phi)$. Sometimes the availability is defined generally in this way with $\theta = E$[time to failure], $\phi = E$ [repair time].

Bibliography

Gray, H. L. and Lewis, T. (1967). *Technometrics*, **9**, 465–471.

Nelson, W. (1968). *Technometrics*, **10**, 594–596.

(QUALITY CONTROL
RELIABILITY)

AVERAGE CRITICAL VALUE METHOD

Geary's [1] average critical value (ACV) method applied to a statistic T used to test the hypothesis $H_0 : \theta = \theta_0$ against the alternative $H_1 : \theta = \theta_1$ determines what the difference between θ_0 to θ_1 should be for $E[T \mid \theta_1]$ to fall on the boundary of the critical region of the test. Test statistics with small differences correspond to tests with high efficiency*. The advantage of this method is that it requires only the calculation of the expected value $E[T \mid \theta_1]$ rather than the distribution of T. (The latter is required for power function* calculations.) Examples are given by Geary [1] and Stuart [2]. Stuart in 1967 showed that the ACV method of gauging the efficiency of tests can usually be represented as an approximation to the use of the asymptotic relative efficiency* of the tests.

General References

[1] Geary, R. C. (1966). *Biometrika*, **53**, 109–119.

[2] Stuart, A. (1967). *Biometrika*, **54**, 308–310.

(EDF STATISTICS)

AVERAGE EXTRA DEFECTIVE LIMIT (AEDL)

The average extra defective limit (AEDL) is a concept introduced by Hillier [1] as a measure of effectiveness of a continuous sampling plan* in adjusting to a process that has gone out of control.

Assume that a process has been operating in control at a quality level p_0 and then instantaneously deteriorates after the mth item to a level p_1, where $0 \leqslant p_0 < p_1 \leqslant 1$. Let D be the number of *uninspected* defectives among the next L items after the mth item is observed. The expected value of D, $E(D)$, is a well-defined quantity for a specific sampling plan. An average extra defective limit is the smallest number denoted by AEDL satisfying

$$E(D) \leqslant \text{AEDL} + L \times A$$

for all possible values of L, p_0, or p_1, where A is the average outgoing quality limit (AOQL)* of the plan.

Equivalently, let

$$X_m = \begin{cases} 1 & \text{if the } m\text{th item is defective but} \\ & \text{not inspected,} \\ 0 & \text{otherwise.} \end{cases}$$

Let the m_0th item be the last item before the shift in quality from p_0 to p_1; then

$$\text{AEDL} = \sup_{(p_0, p_1)} \sup_{L} \sum_{m=m_0+1}^{m_0+L} \left[E(X_m) - A \right].$$

Intuitively, AEDL is the upper limit to the expected number of "extra" defectives that will be left among outgoing items when the process goes out of control regardless of L, p_0, or p_1. Additional interpretations of AEDL, its uses, and methods of computation for continuous sampling plans have been discussed by Hillier [1].

Reference

[1] Hillier, F. S. (1964). *Technometrics*, **6**, 161–178.

[AVERAGE OUTGOING QUALITY
LIMIT (AOQL)
CONTINUOUS SAMPLING PLANS
QUALITY CONTROL]

AVERAGE OUTGOING QUALITY (AOQ)

The definition of this concept suggested by the Standards Committee of ASQC [3] is: "the expected quality of the outgoing product following the use of an acceptance sampling plan* for a given value of incoming product quality." It is basically a ratio of defective items to total items, i.e., the total number of defectives in the lots accepted divided by the total number of items in those lots. Two other formal (but not equivalent) definitions are: (1) the average fraction defective in all lots after rejected lots have been sorted and cleared of defects—this is an average based on practically perfect lots (those sorted) and lots still with fraction of defectives approximately p (it is assumed that lots of stable quality are offered), and (2) the expected fraction of defectives, after substituting good items for bad ones in rejected lots, and in samples taken from accepted lots.

The AOQ serves as a performance measure associated with an (attribute) acceptance sampling plan* when the same sampling plan is used repeatedly.

Wortham and Mogg [4] present formulas for calculating AOQ for nine different ways of carrying out rectifying inspection*. For example, if the defective items are replaced by good ones (thereby returning the lot to its original size N), then

$$\text{AOQ} = \frac{P_a \times p \times (N - n)}{n},$$

where P_a is the probability of acceptance using the given acceptance plan, p the fraction defective, N the lot size, and n the sample size.

References

[1] Burr, I. W. (1953). *Engineering Statistics and Quality Control*, McGraw-Hill, New York.

[2] Juran, J. M. ed. (1951). *Quality Control Handbook*. McGraw-Hill, New York.

[3] Standards Committee of ASQC (1978). Terms, Symbols and Definitions for Acceptance Sampling. *ASQC Standard A3*.

[4] Worthman, A. W. and Mogg, J. W. (1970). *J. Quality Tech.*, **2**(1), 30–31.

[AVERAGE OUTGOING QUALITY LIMIT (AOQL)
OC CURVE
QUALITY CONTROL
SAMPLING PLAN]

AVERAGE OUTGOING QUALITY LIMIT (AOQL)

The current "official" definition of this concept as suggested by the Standards Committee of ASQC [3] reads: "For a given acceptance sampling plan* AOQL is the maximum AOQ* over all possible levels of incoming quality."

Originally, AOQL was defined by Dodge [1] as the upper limit to the percent of defective units that remain in the output after inspection, given that the process is in statistical control (i.e., the proportion of defectives being produced is constant). In other words, it represents the worst average quality the consumer will accept under a particular rectifying inspection* scheme.

References

[1] Dodge, H. F. (1943). *Ann. Math. Statist.*, **14**, 264–279.

[2] Sackrowitz, H. (1975). *J. Quality Tech.*, **7**, 77–80.

[3] Standards Committee of ASQC (1978). Terms, Symbols and Definitions for Acceptance Sampling, *ASQC Standard A3*.

[4] Wald, A. and Wolfowitz, J. (1945). *Ann. Math. Statist.*, **16**, 30–49.

AVERAGE RUN LENGTH (ARL)

The length of time the process must run, on the average, before a control chart* will indicate a shift in the process level* is called the average run length (ARL). It is, of course, desirable that the ARL should be long when no shift has occurred, but short when a shift has occurred.

The ARL is usually measured in terms of the number of consecutive points plotted on the control chart.

Bibliography

Standards Committee of ASQC (1978).
ASQC Standard AL.

(CUMULATIVE SUM CONTROL CHARTS)

AVERAGE SAMPLE NUMBER (ASN)

In a sequential test* $\mathbb{S}$, the final size of the sample (N) required by the test is a random variable. If the sample sequential test $\mathbb{S}$ is carried out repeatedly, N will generally assume different values in successive repetitions of the test. The average amount of sampling per test that would result from the use of $\mathbb{S}$ is measured by the expected value of N and is called the average sampling number (ASN) of the test. If the test relates to the value of a parameter θ, we have formally

$$E[N;\theta] = \sum_{n=1}^{\infty} np(n;\theta),$$

where $p(n;\theta) = \Pr[N = n \mid \theta]$ is the probability of reaching the terminal decision at sample size n. A graph showing $E[N;\theta]$ against various values of θ is called the ASN curve or ASN surface, according as θ is a scalar or a vector, respectively.

Bibliography

Ghosh, B. K. (1970). *Sequential Tests of Statistical Hypotheses*. Addison-Wesley, Reading, Mass.

(SEQUENTIAL ANALYSIS)

AXIAL DISTRIBUTIONS *See* DIRECTIONAL DISTRIBUTIONS

AXIOMS OF PROBABILITY

THE AXIOMATIC METHOD

It is perhaps no coincidence that the axiomatic method in mathematics became prominent somewhat before the "use" theory of meaning became prominent in philosophy. An axiomatic system aims to capture and formalize some way of using language and it sidesteps the difficulties of explicit definitions.

The advantage of the axiomatic method is that theorems can in principle be deduced mathematically from the axioms without new assumptions creeping in surreptitiously, and without necessary philosophical commitment. In other words, the theorems can be proved rigorously according to the usual standards of pure mathematics without involvement in the controversial problems of application to the real world. In practice it is difficult to be totally rigorous, as has been found by philosophers of mathematics. An early example of the axiomatic method is in the geometry of Euclid, although his axioms do not satisfy most modern pure mathematicians. The value of a more precise axiomatic approach was emphasized by David Hilbert near the beginning of the twentieth century. The approach has become a paradigm for pure mathematics, but less so for applied mathematics and for physics because it can lead to rigor mortis (to quote Henry Margenau's joke). Probability theory is both pure and applied*, so that different specialities put more or less emphasis on axiomatic systems.

NOTATION

Many theories of probability have been proposed, and many different notations have been used. In this article we use notations such as $P(E \mid F)$, which can be read as the probability of E given (or assuming, or conditional on) F. Here, depending on the theory or the context, E and F might denote propositions, events*, hypotheses*, scientific theories, or sets, or might even be abstract symbols, such as those in abstract algebra, without ordinary definitions but subject only to some axioms. We can regard $P(\cdot \mid \cdot)$ as a function of two variables, and the domains of E and F are not necessarily identical. The notation $P(E)$ is read "the probability of E" and is used either when F is taken for granted or, in some theories, not as a conditional probability* but as a so-called "absolute probability" in which officially nothing is "given" or "assumed" other than logic and mathematics (were that possible).

When a theory of probability is expressed axiomatically there will usually be axioms satisfied by such symbols as E and F and further axioms satisfied by the "probabilities" themselves. Some theories of probability are formulated as theories of rationality, and then the set of axioms needs to mention either decisions or "utilities" (="desirabilities"). *See* DECISION THEORY; DEGREES OF BELIEF.

ARGUMENTS OF $P(E|F)$

A theory in which E and F denote sets or are abstract symbols can be regarded as a branch of pure mathematics, but propositions, events, and hypotheses are not purely mathematical concepts when they are interpreted as ordinary English words. We shall not try to define the English meanings of events, hypotheses, and theories, but the meaning of "proposition" is especially controversial: see, for example, Gale [11]. Perhaps the best definition is that a proposition is "the meaning of a clear statement." By ruling out unclear statements we are adequately justified in assuming that each proposition is capable of being either true or false, although really there are degrees of meaningfulness because statements can be more or less vague.

Moreover, a statement can be either empirically or mathematically meaningful, a point that is relevant to the choice of axioms. For example, to say that a measurement of a continuous variable lies between 5.25 and 5.35 inches is often empirically meaningful, whereas to say that it is exactly 5.30 inches, with no error at all, is at best mathematically meaningful within an idealized mathematical model. Again, to say that "the limit of the proportion of time that a coin comes up heads is approximately 0.5 in an infinite sequence of tosses" can be fully meaningful only within pure mathematics, because all sequences of tosses in the real world are of finite length.

AXIOMS FOR PROPOSITIONS AND SETS

The *conjunction* of propositions E and F is the proposition E & F and is denoted in this article by EF. The *disjunction* of E and F is denoted by $E \vee F$. This proposition asserts that E or F or both are true. The negation of E is denoted by $\tilde{E}$ or by $\sim E$. If E and F denote the same proposition, then we write $E = F$. (Other notations are in use.)

Some axioms for propositions are:

A1 If E is a proposition, then $\tilde{E}$ is also. This axiom might not be accepted by those who define a proposition as (scientifically) meaningful only if it is refutable if false. This first axiom, if applied to scientific propositions, forces us to the view that a proposition is also scientifically meaningful when it is confirmable if true. There are, however, degrees in these matters: see Good [15, pp. 492–494].

A2 $\sim(\sim E) = E$.

A3 If E and F are both propositions, then so is EF.

A4 *Commutative law*. $EF = FE$.

A5 *Associative law*. $E(FG) = (EF)G$.

A6 *De Morgan's law*. $\sim(EF) = \tilde{E} \vee \tilde{F}$. From this we can prove that the commutative and associative laws apply also to disjunctions.

A7 *Distributive laws*

$$E(F \vee G) = (EF) \vee (EG)$$

and

$$E \vee (FG) = (E \vee F)(E \vee G),$$

of which the second law can be inferred from the first by means of de Morgan's law*.

To these seven axioms, which are essentially the axioms of Boolean algebra, we can append the optional axiom (A8) and perhaps (A9):

A8 The conjunction and disjunction of a countably infinite number of propositions are propositions, with a corresponding "de Morgan law,"

$$\sim(E_1E_2E_3 \cdots) = \tilde{E}_1 \vee \tilde{E}_2 \vee \tilde{E}_3 \vee \cdots .$$

A9 The conjunction and disjunction of any infinite number of propositions are propositions, with yet another "de Morgan law": The negation of the conjunction is the disjunction of the negations.[1]

ORIGINS OF THE AXIOMS

In most theories of probability the probabilities lie in some sense between 0 and 1 and satisfy axioms somewhat resembling in appearance the addition and product axioms, namely:

$$P(A \vee B) = P(A) + P(B)$$

when A and B are mutually exclusive* and

$$P(AB) = P(A) \cdot P(B \mid A).$$

These comments will be clarified in what follows.

The addition and product axioms were known at least implicitly to Fermat and Pascal in 1654 and perhaps to Cardano in the sixteenth century. But it is more convenient to express the axioms explicitly, formally, and completely.

Axioms are seldom slapped down arbitrarily; a system of axioms should be chosen either as a convenient form of other axioms or should be constructed to capture some intuitive ideas about the world or about mathematics, so that the system has some prior justification. The axioms can also be justified by their practical and philosophical implications; e.g., they should not be seen to lead to an irresolvable contradiction. Before the axioms can have practical meaning, some formal rules of application to the real world must be provided. Moreover, in practice a set of axioms and rules of application are still not sufficient: a theory needs to become to some extent a technique if it is to be useful. One needs informal suggestions of how to apply the theory, although these suggestions are not logically essential. In this article no more will be said about such practical suggestions because such matters belong properly to a discussion of the relationship between probability and statistics or between probability and practical decision making. *See* DECISION THEORY; DEGREES OF BELIEF.

The prior justification of a set of axioms must depend on some concept of probability, however vague. One of the earliest concepts of probability was derived from games of chance*, such as those depending on coin spinning, dice throwing, and card drawing. In such games there are some symmetry properties that suggest that some outcomes are at least approximately equally probable, and this is so even for people who have not much idea of what probability means. In the context of such games one might be ready to accept Laplace's* definition of the probability of an event E as k/m, where m is the number of "equally possible" cases (meaning equally probable cases) that could occur, and k is the number of those cases that constitute E. When this "definition" is applicable, it leads to a familiar set of axioms. The main disadvantage of this approach is that a clearly exhaustive set of "equally probable cases" cannot usually be specified with reasonable objectivity in scientific applications. Also, the definition is somewhat circular.

In many experiments or observational circumstances the kind of symmetry required for the direct application of the classical definition of probability is lacking. To get around this difficulty, or for other reasons, a definition in terms of long-run proportional frequency of "successes" was explicitly proposed by Leslie Ellis and Cournot in 1843, and developed in much detail by Venn in 1866. (For these references and further history, see Keynes [21, pp. 92–93]). As usual with simple ideas, "frequentism" had been to some extent foreshadowed long before, e.g., by Aristotle, who said that the probable is what usually happens, or Greek words to that effect, but a self-respecting kudologist would not on that account attribute the theory to Aristotle alone. The frequentist definition is associated with physical probability rather than with logical or subjective (= personal) probability. For discussions of kinds of probability, *see* DEGREES OF BELIEF and its references, and FOUNDATIONS OF PROBABILITY.

It is by no means simple to construct a satisfactory definition of physical probability based on limiting frequencies. Consider, e.g., the following naive approach. By a "trial" we mean an experiment whose outcome is either some event E, or is the negation of E, a "failure" F. For example, a trial might be the tossing of a coin or the throw of a die and E might denote "heads" or "a six." Let an infinite sequence of such trials be performed under "essentially equivalent" conditions. Then the proportion of successes in the first n trials might tend to a limit p when $n \to \infty$. If so, then p might be called the probability of a success.

This naive definition of physical probability by long-run or limiting frequency* has some disadvantages. Even if we admit the possibility of an infinite sequence of trials in the real world, as some kind of approximation, the definition says nothing about whether the sequence of outcomes is in any sense random. A more sophisticated long-run-frequency definition of probability was proposed by von Mises [24] based on the prior notion of a random sequence or irregular Kollektiv*. This approach requires axioms for random sequences and again has severe mathematical and logical difficulties, although it can be presented fairly convincingly to the intuition in terms of generalized decimals [17, and references therein]. The theory of von Mises can be made logically rigorous and then leads to a familiar set of axioms for probability [6, 23]. For a discussion of randomness, with further references, see also Coffa et al. [5] and the article on RANDOMNESS in the present encyclopedia.

An approach to the axioms via sharp absolute probabilities, when there are fewer than 26^M possible mutually exclusive propositions, where $M = 10^{1000}$, is to argue as follows. Suppose that the N mutually exclusive possible propositions $E_1, E_2, \ldots, E_N$ of interest have sharp probabilities approximated to ν places of decimals by $p_1, p_2, \ldots, p_N$, where ν is large, so that $p_i = m_i 10^{-\nu}$, where m_i is a positive integer, and $\sum m_i = 10^\nu$. For each i, take a well-shuffled pack of cards containing m_i equiprobable cards and use it to break E_i into m_i mutually exclusive propositions each of probability $10^{-\nu}$. This leads to 10^ν equally probable propositions and the classical definition can now be used to arrive at a familiar set of axioms. (Compare ref. 12, p. 33, where the argument was expressed somewhat differently.)

An approach that again assumes that probabilities mean something and can be expressed numerically was apparently first suggested by S. N. Bernstein* [3]. It depends on ideas such as that $P((E \vee F) \vee G)$ must equal $P(E \vee (F \vee G))$. It is assumed further that, when E and F are mutually exclusive, then $P(E \vee F)$ is some function of $P(E)$ and $P(F)$. The assumptions lead to functional equations that must be satisfied by probabilities, and these equations can be used to justify the axioms. The idea was developed independently by Schrödinger [32], Barnard and Good [12, pp. 107–108] and especially by R. T. Cox [7, 8]. Cox's assumptions were weakened by Aczél [1]. The approach again leads to a familiar set of axioms, and seems to be the most convincing justification of these axioms for numerical subjective probability among those approaches that makes no reference to decisions or to gambles.

An advantage of bringing decisions or gambles into the discussion is that a prior intuitive concept of probability is then less necessary in arriving at axioms for subjective probabilities. We are then led to a behavioral approach that many people find more convincing than a more purely linguistic approach. With some ingenuity the behavioral approach can be developed to the point where no explicit definition of either probability or utility* is assumed, but only preferences between acts. This approach was adopted by F. P. Ramsey [27], B. de Finetti [9], and L. J. Savage [31]. They assumed that "your" preferences between acts can be completely ordered, and that the preferences satisfy desiderata for rationality that many people find compelling once the complete ordering is granted. These desiderata lead to the conclusion that if you were perfectly rational, you would behave as if you had a set of probabilities (degrees of belief*) satisfying familiar axioms, and a set of utilities, and that you would always prefer the act of maximum expected utility. In this approach the concepts of probability and utility are not separately defined, nor are they taken for granted. In fact, a perfectly rational person might not know the concepts of probability and utility, but these concepts can be used by some one else to describe the rational person. We can *imagine* a doctor or warrior, for example, who always makes the best decisions, although never having heard of probabilities.

THE PURELY MATHEMATICAL APPROACH

Since most of the philosophical approaches lead to somewhat similar formal theories, it is natural for a mathematician to choose a set of axioms based on earlier formalisms. By separating the symbols E, F, etc., from their concrete meanings, the mathematician can avoid philosophical controversies and get on with the job. This approach was adopted by A. N. Kolmogorov* [22], following some earlier writers. His axioms were expressed in the language of sets and measure theory. Borel–Lebesgue measure was introduced at the turn of the century: see, e.g., Carathéodory [4, p. 702]. Before 1890 set theory was not regarded as mathematically respectable, but by 1930 it was regarded as part of the foundation of pure mathematics.

Kolmogorov stated that he agreed with von Mises's frequency interpretation* of probability, but his axioms do not presuppose this interpretation. To begin with, he assumes that there is a set Ω of "elementary events" ω, but the concept of an elementary event requires no definition as long as we are concerned only with the mathematics, and each ω can instead be called a "point" if this helps the imagination. A class $\mathbb{S}$ of subsets of Ω are called "events," not yet to be interpreted in the ordinary sense; and it is assumed that if the subset S is an element of $\mathbb{S}$, then so is its complement $\Omega - S$. Furthermore, it is assumed that if S and T both belong to Ω, then so does the union of S and T. "Events" satisfying these assumptions are said to constitute an *algebra of events** or field of events. Note the similarity to the axioms for propositions given earlier. If the union of any countable infinity of events is also in $\mathbb{S}$, then $\mathbb{S}$ is unfortunately said to be a σ-algebra. We shall soon see why the condition of countability is assumed.

A symbol $P(S)$ is introduced and is called the (absolute) probability of S. It is assumed that $P(S)$ is a real number and lies in the closed interval $[0, 1]$, also that $\Omega \in \mathbb{S}$ and that $P(\Omega) = 1$. Finally, if a countable class of sets $S_1, S_2, S_3, \ldots$ are disjoint, then $P(S_1 \cup S_2 \cup \cdots) = P(S_1) + P(S_2) + \cdots$, where $\cup$ denotes union. This last assumption is called the axiom of complete additivity. A weaker axiom asserts the property only for two sets (which implies the property for any finite number of sets) instead of for a countable infinity of sets. The axiom of complete additivity is the main feature of Kolmogorov's system and makes his system highly reminiscent of Lebesgue measure.

The product axiom is introduced through the back door by defining the conditional probability $P(S \mid T)$ by the quotient $P(S \cap T)/P(T)$ when $P(T) \neq 0$, where $\cap$ denotes the intersection of sets.

The theory is applied by interpreting "events" as meaning physical events.

Kolmogorov's axioms, perhaps better called the measure-theoretic axioms, are the most popular among mathematical statisticians at present. He did not pretend that he had no predecessors, and Rényi [29, p. 55] cites Borel (1909), Lomnicki (1923), Lévy (1925), Steinhaus (1923), and Jordan (1925). As Rényi says, the measure-theoretic approach leads to a rigorous mathematical theory of stochastic processes.

The analogy with Lebesgue measure makes it clear why the axiom of complete additivity is stated only for a countable infinity of sets: the Lebesgue measure of a unit interval is unity, but this measure can hardly be expressed as the sum of the noncountable number of zero measures of the points in the interval. A similar objection has been raised by de Finetti [10, p. 124] against the axiom of complete additivity itself. For consider an infinite sequence in which it is known that there is precisely one "success" but where this success might be anywhere. In the von Mises theory the probability of a success would apparently be zero, and the axiom of complete additivity appears then to lead to the contradiction $1 = \sum 0 = 0$. Perhaps the resolution of this difficulty is to deny that the foregoing sequences count as Kollektivs and say that one models zero probability by a Kollektiv in which the limiting proportional frequency of "successes" is zero. Then there are a noncountable number of such Kollektivs and no paradox arises. (How you could recognize such Kollektivs in practice from finite initial segments is another problem.)

As indicated earlier, from a strictly practical point of view it makes no sense to choose a precise real number at random; in fact, to do so would be like selecting the infinite sequence of its decimal digits. When E and F denote meaningful practical propositions the measure-theoretic approach is not essential, but the approach compensates for this by its valuable mathematical convenience. Then again a price is paid because the mathematics becomes advanced.

The approach in terms of propositions appears more general than in terms of sets, because there need be no concept of an "elementary proposition." It would, however, be possible, at least if the total number of propositions if finite, to define an elementary proposition as a conjunction of propositions that is (a) not strictly impossible while (b) if it can be made any less probable by being conjoined with another proposition, then it becomes impossible. An elementary proposition would be an interpretation of an elementary event ω. In this manner one might be able to subsume the propositional approach under the measure-theory umbrella, but if there were only a finite number of propositions, the axiom of complete additivity would be unnecessary.

AXIOMS EXPRESSED IN TERMS OF CONDITIONAL PROBABILITIES

Since probabilities in practice are always conditional, absolute probabilities do not capture anything concrete, so several writers have proposed sets of axioms stated directly in terms of conditional probabilities; e.g., Wrinch and Jeffreys [33], Keynes [21], Reichenbach [28], Good [12], Popper [26], and Rényi [29]. Some of these writers expressed the axioms in terms of the probabilities of propositions. We give here an example of such a system of axioms based on Good [12, 13]. In these axioms, the symbol H does not necessarily denote a hypothesis, but in many applications it does.

A1 *$P(E \mid F)$ is a real number if F is not self-contradictory.* (A similar caveat applies in the remaining axioms.)

A2 $0 \leqslant P(E \mid F) \leqslant 1$.

A3 *If* $P(EF \mid H) = 0$, *then* $P(E \vee F \mid H) = P(E \mid H) + P(F \mid H)$.

A4 *If H logically implies E* (i.e., if $\overline{H} \vee E$ is a tautology), *then* $P(E \mid H) = 1$ (but not conversely).

A5 (Axiom of equivalence.) *If neither HE nor HF is self-contradictory and HE implies F and HF implies E, then* $P(E \mid H) = P(F \mid H)$.

A6 $P(EF \mid H) = P(E \mid H)P(F \mid EH)$.

A7 $P(H^* \mid H^*) \neq 0$, *where H^* denotes the basic assumptions of logic and pure mathematics.*

A8 $P(E^* \mid H^*) = 0$ *for some proposition* E^*.

A9 (Complete additivity: optional.) *If* $P(E_i E_j \mid H) = 0 (1 < j;\ i, j = 1, 2, \ldots)$, *then* $P(E_1 \vee E_2 \vee \cdots \mid H) = \sum P(E_i \mid H)$.

A10 (The principle of cogent reason: optional: see Keynes [21, p. 56], Russell [30, p. 397], Good [12, p. 37].) *Let ϕ and ψ be propositional functions. Then, for all a and b for which the functions are defined, we have*

$$P(\phi(a) \mid \psi(a)) = P(\phi(b) \mid \psi(b)).$$

For example, the probability of getting 7 hears in a whist hand, given only that the pack contains 13 hearts, is the same if we change "hearts" to "diamonds."

In this theory, the main rule of application is obtained by thinking of the axioms as the workings of a black box* into which judgments of probability inequalities can be plugged and from which discernments of new inequalities can be read. This black-box theory is explained in more detail in the article DEGREES OF BELIEF. Following Keynes [21], who, however, dealt with logical probabilities, Good assumes that (subjective) probabilities are only partially ordered and the use of axioms for sharp probabilities is only a device for expressing the theory in a highly intelligible form. From this form of the theory it is shown by Good [14] that one can derive axioms for the upper* and lower* probabilities themselves. For example, the product axiom splits into six axioms, one of which is

$$P_*(EF \mid H) \leqslant P^*(E \mid H) \cdot P_*(F \mid EH).$$

One can think of upper and lower probabilities as exterior and interior measures, and in a frequency theory they might correspond to upper and lower limits.

If one wishes to talk meaningfully about the probability of a mathematical theorem, as is desirable for the formalizing of "plausible reasoning" (see, e.g., Pólya [25]), then it is necessary as in Good [12, p. 49] to replace the axiom of equivalence by something like A5′. *If at time t you have seen that E and F are equivalent, then* $P_t(E \mid H) = P_t(F \mid H)$ *and* $P_t(H \mid E) = P_t(H \mid F)$, *where the subscript t is self-explanatory.* [Judgments are needed to decide whether $P_t(G \mid K) = P_s(G \mid K)$, where $t \neq s$.] This axiom allows subjective probabilities to vary as time passes, without changing ordinary empirical evidence. This is not, of course, the same as the elementary fact that $P(E \mid GH)$ is not in general equal to $P(E \mid H)$. By allowing probabilities to have the dynamic feature of varying as a consequence of calculations and thinking, such as the variables in FORTRAN, one can say meaningfully and quantitatively that a mathematical theorem conveys information, and one can also solve some otherwise intractable philosophical problems concerning scientific induction. (For example, see Good [16, 18, 19], where these varying probabilities are called "evolving" or "dynamic.")

The theory for partially ordered or comparative probability, as just discussed, extends immediately to a theory of rational behavior, by introducing utilities. For details, see Good [13].

A difficulty in theories of subjective probability, pointed out by Richard Jeffrey [20, p. 154], is that a subjective probability can change as a consequence of an experience that "you" cannot express in words. As a matter of fact, badly remembered experiences cause the same difficulty. Although you might have experiences that you cannot personally express fully in words, you can describe them as experiences that occurred at a certain time, and the meaning of this description can be regarded as a proposition in an extended sense. By allowing the meaning of "proposition" to be extended in this manner, the difficulty seems to be overcome without the need for any new axioms. The difficulty would not be overcome within a theory of logical rather than subjective probability.

A distinctive feature of Jeffrey [20, p. 83] is connected with utilities. Previous theories had led to the conclusion that if preference rankings* are sufficiently extensive, probabilities can be uniquely determined but utilities can be determined only up to linear transformations; that is, if a set of (expected) utilities is given, then each element u of this set can be replaced by $au + b$, where a and b are constants, and this substitution will have no effects on any recommended decisions. In Jeffrey's theory, in which both probabilities and utilities refer to propositions, the probabilities and utilities can undergo a class of transformations, the transformation of the utility being of the form $(au + b)/(cu + d)$. He attributes this result to independent personal communications from Kurt Gödel and Ethan Bolker.

In summary, distinct purposes and distinct philosophies can be associated with distinct systems of axioms of probability, although these systems fortunately have much in common.

NOTE

1. All these axioms and comments are applicable *mutatis mutandis* to sets as well as to propositions. For propositional functions, with "quantifiers" such as "for all" and "there exists," further axioms are necessary, but we shall not labor this point.

References

[1] Aczél, J. (1963). *Ann. Univ. Sci. Budap. Rolando Eötvös Nominatae Sect. Math.*, **6**, 3–11.

[2] Barnard, G. A. (1949). *J. R. Statist. Soc. B*, **11**, 115–139.

[3] Bernstein, S. N. (1917). An attempt at an axiomatic foundation for the calculus of probability (in Russian). *Khar'kov Univ. Kar'kovskoi mat. obshch. Soobshcheniia*, **15**, 209–274. Abstract in German by Bernstein in *Jb. Math.*, **48**, (1920–1921), 596–599.

[4] Carathéodory, C. (1927). *Vorlesungen über Reelle Funktionen*, 2nd ed. Teubner, Leipzig. (Reprint: Chelsea, New York, 1948.)

[5] Coffa, J. A., Good, I. J. and Kyburg, H. E. (1974). *PSA 1972* (Proc. 1972 Bienn. Meet. Philos. Sci. Ass.), K. F. Schaffner and R. S. Cohen, eds. D. Reidel, Dordrecht, pp. 103–149.

[6] Copeland, A. H. (1937). *Trans. Amer. Math. Soc.*, **42**, 333–357.

[7] Cox, R. T. (1946). *Amer. J. Phys.*, **14**, 1–13.

[8] Cox, R. T. (1961). *The Algebra of Probable Inference*. Johns Hopkins University Press, Baltimore, Md.

[9] Finetti, B. de (1937). *Ann. Inst. Henri Poincaré*, **7**, 1–68. English translation in *Studies in Subjective Probability*, H. E. Kyburg and H. E. Smokler, eds. Wiley, New York, 1964, pp. 95–158.

[10] Finetti, B. de (1974). *Theory of Probability*, Vol. 1. Wiley, New York.

[11] Gale, R. M. (1967). In *The Encyclopedia of Philosophy*, Vol. 5, Paul Edwards, ed. Macmillan/The Free Press, New York, pp. 494–505.

[12] Good, I. J. (1950). *Probability and the Weighing of Evidence*. Charles Griffin, London/Hafner, New York.

[13] Good, I. J. (1952). *J. R. Statist. Soc. B*, **14**, 107–114.

[14] Good, I. J. (1962). In *Logic, Methodology, and Philosophy of Science*, E. Nagel, P. Suppes, and A. Tarski, eds. Stanford University Press, Stanford, Calif., pp. 319–329.

[15] Good, I. J. (1962/1966). In *Theories of the Mind*, J. Scher, ed. Glencoe Free Press/Macmillan, New York, pp. 490–518. Misprints corrected in second edition.

[16] Good, I. J. (1968). *Brit. J. Philos. Sci.*, **19**, 123–143.

[17] Good, I. J. (1974). In the symposium cited in ref. 5, pp. 117–135.

[18] Good, I. J. (1975). *Synthése*, **30**, 39–73.

[19] Good, I. J. (1977). In *Machine Intelligence*, Vol. 8, E. W. Elcock and D. Michie, eds. Wiley, New York, pp. 139–150.

[20] Jeffrey, R. (1965). *The Logic of Decision*. McGraw-Hill, New York.

[21] Keynes, J. M. (1921). *A Treatise on Probability*. Macmillan, London (2nd ed., 1929).

[22] Kolmogorov, A. N. (1933). *Grundbegriffe der Wahrscheinlichkeitsrechnung*. Springer-Verlag, Berlin (English translation: Chelsea, New York, 1950).

[23] Martin-Löf, P. (1969). *Theoria*, **35**, 12–37.

[24] Mises, R. von (1919). *Math. Zeit.*, **5**, 52–99.

[25] Pólya, G. (1954). *Mathematics and Plausible Reasoning*, 2 vols. Princeton University Press, Princeton, N. J.

[26] Popper, K. R. (1959). *The Logic of Scientific Discovery*. Hutchinson, London.

[27] Ramsey, F. P. (1926/1931). *The Foundations of Mathematics and Other Logical Essays*. Kegan Paul, London; Harcourt Brace, New York.

[28] Reichenbach, H. (1949). *The Theory of Probability*. University of California Press, Berkeley, Calif.

[29] Rényi, A. (1970). *Foundations of Probability*. Holden-Day, San Francisco.

[30] Russell, B. (1948). *Human Knowledge, Its Scope and Limitations*. Routledge & Kegan Paul, London.

[31] Savage, L. J. (1954). *The Foundations of Statistics*. Wiley, New York. (2nd ed., Dover, New York).

[32] Schrödinger, E. (1947). *Proc. R. Irish Acad.*, **51A**, 51–66, 141–146.

[33] Wrinch, D. and Jeffreys, H. (1919). *Philos. Mag.*, *6th Ser.*, **38**, 715–731.

(CHANCE (II)
DECISION THEORY
DEGREES OF BELIEF
FOUNDATIONS OF PROBABILITY)

I. J. Good

B

BACKWARD DIFFERENCE

A finite difference operator, ∇, defined by

$$\nabla f(x) = f(x) - f(x-1).$$

Symbolically, $\nabla = 1 - E^{-1}$, where E is the displacement operator*. If the differencing interval is h, we define

$$\nabla^h f(x) = f(x) - f(x-h)$$

and

$$\nabla^h \equiv 1 - E^{-h}.$$

Powers of backward operators are defined recursively:

$$\begin{aligned}\nabla^2 f(x) &= \nabla(\nabla f(x)) = \nabla f(x) - \nabla f(x-1)\\ &= f(x) - 2f(x-1) + f(x-2)\end{aligned}$$

and

$$\nabla^n f(x) = \nabla\left(\nabla^{n-1} f(x)\right).$$

Newton's interpolation formula* (which is a finite difference analog of Taylor's series in differential calculus) allows us to evaluate the value of a function at any point x based on the value of the function at x_0 and the values of its successive differences at x_0, where $x = x_0 + uh$ and u is abritrary. In terms of backward differences, Newton's interpolation formula is

$$\begin{aligned}f(x) = f(x_0) &+ \frac{x - x_0}{h}\nabla f(x_0)\\ &+ \frac{(x-x_0)(x-x_{-1})}{2!h^2}\nabla^2 f(x_0)\\ &+ \cdots,\end{aligned}$$

where $x_i = x_0 + ih (i = \pm 2, \ldots)$.

Bibliography

Johnson, N. L. and Kotz, S. (1977). *Urn Models and Their Application*. Wiley, New York.

Jordan, C. (1950). *Calculus of Finite Differences*, 2nd ed. Chelsea, New York.

(FINITE DIFFERENCES
FORWARD DIFFERENCES)

BACKWARD ELIMINATION SELECTION PROCEDURE

This is one of a number of alternative ways of fitting a "best" regression equation* using data on a response and a (usually fairly large) set of predictor* variables. Essentially, the method involves first fitting a least-squares* regression model using all the predictor variables and then testing the regression coefficient* of the predictor variable that provides the smallest "extra sum of squares"* when it is added to the model last. If this coefficient is statistically significant, the procedure is stopped and the full equation is used. Otherwise (and typically), this "worst predictor" is deleted, the equation is refitted by least squares, and the worst predictor of the new equation is checked, and eliminated if the corresponding regression coefficient is not statistically significant*. Predictor variables are deleted in this fashion until the worst predictor produces a statistically significant coefficient, when the procedure stops and the current equation is chosen. (*See* SELECTION PROCEDURES for details on various methods of variable selection.)

Bibliography

Draper, N. R. and Smith, H. (1966). *Applied Regression Analysis*. Wiley, New York (2nd ed., 1981).

(ELIMINATION OF VARIABLES
REGRESSION ANALYSIS)

N. R. DRAPER

BACKWARD-SHIFT OPERATOR *See* BOX–JENKINS MODEL; GRADUATION

BAHADUR EFFICIENCY

Both in theoretical and practical statistical work the choice of a good test procedure for a given hypothesis-testing* problem is of great importance. For finite sample size, the Neyman–Pearson theory* suggests that one select a test based on its power. Various optimality criteria that have been used in the finite-sample-size case include uniformly most powerful*, uniformly most powerful invariant*, uniformly most powerful unbiased*, etc. However, in many hypothesis-testing problems, an optimal *finite* sample test may not exist or may be difficult to determine. In this case an asymptotic approach is often quite useful. Among the asymptotic techniques used for comparing tests, the most frequently discussed approaches are Pitman efficiency* [11], Chernoff efficiency [7] (see below), Hodges–Lehmann efficiency* [9], and Bahadur efficiency [1–3].

Although Bahadur efficiency is a concept of theoretical nature, its basic idea is not difficult to understand. Let $s = (x_1, x_2, \dots)$ be an infinite sequence of independent observations of a normal random variable X with mean θ and variance 1. Imagine that two statisticians A and B are supplied with the sequence $s = (x_1, x_2, \dots)$ for testing $H_0 : \theta = 0$ against $H_1 : \theta = \theta_1$ $(\theta_1 > 0)$. A uses $T_n^{(1)}$ and B uses $T_n^{(2)}$ as test statistics, where $T_n^{(i)}$ is a function of the first n observations $(x_1, x_2, \dots, x_n)$. Suppose that in both cases, H_0 is rejected for large values of $T_n^{(i)}$ $(i = 1, 2)$. The tail probability based on $T_n^{(i)}$ is defined as $P_{\theta=0}(T_n^{(i)} \geqslant T_n^{(i)}(s))$ for $i = 1, 2$, and in practical statistical work, tail probabilities are often computed for the observed values $x_1, \dots, x_n$ in a given experiment. If the tail probability is less than or equal to level α, than H_0 is rejected. Suppose that $\theta = \theta_1$ and that data s is observed. Let N_i be the minimal sample size required for $T_n^{(i)}$ to reject H_0 at the significance level* α for $i = 1, 2$. $\{T_n^{(1)}\}$ is considered to be a better sequence of statistics than $\{T_n^{(2)}\}$if $N_1 < N_2$, since the sample size required by $T_n^{(1)}$ to make the correct decision (rejection of H_0) is less than that of $T_n^{(2)}$. Therefore, it is reasonable to use the ratio N_2/N_1 as a measure of the relative efficiency of $T_n^{(1)}$ to $T_n^{(2)}$. However, the ratio is not very well suited for practical use because it depends on three arguments (α, s, θ_1). Moreover, it is often hard to compute. To avoid these difficulties, let us consider the limit of N_2/N_1 as $\alpha \to 0$. Assume that

$$\epsilon(s, \theta_1) = \lim_{\alpha \to 0} N_2/N_1$$

is a finite number. $\epsilon(s, \theta_1)$ can then be used as a measure of the asymptotic efficiency of $\{T_n^{(1)}\}$ relative to $\{T_n^{(2)}\}$. Note that this method of comparison is stochastic since $\epsilon(s, \theta_1)$ is a random variable. At first sight it seems that $\epsilon(s, \theta_1)$ is not a very useful criterion for comparing tests since $\epsilon(s, \theta_1)$ still depends on two arguments. However, in many cases $\epsilon(s, \theta_1) = \epsilon^*(\theta_1)$ with probability 1. Loosely speaking, $\epsilon^*(\theta_1)$ is the Bahadur efficiency of $\{T_n^{(1)}\}$ relative to $\{T_n^{(2)}\}$ at $\theta = \theta_1$. If $\epsilon^*(\theta_1) > 1$, then $\{T_n^{(1)}\}$ is said to be more efficient. In the next paragraph we define Bahadur efficiency more precisely.

Let $X_1, X_2, \ldots$ be a sequence of independent, identically distributed (i.i.d.) random variables whose common probability distribution is indexed by a parameter θ ranging in a parameter space Θ. For every $1 \leqslant n \leqslant \infty$, let $P_\theta^{(n)}$ denote the probability distribution of $(X_1, \ldots, X_n)$. $P_\theta^{(n)}$ will often be abbreviated to P_θ to simplify the notation. Let $(\mathbb{S}, \mathfrak{Q})$ be the sample space of the sequence of random variables $(X_1, X_2, \ldots)$. Every element s in $\mathbb{S}$ has the form $(x_1, x_2, \ldots)$ and $\mathfrak{Q}$ is a σ-field of subsets of $\mathbb{S}$. We are interested in testing the null hypothesis* $H_0 : \theta \in \Theta_0$ against the alternative hypothesis* $H_1 : \theta \in \Theta_1$, where Θ_0 is a proper subset of Θ and $\Theta_1 = \Theta - \Theta_0$. For each $n = 1, 2, \ldots,$ let $T_n^{(s)}$ be an extended real-valued function such that T_n is $\mathfrak{Q}$-measurable and depends on s only through $(x_1, \ldots, x_n)$. T_n is considered to be a test statistic, large values of T_n being significant. Assume for simplicity, that there exists a distribution function $F_n(t)$ such that $P_\theta(T_n < t) = F_n(t)$ for all $\theta \in \Theta_0$ and for all t, $-\infty \leqslant t \leqslant +\infty$. For every s in $\mathbb{S}$ the tail probability of T_n is defined to be

$$L_n = L_n(s) = 1 - F_n(T_n(s)).$$

Note that L_n depends only on $(x_1, \ldots, x_n)$. Clearly, L_n is a random variable. In typical cases, if $\theta \in \Theta_1$, $L_n \to 0$ exponentially fast (with probability 1 $[P_\theta]$). We shall say that the sequence of test statistics $\{T_n\}$ has the (exact) Bahadur slope $c(\theta)$ if

$$\lim_{n\to\infty} (1/n)\log L_n(s) = -\tfrac{1}{2}c(\theta) \qquad (1)$$

with probability 1 $[P_\theta]$ when θ is the true parameter. Now for a given α, $0 < \alpha < 1$, and a given s, let $N = N(\alpha, s)$ be the smallest integer m such that $L_n(s) < \alpha$ for all $n \geqslant m$ and let $N = \infty$ if no such m exists. Then N is the minimal sample size required for $\{T_n\}$ to reject H_0 when H_1 is true at level α. If (1) holds and $0 < c(\theta) < \infty$, then

$$\lim_{\alpha\to 0} N(\alpha, s)\left[\frac{c(\theta)}{2\log(1/\alpha)}\right] = 1 \qquad (2)$$

with probability 1 $[P_\theta]$. To see this, choose a fixed θ and a fixed s such that

$$0 < c(\theta) < \infty$$

and

$$\lim_{n\to\infty} N^{-1}\log L_n(s) = -c(\theta)/2. \qquad (3)$$

Then $L_n(s) > 0$ for all sufficiently large n and $L_n \to 0$ as $n \to \infty$. It follows that $N < \infty$ for every $\alpha > 0$ and that $N \to \infty$ as $\alpha \to 0$. For all sufficient small α, say that $0 < \alpha < \alpha_1$. We have $L_N < \alpha \leqslant L_{N-1}$. Hence $N^{-1}\log L_N < N^{-1}\log\alpha \leqslant (N-1)N^{-1} \cdot (N-1)^{-1}\log L_{N-1}$. It now follows from (3) that $N^{-1}\log\alpha \to -c(\theta)/2$ as $\alpha \to 0$. This proves (2).

From (2) we can say loosely that when $\theta \in \Theta_1$ the sample size required for $\{T_n\}$ to make a correct decision (rejection of H_0) at level α is approximately $2\log(1/\alpha)(c(\theta))^{-1}$ for small α. The (exact) Bahadur slope $c(\theta)$ can thus be interpreted as a measure of the performance of $\{T_n\}$: For every $\theta \in \Theta_1$, the larger $c(\theta)$ is the faster $\{T_n\}$ makes the correct decision (rejection of H_0). If $\{T_n^{(i)}\}$, $i = 1, 2$, are two sequences of test statistics with (exact) Bahadur slope $0 < c_i(\theta) < \infty$ for all $\theta \in \Theta_1$, then (2) implies that

$$\frac{c_1(\theta)}{c_2(\theta)} = \lim_{\alpha\to 0}\frac{N_2(\alpha, s)}{N_1(\alpha, s)}$$

with probability 1 $[P_\theta]$, where $N_i(\alpha, s)$ is the sample size required for $T_n(i)$ to make the correct decision (rejection of H_0) at level α. This suggests $c_1(\theta)/c_2(\theta)$ as a measure of the asymptotic efficiency of $\{T_n^{(1)}\}$ relative to $\{T_n^{(2)}\}$, when θ is the true parameter. We will define, for every $\theta \in \Theta_1$,

$$\epsilon_{12}(\theta) = c_{1(\theta)}/c_{2(\theta)}$$

as the Bahadur efficiency of $\{T_n^{(1)}\}$ relative to $\{T_n^{(2)}\}$ at θ. If $\epsilon_{12}(\theta) > 1$, the sequence of test statistics $\{T_n^{(1)}\}$ is said to be more efficient than the sequence $\{T_n^{(2)}\}$ in the sense of Bahadur efficiency at θ.

To find the (exact) Bahadur slope of a sequence of test statistics $\{T_n\}$, the following result is often useful. Suppose that for every $\theta \in \Theta_1$

$$n^{-1/2}T_n \to b(\theta) \tag{4}$$

with probability 1 $[P_\theta]$, where $-\infty < b(\theta) < +\infty$, and that

$$n^{-1}\log\left[1 - F_n(n^{1/2}t)\right] \to -f(t) \tag{5}$$

for each t in an open interval I, where f is a continuous function on I, and $\{b(\theta) : \theta \in \Theta_1\} \subset \mathrm{I}$. Then the (exact) Bahadur slope $c(\theta)$ exists and

$$c(\theta) = 2f(b(\theta)) \tag{6}$$

for each $\theta \in \Theta_1$. To establish (6), choose a fixed $\theta \in \Theta_1$ and a fixed s such that

$$n^{-1/2}T_n(s) \to b(\theta)$$

as $n \to \infty$. Let $\epsilon > 0$ be so small that $b(\theta) + \epsilon$ and $b(\theta) - \epsilon$ are in I. Since $F_n(t)$ is nondecreasing in t, $L_n(s) = 1 - F_n(T_n(s))$ and $n^{1/2}(b(\theta) - \epsilon) \leqslant T_n(s) \leqslant n^{1/2}(b(\theta) + \epsilon)$ for sufficiently large n, it follows that $1 - F_n(n^{1/2}(b(\theta) + \epsilon)) \leqslant L_n(s) \leqslant 1 - F_n(n^{1/2} \cdot (b(\theta) - \epsilon))$ for sufficiently large n. From (5), we have $-f(b(\theta) + \epsilon) \leqslant \lim_{n\to\infty} \cdot \inf n^{-1}\log L_n(s) \leqslant \lim_{n\to\infty} \sup n^{-1} \log L_n(s) \leqslant -f(b(\theta) - \epsilon)$. Since f is continuous and ϵ is arbitrary, we conclude that $\lim_{n\to\infty} n^{-1} \cdot \log L_n(s) = -f(b(\theta))$. Hence $2^{-1}c(\theta) = f(b(\theta))$ for each $\theta \in \Theta_1$.

An example is given below to illustrate the computation of the (exact) Bahadur slopes of two sequences of test statistics. Let X_1, $X_2, \dots$ be i.i.d. random variables with normal* $N(\theta, 1)$ distributions* and suppose that $H_0 : \theta = 0$ is to be tested against $H_1 : \theta > 0$. Consider $T_n(1) = n^{-1/2}\sum_{i=1}^n X_i$, $T_n^{(2)} = T_n^{(1)}/S_n$, where $S_n = [\sum_{i=1}^n (X_i - \bar{X})^2/(n-1)]^{1/2}$. H_0 is rejected for large values of $T_n^{(i)}$. Then $T_n^{(i)}$ satisfies (4) with $b_1(\theta) = \theta$ and $b_2(\theta) = \theta$ for every $\theta > 0$. Furthermore, $T_n^{(i)}$ satisfies (5) with $f_1(t) = t^2/2$ and $f_2(t) = (1/2)\log(1 + t^2)$. (For more details, see Bahadur [4].) It follows from (6) that $T_n^{(i)}$ has (exact) Bahadur slope $c_i(\theta)$ and $c_1(\theta) = \theta^2$ and $c_2(\theta) = \log(1 + \theta^2)$ for every $\theta > 0$. Hence the Bahadur efficiency of $\{T_n^{(1)}\}$ relative to $\{T_n^{(2)}\}$ is

$$\epsilon_{12}(\theta) = \theta^2/\log(1 + \theta^2)$$

for every $\theta > 0$. Since $\epsilon_{12}(\theta) > 1$ for all $\theta > O$ $\{T_n^{(1)}\}$ is considered to be a better sequence of test statistics in the sense of Bahadur efficiency.

Since the concepts of Chernoff efficiency and Hodges–Lehmann efficiency are closely related to the concept of Bahadur efficiency, we shall give a very brief introduction to these concepts here. If one wishes to compare statistics $T^{(1)}$ and $T^{(2)}$ for testing hypothesis $H_0 : \theta = \theta_0$, Let $N_i = N_i(\alpha, \beta, \theta_0, \theta)$, $i = 1, 2$, denote sample sizes required for statistical tests based on statistic $T^{(i)}$ to have significance level α and power β for alternative θ. The ratio N_2/N_1 describes in general the relative efficiency of a test based on T_1 compared to T_2. Chernoff's approach is essentially equivalent to considering the limit of N_1/N_2 as α tends to zero and β to 1 at a controlled rate, i.e., $\lim_{\alpha\to 0}(\alpha/(1 - \beta)) = c$, where $0 < c < \infty$, while the Hodges–Lehmann approach takes the limit of N_1/N_2 as $\beta \to 1$.

A sequence of test statistics T_n is said to be asymptotically optimal in the sense of Bahadur efficiency for testing H_0 against H_1 if the (exact) Bahadur slope of $\{T_n\}$ is maximum among all sequences of test statistics for all $\theta \in \Theta_1$. A frequently used method of finding an asymptotically optimal sequence of test statistics is to find an upper bound for the (exact) Bahadur slope. A sequence $\{T_n\}$ is then clearly asymptotically optimal if its (exact) Bahadur slope is equal to this upper bound for every $\theta \in \Theta_1$. Suppose that for each θ the distribution of the single observation X admits a density function $f(x, \theta)$ with respect to a fixed measure μ. For any θ and θ_0 in Θ, let the Kullback–Liebler information* number K be defined by

$$K(\theta, \theta_0) = \int f(x, \theta) \log \frac{f(x, \theta)}{f(x, \theta_0)}\, d\mu.$$

Then $0 \leqslant K \leqslant \infty$ and $K = 0$ if and only if $P_\theta \equiv P_{\theta_0}$. For each $\theta \in \Theta$, let

$$J(\theta) = \inf\{K(\theta, \theta_0) : \theta_0 \in \Theta_0\}.$$

If $c(\theta)$ is the (exact) Bahadur slope of a sequence of test statistics $\{T_n\}$, then

$$c(\theta) \leqslant 2J(\theta)$$

for every $\theta \in \Theta_1$.

Under certain regularity conditions, the (exact) Bahadur slope of the sequence of likelihood ratio test* statistics exists and attains the upper bound $2J(\theta)$ for all $\theta \in \Theta_1$. Hence under these conditions, the likelihood ratio rests are asymptotically optimal in the sense of Bahadur efficiency.

An exceptional fine reference for Bahadur efficiency is the monograph written by Bahadur [4]. It contains an almost exhaustive bibliography on this subject (up to 1970). Groeneboom and Oosterhoff [8] wrote an interesting survey paper on Bahadur efficiency and probabilities of large deviations which contains more recent references. A much more general framework for obtaining asymptotical optimal tests in the sense of Bahadur efficiency was discussed in Bahadur and Raghavachari [5]. In Berk and Brown [6] the notion of Bahadur efficiency was extended to the sequential* case.

References

[1] Bahadur, R. R. (1960). *Ann. Math. Statist.*, **31**, 276–295.

[2] Bahadur, R. R. (1960). *Contributions to Probability and Statistics*. Stanford University Press, Stanford, Calif., pp. 79–88.

[3] Bahadur, R. R. (1960). *Sankhyā*, **22**, 229–252.

[4] Bahadur, R. R. (1971). *Some Limit Theorems in Statistics*. SIAM, Philadelphia.

[5] Bahadur, R. R. and Raghavachari M. (1972). *Proc. 6th Berkeley Symp. Math. Stat. Prob.*, Vol. 1. University of California Press, Berkeley, Calif., pp. 129–152.

[6] Berk, R. H. and Brown, L. D. (1978). *Ann. Statist.*, **6**, 567–581.

[7] Chernoff, H. (1952). *Ann. Math. Statist.*, **23**, 493–507.

[8] Groeneboom, P. and Oosterhoff, J. (1977). *Statist. Neerlandica*, **31**, 1–24.

[9] Hodges, J. L. and Lehmann, E. L. (1956). *Ann. Math. Statist.*, **27**, 324–335.

[10] Pitman, E. J. G. (1948). Nonparametric Statistical Inference. Institute of Statistics, University of North Carolina, Chapel Hill, N. C. (unpublished).

[11] Pitman, E. J. G. (1949). Lecture Notes on Nonparametric Statistical Inferences. Columbia University, New York (unpublished).

(EFFICIENCY, SECOND-ORDER
HYPOTHESIS TESTING
STATISTICAL INFERENCE)

S. K. Perng

BAHADUR–LAZARSFELD EXPANSION

A reparameterization of the formula for a certain multinomial* probability, obtained by Bahadur [1]. Essentially, it is the identity

$$P\left[\bigcap_{j=1}^{m}(X_j - x_j)\right] = \left\{\prod_{j=1}^{m} p_j^{x_j}(1-p_j)^{1-x_j}\right\}$$
$$\times\left[1 + \sum_{j<k}\sum E[Z_jZ_k]z_jz_k + \sum\sum_{j<k<l}\sum E[Z_jZ_kZ_l]z_jz_kz_l + \cdots + E[Z_1Z_2\cdots Z_m]z_1z_2\cdots z_m\right],$$

where $X_1, \ldots, X_m$ are (not necessarily independent) Bernoulli variables (*see* BERNOULLI DISTRIBUTION), with $E[X_j] = p_j$ $(j = 1, \ldots, m)$, $Z_j = (X_j - p_j)/\left(\sqrt{p_j(1-p_j)}\right)$ and $z_j = (x_j - p_j)/\left(\sqrt{p_j(1-p_j)}\right)$.

For one application, see Woodruff, et al. [2].

References

[1] Bahadur, R. R. (1961). In *Studies in Item Analysis and Production*, H. Solomon, ed. Stanford University Press, Stanford, Calif.

[2] Woodruff, H. B., Ritter, G. L., Lowry, S. R., and Isenhour, T. L. (1975). *Technometrics*, **17**, 455–462.

(GENERALIZED MULTINOMIAL
LEXIS
MULTINOMIAL)

BALANCED INCOMPLETE BLOCK DESIGN *See* BALANCING IN EXPERIMENTAL DESIGN; INCOMPLETE BLOCK DESIGNS

BALANCED REPEATED REPLICATIONS

Balanced repeated replications (BRR) is a technique that may be used for the estimation of the sampling errors* (variances) for complex statistics derived from probability sample* designs which make use of both clustering* (*see* CLUSTER SAMPLING)* and stratification*. In such situations, traditional methods of sampling error estimation, based upon the Taylor series expansion*, are often too complex for closed-form expression or too cumbersome for routine calculation.

The general approach of repeated replication was developed at the U.S. Census Bureau (*see* BUREAU OF THE CENSUS, U.S.) [1] from basic replication* concepts [4]. Orthogonal balancing was added later [2, 3]. The BRR method can be briefly described as follows. Assume that we have a stratified sample design* with two primary selections (clusters) from each stratum*. Let S denote the entire sample; let H_i denote the ith half-sample formed by including one of the two primary selections from each of the strata; and let C_i denote the ith complement half-sample, formed by the primary selections in S not in H_i. If k repeated divisions of sample S, into half-samples and complement half-samples, satisfies an orthogonal design* (each stratum represented by a column, each half-sample by a row, the selection of a prespecified primary unit* by $+1$ and the other by -1), the set of half and complement half-samples is said to be balanced.

For some statistic $g(S)$, the BRR estimate of variance is given by

$$\mathrm{var}_{\mathrm{BRR}-S}[g(S)] = (1/2k)\sum_{i=1}^{k}\left[(g(H_i)-g(S))^2 + (g(C_i)-g(S))^2\right].$$

Depending upon the sample design, a finite population correction* term may be added to this formula.

If the amount of computation time required for both half-sample $g(H_i)$ and complement half-sample $g(C_i)$ is excessive, the following approximation may be used:

$$\mathrm{var}_{BRR-H}[g(S)] = (1/k)\sum_{i=1}^{k}[g(H_i)-g(S)]^2.$$

In addition to their use for developing estimates of sampling variance from complex statistics (e.g., regression* and correlation* statistics), BRR methods have been for estimating variances of more simple means, proporitions, and ratios when complex sample weighting procedures are in use. For example, BRR methods may be used when successive marginal weighting procedures have been applied. In this case, the weighting algorithm should be applied to each half and complement half-sample prior to the development of $g(H_i)$ and $g(C_i)$.

When the number of strata is small, estimates of variance based upon fully balanced BRR should be treated as having degrees of freedom equal to the number of strata. This property may be derived in the case of simple linear estimators* and has been demonstrated empiricially in more complex situations.

References

[1] Deming, W. E. (1956). *J. Amer. Statist. Ass.*, **51**, 24–53.

[2] Kish, L. and Frankel, M. R. (1970). *J. Amer. Statist. Ass.*, **65**, 1071–1094.

[3] McCarthy, P. J. (1966). Replication: An Approach to the Analysis of Data from Complex Surveys. *Natl. Center Health Statist. Ser.* 2, **14**, (Washington, D. C.).

[4] Mahalanobis, P. C. (1944). *Phil. Trans. R. Soc. Lond. B*, **231**, 329–451.

(DESIGN AND ANALYSIS OF EXPERIMENTS
REPLICATIONS
SAMPLE SURVEYS
STRATIFIED DESIGNS)

MARTIN R. FRANKEL

BALANCING IN EXPERIMENTAL DESIGN

Assignment of the treatment combinations* to the experimental units* in such a way that a "balanced" or symmetric configuration is obtained. For example, in a two-factor experiment* involving a certain technological process in an accelerated or a steady production, the two factors may be A and B. If the same number of responses are obtained for each A–B combination, we say that the design is balanced. Otherwise, it is unbalanced or the data are missing.

(BLOCKS, BALANCED INCOMPLETE
DESIGN OF EXPERIMENTS)

BALDUCCI HYPOTHESIS

In life tables*, values of l_x (number surviving to age x among l_0 newborn) are given only for discrete values of x (usually at exact years). The Balducci hypothesis is that for purposes of approximation between neighboring tabulated values, l_{x_0} and l_{x_1}, l_x^{-1} may be assumed to be a linear function of x for $x_0 \leqslant x \leqslant x_1$ so that

$$l_x \doteqdot \left[l_{x_0}^{-1} + \frac{x - x_0}{x_1 - x_0} \left(l_{x_1}^{-1} - l_{x_0}^{-1} \right) \right]^{-1}$$

$$\doteqdot \frac{x_1 - x_0}{(x_1 - x_0) l_{x_0}^{-1} + (x - x_0) l_{x_1}^{-1}}.$$

This assumption implies that the force of mortality* ($\mu_x = -l_x^{-1} dl_x/dx$) *increases* with x for $x_0 \leqslant x \leqslant x_1$.

(INTERPOLATION
LIFE TABLES)

BALLOT PROBLEMS

The first ballot theorem was discovered by J. Bertrand in 1887 [5]. He found the following result.

Theorem 1. If in a ballot candidate A scores a votes and candidate B scores b votes where $a \geqslant b$, then the probability that throughout the counting the number of votes registered for A is always greater than the number of votes registered for B is given by

$$P(a,b) = (a - b)/(a + b) \qquad (1)$$

provided that all the possible voting records are equally probable.

$P(a,b)$ can be expressed as $N(a,b)/\binom{a+b}{a}$, where $N(a,b)$ is the number of favorable voting records and $\binom{a+b}{a}$ is the number of possible voting records. Bertrand observed that $N(a,b) = N(a-1,b) + N(a,b-1)$ for $a > b$, $N(a,b) = 0$ for $a \leqslant b$ and deduced (1) from these equations. Also in 1887, D. André [2] proved that

$$N(a,b) = \binom{a+b}{a} - 2\binom{a+b-1}{a} \qquad (2)$$

by using an ingenious method based on the reflection principle*.

Bertrand's result (1) can be traced back to 1708, when A. De Moivre* (1667–1754) solved the following problem of games of chance [9].

Two players, A and B, agree to play a series of games. In each game, independently of the others, either A wins a coin from B with probability p or B wins a coin from A with probability q, where $p > 0$, $q > 0$ and $p + q = 1$. Let us suppose that A has an unlimited number of coins, B has only k coins, and the series ends when B is ruined, i.e., when B loses his last coin. Denote by $\rho(k)$ the duration of games, i.e., the number of games played until B is ruined. The problem is to determine the distribution of $\rho(k)$. De Moivre discovered that

$$\mathbb{P}\{\rho(k) = k + 2j\} = \frac{k}{k+2j} \binom{k+2j}{j} p^{k+j} q^j \qquad (3)$$

for $k \geqslant 1$ and $j \geqslant 0$. De Moivre stated (3) without proof. Formula (3) was proved only in 1773 by P. S. Laplace* (1749–1827) and in 1776 by J. L. Lagrange (1736–1813). In 1802, A. M. Ampère expressed his view that formula (3) is remarkable for its simplicity and elegance.

It is convenient to write

$$L(j,k) = \frac{k}{k+2j}\binom{k+2j}{j} \tag{4}$$

for $j \geqslant 0$, $k \geqslant 1$, $L(0,0) = 1$ and $L(j,0) = 0$ for $j \geqslant 1$. The numbers $L(j,k)$ might appropriately be called De Moivre numbers. They can also be expressed as

$$L(j,k) = \binom{k+2j-1}{j} - \binom{k+2j-1}{j-1} \tag{5}$$

for $j \geqslant 1$, $k \geqslant 0$ and $L(0,k) = 1$ for $k \geqslant 0$. See the following table.

$L(j,k)$

k \ j	0	1	2	3	4	5
0	1	0	0	0	0	0
1	1	1	2	5	14	42
2	1	2	5	14	42	132
3	1	3	9	28	90	297
4	1	4	14	48	165	572
5	1	5	20	75	275	1001

By De Moivre's result, the numbers $L(j,k)$ can be interpreted in the following way: *One can arrange $k+j$ letters A and j letters B in $L(j,k)$ ways so that for every $r = 1, 2, \ldots, k+2j$ among the first r letters there are more A than B.*

Since $N(a,b) = L(b, a-b)$, the result of Bertrand follows from that of De Moivre.

The numbers $L(j,k)$ have appeared in various forms in diverse fields of mathematics.

In 1751, L. Euler (1707–1783) encountered the numbers $L(j,1)$. In 1751, in a letter to Chr. Goldbach, Euler wrote that the number of different ways of dissecting a convex polygon of n sides into $n-2$ triangles by $n-3$ nonintersecting diagonals is

$$D_n = \frac{2 \cdot 6 \cdot 10 \cdots (4n-10)}{2 \cdot 3 \cdot 4 \cdots (n-1)} \tag{6}$$

for $n \geqslant 3$. From De Moivre's result one can deduce that $D_n = L(n-2, 1)$. Euler noticed that $D_n = D_{n-1}(4n-10)/(n-1)$ for $n \geqslant 3$ where $D_2 = 1$; however, apparently, he did not succeed in proving it. In his reply to Euler, Goldbach observed that the numbers $D_n (n \geqslant 2)$ satisfy the equation

$$D_n = \sum_{r=2}^{n-1} D_r D_{n-1-r} \tag{7}$$

for $n \geqslant 3$. It seems that it escaped the attention of both Goldbach and Euler that (7) is easy to prove and that (7) implies (6). Formula (6) was proved only in 1758 by J. A. de Segner. Other proofs and generalizations were given by N. v. Fuss (1793), G. Lamé (1838), E. Catalan (1838; ref. 7), O. Rodrigues (1838), J. Binet (1839), J. A. Grunert (1841), J. Liouville (1843), E. Schröder (1870), T. P. Kirkman (1860), A. Cayley (1890), and others.

In 1838, E. Catalan [7] gave an interesting interpretation of the numbers $L(j,1)$. He proved that the number of ways a product of n factors can be calculated by pairs is

$$P_n = \binom{2n-1}{n-1}\frac{1}{2n-1} \tag{8}$$

for $n \geqslant 2$. By the result of De Moivre, we have $P_n = L(n-1, 1)$.

In 1859, A. Cayley (1821–1895) encountered the numbers $L(j,1)$ in the theory of graphs* [8]. He defined a tree as a connected graph containing no circuits, and a rooted tree as a tree in which one vertex is distinguished as the root of the tree. A planted tree is a tree rooted at an end vertex. The number of edges having an end point in a vertex is called the valence of the vertex. A tree is called trivalent if every vertex has valence 3 except the end vertices, which have valence 1. Cayley proved that the number of trivalent planted trees with $2n$ vertices is given by (8) for $n \geqslant 2$. In 1964, F. Harary, G. Prins, and W. Tutte proved that the number of planted trees with $n+1$ vertices is also given by (8) for $n \geqslant 1$. Other proofs were given by N. G. DeBruijn and B. J. M. Morselt (1967) and D. A. Klarner (1969).

In 1879, W. A. Whitworth demonstrated that one can arrange $j+k$ letters A and j letters B in

$$M(j,k) = \binom{2j+k}{j}\frac{k+1}{j+k+1} \tag{9}$$

ways such that for every $r = 1, 2, \ldots, 2j +$

k among the first r letters there are at least as many A as B. Since, evidently, $L(j,k+1) = (j+k+1)M(j,k)$, (9) follows from (4).

It was already mentioned that in Bertrand's ballot theorem, $N(a,b) = L(b,a-b)$.

In 1897, W. A. Whitworth [32] proved that the number of ways $2n$ points on a circle can be joined in pairs so that the n chords do not intersect each other is $L(n,1)$. This problem has also been studied by A. Errera (1931), J. Touchard (1950), and J. Riordan (1975).

In 1887, É. Barbier [3] generalized Bertrand's result (1) in the following way.

Theorem 2. If in a ballot candidate A scores a votes and candidate B scores b votes, and $a \geqslant \mu b$, where μ is a nonnegative integer, then the probability that throughout the counting the number of votes registered for A is always greater than μ times the number of votes registered for B is given by

$$P(a,b;\ \mu) = (a - b\mu)/(a+b) \qquad (10)$$

provided that all the possible voting records are equally probable.

É. Barbier did not prove (10). This was proved only in 1924 by A. Aeppli [1]. Other proofs were given by A. Dvoretzky and Th. Motzkin (1947), H. D. Grossman (1950), S. G. Mohanty and T. V. Narayana [21], and others.

In 1960, L. Takács [26] generalized Theorem 2 in the following way.

Theorem 3. Let us suppose that a box contains n cards marked $a_1, a_2, \ldots, a_n$, where $a_1, a_2, \ldots, a_n$ are nonnegative integers with sum $a_1 + a_2 + \cdots + a_n = k$, where $k \leqslant n$. We draw all the n cards without replacement. Let us assume that every arrangement has the same probability. The probability that the sum of the first r numbers drawn is less than r for every $r = 1, 2, \ldots, n$ is given by

$$P(n,k) = (n-k)/n. \qquad (11)$$

In 1977, L. Takács [30] pointed out that Theorem 3 can also be deduced from a theorem of G. Hajós [14] which he discovered in 1949 and proposed as a problem. Now there are several proofs for (11), those of J. C. Tanner (1961), M. Dwass (1962), H. Dinges (1963), R. L. Graham (1963), J. L. Mott (1963), S. G. Mohanty (1966), J. G. Wendel (1975), and M. Folledo and I. Vincze (1976).

Theorem 3 has many possible applications in various fields of mathematics, and, in particular, in probability theory and in mathematical statistics. Here are a few examples.

First, an application in mathematical logic. Let $x_1, x_2, x_3, \ldots$ be variables and $f_1, f_2, f_3, \ldots$ be unary, binary, ternary, ... operations. A finite expression such as $f_2(x_1, f_3(x_2, x_3, f_1(x_1)))$ is called a "word" if it represents a meaningful mathematical formula. In 1930, K. Menger and in 1948, D. C. Gerneth gave a necessary and sufficient condition for a finite expression to be a word, or, more generally, a sequence of m finite expressions to be a list of m words. Let us replace in each expression every variable by 0 and every $f_i (i = 1, 2, \ldots)$ by i. Denote by a_i the number of symbols i in the m expressions. A sequence of m expressions is a list of m words if and only if $a_0 = m + \sum_{i=1}^{\infty}(i-1)a_i$ and the sum of the last r numbers is less than r for every $r = 1, 2, \ldots$. In 1960, G. N. Raney found that the number of lists of m words of type $(a_0, a_1, a_2, \ldots)$ with a finite length $a_0 + a_1 + a_2 + \cdots$ is

$$L(m; a_1, a_2, \ldots) = \frac{m}{(m + \sum_{i=1}^{\infty} i a_i)} \times \frac{(a_0 + a_1 + a_2 + \cdots)!}{a_0! a_1! a_2! \cdots}. \qquad (12)$$

Formula (12) is an immediate consequence of (11).

As a second example, let us consider a single-server queue* in the time interval $(0, \infty)$. It is supposed that the server starts working at time $t = 0$ and at this time k $(k \geqslant 0)$ customers are already waiting for service. Denote by $\nu_r (r = 1, 2, \ldots)$ the number of customers arriving during the rth service and by $\sigma(k)$ the number of customers served in the initial busy period. ($\sigma(0) = 0$.) Set $N_r = \nu_1 + \nu_2 + \cdots + \nu_r$ for $r \geqslant 1$ and $N_0 = 0$. If $\nu_1, \nu_2, \ldots, \nu_r$ are mutually independent, identically distributed (i.i.d.) random variables, then

$$\mathbb{P}\{\sigma(k) = n\} = \mathbb{P}\{N_r < r \text{ for } 1 \leqslant r \leqslant n \text{ and } N_n = n - k\} \quad (13)$$

for $0 \leqslant k \leqslant n$. By Theorem 3 one can conclude that

$$\mathbb{P}\{\sigma(k) = n\} = (k/n)\mathbb{P}\{N_n = n - k\} \quad (14)$$

for $0 \leqslant k \leqslant n$. Formula (14) was found by É. Borel [6] in 1942 in the particular case where $\mathbb{P}\{\nu_r = j\} = e^{-\lambda}\lambda^j/j!$ (i.e., Poisson* distributed) for $j \geqslant 0$ and $k = 1$. In the general case the distribution of $\sigma(k)$ was determined by D. G. Kendall [18] in 1951 and L. Takács [25] in 1952. The distribution of $\sigma(k)$ can also be obtained from a result of R. Otter [23] for branching processes* and from a result of J. H. B. Kemperman [17] for random walks*.

Theorem 3 has numerous possible applications in the theory of order statistics*. See Takács [27, 28] for more details. Here only a few results are mentioned. Let

$$\delta_n^+ = \sup_{-\infty < x < \infty} \left[F_n(x) - F(x) \right], \quad (15)$$

where $F_n(x)$ is the empirical distribution function* of a sample of size n whose elements are mutually independent random variables each having the same continuous distribution function $F(x)$. By Theorem 3 we have

$$\mathbb{P}\left\{ \delta_n^+ \leqslant \frac{k}{n} \right\} = 1 - \sum_{j=1}^{n-k} \frac{k}{(n-j)} \binom{n}{j+k} \times \left(\frac{j}{n} \right)^{j+k} \left(1 - \frac{j}{n} \right)^{n-j-k} \quad (16)$$

for $k = 1, 2, \ldots, n$. The probability $\mathbb{P}\{\delta_n^+ \leqslant x\}$ was obtained by N. V. Smirnov [24] in 1944.

Let $\xi_1, \xi_2, \ldots, \xi_m, \eta_1, \ldots, \eta_n$ be mutually independent random variables each having the same continuous distribution function. Denote by $F_m(x)$ and $G_n(x)$ the empirical distribution functions of the samples $(\xi_1, \xi_2, \ldots, \xi_m)$ and $(\eta_1, \eta_2, \ldots, \eta_n)$, respectively. The distribution of the random variable

$$\delta^+(m, n) = \sup_{-\infty < x < \infty} \left[F_m(x) - G_n(x) \right] \quad (17)$$

was found by B. V. Gnedenko and V. S. Koroljuk [12] for $n = m$ and V. S. Koroljuk [19] for $n = mp$, where p is a positive integer. If $n = mp$, where p is a positive integer and k is a nonnegative integer, then by Theorem 3 we obtain that

$$\mathbb{P}\left\{ \delta^+(m,n) \leqslant \frac{k}{n} \right\} = 1 - \frac{1}{\binom{m+n}{m}} \sum_{(k+1)/p \leqslant s \leqslant m} \frac{k+1}{n+k+1-sp} \times \binom{sp+s-k-1}{s} \binom{m+n+k-sp-s}{m-s}. \quad (18)$$

Ballot theorems have been used by S. G. Mohanty and T. V. Narayana [21] in characterizing various sampling plans*. Takács [27] used Theorem 3 in studying the urn model of F. Eggenberger and G. Pólya [10] (see Johnson and Kotz [15].)

The following generalization of Bertrand's ballot problem has attracted a great deal of interest.

Let us assume again that in a ballot candidate A scores a votes and candidate B scores b votes, and all the possible voting records are equally probable. Denote by α_r and β_r the number of votes registered for A and B, respectively, among the first r votes counted. The problem is to determine $P_j(a, b; \mu)$, the probability that the inequality $\alpha_r > \mu\beta_r$ holds for exactly j subscripts $r = 1, 2, \ldots, a + b$.

In the case where $\mu = a/b$ and $(a, b) = 1$, H. D. Grossman (1954) proved that $P_j(a, b; \mu) = 1/(a + b)$ for $j = 0, 1, \ldots, a + b - 1$.

In the case where μ is a positive integer $P_j(a, b; \mu)$ was found in various cases by L. Takács (1962 [26], 1963, 1964), J. Riordan (1964), and O. Engelberg (1964, 1965).

In 1966, M. T. L. Bizley made a conjecture concerning the general form of

$P_j(a,b;\mu)$. Bizley's formula was proved in 1969 by Takács [29].

In 1909, P. A. MacMahon [20] proved a ballot theorem for n candidates. In his theorem it is supposed that in a ballot candidates $A_1, A_2, \ldots, A_n$ score $a_1, a_2, \ldots, a_n$ votes, respectively. Let $a_1 \geqslant a_2 \geqslant \cdots \geqslant a_n$. Denote by $\alpha_1(r), \alpha_2(r), \ldots, \alpha_n(r)$ the number of votes registered for $A_1, A_2, \ldots, A_n$, respectively, among the first r votes counted. If all the possible voting records are equally probable, then

$$\mathbb{P}\{\alpha_1(r) \geqslant \alpha_2(r) \geqslant \cdots \geqslant \alpha_n(r) \text{ for } r = 1, 2, \ldots, a_1 + \cdots + a_n\} = \prod_{1 \leqslant i < j \leqslant n} \left(\frac{a_i - a_j + j - i}{a_i + j - i} \right). \quad (19)$$

We have also

$$\mathbb{P}\{\alpha_1(r) > \alpha_2(r) > \cdots > \alpha_n(r) \; r = 1, 2, \ldots, a_1 + \ldots + a_n\} = \prod_{1 \leqslant i < j \leqslant n} \left(\frac{a_i - a_j}{a_i + a_j} \right). \quad (20)$$

This formula has been proved by H. D. Grossman [13] for $n = 3$ and in another context by R. M. Thrall [31] and R. Srinivasan (1963) for $n \geqslant 3$.

By using a result of S. Karlin and J. McGregor [16], D. E. Barton and C. L. Mallows [4] proved that if $c_1 > c_2 > \cdots > c_n$ are integers, then

$$\mathbb{P}\{\alpha_1(r) + c_1 > \alpha_2(r) + c_2 > \cdots > \alpha_n(r) + c_n \text{ for } r = 1, 2, \ldots, a_1 + \cdots + a_n\} = \det \left[\frac{a_i!}{(a_i + c_i - c_j)!} \right]_{i,j = 1, 2, \ldots, n}, \quad (21)$$

where in the determinant the (i, j)-entry is 0 for $a_i + c_i - c_j < 0$. See also T. V. Narayana [22].

References

[1] Aeppli, A. (1924). Zur Theorie verketteter Wahrscheinlichkeiten. Markoffsche Ketten höherer Ordnung. Dissertation, Die Eidgenössische Technische Hochschule, Zurich.

[2] André, D. (1887). *C. R. Acad. Sci. Paris*, **105**, 436–437.

[3] Barbier, É. (1887). *C. R. Acad. Sci. Paris*, **105**, 407; 440 (errata).

[4] Barton, D. E. and Mallows, C. L. (1965). *Ann. Math. Statist.*, **36**, 236–260.

[5] Bertrand, J. (1887). *C. R. Acad. Sci. Paris*, **105**, 369.

[6] Borel, É. (1942). *C. R. Acad. Sci. Paris*, **214**, 452–456.

[7] Catalan, E. (1838). *J. Math. Pures Appl.*, **3**, 508–516.

[8] Cayley, A. (1859). *Philos. Mag.*, **13**, 172–176; *ibid.*, **18**, 374–378 (1859).

[9] De Moivre, A. (1711). *Philos. Trans. R. Soc.*, **27**, 213–264.

[10] Eggenberger, F. and Pólya, G. (1923). *Zeit. angew. Math. Mech.*, **3**, 279–289.

[11] Fuss, P.-H. (1843). *Correspondance mathématique et physique de quelques célèbres géomètres du XVIIIème siècle*, Vols. 1, 2. St. Petersburg. (Reprinted: *The Sources of Science No.* 35. Johnson Reprint Corporation, New York, 1968.)

[12] Gnedenko, B. V. and Koroljuk, V. S. (1951). *Dokl. Akad. Nauk SSSR*, **80**, 525–528. (in Russian). [English translation: *Select. Transl. Math. Statist. Prob.*, **1**, 13–16 (1961).]

[13] Grossman, H. D. (1952). *Scr. Math.*, **18**, 298–300.

[14] Hajós, G. (1949). *Mat. Lapok*, **1**, 72; *ibid.*, **1**, 152 (1950). [Russian and French translations: *Ibid.*, **7**, 115, 153 (1956).]

[15] Johnson, N. L. and Kotz, S. (1977). *Urn Models and Their Application: An Approach to Modern Discrete Probability Theory*. Wiley, New York.

[16] Karlin, S. and McGregor, J. (1959). *Pacific J. Math.*, **9**, 1141–1164.

[17] Kemperman, J. H. B. (1950). The General One-Dimensional Random Walk with Absorbing Barriers with Applications to Sequential Analysis. Thesis, Excelsior, The Hague.

[18] Kendall, D. G. (1951). *J. R. Statist. Soc. B*, **13**, 151–185.

[19] Koroljuk, V. S. (1955). *Izv. Akad. Nauk SSSR. Ser. Mat.*, **19**, 81–96. [English translation: Select. Transl. Math. Statist. Prob., **4**, 105–121 (1963).]

[20] MacMahon, P. A. (1909). *Philos. Trans. R. Soc. Lond. A*, **209**, 153–175. [Reprinted in *Combinatorics*, G. E. Andrews, ed. MIT Press, Cambridge, Mass., 1978, pp. 1292–1314.]

[21] Mohanty, S. G. and Narayana, T. V. (1961). *Biom. Zeit.*, **3**, 252–258; *Ibid.*, **5**, 8–18 (1963).

[22] Narayana, T. V. (1955). *J. Indian Soc. Agric. Statist.*, **7**, 169–178.

[23] Otter, R. (1949). *Ann. Math. Statist.*, **20**, 206–224.

[24] Smirnov, N. V. (1944). *Uspekhi Mat. Nauk*, **10**, 179–206 (in Russian).

[25] Takács, L. (1955). *Acta Math. Acad. Sci. Hung.*, **6**, 101–129.

[26] Takács, L. (1962). *J. Amer. Statist. Ass.*, **57**, 327–337.

[27] Takács, L. (1967). *Combinatorial Methods in the Theory of Stochastic Processes*. Wiley, New York.

[28] Takács, L. (1970). In *Nonparametric Techniques in Statistical Inference*, M. L. Puri, ed. Cambridge University Press, Cambridge, pp. 359–384.

[29] Takács, L. (1970). *J. Appl. Prob.*, **7**, 114–123.

[30] Takács, L. (1980). *Periodica Math. Hung.*, **11**, 159–160.

[31] Thrall, R. M. (1952). *Mich. Math. J.*, **1**, 81–88.

[32] Whitworth, W. A. (1897). *DCC Exercises Including Hints for the Solution of All the Questions in Choice and Chance*. Deighton Bell, Cambridge. [Reprinted by Hafner, New York, 1965.]

(COMBINATORICS
GRAPH THEORY
QUEUEING THEORY)

LAJOS TAKÁCS

BANACH'S MATCH-BOX PROBLEM

Banach's match-box problem is a classical problem, related to "waiting time" (negative binomial*), discrete distributions, and combinatorial probability theory. (For its history, see Feller [1, p. 166].)

A man carries two boxes of matches, one in his left and one in his right pocket. Initially, they contain N matches each. They are emptied one match at a time—the box to have a match taken from it being chosen each time at random, with each equally likely. Denote by **R** the random variable representing the number of matches remaining in the other box when, for the first time, one of the boxes is found to be empty. The probability distribution of **R** is given by

$$\Pr[R = r] = \binom{2N - r}{N} 2^{-2N+r}$$

and

$$E(R) = 2^{-2N}(2N+1)\binom{2N}{N} - 1$$

$$\approx 2\sqrt{N/\pi} - 1$$

(for N sufficiently large, using Stirling's formula*). The value of $E(R)$ for $N = 50$ is approximately 7.04.

References

[1] Feller, W. (1968). *An Introduction to Probability Theory and Its Applications*, 3rd. ed., Vol. 1. Wiley, New York, pp. 166, 170, 238.

[2] Moran, P. A. P. (1968). *An Introduction to Probability Theory*. Clarendon Press, Oxford, p. 59.

(COMBINATORICS)

BAR CHART

A form of graphical representation applicable to data classified in a number of (usually nonordered) categories. Equal-width rectangular bars are constructed over each category with *height* equal to the frequency or other measurement associated with the category. (This is a widespread representation method in newspapers, magazines, and general publications.)

Example. Energy consumption per capita in millions BTUs in the United States in 1977 was:

Gas	Coal	Petroleum	Other	Total
120	60	170	25	375

The corresponding bar chart is shown in Fig. 1.

(GRAPHICAL REPRESENTATION
OF DATA
HISTOGRAMS
PIE CHART)

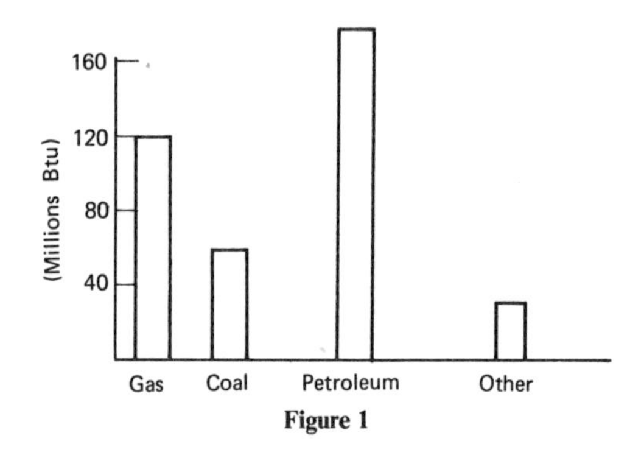

Figure 1

BARLOW – SCHEUER RELIABILITY GROWTH MODEL

A model used to represent probabilities of "inherent" failure and assignable-cause* failure in successive stages of a trial ($i = 1, 2, \ldots, k$) by δ (for all i) and q_i, respectively, with $q_1 \geqslant q_2 \geqslant \cdots \geqslant q_k$. At the ith stage, the probability of survival to the end of the stage, starting from the beginning of the stage, is $(1 - \delta - q_i)$.

If trial outcomes are independent and there are a_i inherent failures, b_i assigned-cause failures and $(n_i - a_i - b_i) - c_i$ survivors among n_i exposed to risk in the ith trial, then the maximum likelihood estimate* of δ is

$$\hat{\delta} = \left(\sum_{i=1}^{k} a_i \right) \Big/ \left(\sum_{i=1}^{k} n_i \right)$$

and (provided that $b_i + c_i \neq 0$) the estimates of $\hat{q}_1, \ldots, \hat{q}_k$ satisfy the equations

$$\hat{q}_i = (1 - \hat{\delta})b_i/(n_i - a_i) \qquad (i = 1, \ldots, k).$$

Read (1971) discusses appropriate procedures when $b_i + c_i = 0$. See also Weinrich and Gross (1978) for a discussion from a Bayesian viewpoint.

References

[1] Barlow, R. E. and Scheuer, E. M. (1966). *Technometrics*, **8**, 53–60.

[2] Read, R. R. (1971). *Technometrics*, **13**, 199–200.

[3] Weinrich, M. C. and Gross, A. J. (1978). *Technometrics*, **20**, 249–254.

(GROWTH MODELS
MULTINOMIAL
RELIABILITY)

BARTLETT'S TEST OF HOMOGENEITY OF VARIANCES

Tests of homogeneity of variances concern the question of whether several (say $k > 2$) populations have the same variance. If the populations are assumed to be approximately normally distributed, then a test due to Bartlett [1] is the standard procedure. This test is frequently called Bartlett's M-test, or the Neyman–Pearson–Bartlett test. Applications arise in the analysis-of-variance* context, principally in testing the equal "within-group" variances assumption. Bartlett's test may also be used to check the homogeneity* of high-order interactions* in factorial experiments* and, in time-series* analysis, to investigate fluctuations in smoothed periodograms*.

METHODOLOGY

Suppose that there are k normally distributed populations whose means μ_j and variances σ_j^2 are unknown, $j = 1, \ldots, k$. Independent random samples are taken, one from each population. Let n_j denote the sample size for the jth population, and $\{X_{ji}\}$, $i = 1, \ldots, n_j$, the sample. Further define $S_j^2 = (1/\nu_j)\sum_{i=1}^{n_j}(X_{ji} - \overline{X}_{j\cdot})^2$, where $\nu_j = n_j - 1$ and $\overline{X}_{j\cdot} = (1/n_j)\sum_{i=1}^{n_j} X_{ji}$. To test $H_0 : \sigma_1^2 = \cdots = \sigma_k^2$ versus $H_1 : \sigma_t^2 \neq \sigma_u^2$ for some $t \neq u$, consider the statistic $L = \prod_{j=1}^{k} (S_j^2)^{\nu_j/\nu} / \sum_{j=1}^{k} (\nu_j/\nu) S_j^2$, where $\nu = \sum_{j=1}^{k} \nu_j$. The Bartlett homogeneity-of-variances test has critical region* $0 < L < A$, where A is determined by the size of the test. In particular, the size α Bartlett critical value, A, is determined by the equation $P_{H_0}\{0 < L < A\} = \alpha$.

Derived from considerations of sufficiency*, the Bartlett test has intuitive appeal. The sample variances S_j^2 are unbiased estimators* of the population variances σ_j^2. The numerator of L is the weighted geometric mean* (GM) of the S_j^2's, where the weights are the relative degrees of freedom. The denominator of L is the corresponding weighted arithmetic mean* (AM) of the S_j^2's. It is wellknown that for positive numbers, the ratio of weighted GM to weighted AM is less than or equal to 1, with equality if and only if the numbers are all equal, (*see* ARITHMETIC MEAN). Consequently, high values of L suggest closeness among the σ_j^2's, and lower values of L suggest greater disparity among the σ_j^2's.

Determination of the critical value, A, needed to carry out the Bartlett procedure is usually accomplished by means of a table. Two types of tables exist: one based on approximate null distributions* of L, and the other based on exact distributions. Comparison with exact results has established that tables [17, pp. 63–66, 204–206; 23] of Hartley's approximation [11] of Bartlett critical values are quite accurate. Furthermore, these tables are quick and easy to use. (A demonstration is offered by Thompson and Merrington [23].) The Hartley approximation supersedes the approximation offered by Bartlett in his original paper [1]. Bartlett actually introduced his procedure interms of the statistic $M = -\nu \ln L = \nu \ln(\sum_{j=1}^{k}(\nu_j/\nu) S_j^2) - \sum_{j=1}^{k}\nu_j \ln S_j^2$, which under H_0 has a limiting χ_{k-1}^2 (chi-squared*) distribution, as each ν_j gets arbitrarily large. To achieve closeness to χ^2 for finite ν_j, Bartlett suggested using the modification $T = M/C$, where $C = 1 + \{[\sum_{j=1}^{k}(1/\nu_j)] - (1/\nu)\} /[3(k-1)]$. A test with size approximately α, then, consists of rejecting H_0 for any T greater than the upper 100α percentage point of the χ_{k-1}^2 distribution.

Exact Bartlett critical values, obtained from the exact null distribution of L, exist in tabular form only for the equal-sample-sizes case $n_1 = \cdots = n_k$ [9, 10]. For the unequal-sample-sizes case, two equivalent approaches requiring the use of a computer are available to provide exact critical values to handle any situation $(\nu_1, \ldots, \nu_k)$. A complicated expression for the exact null density of L can be integrated numerically for this purpose (see Chao and Glaser [5]). Alternatively, an exact expression for the null cumulative distribution function of L involving incomplete beta functions* (*see* BETA DISTRIBUTIONS) can be used (see Glaser [9a]).

PROPERTIES

The Bartlett statistic L is a slight modification of the likelihood ratio* statistic L_1 proposed by Neyman and Pearson [16]: $L_1 = \prod_{j=1}^{k}(\hat{\sigma}_j^2)^{n_j/n} / \sum_{j=1}^{k}(n_j/n)\hat{\sigma}_j^2$, where $n = \sum_{j=1}^{k} n_j$ and $\hat{\sigma}_j^2 = (\nu_j/n_j)S_j^2$, i.e., the maximum likelihood* estimator of σ_j^2. Thus L_1 uses for weights the relative sample sizes rather than the relative degrees of freedom. Although both the L and L_1 tests are consistent* against all alternatives [3], the Bartlett test L is preferable because the L_1 test is biased unless the sample sizes are equal [3], whereas the L test is unbiased for any sample sizes [18]. The competing tests are equivalent in the equal-sample-sizes case, since then $L = L_1$.

In terms of power, the Bartlett test is considered to be the best available performer. No optimum, such as uniformly most powerful unbiased*, test exists for the homogeneity of variances problem. However, Monte Carlo* studies (e.g., refs. 8 and 13) have demonstrated the superiority in power of Bartlett's test (assuming normality) relative to such competitors as tests by Cochran* [6], Hartley* [12], Cadwell [4], Box [2], and the jackknife* method [13]. A challenge to Bartlett superiority is offered in a simulation study by Samiuddin et al. [20], where it is suggested that the procedure of Samiuddin and Atiqullah [19], which employs cube roots of the sample variances, may provide power higher than Bartlett's.

A major objection to use of the Bartlett test is its sensitivity to the assumption of normality* [2]. If the populations are not approximately normally distributed, regardless of the sample sizes used, the actual significance level of the Bartlett procedure can differ greatly from the nominal significance level. In particular, the degree of disparity depends on the kurtosis* of the populations. The real significance level is too low for negative kurtosis (as in uniform distributions) and too high for positive kurtosis (as in double exponential distributions). As a consequence, if the assumption of normality is not reasonable, the use of Bartlett's test is not recommended as a preliminary to standard analysis of variance. (In fact, some would argue that whether normality holds or not, any preliminary homogeneity of variances test is academic, since the analysis-of-variance F-test* is affected little by variance inequalities and is very insensitive to nonnormality [21].) Nevertheless, its very sensitivity to departures from normality allows the Bartlett statistic to perform a dual role as a device for ascertaining normality [2]. To handle the homogeneity of variances problem when normality cannot reasonably be assumed, several tests have been proposed which are robust*; i.e., these tests have true significance level approximately equal to the nominal significance level, regardless of the underlying distributions (see refs. 7, 13, and 14).

AN EXAMPLE

Five brands of men's running shoes are to be compared for rear-foot impact, i.e., the force transferred through the heel of the shoe upon a standard impact. Previous analysis has suggested that the rear-foot impact of a randomly selected shoe is approximately a normally distributed random variable. Samples of 7, 13, 13, 19, and 13 shoes, respectively, are to be taken from the $k = 5$ shoe populations. A size 0.05 Bartlett test for homogeneity of variances is desired. The appropriate approximate critical value, found from Tables 1 and 3 of [23], is $A \doteq 0.849$. (The exact Bartlett critical value, rounded to three decimal places, obtained from the method of Chao and Glaser [5] or Glaser [9a], is 0.849 as well.) Consequently, the hypothesis of equal variances is rejected at the 0.05 level if the value of L computed from the data is less than 0.849. If, for instance, experimentation yields $(s_1^2, \dots, s_5^2) = (10.5, 4.8, 3.6, 8.9, 11.1)$, then the observed value of L is $[(10.5)^{0.1}(4.8)^{0.2}(3.6)^{0.2}(8.9)^{0.3} \cdot (11.1)^{0.2}]/[0.1(10.5) + 0.2(4.8) + 0.2(3.6) + 0.3(8.9) + 0.2(11.1)] = 0.915$, and the hypothesis of equal variances is accepted at the 0.05 level.

References

[1] Bartlett, M. S. (1937). *Proc. R. Soc. Lond. A*, **160**, 268–282.

[2] Box, G. E. P. (1953). *Biometrika*, **40**, 318–335.

[3] Brown, G. W. (1939). *Ann. Math. Statist.*, **10**, 119–128.

[4] Cadwell, J. H. (1953). *Biometrika*, **40**, 336–346.

[5] Chao, M. T. and Glaser, R. E. (1978). *J. Amer. Statist. Ass.*, **73**, 422–426.

[6] Cochran, W. G. (1941). *Ann. Eugen. (Lond.)*, **11**, 47–52.

[7] Draper, N. R. and Hunter, W. G. (1969). *Technometrics*, **11**, 23–40.

[8] Gartside, P. S. (1972). *J. Amer. Statist. Ass.*, **67**, 342–346. (An exhaustive bibliography is included.)

[9] Glaser, R. E. (1976). *J. Amer. Statist. Ass.*, **71**, 488–490. [Tables of exact Bartlett critical values are presented for cases of equal sample sizes $n = 5(1)12$ and $n = 15, 20, 25, 30, 50, 100$, with $k = 3(1)10$ and $\alpha = 0.01, 0.05$, and 0.10.]

[9a] Glaser, R. E. (1980). *Biometrika*, **67**, 53–58.

[10] Harsaae, E. (1969). *Biometrika*, **56**, 273–281. [Tables of exact Bartlett critical values are presented for cases of equal sample sizes $n = 2(1)11$, with $\alpha = 0.001, 0.01, 0.05$, and 0.10, and $k = 3(1)12$ for $n \leqslant 9$ and $k = 3(1)10$ for $n = 10$ and 11.]

[11] Hartley, H. O. (1940). *Biometrika*, **31**, 249–255.

[12] Hartley, H. O. (1950). *Biometrika*, **37**, 308–312.

[13] Layard, M. W. J. (1973). *J. Amer. Statist. Ass.*, **68**, 195–198.

[14] Levene, H. (1960). In *Contributions to Probability and Statistics*, I Olkin, ed. Stanford University Press, Stanford, Calif., pp. 278–292.

[15] Neter, J. and Wasserman, W. (1974). *Applied Linear Statistical Models*. Richard D. Irwin, Homewood, Ill. (A fine presentation and discussion of Bartlett's test is offered in this standard text.)

[16] Neyman, J. and Pearson, E. S. (1931). *Bull. Acad. Pol. Sci. Lett. A*, 460–481.

[17] Pearson, E. S. and Hartley, H. O. (1966). *Biometrika Tables for Statisticians*, 3rd ed., Vol. 1. Cambridge University Press, Cambridge.

[18] Pitman, E. J. G. (1939). *Biometrika*, **31**, 200–215.

[19] Samiuddin, M. and Atiqullah, M. (1976). *Biometrika*, **63**, 206–208.

[20] Samiuddin, M., Hanif, M., and Asad, H. (1978). *Biometrika*, **65**, 218–221.

[21] Scheffé, H. (1959). *The Analysis of Variance*. Wiley, New York.

[22] Snedecor, G. W. and Cochran, W. G. (1967). *Statistical Methods*, 6th ed. Iowa State University Press, Ames, Iowa. (Bartlett's test is given a clear presentation in this standard text.)

[23] Thompson, C. M. and Merrington, M. (1946). *Biometrika*, **33**, 296–301.

(ANALYSIS OF VARIANCE (ANOVA)
COCHRAN'S (TEST) STATISTIC
HOMOGENEITY OF VARIANCES
LIKELIHOOD RATIO TEST
TIME SERIES)

R. E. Glaser

BARTON–DAVID TEST

The Barton–David (B–D) test is a nonparametric (distribution-free*) test used to test the equality of scale parameters of two continuous populations which have a common median*. The data consist of $X_1, X_2, \ldots, X_m$ and $Y_1, Y_2, \ldots, Y_n$. The observations are arranged in increasing order and rank scores are assigned as follows. If $N = n + m$ is even, the scores are $N/2, N/2 - 1, \ldots, 2, 1, 1, 2, \ldots, N/2 - 1, N/2$. If N is odd, the scores are $(N-1)/2, (N-3)/2, \ldots, 1, 0, 1, \ldots, (N-3)/2, (N-1)/2$. The B–D statistic, B_N, is the sum of the ranks assigned to the X's.

The mean and variance of B_N under the null hypothesis* $H_0 : \sigma_x = \sigma_y$ are

$$E(B_N) = \frac{m(N+2)}{4},$$

$$\operatorname{Var}(B_N) = \frac{mn(N^2-4)}{48(N-1)} \quad \text{for } N \text{ even}$$

$$E(B_N) = \frac{m(N^2-1)}{4N},$$

$$\operatorname{Var}(B_N) = \frac{mn(N+1)(N^2+3)}{48N^2} \quad \text{for } N \text{ odd.}$$

For large values of N the statistic $Z = \{B_N - E(B_N \mid H_0)\}/\sqrt{\operatorname{var}(B_N \mid H_0)}$ has an asymptotic standard normal distribution* and may be used to test H_0.

The B–D statistic is linearly related to the Ansari–Bradley statistic*, W, according to $B_N = m[(N+2)/2] - W$, where $[x]$ denotes the greatest integer less than or equal to x. The null distribution of B_N has been tabulated for $N \leqslant 16$ [2]. The W statistic [1] is better known than B_N. *See* ANSARI–BRADLEY *W* STATISTIC.

A more general statistic considered by Barton and David is obtained by summing the ranks of the X's if the ranks 1, 1 are assigned to the Rth and $(R+1)$st observations and increased by 1 monotonically in both directions.

References

[1] Ansari, A. R. and Bradley, R. A. (1960). *Ann. Math. Statist.*, **31**, 1174–1189.

[2] Barton, D. E. and David, F. N. (1958). *Ann. Hum. Genet.*, **22**, 250–257.

(DISTRIBUTION-FREE METHODS)

BENJAMIN S. DURAN

BARYCENTRIC COORDINATES

Introduced by A. F. Möbius in 1827 [2], they are a special case of homogeneous coordinates*.

PHYSICAL MOTIVATION IN TWO DIMENSIONS

Let masses m_1, m_2, and m_3 be concentrated on three noncollinear points A_1, A_2, and A_3. Then in the plane generated by these points a unique point A exists (the centroid or center of gravity of these points) containing mass m such that $m = m_1 + m_2 + m_3$, and the weighted sums of projections of AA_j ($j = 1, 2, 3$; weights m_j/m). Conversely, for any point A—located in the plane of three fixed points A_1, A_2, and A_3—possessing mass $m (m \neq 0)$, there correspond three specific masses m_1, m_2, m_3 ($m_1 + m_2 + m_3 = m$) which, being located at points A_1, A_2, and A_3, respectively, have point A as their center of gravity. The "masses" m_1, m_2, and m_3 may take on positive or negative values. These masses m_1, m_2, m_3 can be viewed as coordinates of the point A possessing mass m and are called barycentric coordinates of point A with respect to A_1, A_2, and A_3.

If the mass m increases k-fold, the coordinates m_1, m_2, and m_3 also increase k-fold. The same point A may have different barycentric coordinates. The point A with mass m possessing barycentric coordinates m_1, m_2, and m_3 may be located inside, outside, or on the boundary of the triangle A_1, A_2, A_3. To normalize, we usually take $m_1 + m_2 + m_3 = 1$.

GENERAL DEFINITION

Let $\mathbf{x}_1, \mathbf{x}_2, \ldots, \mathbf{x}_n$ be linearly independent points (or vectors) in n-dimensional Euclidean space. For any vector (point) $\mathbf{x}$ belonging to E_n, there is only one set of real numbers $b_1, b_2, \ldots, b_n$ which satisfy $\sum_{i=1}^n b_i = 1$ and are such that $\mathbf{x} = b_1\mathbf{x}_1 + \cdots + b_n\mathbf{x}_n$. These numbers $b_1, \ldots, b_n$ are called the barycentric coordinates of the vector $\mathbf{x}$ with respect to simplex* S determined by $\mathbf{x}_1, \ldots, \mathbf{x}_n$. If $\mathbf{x} = b_1\mathbf{x}_1 + b_2\mathbf{x}_2 + b_3\mathbf{x}_3 + \cdots + b_n\mathbf{x}_n$ with the additional requirements $b_1 \geqslant 0$, $b_2 \geqslant 0, \ldots, b_n \geqslant 0$, it is called a *convex linear combination* of $\mathbf{x}_1, \mathbf{x}_2, \ldots, \mathbf{x}_n$. It includes all points belonging to the boundary and the interior of the simplex determined by $\mathbf{x}_1, \mathbf{x}_2, \ldots, \mathbf{x}_n$. Barycentric coordinates are utilized in noncooperative game theory* (see Vorob'yev [4]) and in problems relating to definitions of measures of distances between categories: e.g., in multinomial distributions*. (see Gibbons et al. [1]).

References

[1] Gibbons, J. D., Olkin, I. and Sobel, M. (1977). *Selecting and Ordering Populations*. Wiley, New York.

[2] Möbius, A. F. (1827). *Der baryzentrische Kalkül*. Gesammelte Werke, Vol. 1. Leipzig.

[3] Skibinsky, M. (1976). *Ann. Statist.*, 4, 187–213.

[4] Vorob'yev, N. N. (1977). *Game Theory for System Scientists*, S. Kotz, trans./ed. Springer-Verlag, New York.

BASIS

Any set of linearly independent vectors $\mathbf{v}_1, \mathbf{v}_2, \ldots, \mathbf{v}_n$ such that every vector in a vector space can be written as a linear combination of $\mathbf{v}_1, \ldots, \mathbf{v}_n$ is a *basis* of the space (provided that $v_1, \ldots, v_n$ belong to the space).

In general, there are many different possible bases for a given vector space. The minimum number of members needed for a basis is the *dimension* of the vector space.

Example. $b_1 = \binom{1}{1}$ and $b_2 = \binom{1}{-1}$ is a basis in the two-dimensional Euclidean space R^2. We can express the vector $\binom{2}{1}$ as

$$\begin{pmatrix} 2 \\ 1 \end{pmatrix} = \frac{3}{2}\begin{pmatrix} 1 \\ 1 \end{pmatrix} + \frac{1}{2}\begin{pmatrix} 1 \\ -1 \end{pmatrix}.$$

A basis consisting of *unit* vectors δ_i is called a *standard* or *canonical* basis. A canonical basis of R^n consists of vectors

$$\delta_1 = \begin{bmatrix} 1 \\ 0 \\ \vdots \\ 0 \end{bmatrix}, \quad \delta_2 = \begin{bmatrix} 0 \\ 1 \\ \vdots \\ 0 \end{bmatrix}, \ldots, \delta_n = \begin{bmatrix} 0 \\ 0 \\ 0 \\ \vdots \\ 1 \end{bmatrix}.$$

BASU THEOREMS

The theorems are related to the notions of sufficiency*, ancillarity, and conditional independence*. Let X denote the sample and θ the parameter that completely specifies the sampling distribution P_θ of X. An event E is ancillary* if $P_\theta(E)$ is θ-free, i.e., $P_\theta(E) = P_{\theta'}(E)$ for all $\theta, \theta' \in \Theta$, the parameter space. A statistic $Y = Y(X)$ is ancillary if every Y-event (i.e., a measurable set defined in terms of Y) is ancillary (*see* ANCILLARY STATISTICS). A statistic T is sufficient* if, for every event E, there exists a θ-free version of the conditional probability* function $P_\theta(E \mid T)$ (*see* SUFFICIENCY). The event E is (conditionally) independent of T if, for each $\theta \in \Theta$, the conditional probability function $P_\theta(E \mid T)$ is P_θ-essentially equal to the constant $P_\theta(E)$. The statistic Y is independent of T if every Y-event is independent of T. (Independence is a symmetric relationship between two statistics.)

The theorems originated in the following query. Let X_1, X_2 be independent, identically distributed (i.i.d.) $N(\theta, 1)$ and let $Y = X_1 - X_2$. Clearly, Y is ancillary and, therefore, so also is every measurable function $h(Y)$ of Y. The statistic Y is shift invariant* in the sense that $Y(X_1 + a,\ X_2 + a) = Y(X_1, X_2)$ for all X_1, X_2, and a. It is easy to see that every shift-invariant statistic is a function of Y, and vice versa. Therefore, every shift invariant statistic is ancillary. Is the converse true?

That the answer has to be in the negative is seen as follows. The statistic $T = X_1 + X_2$ is sufficient and is independent of the ancillary statistic $Y = X_1 - X_2$. Let A be an arbitrary T-event and B_1, B_2 be two distinct Y-events such that $P_\theta(B_1) = P_\theta(B_2) = \alpha$, where $0 < \alpha < 1$ is a constant chosen and fixed. Consider the event $E = AB_1 \cup A^cB_2$. The T-events, A, A^c are independent of the Y-events B_1, B_2; therefore,

$$\begin{aligned} P_\theta(E) &= P_\theta(AB_1) + P_\theta(A^cB_2) \\ &= P_\theta(A)P_\theta(B_1) + P_\theta(A^c)P_\theta(B_2) \\ &= \alpha\left[P_\theta(A) + P_\theta(A^c)\right] \\ &= \alpha \qquad \text{for all } \theta. \end{aligned}$$

Thus, E is ancillary even though it is not shift-invariant (not an Y-event). How do we characterize the class of ancillary events in this case?

Consider an arbitrary ancillary event E with $P_\theta(E) \equiv \alpha$. Since $T = X_1 + X_2$ is sufficient, there exists a θ-free version $f(T)$ of the conditional probability function $P_\theta(E \mid T)$ (*see* SUFFICIENCY). Now, $E_\theta f(T) = P_\theta(E) \equiv \alpha$, so $f(T) - \alpha$ is a bounded function of T that has zero mean for each $\theta \in \Theta$. But the statistic T is complete in the sense that no nontrivial function of T can have identically zero mean. Therefore, the event $f(T) \neq \alpha$ is P_θ-null for each $\theta \in \Theta$. In other words, $P_\theta(E \mid T) = P_\theta(E)$ a.s., $[P_\theta]$ for each θ. That is, every ancillary E is independent of $T = X_1 + X_2$. Is the converse true?

Let T be an arbitrary sufficient statistic and let E be independent of T. Let $f(T)$ be a θ-free version of $P_\theta(E \mid T)$. Then, for each $\theta \in \Theta$, $f(T) = P_\theta(E)$ a.s. $[P_\theta]$. If $\mathcal{X}_\theta$ is the set of all sample points for which $f(T) = P_\theta(E)$, then $P_\theta(\mathcal{X}_\theta) = 1$ for all $\theta \in \Theta$. If $P_{\theta_1}(E) \neq P_{\theta_2}(E)$, then the two sets $\mathcal{X}_{\theta_1}$ and $\mathcal{X}_{\theta_2}$ are disjoint and so P_{θ_1} and P_{θ_2} have disjoint supports, which is a contradiction in the present case. Thus the class of ancillary events may be characterized as the class of events that are independent of $X_1 + X_2$.

The "Basu theorems" are direct generalizations of the foregoing results and may be stated as follows.

Theorem 1. Let T be sufficient and boundedly complete*. Then a statistic Y is ancillary only if it is (conditionally) independent of T for each θ.

The measures P_θ and $P_{\theta'}$ are said to overlap if they do not have disjoint supports. The family $\mathcal{P} = \{P_\theta\}$ of measures on a space $\mathcal{X}$ is said to be connected if for all θ, θ' there exists a finite sequence $\theta_1, \theta_2, \dots, \theta_k$ such that every two consecutive members of the sequence $P_\theta, P_{\theta_1}, P_{\theta_2}, \dots, P_{\theta_k}, P_{\theta'}$ overlap. For example, if under P_θ, $-\infty < \theta < \infty$, the random variables $X_1, X_2, \dots, X_n$ are i.i.d. with a common uniform distribution concentrated on the interval $(\theta, \theta + 1)$, then the family $\mathcal{P} = \{P_\theta\}$ is connected even though P_θ and $P_{\theta'}$ do not overlap whenever $|\theta - \theta'| \geqslant 1$.

Theorem 2. Let $\mathcal{P} = \{P_\theta\}$ be connected and T be sufficient. Then Y is ancillary if it is (conditionally) independent of T for each θ.

Neither the condition of bounded completeness (in Theorem 1) nor that of connectedness (in Theorem 2) can be entirely dispensed with. If $T = X$, the whole sample, then it is sufficient. Consider, therefore, a case where a nontrivial ancillary statistic $Y = Y(x)$ exists. Such an Y cannot be independent of $T = X$, because, if it were, then Y has to be independent of itself, which it cannot be unless it is a constant (a trivial ancillary). On the other hand, if $\mathcal{P}$ is not connected, then it is typically true that there exists nonempty proper subsets $\mathcal{X}_0 \subset \mathcal{X}$ and $\Theta_0 \subset \Theta$ of the sample space $\mathcal{X}$ and the parameter space Θ, respectively, such that

$$P_\theta(\mathcal{X}_0) = \begin{cases} 1 & \text{for all } \theta \in \Theta_0 \\ 0 & \text{for all } \theta \in \Theta - \Theta_0 \end{cases}.$$

Koehn and Thomas [7] called such a set $\mathcal{X}_0$ a splitting set. A splitting set (event) $\mathcal{X}_0$ is clearly not ancillary; however, for every $\theta \in \Theta$, it is P_θ-equivalent either to the whole space $\mathcal{X}$ (the sure event) or to the empty set ϕ (the impossible event). Therefore, the non-ancillary event $\mathscr{X}_0$ is independent of every other event E; that is, $\mathscr{X}_0$ is independent of the sufficient statistic $T = X$. Basu [2] gave a pathological example of a statistical model, with a disconnected $\mathscr{P}$, where we have two independent sufficient statistics.

Consider the following three propositions:

(a) T is sufficient.
(b) Y is ancillary.
(c) T and Y are (conditionally) independent for each $\theta \in \Theta$.

Under suitable conditions (a) and (b) together imply (c) (Theorem 1) and (a) and (c) together imply (b) (Theorem 2). The following theorem completes the set.

Theorem 3. If (T, Y) is jointly sufficient, then (b) and (c) imply (a).

Theorem 1 is often used to solve diverse problems involving sampling distributions. The following example illustrates this.

Example. Let $X_1, X_2, \cdots$ be a sequence of mutually independent gamma* variables with shape parameters $\alpha_1, \alpha_2, \ldots,$ (i.e., the PDF* of X_n is $Cx^{\alpha_n - 1}e^{-x}, x > 0,\ n = 1, 2, \ldots$.) Let $T_n = X_1 + X_2 + \cdots + X_n$ and $Y_n = T_n/T_{n+1}, \mathrm{n} = 1, 2, \ldots$. It is easy to show that T_n has a gamma distribution* with shape parameter $\alpha_1 + \alpha_2 + \cdots + \alpha_n$ and that Y_n has a beta distribution* with parameters $\alpha_1 + \alpha_2 + \cdots + \alpha_n$ and α_{n+1}. But it is not clear why $Y_1, Y_2, \ldots, Y_{n-1}, T_n$ have to be mutually independent. This is seen as follows.

Introduce a scale parameter* θ into the joint distribution of $X_1, X_2, \ldots$. That is, suppose that the X_n's are mutually independent and that the PDF of X_n is $c(\theta) \cdot x^{\alpha_n - 1}e^{-x/\theta},\ x > 0, \theta > 0, n = 1, 2, \ldots$. Regard the α_n's as known positive constants and θ as the unknown parameter. With $\mathbf{X}^{(n)} = (X_1, X_2, \ldots, X_n)$ as the sample, $T_n = X_1 + X_2 + \cdots + X_n$ is a complete sufficient statistic*. The vector-valued statistic $\mathbf{Y}^{(n-1)} = (Y_1, Y_2, \ldots, Y_{n-1})$ is scale-invariant. Since θ is a scale parameter, it follows that $\mathbf{Y}^{(n-1)}$ is an ancillary statistic. From Theorem 1 it then follows that, for each n, $\mathbf{Y}^{(n-1)}$ is independent of T_n. Since $(\mathbf{Y}^{(n-1)}, T_n)$ is a function of the first nX_i's, the pair $(\mathbf{Y}^{(n-1)}, T_n)$ is independent of X_{n+1}. Thus $\mathbf{Y}^{(n-1)}, T_n$, and X_{n+1} are mutually independent. It follows at once that $\mathbf{Y}^{(n-1)}$ is independent of $Y_n = T_n/(T_n + X_{n+1})$. Therefore, for each $n \geqslant 2$, the vector $(Y_1, Y_2, \ldots, Y_{n-1})$ is independent of Y_n and this means that the Y_i's are mutually independent.

(Refer to Hogg and Craig [6] for several interesting uses of Theorem 1 in proving results in distribution theory.)

Basu [1] stated and proved Theorem 1 in the generality stated here. At about the same time, Hogg and Craig proved a particular case of the theorem. Basu [1] stated Theorem 2 without the condition of connectedness for $\mathscr{P}$. This has resulted in the theorem being incorrectly stated in several statistical texts. Basu [2] stated and proved Theorem 2 in the general form stated here. Koehn and Thomas [7] noted that the proposition "every Y that is conditionally independent of a sufficient T is ancillary" is true if and only if there do not exist a splitting set as specified above. It turns out, however, that the two notions of connectedness and nonexistence of splitting sets coincide for all typical statistical models.

The Basu theorems are of historical interest because they established a connection between the three apparently unrelated notions of sufficiency, ancillarity, and independence. That the three notions really hang together is not easy to see through if we adopt an orthodox Neyman–Pearson* point of view. However, if we take a Bayesian* view of the matter and regard θ as a random variable with a (prior) probability distribution ξ and the model $\mathscr{P}$ as a specification of the set of conditional distributions of X given θ, then the notions of ancillarity and sufficiency will appear to be manifestations of the notion of conditional independence.

For a model $\mathscr{P}$ and for each prior ξ, consider the joint probability distribution Q_ξ

of the pair (θ, X). A statistic $Y = Y(X)$ is ancillary if its conditional distribution, given θ, is θ-free. In other words, Y is ancillary if, for each joint distribution Q_ξ of (θ, X), the two random variables Y and θ are stochastically independent. A statistic $T = T(X)$ is sufficient if the conditional distribution of X, given θ and T, depends only on T (i.e., the conditional distribution is θ-free). Sufficiency of T may, therefore, be characterized as follows:

Definition. The statistic T is sufficient if, for each Q_ξ, X and θ are conditionally independent given T.

Thus a neo-Bayesian* version of Theorem 1 may be stated as:

Theorem 1(*a*). Suppose that, for each Q_ξ, the variables Y and θ are stochastically independent and also X and θ are conditionally independent given T. Then Y and T are conditionally independent given θ provided that the statistic T is boundedly complete in the sense described earlier.

Refer to Florens and Mouchart [5] for more on the Bayesian insight on the theorems and also to Dawid [4] for a clear exposition on conditional independence as a language of statistics.

References

[1] Basu, D. (1955). *Sankhyā*, **15**, 377.

[2] Basu, D. (1958). *Sankhyā*, **20**, 223.

[3] Basu, D. (1959). *Sankhyā*, **21**, 247.

[4] Dawid, A. P. (1979). *J. R. Statist. Soc. B*, **41**, 1.

[5] Florens, J. P. and Mouchart, M. (1977). Reduction of Bayesian Experiments. *CORE Discuss. Paper 1737.*

[6] Hogg, R. V. and Craig, A. T. (1956). *Sankhyā*, **37**, 209.

[7] Koehn, U. and Thomas D. L. (1975). *Amer. Statist.*, **39**, 40.

(ANCILLARY STATISTICS
CONDITIONAL INFERENCE
STRUCTURAL INFERENCE
SUFFICIENCY)

D. Basu

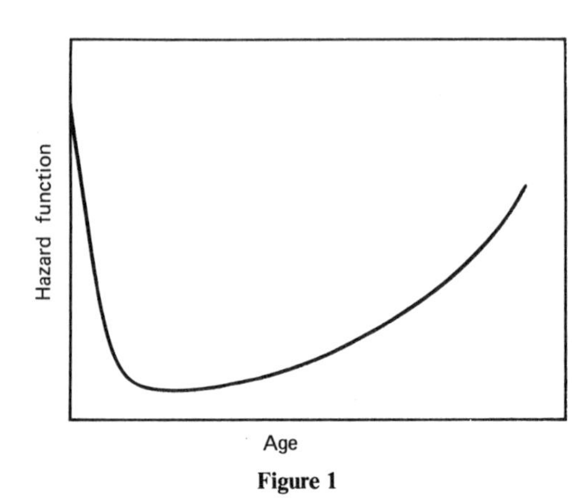

Figure 1

BATHTUB CURVE

For many observed lifetime distributions*, the hazard function* starts (at age 0) at a fairly high level, decreases fairly soon to a minimum, and then starts to increase again, at first slowly but then more and more rapidly. The resultant curve is as shown in Fig. 1.

The name "bathtub curve" is given to curves of this type, although most of them would represent rather uncomfortable bathtubs! Such curves are typical of force of mortality* for human populations. They also apply to some mechanical and electrical products for which there are early failures of substandard specimens, whereas the standard product wears out more slowly but at an increasing rate as time progresses. Of the periods shown in Fig. 1, period I is often called the "infant mortality" period or "break-in" period, and period III is the "wear-out" period.

Bibliography

Bury, A. K. (1975). *Statistical Methods in Applied Science*. Wiley, New York.

(LAWS OF MORTALITY
LIFE TABLES)

BAULE'S EQUATION

Baule's equation is a generalization of Mitscherlich's equation*. It is used, for example, to estimate amounts of two elements in the soil. The equation is given explicitly by

$$Y = A(1 - b_1 r_1^{x_1})(1 - b_2 r_2^{x_2}),$$

where Y represents yield, x_1 and x_2 represent the amounts of the nutrients, A is the theoretical maximum yield, and r_m estimates the efficiency of fertilizer m $(m = 1, 2)$ in the soil. A detailed discussion is given by Patterson [1].

Reference

[1] Patterson, H. D. (1969). *Biometrics*, **25**, 159–162.

BAYESIAN INFERENCE

According to the Bayesian view, all quantities are of two kinds: those known to the person making the inference and those unknown to the person; the former are described by their known values, the uncertainty surrounding the latter being described by a joint probability distribution* for them all. In the context of the usual statistical model with a random variable X having possible distributions indexed by a parameter θ, the data x becomes known to the statistician and the object is to make inferences concerning the unknown parameter; so in the Bayesian approach, the statistician will wish to calculate the probability distribution of θ given $X = x$. Once this is done point estimates* of θ can be calculated; e.g., the mean or mode of this distribution: perform significance tests* of a null value θ_0 by considering the probability that θ equals θ_0: and determine interval estimates* by finding intervals of θ-values containing an assigned probability for θ. If θ is vector-valued, say $\theta = (\theta_1, \theta_2, \ldots, \theta_p)$, and only θ_1 is of interest, then the marginal distribution of θ_1 can be found from the distribution of θ, the remaining values $\theta_2, \ldots, \theta_p$ being regarded as nuisance parameters*. Within the Bayesian framework *all* calculations are performed by the probability calculus, using the probabilities for the unknown quantities.

The statistical model admits, for each θ, a distribution of X. Describe this by a density $p_X(x \mid \theta)$ for $X = x$, given θ, the suffix being omitted when it is clear which random variable is involved. The Bayesian interpretation of this is that it describes the uncertainty about the data, necessarily before it is known, were the parameter value known. The parameter will have its density $p(\theta)$, so that the required distribution of the first paragraph, again described by a density, can be calculated by Bayes' theorem*:

$$p(\theta \mid x) = p(x \mid \theta) p(\theta) / \int p(x \mid \theta) p(\theta)\, d\theta. \tag{1}$$

(The denominator is the unconditional density of X, describing the real uncertainty over the data before it is observed.) It is this basic use of Bayes theorem that gives the method its name. Notice that in addition to the density $p(x \mid \theta)$ used in most statistical situations, the Bayesian view introduces $p(\theta)$. This is called the prior (or a priori) distribution* of θ, prior, that is, to the data being avaialble. Similarly, $p(\theta \mid x)$ is termed the posterior distribution*. The data density and the prior together provide a complete probability specification for the situation from which other uncertainties can be calculated.

As an example, consider a situation in reliability* engineering where n items are put on test, the lifetime x_i of each being supposed exponential* with unknown hazard rate* θ, so that $p(x_i \mid \theta) = \theta e^{-\theta x_i}$ for $x_i \geqslant 0$ and $\theta > 0$, and given θ, the lifetimes of the n items being independent. Suppose that $r (\leqslant n)$ of the items fail at times $x_1, x_2, \ldots, x_r$ and that the remainder of them are withdrawn at times $x_{r+1}, \ldots, x_n$ without having failed (*see* CENSORING). With $X = (x_1, \ldots, x_r;\ x_{r+1}, \ldots, x_n)$, $p(x \mid \theta)$ is $\theta^r e^{=\theta T}$, where $T = \sum_{i=1}^n x_i$, the total time on test. It is technically convenient to suppose that θ has a gamma* density, $p(\theta) = \theta^a e^{-\theta b} b^{a+1} / a!$ for some $b > 0$ and $a > -1$. Then easy calculations in (1) show that

$$p(\theta \mid x) = \theta^{a+r} e^{-\theta(b+T)} \times (b + T)^{a+r+1} / (a + r)!, \tag{2}$$

also gamma. The mean is $(a + r + 1)/(b + T)$, providing an estimate of θ. The probability that the hazard rate exceeds an assigned value θ_1 may be found as the integral of (2) for $\theta \geqslant \theta_1$. If values of the hazard rate exceeding θ_1 are considered dangerous, this probability says how reasonable it is, in the light of the data, that a dangerous situation obtains. Notice that this is a direct probability statement about the unknown quantity of interest (unlike the circumlocutory statements provided by confidence intervals*).

An important deduction can be made from (1). Once the data x are known, θ is the only unknown quantity and $p(\theta \mid x)$ completely describes the uncertainty. But the right-hand side of (1) immediately shows that this depends on the data only through $p(x \mid \theta)$; or, more exactly, through the function $p(x \mid \cdot)$ with argument θ. This is called the likelihood function* (of θ, for $X = x$) and we have the likelihood principle*, which says, in its Bayesian form, that, given the data, all inferences about θ require only the likelihood $p_X(x \mid \cdot)$ for $X = x$. Consider the implications of this for the reliability problem of the preceding paragraph. From (2) we see that r and T, the number of failures and the total time on test, are sufficient* for θ, and that it does not matter how r and T were obtained within the conditions under which (2) was derived. Thus we could have tested for a fixed time t, so that $x_{r+1} = \cdots = x_n = t$, or we could have tested until a preassigned number, r, of failures had been observed; and there are many other possibilities. In all these cases the final inference is the same, and provided by (2), because the likelihood is the same in every one.

Almost all standard, nonBayesian statistical procedures violate the likelihood principle; i.e., they do depend on other aspects of the data besides the likelihood function. This dependence is caused by using integration over x-values to make the inference. Thus the concept of an unbiased estimate*, $t(x)$, demands $\int t(x)p(x \mid \theta)\,dx = \theta$ for all θ; a test of a hypothesis* (*see* HYPOTHESIS TESTING) uses a significance level* $\int_R p(x \mid \theta_0)\,dx$, where R is the region of rejection (critical region*) and θ_0 the "null" value. This integration uses $p_X(\cdot \mid \theta)$ for values other than the observed value, x, thus violating the principle. In many cases it is not obvious which values of X to use for the integration. In the reliability problem we would have one region of integration if T were fixed, and a different one were r to be preselected. The Bayesian argument is often criticized for introducing the prior distribution. It is not often realized that orthodox arguments for inference introduce a set of X-values not needed by the Bayesian. Satisfactory inference scarcely seems possible without using something in addition to the likelihood function, such as other data values or a prior; adherents of the likelihood school having to introduce some device, such as marginal likelihoods*, in order to eliminate nuisance parameters.

There are two fundamental reasons for adopting the Bayesian position. The first is that once the basic step of describing uncertainty through probability is admitted, we have a formal procedure for solving all inference problems; and it is a procedure that works. The procedure, or recipe, runs as follows. First, say what is known; that is, identify the data X. Second, specify the model for the data generation, a family of distributions for x indexed by parameter θ. [This includes nonparametric (distribution-free*) ideas where θ may, e.g., with X a real number, index all continuous distributions on the real line.] Third, specify the uncertainty concerning θ. Fourth, consider what quantity is of interest; usually, this will be a part of θ, so that if $\theta = (\theta_1, \theta_2)$, it will be θ_1, θ_2 being nuisance. Fifth, calculate $p(\theta_1 \mid x)$, the uncertainty of the quantity of interest given the observed value x of the data; this calculation proceeding by Bayes' theorem and subsequent integration of θ_2. This provides a complete inference for θ_1. It is important to recognize that this is a genuine recipe for inference, saying just what has to be done and how to do it; so that there is no need for the introduction of any other principles. For example, a principle that selects a good estimator is not needed, since

$p(\theta_1 \mid x)$ provides everything. Equally, significance levels play virtually no role in Bayesian analysis, although some writers feel they have some use for a Bayesian. The Bayesian recipe is at present the only recipe for inference of wide applicability. It is also important to recognize that the procedure works; there are no examples known to me where the result is unsatisfactory. Indeed, since the procedure is well defined and therefore testable, a single counterexample is all that is needed to make it unacceptable.

The second reason for describing uncertainty in probability terms is mathematical. Using some reasonable requirements of uncertainty as axioms in a mathematical system, it can be proved that such a description exists. The axioms are called those of coherence*, and the Bayesian system is the only coherent one. It therefore follows that other systems are necessarily incoherent and, in some respect, violate simple, axiomatic requirements of uncertainty. It is possible to take any statistical procedure that is non-Bayesian and produce an example where it performs unsatisfactorily. For example, the best, unbiased estimate of θ^2, based on the observation of r successes in n Bernoulli trials* of constant probability θ of success, is $r(r-1)/\{n(n-1)\}$, which is unsound whenever $r = 1$, for to estimate the chance of two future successes as zero when one has been observed is surprising, to say the least.

The concept of probability used in Bayesian inference is not based on frequency considerations; rather, it describes the uncertainty felt by the person making the inference. Some hold that it is objective in the sense that all persons having the same information about θ will have the same probability; others hold a subjective view (*see* SUBJECTIVE PROBABILITY), that $p(\theta)$ reflects the subject's own uncertainty; but neither admits any frequency interpretation. At first sight this seems in conflict with the frequency view that dominates most statistical thinking (*see* FOUNDATIONS OF PROBABILITY), e.g., in the concept of a random sample $x = (x_1, x_2, \ldots, x_n)$ from a population described by θ. There is, however, a simple connection that we illustrate in the context of a Bernoulli sequence of trials*, although the argument is general. The judgment of uncertainty that says a sample is random can be expressed by saying that the order of $x_1, x_2, \ldots, x_n$ is irrelevant. This is captured by saying that $p(x_1, x_2, \ldots, x_n)$ is exchangeable*—the probability here expressing one's uncertainty about the x's. A fundamental theorem due to de Finetti says that if this is true for all n, then this probability must, for the Bernoulli case, be equal to $\int_0^1 \theta^r (1-\theta)^{n-r}\, dP(\theta)$, where $P(\cdot)$ is a distribution function on $[0, 1]$ and r is the number of successes in n trials; furthermore, the limit of r/n as $n \to \infty$ exists and has the distribution described by $P(\cdot)$. Consequently, the exchangeable model has basically the binomial* structure in $\theta^r(1-\theta)^{n-r}$ for fixed θ, this last having the "prior" distribution $P(\cdot)$.

In the case where $P(\cdot)$ has a density $p(\cdot)$, we have the same structure as described earlier, P or p, expressing one's belief about the limiting frequency ratio. Ordinary statistical language would refer to θ as the probability of success for a single trial. In the Bayesian view θ is only a probability in a restricted sense; its proper interpretation is a quantity that is unknown to one and is therefore assessed probabilistically by $p(\theta)$. It is referred to as a chance*, or propensity. It appears, almost incidentally, as a probability since $p(x_i = 1 \mid \theta) = \theta$; that is, were θ known it would be equal to the probability, or the belief (*see* BELIEF FUNCTIONS; DEGREES OF BELIEF), that a trial will result in a success. In the realistic case where θ is unknown, $p(x_i = 1) = \int \theta\, dP(\theta)$, the expectation of θ. Thus the concept of exchangeability provides an important link between Bayesian ideas and the notion of a random sample, a link that is of vital importance if Bayesian ideas are to be applied to the situations of practical importance studied in statistics. Outside the Bernoulli case, a judgment of exchangeability is conveniently described by a random sample, given a parameter θ and a distribution over θ.

There is some advantage in using uncertain quantities that can eventually become known, rather than those, such as, θ, whose value is never known, because they have a reality not possessed by others that makes their uncertainty easier to assess. For this reason it is convenient to express inference rather differently from the manner already mentioned as passing from x to θ, using $p(\theta \mid x)$. Thus when dealing with a random sample $\mathbf{x}^{(n)} = (x_1, x_2, \ldots, x_n)$, it is preferable to consider x_{n+1}, a further value judged exchangeable with the x_i, and to express one's inference through $p(x_{n+1} \mid x^{(n)})$. When x_i is the yield of a variety on the ith plot of an agricultural trial with n plots in all, interest truly centers on what would happen were the variety to be planted on a further plot, $n + 1$. This is a better model of the practical situation than that which introduces parameters such as the "true" yield. In many cases the connection between x_{n+1} and $x^{(n)}$ is more complicated than simple exchangeability, but for that case the de Finetti representation can simplify the calculation. For then $p(x_{n+1} \mid \mathbf{x}^{(n)}) = p(\mathbf{x}^{(n+1)})/p(\mathbf{x}^{(n)})$; and the denominator is $\int \prod_{i=1}^{n} p(x_i \mid \theta)\, dP(\theta)$, the x's, given θ, being a random sample, i.e., independent and identically distributed. A similar expression with $n + 1$ for n holds for the numerator. Alternatively,

$$p(x_{n+1} \mid \mathbf{x}^{(n)}) = \int p(x_{n+1} \mid \theta) p(\theta \mid \mathbf{x}^{(n)})\, d\theta$$

since, given θ, x_{n+1} and $\mathbf{x}^{(n)}$ are independent, whence the inference about x_{n+1} is expressed in terms of that about θ. In the Bernoulli case, $p(x_{n+1} = 1 \mid \theta) = \theta$ and $p(x_{n+1} = 1 \mid x^{(n)})$ is simply the mean of the posterior distribution of θ, $E(\theta \mid x^{(n)})$. In the reliability case above, interests often center on a further item not failing before time s, when it is due for routine inspection. This is $p(x_{n+1} \geqslant s \mid x^{(n)})$ and, since $p(x_{n+1} > s \mid \theta) = e^{-\theta s}$, is $\int e^{-\theta s} p(\theta \mid r, T)\, d\theta$, which, using (2), is easily evaluated to give $[(b + T)/(b + T + s)]^{a+r+1}$. Inferences concerning x_{n+1} are not easily available within a non-Bayesian framework, although tolerance intervals* can be used.

Bayesian inference is closely connected with (Bayesian) decision theory*. By extending the axioms of uncertainty to include further simple postulates concerning one's attitude toward decisions taken under uncertainty, it is possible to prove that the consequences of decisions that are undertaken can be described numerically, just as uncertainty can, the value being termed utility*; and that the best decision is that of maximum expected utility. The extended axioms are also called coherent and the result can be described by saying that the only coherent method of reaching a decision is through maximizing expected utility, the expectation being with respect to the inference distribution previously considered. Other decision methods, like minimax*, are demonstrably incoherent. These ideas can be used in standard statistical problems such as point estimation. These can be modeled by referring to a decision to treat the parameter as having value t. The consequence of so doing when its true value is θ has some utility, $u(t, \theta)$ say, and the best decision (estimate) is that which maximizes $E[u(t, \theta) \mid x] = \int u(t, \theta) p(\theta \mid x)\, d\theta$. It is common to use a loss $l(t, \theta) = u(\theta, \theta) - u(t, \theta)$, equal to the loss in utility through using t rather than θ, and to minimize the expected loss (*see* EXPECTED VALUE). If θ is real and quadratic loss $l(t, \theta) = (t - \theta)^2$ is used, then the best estimate is clearly the posterior mean, having smallest mean square error* for the posterior distribution. (Actually, point estimation rarely occurs as a decision problem and its introduction into statistics is, as we saw above, really unnecessary.) A better, because more practical, example is the decision of whether to reject or accept a batch of items on the basis of a test of n of the items in which r fail and $n - r$ pass. If θ describes the quality of the untested items in the batch (an example of a parameter that can become known), then we have $u_A(\theta)$ and $u_R(\theta)$ as utilities for accepting and rejecting, respectively, a batch of quality θ. With density $p(\theta \mid r, n)$ as our inference concerning θ, the batch is accepted if $\int u_A(\theta) p|\theta|r, n)\, d\theta$ exceeds the corresponding quantity for rejection. For large batches a binomial approximation might be adequate and $p(\theta \mid r, n) \propto \theta^r (1 - \theta)^{n-r} p(\theta)$, where $p(\theta)$ is the prior distribution. It is surprising that relatively little use has been made of Bayesian methods in quality control*, where $p(\theta)$ is not too difficult to obtain. It is important to notice that *every* decision problem concerning θ and data x requires only $p(\theta \mid x)$ as the contribution from the data. Thus the Bayes inference is available for any decision about θ, and similar remarks apply in the alternative inferential form, $p(x_{n+1} \mid x^{(n)})$. This is why inference is conveniently treated as a separate topic, its merit being irrespective of the decision involved.

The quality of a decision depends on the amount of uncertainty present. If decision d has to be chosen from a set D, then without data the best decision is that which maximizes $\int u(d,\theta)p(\theta)\,d\theta$. With data x the criterion becomes $\int u(d,\theta)p(\theta \mid x)\,d\theta$. The difference between these two values is the gain from observing $X = x$. Actually, the gain can be negative; i.e., one can decrease expected utility by having extra data: e.g., when we obtain "surprising" data. Nevertheless, a simple, but important, result is that the *expected* gain is nonnegative: where the expectation is taken over the values of x, having density $p_X(x) = \int p(x \mid \theta)p(\theta)\,d\theta$, the denominator in (1). All relevant data are expected to be of value. The question then arises of whether this gain outweighs the cost, expressed as a loss of utility, in observing X. This leads to the Bayesian solution to the problem of experimental design (*see* DESIGN OF EXPERIMENTS).

Suppose that we are interested in a parameter θ; a similar treatment is available for an observable quantity such as x_{n+1}. Suppose that we can choose an experiment e, from among a class E of experiments, which will lead to data X having density $p(x \mid \theta, e)$. Let the purpose of the inference be to make a choice d of a decision from an available class D. Let $u(d,\theta;x,e)$ be the utility of choosing d when the true value is θ after having obtained data x from experiment e. Often this is of the form $u(d,\theta) - c(x,e)$, where $c(\cdot)$ describes the experimental cost. The best decision, with e performed and x observed, maximizes $\int u(d,\theta;x,e)p(\theta \mid x,e) \cdot d\theta$. Write this maximum $U(x,e)$. Then the best experiment is that which maximizes its expected value $\int U(x,e)p(x \mid e)\,dx$. Notice that the device of maximizing an expected value has been used twice, illustrating the point that all decisions can be solved with the aid of this single principle. This method of experimental design requires decisions d in D. Scientists often feel that experimental design should be separated from the decision aspect, just as inference is separated from it when considering a fixed experiment. This can be done here by using the idea that the purpose of an experiment is to "decide" on the distribution of θ. Formally, one identifies d with the posterior distribution $p(\theta \mid x,e)$ and the utility refers to that of adopting such a distribution when the true value of the parameter is θ. Reasonable restrictions on this utility lead to the concept of information* due to Shannon (*see* INFORMATION THEORY), using $\int p(\theta \mid x,e)\log p(\theta \mid x,e)\,d\theta$, as a criterion for experimental design; and the design chosen maximizes expected information.

The ideas of design explained in the preceding paragraph extend without difficulty to sequential experimentation*. After having observed $x^{(n-1)}$, let the choice be among experiments e_n from a class E_n, consisting in observing X_n. A simple case is where E_n contains only two elements: to stop experimentation or to take a further observation. The same two applications of the principle of maximizing expected utility as used above leads to a solution to the sequential problem. Unfortunately, a serious computational difficulty arises that usually hinders the full operation of this program. It turns out that the choice of e_n cannot be made until e_{n+1} has been selected, which in turns depends on e_{n+2}, and so on. We cannot decide what to do today until we have decided what to do with the tomorrows that today's decisions might bring. Consequently, the calculation has to proceed backwards: from $n+1$ to n, to $n-1$, and back to the start. It often happens that a procedure that looks only one stage ahead at a time gives a reasonable answer, but this is not always so. Only in the case of sequential testing of one simple hypothesis against another, leading to Wald's sequential probability ratio test*, does a complete solution, ignoring boundary effects, exist.

We have seen that in addition to the usual probability $p(x \mid \theta)$ used in a statistical argument, the coherent approach introduces a density for θ. Before Bayesian ideas can be implemented, a way has to be found for determining $p(\theta)$ as the person's initial opinion about θ. [Notice that $p(x \mid \theta)$ is also the person's opinion about x were he or she to know θ, but this is only a difference of interpretation from the frequency view, whereas $p(\theta)$ does not have a meaning in that approach.] In the case of the exponential family*, a conjugate distribution* is often used. In the univariate case the density for the exponential family takes the form

$$p(x \mid \theta) = e^{x\theta} f(x) g(\theta).$$

A member of the conjugate family has the structure

$$p(\theta) = e^{a\theta} g(\theta)^b K(a, b)$$

for some values of a and b that make this a density. The advantage of the conjugate form is that $p(\theta \mid x)$ has the same structure as $p(\theta)$ (*see* CONJUGATE FAMILIES OF DISTRIBUTIONS). For by Bayes' theorem, $p(\theta \mid x) \propto p(x \mid \theta)p(\theta) \propto e^{(a+x)\theta} g(\theta)^{b+1}$, as before but with $a + x$ for a and b increased by 1. Equation (2) illustrates this. In the Bernoulli case where $p(x \mid \theta) = \theta^r (1 - \theta)^{n-r}$, the conjugate family is that of beta distributions*. The idea, then, is for the values of a and b to be selected and calculations to proceed from there. Bayes' original approach did this in the Bernoulli situation, choosing the uniform distribution* in the unit interval, a special beta distribution. These ideas work well if a member of the conjugate family does adequately describe one's opinion, when the mathematics is tractable and results easily obtained, but is otherwise less satisfactory and, in any case, rarely applies outside the exponential family. There are also difficulties in the multivariate case where the conjugate family is too restricted. Another approach tries to replace $p(\theta)$ by a distribution that describes very little initial knowledge of θ so that the major contribution to the inferential $p(\theta \mid x)$ comes from the likelihood $p(x \mid \theta)$, and the data "speak for themselves." For example, small values of b above are clearly indicated since b increases with each observation. Such distributions may be described as "objective," as explained above, and might be used by common agreement among all scientists. Unfortunately, many of these distributions lead to curious paradoxes, usually caused by the resulting objective distributions being improper*, i.e., not having a finite integral, although the work on eliminating these continues. Another approach, using hyperparameters, will be mentioned later.

The methods just mentioned do not address themselves to the fundamental problem of describing someone's uncertainty about a set of quantities, a description which, according to the coherent argument, must be probabilistic. For simplicity, take the case of an uncertain quantity which takes only one of two values, conventionally 1 and 0, such as an event that can either occur or not. How can one assess the probability that the event will occur? One way of doing this is through a scoring rule which assigns a score $\phi(u)$ if the event occurs after having been assigned probability u, and $\phi(1 - u)$ if it does not, since the nonoccurrence will have probability $(1 - u)$. If p is the probability, then the expected score, if u is assigned, is $\phi(u)p + \phi(1 - u)(1 - p)$. If this is identified with the individual's expected utility, or more usually, its negative when the score is a penalty, this should be a minimum when u is p, since otherwise one will be motivated to assign u a value other than p. For this to happen $\phi'(p)p = \phi'(1 - p)(1 - p)$ and a rule with this property is called proper. $\phi(u) = (1 - u)^2$ is an example of such a rule; $\phi(u) = u$ is improper and leads always to stating 0 or 1. Another proper rule is $\phi(u) = \log u$ and this leads again to Shannon's information. Proper scoring rules have been used in weather forecasting. They should be of value in the forecasting of other time series.

We consider how Bayesian ideas apply to the popular statistical situations. Easily the most widely used one in statistics is the general linear model*, in which the data x have expectations that are known linear functions of unknown parameters θ, with the associated concepts of least squares* and the analysis of variance*. The linear relation $E(x) = A\theta$ describes one aspect of $p(x \mid \theta)$. It is usual to complete the model by supposing $p(x \mid \theta)$ to be normal* with a known, or partially known, variance–covariance matrix, although there is a purely linear Bayesian theory which assumes that $E(\theta \mid x)$ is also linear and parallels the classical Gauss–Markov theory*. To proceed with the coherent analysis it is necessary to assess $p(\theta)$. Let us consider the case where $x = x^{(n)}$ and $E(x_i \mid \theta) = \theta_i$, where, for given θ and σ^2, the x_i are uncorrelated with equal variance σ^2. This is a canonical form for the linear model. Now it will typically happen that the θ's are related; for example, θ_i may be the true yield of the ith variety, and all yields will increase under favorable conditions. Generally, we may regard the θ's as exchangeable and, more restrictively but in the spirit of the linear model, suppose them to be a random sample from $N(\mu, \tau^2)$ with τ^2 known. Then μ, corresponding to some average yield over varieties, can itself be assigned a distribution. Hence the distribution of θ is described in two stages: for fixed μ, and then over μ. μ is an example of a hyperparameter, describing the distribution of θ just as θ describes that of x, its role here being to exhibit correlation between the θ_i. It turns out that the inference $p(\theta \mid x)$ is insensitive to the distribution of μ provided that its variance is large and that then the inference is normal with means $E(\theta_i \mid x) = (\tau^2 x_i + \sigma^2 x_.)/(\tau^2 + \sigma^2)$, where $x_. = n^{-1}\sum x_i$, the overall mean. The usual least-squares estimate* is x_i, obtained here only as $\tau \to \infty$. The effect of the prior distribution, which recognizes the similarity between the θ's, is to "shrink" the means of the posterior distribution toward the overall mean $x_.$, having most effect on the extreme values of x_i. The argument can be extended to the case where σ^2 and τ^2 are themselves unknown. Notice that this canonical form is essentially the problem of estimating the mean θ of a multivariate normal distribution*. The Bayesian approach suggests that the sample mean may not often be a sensible estimate. Similar results apply to the general, linear regression* problem where the largest and the smallest of the sample regression coefficients are typically reduced in value in a procedure closely akin to ridge regression*. One interesting feature of these results is that the least-squares estimate is replaced by another. Even in the sampling-theory view, the least-squares estimate is inadmissible* except when only a few parameters are involved, so that the least-squares procedure needs a complete rethink. Closely related to least squares is the procedure of the analysis of variance with its series of significance tests and various estimates of error. We have remarked above that significance tests, with their tail-area probabilities obtained by integration, play no role in coherent analyses, although it is always possible to calculate the probability of a null hypothesis* given the data. This can be done in the case of the linear model, with the result that one obtains procedures that, for any sample size, can closely parallel the routine of significance tests but which behave quite differently as the sample size changes. For example, let $x^{(n)}$ be a random sample from $N(\theta, 1)$: then a conventional significance test rejects the hypothesis that $\theta = 0$ if $\bar{x}_n^2 > c_1 n^{-1}$, where $\bar{x}_n$ is the sample mean and c_1 is a constant, depending on the chosen level. The Bayes procedure shows that under reasonable assumptions, which include a nonzero prior value for $p(\theta = 0)$, $p(\theta = 0 \mid x^{(n)}) \leqslant \alpha$ when $\bar{x}^2 > c_2 n^{-1} + n^{-1}\log n$, where $c_2 = c_2(\alpha)$. Clearly, for any n, these two inequalities can be made to match for suitable choices of c_1 and c_2, but the agreement cannot be complete for all n. This is an example of coherence between results for different sample sizes, a concept omitted from orthodox statistical theory.

Bayesian analysis of variance does not need significance tests. We illustrate with a two-way, balanced design* (*see* BALANCING IN EXPERIMENTAL DESIGN) with m rows, n columns, r replicates per cell, and observations $x_{ijk}(i = 1, \ldots, m;\ j = 1, \ldots, n;\ k = 1, \ldots, r)$ having $E(x_{ijk}) = \theta_{ij}$. Then using ideas of exchangeability between rows and between columns, in extension of the ideas just discussed, the posterior mean $E(\theta_{ij} \mid x)$ is of the form, using the usual dot notation for sample means,

$$w_I(x_{ij.} - x_{i..} - x_{.j.} + x_{...}) + w_R(x_{i..} - x_{...}) + w_C(x_{.j.} - x_{...}) + wx_{...},$$

where the w's are weights that can be calculated from the data. [The earlier mean was of this form. $w(x_i - x_.) + (1 - w)x_.$] If the interaction* is thought, in the light of the data, to be small, then w_I will be small and the interaction, $(x_{ij.} - x_{i..} - x_{ij.} + x_{...})$'s, contribution to the mean will be small. In this way, tests are avoided and a main effect* or interaction included by an amount that is dictated by the weights.

The Bayesian method, or recipe, or paradigm, is complete and provides a tool for all inference (and decision) problems. Bayesian inference is not a branch of statistics: it is a new way of looking at the whole of statistics. Its potential is enormous, but its implementation requires extensive calculations, principally in the performance of numerical integration (to remove nuisance parameters and to find expectations) and maximization (to select optimal decisions). Suitable computer packages to do these calculations, and to interrogate the scientist on his or her knowledge about the situation and convert it into probability forms, have just begun to be written. Until they are more widely available, applications to elaborate sets of data are bound to be limited.

Bibliography

Box, George E. P. and Tiao, George C. (1973). *Bayesian Inference in Statistical Analysis.* Addison-Wesley, Reading, Mass. (An excellent account of standard statistical models from a Bayesian viewpoint.)

de Finetti, Bruno (1974–1975). *Theory of Probability. A Critical Introductory Treatment* (translated from the Italian by Antonio Machi and Adrian Smith), 2 vols. Wiley, London. (An outstanding book on subjective probability; no one could claim to understand this topic without having read it.)

DeGroot, Morris H. (1970). *Optimal Statistical Decisions.* McGraw-Hill, New York. (In addition to a Bayesian treatment of statistical models, particularly sequential experimentation, this book contains a fine account of the axiomatic treatment. Excellent bibliography.)

Jeffreys, Harold (1961). *Theory of Probability*, 3rd ed. Clarendon Press, Oxford. (An eminent geophysicist writing originally about the theory and practice of statistics. The first edition appeared in 1939, but it is still fresher than many later books.)

Lindley, D. V. (1971). *Bayesian Statistics: A Review.* SIAM, Philadelphia. (A review in outline of the whole field, with an extensive bibliography.)

Novick, Melvin R. and Jackson, Paul H. (1974). *Statistical Methods for Educational and Psychological Research.* McGraw-Hill, New York. (Perhaps the best book on statistical methods from a Bayesian view.)

Raiffa, Howard and Schlaifer, Robert (1961). Applied Statistical Decision Theory. Harvard University, Boston. (Bayesian decision theory with extensive technical developments.)

Savage, Leonard J. (1972). *The Foundations of Statistics*, 2nd ed. Dover, New York. [The first edition (1954) started the modern Bayesian movement.]

Zellner, Arnold (1971). *An Introduction to Bayesian Inference in Econometrics.* Wiley. New York. (A comprehensive account of linear econometric models using objective priors.)

(BAYES' THEOREM
COHERENCE
CONJUGATE FAMILIES OF DISTRIBUTIONS
DECISION THEORY
DEGREES OF BELIEF
FIDUCIAL INFERENCE
FOUNDATIONS OF PROBABILITY
STATISTICAL INFERENCE)

D. V. LINDLEY

BAYES' RISK *See* DECISION THEORY

BAYES' THEOREM

This theorem is obtained from the relationships

$$\Pr[E \mid H_1]\Pr[H_1] = \Pr[E \cap H_1] = \Pr[H_1 \mid E]\Pr[E]$$

and (if $\sum_{i=1}^{k} \Pr[H_i] = 1$)

$$\Pr[E] = \sum_{i=1}^{k} \Pr[H_i]\Pr[E \mid H_i].$$

It is usually stated in the form

$$\Pr[H_j \mid E] = \frac{\Pr[H_j]\Pr[E \mid H_j]}{\sum_{i=1}^{k} \Pr[H_i]\Pr[E \mid H_i]}.$$

In applications, $H_1, \ldots, H_k$ usually represent possible "states of nature" ("hypotheses") and E an observed event. The formula then provides a value for the probability of H_j given that E has been observed—the posterior probability*. Calculation of this value requires not only a knowledge of the probabilities $\Pr[E \mid H_i]$, but also of the prior (a priori) probabilities* $\Pr[H_i]$.

Another form of Bayes's theorem relates the posterior distribution of a parameter $\boldsymbol{\theta}$, given observed values $\mathbf{x}$ to its prior distribution*. When $\mathbf{x}$ has a PDF $f(\mathbf{x} \mid \boldsymbol{\theta})$, and $p(\boldsymbol{\theta})$ is the prior PDF of $\boldsymbol{\theta}$, the posterior PDF of $\boldsymbol{\theta}$ is

$$p(\boldsymbol{\theta} \mid \mathbf{x}) = \frac{p(\boldsymbol{\theta}) f(\mathbf{x} \mid \boldsymbol{\theta})}{\int \cdots \int p(\boldsymbol{\theta}) f(\mathbf{x} \mid \boldsymbol{\theta})\, d\boldsymbol{\theta}}$$

($\boldsymbol{\theta}$ and/or $\mathbf{x}$ may be vectors; the integral is over the whole range of variation of $\boldsymbol{\theta}$).

Considerable controversy has existed, and still continues, on the possibility of using Bayes' theorem when—as is often the case—the values to be used for the prior probabilities are not clearly established.

More detailed discussion of these matters will be found in the entries listed below.

(BAYESIAN INFERENCE
DECISION THEORY
FIDUCIAL INFERENCE
FIDUCIAL PROBABILITY)

BECHHOFER TEST

For a fixed-effects ANOVA model*, Bechhofer [1] developed tables that enable us to determine the necessary sample size n so that with probability $1 - \alpha$, the highest (lowest) sample factor-level mean is indeed from the factor level with the highest (lowest) population mean. We need to specify the probability $1 - \alpha$, the standard deviation σ of the residual errors*, and the smallest difference λ between the highest (lowest) and second highest (second lowest) factor level means which "ought" to be detected with this probability. Bechhofer's tables assume that equal sample sizes (n) are used for all factor levels.

Example. For $r = 4$ (the number of populations or factor levels), $\sigma = 2, \lambda = 1$ and $1 - \alpha = 0.90$, the entries in Bechhofer's tables are $\lambda\sqrt{n}/\sigma$, and for this particular choice of the parameters r and $(1 - \alpha)$, the corresponding value of $\lambda\sqrt{n}/\sigma$ is 2.4516, which yields $\sqrt{n} = 4.90$ or $n = 24$. [The range of r is 2(1)10 and $1 - \alpha = 0.90$, 0.95, and 0.99.]

References

[1] Bechhofer, R. E. (1954). *Ann. Math. Statist.*, **25**, 16–39.

[2] Neter, J. and Wasserman, W. (1974). *Applied Linear Statistical Models*, 2nd ed. Richard D. Irwin, Homewood, Ill.

(GENERAL LINEAR MODEL
ONE-WAY CLASSIFICIATION)

BEHRENS–FISHER PROBLEM

The Behrens–Fisher problem is that of testing whether the means of two normal populations are the same, without making any assumption about the variances. It is also referred to as the two-means problems and as Behrens' problem. An essentially equivalent problem is that of finding an interval estimate* for the difference between the population means.

Suppose that the first of the normal populations has mean μ_1 and variance σ_1^2 and that a sample of size n_1 yields sample mean $\overline{X}_1$ and sample variance s_1^2. We will use corresponding notation for the second sample, but with subscript 2.

If the population variances are assumed to be identical (at least under the null hypothesis* of no difference between means), then the quantity

$$\frac{(\bar{X}_1 - \bar{X}_2) - (\mu_1 - \mu_2)}{\left[s_1^2(n_1 - 1) + s_2^2(n_2 - 1)\right]^{1/2}} \times \left[\frac{(n_1 + n_2 - 2)n_1 n_2}{n_1 + n_2}\right]^{1/2} \quad (1)$$

has a t-distribution* with $n_1 + n_2 - 2$ degrees of freedom. Confidence intervals* for $\mu_1 - \mu_2$ and tests for the hypothesis $\mu_1 - \mu_2 = 0$ based on this fact are widely used. The Behren–Fisher problem allows the assumption of equal population variances to be relaxed.

The Behrens–Fisher problem has considerable theoretical importance as a situation where fiducial (*see* FIDUCIAL INFERENCE) and Neyman–Pearson* schools of thought give different results. Kendall and Stuart [12] discuss it extensively in this role. Lee and Gurland [13] and Scheffé [21] provide useful reviews of the various solutions that have been proposed.

BEHRENS' SOLUTION

Behrens [3] was the first to offer a solution to the Behrens-Fisher problem. However, the basis of his solution was not made completely clear. Fisher [6] showed that Behrens' solution could be derived using his theory of statistical inference called fiducial probability*. Both Fisher and Yates [9] and Isaacs et al. [10] give tables of significance points. Patil [16] gives a useful approximation.

A short fiducial derivation of the test starts by noting that, for fixed μ_1, μ_2, σ_1^2, and $\sigma_2^2, t_1 = (\bar{X}_1 - \mu_1)n_1^{1/2}/s_1$ has a t-distribution with $n_1 - 1$ degrees of freedom and $t_2 = (\bar{X}_2 - \mu_2)n_2^{1/2}/s_2$ has an independent t-distribution with $n_2 - 1$ degrees of freedom. If we treat these distributions as probability statements about μ_1 and μ_2 given $\bar{X}_1, \bar{X}_2, s_1^2$, and s_2^2, then a little manipulation shows that

$$d = \frac{(\bar{X}_1 - \bar{X}_2) - (\mu_1 - \mu_2)}{\left(s_1^2/n_1 + s_2^2/n_2\right)^{1/2}} = t_1 \sin R - t_2 \cos R, \quad (2)$$

where $\tan R = (s_1/s_2)(n_2/n_1)^{1/2}$. Thus the distribution of d is that of a mixture* of two t-distributions.

The controversial step in this derivation is what Fisher called logical inversion. Here we know the distribution of t_1 and t_2 and it describes the distribution of $\bar{X}_1, s_1, \bar{X}_2$, and s_2 for given μ_1, σ_1^2, μ_2, and σ_2^2 and somehow proceed to the conclusion that the same distribution of t_1 and t_2 describes a distribution of the parameters for given values of the observable random variables.

Jeffreys [11] pointed out that a Bayesian calculation based on a $(\sigma_1^2\sigma_2^2)^{-1}$ improper prior density for μ_1, μ_2, σ_1^2, and σ_2^2 yields a solution that is numerically the same, although the interpretation is different. This derivation proceeds by finding the posterior distribution* of μ_1, σ_1^2, μ_2, and σ_2^2 for given $\bar{X}_1, s_1^2, \bar{X}_2$, and s_2^2 for a $(\sigma_1^2\sigma_2^2)^{-1}$ improper prior density. Integrating out σ_1^2 and σ_2^2 shows that the marginal posterior distribution of μ_1 and μ_2 may be described by the t-distributions of the statistics t_1 and t_2 given above. Hence the marginal* posterior distribution of d is that of a mixture* of t_1 and t_2 as in (2). The controversial step in this derivation is the assumption of the prior density.

WELCH–ASPIN SOLUTION

Welch [22] derived a test for which the probability of error was very nearly the nominal value throughout the parameter space. It is based on the same statistic, d, as in Behrens' test but uses different significance points. Asymptotic calculations of Welch [22] were extended by Aspin [1] and the significance points are available in the *Biometrika Tables* [17, Table 11, p. 136].

PRACTICAL COMPARISON OF SOLUTIONS

For sample sizes greater than about 10, the differences between the various proposed solutions are generally much less their differences from the t-test of (1), so the use of any one of them is better than using the t-test, unless the assumption of equal variances is warranted. The tests of McCullough-Banerjee and Cochran and Cox may be regarded as *approximations* to the Behrens–Fisher solution. Most of the other tests discussed by Lee and Gurland [13] may be regarded as approximations to the Welch–Aspin test.

Example. For $n_1 = 13$, $n_2 = 25$, suppose that $s_1^2 = 2.0$, $s_2^2 = 1.0$, and $\bar{X}_1 - \bar{X}_2 = 0.9$. From (2), $d = 2.044$. The 5% significance point is 2.15 for the Behrens–Fisher test, 2.16 for both the McCullough-Banerjee and Cochran and Cox tests, 2.11 for the Welch–Aspin test, and 2.10 for Welch's approximate t-solution. We would not reject the hypothesis of equal population means at the 5% level.

The t-statistic from (1) takes the value 2.27, which is significant at the 5% level, the significance point being 2.03.

If the sample variances were $s_1^2 = 1$ and $s_2^2 = 1.5$, then the t-statistic would still be 2.27. However, the d-statistic would be changed to 2.43, which would be considered significant by the various tests for the Behrens–Fisher problem. (The significance points for all these tests vary with the ratio s_1^2/s_2^2, but are similar to their previously quoted values for this particular case.)

It should be noted that the hypothesis of equal population variances would not be rejected by the usual F-test for either set of sample variances.

CRITICISM OF BEHRENS' SOLUTION

Bartlett [2] noted that the probability of error* for Behrens' test was not always the same as the nominal significance level*. This set the scene for a major controversy between Neymen's theory of confidence intervals and Fisher's theory of fiducial inference. Fisher [7] argued that the probability of including the true value is not an important property, while the Neyman–Pearson school of thought has maintained that it is important, and supported the search for similar*, or approximately similar tests*.

SIMILAR TESTS

A test is said to be similar if its error probability is constant throughout the parameter space. Wilks [23] suggested that similar tests for the Behrens–Fisher problem would be difficult to find. Linnik [14, 15] has proved that no similar tests having certain desirable properties exist. The Welch–Aspin solution is approximately similar for moderate sample sizes (see Lee and Gurland [13] for references). Behrens' solution seems to have error rate not greater than the normal value (see Robinson [20]).

The solutions to the Behrens–Fisher problem not discussed here are generally attempts to find approximately similar tests.

CRITICISM OF THE WELCH–ASPIN SOLUTION

Fisher [8] criticized Welch's test in a way that amounts to showing the existence of a negatively biased relevant subset in the sense of Buehler [4]. In our notation, independently of s_1/s_2, $U = (n_1 - 1)(s_1^2/\sigma_1^2) + (n_2 - 1)(s_2^2/\sigma_2^2)$ is distributed as chi-square* with $n_1 + n_2 - 2$ degrees of freedom and $(\bar{X}_1 - \mu_1) + (\bar{X}_2 - \mu_2)$ is distributed normally with mean zero and variance $\sigma_1^2/n_1 + \sigma_2^2/n_2$. It follows that

$$d\sqrt{\frac{s_1^2/n_1 + s_2^2/n_2}{\sigma_1^2/n_1 + \sigma_2^2/n_2} \frac{n_1 + n_2 - 2}{U}} \tag{3}$$

has a t-distribution with $n_1 + n_2 - 2$ degrees of freedom given s_1/s_2, μ_1, μ_2, σ_1^2, and σ_2^2. When $s_1^2/s_2^2 = 1$, $n_1 = n_2$, and $\sigma_1^2/\sigma_2^2 = w$, expression (3) simplifies to

$$d\sqrt{\frac{4}{2 + 1/w + w}} .$$

Now $1/w + w \geqslant 2$, so

$$\sqrt{\frac{4}{2 + 1/w + w}} \leqslant 1$$

and so

$$\Pr\left[|d| > a \mid s_1/s_2 = 1\right] \geqslant \Pr\left[|t_{(n_1+n_2-2)}| > a\right]$$

for all $a > 0$, where $t_{(n_1+n_2-2)}$ denotes a variable having a t-distribution with $n_1 + n_2 - 2$ degrees of freedom. In particular, taking $n_1 = n_2 = 7$ and $a = 1.74$ (from the *Biometrika Tables*),

$$\Pr[|d| > 1.74 \mid s_1/s_2 = 1] \geqslant \Pr[t_{(12)} > 1.74] > 0.1.$$

Thus the set where $s_1/s_2 = 1$ is a relevant subset where confidence intervals based on the Welch-Aspin test cover the true value of $\mu_1 - \mu_2$ less often than the nominal confidence level suggests.

Robinson [20] has shown that negatively biased relevant subsets do not exist for Behrens' solution. Positively biased relevant subset similar to that of Buehler and Fedderson [5] for the t-distribution presumably exist, but Robinson [20] argues that this is less important.

CURRENT STATUS OF THE BEHRENS–FISHER PROBLEM

Within the Neyman–Pearson school of thought, the Welch–Aspin test is regarded as satisfactory except that its derivation using asymptotic series makes calculation of significance points for low numbers of degrees of freedom quite difficult. There is still some interest in other approximately similar tests. Other schools of thought tend to regard the Behrens–Fisher problem as an example when the Neyman–Pearson school's quest for a similar test has led it astray.

Fisher's [8] criticism of the Welch–Aspin test and Robinson's [20] support of Behrens' test have made the Behrens–Fisher problem a focal point for discussion of the importance of relevant subsets.

Pedersen [18] and Fisher [8] suggest that the choice of a reference set—the set relative to which a quoted level of confidence applies—is the major issue illustrated by the Behrens–Fisher problem.

It seems likely that the Behrens–Fisher problem will be quoted as an important example in the discussion of rival theories as long as disagreements about the foundations of statistics persist.

Further Reading

The reviews of Scheffé [21] and Lee and Gurland [13] provide an introduction to practical solutions to the Behrens–Fisher problem. Readers interested in the role of the Behrens–Fisher problem in the foundations of statistics should become familiar with the theories of confidence intervals and fiducial inference and the concepts of relevant subsets and similar regions, then look at Bartlett [2], Fisher [8], Linnik [15], Robinson [20], and Pedersen [18].

In comparing various calculations concerning Behrens' solution, it is important to note whether exact significance points or asymptotic approximations have been used.

References

[1] Aspin, A. A. (1948). *Biometrika*, **35**, 88–96.

[2] Bartlett, M. S. (1936). *Proc. Camb. Philos. Soc.*, **32**, 560–566.

[3] Behrens, B. V. (1929). *Landwirtsch. Jb.*, **68**, 807–837.

[4] Buehler, R. J. (1959). *Ann. Math. Statist.*, **30**, 845–867.

[5] Buehler, R. J. and Fedderson, A. P. (1963). *Ann. Math. Statist.*, **34**, 1098–1100.

[6] Fisher, R. A. (1935). *Ann. Eugen.* (*Lond.*), **6**, 391–398.

[7] Fisher, R. A. (1939). *Ann. Math. Statist.*, **10**, 383–388.

[8] Fisher, R. A. (1956). *J. R. Statist. Soc. B*, **18**, 56–60. (There are several replies to this article in the same volume.)

[9] Fisher, R. A. and Yates, F. (1957). *Statistical Tables for Biological, Agricultural and Medical Research*, 5th ed. Oliver & Boyd, Edinburgh.

[10] Isaacs, G. L., Christ, D. E., Novick, M. R., and Jackson, P. H. (1974). *Tables for Bayesian Statisticians*. Iowa State University Press, Ames, Iowa.

[11] Jeffreys, H. (1940). *Ann. Eugen.* (*Lond.*), **10**, 48–51.

[12] Kendall, M. G. and Stuart, A. (1967). *The Advanced Theory of Statistics*, Vol. 2: *Inference and Relationship*, 2nd ed. Charles Griffin, London.

[13] Lee, A. F. S. and Gurland, J. (1975). *J. Amer. Statist. Ass.*, **70**, 933–941.

[14] Linnik, Yu. V. (1966). *Sankhyā A*, **28**, 15–24.

[15] Linnik, Yu. V. (1968). *Statistical Problems with Nuisance Parameters*, Vol. 20. (Amer. Math. Soc. Transl. Math. Monogr.; translated from the Russian.) American Mathematical Society, Providence, R.I.

[16] Patil, V. H. (1965). *Biometrika*, **52**, 267–271.

[17] Pearson, E. S. and Hartley, H. O., eds. (1954). *Biometrika Tables for Statisticians*, Vol. 1. Cambridge University Press, Cambridge.

[18] Pedersen, J. G. (1978). *Int. Statist. Rev.*, **46**, 147–170.

[19] Rahman, M. and Saleh, A. K. M. E. (1974). *J. R. Statist. Soc. B*, **36**, 54–60. (Not recommended; note corrigendum on p. 466 of same volume.)

[20] Robinson, G. K. (1976). *Ann. Statist.*, **4**, 963–971.

[21] Scheffé, H. (1970). *J. Amer. Statist. Ass.*, **65**, 1501–1508.

[22] Welch, B. L. (1947). *Biometrika*, **34**, 28–35.

[23] Wilks, S. S. (1940). *Ann. Math. Statist.*, **11**, 475–476.

(FIDUCIAL DISTRIBUTIONS
FIDUCIAL INFERENCE
SIMILAR TESTS
STUDENT'S *t*-TESTS)

G. K. ROBINSON

BEHRENS' PROBLEM *See* BEHRENS–FISHER PROBLEM

BELIEF FUNCTIONS

The theory of belief functions is a generalization of the Bayesian theory of subjective probability*. It generalizes the Bayesian theory in two ways. First, it permits nonadditive degrees of belief*. If, for example, there is no evidence for a proposition A and yet only inconclusive evidence for its negation $\bar{A}$, then the theory permits us to assign the degree of belief zero to A while assigning a degree of belief less than one to $\bar{A}$. Secondly, it generalizes the Bayesian rule of conditioning to "Dempster's rule of combination," a rule for combining degrees of belief based on one body of evidence with degrees of belief based on another body of evidence.

If Ω denotes the set of possible answers to some question, then a "belief function over Ω" assigns to each subset A of Ω a number $\mathrm{Bel}[A]$ that is interpreted as one's degree of belief that the correct answer is in A. Although belief functions are not required to be additive (they may fail to satisfy $\mathrm{Bel}[A] + \mathrm{Bel}[\bar{A}] = 1$), they are required to have a weaker property called "monotonicity of order ∞." This property was first studied in the early 1950s by Gustave Choquet.

In addition to their general interest as a means for quantifying the weight of evidence*, belief functions are also of interest because of the statistical methods that use them. Some of these methods can be understood as generalizations of Bayesian methods (*see* NONADDITIVE PROBABILITIES).

Bibliography

Choquet, G. (1954). *Ann. Inst. Fourier*, **5**, 131–295.

Dempster, A. P. (1968). *J. R. Statist. Soc. B*, **30**, 205–247.

Shafer, G. (1976). *A Mathematical Theory of Evidence*. Princeton University Press, Princeton, N.J.

(DEGREE OF BELIEF
NONADDITIVE PROBABILITIES
SUBJECTIVE PROBABILITIES)

G. SHAFER

BELL–DOKSUM TESTS

These tests are obtained from Kruskal–Wallis tests* by replacing ranks by values of ordered unit normal* random variables. The data consist of k random samples of size n_i, $i = 1, \ldots, k$. Denote the ith sample of size n_i by $X_{i1}, X_{i2}, \ldots, X_{in_i}$. Let $N = \sum_{i=1}^{k} n_i$. Rank all N values from rank 1 to N and let $R(X_{ij})$ denote the rank of X_{ij}. Next, draw a group of N numbers from a table of random normal deviates* (or normal scores*). The rth smallest of these numbers is assigned to X_{ij}, which has rank r in the original data. Symbolically, if $Z(r)$ is the rth smallest random deviate, then $Z[R(X_{ij})]$ is assigned to X_{ij} (the so-called random normal scores transformation). Next compute

$$Z_i = \frac{1}{n_i} \sum_{j=1}^{N} Z(R(X_{ij})) \quad \text{for} \quad i = 1, \dots, k$$

and

$$\bar{Z} = \frac{1}{N} \sum_{r=1}^{N} Z(r).$$

The null hypothesis is that all k *population distribution functions are the same*, the alternative being that they are not identical. The test statistic is

$$T = \sum_{i=1}^{k} n_i (Z_i - \bar{Z})^2.$$

The null hypothesis is rejected at level α if the observed value of T exceeds the $(1 - \alpha)$ quantile of χ^2_{k-1} (chi-squared* variable with $k - 1$ degrees of freedom).

For assumptions concerning the applicability of this test and the procedure for handling ties among X_{ij}, see Bell and Doksum [1] and Conover [3, pp. 284–287]. It has been observed that for certain sets of data the Bell–Doksum test may have a slightly higher power than the Kruskal–Wallis test*, although the former test has the unnatural characteristic that after the data are collected, an element of chance is introduced by the choice of random normal deviates, completely unrelated to the data. This experiment is permitted to influence the test and therefore the decision. On the other hand, as a consequence of the transformation, powerful t-* and F-tests* for the equality of means can be applied, without needing any assumption on the form of population distribution except continuity. (Independence and homogeneity are still required.)

General References

[1] Bell, C. B. and Doksum, K. A. (1965). *Ann. Math. Statist.*, **36**, 203–214.

[2] Bradley, J. V. (1968). *Distribution-Free Statistical Tests*, Prentice-Hall, Englewood Cliffs, N.J.

[3] Conover, W. J. (1971). *Practical Non-parametric Statistics*, Wiley, New York.

(KRUSKAL–WALLIS TESTS)

BELLMAN–HARRIS PROCESS *See* BRANCHING PROCESSES

BELL POLYNOMIALS

These polynomials represent expressions for derivatives of composite functions. The equations

$$D\{f(g(x))\} = f'(g(x)) \cdot g'(x),$$

$$D^2\{f(g(x))\} = f'(g(x)) \cdot \{g'(x)\}^2 + f'(g(x)) \cdot g''(x),$$

etc., can be written formally as

$$Df(g(x)) = f_1 g_1,$$

$$D^2 f(g(x)) = f_2 g_1^2 + f_1 g_2,$$

etc.

The quantities on the right-hand side are called (by Riordan [2, 3]) *Bell polynomials* (see Bell [1]) of order $1, 2, \dots$. The Bell polynomial of order r is

$$Y_r = r! \sum \cdots \sum f_k \prod_{j=1}^{n} \left(\frac{g_j^{k_j}}{j! k_j!} \right),$$

where the summation is over all sets of nonnegative integer values $k_1, k_2, \dots, k_n$ such that $k_1 + 2k_2 + \cdots + nk_n = n$.

Interpreting f^k as f_k, and Y^k by Y_k, the Bell polynomials are generated by the identity

$$\exp(sY) = \exp\left(f \sum_{j=1}^{\infty} \frac{s^j g_j}{j!} \right).$$

Bell polynomials are useful in the analysis of Lagrange distributions*.

For further details, refer to Riordan [2, 3].

References

[1] Bell, E. T. (1927). *Ann. Math.*, **29**, 38–46.

[2] Riordan, J. (1958). *An Introduction to Combinatorial Analysis*. Wiley, New York.

[3] Riordan, J. (1968). *Combinatorial Identities*. Krieger, Huntington, N.Y.

BENFORD'S LAW *See* FIRST-DIGIT PROBLEM

BENNETT'S BIVARIATE SIGN TEST

For the problem of testing whether a univariate distribution has a specified median*, the sign test* holds a prominent role in regard to historical importance, simplicity, versatility, and minimal assumptions about the shape of the distribution. Bennett [1] proposed an extension of the sign test for testing the null hypothesis* that a bivariate distribution has specified marginal medians. Let (X_i, Y_i), $i = 1, \dots, n$ be a random sample from a bivariate population with an unspecified continuous distribution function $F(x - \mu, y - \nu)$ whose marginal medians are denoted by μ and ν and are assumed to be unique. Bennett's test is concerned with testing $H_0 : (\mu, \nu) = (0, 0)$ vs. $H_1 : (\mu, \nu) \neq (0, 0)$, with no assumption made on the shape or association of the bivariate distribution. A typical context of application is a matched-pair* experiment where two, possibly associated, characteristics are observed simultaneously, and (X_i, Y_i) is identified as the vector of paired differences for the ith pair.

Consider the Euclidean plane divided in four quadrants with the origin taken at the median point specified under H_0. Denoting the respective quadrant probabilities by p_1, p_2, p_3, and p_4, an equivalent formulation of the null hypothesis is then $H_0 : p_1 = p_3, p_2 = p_4$. Bennett's test is basically structured in the frame of Pearson's chi-square statistic* for goodness of fit* based on the quadrant frequency counts. Specifically, the test statistic has the form $B = \sum_{j=1}^{4} (n_j - n\hat{p}_j)^2 / n\hat{p}_j$, where n_j denotes the number of (X_i, Y_i)'s in the jth quadrant, and $\hat{p}_j$, $j = 1, \dots, 4$, are the maximum likelihood* estimates of the p_j's based on the multinomial distribution* of $(n_1, \dots, n_4)$ with cell probabilities constrained by $p_1 = p_3$ and $p_2 = p_4$. These estimates are $\hat{p}_1 = \hat{p}_3 = (n_1 + n_3)/2n$, $\hat{p}_2 = \hat{p}_4 = (n_2 + n_4)/2n$, and, after substitution, the expression for B simplifies to

$$B = \frac{(n_1 - n_3)^2}{n_1 + n_3} + \frac{(n_2 - n_4)^2}{n_2 + n_4},$$

which is the test statistic of Bennett's bivariate sign test. Noting that the asymptotic null distribution of B is central chi-squared* with 2 degrees of freedom, the level α critical region* is set as $B \geqslant \chi^2_{2,\alpha}$, where $\chi^2_{2,\alpha}$ is the upper α-point of the chi-squared distribution with 2 degrees of freedom. Evidently, usefulness of the test is limited to large samples. In fact, the exact distribution of B involves the unknown parameter p_1. While the univariate sign test is strictly distribution-free*, Bennett's bivariate extension is only asymptotically distribution-free.

Bhattacharyya [2] studied the Pitman efficiency* of B under local translation alternatives $F(x - \delta_1 n^{-1/2}, y - \delta_2 n^{-1/2})$ and some smoothness conditions on F. In particular, when $F(x, y)$ is diagonally symmetric with a covariance matrix $\mathbf{\Sigma}$, the asymptotic efficiency of B relative to the optimum normal theory (Hotelling's) T^2*-test is given by $e(B : T^2) = \boldsymbol{\delta}'\boldsymbol{\tau}^{-1}\boldsymbol{\delta} / \boldsymbol{\delta}'\mathbf{\Sigma}^{-1}\boldsymbol{\delta}$, where $\boldsymbol{\delta}' = (\delta_1, \delta_2)$,

$$\boldsymbol{\tau} = \frac{1}{4\alpha_1^2\alpha_2^2} \begin{bmatrix} \alpha_2^2 & \alpha_1\alpha_2(4p_1 - 1) \\ \alpha_1\alpha_2(4p_1 - 1) & \alpha_1^2 \end{bmatrix}$$

and $\alpha_1 = f_1(0)$, $\alpha_2 = f_2(0)$ are the respective marginal pdf's evaluated at 0. When X and Y are independent normal, $e(B : T^2) = 2/\pi$, which is also the Pitman efficiency of the univariate sign test relative to the t-test. With a scale contamination* model of the form $F(x, y) = (1 - \epsilon)G_1(x, y) + \epsilon G_2(x/\sigma_1, y/\sigma_2)$, the efficiency can be arbitrarily large with increasing σ_1 or σ_2 for any fixed $0 < \epsilon < 1$. On the other hand, the efficiency could approach 0 when the bivariate distribution tends to be singular.

A strictly distribution-free bivariate sign test is proposed by Chatterjee [4] through a conditioning approach to remove the nuisance parameter* p_1. His test statistic is equivalent to B, but the rejection region is determined from the fact that conditionally, given $(n_1 + n_3)$, the counts n_1 and n_2 are independent binomial* variables. This conditional version of Bennett's test is unbiased* and consistent* under a wide class of alternatives and has the same Pitman efficiency as the unconditional test.

Two other bivariate extensions of the sign test are due to Hodges [5] and Blumen [3]. These tests are exact and share a property of Hotelling's T^{2*} in that they are invariant under nonsingular linear transformations of the observations (X_i, Y_i). On the other hand, Bennett's test is invariant under transformations of the form $X_i' = h_1(X_i), Y_i' = h_2(Y_i)$, where h_1 and h_2 are monotone increasing and zero-preserving functions. Bennett [1] also discusses a general multivariate extension of the sign test along the same lines as the bivariate test. Such extensions of Hodges' and Blumen's tests do not appear to be tractable. For additional discussions and relevant references on multivariate sign tests, see Puri and Sen [6].

References

[1] Bennett, B. M. (1962). *J. R. Statist. Soc. B*, **24**, 159–161.

[2] Bhattacharyya, G. K. (1966). *J. R. Statist. Soc. B*, **28**, 146–149.

[3] Blumen, I. (1958). *J. Amer. Statist. Ass.*, *53, 448–456.*

[4] Chatterjee, S. K. (1966). *Ann. Math. Statist.*, **37**, 1771–1782.

[5] Hodges, J. L., Jr. (1955). *Ann. Math. Statist.*, **26**, 523–527.

[6] Puri, M. L. and Sen, P. K. (1971). *Nonparametric Methods in Multivariate Analysis*. Wiley, New York.

(DISTRIBUTION-FREE METHODS
SIGN TESTS)

G. K. BHATTACHARYYA

BERGE INEQUALITY

A two-dimensional probability inequality of the Chebyshev* type. Given a two-dimensional random variable $X = (X_1, X_2)$ with

$$\mu_i = E[X_i], \qquad \text{var}(X_i) = \sigma_i^2 \qquad (i = 1, 2)$$

$$\text{cov}(X_1, X_2) = \sigma_{12} \qquad \text{and} \quad \rho = \frac{\sigma_{12}}{\sigma_1 \sigma_2},$$

then

$$\Pr\left(\max\left\{\frac{|X_1 - \mu_1|}{\sigma_1}, \frac{|X_2 - \mu_2|}{\sigma_2}\right\} \geqslant \lambda\right) \leqslant \frac{1 + \sqrt{1 - \rho^2}}{\lambda^2};$$

also,

$$\Pr\left(\text{either } \frac{|X_1 - \mu_1|}{\sigma_1} \geqslant \lambda \text{ or } \frac{|X_2 - \mu_2|}{\sigma_2} \geqslant \lambda\right) \leqslant \frac{1 + \sqrt{1 - \rho^2}}{\lambda^2}$$

for any $\lambda > 0$.

This inequality bounds the probability of falling outside a rectangle centered at the means of a bivariate distribution; it uses a measure of the dependence* between the random variables.

Bibliography

Berge, P. O. (1938). *Biometrika*, **29**, 405–406.

Karlin, S. and Studden, W. J. (1966). *Tchebycheff Systems with Applications in Analysis and Statistics*. Wiley, New York.

Leser, C. E. V. (1941). *Biometrika*, **32**, 284–293.

Savage, I. R. (1961). *J. Res. Natl. Bur. Stand.*, **65B**(3), 211–222.

(BIRNBAUM–RAYMOND–
ZUCKERMAN INEQUALITY
MULTIVARIATE CHEBYSHEV–TYPE
INEQUALITIES)

BERKSON'S MINIMUM LOGIT χ^2 PROCEDURE

A special form of dose–response curve* analysis, in which it is supposed that the probability of response to a dosage x is a logistic function*, of form

$$P(x) = [1 + \exp(-\alpha - \beta x)]^{-1}.$$

If dosages x_i are applied to n_i individuals, of whom d_i exhibit response, for $i = 1, \ldots, k$ we have data from which estimates of α and β have to be derived. An alternative to the method of maximum likelihood* (described in the entry BIOASSAY) is to minimize the quantity ("logit χ^2")

$$\sum_{i=1}^{k} \frac{[d_i - n_i P(x_i)]^2}{n_i P(x_i)[1 - P(x_i)]} = \sum_{i=1}^{k} e^{\alpha+\beta x_i}[d_i(1 + e^{-\alpha-\beta x_i}) - n_i]^2 n_i^{-1}$$

with respect to α and β. Berkson [1] presents evidence that these estimators can have a smaller mean square error than the maximum likelihood estimators in small samples.

Reference

[1] Berkson, J. B. (1955). *J. Amer. Statist. Ass.*, **50**, 130–162.

(LOGISTIC DISTRIBUTION
LOGIT
QUANTAL RESPONSE)

BERKSON'S MINIMUM NORMIT χ^2 PROCEDURE

A similar procedure to Berkson's minimum logit χ^2 procedure* except that the tolerance distribution* is supposed to be normal. It has similar properties vis-à-vis maximum likelihood estimators*.

Bibliography

Berskon, J. B. (1955). *J. Amer. Statist. Ass.*, **50**, 529–549.

Berkson, J. B. (1957). *Biometrika*, **44**. 411–435.

(PROBIT ANALYSIS
QUANTAL RESPONSE)

BERKSON'S 2*n* RULE

Consider a random variable R with a binomial distribution* (n, p). Berkson's estimator [1,2]—the so-called $2n$ rule—of the logit* $\ln(p/q)$, where $q = 1 - p$, is

$$L_1(R) = \begin{cases} -\ln(2n-1) & (R = 0) \\ \ln\{R/(n-R)\} & (R = 1, 2, \ldots, n-1) \\ \ln(2n-1) & (R = n). \end{cases}$$

The estimator is unbiased* only for $p = \frac{1}{2}$. Gart and Zweifel [3] show that the bias of Berkson's estimator is usually larger than those of both the Haldane–Anscombe* and Tukey* estimators.

References

[1] Berkson, J. (1944). *J. Amer. Statist. Ass.*, **39**, 357–365.

[2] Berkson, J. (1953). *J. Amer. Statist. Ass.*, **48**, 565–599.

[3] Gart, J. J. and Zweifel, J. R. (1967). *Biometrika*, **54**, 181–187.

BERNOULLI DISTRIBUTION

A random variable X possesses a Bernoulli distribution with parameter p $(0 < p < 1)$ if

$$\Pr\{X = 1\} = p \quad \text{and} \quad \Pr\{X = 0\} = 1 - p.$$

The corresponding CDF is

$$F(x) = \begin{cases} 0 & (x < 0) \\ 1 - p & (0 \leqslant x < 1) \\ 1 & (x \geqslant 1) \end{cases}$$

and the characteristic function* is

$$\phi_x(t) = 1 + p(e^{it} - 1).$$

The moments of this distribution are $\mu_k' = E(X^k) = p \quad (k = 1, 2, \ldots), \operatorname{var}(X) = p(1-p)$.

The Bernoulli distribution is basic in probability theory and statistics, being a model of any random experiment with outcomes belonging to two mutually disjoint classes.

(BERNOULLI NUMBERS
BERNOULLIS, THE
BINOMIAL DISTRIBUTION
LAWS OF LARGE NUMBERS)

BERNOULLI NUMBERS

Bernollui numbers $B_r (r = 0, 1, \ldots)$ (first introduced by J. Bernoulli [2]) are used in numerical integration formulas, in the calculus of finite differences*, and also occur in combinatorial-probabilistic problems. They can be defined in various ways.

1. The original definition given by Bernoulli is

$$\sum_{\mu=0}^{m-1} \mu^k = \frac{1}{k+1} \sum_{r=0}^{k} \binom{k+1}{r} B_r m^{k+1-r}$$

for $k = 0, 1, 2$ and $m = 1, 2 \dots$.

2. For even subscripts,

$$B_{2n} = (-1)^{n-1} \frac{2(2n)!}{(2\pi)^{2n}} \sum_{m=1}^{\infty} \frac{1}{m^{2n}} \qquad (n \geqslant 1).$$

For odd subscripts (except for B_1) the Bernoulli numbers are all zero. The values of the first several Bernoulli numbers are $B_0 = 1$, $B_1 = -\frac{1}{2}$, $B_2 = \frac{1}{6}$, $B_4 = -\frac{1}{30}$, $B_6 = \frac{1}{42}$, $B_8 = -\frac{1}{30}$, $B_{10} = \frac{5}{66}$, $B_{12} = -\frac{691}{2730}$, and $B_{14} = \frac{7}{6}$.

3. Alternative definitions are

$$B_{2n} = (-1)^{n-1} 4n \int_0^{\infty} \frac{x^{2n-1}}{e^{2\pi x} - 1}\, dx \qquad (n \geqslant 1)$$

or indirectly via the expansion

$$u/(e^u - 1) = \sum_{n=0}^{\infty} B_n u^n / n! \qquad (0 < |u| < 2\pi),$$

which is often used.

Among the useful properties of these numbers we mention

$$\sum_{\nu=0}^{n-1} \binom{n}{\nu} B_\nu = 0 \qquad (n \geqslant 2).$$

These numbers also occur in the Euler–Maclaurin formula:

$$\int_{x_0}^{x_m} f(x)\, dx = h(f_0/2 + f_1 + f_2 + \cdots + f_{n-1} + f_m/2) - \sum_{2}^{\infty} B_\nu h^\nu \left(f_m^{(\nu-1)} - f_0^{(\nu-1)}\right)/\nu,$$

where $f_i = f(x_i) = f(x_0 + ih)$, as well as in the expansion of the complex-valued trigonometric functions $\cot Z$ and $\tan Z$.

Useful bounds on these numbers are

$$\frac{2 \cdot (2n)!}{(2\pi)^{2n}} < (-1)^{n-1} B_{2n} \leqslant \frac{(\pi^2/3)(2n)!}{(2\pi)^{2n}} \qquad (n \geqslant 1),$$

which shows that $(-1)^{n-1} B_{2n} \to \infty$ as $n \uparrow \infty$. Tables of Bernoulli numbers are presented in Abramowitz and Stegun [1].

References

[1] Abramowitz, M. and Stegun, I. A. (1964). Handbook of Mathematical Functions. *Natl. Bur. Stand. U. S. Appl. Math. Ser. 55* (Washington, D.C.).

[2] Bernoulli, J. (1713). *Ars. Conjectandi.*

[3] Davis, H. T. (1935). *Tables of the Higher Mathematical Functions.* Principia Press, Bloomington, Ind.

(BERNOULLI POLYNOMIALS
BERNOULLIS, THE)

BERNOULLI POLYNOMIALS

The rth Bernoulli polynomial $B_r(y)$ is defined by

$$x e^{yx} (e^x - 1)^{-1} = \sum_{r=0}^{\infty} B_r(y) \frac{x^r}{r!}.$$

$B_r(0)$ is called the rth Bernoulli number* and written B_r.

The Bernoulli polynomials of order n are defined by

$$x^n e^{yx} (e^x - 1)^{-n} = \sum_{r=0}^{\infty} B_r^{(n)}(y) \frac{x^r}{r!}.$$

The following relations are of interest:

$$B_n^{(n+1)}(x) = x^{(n)}$$

$$B_r^{(n)}(x) \simeq (B^{(n)} + x)^r,$$

where, after expansion, $\{B^{(n)}\}^j$ is replaced by $B_j^{(n)}$.

BERNOULLIS, THE

Bernoulli is one of the most illustrious names in the history of mathematics. At

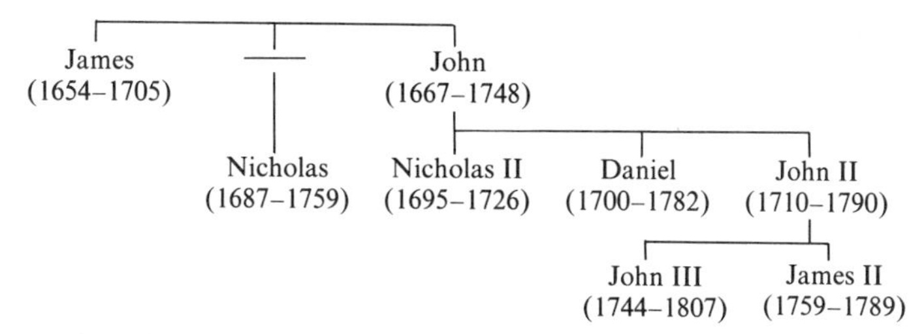

Figure 1. The mathematicians Bernoulli. The Roman numerals have been added to their names by historians. *James*, *John*, and *Nicholas* are English names. They become *Jakob*, *Johann*, and *Nikolaus* in German, and *Jacques*, *Jean*, and *Nicolas* in French.

least eight distinguished mathematicians bore the name (see Fig. 1), and three of these, the brothers James and John and John's son Daniel, were in the first rank of the mathematicians of their time. All the mathematicians Bernoulli were descendants of James' and John's father Nicholas Bernoulli, a prominent merchant in Basel, Switzerland. And all considered Basel their home.

Five of the Bernoullis made contributions to the theory of probability: James and John, their nephew Nicholas, Daniel, and John's grandson John III. The major contributions were those of James, Nicholas, and Daniel.

JAMES AND JOHN

James Bernoulli, the eldest of four brothers, was meant by his father to be a theologian. John, the third of the brothers, was meant to be a merchant. Both studied mathematics secretly and against their father's will.

James completed a degree in theology at the University of Basel in 1676. He then left Basel for several years of travel, during which he tutored and taught in several parts of France and Switzerland and further pursued his studies, especially Cartesian philosophy. After his return to Basel in 1680, he turned more to mathematics. In 1681, at a time when his fellow theologians were proclaiming a recent spectacular comet a sign of God's anger, he published a mathematical theory of the motion of comets. And he devoted a second educational journey, to Holland and England in 1681–1682, to visits with mathematicians. He returned to the University of Basel to lecture in experimental physics in 1683, and he obtained the university's chair of mathematics in 1687.

John, too, disappointed his father. He proved unsuited for a business career and obtained permission to enroll at the university when James began teaching there. His father now wanted him to study medicine, and he did so after completing his master of arts degree in 1685. But James secretly tutored him in mathematics, and about the time that James obtained his chair in mathematics, the two began to puzzle out the cryptic version of the infinitesimal calculus that G. W. Leibniz (1646–1716) published in 1684. They were the first to master Leibniz' method, and it was through the brilliant papers that they and Leibniz published from 1690 on that the calculus became known to the learned world.

As their mathematical powers developed, the two pugnacious brothers became bitter rivals. By the late 1690s, after John had obtained his own position in Groningen, the Netherlands, they were publicly trading insults and challenging each other with mathematical problems. It is unfortunate, perhaps, that these problems were so interesting, for memory of them has kept the brothers' quarrels alive even after their own lifetimes.

It is difficult to compare the accomplishments of the two brothers. The slower and more careful James bested his brother in their public exchanges, but John ultimately made the greater contribution to the calculus. John inherited his brother's chair at

Basel after his death in 1705, and remained active for 30 years thereafter. After Newton's death in 1727, he was unchallenged as the leading mathematician of all Europe.

In the field of probability, it is certainly James who was the giant. His philosophical training, together with his deep mathematical intuition, uniquely fitted him to attack the conceptual problems involved in applying the theory of games of chance*, as it had been developed by his predecessors Pascal (1623–1662), Fermat (1601–1665), and Huygens* (1629–1695), to problems of probability and evidence. And his struggle with these conceptual problems led him to the first limit theorem* of probability—the law of large numbers*.

James' great treatise on probability, *Ars Conjectandi* or the "Art of Conjecture," has four parts. The first three parts are in the tradition of his predecessors; he studies Huygens' pioneering book (*De ratiociniis in ludo aleae*, 1657), develops new combinatorial tools, and applies these to new problems. But the fourth part develops a new theory of probability—a theory that uses Huygens' ideas together with the law of large numbers to go beyond games of chance to problems of evidence and practical decision. (See below for a brief explanation of James' theory.)

James' mathematical diary shows that he had begun to work on probability in the 1680s and that he had proven the law of large numbers by 1689. Thus his completion of *Ars conjectandi* was impeded by ill helath, and after his death its publication was delayed by his wife and son's fears that an editor might treat it unfairly. It was finally published in 1713.

NICHOLAS

Nicholas Bernoulli studied mathematics under both his uncles. In 1709 he obtained a law degree with a dissertation that applied some of the ideas of *Ars conjectandi* to problems of law. Later, he was the editor who finally published *Ars conjectandi*. His first academic position was at Padua, Italy, but from 1722 on he held chairs at Basel, first in logic and then in law.

Nicholas was a gifted mathematician, but he published very little. Most of his achievements are hidden in the some 500 letters he wrote to other mathematicians. His correspondents included Leibniz, Euler (1707–1783), De Moivre* (1667–1754), and Montmort (1678–1719).

He is especially remembered for his letters to Montmort, where he generalized many of the results of Montmort's 1708 book *Essay d'analyse sur les jeux de hazard*. Montmort published most of his correspondence in the second edition of the book, in 1713; he and Nicholas had become so close that Nicholas stayed in Montmort's home for several months helping him prepare that edition.

One of the topics in the correspondence between Nicholas and Montmort was the excess of boys over girls in the christening records in London. This excess had been pointed out around 1711 by John Arbuthnot, who observed that it was too persistent to attribute to chance*. Arbuthnot saw this as evidence of divine intervention, but Nicholas, mindful of James' law of large numbers, explained the excess by saying that the true odds for a boy are closer to 18 : 17 than to 50 : 50.

DANIEL

Daniel Bernoulli was born during his father John's stay in Groningen, but he was educated at Basel. John made Daniel repeat his own youthful experience—first he tried to make him a merchant, and then he directed him toward medicine. Daniel's doctoral dissertation, completed in 1721, was on the mechanics of breathing.

In 1725, Daniel and his beloved older brother Nicholas II, who had tutored Daniel in mathematics, both obtained academic posts in St. Petersburg, Russia. Daniel was saddened by his brother's sudden death the following year, but he remained in St. Petersburg for eight years, during which he was very productive. In 1733, he returned to Basel, where he successively held chairs in anatomy and botany, in physiology, and in physics.

Daniel was a prolific and original mathematician, especially in physics. His greatest work was his treatise on hydrodynamics, written while he was in St. Petersburg. He also made major contributions to the mechanics of flexible and elastic bodies. It is an indication of his stature, perhaps, that his competitive father was often antagonistic toward him.

Daniel made several contributions to probability and statistics. The first and most influential was his *Specimen theoriae novae de mensura sortis* (1738), in which he developed the idea of utility*. His inspiration for this paper was a problem posed to him by his cousin Nicholas: how to determine the fair price of a game with infinite mathematical expectation*. We usually think of the expectation as the game's fair price, but as Nicholas pointed out, it is hardly fair to pay an infinite amount when one is certain to get only a finite amount back. Daniel's solution was to replace the monetary value of the payoff by its utility to the player, which may have a finite expectation. Since Daniel's paper was published in the St. Petersburg academy's journal, Nicholas' problem has become known as the St. Petersburg Paradox*.

Daniel is also remembered for his spirited statistical defense of inoculation for smallpox and for his statistical test of randomness* for the planetary orbits. And in several papers written around 1770, he participated in the effort to develop a probabilitistic theory of errors* that could be applied to astronomical observations. One of these papers is notable, in retrospect, for its use of what we now call the method of maximum likelihood*. Daniel did not develop this idea extensively, and his work was soon superseded by Laplace's* Bayesian theory.

JAMES' THEORY OF PROBABILITY

In the long tradition of philosophical and theological thought to which James Bernoulli was heir, the idea of probability was not closely tied to the idea of chance. Pascal, Fermat, and Huygens did not even use the word *probability* in their writings on chance [*see* CHANCE (I); CHANCE (II)]; probability, as these scholars knew, was an attribute of opinion, a product of argument or authority. The theory that James set forth in Part IV of *Ars conjectandi* was an attempt to bridge this gap. It was an attempt to apply the new theory of games of chance to probability while preserving the idea that probability is based on argument.

James' theory had two parts: his law of large numbers, which he thought could be used to assess the strength of an argument, and his rules for combining arguments.

Here is one of the examples of practical reasoning that made James see the need for his law of large numbers. Titius is marrying Caja, and an agreement is to be made concerning the division of her estate between him and their children in case she dies before him. But the size of the estate will depend on whether their own fathers are still living. Titius considers two possible agreements: one specifies that two-thirds of the estate will fall to him in any case; the other varies the proportion according to which fathers are still living. Which agreement is most advantageous to Titius? As James saw, the answer depends on how probable it is that Caja will outlive one or both of the fathers. But we cannot evaluate the probabilities a priori*, for we cannot count the equally possible ways in which Caja might or might not outlive the fathers. James thought there must be some number r of ways in which she might outlive them and some number s in which she might not. But since we do not fully understand the possible causes of death for Caja or the fathers, we cannot find r and s so as to calculate the probability $r/r + s$.

But it should be possible, James reasoned, to evaluate this probability *a posteriori*, from outcomes observed in many similar cases. The numbers r and s should be the same for other young women in similar circumstances. So in large number n of such cases, there will be a total of $(r + s)^n$ equally possible ways things might turn out: r^n ways for all the young women to outlive their fathers and fathers-in-law, $nr^{n-1}s$ ways for all but one to, and so on. Studying this binomial expansion, James realized that most of the $(r+s)^n$ equally possible ways involve a number k of the n young women outliving their fathers and fathers-in-law such that k/n is close to $r/(r+s)$. So if we actually observe n such young women, the observed frequency k/n will probably be close to $r/(r+s)$. As James proved, we can always choose n large enough for it to be morally certain (he thought we should decide on some high value of probability to call "moral certainty") that k/n will be close enough to $r/(r+s)$ that the two are identical for our practical purposes. This is the theorem we now call "the weak law of large numbers"*.

There remains the problem of combination. In the preceding example there may be no need to combine the probabilities based on mortality* statistics with probabilities based on other arguments. But in general, James believed, probability will depend on the combination of many arguments. Among the examples he used to make his point was the following very traditional one, which goes back to Cicero (106–43 B.C.) and Quintilian (ca A.D. 35–96). Titius is found dead on the road and Maevius is accused of committing the murder. There are several arguments against him. Maevius hated Titius, and they had been quarreling the day before; Maevius traveled over the road that day; a blood-stained sword is found in his house; and he turns pale and answers apprehensively when he is questioned. Perhaps none of these arguments alone would make Maevius' guilt probable. But together they make it almost certain.

As James saw it, his theorem should enable us to assess probabilities based on each argument. By observing, for example, other instances where a blood-stained weapon is found in a suspect's house, we should be able to assess the probability of Maevius' guilt based on that argument alone. But after thus assessing probabilities based on each argument alone, we should use his rules, which were derived from the theory of games of chance, to combine these probabilities. Only then do we have a probability based on the total evidence. The rules of combination James gave were imperfect, but they are interesting even today. Some of them explicitly recognize the possible nonadditivity of probability (*see* NONADDITIVE PROBABILITIES).

James' theory was not, of course, an adequate theory of statistical inference*. He suggested we find out how many observations are required for moral certainty and then make that many. He did not recognize, perhaps because he had no opportunity to apply his theory to actual numerical data, that we cannot always make as many observations as we would like. Statistical methods for dealing with limited observations did not emerge until more than 50 years after James' death, in response to the need to combine discrepant astronomical observations. *See* LEAST SQUARES.

Literature

The entries on the Bernoullis in the *Dictionary of Scientific Biography* include excellent descriptions of their contributions to mathematics and mechanics and extensive bibliographies. Detailed accounts of their contributions to probability can be found in Issac Todhunter's *History of the Theory of Probability* (1865; reprinted by Chelsea, New York, in 1965) and in L. E. Maistrov's *Probability Theory: A Historical Sketch* (1974, Academic Press, New York). James' work is also discussed by F. N. David in *Games, Gods and Gambling* (1962, Hafner, New York) and by Ian Hacking in *The Emergence of Probability* (1975, Cambridge University Press, Cambridge).

In 1935, the Bernoulli Commission was created in Basel for the purpose of publishing a new edition of the works of the Bernoullis. The plan is to publish everything left by the three great Bernoullis and a selection of the work of the five lesser ones. So far three volumes have appeared, one volume of John's letters and two volumes of James' work (*Der Briefwechsel von Johann Bernoulli*,

Vol. 1, 1955, Otto Spiess, ed.; *Die Werke von Jakob Bernoulli*, Vol. 1, 1969, J. O. Fleckenstein, ed., and Vol. 3, 1975, B. L. van der Waerden, ed.; all published in Basel by Birkhaüser Verlag). The volume edited by van der Waerden includes most of James' work on probability, including the parts of his previously unpublished diary bearing on probability. (It is in this diary that he discusses the problem of Caja's marriage contract.) It also includes Nicholas' dissertation, John de Witt's treatise on annuities, and extensive commentary, in German, on James' work and on most of the Bernoullis' correspondence on probability. [For further information on the new Bernoulli edition, see Clifford Truesdell's reviews in *ISIS*: 1958 (**49**:54–62) and 1973 (**64**:112–114).]

James' *Ars Conjectandi* was reprinted in the original Latin in 1968 by the Belgian publishing house Culture et Civilisation. A German translation was published in 1899: *Wahrscheinlichkeitsrechnung* (2 vols., Engelman, Leipzig). Unfortunately, there is still no definitive and widely available English translation, but a rough translation of Part IV is available from the National Technical Information Service, 5285 Port Royal Road, Springfield, VA 22161. (*Translations from James Bernoulli*, by Bing Sung, Harvard University Technical Report, 1966; the order number is AD-631 452.) For a faithful account in English of James' proof of his theorem, see pp. 96–101 of J. V. Uspensky's *Introduction to Mathematical Probability* (1937, McGraw-Hill, New York). For a discussion of his ideas on the combination of arguments, see G. Shafer's "Non-additive Probabilities in the Work of Bernoulli and Lambert," *Archive for History of Exact Sciences*, 1978 (**19**:309–370), and for a discussion of the relation of his ideas to classical rhetoric, see D. Garber and S. Zabell's "On the Emergence of Probability" in the same journal, 1979 (**21**:33–54).

An English translation of Daniel's *De mensura sortis* was published in *Econometrica* in 1954 (**22**:23–36). In 1967, Gregg Press (Farnborough, Hampshire, England) reprinted this translation, together with an earlier German translation and commentary, under the original Latin title. L. J. Savage discussed the relation between Daniel's concept of utility and contemporary concepts in Sec. 5.6 of *The Foundations of Statistics* (1972, Dover, New York). Daniel's essays on inoculation appeared in French in the *Mémoires* of the Paris academy (1760, pp. 1–45) and in *Mercure de France* (June 1760). An English translation of the paper in which he used maximum likelihood* was published in *Biometrika* in 1958 (**45**:293–315) and reprinted in Vol. 1 of *Studies in the History of Statistics and Probability* (E. S. Pearson and M. G. Kendall, eds.; Charles Griffin, London, 1970). Articles by O. B. Sheynin on his work are included in Vol. 2 of these *Studies* (Kendall and Plackett, eds.; 1977).

(BINOMIAL DISTRIBUTION
COMPETING RISKS
GAMES OF CHANCE
MAXIMUM LIKELIHOOD
NONADDITIVE PROBABILITIES
TESTS OF RANDOMNESS
UTILITY)

G. SHAFER

BERNOULLI SOCIETY

The Bernoulli Society for Mathematical Statistics and Probability is the international learned society in these fields, and it is the only one. Its objective is the advancement of probability and mathematical statistics in the widest sense through international contacts and international cooperation. Thus theoretical research, applications, dissemination of knowledge inside and outside the circle of the workers in the field, and development of teaching methods in the subject on all levels are all covered.

The Bernoulli Society has only individual members. Membership is open to anybody interested in its fields. In order to emphasize interaction with other branches of statistics such as official (governmental) statistics*, survey sampling*, and statistical computing, the Bernoulli Society is organized as a Section of the International Statistical Institute* (ISI). It is using the facilities of the ISI office in The Hague (the Netherlands) for its administration, and its ordinary meetings are being held biennially conjointly with the ordinary sessions of the ISI.

The statutes foresee, in addition to the general assembly composed of all members, an executive committee (President, elected for 2 years; president-elect; scientific secretary; treasurer; executive secretary of the ISI) and a council, consisting of 12 members elected for 4 years, and the members of the executive committee.

The Bernoulli Society was founded in its present form in 1975 and succeeds the International Association for Statistics in the Physical Sciences (IASPS), which originated in 1963. The following were the presidents of IASPS and the Bernoulli Society: Kitagawa (Japan), Bartlett (United Kingdom), Kolmogorov* (USSR), Neyman* (United States), Schmetterer (Austria), D. G. Kendall (United Kingdom), Blackwell (United States), Krickeberg (France), D. R. Cox (United Kingdom); the president-elect during 1979–1981 is P. Révész (Hungary). Some of the activities of the Bernoulli Society are being conducted by standing committees devoted to a particular subject area or geographical region.

To reach its objective, the Bernoulli Society is active in the following ways: first by organizing its own international meetings and joint meetings with other bodies, or by sponsoring such meetings. Thus the Committee on Statistics in the Physical Sciences, which continues the work of the former IASPS, is involved in meetings in its field of application. The Committee for Conferences on Stochastic Processes*, which is now also a standing committee of the Society, has been organizing annual meetings in its own domain on four continents since 1971. The European Regional Committee of the Society and its forerunner has been arranging the European Meetings of Statisticians since 1962. Similar regional committees in Latin America, in East Asia, and in the Pacific were recently formed. The Bernoulli Society also attends to relevant parts of the program for the biennial sessions of the ISI.

Second, the Bernoulli Society supports and sponsors relevant publications. In particular, it takes part in the "International Statistical Review"*. All members receive the "Review" and the newsletter of the ISI, "International Statistical Information," which contains, among other items, complete and early announcements of all meetings of interest to statisticians and probabilists. Members also receive the ISI *International Statistical Education Newsletter* and *Short Book Reviews*, which provides a rapid reviewing service. Many professional journals grant reduced rates to members of the Society.

The Bernoulli Society actively seeks contacts with the users of mathematical statistics and probability on an international basis. It is trying to eliminate existing barriers between statisticians or probabilists working in academic institutions and those employed by agricultural, medical, meteorological, industrial, and similar institutions.

One of the main concerns of the Bernoulli Society has been to overcome obstacles posed by geographical and economical conditions, currency problems, or political restrictions. For example, the membership dues may in principle be paid in national currency in all countries whose currency is not convertible for this purpose. The Society feels a very strong responsibility toward advancing its science in the wide sense defined by the opening paragraph, especially in those countries where mathematical statistics and probability is still relatively weak. It is doing this by intensifying its work on a regional basis and by offering the services of its members in various forms: lectures, courses, documentation, and specialized consulting. Its participation in the work of the ISI on statistical education* is seen to a large extent from this angle.

On June 1, 1981, the Bernoulli Society had 936 members in 57 countries. The membership dues are at present 35 Swiss francs per year, and half of this for members of age less than 30 years.

K. KRICKEBERG

BERNOULLI'S THEOREM

Bernoulli's theorem is a simple form of the weak law of large numbers* which follows from Chebyshev's inequality*. If S_n represents the number of successes (or occurrences) in n independent trials of an event with probability p, and if $\epsilon > 0$, then

$$\Pr\left[|n^{-1}S_n - p| \geqslant \epsilon\right] \to 0 \qquad \text{as } n \to \infty$$

or

$$\Pr\left[|n^{-1}S_n - p| < \epsilon\right] \to 1 \qquad \text{as } n \to \infty.$$

Equivalently, $n^{-1}S_n \xrightarrow{p} p$ as $n \to \infty$.

The last equation means that the relative frequency of successes approaches *in probability* the probability of success as the number of independent trials increases indefinitely.

Bibliography

Rényi, A. (1970). *Probability Theory*. North-Holland, Amsterdam.

(LAWS OF LARGE NUMBERS)

BERNOULLI TRIALS

Named after the Swiss mathematician J. Bernoulli (*see* BERNOULLIS, THE), Bernoulli trials are repeated trials of an experiment that obey the following conditions:

1. Each trial yields one of two outcomes [often called *success* (S) and *failure* (F)].
2. For each trial the probability of success is the same [usually denoted by $p = P(S)$, and the probability of failure is denoted by $q = 1 - p$].
3. The trials are independent; the probability of success in a trial does not change given any amount of information about the outcomes of other trials.

The simplest example of Bernoulli trials is the model for tossing a fair coin, where the occurrences head and tail are labeled S and F, respectively (here $p = q = \frac{1}{2}$).

(BINOMIAL DISTRIBUTION)

BERNSTEIN, SERGEI NATANOVICH

Born: March 5 (n.s.), 1880, in Odessa (a port on the Black Sea in the Ukraine).

Died: October 26, 1968, in Moscow, USSR.

Contributed to: theory and application of differential equations, function approximation theory, probability.

Bernstein received his mathematical education in Paris on completing high school in Odessa (where his father was a doctor and university lecturer) in 1898, and also studied at Göttingen. He defended a doctoral dissertation at the Sorbonne at age 24 in 1904, and another in Kharkov in 1913. From 1908 to 1933 he taught at Kharkov University, and from 1933 he worked at the Mathematical Institute of the USSR Academy of Sciences, also teaching at the University and Polytechnic Institute, in Leningrad, and continued to work at the Mathematical Institute in Moscow from 1943. He was much honored, including foreign membership of the Paris Academy of Sciences, and membership of the USSR and Ukrainian Academies. In his general mathematical work he united the traditions of the St. Petersburg "school", founded by P. L. Chebyshev* with those of modern western European thinking. The scope of his probabilistic work was in general ahead of its time, and his writings, including his textbook [4], which first appeared in 1927, were largely responsible for determining the course of development of this subject area in the USSR.

Bernstein's early publications of a probabilistic nature have a heavily analytical character: a constructive proof, using the sequence of what came to be known as Bernstein polynomials* and Bernoulli's law of large numbers*, of Weierstrass's uniform approximation theorem, and a consideration of the accuracy of approximation of the binomial distribution* by the normal distribution*, a problem to which he was to return several times. These early writings include an interesting attempt (1917) at the axiomatization* of probability theory (see ref. 5, pp. 10–60, and refs. 6 and 8).

One of the most significant areas of Bernstein's creativity, however, was a reexamination in a new light of the main existing theorems of probability theory of his times. For example, if $X_1, X_2, \ldots$ are independent random variables with finite variance, adjusted to have zero mean, and $S_n = \sum_{i=1}^{n} X_i$, then Chebyshev's inequality* may be written

$$P\{|S_n| < tB_n\} > 1 - t^{-2},$$

where $B_n = \operatorname{var} S_n$, and Kolmogorov's inequality* strengthens this result by replacing $|S_n|$ by $\sup_{1 \leqslant k \leqslant n} |S_k|$. Bernstein succeeded in raising the lower bound $1 - t^{-2}$ to $1 - 2e^{-t^2}$ in both, under additional assumptions on the X_k, this variant assuming the name Bernstein's inequality* and even more significantly showed the refinements to hold where the X_k are no longer necessarily independent but form what is now termed a martingale-difference sequence, so that $\{S_n\}$ is a martingale*. Another direction taken by Bernstein within this area was to generalize the conditions of Liapunov* for the applicability of the central limit theorem* for sums of random variables. In 1922, Bernstein proved this under conditions which, when specialized to the same setting, are equivalent to those of Lindeberg, whose paper appeared in the same year; a fundamental and perhaps his best-known paper [3] extended this work to sums of "weakly dependent" random variables X_k and Markov chains*; he later proved it for martingales $\{S_n\}$. (He speaks of "sums of dependent variables, having mutually almost null regression," since for a martingale difference sequence $E[X_{k+1} \mid X_k, X_{k-1}, \ldots, X_1] = 0$.) A group of papers, and Appendix 6 of [4], deal in essence with the weak convergence of a discrete-time stochastic process* to one in continuous time whose probability distribution satisfies a diffusion equation, together with an examination of the boundary theory of such an equation, anticipating later extensive development of this subject matter. Little known also are surprisingly advanced mathematical investigations in population genetics (e.g., ref 2), including a synthesis of Mendelian inheritance and Galtonian "laws" of inheritance. In addition, Bernstein took a keen interest in the methodology of the teaching of mathematics at secondary and tertiary levels.

In the years 1952–1964 he devoted much time to the editing and publication of the four-volume collection of his works [5], which contains commentaries by his students and experts in various fields.

References

[1] Alexandrov, P. S., Akhiezer, N. I., Gnedenko, B. V., and Kolmogorov, A. N. (1969). *Uspekhi Mat. Nauk*, **24**, 211–218 (in Russian). (Obituary; emphasis is mathematical rather than probabilistic, but interesting commentary on Bernstein's population mathematics.)

[2] Bernstein, S. N. (1924). *Uch. Zap. Nauchno-Issled. Kafedr Ukr., Otd. Mat.*, **1**, 83–115 (in Russian). [Also in ref. 5, pp. 80–107; in part (Chap. 1) in English translation (by Emma Lehner) in *Ann. Math. Statist.*, **13**, 53–61 (1942).]

[3] Bernstein, S. N. (1926). *Math. Ann.*, **97**, 1–59. (Published in 1944 in Russian translation, in *Uspekhi Mat. Nauk*, **10**, 65–114. Also in ref. 5, pp. 121–176.)

[4] Bernstein, S. N. (1946). *Teoriya veroiatnostei*, 4th ed. Gostehizdat, Moscow-Leningrad. (1st ed.: 1927; 2nd and 3rd eds.: 1934. Portions on nonhomogeneous Markov chains—pp. 203–213, 465–484 reprinted in ref. 5, pp. 455–483.)

[5] Bernstein, S. N. (1964). *Sobranie sochineniy*, Vol. 4: *Teoriya veroiatnostei i matematicheskaia statistika* [1911–1946]. Izd. Nauka, Moscow. [This fourth volume of Bernstein's collected works contains Russian language versions of most of his probabilistic work. Volume 1 (1952) and Vol. 2 (1954) deal with constructive function theory, and Vol. 3 (1960) with differential equations.]

[6] Kolmogorov, A. N. and Sarmanov, O. V. (1960). *Teor. Veroyatn. ee primen.*, **5**, 215–221 (in Russian). (Division of Bernstein's writings into groups, with commentary; complete listing of his probabilistic writings up to ref. 5; full-page portrait.)

[7] Kolmogorov, A. N., Linnik, Yu. V., Prokhorov, Yu. V., and Sarmanov, O. V. (1969). Sergei Natanovich Bernstein. *Teor. veroyatn. ee primen.*, **14**, 113–121 (in Russian). (Obituary, with good detail on biography and probabilistic work, and a photograph on p. 112.)

[8] Maistrov, L. E. (1974). *Probability Theory. A Historical Sketch*. Academic Press, New York, pp. 250–252. [Translated and edited from the Russian edition (1967) by S. Kotz.]

(AXIOMS OF PROBABILITY
BERNSTEIN POLYNOMIALS
BERNSTEIN'S INEQUALITY
CHEBYSHEV'S INEQUALITY
CONVERGENCE OF SEQUENCES OF RANDOM VARIABLES
DIFFUSION PROCESSES
GENETICS, STATISTICS IN
KOLMOGOROV'S INEQUALITY
LAWS OF LARGE NUMBERS
LIMIT THEOREMS, CENTRAL
MARKOV PROCESSES
MARTINGALES)

E. SENETA

BERNSTEIN LEMMA

Let X_n, Y_n, and Z_n be random variables such that $X_n = Y_n + Z_n$, where Y_n and Z_n have some joint distribution. If Y_n has a distribution converging as $n \to \infty$ to a proper distribution $F(\cdot)$ with nonzero variance and $\text{var}(Z_n) \to 0$, then the distribution of X_n also converges to $F(\cdot)$. (The variables Y_n and Z_n need not be independent.) This result is useful for proving theorems on convergence in distributions of random variables, without using characteristic functions*. It is used in problems related to asymptotic distributions* of test statistics.

Bibliography

Bernstein, S. (1927). *Math. Ann.*, **97**, 1–59.

Moran, P. A. P. (1968). *An Introduction to Probability Theory*. Clarendon Press, Oxford, p. 345.

For a recent application, see:

Serfling, R. L. (1969). Probability Inequality and Convergence Properties of Sums of Multiplicative Random Variables. *Statistics Report M151*, Florida State University, Tallahassee, Fla.

(CONVERGENCE OF SEQUENCES OF RANDOM VARIABLES)

BERNSTEIN POLYNOMIALS

Polynomials useful in approximation and interpolation theory. For a function $f(x)$ defined on the closed interval $[0, 1]$, the nth Bernstein polynomial is

$$B_n(x) = \sum_{k=0}^{\infty} f\left(\frac{k}{n}\right)\binom{n}{k} x^k (1-x)^{n-k}$$

as $n \to \infty$. The polynomial converges *uniformly* to the function $f(x)$. More precisely, [2],

$$|B_n(x) - f(x)| \leqslant \tfrac{5}{4}\omega(f, 1/\sqrt{n}),$$

where the modulus of continuity, $\omega(f; \delta)$, is defined by

$$\omega(f;\delta) = \sup_{|x-y| \leqslant \delta} |f(x) - f(y)|$$

$$= \sup_{|h| \leqslant \delta} |f(x+h) - f(x)|$$

$(x, y;\ x + h \in [0, 1])$. Also if the derivative is finite, then $\lim_{n\to\infty} B_n'(x) = f'(x)$ and convergence is uniform for a continuously differentiable $f(x)$. Applications of Bernstein's polynomials are for proving Weierstrass' approximation theorem using the weak law of large numbers and for density function estimation [4].

References

[1] Lorentz, G. G. (1953). Bernstein Polynomials. *Mathematical Expositions No*. 8. University of Toronto Press, Toronto.

[2] Popoviciu, T. (1935). *Mathematica* (*Cluj*), **10**, 49–54.

[3] Rényi, A. (1970). *Probability Theory*. North-Holland, Amsterdam, p. 165.

[4] Vitale, R. A. (1975). *Stochastic Processes*, **2**, 87–100.

BERNSTEIN'S INEQUALITY

A refinement of Chebyshev's inequality* published in 1926.

If $X_1, X_2, \ldots, X_n$ are independent random variables with $E(X_i) = 0$, then

$$\Pr[|X_1 + X_2 + \ldots + X_n| \leqslant \theta] > 1 - 2\exp\left[-\theta^2(2B_n + c\theta)^{-1}\right],$$

where B_n is the variance of the sum $S = X_1 + \cdots + X_n$ and c is a constant. (Chebyshev's inequality states: $\Pr[|X_1 + \cdots + X_n| \leqslant t\sqrt{B_n}\,] > 1 - 1/t^2$.) A useful form of Bernstein's inequality, quoted by Uspensky [3], is

$$\Pr[|X - \mu| \geqslant \lambda] \leqslant 2\exp\left[-\lambda^2\left(2\sigma^2 + \tfrac{2}{3}m\lambda\right)^{-1}\right]$$

for any $\lambda > 0$, where $\mu_i = EX_i$, $\mu = \sum_{i=1}^n \mu_i$; $\sigma_i^2 = \mathrm{var}(X_i)$, $\sigma^2 = \sum_1^n \sigma_i^2$, $X = \sum_1^n X_i$, X_i is independent of X_j for $i \neq j = 1, \ldots, n$, and m is a constant such that $\Pr[|X_i - \mu_i| > m] = 0$. That is, with probability 1, the maximal deviation of a random variable X_i from its mean μ_i does not exceed m. This condition also bounds all the central moments.

For additional information, see refs. 1 and 2.

References

[1] Renyi, A. (1970). *Probability Theory*. North-Holland, Amsterdam, pp. 284–386.

[2] Savage, I. R. (1961). *J. Res. Natl. Bur. Stand.*, **65B**, (3), 211–222.

[3] Uspensky, J. V. (1937). *Introduction to Mathematical Probability*. McGraw-Hill, New York, p. 205.

(CHEBYSHEV'S INEQUALITY)

BERNSTEIN'S THEOREM *See* DAMAGE MODELS

BERRY–ESSÉEN INEQUALITY *See* ASYMPTOTIC NORMALITY

BERRY–ESSÉEN THEOREM *See* ASYMPTOTIC NORMALITY

BERTRAND'S LEMMA

If in a ballot, two candidates A and B obtain a and b votes, respectively, with $a > b$, then the probability that A leads B throughout the voting, on the assumption that the votes are counted in random order, is $(a - b)/(a + b)$. The classical solution is based on the reflection principle*.

Bibliography

Feller, W. (1968). *An Introduction to Probability Theory*, 3rd ed., Vol. 1. Wiley, New York, p. 69.

Takács, L. (1967). *Combinatorial Methods in the Theory of Stochastic Processes*. Wiley, New York.

Whitworth, W. A. (1901). *Choice and Chance*, 5th ed. London. (Reprinted by G. E. Stechert, New York, 1942.)

(BALLOT PROBLEMS)

BERTRAND'S THEORY OF HYPOTHESIS TESTING *See* FREQUENCY INTERPRETATION IN PROBABILITY; HYPOTHESIS TESTING; STATISTICAL INFERENCE

BESSEL FUNCTION DISTRIBUTIONS

A system of distributions developing from the work of McKay [3]. Bhattacharyya [1] showed that these could be obtained as the distributions of linear functions of independent chi-squared* variables X_1, X_2 with common degrees of freedom ν.

The density function of $Y = a_1X_1 + a_2X_2$ (with $a_1 > a_2 > 0$) is

$$\frac{(c^2 - 1)^{m+1/2}}{\pi^{1/2}2^m b^{m+1}\Gamma(m + \frac{1}{2})}\, y^m e^{-cy/b} I_m(y/b) \qquad (y > 0),$$

where $b = 4a_1a_2(a_1 - a_2)^{-1}$; $c = (a_1 + a_2)/(a_1 - a_2)$; $m = 2\nu + 1$ and

$$I_m(x) = \left(\tfrac{1}{2}x\right)^m \sum_{j=0}^{\infty} \frac{(x/2)^{2j}}{j!\Gamma(m + j + 1)}$$

is a modified Bessel function of the first kind.

The density function of $Y = a_1X_1 - a_2X_2$ ($a_1 > 0$, $a_2 > 0$) is

$$\frac{(1 - c'^2)^{m+1/2}}{\pi^{1/2}2^m b'^{m+1}\Gamma(m + \frac{1}{2})} \cdot |g|^m e^{-cz/b} K_m(|z/b|),$$

where $b' = 4a_1a_2/(a_1 + a_2)^{-1}$, $c' = (a_2 - a_1)/(a_1 + a_2)$ and

$$K_m(x) = \frac{1}{2}\,\frac{I_{-m}(x) - I_m(x)}{\sin \nu\pi}$$

is a modified Bessel function of the second kind.

The curves of this system are rather troublesome to handle numerically and mathematically, mainly because of the functions $I_m(\cdot)$ and (especially) $K_m(\cdot)$, and they are not at all widely used.

The function $I_m(\cdot)$ also appears in the distribution of the difference between two independent Poisson* variables W_1, W_2 with expected values θ_1, θ_2, respectively. We have

$$\Pr[W_1 - W_2 = w] = e^{-(\theta_1 - \theta_2)}(\theta_1/\theta_2)^{w/2} \times I_{w/2}\left(2\sqrt{\theta_1\theta_2}\right).$$

References

[1] Bhattacharyya, B. C. (1942). *Sankhyā*, **6**, 175–182.

[2] Johnson, N. L. and Kotz, S. (1970). *Distributions in Statistics: Continuous Univariate Distributions*, Vol. 1. Wiley, New York, Chap. 12.

[3] McKay, A. T. (1932). *Biometrika*, **24**, 39–44.

BESSEL FUNCTIONS

Introduced by F. W. Bessel (1784–1846), these functions are also called cylinder functions.

Bessel functions of various kinds appear in several statistical distribution formulas. For example: distribution product of normal variables* [4], extreme values* in samples from a Cauchy distribution* [7], distribution of quadratic forms [5], asymptotic distribution of range* [6], noncentral χ^2 distribution* [11], moments of mean differences in samples from a Poisson distribution* [8], distribution of the Cramér–von Mises statistic* [2], distribution of the covariance of two normal correlated variables, directional distributions* of various kinds [9], and distributions of random variables on a cylinder [10] involve Bessel functions. (*See* BESSEL DISTRIBUTION.)

BESSEL FUNCTIONS OF THE FIRST AND SECOND KINDS

Bessel functions of the *first* kind of order ν and argument z are defined by

$$J_\nu(z) = \sum_{r=0}^{\infty} \frac{(-1)^r (z/2)^{\nu+2r}}{r!\Gamma(\nu + r + 1)}.$$

Similarly,

$$J_{-\nu}(z) = \sum_{r=0}^{\infty} \frac{(-1)^r (z/2)^{-\nu+2r}}{r!\Gamma(-\nu + r + 1)}.$$

These are independent solutions for the *Bessel equation*. If n is an integer, we have

$$J_{-n}(z) = (-1)^n J_n(z).$$

The identity

$$\exp\left[\tfrac{1}{2}z(t - t^{-1})\right] = \sum_{n=-\infty}^{\infty} t^n J_n(z)$$

gives the *generating function** of the Bessel functions $J_n(z)$, which are also called the Bessel coefficients.

Bessel functions of the *second* kind of order ν and argument z are defined by

$$Y_\nu(z) = \frac{(\cos \nu\pi)J_\nu(z) - J_{-\nu}(z)}{\sin \nu\pi}.$$

If $\nu = n$ is an integer, we also have

$$Y_{-n}(z) = (-1)^n Y_n(z).$$

SPECIAL CASES OF BESSEL FUNCTIONS OF THE FIRST AND SECOND KINDS

(a) $$J_{1/2}(x) = (\pi x/2)^{-1/2}\sin x;$$

$$J_{-1/2}(x) = (\pi x/2)^{-1/2}\cos x;$$

$$J_{3/2}(x) = (\pi x/2)^{-1/2}\left[(\sin x/x) - \cos x\right],$$

and in general

$$J_{\pm n+1/2}(x) = \left(\frac{2}{\pi}\right)^{1/2} x^{n+1/2} \times \left(\mp \frac{d}{x\,dx}\right)^n \left(\frac{\sin x}{x}\right)$$

are expressible in a finite number of terms involving sines, cosines, and powers of z and are sometimes referred to as spherical Bessel functions.

(b) $$J_0(x) = 1 - \frac{x^2}{2^2(1!)^2} + \frac{x^4}{2^4(2!)^2} - \frac{x^6}{2^6(3!)^2} + \cdots$$

$$J_1(x) = \frac{x}{2} - \frac{x^3}{2^3 1!2!} + \frac{x^5}{2^5 2!3!} - \cdots.$$

For small x, these series show that $J_0(x) \simeq 1$ and $J_1(x) \simeq x/2$. Similarly, it can be shown that $Y_0(x) = (2/\pi)\{\gamma + \log(x/2)\}$ and $Y_1(x) \simeq -2/(\pi x)$, where $\gamma = 0.5772$ is *Euler's constant**.

Recurrence formulas:

$$J_{\nu+1}(z) = \frac{\nu}{z} J_\nu(z) - J_\nu'(z),$$

$$J_{\nu-1}(z) = \frac{\nu}{z} J_\nu(z) + J_\nu'(z),$$

$$J_{\nu-1}(z) + J_{\nu+1}(z) = \frac{2\nu}{z} J_\nu(z),$$

$$J_{\nu-1}(z) - J_{\nu+1}(z) = 2J_\nu'(z),$$

and the special case

$$J_1(z) = -J_0'(z),$$

which is an identity of great utility.

BESSEL FUNCTIONS OF THE THIRD KIND

Bessel functions of the *third* kind, also known as *Hankel functions* are defined

$$H_\nu^{(1)}(z) = J_\nu(z) + iY_\nu(z)$$

and

$$H_\nu^{(2)}(z) = J_\nu(z) - iY_\nu(z).$$

These functions are *also* independent solutions of the Bessel equation. The identity

$$H_{n+(1/2)}^{(1)}(z) = J_{n+(1/2)}(z) + i(-1)^{n+1} J_{-n-(1/2)}(z)$$

is sometimes useful.

MODIFIED BESSEL FUNCTIONS

The function $e^{-\nu\pi i/2} J_\nu(iz)$ is a real function of z which is a solution of the equation

$$z^2 \frac{d^2w}{dz^2} + z\frac{dw}{dz} - (z^2 + \nu^2)w = 0$$

called the *modified Bessel function* of the *first* kind and denoted by $I_\nu(z)$. The infinite series representation of $I_\nu(z)$ is

$$I_\nu(z) = \sum_{r=0}^{\infty} \frac{(z/2)^{\nu+2r}}{r!\Gamma(\nu + r + 1)}.$$

Observe that $I_{-n}(z) = I_n(z)$ for a *positive* integer n. The function

$$K_\nu(z) = \frac{\pi}{\sin(\nu\pi)} \{I_{-\nu}(z) - I_\nu(z)\}$$

when ν is not an integer is called the *modified Bessel function* of the *second* kind. [Note that $K_\nu(z) = K_\nu(-z)$.] Some particular cases of modified Bessel functions are of importance:

1. $$I_0(x) = 1 + \frac{x^2}{2^2(1!)^2} + \frac{x^4}{2^4(2!)^2} + \frac{x^6}{2^6(3!)^2} + \cdots$$

$$K_0(x) = -\left(\gamma + \log_e \frac{x}{2}\right) I_0(x) + \sum_{r=1}^{\infty} \frac{(x/2)^{2r}}{r!r!} \times \left(1 + \frac{1}{2} + \frac{1}{3} + \cdots + \frac{1}{r}\right).$$

Also,

$$I_1(x) = \frac{x}{2} + \frac{x^3}{2^3 1!2!} + \frac{x^5}{2^5 2!3!} + \cdots$$

and

$$K_1(x) = \left(\gamma + \log_e \frac{x}{2}\right) I_1(x) + \frac{1}{x} - \frac{1}{2}\sum_{r=0}^{\infty} \frac{(x/2)^{2r+1}}{r!(r+1)!} \times \left\{2\left(1 + \frac{1}{2} + \cdots + \frac{1}{r}\right) + \frac{1}{r+1}\right\},$$

which shows that

$$I_0(x) \simeq 1, \quad K_0(x) \simeq -\gamma - \log_e(x/2),$$
$$I_1(x) \simeq (x/2), \quad K_1(x) = 1/x.$$

2. Recurrence relations:

$$-I_{\nu+1}(z) = \frac{\nu}{z} I_\nu(z) - I'_\nu(z)$$
$$I_{\nu-1}(z) = \frac{\nu}{z} I_\nu(z) + I'_\nu(z)$$
$$I_{\nu-1}(z) - I_{\nu+1}(z) = \frac{2\nu}{z} I_\nu(z)$$
$$I_{\nu-1}(z) + I_{\nu+1}(z) = 2I'_\nu(z)$$
$$-K_{\nu+1}(z) = -\frac{\nu}{z} K_\nu(z) + K'_\nu(z)$$
$$-K_{\nu-1}(z) = \frac{\nu}{z} K_\nu(z) + K'_\nu(z)$$
$$K_{\nu-1}(z) - K_{\nu+1}(z) = -\frac{2\nu}{z} K_\nu(z)$$
$$K_{\nu-1}(z) + K_{\nu+1}(z) = 2K'_\nu(z)$$

are worth noting.

Modified functions $I_n(x), K_\nu(x)$ bear to the exponential functions similar relations to those which the functions $J_\nu(x)$, $Y_\nu(x)$ bear to the trigonometric functions; modified functions have *no* zeros for real values of x. [Note that $I_1(z) = I'_0(z)$ and $K_1(z) = -K'_0(z)$; the latter is similar to the corresponding relation for $J_n(z)$.]

Graphs of Bessel functions $J_i(x)$, $Y_i(x)$, $I_i(x)$, and $K_i(x)$ for $i = 0, 1$ are presented in Fig. 1.

Tabulated values of $J_n(x)$, $Y_n(x)$, $I_n(x)$, and $K_n(x)$ for $n = 0(1)20$ are presented in the *British Association Mathematical Tables* [3] and in Abramowitz and Stegun [1].

References

[1] Abramowitz, M. and Stegun, I. A. (1964). Handbook of Mathematical Functions. *Natl. Bur. Stand. Appl. Math Ser. 55* (Washington, D. C.).

[2] Anderson, T. W. and Darling, D. A. (1952). *Ann. Math. Statist.*, **23**, 193–212.

[3] *British Association Mathematical Tables* (1952). Vol. 10.

[4] Craig, C. C. (1936). *Ann. Math. Statist.*, **7**, 1–15.

[5] Grad, A. and Solomon, H. (1955). *Ann. Math. Statist.*, **26**, 464–477.

[6] Gumbel, E. J. (1947). *Ann. Math. Statist.*, **18**, 384–412.

[7] Gumbel, E. J. and Keeney, R. D. (1950). *Ann. Math. Statist.*, **21**, 523–538.

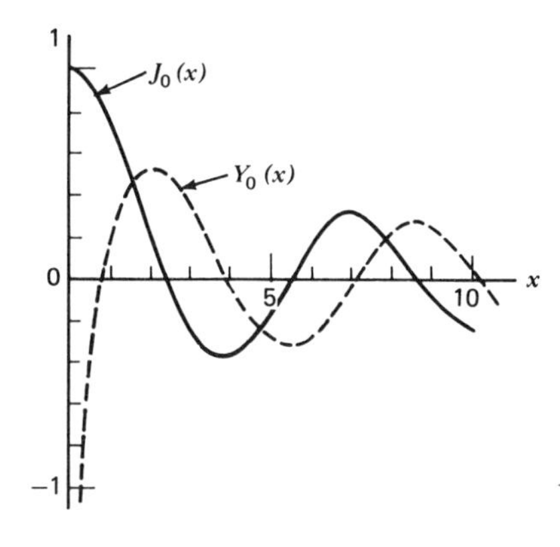

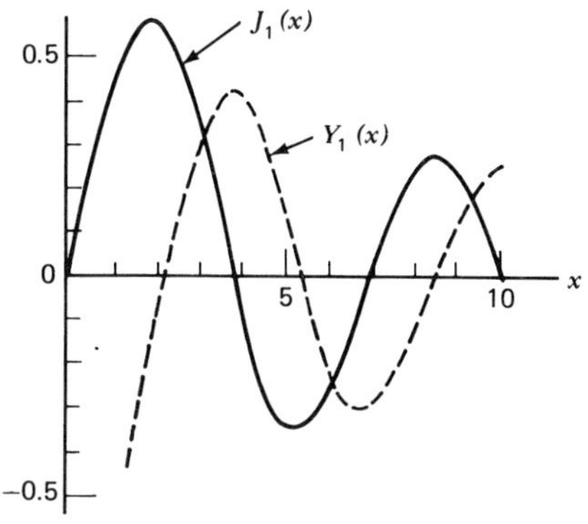

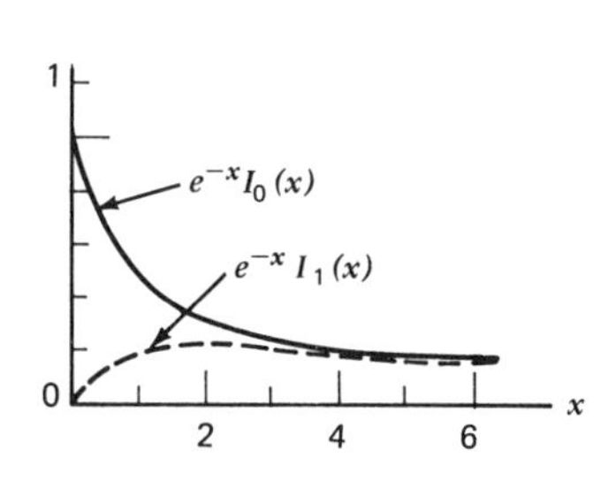

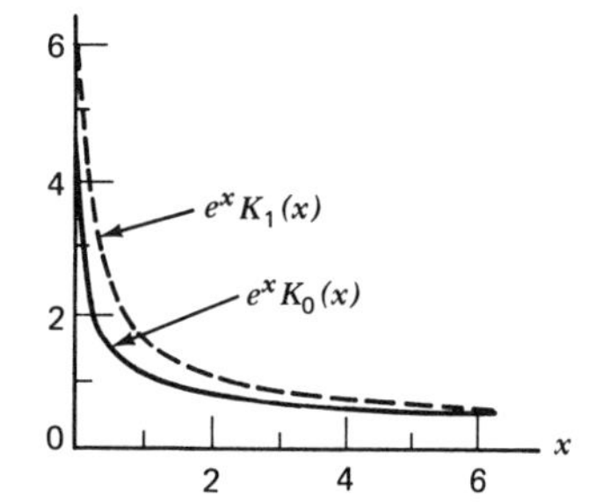

Figure 1

[8] Katti, S. K. (1960). *Ann. Math. Statist.*, **31**, 78–85.

[9] Mardia, K. V. (1975). *J. R. Statist. Soc. B*, **37**, 349–393.

[10] Mardia, K. V. and Sutton, T. W. (1978). *J. R. Statist. Soc. B*, **40**, 229–233.

[11] Ruben, H. (1960). *Ann. Math. Statist.*, **31**, 598–618.

Bibliography

See the following works, as well as the references just given, for more information on the topic of Bessel functions.

Erdelyi, A., ed. (1953). *Higher Transcendental Functions*, Vol. 2. McGraw-Hill, New York.

Luke, Y. L. (1976). *Mathematical Functions and Their Approximations*. Academic Press, New York.

McLachlan, N. W. (1955). *Bessel Functions for Engineers*. Oxford University Press, New York.

Relton, F. E. (1946). *Applied Bessel Functions*, Blackie, London.

Traner, C. J. (1968). *Bessel Functions with Some Physical Applications*. English University Press, London.

Watson, G. N. (1966). *Theory of Bessel Functions. Cambridge University Press, Cambridge. (Paperback edition; reprint of the 1944 edition.)*

BESSEL PROCESS *See* BROWNIAN MOTION

BESSEL'S INTERPOLATION FORMULA

Suppose that a function $f(\cdot)$ is tabulated at points h units apart. We want to evaluate $f(t)$. We first find the interval t_0 to $t_0 + h$ containing t, and define $y_{(t-t_0)/h} = f(t)$. Putting $(t - t_0)/h = x$ and $v = x - \frac{1}{2}$, Bessel's interpolation formula is

$$\begin{aligned} y_x &= P_{2n+1}(x) \\ &= \frac{y_0 + y_1}{2} + v\Delta y_0 + \frac{v^2 - \frac{1}{4}\Delta^2 y_{-1} + \Delta^2 y_0}{2} \\ &\quad + \frac{v(v^2 - \frac{1}{4})}{3!}\Delta^3 y_{-1} + \frac{(v^2 - \frac{1}{4})(v^2 - \frac{9}{4})}{4!} \\ &\quad \times \frac{\Delta^4 y_{-2} + \Delta^2 y_{-1}}{2} + \cdots \\ &\quad + \left[(2n+1)!\right]^{-1} v(v^2 - \tfrac{1}{4})(v^2 - \tfrac{9}{4}) \cdots \\ &\quad \times \left[v^2 - (2n-1)^2/4\right]\Delta^{2n+1} y_{-n}, \end{aligned}$$

where Δ^r is the rth-order forward difference*. Generally, Bessel's formula is used for interpolation near the midpoint between successive tabulated values (e.g., for $|v| < \frac{1}{4}$).

Bibliography

Chakravarti, I. M., Laha, R. G., and Roy, J. (1967). *Handbook of Methods of Applied Statistics*, Vol 1. Wiley, New York.

(INTERPOLATION
NEWTON'S FORMULA)

BETA COEFFICIENTS

In writing the model for multiple linear regression*, it is customary to use the Greek symbol β for the coefficients of the independent (predictor) variables. This notation is so widely used that the term "beta coefficients" causes little confusion.

(REGRESSION COEFFICIENTS)

BETA DISTRIBUTIONS

The *standard* beta distribution [Beta(a, b)] form of two-parameter family of continuous distributions with density functions

$$\{B(a,b)\}^{-1}x^{a-1}(1-x)^{b-1} \qquad (0 < x < 1),$$

where $B(a,b)$ is a beta function ($B(a,b) = \int_0^1 t^{a-1}(1-t)^{b-1}\,dt$). The parameters are $a(> 0)$ and $b(> 0)$. If X has a standard beta distribution, then $Y = \alpha + \beta X$, with $\beta > 0$, has a beta distribution over the range α to $(\alpha + \beta)$, with the same parameter values as X.

The rth moment of X is

$$\begin{aligned} E[X] &= B(a+r,b)/B(a,b) \\ &= a^{[r]}/(a+b)^{[r]} \end{aligned}$$

if r is an integer [$a^{[r]} = a(a+1)\cdots(a+r-1)$]. The expected value of X is $a/(a+b)$; the variance is $ab(a+b)^{-2}(a+b+1)^{-1}$.

The distribution of Beta(b,a) is the same as that of $\{1 - \text{Beta}(a,b)\}$. This is sometimes referred to as quasi-symmetry*.

Beta distributions arise in statistical theory as the distribution (Beta(α_1, α_2)) of the ratio $Y_1/(Y_1 + Y_2)$, where Y_1, Y_2 are independent standard gamma variables with parameters α_1, α_2. As a particular case of this, if $X_1, \ldots, X_n$ are independent normal* $N(\mu, \sigma^2)$ variables, then

$$\frac{1}{n-1}(X_i - \overline{X})^2 / \sum_{j=1}^{n} (X_j - X)^2$$

(with $\overline{X} = n^{-1}\sum_{j=1}^{n} X_j$) is distributed as Beta $(\frac{1}{2}, \frac{1}{2}(n-1))$. It is also used as a prior density* in Bayesian inference*. Beta distributions are members (Types I and II) of the Pearson system* of frequency curves.

The four-parameter beta distribution with density of the form

$$\frac{1}{B(p,q)} \frac{(y-a)^{p-1}(b-y)^{q-1}}{(b-a)^{p+q-1}} \qquad (a \leqslant y \leqslant b)$$

with $p > 0$, $q > 0$, was studied in detail by Whitby [2].

For additional information, see ref. 1.

References

[1] Johnson, N. L. and Kotz, S. (1970). *Continuous Univariate Distributions*, Vol. 2. Wiley, New York, Chap. 23.

[2] Whitby, O. (1972). Ph.D. dissertation, Stanford University.

[APPROXIMATIONS TO DISTRIBUTIONS
ARC-SINE DISTRIBUTION
F-DISTRIBUTION
PRIOR DENSITY
RECTANGULAR (UNIFORM) DISTRIBUTION]

BHATE'S CONJECTURE

Bhate [2] conjectured that the average sample number (ASN)* of a sequential probability ratio test (SPRT)* discriminating between two hypotheses, H_0 and H_1, can be approximated by the solution for n of the equation

$$E[L_N] = E[L_n]$$

where L_n denotes the likelihood ratio* used in the SPRT test, when the sample size is equal to n and N is the sample size of the SPRT. N, of course, corresponds to a random variable. The expectation on the left-hand side involves the joint distribution of sample size N and of the random variables $X_1, X_2, \ldots$ corresponding to observations; that on the right-hand side depends only on the latter.

When the random variables X_1, $X_2, \ldots$ are independent and identically distributed, the conjecture leads to Wald's approximate ASN formula*.

It has been shown [1] that Bhate's conjecture is valid in a certain asymptotic sense, when the ASN is large, for a broad class of sequential tests*.

References

[1] Berk, R. H. (1973). *Ann. Statist.*, **1**, 1126–1138.

[2] Bhate, D. H. (1955). Ph.D. dissertation, University of London, Chap. VI.

(AVERAGE SAMPLE NUMBER
SEQUENTIAL ANALYSIS
SEQUENTIAL PROBABILITY RATIO TEST)

BHATTACHARYYA BOUNDS

A set of lower bounds for the variance of an unbiased estimator* T of a parameter θ based on a random sample of size n. (The Cramér–Rao lower bound* is a special case of these bounds.) The derivation of the kth bound involves the determination of the inverse of the Bhattacharyya matrix of $u_i = L^{(i)}(\theta)/L(\theta)$, $i = 1, \ldots, k$, where $L(\theta)$ is the likelihood function for θ and $L^{(i)}(\theta)$ is its ith derivative with respect to θ. For the case of the exponential family of distributions*, Bhattacharyya bounds converge to the variance of the best unbiased estimator [2].

For additional information, see refs. 1, 3, and 4.

References

[1] Bhattacharyya, A. (1946). *Sankhyā*, **8**, 1–14, 201–218, 315–328.

[2] Blight, B. J. N. and Rao, P. V. (1974). *Biometrika*, **61**, 137–142.

[3] Fend, A. V. (1959). *Ann. Math. Statist.*, **30**, 381–388.

[4] Seth, G. R. (1949). *Ann. Math. Statist.*, **20**, 1–27.

(CRAMÉR–RAO BOUNDS)

BIAS

BIAS OF AN ESTIMATOR

The statistical meaning of bias is quite precise. It is defined as the difference between the expected value* of a statistic and the value of a population parameter. It thus depends on both the statistic and the parameter. It is relevant when the statistic is to be used as an estimator* of the parameter. (See below for a formal definition.)

Sources of bias fall into three main categories. The most prominently studied in statistical literature is the statistical (technical) bias due to the specific algebraic form of an estimator. The other two categories are selection bias and measurement error. Williams [2] provides an overview of this topic.

Statistical Bias

Consider a statistic T which estimates a parameter θ. If $E_\theta(T) = \theta + b(\theta)$, where E_θ indicates that the expectation of T is computed under the assumption that the true parameter value is θ, the quantity $b(\theta) = E_\theta(T) - \theta$ is the *bias* of the statistic T, regarded as an estimator of θ. If $b(\theta) \equiv 0$ for all values of θ, T is called an *unbiased* estimator of θ.

Selection Bias

This kind of bias can arise when a sample is not drawn according to the prearranged specifications. Population units are included in the sample with probabilities other than those specified in the sampling design*. As a typical example, a design might specify an equal probability of selection for each of a number of households, but in fact, households with children may appear more often than do those without children. If no adjustment is applied, estimates based on the sample are likely to be *biased*. In other words, selection bias occurs if the real probabilities of selection are different from the intended probabilities, and the real probabilities are not known while the incorrect design probabilities are used in the estimator. An extreme case of selection bias occurs when the chance of selection is zero for some subgroup of the population. Then the sampled population is not the same as the target population*.

Measurement Error*

Conceptually, measurement error is easy to understand, but it is often difficult to assess in practice because it can arise in many subtle ways.

A measuring device can malfunction; the device can be mechanical or human and the errors may be generated completely randomly, mechanically, systematically, or even deliberately. Imprecise definitions also cause measurement problems; another major potential source of measurement bias occurs in the handling of the data. The effect of a single mispunched card may not be trivial. Various suggestions available in the literature to combat measurement bias include: (1) to subsample the data and remeasure the subset of observations; (2) to assume that the errors made are randomly distributed (averaging to zero); and (3) to utilize the randomized response techniques* introduced by Warner [1]; [4] to take "parallel" (similar) samples.

BIAS OF A TEST

A statistical test is said to be *biased* if the probability of rejecting the hypothesis tested

(H_0), when H_0 is valid—the significance level*—is greater than the probability of rejection when some other hypothesis (H, say) is valid. The property of *biasedness* is relative to the set of alternative hypotheses considered (i.e., the set from which H may be chosen). In fact, it is always possible to construct an artificial hypothesis, $\tilde{H}$ say, different from H_0, such that the probability of rejection is zero—simply by excluding the possibility, under $\tilde{H}$, of getting a sample point in the critical region*. (This does not constitute bias unless H is included in the set of alternative hypotheses for which the test is intended to be appropriate.)

References

[1] Warner, S. L. (1965). *J. Amer. Statist. Ass.*, **60**, 63–69.

[2] Williams, B. (1978). *A Sampler on Sampling*. Wiley, New York, Chap. 6.

(ESTIMATION, POINT
HYPOTHESIS TESTING
POWER
SIGNIFICANCE LEVEL)

BIAS, ORDER OF

Let $\hat{\theta}_1$ and $\hat{\theta}_2$ be two estimators* of θ defined on a sample of size n with nonnegative biases*

$$B_1(n,\theta) = E[\hat{\theta}_1 - \theta] \neq 0$$

and

$$B_2(n,\theta) = E[\hat{\theta}_2 - \theta] \neq 0.$$

If

$$\left| \lim_{n\to\infty} \frac{B_1(n,\theta)}{B_2(n,\theta)} \right| = 1 \qquad (1)$$

$\hat{\theta}_1$ and $\hat{\theta}_2$ are said to be *same order* bias estimators of θ. This concept is sometimes abbreviated to $\hat{\theta}_1$ S.O.B.E. $\hat{\theta}_2$. If the absolute value of the limit in (1) is greater than zero and less than 1, $\hat{\theta}_1$ is said to be a *better same order bias estimator* than $\hat{\theta}_2$ (abbreviated as $\hat{\theta}_1$ B.S.O.B.E. $\hat{\theta}_2$). If, without restricting $B_1(n, \theta) \neq 0$,

$$\lim_{n\to\infty} \frac{B_1(n,\theta)}{B_2(n,\theta)} = 0,$$

we then say that $\hat{\theta}_1$ is a lower-order bias estimator than $\hat{\theta}_2$ ($\hat{\theta}_1$ L.O.B.E. $\hat{\theta}_2$). These concepts are useful in bias reduction techniques of estimators.

Bibliography

Gray, H. L. and Schucany, W. R. (1972). *The Generalized Jackknife Statistic*. Marcel Dekker, New York.

(ESTIMATION POINT
JACKKNIFING)

BIENAYMÉ, IRENÉE-JULES

Born: August 28, 1796, in Paris, France.

Died: October 19, 1878, in Paris, France.

Contributed to: probability, mathematical statistics, demography, social statistics.

Bienaymé's life initially followed a somewhat erratic course with regard to scientific pursuits, partly because of the times through which he lived. The École Polytechnique, in which he enrolled in 1815, was dissolved in 1816, because of the Napoleonic sympathies of its students. Subsequently, he worked as a translator for journals, and in 1818 became lecturer in mathematics at the military academy at St. Cyr, leaving in 1820 to enter the Administration of Finances. Here he became Inspector, and, in 1834, Inspector General, but because of the revolution of 1848, retired to devote all his energies to scientific work.

Up to this time he had been active in the affairs of the Société Philomatique de Paris and his contributions to its meetings were reported in the now-obscure scientific newspaper-journal *L'Institut, Paris*, being reprinted at the end of the year of their appearance in the collections *Procès-Verbaux* *de la Société Philomatique de Paris—Extraits*. The most startling of Bienaymé's contributions to probability occurs in this context, where he gives a completely correct statement of the criticality theorem* for simple branching processes*. His communication [2], which may have been stimulated by work of L. F. B. de Châteauneuf (1776–1856), precedes the partly correct statement of F. Galton* and H. W. Watson by some 30 years, and the first subsequent completely correct one by over 80 years [5].

Bienaymé began to publish only in 1829, and his early (Civil Service period) writings lean to demography and actuarial matters. For example, a major one [1] discusses the then used *life tables** of A. Deparcieux and E. E. Duvillard, with the object of presenting overwhelming evidence against continued use of the Duvillard table, since its predicted mortality rates were heavier than appropriate. These interests persisted in Bienaymé's work, for despite his retirement, Bienaymé had considerable influence as a statistical expert in the government of Napoleon III, being praised in a report to the Senate in 1864 for his actuarial work in connection with the creation of a retirement fund. He was elected to the Academy of Sciences in 1852.

Even those of his papers with primarily sociological or demographic intent sometimes involve significant methodological contributions to probability and mathematical statistics. Bienaymé was active in three general areas: the stability and dispersion theory* of statistical trials (before his retirement), theory associated with linear least squares* (after his retirement), and limit theorems*.

In dispersion theory (which later came to be associated with the names of W. Lexis*, L. Bortkiewicz*, and A. A. Chuprov*) Bienaymé introduced a physically motivated principle of *durée des causes*, under the operation of which the proportion of successes in a sequence of trials exhibits more variability than in the homogeneous case of all Bernoulli trials*; and which might therefore be used to explain such observed variability. Bienaymé also manifested an understanding of the concept of a sufficient statistic*.

Laplace's treatise [9] acted as a basis for some of Bienaymé's best work. In the area of least squares in particular, he is concerned with generalizing and defending Laplacian positions [3, 4]. Reference 4 contains, *interalia*, the general Bienaymé–Chebyshev inequality (*see* CHEBYSHEV INEQUALITY). $\Pr[|\bar{X} - EX| \geqslant \epsilon] \leqslant \operatorname{Var} X/(\epsilon^2 n)$ (proved by the simple reasoning still used today), which is used to deduce a weak law of large numbers*. P. L. Chebyshev* obtained the inequality in 1867 (in a much more restricted setting and with a more difficult proof), in a paper published simultaneously in Russian, and in French juxtaposed to a reprinting of Bienaymé's paper. Later, Chebyshev gave Bienaymé credit for arriving at the inequality via the "method of moments"*, whose discovery he ascribes to Bienaymé. The slightly earlier paper [3], partly on the basis of which Bienaymé was elected to the Academy, contains, again as an incidental result, the deduction of an almost final form of the continuous chi-square* density, with n degrees of freedom, for the sum of squares of n independently and identically distributed normal $N(0, 1)$ random variables.

A limit theorem reobtained in 1919 by R. von Mises [10] (who regarded it as a *Fundamentalsatz* of the same significance as the central limit theorem*), was actually proved by Bienaymé in 1838. It asserts the following: If the random variables W_i, $i = 1, \ldots, m$ ($\sum W_i = 1$) have a joint Dirichlet distribution* with parameters $x_1, x_2, \ldots, x_m$ then, as $n = \sum x_i \to \infty$ with $r = x_i/n = \text{constant}(i) > 0$, the limiting standardized distribution of $V = \sum \gamma_i W_i$, is $N(0, 1)$. Late in his life, he constructed a simple combinatorial "runs up and down" test* of randomness of a series of observations on a continuously varying quantity, based on the number of local maxima and minima in the series. In particular he stated that the number of intervals between extrema ("runs up and down") in a sequence of n observations is, under assumption of randomness* of sample from a continuous distribution, approximately normally distributed (for large n) about a mean of $(2n - 1)/3$ with variance $(16n - 29)/90$.

Bienaymé was far ahead of his time in the depth of his statistical ideas and, as a result, has been largely ignored in the literature, although he was an important figure in nineteenth-century statistics. He was a correspondent of L. A. J. Quetelet* and a friend of A. A. Cournot* and of Chebyshev. However, his papers were verbose, his mathematics was laconic, and he had a penchant for controversy. He invalidly criticized the law of large numbers of S. D. Poisson*, and no sooner was he in the Academy than he engaged in a furious controversy with A. L. Cauchy*.

Bienaymé's life and scientific contributions, within a general framework of nineteenth century statistics, have been extensively described and documented by Heyde and Seneta [6]. Reference 8 provides information on Bienaymé's contributions to linear least squares, and ref. 7 expands on ref. 5 in relation to the discovery of the criticality theorem.

References

[1] Bienaymé, I. J. (1837). *Ann. Hyg. Paris*, **18**, 177–218.

[2] Bienaymé, I. J. (1845). *Soc. Philom. Paris, Extraits, Ser. 5*, 37–39. (Also in *L'Institut, Paris*, **589**, 131–132; and reprinted in ref. 7.)

[3] Bienaymé, I. J. (1852). *Liouville's J. Math. Pures Appl.*, **17** (1), 33–78. [Also in *Mém. Pres. Acad. Sci. Inst. Fr.*, **15** (2), 615–663 (1858).]

[4] Bienaymé, I. J. (1853). *C. R. Acad. Sci. Paris*, **37**, 309–324. [Also in *Liouville's J. Math. Pures Appl.*, **12** (2) 158–176 (1867).]

[5] Heyde, C. C. and Seneta, E. (1972). *Biometrika*, **59**, 680–683.

[6] Heyde, C. C. and Seneta, E. (1977). *I. J. Bienaymé: Statistical Theory Anticipated*. Springer-Verlag, New York. (The basic modern source on Bienaymé.)

[7] Kendall, D. G. (1975). *Bull. Lond. Math. Soc.*, **7**, 225–253.

[8] Lancaster, H. O. (1966). *Aust. J. Statist.*, **8**, 117–126.

[9] Laplace, P. S. de (1812). *Théorie analytique des probabilités*. V. Courcier, Paris.

[10] Mises, R. von (1964). *Mathematical Theory of Probability and Statistics*, Hilda Geiringer, ed. Academic Press, New York, pp. 352–357.

(BRANCHING PROCESSES
CHEBYSHEV, PAFNUTY LVOVICH
CHEBYSHEV INEQUALITY
CHI-SQUARE DISTRIBUTION
CRITICALITY THEOREM
DEMOGRAPHY
DISPERSION THEORY
LAWS OF LARGE NUMBERS
LEAST SQUARES
LIFE TABLES
LIMIT THEOREMS
METHOD OF MOMENTS
RANDOMNESS, TESTS FOR
SUFFICIENT STATISTICS]

E. SENETA

BIENAYMÉ–CHEBYSHEV INEQUALITY *See* CHEBYSHEV INEQUALITY

BIGEOMETRY *See* BIPLOT

BIMODAL *See* BIPLOT

BINARY DATA

These are data arising when measurement can give only one of two values. Conventionally these are represented by 0 and 1, but they need not even be numerical. Survival for a specified period, success or failure in a specified task, sex, are all examples of sources of binary data. The *number* of 1's obtained from a set of n measurements, of course, is a numerical variable, and appropriate statistical analysis can be applied to it. In particular, it is often supposed, as a first approximation, that successive observations are independent and the probability of a 1 (or a 0) is constant (*see* BERNOULLI TRIALS). The number of 1's would then have a binomial distribution*. More elaborate models are, of course, available (*see* LEXIS SERIES,

POISSON SERIES, SUBNORMAL AND SUPERNORMAL DISPERSION).

If measurements on concomitant variables* $X_1, \ldots, X_k$ are available, dependence of $\Pr[X = 1]$ on values of the X's is often represented in one of the following ways:

1. *Linear model*: $\Pr[X = 1] =$ linear function of X's
2. *Log-linear model*: $\log\{\Pr[X = 1]\}$ = linear function of X's
3. *Log-logistic model*:

$$\log\left\{\frac{\Pr[X=1]}{\Pr[X=0]}\right\} = \text{linear function of} X\text{'s}$$

Although models 1 and 2 (especially Model 1) may have greater immediate appeal, being rather simpler to understand, model 3 has the advantage that as $\Pr[X = 1]$ increases from 0 to 1, the left-hand side increases from $-\infty$ to $+\infty$, and so covers all possible values of the linear function on the right-hand side. However, models 1 and 2 may give good approximations for those ranges of values of the X's which are of practical importance. Provided that risks attendant on extrapolation* are recognized and avoided, such models can be very useful.

Of course, it is possible that some (or all) of the concomitant ("explanatory") variables may be binary. Analysis of multivariate binary data has attracted increasing attention recently (e.g., Cox [3], Ott and Kronmal [5]).

It is common practice to analyze observed proportions of 1's as if the residual variation in the models were normal. For sufficiently large samples this is not likely to be misleading, although some attention should be paid to using appropriate weights (e.g., in least-squares* equations). Lindsey [4] explains how to use direct likelihood* approaches (for model 3 and by analogy for other models).

Binary data can present themselves in many different forms—*see* e.g., CATEGORICAL DATA, CONTINGENCY TABLES, DICHOTOMY, PAIRED COMPARISONS, QUANTAL RESPONSE, and entries on various discrete distributions. There is a very useful collection of methods of analysis appropriate to such data in Cox [2]. There are further interesting examples, and technical developments in Bishop et al. [1].

References

[1] Bishop, Y. M. M., Fienberg, S. E., and Holland, P. W. (1975). *Discrete Multivariate Analysis: Theory and Practice*. MIT Press, Cambridge, Mass.

[2] Cox, D. R. (1970). *Analysis of Binary Data*. Methuen, London.

[3] Cox, D. R. (1972). *Appl. Statist.*, **21**, 113–120.

[4] Lindsey, J. K. (1975). *Appl. Statist.*, **24**, 1–16.

[5] Ott, J. and Kronmal, R. A. (1976). *J. Amer. Statist. Ass.*, **71**, 391–399.

(CATEGORICAL DATA
CONTINGENCY TABLES
PATTERN RECOGNITION)

BINARY SEQUENCE

A sequence, each element of which has one of only two possible values—usually 0 or 1. Such a sequence can represent the results of a series of trials at each of which the occurrence or nonoccurrence of an event—E, say—is observed, with 0 representing nonoccurrence and 1 representing occurrence. The resulting sequence is an example of binary data*. Binary sequences also arise in nonstatistical contexts: e.g., in the representation of a number of the binary scale.

(BERNOULLI TRIALS
BINARY DATA)

BINARY STRING *See* ALGORITHMIC INFORMATION THEORY

BINARY STRING TECHNIQUE *See* EDITING STATISTICAL DATA

BINGHAM DISTRIBUTION *See* DIRECTIONAL DISTRIBUTIONS

BINOMIAL DISTRIBUTION

This important discrete distribution* is commonly used to model observed counts of events. It is based on the assumption that the counts can be represented as the results of a sequence of independent Bernoulli trials*. If the probability of observing an event E is p for each of n such trials, the probability that E will be observed in exactly x of the trials is

$$p_x = \binom{n}{x} p^x (1-p)^{n-x} \qquad (x = 0, 1, \ldots, n).$$

The distribution defined by

$$\Pr[X = x] = P_x \qquad (x = 0, 1, \ldots, n)$$

is called a *binomial distribution with parameters n and p*. The name comes from the fact that the binomial expansion of $(p + \overline{1-p})^n$ is $P_0 + P_1 + \cdots + P_n$.

It is customary to use the symbol q to represent the quantity $(1 - p)$. This leads to some simplification of formulae; it does not, of course, increase the number of parameters beyond two.

The rth factorial moment* is

$$E[X^{(r)}] = E[X(X-l) \cdots (X-r+1)] = n^{(r)} p^r.$$

The expected value is np; the standard deviation is $\sqrt{(npq)}$. The mean deviation* is

$$n \binom{n-1}{[np]} p^{[np]+1} q^{n-[np]}$$

($[np]$ = integer part of np).

The ratio of successive P_x's is

$$\frac{P_{x+1}}{P_x} = \frac{n-x}{x+1} \frac{p}{q},$$

from which it can be seen that as x increases from zero, P_x increases until $x = [(n+1)p]$. Thereafter P_x decreases, so that there is a mode at $x = [(n+1)p]$. If $(n+1)p < 1$, the mode is at zero, and $P_{x+1}/P_x < 1$ for all $x = 0, 1, \ldots$. If $(n+1)p$ is an integer, there are two equal maximum values of P_x at $x = (n+1)p - 1$ and $x = (n+1)p$.

Numerical values of P_x, for selected values of n and p, are available in published tables (see, e.g., Johnson and Kotz [1]). Although direct calculation of P_x is feasible, calculations become very heavy if n is large. There are very good approximations which may be used when n is large.

If $\min(np, n(1-p)) > 10$, the normal approximation*

$$\Pr[X \leqslant x] \doteqdot \Phi\left(\frac{x + \frac{1}{2} - np}{(npq)^{1/2}} \right),$$

where $\Phi(\cdot)$ is the unit normal distribution function, incorporating the continuity correction*, $\frac{1}{2}$, is quite good. Better approximations can usually be obtained from formulas developed by Molenaar [3]. Particular mention may be made of the formula

$$\Pr[X \leqslant x] \doteqdot \Phi\left([(nx+3)q]^{1/2} - [(4n - 4x - 1)p]^{1/2} \right).$$

When $np < 10$ and p is small (less than 0.1, say) the Poisson* approximation

$$\Pr[X \leqslant x] \doteqdot e^{-np} \sum_{j=0}^{x} \frac{(np)^j}{j!}$$

can be used. If $n(1-p) < 10$ and p is large (greater than 0.9, say) the Poisson approximation may be applied to the distribution of $Y = n - X$.

Littlewood has published an exhaustive study of approximations to sums of binomial probabilities [2].

References

[1] Johnson, N. L. and Kotz, S. (1969). *Distributions in Statistics: Discrete Distributions*. Wiley, New York, Chap. 3.

[2] Littlewood, J. E. (1969). *Adv. Appl. Prob.*, **1**, 43–72.

[3] Molenaar, W. (1970). *Approximations to the Poisson, Binomial and Hypergeometric Distribution Functions*. Mathematical Centre, Tracts 31, Amsterdam: Mathematisch Centrum.

(BERNOULLI TRIALS
LAW OF LARGE NUMBERS
LEXIS RATIO
LIMIT THEOREMS
LIMIT THEOREMS, CENTRAL
NEGATIVE BINOMIAL DISTRIBUTIONS
POISSON DISTRIBUTIONS)

BINOMIAL MOMENT *See* FACTORIAL MOMENT

BINOMIAL PROBABILITY PAPER

This is a graph paper with both scales proportional to square roots. It was introduced by Mosteller and Tukey [2]. The paper is used by plotting the nominal sample point $(n - x, x)$ (where x is the number of occurrences of an event in n binomial trials). The actual coordinate distances are $(\sqrt{n - x}, \sqrt{x})$, so the point lies somewhere on a quarter-circle of radius $\sqrt{n}$ with center at the origin. The angle between the line connecting the origin with the point and the horizontal axis is $\arcsin \sqrt{x/n}$. Repeated random samples of size n will result in sample points at different points on the circumference of the circle. The standard error*, measured around the circumference, will be about $\frac{1}{2}$. It is a graphical representation of the *angular transformation**.

For additional information, see ref. 1.

References

[1] Hald, A. (1952). *Statistical Theory with Engineering Applications*, Wiley, New York.

[2] Mosteller, F. and Tukey, J. W. (1949). *J. Amer. Statist. Ass.*, **44**, 174–212.

(ANGULAR TRANSFORMATION
BINOMIAL DISTRIBUTION)

BINOMIAL PROPORTION

The parameter p in the binomial distribution*

$$\Pr[X = x] = \binom{n}{x} p^x (1 - p)^{n-x} \qquad (x = 0, 1, \ldots, n).$$

Sometimes p is called the *population* binomial proportion, to distinguish it from the *sample* binomial proportion X/n.

BINOMIAL TEST

A test of the value of a (population) binomial proportion*. It is supposed that we have an observed value which can be represented by a random variable R that has a binomial distribution*.

$$\Pr[R = r] = \binom{n}{r} p^r (1 - p)^{n-r} \qquad (r = 0, 1, \ldots, n)$$

n is supposed known; the hypothesis tested relates to the value of p.

The sample size (n) corresponds to the number of trials. For example, if $H_0 : p = \frac{1}{2}$ and the alternative hypotheses* are $H_1 : p > \frac{1}{2}$ with $n = 25$, and the significance level* is $\alpha = 0.05$, the test rejects H_0 when $R > r_0$, where r_0 is usually chosen as the least integer for which the cumulative probability $\Pr[R \leqslant r_0] = \sum_{r=0}^{r_0} \binom{n}{r} p^r (1 - p)^{n-r}$ exceeds $1 - \alpha$. Sometimes r_0 is chosen to make this quantity as close to α as possible.

In the example above, when the hypothesis H_0 is valid,

$$\Pr[R > 16] = \sum_{r=17}^{25} \binom{n}{r} p^r (1 - p)^{n-r} = \sum_{r=17}^{25} \binom{25}{r} \left(\frac{1}{2}\right)^r \left(\frac{1}{2}\right)^{25-r} = 0.0530$$

and $\Pr[R > 17] = 0.0217$. The critical region* $R > 16$ gives the level of significance $\alpha = 0.0530$.

(BINOMIAL DISTRIBUTION
HYPOTHESIS TESTING)

BIOASSAY, STATISTICAL METHODS IN

Bioassay refers to the process of evaluating the potency of a stimulus by analyzing the responses it produces in biological organisms. Examples of a stimulus are a drug, a hormone, radiation, an environmental effect, and various forms of toxicants. Examples of biological organisms are experimental animals, human volunteers, living tissues, and bacteria.

When a new drug is introduced, we are often interested in how it compares with a standard drug. One means of approaching this problem is to use the two drugs on living organisms from a common population. On the other hand, when an insecticide is being considered for use on an unfamiliar insect population, we may be more interested in the tolerance* of the insect population to the insecticide than in the relative strength of the insecticide compared to another. These are examples of two basic problems in bioassay. One is the evaluation of the relative potency of an unknown drug to a standard. The other is the estimation of the stimulus response.

The response to a stimulus may often be classified as quantitative or quantal. If the response can be measured, as would be the case if we were studying weight changes following the use of a vitamin, it is quantitative. On the other hand, if the response is all or nothing, as would be the case of death or survival, it is quantal (*see* BINARY DATA).

The development of statistical methods in bioassay has paralleled the development of statistics in general. Not only are the familiar techniques of regression analysis* and maximum likelihood* extensively used in bioassay, but a number of statistical problems have originated in bioassay and have stimulated the general development of statistics. The companion books by Finney [14, 16], which are now in their third edition, have contributed greatly toward unifying the field and introducing many biomedical researchers, as well as seasoned statisticians, to current practices in bioassay. Reference 14 deals primarily with the probit model* for quantal responses, and ref. 16 with bioassay more generally, including recent developments in the field.

One of the earliest applications of the normal distribution* and normal deviate* to quantal responses was Fechner's [12] psychophysical experiment on human sensitivity to various physical stimuli. Probit analysis*, which is based on normal deviates, was widely used for estimating quantal response curves in the late 1920s and early 1930s and put into its present form by R. A. Fisher* [17]. During this period the need for standardizing various drugs, hormones, and toxicants brought about statistical methods for estimating relative potencies [22]. The multitude of competing methods for estimating parameters of quantal response curves were unified under the theory of RBAN estimation* by Neyman [25]. A number of sequential designs* for quantal responses were proposed in the 1950s and 1960s [9] and a limited number of Bayesian methods were introduced in the 1970s [18]. The extensive use of animals for measuring concentrations of hormones and enzymes have been recently replaced by radioimmunoassay and related procedures known for their high precision [27].

ESTIMATION OF RELATIVE POTENCY FROM QUANTITATIVE RESPONSES

Consider a test drug T being compared to a standard drug S. T and S are said to be similar if there is a constant $\rho > 0$ such that the distribution of the response to dose z_2 of the test drug is equivalent to that of dose $z_1 = \rho z_2$ of the standard drug. When the drugs are similar, ρ is called the relative potency of T to S.

In many dose–response experiments the quantitative response Y is linearly related to the log dose, $x = \log z$, and we can assume the model

$$E[Y|x] = \alpha + \beta x, \tag{1}$$

where α and β are unknown real-valued parameters (*see* REGRESSION). If S and T are similar, then the responses Y_1 and Y_2 to S and T, respectively, are related by

$$\begin{aligned} E[Y_1|x] &= \alpha + \beta x \\ E[Y_2|x] &= \alpha + \beta \log \rho + \beta x \end{aligned} \tag{2}$$

for some α, β, and ρ. This follows from the assumption that dose z of T is equivalent to dose ρz of S. The estimation of ρ based on this model is called a parallel-line assay, in contrast to a slope-ratio assay [16], which is based on a model where the regression is linear with respect to $x = z^{\lambda}$, for some $\lambda \neq 0$.

Given the intercepts $\alpha_1 = \alpha$ and $\alpha_2 = \alpha + \beta \log \rho$ and the common slope β, the horizontal distance between the regression lines (2) is $\log \rho = (\alpha_2 - \alpha_1)/\beta$. Thus if one has estimates a_1, a_2, and b of α_1, α_2, and β, he or she may estimate $\log\rho$ by $M = (a_2 - a_1)/b$ and ρ by $\log^{-1}(M)$.

A typical experiment for a parallel-line assay consists of a series of n_1 observations $(Y_{11}, \ldots, Y_{1n_1})$ at log doses $(x_{11}, \ldots, x_{1n_1})$ of the standard and n_2 observations $(Y_{21}, \ldots, Y_{2n_2})$ at log doses $(x_{21}, \ldots, x_{2n_2})$ of the unknown. In addition to the linearity assumption, suppose that the observations are independent and normally distributed with common variance σ^2. Then α_1, α_2, β and σ^2 may be estimated by the standard least-squares* method, and a γ-level confidence interval* for $\log\rho$ is given by

$$(\bar{x}_1 - \bar{x}_2) + (1-g)^{-1} \times \left\{ M - \bar{x}_1 + \bar{x}_2 \pm \frac{st}{b} \times \left[(1-g)\left(\frac{1}{n_1} + \frac{1}{n_2}\right) + \frac{(M - \bar{x}_1 + \bar{x}_2)^2}{\sum_{i=1}^{2}\sum_{j=1}^{n_i}(x_{ij} - \bar{x}_i)^2} \right]^{1/2} \right\}, \quad (3)$$

where

$$\bar{x}_i = \sum_{j=1}^{n_i} x_{ij}/n_i,$$

$$s^2 = \sum_{i=1}^{2}\sum_{j=1}^{n_i} (Y_{ij} - a_i - b_i x_{ij})^2/(n_1 + n_2 - 3),$$

$$g = \frac{t^2 s^2}{b^2 \sum_{i=1}^{2}\sum_{j=1}^{n_i}(x_{ij} - \bar{x}_i)^2},$$

and t is the $(1+\gamma)/2$ quantile of the t-distribution* with $n_1 + n_2 - 3$ d.f.

The expression (3) follows from the celebrated Fieller's theorem* [13], which has been widely used in deriving confidence intervals for ratios of two parameters whose estimators have a bivariate normal distribution*. We note that small values of g indicate a significant departure of β from 0. However, large values of g, (i.e., $g > 1$) indicate that β is not significantly different from 0 and the confidence region is the complement of the interval defined by (3) and is of little practical value.

Table 1 Uterine Weights in Coded Units

Standard Dose (mg)			Unknown Dose (mg)	
0.2	0.3	0.4	1.0	2.5
73	77	118	79	101
69	93	85	87	86
71	116	105	71	105
91	78	76	78	111
80	87	101	92	102
110	86		92	107
	101			102
	104			112

Source. Brownlee [7].

As an example, consider an estrogen hormone assay illustrated by Brownlee [7] using data derived from Emmons [11]. A total of 33 rats were assigned to three levels of the standard and two levels of the unknown. The response variable Y is some linear function of the logarithm of the weights of the rats' uteri and the independent variable x the logarithm of 10 times the dose. (See Table 1.)

In this example the estimated variance about the regression lines is $s^2 = 153.877$ and the estimated regression coefficients* are $a_1 = 67.7144$, $a_2 = 32.8817$, and $b = 50.3199$. Thus the estimated $\log\rho$ is $M = -0.6722$. Moreover, $g = 0.307316$ and the 95% confidence interval for $\log\rho$ based on (3) is $(-0.8756, -0.4372)$. Thus the estimate of ρ is 0.213 and its 95% confidence interval is (0.133, 0.365). For computational details and tests for the validity of assumptions, see Brownlee [7, pp. 352–358].

QUANTAL RESPONSE MODELS

When the response to a drug (or other stimulus) is quantal, it is often reasonable and convenient to assume that each member of the population has a tolerance level such that the member responds to any dose greater than this level and does not to any lesser dose. This gives rise to the concept of a continuous tolerance distribution*, F, which has the characteristics of a CDF. In particular, the probability of a response from a randomly chosen member at dose x (usually measured in log units) is $F(x)$ and the probability of nonresponse is $(1 - F(x))$. In practice, this model is sometimes generalized to include an unknown proportion C of the population consisting of those who will respond in the absence of any stimulus and an unknown proportion D consisting of those that are immune to the stimulus and will not respond to any finite dose. When the responses are deaths and survivals, the dose which is lethal to 50% of the population is called LD50. More generally ED05, ED50, and ED90 denote doses that affect 5, 50 and 90% of the population.

Relative to a suitable transformation of the dose, such as the logarithm, the tolerance distribution is often approximated by a normal* or logistic* distribution, i.e.,

$$F(x) = \Phi(\alpha + \beta x) \tag{4}$$

or

$$F(X) = \psi(\alpha + \beta x), \tag{5}$$

where

$$\Phi(t) = \int_{-\infty}^{t} \frac{1}{\sqrt{2\pi}} \exp(-w^2/2)\, dw$$

$$\psi(t) = \left[1 + \exp(-t)\right]^{-1},$$

$-\infty < t < \infty$, and α and β are unknown parameters $-\infty < \alpha < \infty, 0 < \beta < \infty$. Given the probability of response P, the probit* is defined for the normal (probit model) by the value of y such that $P = \Phi(y - 5)$ and the logit* for the logistic (logit model) by $y = \ln[P/(1 - P)]$. Berkson [4] promoted the use of the logit model because of its similarity to the normal and numerical tractability. In view of the recent development of computational facilities, the numerical convenience is usually negligible and the choice between the two is generally not crucial. Prentice [26] has introduced other distributions, including those that are skewed and may be better suited for modeling extreme percent points such as the ED99.

ESTIMATION OF QUANTAL RESPONSE CURVES

A typical experiment for quantal responses consists of taking n_i independent observations at log dose x_i and observing the frequency of responses r_i, $i = 1, \ldots, k$, where the number, k, of dose levels is generally at least two, and more than two when the model is being tested. If we assume that the tolerance distribution $F(x \mid \theta)$ belongs to a family of distributions with parameter θ, the log likelihood* function is given by

$$\sum_{i=1}^{k} \{r_i \log F(x_i \mid \theta) + (n_i - r_i) \log\left[1 - F(x_i \mid \theta)\right]\}.$$

For most of the commonly used models, including the probit and logit, the maximum likelihood estimators of θ do not have explicit expressions and estimates must be found numerically by iterative schemes such as the Newton–Raphson*. The probit method originally proposed by Fisher [17] is still a widely used method for finding such estimates and has been described in detail by Finney [14].

There are a number of other estimators, such as the minimum chi-square* and weighted least squares* which belong to the RBAN family [25] and hence have the same asymptotic efficiency* as the maximum likelihood estimator*. Their relative merits for small samples have not been studied extensively and the choice among them is usually a matter of convenience.

For the probit model, with x the log dose, log ED50 is estimated by $\hat{\mu} = a/b$, where (a, b) is the maximum likelihood estimate of (α, β). The γ-level confidence interval for log ED50, valid for large samples, is given by

$$\hat{\mu} + \frac{g}{1-g}(\hat{\mu} - \bar{x}) \pm \frac{K}{b(1-g)} \times \left[\frac{1-g}{\sum_{i=1}^{k} n_i w_i} + \frac{(\hat{\mu} - \bar{x})^2}{\sum n_i w_i (x_i - \bar{x})^2}\right]^{1/2},$$

where

$$\bar{x} = \sum_{i=1}^{k} n_i w_i x_i / \sum_{i=1}^{k} n_i w_i,$$

$$w_i = z_i^2 / P_i Q_i,$$

$$z_i = (2\pi)^{1/2} \exp\left[-\tfrac{1}{2}(a + bx_i)^2\right],$$

$$g = \frac{K^2}{b^2 \sum_{i=1}^{k} n_i w_i (x_i - \bar{x})^2},$$

and K is the $(1 + \gamma)/2$ quantile of the standard normal distribution. As in the case of relative potency, the interval should be used only when $g < 1$.

Among the several nonparametric (distribution-free*) estimators of the ED50 that have been used, the Spearman–Kärber* estimator [23, 28] is not only quite simple to use but is considered quite efficient [5]. If the levels are ordered such that $x_1 < \cdots < x_k$, this estimator is defined by

$$\tilde{\mu} = \sum_{i=1}^{k-1} (p_{i+1} - p_i)(x_i + x_{i+1})/2,$$

provided that $p_1 = 0$ and $p_k = 1$, where $p_i = r_i/n_i$, $i = 1, \ldots, k$. If $p_1 > 0$, then an extra level is added below x_1, where no responses are assumed to occur. Similarly, if $p_k < 1$, an extra level is added above x_k, where responses only are assumed to occur. When the levels are equally spaced with interval d, the variance of $\tilde{\mu}$ is estimated by

$$\text{var}(\tilde{\mu}) = d^2 \sum p_i(1 - p_i)/(n_i - 1),$$

provided that $n_i \geqslant 2, i = 1, \ldots, k$.

The maximum likelihood and Spearman–Kärber methods may be illustrated using data from Finney [14] on the toxicity of quinidine to frogs, given in Table 2. The dose was measured in units of 10^{-2} ml per gram of body weight. If the probit model with respect to log dose is assumed, maximum likelihood estimates may be computed using a SAS program [3]. The resulting maximum likelihood estimates are $a = -1.313$, $b = 4.318$, and log LD50 = 1.462. The 95% confidence interval for log LD50 is (1.398, 1.532). On the other hand, the Spearman–Kärber estimates are $\tilde{\mu} = 1.449$ and $\text{var}(\tilde{\mu}) = 9.334 \times 10^{-4}$. (The estimate here uses an additional level at log dose = 1.879, where $p_i = 1$ is assumed.)

Table 2 Toxicity of Quinidine to Frogs

Log Dose	n	r	p
1.0000	26	0	0.0000
1.1761	24	2	0.0833
1.3522	23	9	0.3913
1.5276	24	17	0.7083
1.7033	26	20	0.7692

Source. Finney [14].

There are a few nonparametric procedures for estimating the entire tolerance distribution F. One approach is to use the isotone regression* method for estimating ordered binomial parameters [2]. Under this method the estimates of F at the points $x_1 < \cdots < x_k$ are given by

$$\hat{F}(x_i) = \min_{i \leqslant v \leqslant k} \max_{l \leqslant u \leqslant i} \left(\sum_{\nu=u}^{v} r_\nu / \sum_{\nu=u}^{v} n_\nu \right),$$

$i = 1, \ldots, k$. Between these points and outside the interval $[x_1, x_k]$ the estimate $\hat{F}$ may be defined arbitrarily, subject to the constraint that $\hat{F}$ be nondecreasing.

Another parametric approach, which is Bayesian*, assumes that F is a random distribution function, whose distribution is defined by a Dirichlet process*, and uses the Bayes estimate of F with respect to a suitable loss function. Antoniak [1] gives a theoretical discussion of the Dirichlet prior and resulting posterior, as well as references to related works.

ESTIMATION OF RELATIVE POTENCY FROM QUANTAL RESPONSES

When quantal responses satisfy the probit or logit model with respect to log dose and the drugs are similar, a parallel-line assay may be performed to estimate the relative potency of an unknown to a standard. Under the probit model, for example, the condition of similarity requires that the probabilities of response at log dose x must satisfy

$$F_1(x) = \Phi(\alpha_1 + \beta x),$$

$$F_2(x) = \Phi(\alpha_2 + \beta x),$$

for the standard and unknown, respectively, for some parameters α_1, α_2, and $\beta > 0$. In this case, $\alpha_2 = \alpha_1 + \beta \log \rho$ or $\log \rho = (\alpha_2 - \alpha_1)/\beta$.

If we have two independent series of independent quantal response observations, one for the standard and the other for the unknown, the joint likelihood function may be formed and the maximum likelihood estimates a_1, a_2, and b of α_1, α_2, and β may be computed by the Newton–Raphson or similar method. These estimates may be used to estimate $\log \rho$ by $M = (a_2 - a_1)/b$. Use of Fieller's theorem gives an approximate confidence interval for $\log \rho$ and ρ. Computational details and numerical illustrations are given in Finney [16].

DESIGN OF THE EXPERIMENT

The design of the experiment* for bioassay involves the selection of dose levels and the allocation of the living organisms to these levels in order to obtain experimental results that can be used to answer some predetermined question.

In estimating the relative potency, the objective may be to select the design that minimizes the width of the confidence interval* for a given confidence level and total sample size. If a parallel-line assay is being considered, it is important to have a rough estimate of ρ so that doses chosen for the unknown will give results comparable to those of the standard. One recommended approach is to use the same number of dose levels, usually between 3 and 6, for both the unknown and standard and the same number of organisms at each level. For each level of the standard, the level of the unknown should be chosen so that the predicted outcome will be like that of the standard. Although there are certain advantages to using dose levels that cover a wide range for qualitative responses, this does not hold for quantal responses since responses at extreme levels are quite predictable (being all responses or all nonresponses) and little information is gained. (See Finney [16] for more detailed instructions.)

In estimating the ED50 from quantal responses, again the objective may be to minimize the width of the confidence interval. For both the probit and Spearman–Kärber methods, it is generally recommended that an equal number of observations be taken at three to six equally spaced levels which are between 0.5 and 2 standard deviations (of F) apart. The specific recommendations vary according to the number of levels to be used, total sample size available, and previous information. (See Finney [14] for the probit method and Brown [6] for the Spearman–Kärber.)

A number of sequential* designs have been proposed as a means of allocating N experimental units to different levels as information becomes available. The Robbins–Monro* process [9] and up-and-down method* [10], together with their variations, are the most intensively studied sequential procedures for estimating the ED50. For the up-and-down method, one observation at a time is taken, starting at some initial level x_0 and successively at levels $x_{i+1} = x_i \pm d$, where d is the step size chosen to be close to the standard deviation of F, a + used if the ith observation is a nonresponse and a − if it is a response. Under this scheme, the experimental dose levels tend to fluctuate about the ED50 and the average, $\sum_{i=2}^{N+1} x_i/N$, is used to estimate the ED50. For the probit model this estimator approximates the maximum likelihood estimator and has better efficiency than do the fixed sample designs [8, 10]. The up-and-down method, when modified by using several observations at a time, not only reduces the number of trials but has been shown to provide even greater efficiency [29].

The selection of dose levels for estimating ED50 and extreme percent points, such as the ED90, of F may also be based on Bayesian principles. Freeman [18] has proposed a sequential method for estimating the ED50 of a one-parameter logit model, where the slope β is assumed known, and Tsutakawa [30] has proposed nonsequential designs for logit models, with unknown slope. These methods depend on the explicit and formal use of a prior distribution of the parameters and are aimed at minimizing the posterior variance of the percent point of interest.

RELATED AREAS

Some of the techniques that have been described above have been modified and extended to more complex problems. We will briefly describe some of these where further work can be expected.

The analysis of quantal responses to combinations or mixtures of two or more drugs introduces many additional problems, since we must not only consider the relative proportion of the drugs but also the interaction of the drugs. When there are two drugs and one is effectively a dilution of the other or they act independently, it is not difficult to extend the models for single drugs to those for a mixture. However, when the drugs interact, either antagonistically or synergistically, the model building becomes considerably more complex. Hewlett and Plackett [21] have discussed different modes in which the effect of the mixture of drugs may depend on the amount of drugs reaching the site of the action.

In many quantal response studies the time of response is an important variable. For example, in a carcinogenic experiment animals are often exposed to different doses of a carcinogen and kept under observation until they develop a tumor, die, or the experiment is terminated. For a discussion of such experiments and related references, see Hartley and Sielken [20].

Statistical techniques for bioassay have been used in radioimmunoassay and related techniques for measuring minute concentrations of hormones and enzymes. Radioactivity counts, resulting from antigen–antibody reactions, are observed at different doses of a ligand in order to estimate a dose–response curve. Under appropriate transformations, the counts are often related to the dose by a logistic model with unknown asymptotes (generally different from 0 and 1). For transformations, weighted least-squares methods*, and references, see Rodbard and Hutt [27] and Finney [15].

The probit and logit models have also been applied to mental testing and latent trait analysis, where human subjects with different mental abilities respond to questions of different degrees of difficulty [24]. When the responses are classified correct or incorrect, the probability of a correct answer to a particular question usually depends on the ability of the subject and can often be approximated by one of these models. In such cases, the ability of each subject can be estimated by using the joint response to several questions. See Hambleton and Cook [19] for a review of this area and related references.

References

[1] Antoniak, C. E. (1974). *Ann. Statist.* **2**, 1152–1174.

[2] Ayer, M., Brunk, H. D., Ewing, G. M., Reid. W. T., and Silverman, E. (1955). *Ann. Math. Statist.* **26**, 641–647.

[3] Barr, A. J., Goodnight, J. H., Sall, J. P., and Helwig, J. T. (1976). *A User's Guide to SAS76.* SAS Institute, Raleigh, N. C.

[4] Berkson, J. (1944). *J. Amer. Statist. Ass.* **39**, 357–365.

[5] Brown, B. W., Jr. (1961). *Biometrika*, **48**, 293–302.

[6] Brown, B. W., Jr. (1966). *Biometrics*, **22**, 322–329.

[7] Brownlee, K. A. (1965). *Statistical Theory and Methodology in Science and Engineering*, 2nd ed. Wiley, New York.

[8] Brownlee, K. A., Hodges, J. L., Jr., and Rosenblatt, M. (1953). *J. Amer. Statist. Ass.*, **48**, 262–277.

[9] Cochran, W. G. and Davis, M. (1965). *J. R. Statist. Soc. B*, **27**, 28–44.

[10] Dixon, W. J. and Mood, A. M. (1948). *J Amer. Statist. Ass.*, **43**, 109–126.

[11] Emmons, C. W. (1948). *Principles of Biological Assay.* Chapman & Hall, London.

[12] Fechner, G. T. (1860). *Elemente der Psychophsysik*, Breitkopf und Hartel, Leipzig. (Translated into English in 1966 by H. E. Adler, Holt, Rinehart and Winston, New York.)

[13] Fieller, E. C. (1940). *J. R. Statist. Soc. Suppl.*, **7**, 1–64.

[14] Finney, D. J. (1971). *Probit Analysis*, 3rd ed. Cambridge University Press, Cambridge.

[15] Finney, D. J. (1976). *Biometrics*, **32**, 721–740.

[16] Finney, D. J. (1978). *Statistical Methods in Biological Assay*, 3rd ed. Macmillan, New York.

[17] Fisher, R. A. (1935). *Ann. Appl. Biol.*, **22**, 134–167.

[18] Freeman, P. R. (1970). *Biometrika*, **57**, 79–89.
[19] Hambleton, R. K. and Cook, L. L. (1977). *J. Educ. Meas.*, **14**, 75–96.
[20] Hartley, H. O. and Sielken, R. L., Jr. (1977). *Biometrics*, **33**, 1–30.
[21] Hewlett, P. S. and Plackett, R. L. (1964). *Biometrics*, **20**, 566–575.
[22] Irwin, J. O. (1937). *J. R. Statist. Soc. Suppl.*, **4**, 1–48.
[23] Kärber, G. (1931). *Arch. exp. Pathol. Pharmakol.*, **162**, 480–487.
[24] Lord, F. M. and Novick, M. R. (1968). *Statistical Theories of Mental Test Scores.* Addison-Wesley, Reading, Mass.
[25] Neyman, J. (1949). *Proc. Berkeley Symp. Math. Statist. Prob.*, University of California Press, Berkeley, Calif., pp. 239–273.
[26] Prentice, R. L. (1976). *Biometrics*, **32**, 761–768.
[27] Rodbard, D. and Hutt, D. M. (1974). *Radioimmunoassay and Related Procedures in Medicine*, Vol. 1. International Atomic Energy Agency, Vienna, pp. 165–192.
[28] Spearman, C. (1908). *Br. J. Psychol.*, **2**, 227–242.
[29] Tsutakawa, R. K. (1967). *J. Amer. Statist. Ass.*, **62**, 842–856.
[30] Tsutakawa, R. K. (1980). *Appl. Statist.*, **29**, 25–33.

(FIELLER'S THEOREM
ISOTONIC REGRESSION
LOGIT
PROBIT
QUANTAL RESPONSE
RANKIT
REGRESSION ANALYSIS)

ROBERT K. TSUTAKAWA

BIOLOGY *See* STATISTICS IN BIOLOGY

BIOMETRICAL JOURNAL

Biometrical Journal (Biom. J.)—subtitled "Journal of Mathematical Methods in Biosciences"—is published at the Central Institute of Mathematics and Mechanics of the Academy of Sciences of the German Democratic Republic by Erna Weber (Berlin), Chief Editor: Heinz Ahrens (Berlin); and Klaus Bellmann (Berlin), Akademie-Verlag, Berlin.

Editorial Board: J. Adam (Halle), G. Enderlein (Berlin), D. J. Finney (Edinburgh), D. Firescu (Bucharest), B. V. Gnedenko (Moscow), J. Hartung (Hannover), S. Koller (Mainz), H. L. Le Roy (Zurich), C. C. Li (Pittsburgh), G. A. Lienert (Düsseldorf), A. Linder (Geneva), L. Martin (Brussels), W. Oehmisch (Berlin), T. Postelnicu (Bucharest), Dr. Rasch (Rostock), V. A. Ratner (Novosibirsk), J.-H. Scharf (Halle), L. Schmetterer (Vienna), P. Tautu (Heidelberg), and R. Trommer (Eberswalde).

The *Biometric Journal* was founded as *Biometrische Zeitschrift* by Ottokar Heinisch and Maria Pia Geppert in 1959 (four issues per year). The first issue was published on January 23, 1959. Beginning with Volume 9 (1967), No. 1, the journal was published by Maria Pia Geppert and Erna Weber. The publication frequency was changed to six issues per year with the beginning of 1969 (Vol. 11), and the publisher was Erna Weber (Berlin). A new graphic concept for the cover page was introduced with the first issue of Vol. 15 (1973), when the publication frequency was changed to the current eight issues per year, with a total of 832 pages per volume.

The title *Biometrical Journal* was introduced in 1977 (Vol. 19). Part of the edition was published by Academic Press, London and New York, in 1977 and 1978. From 1979 on, publication of part of the edition was transferred to Chemie International, Inc., New York. The *Biometrical Journal* is an international journal that is independent of any region of the Biometric Society. It was founded for the purpose of publishing contributions on the mathematical penetration of biosciences.

The scope of the journal covers papers on new theoretical aspects of mathematics and its application to biological sciences in the widest sense (including biology, medicine, agriculture, and forestry), or on the application of known mathematical and statistical methods to new areas. These may be methods of mathematical statistics and approaches to mathematicocybernetic model construction of biological systems with due consideration to electronic data processing.

Accepted for publication are original papers, summary reports on the latest developments in the areas cited above, proceedings, and book reviews. Manuscripts should be submitted preferably in English, but papers in German, French, and Russian are also accepted. Papers are chosen for inclusion based on the expert opinion of the chief editor or other members of the editorial board. The editorial policy is to maintain a careful balance between papers representing good applications of well-established methods, and papers containing new procedures and methods of probability theory and mathematical statistics for biosciences.

The address of the *Journal's* editorial office is: Institut für Mathematik der AdW der DDR, Redaktion *Biometrical Journal*, Mohrenstrasse 39, DDR-1080 Berlin. The address of the *Journal's* business office is: *Biometrical Journal*, Akademie-Verlag, Leipziger Strasse 3-4, DDR-1080 Berlin.

H. AHRENS

BIOMETRIC FUNCTIONS

In biometrics, actuarial science*, and demography*, one wishes to study a nonnegative random variable X, the lifetime* of a subject selected at random from a population of interest.

The natural quantities of interest in the statistical analysis of lifetime data are $S(x)$, the probability that an individual survives longer than age x, and $e(x)$, the expected remaining lifetime of an individual who has attained age x. Consistent with the definition of lifetime, the age of an individual is either the chronological age or the time since entering the study as the case may be. Mathematically, these are defined in terms of $F(x)$, the cumulative distribution function* of the lifetime X, by the equations

$$S(x) = \Pr[X > x] = 1 - F(x) \quad (1)$$

$$e(x) = E[X - x \mid X > x] = \int_x^{\infty} S(v)\,dv / S(x) \quad (2)$$

for any x such that $S(x) > 0$. Note that

$$S(x) = e(0)e^{-1}(x)\exp\left[-\int_0^x e^{-1}(v)\,dv\right]. \quad (3)$$

The functions $S(x)$ and $e(x)$ are termed *biometric functions* (BF). Generally speaking, any function arising in probability modeling of lifetimes may be regarded as a BF. Other examples include $\mu(x)$, the force of mortality* (or intensity function), and $q(x, y)$, the conditional probability of dying in age interval (x, y) given that the subject is alive at age x. The defining equations are

$$\mu(x) = F'(x)/S(x) = -(d/dx)\log S(x), \quad (4)$$

$$q(x, y) = [S(x) - S(y)]/S(x), \quad (x \leqslant y). \quad (5)$$

The quantities $S(x)$, $e(x)$, $\mu(x)$, and $q(x, y)$ are mathematically equivalent [9] in the sense that each expression can be derived from any of the others as illustrated by (1) to (5). In actuarial science $e(x)$ is denoted by e_x and is termed the life expectancy at age x.

Use of the term "biometric function " can be traced back to a paper by Lotka [7], who called the birthrate B_t of a population at time t a BF. B_t and $S(x)$ are related by

$$m_t = \int_0^{\infty} B_{t-x} S(x)\,dx,$$

where m_t is the population size at time t. Chiang [2] refers to the statistical estimators of $e(x)$, $S(x)$, and other life-table functions as BF.

In the engineering, physical, and management sciences, the study of lifetime distributions is part of reliability theory*. In this context the functions $\mu(x)$ and $e(x)$ and $S(x)$ are usually called the hazard function*, the mean residual life*, and the survival function*, respectively.

STATISTICAL INFERENCE

A general estimation technique for these biometric functions is the construction of life tables*. This method produces nonparametric estimates of $S(x)$, $e(x)$, and $q(x, y)$ for a sequence of ages $0, x_1, x_2, \ldots, x_w$ [3]. For nonparametric estimates of $\mu(x)$, *see* HAZARD RATES.

In the parametric approach, Gompertz in 1825 modeled the force of mortality by $\mu(x) = \theta c^x$ for appropriate values of θ and c. Other models considered in the literature include the exponential*, Weibull*, gamma*, log normal*, and extreme value* distributions.

CENSORED DATA

In practice, some lifetime observations may not be completed. Such data are called censored*. The censored lifetimes complicate statistical analyses, but they cannot be ignored without introducing bias into the estimates of the BF.

The statistical treatment of the censored data varies with types of censorship. For example, in the case of right censorship* (i.e., instead of the lifetime X, one observes a random variable Y and the fact that $X > Y$), the survival function can be estimated using life-table methods or the Kaplan–Meier product limit (PL) estimator*. The PL estimator is nonparametric and generalizes the usual empirical distribution function* for uncensored data. Some asymptotic properties of the PL estimator have been rigorously established by Breslow and Crowley [1]. Nonparametric estimation of $e(x)$ has been investigated by Yang [8, 9]. Various parametric models for censored data have also been analyzed [5, 6]. Cox [4] has proposed a regression model for inference on biometric functions in the presence of covariates. *See* COX REGRESSION MODEL.

COMPETING RISKS

In a competing risk model*, an individual is exposed to several possible causes of death (called risks). One wishes to study, in the presence of all risks, the force of mortality, the life expectancy, and the survival function (the so-called crude survival probability). One also wishes to study these BF pertaining to a particular risk when some or all other risks have been eliminated. The survival function corresponding to the elimination of all but one risk is termed the net survival function.

References

[1] Breslow, N. and Crowley, J. (1974). *Ann. Statist.*, **2**(3), 437–453.

[2] Chiang, C. L. (1960). *Biometrics*, **16**, 618–635.

[3] Chiang, C. L. (1968). *Introduction of Stochastic Processes in Biostatistics*. Wiley, New York.

[4] Cox, D. R. (1972). *J. R. Statist. Soc. B*, **34**, 187–220.

[5] David, H. A. and Moeschberger, M. L. (1978). *The Theory of Competing Risks*. Charles Griffin, London.

[6] Kalbfleisch, J. D. and Prentice, R. L. (1980). *The Statistical Analysis of Failure Time Data*. Wiley, New York.

[7] Lotka, A. J. (1929). *Proc. Natl. Acad. Sci. USA*, **15**, 793–798.

[8] Yang, G. L. (1977). *Stoch. Processes Appl.*, **6**, 33–39.

[9] Yang, G. L. (1978). *Ann. Statist.*, **6**, 112–116.

(COMPETING RISKS
HAZARD RATES
KAPLAN–MEIER ESTIMATOR
LIFE TABLES
RELIABILITY THEORY)

GRACE L. YANG

BIOMETRICS

Biometrics is the journal of the Biometric Society* and publication was initiated in 1945 as the *Biometrics Bulletin*, under the auspices of the Biometrics Section of the American Statistical Association. The founding editor of the journal was Gertrude M. Cox, who served in this capacity until 1955. During this time, the name was changed to *Biometrics* in 1947 and it became the official publication of the Biometric Society in 1950. Cox was succeeded by John W. Hopkins (1955–1957), Ralph A. Bradley (1957–1962),

Michael R. Sampford (1962–1967), Herbert A. David (1967–1972), Franklin A. Graybill (1972–1975), and Foster B. Cady (1975–1979). The current editorial address is: Professor Peter Armitage, Department of Biomathematics, University of Oxford, Pusey Street, Oxford OX1 2JZ, United Kingdom.

The general objectives of the journal are to promote and extend the use of mathematical and statistical methods in pure and applied biological sciences, by describing and exemplifying developments in these methods and their applications in a form readily usable by experimental scientists. It is also intended to provide a medium for exchange of ideas between experimenters and those concerned primarily with the analysis and development of statistical methodology.

Published papers may deal with (1) statistical methodology applied to specific biological contexts, (2) topics in mathematical biology, and (3) statistical principles and methodology of general applicability for readers interested in both the biological and the statistical motivation and formulation of a problem. In addition to regular papers, the journal has sections for shorter communications and book reviews, together with a consultant's forum which includes queries and responses. Expository articles are encouraged through invitation, and special issues have centered on analysis of variance* (1947), components of variance* (1951), analysis of covariance* (1957), and multivariate methods* (1972). Memorial issues to R. A. Fisher* (1964), G. W. Snedecor (1975), and G. M. Cox (1979) have also appeared.

Currently *Biometrics* is being published annually in four parts. Volume 35 (1979) comprised 917 pages and contained the editor's 5-year report.

(BIOMETRIC SOCIETY)

FOSTER B. CADY

BIOMETRIC SOCIETY

The Biometric Society is an international society devoted to the mathematical and statistical aspects of biology. The society was founded at the Marine Biological Laboratory, Woods Hole, Massachusetts, in September 1947 during a conference, called the First International Biometric Conference, arranged for that purpose.

It was envisaged that the society would hold international meetings in biometry in different countries from time to time—as it turned out, about every second year. These meetings have served to bring together biometricians from all over the world. At each conference, attendance continues to increase both in terms of numbers of participants and countries represented. Places and dates of international conferences and symposia are listed in Table 1.

Since 1952 the Biometric Society has been the Biometric Section of the International Union of Biological Sciences (IUBS), which is a member of the International Council of Scientific Unions. The IUBS have given the society grants and loans for conferences and symposia.

It was planned that as membership grew, regions would be formed which would, in addition, hold their own meetings and conduct their own programs. In the constitution adopted at the Woods Hole meeting, the officers specified for the society were a president, a vice-president representing each region, a secretary, and a treasurer, plus re-

Table 1 International Meetings

Date	Place
	Conferences
1947	Woods Hole, Mass., United States
1949	Geneva, Switzerland
1953	Bellagio, Italy
1958	Ottawa, Canada
1963	Cambridge, England
1967	Sydney, Australia
1970	Hannover, Germany
1974	Constanza, Romania
1976	Boston, Mass., United States
1979	Guaruja, Brazil
	Symposia
1951	Indian Statistics Institute, Calcutta, India
1955	Campinas, Brazil
1960	Leiden, The Netherlands
1964	Berne, Switzerland

gional secretaries and treasurers. There was also a council composed of the principal officers plus members elected to represent the geographic areas and fields of activity in the society.

At the second meeting of the council, four regions were set up: British, Indian, Western American (including western Canada and western Mexico), and Eastern American (including eastern Canada). As soon as seemed desirable, five other regions were to be activated: Scandinavian, Benelux, Australian, French, and Russian.

As of April 1, 1979, there were 13 regions and 8 groups. The regions and memberships were: ANed, Netherlands (87); AR, Australasian (190); BR, British (355); DR, German (Federal Republic) (420); ENAR, Eastern North American (2,114); JR, Japanese (106); RBe, Belgian (62); RBras, Brazilian (164); RF, French (141); RGDR, German Democratic Republic (54); RItl, Italian (73); ROeS, Austro-Swiss(198); and WNAR, Western North American (436). Of these, the Japanese was the newest, having changed from being a group in 1979. The groups were: GDe, Denmark (23); GHu, Hungary (15), GInd, India (18); GMex, Mexico (41); GNo, Norway (13); GRo, Romania (22); GSd, Sweden (35); and NGi, Indonesia (18). There were also 95 members-at-large, bringing the total membership to 4,680. This does not include an additional 2,400 nonmember subscribers to *Biometrics**.

Apart from international conferences to ensure communication among regions, local meetings are held more frequently so that members can exchange ideas with others in their regions.

The society is essentially open to anyone interested in membership. Currently, the conditions for membership are sponsorship by one member and approval by the appropriate secretary.

The society has, in addition to Members, Sustaining Members, Associate Members, and Student Members. A few distinguished members have by vote been awarded Honorary Membership. Members so honored are R. A. Fisher*, C. I. Bliss, G. M. Cox, G. Barbensi, G. W. Snedecor*, F. Yates, W. G. Cochran, and Erna Weber.

The constitution of the society and of each region are published in the 1975 Membership Directory, together with the full listings of officers, both international and regional, since the founding of the society.

The origins of the society can be traced to the work and influence of R. A. Fisher in the 1920s and 1930s. The new techniques for collecting and analyzing data that he produced in a steady stream found their readiest applications in biological problems. In 1938, the members of the American Statistical Association who had biometric interests formed a Biometric Section of the association, primarily to ensure that the annual meetings of the association would have some sessions devoted to topics in biometry. From 1941 on, the section also arranged programs during the meetings of some of the principal biological societies in the United States, to foster contacts between statisticians and biologists with common interests.

In 1945, with World War II drawing to a close, two important developments occurred. The Biometric Section, with the support of the association, started a small publication, the *Biometrics Bulletin*, with Gertrude Cox as editor, which initially appeared six times yearly. In the same year the society began to give serious consideration to the formation of an autonomous Biometry Society. Further discussions of this issue were held with Fisher in the summer of 1946 when he visited North Carolina. Planning for the Woods Hole conference began in April 1947. At that time, the *Biometrics Bulletin*, under the new name *Biometrics**, became the journal of the society.

It was hoped by some of its founders that the society would be a genuinely cooperative venture between biologists and statisticians. To quote some excerpts (in a rough translation) from the remarks by the biologist George Peissier in closing the Woods Hole conference:

> Our Society should not be a new association of statisticians, nor a new association of biologists. . . . In the Biometric Society,

> we must work together to advance biology by means of statistics. ... Now there are very few who have had the good fortune to benefit, from their youth on, from the exposure to both mathematics and biology that seems essential to a full understanding of the science that we follow.
>
> We must therefore—and this should be our primary objective—teach one another, the statistician learning from the biologist what his major problems are, the biologist learning from the statistician the difficult art of stating a quantitative problem correctly. Next we must recruit future biometricians from among young people whom we must train, better than we have been trained, in this exciting but severe discipline.

This goal has been pursued by successive presidents from different countries, each serving a 2-year term. They have come from a variety of regions: British, French, Eastern North American, Australasian, Belgian, Italian, German Democratic Republic, and Austro-Swiss, reflecting the international membership. Table 2 presents the distinguished list of names.

The society would not flourish without the dedicated work of its officers. The first secretary, C. I. Bliss, served from 1948 to 1955 and recorded the first 10 years of growth of the society (*Biometrics*, 1958, **14**, 309–329).

Table 2 Presidents of the Society

Term	President
1948–1949	R. A. Fisher
1950–1951	A. Linder
1952–1953	G. Darmois
1954–1955	W. G. Cochran
1956–1957	E. A. Cornish
1958–1959	C. H. Goulden
1960–1961	L. Martin
1962–1963	C. I. Bliss
1964–1965	D. J. Finney
1966–1967	L. L. Cavalli-Sforza
1968–1969	G. M. Cox
1970–1971	B. Schneider
1972–1973	P. Armitage
1974–1975	C. R. Rao
1976–1977	A. L. Le Roy
1978–1979	J. A. Nelder

He was active in society affairs until his death in 1979. The first editor, G. M. Cox, served from 1948 to 1955 and was active until her death in 1978.

In recent years the size of the society has warranted a business manager, who takes care of international matters, including the membership lists. The editing of the society's journal *Biometrics* is still a voluntary contribution to the society.

The business office of the society since December 1978 has been located at 806 15th Street, N.W., Suite 621, Washington, DC 20005, and is run by the business manager, Elsie Thull.

WILLIAM G. COCHRAN
YVONNE M. M. BISHOP

BIOMETRIKA

Biometrika contains about 100 papers per year on theoretical statistics. The journal is published by the Biometrika Trust in a volume of 600 to 700 pages annually in three parts, issued in April, August, and December. The circulation is about 3500.

In general, the papers contain new statistical theory or methods which are capable of a useful application. Purely mathematical papers are excluded, as are applications of known theory. The editor tries to maintain a balance among the various branches of statistics; see Fig. 1, showing the contents list for the third part of the 1978 volume. Table 1 shows the countries of residence of authors of papers published in 1979.

A few review papers have been published and also a series on the history of probability and statistics. There is a Miscellanea section for shorter contributions, and another section for brief amendments and corrections to published work. Book reviews are no longer included. No advertisements are carried except for publications of the Biometrika Trust.

At the time of writing (1979), 350 to 400 papers are submitted a year, compared with about 250 in 1965. Thus the proportion ac-

Figure 1 Contents page from *Biometrika*, Vol. 65, No. 3 (1978).

Table 1 Country of Residence of Authors of the 102 Papers Published in 1979

Country	Number of Authors	Country	Number of Authors
United States	62	Israel	2
United Kingdom	34	Brazil	1
Australia	20	Denmark	1
Canada	10	Italy	1
Hong Kong	4	Norway	1
Japan	4	Pakistan	1
India	3	Total	144

cepted at present is between one-fourth and one-third. The editor deals directly with a number, and about eight associate editors each handle 30 to 50 papers per year. They act as referees themselves or send a paper to one or more readers for assessment. All correspondence with authors is done through the editorial office, and referees and associate editors act anonymously; associate editors see all correspondence concerning papers they have dealt with. The editor aims to get reports to authors within 2 months of submission. Referees receive modest payment for their work.

Publication takes 5 to 10 months from the date of final submission of an acceptable manuscript. Authors are asked to check the first proof of their papers. After publication, authors receive 50 free reprints (joint authors 25 copies each); they are also able to order extra copies when they return their proofs.

HISTORY

Biometrika was founded by Karl Pearson* and W. F. R. Weldon in consultation with Francis Galton*; the first issue appeared in October 1901. From 1906 on, Pearson assumed entire editorial responsibility, and since then the editorship has changed only twice; E. S. Pearson (Karl Pearson's son) acted from 1936 to 1965, and D. R. Cox from 1966.

The journal's origin was due in part to the Royal Society's request that, in papers submitted for publication, mathematics be kept apart from biological applications.

The editorial in 1901 stated that the journal would include:

> (a) memoirs on variation, inheritance, and selection in Animals and Plants, based on the examination of statistically large numbers of specimens . . . ;
> (b) those developments of statistical theory which are applicable to biological problems;
> (c) numerical tables and graphical solutions tending to reduce the labour of statistical arithmetic.

The early volumes contained many diagrams, photographs, and tables of measurements of parts of the human body, animals, and plants. Over the years the amount of biological and anthropological work has become less, and applications of theory have not been restricted to biological problems. Publication of tables has continued, although on a reduced scale.

The number of pages published over the years is indicated in Table 2. The average length of Miscellanea papers has remained more or less constant: $2\frac{1}{2}$ to $3\frac{1}{2}$ pages per paper. However, main-section papers have become a lot shorter.

There were in general two journal issues per year up to 1967 and have been three per year since. World Wars I and II resulted not only in fewer papers and pages per year but also in the frequency of publications falling to one issue per year. However, no year passed without publication, although Volume 33 covering the years 1943–1946, contained only 362 pages.

COMPARISON WITH OTHER STATISTICAL JOURNALS

*Journal of the Royal Statistical Society**, Series B, and *Biometrics** contain material very similar to that in *Biometrika*, although *Biometrics* lays more stress on biological applications and less on theoretical novelty; dur-

Table 2 Variation in the Size of *Biometrika* and the Length of Papers Published, 1905–1979

Year	Pages per Year	Issues per Year	Papers: Main	Papers: Miscellanea	Average Paper* Length[a] (pages)
1905	384	2	13	3	28
1920	132	1	7	0	19
1935	471	2	18	6	25
1945	85	1	8	4	9
1950	454	2	41	15	10
1965	675	2	45	30	13
1979	689	3	76	26	8

[a]Main section.

ing 1978, *J. R. Statist. Soc. B* published 375 pages and *Biometrics* 764 pages. Their formats are smaller: the page areas of both are 81% that of *Biometrika*. They have narrower margins and this, together with the fact that *Biometrika* uses a larger typeface (11-point Modern for main-section papers), results in *J. R. Statist. Soc. B*'s printing about 20% more words per page, and *Biometrics* about 8% more.

MANAGEMENT

In the early days, a guarantee fund was provided by the founders and their friends. After Weldon's death in 1906, the practical management was in Karl Pearson's hands. Circulation dropped during World War I, and Pearson and Mrs. Weldon contributed money to help keep the journal going. *Biometrika* has never been run for financial profit. Auxiliary publications, such as books of tables, have been issued since 1914, notably *Biometrika Tables for Statisticians*, two volumes, edited by E. S. Pearson and H. O. Hartley. In 1935, a trust was set up by Karl Pearson with at least five trustees, responsible for financial matters and the issuance of back numbers, auxiliary publications, and so on. The trust also holds the copyright of *Biometrika* and its associated publications. The present chairman of the trustees is J. Durbin, and the business manager is R. F. Galbraith.

Subscriptions are dealt with at the business office: *Biometrika* Office, University College London, Gower Street, London WC1E 6BT, England. Papers for consideration for publication should be sent to: Professor D. R. Cox, Editor, *Biometrika*, Department of Mathematics, Imperial College, London SW7, England.

BRENDA SOWAN

BIOMETRISCHE ZEITSCHRIFT *See* BIOMETRICAL JOURNAL

BIOMETRY *See* BIOSTATISTICS

BIOSTATISTICS

Biostatistics is that branch of science which applies statistical methods to biological problems, the common prefix being derived from the Greek word *bios*, meaning life.

The first major applications started in the middle of the seventeenth century when Sir William Petty and John Graunt conceived new and creative methods to analyze the London Bills of Mortality. Petty and Graunt essentially invented the field of *vital statistics** by studying the reported christenings and causes of death, and proposing measures of what they called "political arithmetick." Graunt recognized problems of inference when there has been inaccurate reporting of causes of death; he created methods of estimating mortality rates by age when age was not even recorded on the death certificate; and he devised estimates of birthrates, as well as a method to estimate the population from birthrates and other ingenious techniques for interpreting the data in the records of christenings and burials. Sir William Petty developed an enumeration schedule for a population census, proposed a centralized statistical department in the government, conceived the idea of life expectancy before Halley developed the first actual life table*, and proposed clever and original ideas on how to estimate population sizes. For further details, see the monograph by Greenwood [29].

Today, vital statistics is generally restricted by definition to the statistics of births, deaths, marriages, and divorces, and thus the term has a current connotation considerably more limited than "biostatistics," despite its derivation from the same root in its Latin form *vita*. Biometry or biometrics is another term closely identified with biostatistics but also more restricted in scope. The biostatistician must deal not only with biometrical techniques used in the design and analysis of experiments but also with some sociometric and psychometric procedures plus most of the methods used by demographers. Thus, the biostatistician works closely not only with the biological researcher but also with the epidemiologist, survey researcher, local community planner, state and national health policy analyst, and those government officials concerned with developing procedures for registering births, deaths, marriages, divorces, abortions, morbidity reports, the description of populations by sample surveys and census enumeration, and with health regulatory agencies.

FIELDS OF APPLICATION OR AREAS OF CONCERN

The biostatistician differs from the traditional statistician in that he or she is confronted by a wider range of problems dealing with all the phenomena that affect people's physical, social, and mental well-being. These phenomena consist of our relationship to other human beings, to animals and microbes, to plants, and to the physical and chemical elements in the environment. In dealing with these problems the biostatistician encounters theoretical difficulties, such as analyzing autocorrelated data in time series, in addition to practical and applied problems, such as working with accountants and economists to calculate costs versus benefits in evaluating the efficiency of a health program.

This means that the biostatistician must have familiarity with the concepts, goals, and specialized techniques of numerous fields beyond what might be considered a standard knowledge of statistics and probability. Some of these fields and areas of concern are mentioned here briefly and the remainder of this article will comment on a few of them at greater length.

Statistical Genetics

After the early developments in vital statistics, the field of *statistical genetics** was the next area that benefited most from the new ideas emerging in statistics. Any discussion of biostatistics and biometry would be incomplete without the names of Charles Darwin (1809–1882), Francis Galton* (1822–1911), Karl Pearson (1857–1936), and Ronald A. Fisher* (1890–1962).

Galton is responsible for the use of the term "regression"* when he observed that sons regressed linearly on their fathers with respect to stature. His thesis was to call the phenomenon a "regression to mediocrity" because children deviated less from the mean height of all children than the amount their fathers deviated from the mean height of all fathers. This bivariate normal distribution* gave rise to the measurement of the association by the coefficient of (product-moment) correlation* in 1897 by Karl Pearson* and to many other contributions by him. He is also generally credited with the creation of the new discipline of biometry and established with Walter F. R. Weldon and C. B. Davenport, in consultation with Galton, a new journal called *Biometrika** to provide for study of these problems. The journal has been in continuous publication since 1901, and after an unsigned editorial presumably written by Pearson, had as its first paper an article entitled "Biometry" by Francis Galton. The journal is still a highly regarded source for communications in biometry. Fisher's major contributions were to genetics and statistical theory, and he published the genetical theory of natural selection in 1930. This landmark book, plus earlier and later publications, represented attempts by Fisher to give quantitative form to Darwin's views and a statistical theory of evolution.

For more detailed discussion of current statistical problems in genetics, readers are referred to GENETICS, STATISTICS IN. For a history of early developments in statistical genetics, see Norton [48]. For biographical accounts of the statistical geneticists, see the appropriate entries in Kruskal and Tanur [35], and for a stimulating account of the life of Fisher, the biography by his daughter, Joan Fisher Box [4], is unrivaled (*see also* R. A. FISHER).

Bioassay

*Bioassay** techniques cover the use of special transformations such as probits and logits, as well as the application of regression to the estimation of dosages that are p percent effective within stated confidence limits. There are also problems in measuring relative potency, slope-ratio assays, and quantal responses* vis-à-vis tolerance distributions*. The reader interested in this subject is well advised to consult BIOASSAY and a standard textbook such as Finney [18].

Demography

A knowledge of *demography** which includes traditional vital statistics, rates and ratios, life tables, competing risks, actuarial statistics, and census enumeration techniques, is necessary in biostatistics. In this category, many tabulations of data will consist of a time series of events or rates classified by age. For the appropriate analysis of such data, reference should be made to cohort analysis techniques collected in a monograph by Hastings and Berry [31]. For further details in this broad area, *see* ACTUARIAL STATISTICS, LIFE TABLES, COMPETING RISKS, and DEMOGRAPHY, as well as Linder and Grove [37] and the book by Spiegelman [52].

Epidemiology

Some knowledge is required about the measurement of disease, including false-negative and false-positive results, so that sensitivity and specificity of a diagnostic test can be estimated, as well as survey results used to estimate the true incidence and prevalence of disease. It is necessary to have knowledge of epidemic theory and the use of deterministic and stochastic models [1]. Fundamental to this whole field of application is an understanding of causality and association [3, 24, 36].

In the case where clinical trials can be conducted, two groups of persons, one "treated" and the other "untreated," are observed over a period of time with respect to attack by or relief from the disease that is the object of the study. Here the biostatistician must know how to develop a protocol [17], how to randomize (*see* RANDOMIZATION), use double-blind techniques, and combine multiple-response variables into a multivariate analysis*. If several medical centers are involved, it is important to know how to operate a statistical coordinating center for collaborative clinical trials (see refs. 20 and 23, and CLINICAL TRIALS).

In situations where moral concerns prohibit a comparative experiment, such as in the study of whether exposure of a woman during pregnancy to infection by German measles (rubella) causes congenital malformations, it is necessary to know how to conduct retrospective case-control studies and measure the relative risk caused by exposure. In fact, with the sole exception of clinical trials, almost all the statistical research in epidemiology is retrospective in nature. That is, the research is ex post facto because the investigators seek to describe and analyze a series of events that are customarily a rather sudden, unusual, and significant increase in the incidence of disease.

The so-called case-control study is the most common procedure used to investigate an epidemic or unusual increase in disease. By this approach, a special group, frequently a 100% sample of available cases, is studied in detail to ascertain whether there were one or more common factors to which the members of the group were exposed. The exposure might be a drug, a food, or an environmental factor. A comparable group of noncases, frequently called controls or compeers or referents, is also selected at random in order to determine whether its members had the same, less, or more exposure to the suspected factor(s).

In the typical design of such studies, the data are presented in a 2×2 contingency table* of the following form, wherein a, b, c, and d are category frequencies.

Factor F	Cases	Compeers	Total
Exposed	a	b	$a + b$
Nonexposed	c	d	$c + d$
Total	$a + c$	$b + d$	N

If the proportion $a/(a + c)$ is significantly greater than $b/(b + d)$, one can safely assume that factor F is associated in some way with the occurrence of the event. The test of significance* to validate this may be the common χ^2 with 1 degree of freedom.

Owing to the fact that the design is retrospective, the comparable groups are cases and compeers, *not* exposed and nonexposed. Thus one cannot calculate the rates of disease as simply $a/(a + b)$ and $c/(c + d)$ in order to divide the former by the latter to derive a measure of relative risk associated with factor F. Although other researchers in genetics had previously used a solution similar to his, it was Cornfield [10] who demonstrated clearly that an estimate of relative risk is obtainable from the ratio of cross-products, ad/bc. [If $a/(a + c)$ is designated as p_1, and $b/(b + d)$ is designated as p_2, the relative risk is equivalently estimated as $p_1(1 - p_2)/p_2(1 - p_1)$.] The ratio of cross-products is commonly referred to as the *odds ratio**, motivated by the comparison of exposed-to-nonexposed "odds" in the two groups, $a : c$ and $b : d$.

Cornfield clearly emphasized that the validity of such estimation is contingent upon the fulfillment of three assumptions:

1. The rate of disease in the community must be comparatively small, say in the order of magnitude of 0.001 or less, relative to both the proportion of exposed cases and the proportion of nonexposed persons in the nonattacked population.
2. The $(a + c)$ cases must represent a random, unbiased sample* of all cases of the disease.
3. The $(b + d)$ controls must represent a random, unbiased sample of all noncases of the disease.

In actual practice, fulfillment of the first assumption is usually easily attainable, and any minor deviation from it causes no serious distortion in the results. The remaining two assumptions, however, are extremely difficult, if not actually impossible, to satisfy. Failure can cause considerable bias in the results, and is the basis of disagreement among both biostatisticians and epidemiologists. For instance, the detection of cases, referred to as ascertainment by R. A. Fisher, may be biased because cases are selected in a large medical referral center which is not representative of all cases in the community. In addition to being certain that the cases have all been diagnosed properly, the biostatistician must check to be sure that the cases were not selected because of distinguishing attributes such as socioeconomic status, location with respect to the center, race, sex, medical care previously received, or even whether the cases had close relatives with a similar disease and sought special diagnostic attention. The controls are sometimes chosen to be persons in the same hospital with a different diagnosis, or neighbors in the community. The biostatistician has to determine whether they are comparable in such factors as age, race, sex, severity (stage or grade) of the disease, and many other variables that tend to confound a just and fair comparison on factor F alone.

Three statistical problems are mentioned as a result of this type of study.

1. How to select the cases and controls? The literature on this is voluminous, but a few references may be mentioned [9, 32, 33].
2. In selecting the compeers, is it worthwhile to try to pair one or more controls to each case based upon certain characteristics which influence the probability of disease so that factor F will be the primary residual influence? References to this are found in Cochran [8] and McKinlay [38, 39], among others.
3. After selecting cases and controls, it is difficult to estimate the influence of all factors other than F. These other variables are referred to as confounding variables, and they need to be adjusted or accounted for so as to enable a valid comparison to be made on factor F. Historically, the procedure of adjusted or standardized rates is one way of achieving this goal. Another common statistical procedure is the use of covariance analysis [*see* ANALYSIS OF COVARIANCE (ANOCOVA)] [22]. More elaborate statistical procedures for dealing with confounding* may be found in Miettinen [42–45], Rothman [51], and many others.

An excellent introductory reference to this class of problems is the monograph of Fleiss [19] and a review paper by Walter [54].

Clinical Trials

In the case of *clinical trials**, a whole host of special problems arise for which the biostatistician has had to develop special techniques. One of these is the detection of unexpected and untoward rare effects of drugs with the consequent need to terminate a trial early. Moreover, when data are demonstrating a trend earlier than expected, there is also a need to end the accession of patients and to stop further treatment with what may be an inferior regimen. This means the biostatistician must be familiar with the problems of multiple examinations of data [40, 41], multiple comparisons* [46, 53], and other adjustment procedures made necessary by the ex post facto dredging of data. An excellent pair of references on randomized clinical trials is the set of articles by Peto et al. [49, 50] and one by Byar et al. [5]. *See also* CLINICAL TRIALS. For the ethical problems involved in conducting clinical trials, the reader is urged to read Gilbert et al. [21] and Courand [12].

Confidentiality and Privacy of Records

The biostatistician is continually confronted with the demands of *confidentiality* and *privacy* of records in dealing with health and medical records. This subject begins with the controversial area of what constitutes informed consent when a patient is a minor, ill, comatose, or otherwise incompetent. The use of informed consent and randomization may present a conflict as far as a patient is concerned, and other design techniques may have to be used to overcome this problem [55].

Safeguarding of Computer Data

Related to the problem of privacy and confidentiality of data is how to safeguard these attributes when data are stored in *computers*, or records are to be linked from multiple sources. The linkage of records regarding an individual or an event requires the knowledge of computers, computer languages, and programming, as well as means for protecting identification of information stored in computers [6, 14, 15]. Of course, knowledge of computers is necessary in general because the tabulation of large volumes of data and analysis by current techniques could not be carried out if the methods in vogue 25 years ago, such as punch-card machines and desk calculators, were still relied upon to carry out the necessary mechanical procedures. Reference should be made to COMPUTERS IN STATISTICS*.

Nature of Human Surveys

The nature of the field of inquiry in *human surveys* involving personal knowledge, attitudes, and behavior confront the biostatistician with a challenge akin to that of the sociologist. The goal is to obtain cooperation and a truthful response when the question(s) to be asked may be highly personal, sensitive, or stigmatizing in nature. Biostatisticians and sociologists have developed techniques to maximize cooperation with incentives of various kinds, including even monetary payment, as well as devices to assure anonymity. The use of response cards, telephone surveys, and a technique known as randomized response* have all shown promise in this field [27, 28].

Censoring of Observations

The *censoring* of observations* in statistics is not a matter related to the sensitivity of highly classified data but rather is the purposeful or accidental blurring of an actual value for other reasons. For instance, in studies of life expectancy subsequent to an event such as exposure to a carcinogenic agent, the limited amount of time available for observation may require the experimenter to terminate the experiment after 1 or 2 years. Those individuals who have survived up to that point have had their *actual* life expectancy "censored,"and the appropriate analysis must consider such observations in estimating parameter values. Similarly, the speed of some reactions may be so rapid, or the recording of data so crude at the beginning of the period of observation, that a certain number of early observations are censored initially. Thus, in deaths of infants under 1 day old, some countries do not record the actual number of hours lived under 24. For most of these problems the use of order statistics* has been the preferred solution to the problem of calculating unbiased estimates of the distribution parameters. (See refs. 16 and 26, and ORDER STATISTICS.)

Community Diagnosis

The biostatistician collaborates with health planning personnel to establish bench marks that describe the health status of a community so as to earmark places where greater attention and/or funds should be directed. The first step in community diagnosis is to study the population in terms of its magnitude and distribution by attributes such as age, sex, ethnicity, occupation, residence, and other factors that are related either to health or to the ability to obtain needed health services. Special studies may also be made of the community itself with respect to special environmental factors (industries and occupations, climate, pollution, etc.) and availability of health facilities, resources, and personnel. This information is combined with vital statistics and morbidity data to ascertain health problems characteristic of the community. For example, an unusually high birth rate may signify the need for a program of family planning services or it may simply be a fact caused by the peculiar age and sex distribution of the population. An excessive rate of lung cancer may suggest the need for an antismoking campaign or an investigation as to whether air pollution or a special industry, like the manufacture of products containing asbestos, may be involved.

In making judgments regarding the health status of a community, the biostatistician must be aware of the many possible comparisons that might be drawn. The age-adjusted death rate in a county, for example, might be compared to that of an adjacent county, to the rate of the entire state or nation, to the lowest county rate in the area, to the rate of a county or counties similar in the composition of the population of the study county, or simply to the trend of the rates for that county over the past 10 to 20 years.

Finally, community diagnosis would be

incomplete without an attempt to study the effectiveness of treatment efforts. The biostatistician collaborates with health service providers to evaluate the effectiveness of any program instituted to improve the status of the community. With the assistance of cost-accounting specialists, the benefits (and any undesirable effects) of the program are compared with the costs in terms of funds and personnel so as to balance the relative weights of these items from a societal point of view. There are many references for studying this aspect in greater detail, and two helpful ones are Caro [7] and Greenberg [25].

FUTURE TRENDS

There are two areas in which the biostatistician has been playing a leading role lately, and these cut across many kinds of applications and problems. It is highly likely that considerable research in methodology will continue to be devoted to these two special areas of concern. The first of these areas might be called modeling.

Mathematical Models

The relationship between a set of independent variables and the dependent or response variable(s) is usually referred to as the mathematical model. This model may take the form of a standard multiple regression* analysis with a single response variable, a surface, or multiple response variables as in multivariate analysis.

It is generally assumed that the technical specialist with substantive knowledge of the field of application (epidemiology, toxicology, pharmacology, radiology, genetics, etc.) will play a crucial role in determining the model or relationship between a set of independent variables and the response variable(s). In actual practice, however, the biostatistician is the one who finally selects the specific model that establishes this functional relationship and then attempts to measure the strength or influence of the independent variables therein. Moreover, the biostatistician is often expected to contribute strongly to the decision as to whether the relationship is a causal one or merely one of association and correlation.

For example, in the measurement of the carcinogenicity of a food additive or drug, questions may arise as to whether a substance can be judged harmful if it "accelerates" the appearance of a tumor even though it does not increase the incidence of the abnormal growth. In general, the answer to this question is in the affirmative provided that there can be unequivocally demonstrated a dosage–response relationship between the substance and the tumor—i.e., the more of the suspected compound that is given or exposed to the test animal, the greater is the probability or likelihood that the tumor will occur earlier.

In the case of bioassay procedures, the biostatistician may be called upon to decide which sigmoid or S-shaped curve to use in order to relate the dosage to the response. This problem is more than simply a decision between the integrated normal curve (or probit*) vis-à-vis the logit function [i.e., $\log_e(p/(1-p))$, where $0 < p < 1$]. In the case of harmful or toxic substances or low-level irradiation (as explained in the subsequent section), it is a question as to whether even a regression relationship can be assumed for purposes of extrapolation.

In many data sets that arise in biostatistics the data are in the form of contingency tables* arising from frequency counts in cells created as a result of an n-way cross-classification of the population. The variable being measured is frequently on a nominal scale, i.e., the categories are simply the set of names corresponding to an attribute such as occupation, place of residence, cause of death, religion, or sex. In a few instances the categories of classification may be ordered or ranked on a scale whose orientation is clear but whose spacings are not known. For example, socioeconomic class may be classified as low, medium, or high; or birth order may be determined as first, second, third, and so on. In those special cases where the number of possible categories for an attribute is limited to two, the data are referred to as *dichotomous**. Thus one can have dichotomous information on sex disaggregated by male, female; or data where simply the presence or absence of a factor is noted and a dummy variable is created such that $1 = \text{yes}$ and $0 = \text{no}$.

If we consider one kind of contingency table encountered frequently in surveys, we may have as cell entries the annual mean number of doctor visits, illnesses, or days of disability. The number of respondents in each cell may be known but the relationship of the attributes in the classification scheme to the dependent variable is not. The aim of the analysis is to study how the different subclasses of the categories relate to the response variable.

For instance, let us consider that the number of doctor visits for a given age group has been averaged in each cell, and the number of classification variables is three: location ($L =$ urban, rural), sex ($S =$ male, female), and highest level of education ($E =$ no high school, high school, college, postcollege). This means that there are $2 \times 2 \times 4 = 16$ cells, and the response variable, R, can be a function of all 16 parameters.

Instead of having to contend with 16 parameters, a general linear model* might consist of

$$R = \lambda + \lambda_L + \lambda_S + \lambda_E + \lambda_{LS} + \lambda_{LE} + \lambda_{SE} + \lambda_{LSE},$$

where $\lambda =$ general mean,

$\lambda_L, \lambda_S, \lambda_E =$ main effects of location, sex, and education, respectively

$\lambda_{LS}, \lambda_{LE}, \lambda_{SE} =$ the first-order interactions, each the simultaneous effect of two factors shown

$\lambda_{LSE} =$ a second-order interaction, or the simultaneous effect of all three factors

In fitting the parameters in this model, it is highly unlikely that the general mean will itself provide an adequate fit. By adding the main effects, one can ascertain whether an adequate fit has been obtained. If it doesn't, one proceeds to the next level and introduces whichever of the first-order interactions seem necessary. Finally, and only if required, the highest-order interaction would be brought into the final model. This hierarchical approach is similar to that of a stepwise regression* analysis and can be carried out either forward or backward.

Now, if the model is restructured so that one may consider the effects as a linear sum for the *logarithm* of R, then the model is referred to as multiplicative, because effects are being multiplied. Thus, if

$$\log_e R = \lambda + \lambda_L + \lambda_S + \lambda_E + \lambda_{LS} + \lambda_{LE} + \lambda_{SE} + \lambda_{LSE},$$

we can define a new set of parameters such that $\lambda = \log_e \lambda'$, whence

$$\log_e R = \log\lambda' + \log\lambda'_L + \log\lambda'_S + \log\lambda'_E + \log\lambda'_{LS} + \log\lambda'_{LE} + \log\lambda'_{SE} + \log\lambda'_{LSE}$$

or

$$\log_e R = \log(\lambda' \cdot \lambda'_L \cdot \lambda'_S \cdot \lambda'_E \cdot \lambda'_{LS} \cdot \lambda'_{LE} \cdot \lambda'_{SE} \cdot \lambda'_{LSE}).$$

Taking antilogarithms results in a form that shows why the model is referred to as a multiplicative model:

$$R = \lambda' \cdot \lambda'_L \cdot \lambda'_S \cdot \lambda'_E \cdot \lambda'_{LS} \cdot \lambda'_{LE} \cdot \lambda'_{SE} \cdot \lambda'_{LSE}.$$

This form of relationship is referred to as a *log-linear model**, and the predicted response value must always be positive.

In the case where the response variable is dichotomous yes–no, the mean values in the cell entries are really proportions. In the case of proportions, one may also use a logit-linear model not only to assure that the predicted values of p will lie between zero and 1 but also to help obtain a better fit.

A frequently encountered multiway contingency table consists of s samples from s multinomial distributions having r categories of response. There are then counterparts to the linear models and log-linear models discussed earlier, but the problem of estimation of the parameters involves choices between ordinary least squares*, weighted least squares*, maximum likelihood*, and minimum χ^{2*}. The complexity of these considerations is beyond the scope of this section, but the reader will find that refs. 2, 13, 30, and 47 fairly well summarize the state of the art.

Detection of Hazardous Substances

With the successful conquest of most of the infectious diseases that have plagued mankind throughout history, health authorities have been concentrating recently upon two chronic diseases whose etiology is yet to be determined: cardiovascular disease and cancer. In both cases, there is no disagreement with the thesis that heredity exercises a determining influence, but the role of the environment in causing many cases is also unquestioned. Attempts to measure the harm that may be caused by potentially hazardous substances in the environment, principally with respect to these two diseases, represent the greatest challenge to the biostatistician today. The benefit to society when successes are obtained make this area of research rewarding emotionally and scientifically despite the exceptional complexities involved.

The number of factors included under the rubric of environment which have a human impact is probably infinite. In addition to food, air, and water, environment includes everything that is not included under the genetics label. Thus in addition to known or unknown chemical and physical substances, there are such variables as exercise, noise, tension, and stress, plus all the psychosocial elements that affect people.

When one starts to limit study to hazardous substances by themselves, say cigarette smoking and exposure to asbestos, the interaction of these two factors and of these two with stress, sleep, use of alcohol, and psychosocial elements soon points to the impracticality of drawing too constraining a line around the hazardous substances. Thus if one accepts the assertion that an overwhelming majority of cancers are environmentally induced, and that of these perhaps 5 to 10% are occupation-related, any good inquiry into the incriminating substance(s) in the workplace cannot overlook the personal characteristics and habits of the workers themselves.

This highlights the first two difficulties in measuring the health importance of hazardous substances: that the list of substances and important factors is substantially great if not infinite, and that these factors have interactions or synergistic reactions that may be more important than the main effects themselves. (The effect of cigarette smoking and asbestos exposure referred to a moment ago is a perfect example of how the two factors in combination are much more important than the addition of the two by themselves in promoting lung cancer.)

Some of the other difficulties in studying the hazardous substances is that they are frequently available only in low doses and administered over a long period of time. The low dosage creates many problems of its own. For example, there is the question of how reliable the measurement is, especially when many of the estimates of exposure have to be retrospective or ex post facto in nature. Even if prospective in time, a sampling of the air in an environment requires a suitable model of air circulation so as to know whether to collect samples at different locations, at different times of the day, indoors or outdoors, with or without sunlight and wind, and so on. Furthermore, total exposure over an n-hour period may be more or less important than a peak period of exposure during a short period of time.

Determination of the impact of low doses of hazardous substances is especially complex because of other peculiarities. Experiments on human beings are, of course, out of the question, so reliance must be placed on accidental exposure of human beings, long-term exposure retrospectively, or the effect upon animals. Since low doses are likely to cause small effects, the results are extremely difficult to detect and usually require large numbers of animals plus extensive and expensive examinations of many organs. Since the number of animals required might be prohibitive anyway, reliance is often placed upon artificially high doses.

This confuses the picture further because one needs to know what model to use in projecting by extrapolation from high doses the effect at low dose levels, as well as what rationale for judging how valuable it is to measure the possible effects on human beings from data on the health of animals, usually mice and other rodents. The models used to extrapolate from high dose to low dose, especially in problems involving radiation, include the one-hit, two-hit, probit*, and a variety of empirical models. This subject gets still more complicated when assumptions are made about threshold levels and rates of neutralization or detoxification [11].

There is here also the effect of confounding variables, which were referred to earlier, when discussing the selection of epidemiological cases and controls, as a special field of application.

There is also the entire question of risk versus benefit, which is an important consideration in determining public policy. As an example of how legislation in the United States which fails to consider both risk and benefit can adversely affect public policy, the Delaney Amendment to the Food Additive Amendments of 1958 (P.L. 85–959) can be cited. The Delaney clause stated that any food additive that is capable of producing cancer in animals or human beings is assumed harmful to human beings and must be banned by the Food and Drug Administration regardless of any benefit. Although the methodology for studying risks vis-à-vis benefits is in a primitive state, the present law does not encompass the possibility that the risk might be miniscule and the benefit substantial. Such has probably been the situation regarding the use of saccharin. Another drawback with the Delaney clause is that certain food additives that have a low carcinogenic effect, such as nitrites, might be found naturally in other foods. They have to be banned as a food additive, even though the benefits might outweigh the disadvantages and even though other foods already possess the particular substance.

This has been a brief overview of a most important area. Readers who are interested in this subject are urged to examine the technical report by Hunter and Crowley [34].

EDUCATION AND EMPLOYMENT OF BIOSTATISTICIANS

What does it take to make a biostatistician? To this question there are a host of responses but no firm, positive answer. This is not surprising in view of the fact that two other quite different occupations that have been studied probably more than any other during the past 40 years—airplane piloting and medicine—still have a long way to go before most of the answers are known regarding the optimal selection of applicants and the most effective method of education.

Undoubtedly, one of the most important characteristics in the outstanding biostatistician is the individual himself or herself. Essential attributes are an inquisitive curiosity or "burning yearning for learning," a constancy of purpose, an ability to think quantitatively, an interest in applying statistical methods to biological problems, and probably a personality or mental disposition that encourages close working relationships with collaborators from many fields.

The field of biostatistics has many avenues of access for entry purposes. Although most persons probably enter from a traditional mathematical or mathematical–statistical background, many others have come from an original interest in biomedical fields, sociology, psychology, engineering, and computer sciences. There is no unique or assured pathway to biostatistics; the right person can approach it from whatever direction maximizes his or her own potential.

The most frequently used institutions for educating biostatisticians are the departments of biostatistics and statistics. These departments should be located in a university setting where there are, first, an academic health center or, at the very least, a medical school and hospital. There must also be strong units in the remainder of the university concerned with the teaching of graduate students in mathematics, probability and statistics, and computer sciences. More than the structural units in the university, however, there must be a pattern or tradition of close working relationships between the biostatistics faculty and those from the other entities. Training at the doctoral level will not be truly meaningful unless the biostatistical faculty are actively engaged in and publish work on applied as well as theoretical research concerning statistical methods.

Merely because the university has, say, an accredited school of public health with a department of biostatistics is no guarantee that it is a good one. (Moreover, biostatisticians can be, and have been, educated in statistical departments other than those found in schools of public health.) Students seeking training in biostatistics would be well advised to be certain that the teaching faculty are engaged in both applied and methodological research, and most important, that there is a close affiliation or working relationship with a medical unit of some kind. Unless such affiliation exists, it will be more difficult to get exposure to good experience involving clinical data, clinical trials, epidemiology, hospital and clinic studies, and health services research.

The doctoral training might consist of either the program for the traditional academic Ph.D. degree or a program that is professionally oriented, such as one directed to the Doctor of Hygiene or Doctor of Public Health degree. The Ph.D. is usually intended for those persons planning careers in an academic or research setting where emphasis is on developing statistical methodology to solve important biological and public health problems. The related doctoral dissertations are most often published in statistical journals.

The professional doctoral degree is usually intended for persons who plan careers in government or industry and whose emphasis is on service to persons seeking statistical advice. The nature of the doctoral dissertation is frequently the new application of statistical concepts to important public health or biological problems. What is novel in the dissertation is the application of a known statistical technique to solve an important health problem. This type of dissertation is usually published in biological and public health journals.

Training at the master's degree level for persons interested in beginning and intermediate-level positions in biostatistics is determined and evaluated by approximately the same guidelines as the foregoing. The main difference, perhaps, is that the criterion of an active ongoing research program is not as stringent. Instead, emphasis should be placed on good teaching and other pedagogical processes that will enable students to learn practical techniques at the same time that a few of them are stimulated to pursue advanced training at the doctoral level.

Employers of biostatisticians have ranged from local, state, and federal government to industry and academic institutions. Each employer will require a different set of areas of knowledge over and above the general field of statistics and probability. For example, one type of government biostatistician may be required to be an expert in registration of vital statistics, demography, survey research, and the special problems associated with confidentiality. The person working as a biostatistician in a pharmaceutical firm may require special training in bioassay techniques, mathematical modeling, and those aspects of clinical trials related to the study of new drugs and their toxicity, dosage, and effectiveness, in order to collaborate in preparing applications for a new drug to be approved by a regulatory agency such as the Food and Drug Administration, U.S. Public Health Service.

It is difficult to know even approximately how many persons there are in the United States who would classify themselves as biostatisticians. There is no one professional organization designed for affiliation of persons who are primarily biostatisticians. A rough guess would be that one-fifth of the 15,000 statisticians listed in the 1978 Directory published by the American Statistical Association* are strongly interested in biostatistical problems, and, of these, about one-half would be individuals who classify themselves primarily as biostatisticians. An international Biometric Society* was established in 1947, and in 1950 it assumed responsibility for publication of the journal *Biometrics**, which had had its inception as the *Biometrics Bulletin* in 1945 under the aegis of the American Statistical Association.

The reader interested in further readings about biostatistics is recommended to consult especially the journals *Biometrika** and *Biometrics**.

References

The citations referred to in the article are presented in alphabetical order. For convenience, they have been classified into one of seven categories shown at the end of the reference and coded according to the following scheme.

A: historical and biographical
B: epidemiology: models and causality
C: epidemiology: relative risk
D: edipemiology: clinical trials
E: demography and community diagnosis
F: surveys, privacy and confidentiality
G: general biometry

[1] Bailey, N. T. J. (1957). *The Mathematical Theory of Epidemics*. Charles Griffin, London/Hafner, New York. (B)

[2] Bishop, Y. M. M., Fienberg, S. E., and Holland, P. W. (1975). *Discrete Multivariate Analysis*. MIT Press, Cambridge, Mass. (G)

[3] Blalock, H. C., Jr. (1964). *Causal Inference in Nonexperimental Research*. University of North Carolina Press, Chapel Hill, N.C. (B)

[4] Box, J. F. (1978). *R. A. Fisher: The Life of a Scientist*. Wiley, New York. (A)

[5] Byar, D. P., Simon, R. M., Friedewald, W. T., Schlesselman, J. J., DeMets, D. L., Ellenberg, J. H., Gail, M. H., and Ware, J. H. (1976). *N. Engl. J. Med.*, **295**, 74–80. (D)

[6] Campbell, D. T., Baruch, R. F., Schwartz, R. D., and Steinberg, J. (1974). *Confidentiality—Preserving Modes of Access to Files and to Interfile Exchange for Useful Statistical Analysis*. Report of the National Research Council Committee on Federal Agency Evaluation Research. (F)

[7] Caro, F. G., ed. (1971). *Readings in Evaluation Research*. Russell Sage Foundation, New York. (E)

[8] Cochran, W. G. (1953). *Amer. J. Public Health*, **43**, 684–691. (C)

[9] Cochran, W. G. (1965). *J. R. Statist. Soc. A*, **128**, 234–255. (C)

[10] Cornfield, J. (1951). *J. Natl. Cancer Inst.*, **11**, 1269–1275. (C)

[11] Cornfield, J. (1977). *Science*, **198**, 693–699. (G)

[12] Courand, A. (1977). *Science*, **198**, 699–705. (D)

[13] Cox, D. R. (1970). *The Analysis of Binary Data*. Methuen, London. (G)

[14] Dalenius, T. (1974). *Statist. Tidskr.* **3**, 213–225. (F)

[15] Dalenius, T. (1977). *J. Statist. Plan. Infer.*, **1**, 73–86. (F)

[16] David, H. A. (1970). *Order Statistics*. Wiley, New York. (G)

[17] Ederer, F. (1979). *Amer. Statist.*, **33**, 116–119. (D)

[18] Finney, D. J. (1964). *Statistical Method in Biological Assay*, 2nd ed. Hafner, New York. (G)

[19] Fleiss, J. L. (1973). *Statistical Methods for Rates and Proportions*. Wiley, New York. (C)

[20] George, S. L. (1976). *Proc. 9th Int. Biom. Conf.*, (Boston), **1**, 227–244. (D)

[21] Gilbert, J. P., McPeek, B., and Mosteller, F. (1977). *Science*, **198**, 684–689. (D)

[22] Greenberg, B. G. (1953). *Amer. J. Public Health*, **43**, 692–699. (C)

[23] Greenberg, B. G. (1959). *Amer. Statist.*, **13**(3), 13–17, 28. (D)

[24] Greenberg, B. G. (1969). *J. Amer. Statist. Ass.*, **64**, 739–758. (B)

[25] Greenberg, B. G. (1974). *Medikon*, **6/7**, 32–35. (E)

[26] Greenberg, B. G. and Sarhan, A. E. (1959). *Amer. J. Public Health*, **49**, 634–643. (G)

[27] Greenberg, B. G. and Sirken, M. (1977). *Validity Problems, Advances in Health Survey Research Methods*. National Center of Health Services Research, Research Proceedings Series, DHEW Publication No. (HRA) 77-3154, pp. 24–31. (F)

[28] Greenberg, B. G. and Abernathy, J. R., and Horvitz, D. G. (1970). *Milbank Mem. Fund Quart.*, **48**, 39–55. (F)

[29] Greenwood, M. (1948). *Medical Statistics from Graunt to Farr*. Cambridge University Press, Cambridge. (A)

[30] Grizzle, J. E., Starmer, C. F., and Koch, G. G. (1969). *Biometrics*, **25**, 489–503. (G)

[31] Hastings, D. W. and Berry, L. G., eds. (1979). *Cohort Analysis: A Collection of Interdisciplinary Readings*. Scripps Foundation for Research in Population Problems, Oxford, Ohio. (E)

[32] Horwitz, R. I. and Feinstein, A. R. (1978). *N. Engl. J. Med.*, **299**, 1089–1094. (C)

[33] Hulka, B. S., Hogue, C. J. R., and Greenberg, B. G. (1978). *Amer. J. Epidemiol.*, **107**, 267–276. (C)

[34] Hunter, W. G., and Crowley, J. J. (1979). Hazardous Substances, the Environment and Public Health: A Statistical Overview. *Wisconsin Clinical Cancer Center Tech. Rep. No. 4*, University of Wisconsin, Madison, Wis. (G)

[35] Kruskal, W. and Tanur, J. M., eds. (1978). *International Encyclopedia of Statistics.* Free Press, New York. (A)

[36] Lave, L. B. and Seskin, E. P. (1979). *Amer. Sci.*, **67**, 178–186. (B)

[37] Linder, F. E. and Grove, R. D. (1947). *Vital Statistics Rates in the United States, 1900–1940.* U.S. Government Printing Office, Washington, D.C. (See especially, Chaps. 3 and 4.) (E)

[38] McKinlay, S. M. (1975). *J. Amer. Statist. Ass.*, **70**, 859–864. (C)

[39] McKinlay, S. M. (1977). *Biometrics*, **33**, 725–735. (C)

[40] McPherson, K. (1974). *N. Engl. J. Med.*, **290**, 501–502. (D)

[41] McPherson, C. K. and Armitage, P. (1971). *J. R. Statist. Soc. A*, **134**, 15–25. (D)

[42] Miettinen, O. S. (1970). *Biometrics*, **26**, 75–86. (C)

[43] Miettinen, O. S. (1970). *Amer. J. Epidemiol.*, **91**, 111–118. (C)

[44] Miettinen, O. S. (1972). *Amer. J. Epidemiol.*, **96**, 168–172. (C)

[45] Miettinen, O. S. (1974). *Amer. J. Epidemiol.*, **100**, 350–353. (C)

[46] Miller, R. G. (1966). *Simultaneous Statistical Inference*. McGraw-Hill, New York. (G)

[47] Nelder, J. A. and Wedderburn, R. W. M. (1972). *J. R. Statist. Soc. A*, **135**, 370–384. (G)

[48] Norton, B. J. (1976). *Proc. 9th Int. Biom. Conf.* (Boston), **1**, 357–376. (A)

[49] Peto, R., Pike, M. C., Armitage, P., Breslow, N. E., Cox, D. R., Howard, S. V., Mantel, N., McPherson, K., Peto, J., and Smith, P. G. (1976). *Br. J. Cancer*, **34**, 585–612. (D)

[50] Peto, R., Pike, M. C., Armitage, P., Breslow, N. E., Cox, D. R., Howard, S. V., Mantel, N., McPherson, K., Peto, J., and Smith, P. G. (1977). *Br. J. Cancer*, **35**, 1–39. (D)

[51] Rothman, K. J. (1976). *Amer. J. Epidemiol.*, **103**, 506–511. (C)

[52] Spiegelman, M. (1968). *Introduction to Demography*, Harvard University Press, Cambridge, Mass. (E)

[53] Tukey, J. W. (1977). *Science*, **198**, 679–684. (D)

[54] Walter, S. D. (1976). *Biometrics*, **32**, 829 – 849. (C)

[55] Zelen, M. (1979). *N. Engl. J. Med.*, **300**, 1242–1245. (D)

BERNARD G. GREENBERG

BIPLOT

DEFINITION

A *biplot* is a graphical display of a matrix $Y_{n\times m}$ by means of markers $\mathbf{a}_1, \ldots, \mathbf{a}_n$ for its rows and $\mathbf{b}_1, \ldots, \mathbf{b}_m$ for its columns, such that inner product $\mathbf{a}_i'\mathbf{b}_v$ represents element $y_{i,v}$ ($i = 1, \ldots, n$; $v = 1, \ldots, m$).[1] The prefix "bi" indicates that this is a joint display of the rows and columns; in that, biplots differ from most other plots which display only rows or only columns.

(A geometric discussion for the mathematical reader is given in the concluding section.)

USES

Biplots are useful for visual inspection of data matrices, allowing the eye to pick up patterns, regularities, and outliers. Multivariate data on single batches can be biplotted in a manner analogous to principal component analysis*, multiple sample data in analogy with MANOVA*, and discriminant analysis*. Biplots are also available for canonical correlations* and contingency tables* and can incorporate approximate graphical significance tests. They can be used to diagnose models, such as Tukey's degree of freedom for nonadditivity* or a harmonic series*.

A SIMPLE EXAMPLE

A biplot of the matrix Y of Table 1 is shown in Fig. 1, with row (tribe) markers $\mathbf{a}_1, \mathbf{a}_2, \ldots, \mathbf{a}_5$ displayed as points and column (characteristics) markers $\mathbf{b}_1, \mathbf{b}_2, \mathbf{b}_3$ displayed as arrows from the origin.

Table 1 Demographic Characteristics of Some American Indian Tribes

Tribe	Median Years of Schooling	Percentage below Poverty Line	Economic Index[a]
Shoshones	10.3	29.0	9.08
Apaches	8.9	46.8	10.02
Sioux	10.2	46.3	10.75
Navajos	5.4	60.2	9.26
Hopis	11.3	44.7	11.25

Source. Kunitz [8].
[a]0.6 median school year + 0.1% poverty.

The biplot represents each element $y_{i,v}$ geometrically. This is illustrated as follows for element $y_{1,2}$: A perpendicular is dropped from point $\mathbf{a}_1$ onto arrow $\mathbf{b}_2$ (or onto the straight line through $\mathbf{b}_2$). The distance from the origin to the foot P of this perpendicular is measured as 2.3 and this is multiplied by 12.6, the length of arrow $\mathbf{b}_2$. The product $2.3 \times 12.6 = 29.0$ corresponds to value $y_{1,2}$. This construction is the *inner product* $\mathbf{a}_1'\mathbf{b}_2 \Rightarrow y_{1,2}$.

The **a**-scatter reflects the tribes. The cluster of three points shows that Apaches, Sioux, and Hopis were quite similar in the characteristics displayed here, whereas Shoshones and Navajos differed from them in opposite directions.

Linear combinations of rows of Y can be represented on the biplot. Thus the average of rows $\mathbf{y}_1'$ and $\mathbf{y}_4'$ can be represented by a point midway between $\mathbf{a}_1$ and $\mathbf{a}_4$. This would be close to the cluster of the other three points. Evidently, the Shoshone–Navajo average was similar to the other three tribes. Similarly, the Navajo–Apache difference $\mathbf{a}_4 - \mathbf{a}_2$ can be constructed as a point in the same direction and distance from the origin as $\mathbf{a}_4$ is from $\mathbf{a}_2$.

The configuration of arrows reflects the magnitude and relations of the characteristics. The longer arrow $\mathbf{b}_2$ shows that the numbers in the second column are larger than those in the other columns. The near collinearity of arrows $\mathbf{b}_1$ and $\mathbf{b}_3$ reflects the

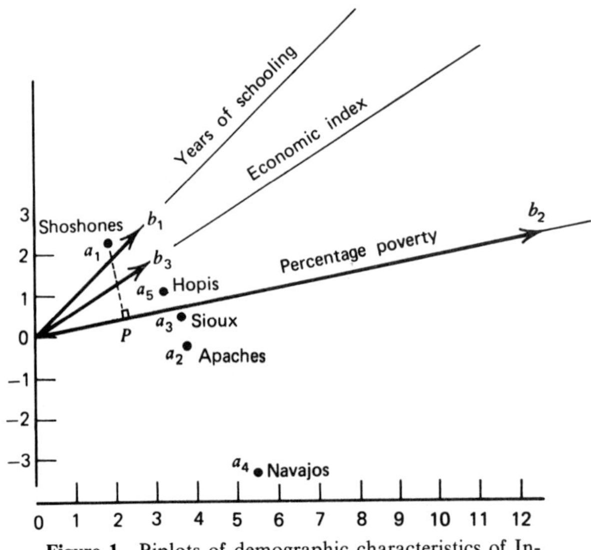

Figure 1 Biplots of demographic characteristics of Indian tribes.

rough proportionality of the first and third columns.

Combinations of columns may be represented just like combinations of rows. As an example, a new variable defined as (economic index) – (median years of schooling) can be represented by an arrow in the direction from $\mathbf{b}_1$ to $\mathbf{b}_3$. This is very much the same direction as the Navajo–Apache difference $\mathbf{a}_4 - \mathbf{a}_2$. Evidently, these two tribes differed most strongly on that variable. The special feature of the biplot is that it displays such interrelations between the rows and columns.

APPROXIMATION AND COMPUTATION

In practical applications, biplot display is approximative because matrices are usually of rank greater than 2. Inner-product representation relation $\mathbf{a}_i'\mathbf{b}_v \Rightarrow y_{i,v}$ can be written $AB' \Rightarrow Y$, where the **a**'s and **b**'s are rows of A and B, respectively. But this AB'-biplot can represent Y well only if the rank 2 product AB' is close to Y. One may obtain AB' by least squares* through the Householder–Young theorem (see below) [1]. Algorithms are also available for weighted least-squares* fitting [6] and for adaptive fits [9]. (If either A or B is given from some prior considerations, the other factor may be obtained by linear least squares.)

Adding another dimension to the biplot display may yield a closer approximation. Such a three-dimensional bimodel can be partly represented by a number of planar projections but is most useful when displayed on a CRT with a fast rotation-projection algorithm [12].

PRINCIPAL COMPONENT* BIPLOTS AND AN EXAMPLE

A special type of biplot is suitable for multivariate data matrices in which i indexes the n units of a batch and v the m variables, the observations $y_{i,v}$ being measured from the batch means. These *principal component biplots* [3] are fitted by least squares as follows. Solve $Y'Y\mathbf{q}_\alpha = \lambda_\alpha^2\mathbf{q}_\alpha$ subject to $\mathbf{q}_\alpha'\mathbf{q}_\alpha = 1$ for the largest two roots $\lambda_1^2 \geqslant \lambda_2^2$; compute $\mathbf{p}_\alpha = Y\mathbf{q}_\alpha\lambda_\alpha^{-1}$ and define $G = (\mathbf{p}_1, \mathbf{p}_2)$, $H = (\lambda_1\mathbf{q}_1, \lambda_2\mathbf{q}_2)$, $J = (\lambda_1\mathbf{p}_1, \lambda_2\mathbf{p}_2)$, and $K = (\mathbf{q}_1, \mathbf{q}_2)$. This gives the GH'- and the JK'-biplots, both of which approximate Y to the extent of $(\lambda_1^2 + \lambda_2^2)/\sum_{\alpha=1}^{y}\lambda_\alpha^2$. The GH'-biplot gives an even better fit to the variables (columns) *configuration* (i.e., the inner products $Y'Y$) in that $h_v'h_w \Rightarrow y_{(v)}'y_{(w)}$, which is n times the (v,w)-covariance; but its representation $\|\mathbf{g}_i - \mathbf{g}_e\| \Rightarrow \sqrt{(\mathbf{y}_i - \mathbf{y}_e)' \cdot (Y'Y)^{-1}(\mathbf{y}_i - \mathbf{y}_e)}$ of the standardized interunit distances is poorer. Conversely, the JK'-biplot represents the interunit *distances* very well, but not the configuration of the variables. (See Table 2.)

Figure 2 displays a GH'-biplot of 27 European countries' per capita consumption of protein from nine sources, measured from the European means [11].

1. Lengths of **h**-arrows are roughly proportional to the standard deviations* of the variables (sources) represented. Cosines

Table 2 Goodness of Fit of Biplot Approximations

Biplot Approximation		Goodness of
GH'	JK'	Fit[a]
$\mathbf{g}_i'\mathbf{h}_v \Rightarrow y_{i,v}$	$\mathbf{j}_i'k_v \Rightarrow y_{i,v}$	$(\lambda_1^2 + \lambda_2^2)/\sum_{\alpha=1}^{y}\lambda_2^2$
$\mathbf{g}_i'\mathbf{g}_e \Rightarrow \mathbf{y}_i'(Y'Y)^{-1}y_e$	$\mathbf{k}_v'\mathbf{k}_w \Rightarrow y_{(v)}'(YY')^{-1}y_{(w)}$	$2/y$
$\mathbf{h}_v'\mathbf{h}_w \Rightarrow \mathbf{y}_{(v)}'y_{(w)}$	$\mathbf{j}_i'\mathbf{j}_e \Rightarrow \mathbf{y}_i'\mathbf{y}_e$	$(\lambda_1^4 + \lambda_2^4)/\sum_{\alpha=1}^{y}\lambda_\alpha^4$

[a] Goodness of fit

$$= 1 - \frac{\text{sum of squares of residuals from approximation}}{\text{sum of squares of data}}$$

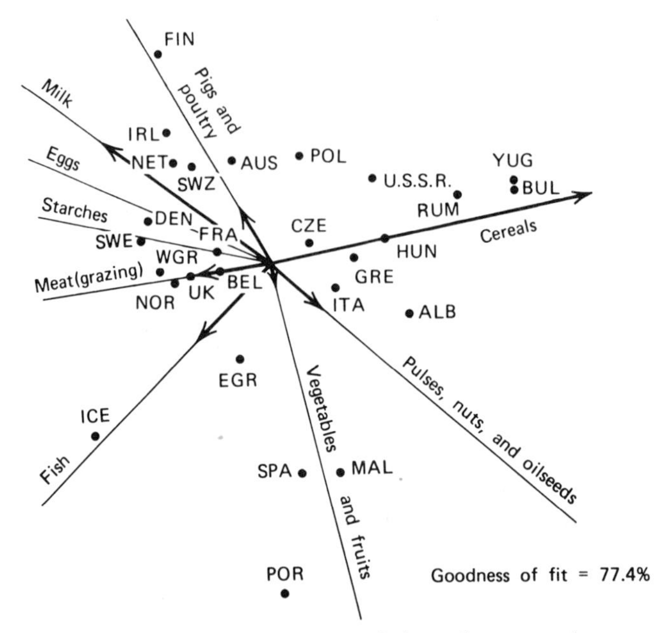

Figure 2 *GH*-biplot of European countries' protein consumption.

of angles between **h**-vectors approximate the correlations* between the sources represented.

Cereal protein consumption had by far the highest variability, followed by milk. Eggs and starches had little variability. Cereal protein consumption was negatively correlated with meat proteins, (180° angle $\Rightarrow -1$ correlation). Vegetables and fruit appeared uncorrelated with either of the above, but somewhat with pulses, nuts, and oilseeds and with fish.

2. Distances between **g**-points represent standardized intercountry differences in protein consumption. Points close together represent countries with similar patterns.

   There was a clear geographical division—the eastern countries to the right, the northwestern ones to the left, and the southwestern ones at the bottom. Geographical contiguity appears to have been related to similarity of protein consumption. (Only Iceland and Finland were exceptional.)

3. Inner products $\mathbf{g}_i'\mathbf{h}_v$ represent the actual v-source protein consumption in country i (as measured from the European mean). If $\mathbf{g}_i$ is in (opposite to) the direction of $\mathbf{h}_v$, country i consumes more (less) of protein v than the European average.

Eastern European countries are clustered along the arrow for cereals and opposite those for animal proteins. Evidently, Eastern Europeans consumed more cereals and fewer animal proteins than average Europeans did. Northwestern Europeans have the reverse pattern. Southwestern Europeans are seen to have consumed large amounts of vegetables and fruit, of pulses, nuts, and oilseeds, and of fish. (Iceland and Finland showed unusually large consumption of fish and milk, respectively.)

The biplot's special feature is that it shows the relation of the variables' configuration—**h**-arrows—to the units' scatter—**g**-points—and thus indicates on what variables the clusters of units differ.

A DIAGNOSTIC EXAMPLE

A biplot can display data from a three-way and higher-order layout and can be used to diagnose a model. Figure 3 displays data from a three-factor, two-replication experi-

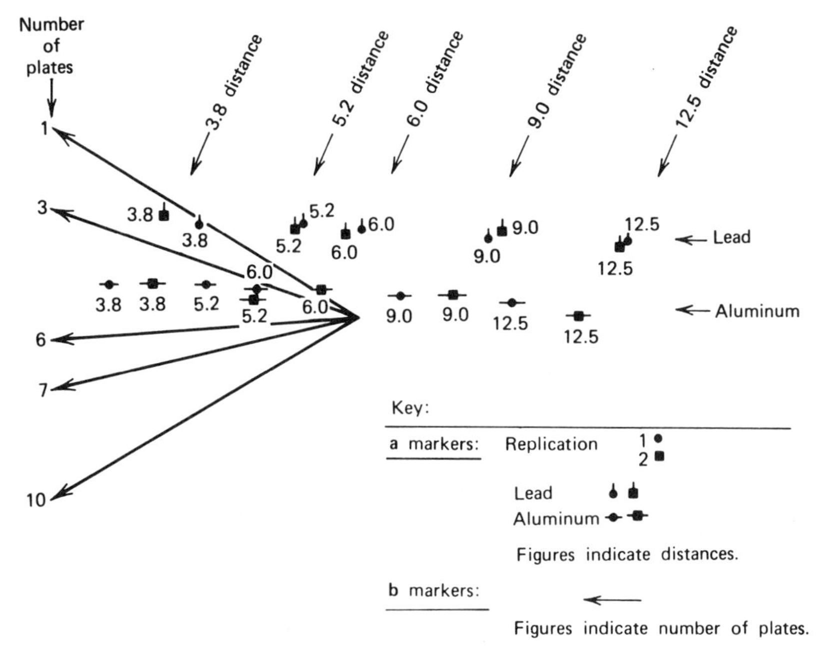

Figure 3 Biplot of gamma-radiation data.

ment (*see* DESIGN OF EXPERIMENTS) on absorption of gamma radiation. The factors metal, distance, and replication were cross-classified within the rows of Y and are jointly represented by **a**'s; **b**'s represent number of plates, the column factor.

Figure 3 shows collinearity of the **b** markers as well as of the **a** markers for each metal. One may apply the diagnostic rules from Table 3. When these patterns hold for only a part of the row markers or the column markers, the model may be diagnosed for the appropriate submatrix. These rules suggest two separate *additive* models, one for lead and another for aluminum. Both replicates fit the models, but whereas lead replicates hardly differ, those for aluminum do.

The order of markers allows one to detail the diagnosis further. The **b** markers show a trend proportional to the number of plates on which radiation is seen to depend linearly. For each metal, the **a** markers have a trend that indicates quadratic dependence on distance. The two metals' **a**'s are close to two parallel lines. This indicates null metal × distance interaction*. Hence the biplot suggests diagnosis of a model

$$\text{radiation}_{(\text{plts, dist, metal})} = \mu + \alpha(\text{dist}) + \beta\,(\text{dist})^2 + \gamma_{(\text{metal})}(\text{plts}).$$

Least-squares fitting confirmed this as a good model [2].

Table 3 Biplot Diagnostic Rules

Row Markers	Column Markers	The Model is:	
Collinear	—	$y_{i,v} = \beta_v + \alpha_i \beta_v^*$	Columns' Regression
—	Collinear	$y_{i,v} = \alpha_i + \alpha_i^* \beta_v$	Rows' Regression
Collinear	Collinear	$y_{i,v} = \mu + \alpha_i \beta_v$	Concurrent
Collinear at 90° to	Collinear	$y_{i,v} = \alpha_i + \beta_v$	Additive

Source. Bradu and Gabriel [2].

Nonlinear models may also be revealed. Thus an elliptic structure of **b**'s in three dimensions corresponds to a harmonic model for the rows of Y [10].

A MULTIVARIATE ANALYSIS-OF-VARIANCE* (MANOVA) EXAMPLE

Data of the first randomized Israeli rainmaking experiment [4] are displayed in Fig. 4. The days were classified into four samples by the area allocated to seeding (north or center) and by whether seeding actually took place (suitable clouds available). There also was a fifth sample of preexperimental days. The variables were daily precipitation in the following four areas: north, center, a buffer zone between these two alternate target areas, and south.

A JK'-biplot was fitted to deviations from the overall mean, weighted by sample sizes n_i and by the inverse of the "within" variance–covariance* matrix. Around each sample marker $\mathbf{j}_i$, a circle J_i was drawn with radius $[n_i\theta(1-\theta)/2]^{1/2}$, where θ is the upper 5% point of the maximum characteristic root distribution*.

1. Each **k**-vector represents "within" variation of one area's rainfall and the cosines approximate the correlations. **k**-vectors for linear combinations of variables can be constructed. Thus $\mathbf{k}_{N-C} = \mathbf{k}_N - \mathbf{k}_C$ is shown; it displays the "within" variation of the north–center rainfall differences.
2. The distance between **j** markers displays the dissimilarity of the samples with respect to rainfall in the four areas. (It is the Mahalanobis distance* of the planar approximation.)

   The two unseeded samples cluster together, with the nonseeded preexperimental sample between them and the two seeded samples. This reflects the division of days into those suitable for seeding and those not suitable, and the preexperimental days, which included both types. The seeded samples differed more than the unseeded; this may have been due to seeding.
3. Approximate tests are: "Declare any two samples significantly different if their J circles do not overlap."

   Only circles J_2 and J_4 overlap. All samples are therefore declared "significantly" different except the two unseeded samples.
4. For a rough test of whether two samples differ significantly on a particular vari-

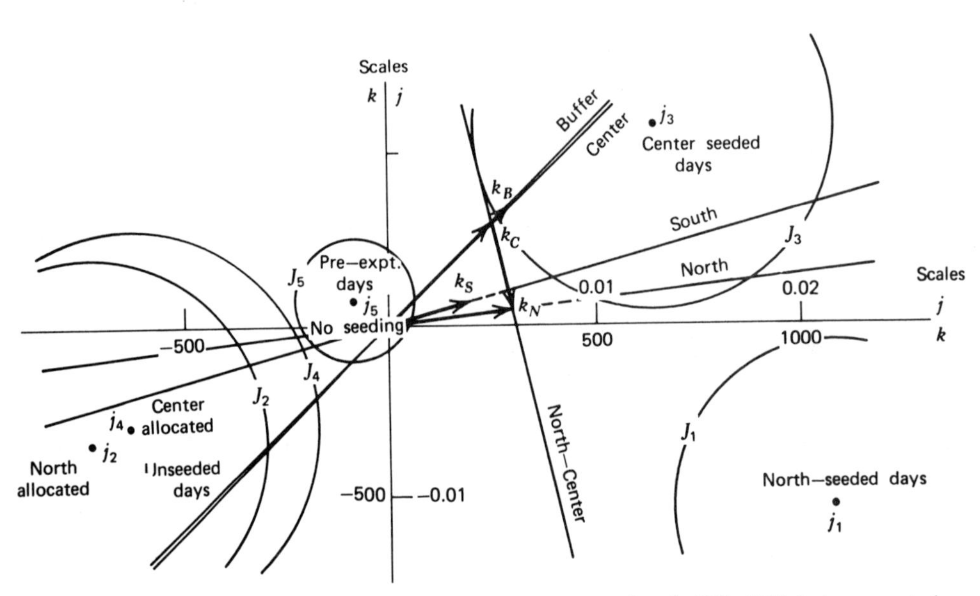

Figure 4 MONOVA biplot of results of rainmaking experiment, Israel, 1961–1967 (rain amounts in millimeters).

able v, one may project the corresponding two circles onto the $\mathbf{k}_v$-vector.

North-seeded days are seen to have had a "significantly" larger north–center rainfall difference than center–seeded days, although these two samples of days did not differ "significantly" on the rainfall in any single one of the areas.

This MANOVA biplot allowed graphic approximation of complex data and simultaneous significance testing*, and it provides a picture of the multivariate structure of the data. It would have been difficult to obtain these by purely analytical methods. We must caution against interpreting "significance" literally in this particular example, as rainfall data are far from Gaussian* and observations are dependent from day to day.

RELATION TO OTHER TECHNIQUES

GH'- and JK'-biplots are complementary to principal component analysis* in that they display two (or three, for bimodels) dimensions simultaneously. The MANOVA* biplot displays samples along discriminant function*, or canonical variate* axes. Biplots of canonical correlations* or contingency tables* also complement rather than supplant standard analyses.

Multidimensional scaling* methods differ from classical multivariate analyses and the related biplots in using nonmetric methods of fitting [7]. This makes them more flexible for the purpose they are designed for, but less amenable to joint representation of rows and columns. However, a multidimensional scaling model for the variables (or units) can be supplemented by a superimposed biplot display of the units (or variables) [5].

The relation to factor analysis* is less clear, although rotational techniques may be applied to the biplot.

RELATION TO MULTIVARIATE GEOMETRY

Any matrix $Y_{(n\times m)}$ can be factored as $Y = A_{(n\times r)}(B_{(m\times r)})'$ for any $r \geqslant y$. A display of $\mathbf{a}_1, \ldots, \mathbf{a}_n$ and $\mathbf{b}_1, \ldots, \mathbf{b}_m$ in E_r is called the *AB'-bigeometry* of Y and represents the matrix elementwise by inner products $y_{i,v} = \mathbf{a}_i'\mathbf{b}_v$. The factorization is not unique and neither is the bigeometry of Y [3].

Two bigeometries of particular interest are the *columns geometry*—much used for representing the configuration $Y'Y$ of a general linear model*—and the *rows geometry*—commonly known, for $m = 2$, as the scatter plot. The former is used for representation of the variance–covariance configuration of the columns of Y, whereas the latter displays interrow distances.

The columns geometry corresponds to the IY-bigeometry with $r = n$, in which the columns of Y are displayed as $\mathbf{b}_v = \mathbf{y}_{(v)}$, and the rows as $\mathbf{a}_i' = \mathbf{u}_i'$,—unit-length vectors along the n orthogonal axes of reference of E_n. The inner products are $\mathbf{a}_i'\mathbf{b}_v = \mathbf{u}_i'\mathbf{y}_{(v)} = y_{i,v}$, and the column configuration is preserved by the $\mathbf{b}$'s, i.e., $\mathbf{b}_v'\mathbf{b}_w = \mathbf{y}_{(v)}'\mathbf{y}_{(w)}$.

The rows geometry corresponds to the YI-bigeometry, with $r = m$, in which the rows are displayed as $\mathbf{a}_i' = \mathbf{y}_i'$, and the columns as $\mathbf{b}_v = \mathbf{u}_v$—unit-length vectors along the orthogonal axes of reference of E_m. Again, the inner products are $\mathbf{a}_i'\mathbf{b}_b = \mathbf{y}_i'\mathbf{u}_v = y_{i,v}$, but here interrow distances are preserved by the $\mathbf{a}$'s display—$\|\mathbf{a}_i - \mathbf{a}_e\| = \|\mathbf{y}_i - \mathbf{y}_e\|$.

The columns' configuration does not change if it is rotated into another set of orthogonal axes*. Thus if $\tilde{P}_{(n\times m)}$ satisfies $\tilde{P}\tilde{P}' = \tilde{P}'\tilde{P} = I_n$, then the rotated $(\tilde{P}')(\tilde{P}Y)$-bigeometry has the same configuration $(\tilde{P}Y)'\cdot(\tilde{P}Y) = Y'Y$ as the IY-bigeometry. One such rotation is that obtained from the singular-value decomposition:

$$\begin{aligned}
Y &= \tilde{P}'_{(n\times m)}\tilde{\Lambda}_{(n\times m)}\tilde{Q}_{(m\times m)} \\
&= \left(P'_{(n\times y)}, P^{*\prime}_{(n\times(n-y))}\right) \\
&\quad\times \begin{bmatrix} \Lambda_{(y\times y)} & 0 \\ & \\ 0 & 0 \end{bmatrix} \\
&\quad\times \begin{pmatrix} Q_{(y\times m)} \\ Q^*_{((m-y)\times m)} \end{pmatrix} \\
&= P'\Lambda Q \\
&= \sum_{\alpha=1}^{y} \lambda_\alpha \mathbf{p}_\alpha \mathbf{q}_\alpha',
\end{aligned}$$

which would yield the $(\tilde{P}')(\tilde{\Lambda}\tilde{Q})$-bigeometry. Since only the first y rows of Q are nonzero, this may be represented in E_y by the GH'-bigeometry with $G = P'$ and $H' = \Lambda Q$.

Omission of the last $(n - y)$-axes is equivalent to projection onto what are known as the first *y principal axes* for columns*.

The vectors $\mathbf{h} = (\lambda_1 q_{1,v}, \ldots, \lambda_y q_{y,v})'$ have the same configuration $\mathbf{h}_v'\mathbf{h}_w = \mathbf{y}_{(v)}'\mathbf{y}_{(w)}$, as the columns of Y. Their elements $h_{v,\alpha} = \lambda_\alpha q_{\alpha,v}$ are known as the *loadings** of column v onto principal axis α. The vectors $\mathbf{g}_i' = (p_{1,i}, \ldots, p_{y,i})$, are projections of unit length $\tilde{\mathbf{p}}_{(i)}$'s onto these *principal axes*. Since $\mathbf{g}_i'\mathbf{g}_e = \mathbf{y}_i'(Y'Y)^{-1}\mathbf{y}_e$, the $\mathbf{g}$'s display the scatter of the rows of Y in a standardized form.

The rows scatter will similarly be preserved if the YI-bigeometry is rotated into another set of axes. Choice of rotation $\tilde{Q}$ from the singular-value decomposition, and projection into the y-dimensional subspace spanned by what are known as the *principal axes for rows*, yields the JK'-bigeometry with $J = P'\Lambda$ and $K' = Q$. Vectors $\mathbf{k}_v = (q_{1,v}, \ldots, q_{y,v})$, are projections of unit length $\tilde{q}_{(v)}$'s onto these axes. They display the column configuration of Y in standardized form $\mathbf{k}_v'\mathbf{k}_w = \mathbf{y}_{(v)}'(YY')^{-}\mathbf{y}_{(w)}$. The vectors $\mathbf{j}_i' = (\lambda_1 p_{1,i}, \ldots, \lambda_y p_{y,i})$ preserve the row scatter in that $\mathbf{j}_i'\mathbf{j}_e = \mathbf{y}_i'\mathbf{y}_e$. Their elements $j_{i,\alpha} = \lambda_\alpha p_{\alpha,i}$ are known as row i's *principal coordinates** with respect to axis α, or αth *principal component scores**.

The rows geometry and the columns geometry, regarded, respectively, as the JK' and the GH' bigeometries, are seen to differ only by reciprocal scalings along their principal axes, i.e., $j_{i,\alpha}\lambda_\alpha^{-1} = g_{i,\alpha}$ and $k_{v,\alpha}\lambda_\alpha = h_{v,\alpha}$. The Y rows' scatter, as displayed by the $\mathbf{j}$'s, is scaled down to the standardized scatter displayed by the $\mathbf{g}$'s, whereas the standardized Y columns' configuration displayed by the $\mathbf{k}$'s is scaled up to the original configuration displayed by the $\mathbf{h}$'s. The relation between the two geometries is evident when they are regarded as bigeometries.

For $y = 2$ or 3 the bigeometry reduces to a *biplot* or *bimodel*, in the plane or in three dimensions, respectively. When y is larger, the bigeometry may be projected onto a plane to obtain an *approximate biplot* (and similarly for an approximate bimodel). Projection can be obtained by omitting all but the first two (or three) coordinates of the $\mathbf{a}$'s and $\mathbf{b}$'s. In case of the GH' and JK' bigeometries, this leads to least-squares approximation of rank 2 (or 3). (This is the Householder–Young theorem.) The GH'-biplot thus displays vectors $\mathbf{g}_i' = (p_{1,i}, p_{2,i})$ and $\mathbf{h}_v = (\lambda_1 q_{1,v}, \lambda_2 q_{2,v})$, and the JK'-biplot displays $\mathbf{j}_i' = (\lambda_1 p_{1,i}, \lambda_2 p_{2,i})$ and $\mathbf{k}_v = (q_{1,v}, q_{2,v})'$. (For brevity, we use the same notation for the y-dimensional $\mathbf{g}, \mathbf{h}, \mathbf{j}$, and $\mathbf{k}$ for their two-dimensional biplot projections.) These biplots serve as approximate displays of Y with properties similar to those of the corresponding bigeometries.

NOTE

1. A matrix is denoted by an uppercase letter; its (i, v)th element is subscripted i, v; its ith row and vth column are set in boldface type and subscripted i and (v), respectively; $(\ )^{-1}$ indicates a generalized inverse. $\|\mathbf{x}\|$ denotes the length of $\mathbf{x}$. The row vectors of identity matrix I_m are denoted $\mathbf{u}_i (i = 1, \ldots, m)$. The notation $\mathbf{a}_i'\mathbf{b}_v \Rightarrow y_{i,v}$ means "$\mathbf{a}_i'\mathbf{b}_v$ represents $y_{i,v}$."

References

[References 2 and 3 provide useful introductions to the biplot and its diagnostic uses. References 4, 5, and 6 give additional applications.]

[1] BIPLOT computer program available from Division of Biostatistics, University of Rochester, Rochester, NY 14642.

[2] Bradu, D. and Gabriel, K. R. (1978). *Technometrics*, **20**, 47–68.

[3] Gabriel, K. R. (1971). *Biometrika* **58**, 453–467.

[4] Gabriel, K. R. (1971). *J. Appl. Meteorol.*, **11**, 1071–1077.

[5] Gabriel, K. R. (1978). In *Theory Construction and Data Analysis in the Behavioral Sciences*, S. Shye, ed. Jossey-Bass, San Francisco, pp. 350–370.

[6] Gabriel, K. R. and Zamir, S. (1979). *Technometrics*, **21**, 489–498.

[7] Kruskal, J. B. (1978). *International Encyclopedia of Statistics*, W. H. Kruskal and J. M. Tanur, eds. Free Press, New York. pp. 307–330.

[8] Kunitz, S. J. (1976). *Hum. Biol.*, **48**, 361–377.

[9] McNeil, D. R. and Tukey, J. W. (1975). *Biometrics*, **31**, 487–510.

[10] Tsianco, M. and Gabriel, K. R. (1981). Biplot diagnosis of harmonic models for meteorological data. (To be published.)

[11] Weber, A. (1973). Agrarpolitik in Spannungsfeld der internationalen Ernährungspolitik. Institut für Agrarpolitik und Marktlehre, Kiel (mimeographed).

[12] Weber, E. and Berger, J. (1978). *COMSTAT 78–Proceedings*, L. C. A. Corsten and J. Hermans, eds. Physica Verlag, Vienna. pp. 331–337.

(ANDREWS' FUNCTION PLOT
FACTOR ANALYSIS
GRAPHICAL REPRESENTATION OF DATA
PRINCIPAL COMPONENT ANALYSIS)

K. RUBEN GABRIEL

BIRNBAUM–HALL TEST

A k-sample analog of the two-sided Kolmogorov–Smirnov test* [1]. The available theory and tables are at present (1979) restricted to the case $k = 3$ and equal sample sizes. An elementary discussion of this test, with a numerical example, is given in Conover [2, pp. 317–319, tables on p. 402].

References

[1] Birnbaum, Z. W. and Hall, R. A. (1960). *Ann. Math. Statist.*, **31**, 710–720.

[2] Conover, W. J. (1971). *Practical Non-parametric Statistics*, Wiley, New York.

BIRNBAUM–RAYMOND–ZUCKERMAN INEQUALITY

A probability inequality of the Chebyshev* type. Let X_i be independent random variables with common variance, $\mu_i = EX_i$ and $\sigma^2 = \text{var}(X_i)$, $i = 1, \ldots, n$. Then for even n

$$\Pr\left(\sum_{i=1}^{n}(X_i - \mu_i)^2 \geqslant \lambda^2\right) \leqslant \begin{cases} 1 & \text{if } \lambda^2 \leqslant n\sigma^2 \\ \dfrac{n\sigma^2}{2\lambda^2 - n\sigma^2} & \text{if } n\sigma^2 \leqslant \lambda^2 \leqslant \dfrac{n\sigma^2}{4}(3+\sqrt{5}) \\ \dfrac{n\sigma^2}{\lambda^2}\left(1 - \dfrac{n\sigma^2}{4\lambda^2}\right) & \text{if } \dfrac{n\sigma^2}{4}(3+\sqrt{5}) \leqslant \lambda^2, \end{cases}$$

and for odd n

$$\Pr\left(\sum_{i=1}^{n}(X_i - \mu_i)^2 \geqslant \lambda^2\right) \leqslant \begin{cases} 1 \quad \text{if } \lambda^2 \leqslant n\sigma^2 \\ \dfrac{(n+1)\sigma^2}{2\lambda^2 - (n-1)\sigma^2} \\ \qquad \text{if } n\sigma^2 \leqslant \lambda^2 \leqslant \dfrac{\sigma^2}{4}\left(3n + 1 + \sqrt{5n^2 + 6n + 5}\right) \\ \dfrac{n\sigma^2}{\lambda^2} - \dfrac{(n^2-1)}{4}\dfrac{\sigma^4}{\lambda^4} \\ \qquad \text{if } \dfrac{\sigma^2}{4}\left(3n + 1 + \sqrt{5n^2 + 6n + 5}\right) \leqslant \lambda^2. \end{cases}$$

This inequality provides an upper bound on the probability of the sample point falling outside a hypersphere centered at the population mean. It has applications to bombing and other aiming problems. Although the random variables are assumed to be independent, it is a *multidimensional* inequality in the sense that it gives a bound on the probability of a multidimensional set. (Compare with Berge's inequality*.) As n becomes large, the results for odd and even integers n approach each other.

Bibliography

Birnbaum, Z. W., Raymond, J., and Zuckerman, H. S. (1947). *Ann. Math. Statist.*, **18**, 70–79.

Savage, I.R. (1961). *J. Res. Natl. Bur. Stand.* **65B**(3), 211–222.

(BERGE INEQUALITY
MULTIVARIATE CHEBYSHEV-TYPE INEQUALITY)

BIRNBAUM'S THEOREM *See* ANCILLARY STATISTICS

BIRTH-AND-DEATH PROCESSES

It would seem that the name "birth-and-death process" should apply to any stochastic process* that models the development of a population, the individual members of which are subject to the vicissitudes of births and deaths. It is usual, however, for the name to be used with reference to a very specific class of stationary Markov* chains in continuous time; this article is restricted to a discussion of this class.

We begin by describing what we shall call a *generalized* birth-and-death process; this terminology is convenient but not universal, and some authors refer to these "generalized" processes simply as "birth-and-death" processes.

Let $N(t)$ be the size of the population at time t, i.e., the number of *live* individuals at that time; assume that $N(T)$ is, as mentioned above, a Markov chain with stationary transition probabilities*:

$$P_{ij}(s) = \Pr[N(t+s) = j \mid N(t) = i], \quad (1)$$

where, for $s > 0, i = 0, 1, 2, \ldots,$

$$P_{i,i+1}(s) = \alpha_i s + o(s),$$

$$P_{i,i}(s) = 1 - (\alpha_i + \beta_i)s + o(s),$$

and, provided that $i \neq 0$,

$$P_{i,i-1}(s) = \beta_i s + o(s).$$

Here the numbers α_i and β_i are nonnegative constants that may be referred to as the birthrates and deathrates, respectively.

Unfortunately, there are technical mathematical difficulties if the $\{\alpha_i\}$ sequence grows too rapidly with increasing i [causing the $N(t)$-process to "explode" to arbitrarily large values]; the remarks of this article assume that the birthrates, $\{\alpha_i\}$, grow no more rapidly than a polynomial in i. There are few applications where such an assumption is violated; indeed, a large part of the interesting ones need the birthrates to grow no faster than a linear function of i.

If we set $P_j(t) = \Pr[N(t) = j]$, then it is possible to show these $P_j(t)$ satisfy the system of differential equations

$$\frac{dP_j(t)}{dt} = \alpha_{j-1}P_{j-1}(t) - (\alpha_j + \beta_j)P_j(t) + \beta_{j+1}P_{j+1}(t) \quad (2)$$

$(j = 0, 1, 2, \ldots)$, where we introduce a constant $\beta_0 = 0$, for ease of writing.

Thus, in theory, if we know the probability distribution of $N(0)$, say, we can derive the probability distribution of $N(t)$ for any $t > 0$. Unfortunately, it is impossible to obtain usable solutions of the system (2) for quite arbitrary birth and death rates and, indeed, the class of processes for which some sort of explicit solution to (2) may be exhibited is decidedly circumscribed; we shall discuss specific examples of such tractable processes below.

Nonetheless, if the birth and death rates are such that a steady-state (stationary) solution* to (2) exists, then that stationary solution is easily calculated for quite general processes. If we set

$$S = 1 + \frac{\alpha_0}{\beta_1} + \frac{\alpha_0\alpha_1}{\beta_1\beta_2} + \frac{\alpha_0\alpha_1\alpha_2}{\beta_1\beta_2\beta_3} + \cdots \quad (3)$$

then the finiteness of S ensures the existence of a steady-state solution $P_j(t) = \pi_j$, say $(j = 0, 1, \ldots)$, where every $\pi_j \geqslant 0$ and $\sum_0^\infty \pi_j = 1$. Indeed, it is easy to discover that

$$\pi_j = S^{-1}\left(\frac{\alpha_0\alpha_1 \cdots \alpha_{j-1}}{\beta_1\beta_2 \cdots \beta_j}\right) \qquad (j \geqslant 1) \quad (4)$$

and $\pi_0 = S^{-1}$.

A consequence of (4) is the revelation that a very large number of different mechanisms can lead to the same steady-state solution to (2). Thus the practical observation of a certain probability distribution (e.g., the Poisson*) in, shall we say, a biological context, can offer, at best, only partial confirmation of some hypothesis about the mechanism of that biological process.

When $S < \infty$, one can infer from the general theory of Markov chains* in continuous time that, whatever the initial probability distribution $\{P_j(0)\}_{j=0}^\infty$, $P_j(t) \to \pi_j$ as $t \to \infty$. This result demonstrates the supreme importance, from the viewpoint of applications, of solution (4) to equations (2) of the generalized birth-and-death process.

If specific assumptions are made about the $\{\alpha_i\}$ and $\{\beta_i\}$ as functions of i, then it is sometimes possible to determine useful information about the "transient" probabilities $P_j(t)$. The most useful tool here is the probability generating function* (PGF):

$$\Pi(z,t) = \sum_{j=0}^{\infty} z^j P_j(t), \quad (5)$$

whose use in the present context was

championed (although not invented) by M. S. Bartlett (see, e.g., Bartlett [1]). In (5) the dummy variable z may be supposed to be real, $0 \leqslant z \leqslant 1$.

Let us examine briefly certain interesting examples where this generating function technique is successful.

"PURE" BIRTH-AND-DEATH PROCESS

In this example we imagine that for all $j \geqslant 0$, $\alpha_j = \lambda j$, and $\beta_j = \mu j$, where $\lambda > 0$ and $\mu > 0$ are two constants. The intuitive idea is that if $N(t) = j$, then, in the short time interval $(t, t + \delta t)$, each individual has a small probability $\lambda \delta t + o(\delta t)$ of splitting into two individuals (a "birth") and a small independent probability $\mu \delta t + o(\delta t)$ of dying. It can be shown that the infinite set of equations (2) reduce, with the introduction of the PGF, to the single equation

$$\frac{\partial}{\partial t} \Pi(z,t) = (\lambda z - \mu)(z - 1) \frac{\partial}{\partial z} \Pi(z,t). \tag{6}$$

This partial differential equation is a special example of what is often, in the literature devoted to applied probability, called the Lagrange equation; fortunately, a method of attacking such equations has long been known. It yields (if $\lambda \neq \mu$) the solution

$$\Pi(t,z) = \Psi\left[\frac{\mu(z-1) - (\lambda z - \mu)e^{-(\lambda-\mu)t}}{\lambda(z-1) - (\lambda z - \mu)e^{-(\lambda-\mu)t}}\right], \tag{7}$$

where we have written $\Psi(z)$ for the initial PGF $\Pi(z, 0)$, which we assume to be known. The case $\lambda = \mu$ presents no special difficulty, but must be treated specially.

In many cases formula (7) can be expanded as a power series in z to yield the desired probabilities $P_j(t)$.

Even in intractable problems where such expansion is impracticable, one can still obtain useful information by falling back on the fact that repeated differentiations, followed by the substitution $z = 1$, yield factorial moments* of $N(t)$.

Notice that if $\lambda > \mu$, then, as $t \to \infty$,

$$\Pi(z,t) \to \Psi(\mu/\lambda);$$

while if $\lambda < \mu$,

$$\Pi(z,t) \to \Psi(1) = 1.$$

From the first result, by examining coefficients of z^n, we can infer that, as $t \to \infty$, $P_j(t) \to 0$ if $j \geqslant 1$ and $P_0(t) \to \Psi(\mu/\lambda)$. Thus we may deduce that if the birthrate exceeds the deathrate, there is a probability $\Psi(\mu/\lambda)$ that the population will ultimately become extinct; but if that extinction does not occur, the population will become arbitrarily large. The latter deduction follows by observing that if K is any large number, then $P\{1 \leqslant N(t) \leqslant K\}$ tends to zero as $t \to \infty$. In a similar way we see that the second result implies that if the deathrate exceeds the birthrate then the ultimate extinction has a probability 1 of occurring.

A possible modification of this pure birth-and-death process is to suppose that at a birth, instead of there being exactly one "offspring," there appears a random number ξ, say, of offspring; one may assume that the probabilities

$$\Pr[\xi = r] = g_r(t) \qquad (r = 0, 1, 2, \ldots)$$

depend on the time t at which the births occur. If one introduces the PGF

$$\Gamma(z,t) = \sum_{r=0}^{\infty} z^r g_r(t),$$

it can be shown that (6) is to be replaced by

$$\frac{\partial \pi}{\partial t} = \{\lambda[\Gamma(z,t) - z] + \mu[z - 1]\} \frac{\partial \pi}{\partial z}.$$

The tractability of this equation depends very much on assumptions made about $\Gamma(z,t)$. We refer to the references cited at the end of this article for further details.

Another possible, and interesting, modification of the pure birth-and-death model arises if we suppose λ and μ to be dependent on t, the time. This yields the so-called *nonhomogeneous* pure birth-and-death process. It is governed by an equation identical to (6) except that λ and μ should now become $\lambda(t)$ and $\mu(t)$. This model was fully explored by Kendall [4], who showed, for instance, that if $\Pi(z,0) = z$ (corresponding to exactly *one* initial member of the population), then

$$\Pi(z,t) = 1 + \frac{1}{e^{\rho}(z-1)^{-1} - \int_0^t \lambda e^{\rho}\, dt}, \quad (8)$$

where

$$\rho \equiv \rho(t) \equiv \int_0^t (\mu - \lambda)\, dt.$$

BIRTH–DEATH–IMMIGRATION PROCESS

This process is similar to (a) except that (in addition to the births and deaths) in every small time interval there is an independent probability $\nu\delta t + o(\delta t)$, where $\nu > 0$ is the constant "immigration" rate, of an addition to the population from outside the system being studied. The differential equations (2) can again be summed up in a single equation, such as (6), involving the PGF:

$$\frac{\partial \Pi}{\partial t} = (\lambda z - \mu)(z-1)\frac{\partial \Pi}{\partial z} + \nu(z-1)\Pi. \quad (9)$$

This is another partial differential equation of the Lagrange kind mentioned earlier and, once again, textbook methods are adequate and yield the solution (for $\lambda \neq \mu$):

$$\Pi(z,t) = A(z,t)B(z,t), \quad (10)$$

where $B(z,t)$ is the function displayed in (7) as the PGF associated with the pure-birth process and involves $\Psi(z)$, the assumed initial PGF, while $A(z,t)$ is given by

$$A(z,t) = \left[1 - \frac{\lambda(z-1)(1-e^{(\lambda-\mu)t})}{(\mu - \lambda)}\right]^{-\nu/\lambda}. \quad (11)$$

Thus $A(z,t)$ is the PGF of $N_I(t)$, say, that part of the total population $N(t)$ that has descended from immigrants. As is clear from the considerable amount of independence assumed in our model, if we write $N_0(t)$ for those in the population at time t who have descended from the original population, then $N(t) = N_0(t) + N_I(t)$, where, as (10) shows, N_0 and N_I are independent.

If $\lambda < \mu$, we see that

$$\Pi(z,t) \to \left(\frac{\mu}{\mu - \lambda} - \frac{\lambda z}{\mu - \lambda}\right)^{-\nu/\lambda}.$$

The PGF on the right-hand side of this result is that of a negative-binomial distribution*. Thus we see that, with the passage of time, the distribution of $N(t)$ approaches more and more closely a particular negative-binomial distribution, whose parameters depend on μ, λ, and ν.

IMMIGRATION–DEATH PROCESS

This model arises in several fields of application; it is really a special case of the birth–death–immigration process with $\lambda = 0$. It is possible to deduce the appropriate PGF by letting $\lambda \to 0$ in (10), (7), and (11). One finds that

$$\Pi(z,t) = e^{(\nu/\mu)(z-1)(1-e^{\mu t})} \cdot \Psi(1 + [z-1]e^{-\mu t}), \quad (12)$$

where, as before, we write $\Psi(z)$ for the initial $\Pi(z,0)$, assumed known. Notice that if the population contains, initially, zero particles [so that $\Psi(z) \equiv 1$], then $\Pi(z,t)$ has the form of the PGF of a Poisson* distribution; thus, in this case, $N(t)$ always has a familiar Poisson distribution with mean value

$$\mathcal{E}N(t) = \frac{\nu}{\mu}(1 - e^{\mu t}). \quad (13)$$

In any case, whatever $\Psi(z)$ may be, as $t \to \infty$

$$\Pi(z,t) \to e^{(\nu/\mu)(z-1)}.$$

The limit function is the PGF of a Poisson distribution with mean ν/μ. Thus for large t, the distribution of $N(t)$ will always be close to a Poisson distribution (with mean value ν/μ).

SIMPLE QUEUE PROCESS

This model would seem, from its specifications, to be the simplest of all possible birth-

and-death processes. It stems from the basic assumptions

$$\alpha_n = \lambda > 0 \qquad \text{for } n = 0, 1, 2, \ldots$$
$$\beta_n = \mu > 0 \qquad \text{for } n = 1, 2, 3, \ldots .$$

The corresponding equation for $\Pi(z, t)$ here becomes

$$\frac{\partial \Pi}{\partial t} = \left(\lambda z - \lambda - \mu + \frac{\mu}{z}\right)\Pi + \mu\left(1 - \frac{1}{z}\right)P_0(t). \qquad (14)$$

The partial differential equations previously encountered for $\Pi(z, t)$ have involved no unknown function other than $\Pi(z, t)$ itself and they have thus been amenable to conventional textbook attack. The present equation (14) raises an awkward difficulty in that it involves the unknown $P_0(t)$ as well as $\Pi(z, t)$. Special methods are needed to solve this problem; we cannot go into them here (but see the article on queueing theory*, which will also explain the choice of name for the present special birth-and-death process).

We here end our discussion of certain special birth-and-death processes and turn to some related matters. The useful notion of the *characteristic functional* has been discussed by several authors, but the most useful results are due to Bartlett and Kendall [2]. For a population initially of unit size, we may define it as the expectation

$$\mathcal{C}\left[\omega, \theta(x); t\right] = \mathcal{E}\left\{\omega^l \exp\left[i \sum_t \theta(x)\right]\right\},$$

Where l is a "marker" variable equal to unity throughout the life of the initial ancestor and zero after his or her death, and the summation $\sum_t$ is over the ages of all individuals (other than the initial ancestor) alive at t (*see* CHARACTERISTIC FUNCTIONS). The function $\theta(x)$ is purely arbitrary and by skillful choices of $\theta(x)$ one can learn a great deal about the development of the birth-and-death process. It is gratifying that for many specific birth-and-death processes the characteristic functional can be explicitly determined. For the pure birth-and-death process, Kendall [5] showed that

$$\mathcal{C}[\omega, \theta(x); t] = 1 + \left\{(\omega - 1)e^{-\mu t} + \int_0^t (e^{i\theta(x)} - 1)g(x, t)\,dx\right\} \times \left\{1 - \int_0^t (e^{i\theta(x)} - 1)h(x, t)\,dx\right\}^{-1}$$

where

$$g(x, t) = \lambda e^{-\mu x} e^{(\lambda - \mu)(t - x)}$$

and

$$h(x, t) = \frac{\lambda}{\lambda - \mu} e^{-\mu x}\left[\lambda e^{(\lambda - \mu)(t - x)} - \mu\right].$$

No discussion, however short, of birth-and-death processes would be complete without some reference to the original and profound approach to the general problem adopted by Karlin and McGregor [3], who relate the task of solving equations [2] to the solution of a Stieltjes moment problem*. Their methods require extensive discussion for a complete exposition and we must settle here for the slightest of sketches. They develop a sequence of polynomials $Q_n(x)$, $n = 0, 1, \ldots,$ by taking $Q_0(x) \equiv 1$ and then requiring

$$-xQ_0(x) = -(\alpha_0 + \beta_0)\,Q_0(x) + \alpha_0\,Q_1(x),$$

while, for $n \geqslant 1$,

$$-xQ_n(x) = \alpha_n Q_{n-1}(x) - (\alpha_n + \beta_n)\,Q_n(x) + \alpha_n Q_{n+1}(x).$$

They define $\pi_0 = 1$ and, for $n \geqslant 1$,

$$\pi_n = \frac{\alpha_0 \alpha_1 \cdots \alpha_{n-1}}{\beta_1 \beta_2 \cdots \beta_n}.$$

They are then able to show the existence of a weight function $\Psi(x)$ on $[0, \infty)$ such that the transition probabilities $P_{ij}(t)$ of the general birth-and-death process are given by equations

$$P_{ij}(t) = \pi_j \int_0^\infty e^{-xt} Q_i(x) Q_j(x)\,d\Psi(x), \qquad (15)$$

where

$$\int_0^\infty \{Q_j(x)\}^2\,d\Psi(x) = \frac{1}{\pi_j}.$$

Remarkably, for many specific birth-and-death processes of practical interest, the weight function $\Psi(x)$ has been explicitly determined and general formulas derived for the Q-polynomials. The integral representations (15) then give considerable insight into the temporal development of the process.

There are many possible generalizations of birth-and-death processes that we have no room to discuss in this, necessarily brief, general article. One important inquiry allows the particles comprising the population to be of various "sexes" and the birthrate to depend in some way on the relative numbers of particles of different sexes. Another inquiry is concerned with "variable generation time"; a parent particle's ability to spawn an offspring depends on the age of the parent. A birth-and-death process in which overcrowding has a deterrent effect on births is the *logistic* process*; this is notoriously difficult to treat analytically and merits a separate article.

References

[1] Bartlett, M. S. (1955). *An Introduction to Stochastic Processes*. Cambridge University Press, Cambridge. (This book gives a brief and authoritative discussion of, particularly, those aspects of birth-and-death processes on which Bartlett did pioneering work. However, the style of the book is taciturn and some readers may find it difficult.)

[2] Bartlett, M. S. and Kendall, D. G. (1951). *Proc. Camb. Philos. Soc.*, **47**, 65–76.

[3] Karlin, S. and McGregor, J. L. (1957). *Trans. Amer. Math. Soc.*, **85**, 489–546.

[4] Kendall, D. G. (1948). *Ann. Math. Statist.*, **19**, 1–15.

[5] Kendall, D. G. (1950). *J. R. Statist. Soc. B*, **12**, 278–285.

Bibliography

See the following works, as well as the references just given, for more information on the topic of birth and death processes.

Bharucha-Reid, A. T. (1960). *Elements of the Theory of Markov Processes and Their Applications*. McGraw-Hill, New York.

Bhat, U. N. (1972). *Elements of Applied Stochastic Processes*. Wiley, New York. (See, especially, Chap. 7.)

Cox, D. R. and Miller, M. D. (1956). *The Theory of Stochastic Processes*. Wiley, New York.

Hoel, P. G., Port, S. C., and Stone, C. J. (1972). *An Introduction to Stochastic Processes*. Houghton Mifflin, Boston. (Chapter 3 is an accessible and clear introduction to the sort of random processes discussed in this article.)

Karlin, S. and Taylor, H. M. (1975). *A First Course in Stochastic Processes*, 2nd ed. Academic Press, New York. (Chapter 4 is a good introductory account containing a wealth of examples in applied fields.)

Moyal, J. E., Bartlett, M. S., and Kendall, D. G. (1949). *J. R. Statist. Soc. B*, **11**, 150–282. (A most valuable source of information touching on many of the points raised in this entry.)

(APPLIED PROBABILITY
BRANCHING PROCESSES
GALTON–WATSON PROCESS
MARKOV PROCESSES
POPULATION GROWTH MODELS
QUEUEING THEORY)

WALTER L. SMITH

BISERIAL CORRELATION

Biserial correlation refers to an association (*see* ASSOCIATION, MEASURES OF) between a random variable X which takes on only two values (for convenience 0 and 1), and a random variable Y measured on a continuum. Choice of a parameter to measure such association, and a statistic to estimate and test the parameter, depend on the conceptualization of the nature of the (X, Y)-population. Common forms are the point biserial, the biserial, forms that here are termed the modified biserial and the rank biserial correlation coefficient.

For example, in the Appendix are data representing a sample used to test the hypothesis that expectant mothers with extreme lengths of labor (either very short or very long) are more likely to be administered analgesia during labor. In this case X refers to use or nonuse of analgesia, Y is the absolute deviation* of length of labor from 6 hours (median labor).

POINT BISERIAL CORRELATION

The point biserial correlation coefficient is probably the earliest statistical approach to this problem because of its close relationship both to the product-moment correlation coefficient* and to the two-sample t-test*.

If it is assumed that the distributions of Y, conditional on $X = 0$ and 1, are normal with different means but with a common variance, the product moment correlation coefficient* between X and Y (ρ_{pb}) is estimated by the *point* biserial correlation coefficient

$$r_{pb} = (pq)^{1/2}(\bar{y}_1 - \bar{y}_0)/s_y,$$

where $(x_1, y_1), (x_2, y_2), \ldots, (x_n, y_n)$ is a sample from the (X, Y) population, $\bar{y}_1$ and $\bar{y}_0$ are the mean y-values of observations having $x_i = 1$ and $x_i = 0$, respectively; s_y^2 is the sample variance of Y; and p is the proportion of the X-sample with $x_i = 1$, $(q = 1 - p)$.

The statistic t may be used to test the null hypothesis* that $\rho_{pb} = 0$, where

$$t = (n-2)^{1/2} r_{pb}\left(1 - r_{pb}^2\right)^{-1/2},$$

and t is distributed as Student's t* with $n - 2$ degrees of freedom. This test is equivalent to a two-sample t-test of the null hypothesis that the mean of Y values with $X = 1$ equals that with $X = 0$. When $\rho_{pb} \neq 0$, the statistic t, where

$$t = (n-2)^{1/2}(r_{pb} - \rho_{pb}) \cdot \left[\left(1 - r_{pb}^2\right)\left(1 - \rho_{pb}^2\right)\right]^{-1/2}$$

is asymptotically normally distributed (*see* ASYMPTOTIC NORMALITY) with mean zero and variance

$$1 + \rho_{pb}^2 \lambda/4,$$

where λ is the kurtosis* of the X-distribution:

$$\lambda = (1 - 6PQ)/4PQ$$

and $P = 1 - Q = \Pr[X = 1]$.

This distribution theory is generally robust with respect to departures from the assumption of the normal conditional distributions* of Y, but sensitive to departures from the assumption of equal variances.

In the numerical example presented in the Appendix, the mean Y-value for subjects with $X = 1$ was 5.31 hours, with $X = 0$ was 2.48 hours, and $s_y = 3.86$ hours. Of the 34 subjects, a proportion $p = 0.382$ had $X = 1$. Hence

$$r_{pb} = (0.382 \times 0.618)^{1/2}(5.31 - 2.48)/3.86 = 0.36$$

$$t_{32} = (32)^{1/2} \times 0.36(1 - 0.36^2)^{1/2} = 2.2 \qquad (p < 0.05).$$

Statistical papers related to the use of r_{pb} appear in the same era (1900–1920) as those introducing the biserial correlation. Development of the latter approach was motivated by the fact that underlying the dichotomy* X there often exists an unobservable random variable Z measured on a continuum, with $X = 1$ if Z exceeds some unknown threshold value and $X = 0$ otherwise. In the illustration, for example, Z might represent the physician's perception of the difficulty of labor. There is no loss in generality entailed in assuming that $E(Z) = 0$ and $\text{var}(Z) = 1$. The parameter of interest in this approach is some correlation coefficient between Z and Y, the estimation of which is based on the sample from the (X, Y)-population.

BISERIAL CORRELATION

If (Z, Y) is assumed to have a bivariate normal distribution* with correlation coefficient ρ_b, a consistent estimator* of ρ_b based on the sample from the (X, Y) population is the biserial correlation coefficient r_b:

$$r_b = pq(\bar{y}_1 - \bar{y}_0)/(s_y u),$$

where

$$u = (2\pi)^{-1/2} e^{-h^2/2}$$

and h is defined by

$$\Pr[Z \geqslant h] = p,$$

with Z as a standard normal variate.

The statistic $z(2r_b/\sqrt{5})$, where z is Fisher's z-transformation* ($z(r) = 0.5\ln[(1+r)/(1-r)]$) is asymptotically normally distributed with mean $z(2\rho_b/\sqrt{5})$ and variance $5/(4n)$ when ρ_b is near zero.

The distribution theory of r_b is sensitive to departures from the assumption of bivariate normality* of (Z, Y). Under such circumstances, the magnitude of $|r_b|$ is often found to exceed 1, although by definition ρ_b cannot exceed 1. Even when the assumptions are met, the asymptotic efficiency of $|r_b|$ as an estimator of ρ_b approaches zero as $|\rho_b|$ approaches 1.

In the illustrative data, since $p = 0.382$, $h = 0.300$, and $u = 0.381$. Thus

$$r_b = (0.382 \times 0.618)(5.31 - 2.48)/\{(3.86)(0.381)\}$$
$$= 0.45.$$

Although both point biserial and biserial correlation coefficients were known and used in 1920, knowledge of the nonnull distribution theory and general statistical properties of these statistics were not developed until the 1950s with Robert F. Tate's work [4, 5]. In the long interim period (1920–1950), there were a number of papers dealing with computation methods. This signaled that, even lacking theoretical base, these approaches were being utilized. The awakening of interest in theoretical development in 1950–1960, however, is marked by an unfortunate series of papers that confused the conditional distribution* theory of r_{pb} with its unconditional distribution. On this basis procedures were evolved and published for estimating and testing point biserial correlation coefficients.

Tate's study corrected these problems with r_{pb}, and in addition suggested the lack of robustness* of r_b and stimulated interest in more robust approaches. The first of these was actually introduced by Brogden in 1949 [1], but further interest does not appear until the late 1950s.

MODIFIED BISERIAL CORRELATION

If one weakens the foregoing assumptions by assuming only a monotonic relationship between Y and Z, then the product moment correlation coefficient between Y and Z is ρ_{mb}, where

$$\rho_{mb} = \frac{E(\bar{y}_1) - E(\bar{y}_0)}{\sigma_Y[E(Z \mid Z \geqslant h) - E(Z \mid Z < h)]},$$

where σ_Y^2 is the population variance of Y. A consistent* estimator would be of the form

$$r_{mb} = (\bar{y}_1 - \bar{y}_0)/D.$$

There have been several suggestions as to the choice of the statistic D. The one suggested by Brogden [1] requires that the Y-sample be rank-ordered:

$$y_1 > y_2 > y_3 > \cdots > y_n,$$

with

$$D_m = \left(\sum_{i=1}^{m} y_i/m\right) - \left(\sum_{i=m+1}^{n} y_i/(n-m)\right),$$

where m is an integer between 1 and the sample size n. If $\bar{y}_1 \geqslant \bar{y}_0$, let $D = D_{np}$, and if $\bar{y}_1 < \bar{y}_0$, let $D = D_{nq}$. Such a choice guarantees that if $|\rho_{mb}| = 1$, its estimator r_{mb} will equal ρ_{mb} with probability 1. Empirical study indicates that r_{mb} is preferable to r_b for estimation of ρ_b when $|\rho_b|$ is large, even if the more stringent assumptions underlying the definition of the biserial correlation coefficient, ρ_b, are met.

In the illustrative data, $np = 13$ and

$$D_m = \text{(average of 13 most extreme labors)}$$
$$- \text{(average of 21 least extreme labors)}$$
$$= 7.08 - 1.38 = 5.70.$$

Thus $r_{mb} = (5.31 - 2.48)/5.70 = 0.50$.

RANK BISERIAL CORRELATION

Another modification of the biserial approach is to focus attention on the *rank* correlation coefficient* between Y and Z, rather than the product moment correlation coefficient between Y and Z (as for r_b) or the product moment correlation coefficient between X and Y (as for r_{pb}). This association would then include not just linear relationships, but any monotonic relationship between Y and Z. The rank correlation coefficient (either Spearman's* or Kendall's* forms) between Y and Z, as estimated from observation of X and Y, is simply

$$r_{rb} = 2(\bar{R}_1 - \bar{R}_0)/n,$$

where $\bar{R}_1$ and $\bar{R}_0$ are the average ranks of the y-values associated with $x_i = 1$ and $x_i = 0$, respectively.

The test of the null hypothesis that $\rho_{rb} = 0$ is equivalent to a Mann–Whitney test* applied to the groups with $x_i = 1$ and $x_i = 0$, respectively.

In the illustrative data, $\bar{R}_1 = 21.15$, $\bar{R}_0 = 13.76$, and

$$r_{rb} = 2(21.15 - 13.76)/34 = 0.43.$$

At present, although several forms of robust or nonparametric (distribution-free*) biserial correlation coefficients have been proposed, little is known of the distribution theory or any of their statistical properties. In particular, little is known of the advantages or disadvantages these forms may have over the classical forms.

In the illustration, the fact that the conditional distributions of Y when $X = 0$ and 1 appear nonnormal, and that the sample variances of these distributions differ widely, would militate against use of r_{pb}. The fact that the marginal distribution of Y is nonnormal warns against the use of r_b. Whether r_{mb} or r_{rb} is preferred remains an open question.

References

[1] Brogden, H. E. (1949). *Psychometrika*, **14**, 169–182.

[2] Cureton, E. E. (1956). *Psychometrika*, **21**, 287–290. (A brief and simple definition with an illustration of its use.)

[3] Lord, F. M. (1963). *Psychometrika*, **28**, 81–85. (Introduces a new form of modified biserial, reviews earlier suggested forms, and presents a brief empirical comparison of one such form with the biserial.)

[4] Tate, R. F. (1954). *Ann. Math. Statist.*, **25**, 603–607.

[5] Tate, R. F. (1955). *Biometrika*, **42**, 205–216. (The level of Tate's two papers requires a degree of mathematical facility. However, the exposition is clear and the results are basic to an understanding of the biserial and point biserial coefficients.)

(ASSOCIATION, MEASURES OF
CORRELATION
KENDALL'S τ
SPEARMAN RANK CORRELATION COEFFICIENT

APPENDIX

X (Analgesia)[a]	Rank	Y |Labor-6|	X (Analgesia)[a]	Rank	Y |Labor-6|
Y	33	14.8	Y	16	2.3
N	32	13.8	Y	15	2.1
Y	31	12.4	N	13.5	1.7
Y	30	10.1	N	13.5	1.7
Y	29	7.1	N	12	1.5
Y	28	6.1	N	10.5	1.3
N	27	5.8	N	10.5	1.3
Y	26	4.6	N	8.5	1.2
N	25	4.3	N	8.5	1.2
N	24	3.5	N	7	1.1
N	23	3.3	Y	6	0.8
Y	22	3.2	N	5	0.7
Y	21	3.0	N	4	0.6
N	19.5	2.8	N	3	0.5
N	19.5	2.8	N	2.5	0.2
N	18	2.5	N	2.5	0.2
Y	17	2.4	Y	1	0.1

[a]Y, yes; N, no.

TETRACHORIC CORRELATION COEFFICIENT)

HELENA CHMURA KRAEMER

BIT (BINARY DIGIT) *See* ENTROPY

BITTERLICH'S ANGLE-COUNT METHOD

A technique for estimating the proportion of the area of a forest that is actually covered by the bases of trees. An observer goes to each of a number of points in the forest, chosen either randomly or systematically and counts the number of trees that subtend at that point an angle greater than some fixed critical angle 2α. (This can be done either with a simple rod and crosspiece or with a special instrument known as the relascope.) The major drawback of this method is that a tree may be missed by an observer if it falls behind another one. Holgate [2] presents a statistical analysis of this estimator, derives its sampling distribution both under random and regular spatial distribution of the trees, and compares it with other more conventional methods based on *sample plots**.

For additional information, see refs. 1 and 3.

References

[1] Bitterlich, W. (1948). *Allg. Forst Holzwirtsch. Zeit.*, **59**, 4–5.

[2] Holgate, P. (1967). *Biometrika*, **54**, 615–623.

[3] Kuusela, K. (1966). *Commun. Inst. For. Fenn.*, **61**(2), 32 pp. [This survey paper (in a relatively obscure publication) contains a detailed and lucid summary of the method.]

(FORESTRY, STATISTICS IN)

BIVARIATE DISTRIBUTION *See* MULTIVARIATE DISTRIBUTION

BIVARIATE NORMAL DISTRIBUTION

This is a special case of the multinormal distribution*, for two variables. It has a position of considerable historical importance (*see* F. Galton*).

The *standard* bivariate normal distribution is a joint distribution of two standardized normal variables* X_1, X_2 with correlation coefficient* ρ. It has density function

$$\left\{2\pi\sqrt{(1-\rho^2)}\right\}^{-1}\exp\left[-\tfrac{1}{2}(1-\rho^2)^{-1}\times\left(x_1^2-2\rho x_1x_2+x_2^2\right)\right]. \quad (1)$$

The regression* of either variable on the other is linear; in fact, the distribution of X_2, given $X_1 = x_1$, is normal with expected value ρx_1 and standard deviation $\sqrt{(1-\rho^2)}$—a similar result holds for the distribution of X_1, given $X_2 = x_2$.

The density is constant on the ellipse

$$x_1^2 - 2\rho x_1x_2 + x_2^2 = K \qquad \text{(a constant)}. \quad (2)$$

The distribution of $(X_1^2 - 2\rho X_1X_2 + X_2^2)(1-\rho^2)^{-1}$ is chi-square* with 2 degrees of freedom (which is a special case of the exponential* distribution), so the integral of (1) over the interior of the ellipse (2) is

$$\Pr\left[\chi_2^2 \leqslant K(1-\rho^2)\right] = 1 - \exp\left[-\tfrac{1}{2}K(1-\rho^2)\right].$$

Extensive tables of integrals of (1) over various regions of the (x_1, x_2) space have been published (*see* Johnson and Kotz [1]). These include values of

(a) $\Pr[(X_1 > h) \cap (X_2 > k)] = L(h,k;\rho)$.

(b) $V(h,k) = \Pr[(0 < X_1 < h) \cap (0 < X_2 < k) \cap (X_1 > X_2)]$.
[This is the integral of (1) over triangle *OPH* in Fig. 1.]

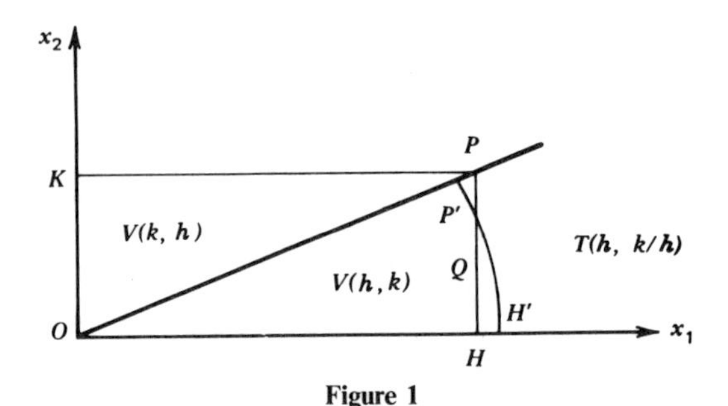

Figure 1

(c) $\Pr[\sigma_1^2 X_1^2 + \sigma_2^2 X_2^2 < K]$.

(d) $\Pr[(X_1 + \xi_2)^2 + (X_2 + \xi_2)^2 < K]$.

The *general* bivariate normal distribution is obtained by the transformation

$$Y_i = \eta_i + \sigma_i X_i \qquad (i = 1, 2; \sigma_i > 0).$$

The variables Y_1, Y_2 are each normally distributed; their correlation coefficient is ρ, and $E[Y_i] = \eta_i$; $\text{var}(Y_i) = \sigma_i^2 (i = 1, 2)$.

Relations such as

$$\begin{aligned}\Pr[(Y_1 > a_1) \cap (Y_2 > a_2)] &= \Pr\left[\left(X_1 > \frac{a_1 - \eta_1}{\sigma_1}\right) \cap \left(X_2 > \frac{a_2 - \eta_2}{\sigma_2}\right)\right] \\ &= L\left(\frac{a_1 - \eta_1}{\sigma_1}, \frac{a_2 - \eta_2}{\sigma_2}; \rho\right)\end{aligned}$$

make it possible to use tables of the standard bivariate normal in evaluating probabilities for general bivariate normal distributions.

Two normally distributed variables do not necessarily have a joint bivariate normal distribution (*see*, e.g., FARLIE–GUMBEL–MORGENSTERN DISTRIBUTIONS). However, there are several other sets of conditions that do have this consequence (*see* CHARACTERIZATION OF DISTRIBUTIONS).

Reference

[1] Johnson, N. L. and Kotz, S. (1972). *Distributions in Statistics: Continuous Multivariate Distributions*, Wiley, New York, Chap. 36.

(BISERIAL CORRELATION
LIMIT THEOREMS, CENTRAL
MULTINORMAL DISTRIBUTION
REGRESSION, LINEAR
TETRACHORIC CORRELATION)

BIVARIATE RANK SUM TEST (W)

A test—introduced by Chatterjee and Sen [2]—of identity of two bivariate distributions on two sets of pairs of sample values $(X_{1j}, X_{2j})(j = 1, \ldots, n_1)$—the first sample—and $(X_{1,n_1+j}, X_{2,n_1+j})(j = 1, \ldots, n_2)$—the second sample. The test is carried out by:

1. Ranking all $(n_1 + n_2)$ X_1's and all $(n_1 + n_2)$ X_2's separately—denote the ranks of X_{1j}, X_{2j} by

   R_{1j}, R_{2j} for $j = 1, \ldots, n_1$ for the first sample,

   R_{1,n_1+j}, R_{2,n_1+j} for $j = 1, \ldots, n_2$ for the second sample.

2. Calculating

   $$W = (1 - q^2)^{-1}(W_1^2 - 2qW_1W_2 + W_2^2)$$

   where

   $$W_h = \sum_{j=1}^{n_1} R_{hj} - \frac{1}{2} m(n_1 + n_2) \quad (h = 1, 2)$$

   (*see* WILCOXON TEST*) and

   $$q = 12\left[(n_1 + n_2)^3 - n_1 - n_2\right]^{-1} \times \sum_{j=1}^{n_1} R_{1j}\left[R_{2j} - \frac{1}{2}(n_1 + n_2 + 1)\right]$$

   (*see* RANK CORRELATION*) *for each possible partitioning* of (R_{1j}, R_{2j}) into sets of n_1 and n_2.

3. The resultant $\binom{n_1+n_2}{n_1}$ values of W constitute a randomization distribution* against which the observed value can be judged. (Bhattacharyya et al. [1] compare this test with Hotelling's T^2*.)

References

[1] Bhattacharyya, G. K., Johnson, R. L., and Neave, H. R. (1971). *Technometrics*, **13**, 191–198.

[2] Chatterjee, S. K. and Sen, P. K. (1964). *Calcutta Statist. Ass. Bull.*, **13**, 18–58.

(DISTRIBUTION-FREE METHODS
PERMUTATION TESTS
RANDOMIZATION MODEL
RANK SUM TESTS)

BLACK BOX

A term borrowed from physics, and introduced into statistical theory by D. A. S. Fraser [1] to describe features common to most techniques of statistical inference*. The following quotations from Fraser [2] describe the concept. "The black box has an input variable θ receiving any value in Ω, it has an output variable x producing values in X, and it has behavioral characteristics $f(x; \theta)$ describing the frequency distribution* of output values x for any input θ. . . . " "The black box has *complete behavioral characteristics*—provided θ is treated as output." *But* "an investigator . . . needs the black box with the opposite direction of throughput; in by means of x, out by means of θ."

"The decision theory model introduces an additional black box . . . The Bayesian method also introduces an additional black box."

In Fraser [2] the distinction between the two additional black boxes is explained.

References

[1] Fraser, D. A. S. (1968). *The Structure of Inference*. Wiley, New York.

[2] Fraser, D. A. S. (1968). *Technometrics*, **10**, 219–229. (A concise description of the concept.)

(BAYESIAN INFERENCE
DECISION THEORY
ESTIMATION
HYPOTHESIS TESTING
PRIOR PROBABILITY)

BLACKJACK

Blackjack or "twenty-one" is a popular casino card game attracting some 24 million gamblers wagering over $15 billion annually. Its origin is unknown; however, the most recent version can be traced to the French game vingt-et-un. The game was introduced to the United States as "twenty one" in the early 1900s, but the name "blackjack" did not evolve until 1912, when a gambling house in Evansville, Indiana, offered a bonus if the first two cards dealt were the ace of spades and either of the black jacks. The name persisted as the game spread throughout the United States in the 1920s. (See Scarne [7] for a complete historical background.) The game continues to gain in popularity. Las Vegas casinos in the fiscal year 1977–1978 had at least 3,670 tables in operation. These tables generated a median win per unit ranging from $106,720 for casinos with $2 to 5 million total gaming revenue to $321,720 for casinos with an excess of $20 million. These and similar statistics are available in the *Nevada Gaming Abstract*, which may be obtained by writing to the State Gaming Control Board, Las Vegas, Nevada.

Under casino rules, the dealer has a rigid line of play. With 16 or fewer points, the dealer *must* draw. With 17 or more, the dealer *must* stand (not draw). An ace counts 11 points unless the total count exceeds 21, in which case the ace counts 1.

Bets are made, then two cards are dealt to each player, including the dealer. Players exercise their options, basically to draw or stand, starting with the player to the immediate left of the dealer. If a player goes bust (exceeds 21), he or she loses regardless of the dealer's actions. This rule puts the odds in the casino's favor because both player and dealer simultaneously go bust about 7% of the time.

The player's odds are somewhat improved because the dealer pays out a 3 : 2 ratio when the player has "blackjack," an ace and a 10-point card as the first two cards dealt. An exception occurs when the dealer also has "blackjack"; in this case no money exchanges hands. Baldwin et al. [1] were the first to publish an analytical strategy for the player. The use of their method increases the player's probability of winning to 49.5%. It was E. O. Thorp's best-selling book *Beat the Dealer* [8] that terrified casino owners along the Las Vegas strip and brought joy to studious gamblers. As a result of Thorp's classic, many casinos shifted from a single deck to a four-deck "shoe" or resorted to shuffling a single deck after each hand. Many players used Thorp's counting methods to improve their play. In addition, Thorp's work stimulated the creation of new computer evaluations to give more accurate estimates of probabilities associated with the "correct" line of play. Analytical computations are too cumbersome to use because of the myriad of combinations involved. A more formal development, including some analytical computations, can be found in Griffin [3]. Julian Braun [6, 8] has devised some useful computer algorithms, and Manson et al. [4] have made successful computer applications for the four-deck game to provide further sources of favorable strategy. On the other hand, the literature abounds with misleading schemes. (See Wilson [10] for an informative list.)

Casino operators turn away customers who use Thorp's methods and consider "counters" to be "cheaters." Ken Uston, a gambler who has managed to outwit casino owners many times by such means as disguise, had made blackjack a profitable vocation for the past 10 years. According to Uston, casinos are depriving him of a living. He has introduced several court cases [9], which are still pending. The courts will decide whether counting cards in blackjack is legal.

An important concept used by counters is that of a favorable or "rich" deck. A rich deck contains a large number of high-value cards (aces and 10-point cards). When the deck is rich, the counter gains an advantage because he or she can increase the size of a bet and optimally exercise several options with the standard playing procedures. The options include:

1. Doubling down; after two cards are drawn, a bet is doubled and only one more card is dealt.
2. Pair splitting; if the first two cards make a pair whose indices show the same value, the cards may be split into two separate hands.
3. Insurance; if the dealer has an ace showing, the player may wager up to half his or her bet that the dealer has blackjack.

Because few players are familiar with scientific strategies, dealers should expect an approximate 3% advantage; however, returns typically run over 20%. This hefty percentage may be accounted for by both amateur players and the casino atmosphere—in which player concentration is diminished by free drinks from scantily dressed waitresses and by continuous long hours at uninterrupted play. Another possible reason for returns over 20% is the wrong stopping policy used by the majority of "average" players. They stop playing only when they are broke or nearly broke; just a few have the willpower to stop when winning.

References

[1] Baldwin, R. R., Cantey, W. E., Maisel, H., and McDermott, J. P. (1956). *J. Amer. Stat. Ass.*, **51**, 429–439. (For the researcher or the serious player.)

[2] Epstein, R. A. (1977). *The Theory of Gambling and Statistical Logic*. Academic Press, New York. (Excellent at all levels; contains techniques for computing probabilities in blackjack and other games as well.)

[3] Griffin, P. (1979). *The Theory of Blackjack (The Complete Card Counter's Guide to the Casino Game of "21")*. GBC Press (Gambler's Book Club), Las Vegas, Nev. (Thorp's book [8] should be read as a prerequisite for complete understanding. Some new results that can benefit all levels of players.)

[4] Manson, A. R., Barr, A. J., and Goodnight, J. A. (1975). *Amer. Statist.*, **29**(2), 84–88. (Interesting tables that can be used by all level players.)

[5] Morehead, A., Frey, R., and Mott Smith, G. (1964). *The New Complete Hoyle*. Doubleday (Garden City Books), Garden City, NY. (For the beginner who knows little about cards.)

[6] Revere, L. (1973). *Playing Blackjack as a Business*. Paul & Mann, Las Vegas, Nev. (For those interested in computer simulations and for serious players. Contains Braun's computer calculations of exact strategies.)

[7] Scarne, J. (1978). *Scarne's Guide to Casino Gambling*. Simon and Schuster, New York. (Some misleading information concerning player strategy; good for those interested in historical background and basic rules.)

[8] Thorp, E. O. (1966). *Beat the Dealer*. Random House, New York. (Excellent; a classic; readable at all levels. The novice should read ref. 2 first. Contains Braun's computer calculations of exact strategies.)

[9] Uston, K. and Rapoport, R. (1977). *The Big Player*. Holt, Rinehart and Winston, New York. (Strategy; for the serious player.)

[10] Wilson, A. N. (1965). *The Casino Gambler's Guide*. Harper & Row, New York. (Well written and entertaining; a must for all players as well as those interested in computer simulations.)

(GAMBLING, STATISTICS IN
GAMES OF CHANCE
SAMPLING WITHOUT
REPLACEMENT)

A. GOLDMAN

BLOCKING *See* BLOCKS, RANDOMIZED COMPLETE

BLOCKS, BALANCED INCOMPLETE

When R. A. Fisher* introduced randomized complete block designs (*see* BLOCKS, RANDOMIZED COMPLETE, which should be read in conjunction with this discussion), he had in mind experiments such as agricultural field trials, in which block sizes were moderately flexible. In these studies, the requirement that the common block size equal the number of treatments under study created no difficulty for the experimenter. In many areas of research, however, the choice of block size is severely limited. Yates [14] mentions the restrictions on block size in studies where animal litters or monozygotic human twins form the blocks. Scheffé [13, p. 160], citing examples from consumer product testing, alludes to the limitations on block size imposed by an automobile's having only four wheels, or by an individual's wearing only two shoes at a time. He also comments that in certain cases the block size could theoretically be set large enough to accommodate all t treatments of interest, but one might then lose both precision and accuracy in assessing treatment differences because a large block size would contribute to measurement instability. To illustrate this point, consider the taste testing and rating of wines and spirits. Expert tasters find it increasingly difficult to evaluate treatments in a meaningful fashion as the number of tastings per session increases ("palate paralysis," or simply "paralysis" is sometimes used to describe the phenomenon associated in these studies with block sizes that are too large).

Yates [14] perceived the need in many studies for a common block size k that was smaller than the number of treatments t, so he relaxed the randomized complete block design requirement that k equal t. Instead, he imposed the weaker requirement that each *pair* of treatments in a blocked experiment occur together in λ blocks. Designs satisfying this restriction, originally labeled by Yates symmetrical incomplete randomized block arrangements, are more commonly known today as *balanced incomplete block designs*, the need for randomization* being understood. Yates illustrated his new class of designs with a design for six treatments, say a, b, c, d, e, and f, to be studied in 10 blocks of size 3:

1:	$a\ b\ c$	6:	$b\ c\ f$
2:	$a\ b\ d$	7:	$b\ d\ e$
3:	$a\ c\ e$	8:	$b\ e\ f$
4:	$a\ d\ f$	9:	$c\ d\ e$
5:	$a\ e\ f$	10:	$c\ d\ f$

For this design, it is easily verified that $\lambda = 2$.

For the implementation in practice of a balanced incomplete block design, Cochran and Cox [4] list the following steps that require randomization:

1. Randomly assign the blocks of a design plan as given in a reference work to the groupings of experimental units.
2. Randomly assign the letters (or numbers, which some authors substitute for letters) of each block to the experimental units associated with that block in step 1.
3. Randomly assign treatments to the letters or numbers.

The number of blocks b and the number of replications r per treatment, together with t, k, and λ, serve as a set of five integers to index the class of balanced incomplete block designs. Two easily verified restrictions on the class are:

$$rt = bk \quad \text{and} \quad r(k-1) = \lambda(t-1).$$

Not so easily seen is the further restriction (Fisher's inequality*) that $b \geqslant t$ or, equivalently, $k \leqslant r$ [8, p. 220].

For a given k and t, with $k < t$, there always exists a balanced incomplete block design; it may be formed by listing all subsets of the numbers 1 to t that are of size k. This subclass of balanced incomplete block designs is referred to as the unreduced class, and is described further by the following equations:

$$b = \binom{t}{k}, \quad r = \binom{t-1}{k-1}, \quad \text{and}$$

$$\lambda = \binom{t-2}{k-2}.$$

Unfortunately, even for moderate values of k and t, these unreduced designs frequently require too many blocks for most studies; e.g., for $t = 8$ and $k = 4$, one would need 70 blocks and a total of 280 experimental units.

The combinatorial problems* encountered in studying the class of balanced incomplete block designs have long challenged both statisticians and mathematicians. A detailed discussion of this research on design existence and construction for given (t, k, b, r, λ) is found in John [8, Sec. 13]. From the standpoint of practice, the important results of this research are extensive catalogs of balanced incomplete block designs, such as those found in Fisher and Yates [7, Tables XVIII and XIX] and Cochran and Cox [4, pp. 469–482].

The parametric model commonly adopted for the analysis of data from a balanced incomplete block design is identifical in form to the model assumed for a randomized complete block design*:

$$y_{ij} = \mu + \tau_i + \beta_j + \epsilon_{ij}. \tag{1}$$

Here y_{ij} is the response observed if treatment i has been applied to a randomly chosen experimental unit in block j, μ is an overall mean, τ_i is the contribution of the ith treatment, β_j is the contribution of the jth block and ϵ_{ij} is an observational error. The $\{\epsilon_{ij}\}$ are assumed to be independent, each with mean zero and variance σ^2.

In contrast to the analysis of randomized complete block designs, there is a difference in analysis for balanced incomplete block designs depending upon whether blocks are considered to be fixed or random. Cochran and Cox [4, p. 383] comment that the random assumptions made for block contributions are very reasonable in certain contexts, but that these "assumptions cannot be taken for granted, and with certain types of data we might not wish to make them." Whether block contributions are assumed to be fixed or random, the corresponding analysis of a balanced incomplete block design is more complex than is the analysis for a randomized complete block design. This increased complexity is attributable to the lack of orthogonality* of treatments and blocks.

For the case of fixed block effects, the least-squares* normal equations* are of the form

$$rt\hat{\mu} + r \sum_{i=1}^{t} \hat{\tau}_i + k \sum_{j=1}^{b} \hat{\beta}_j = S$$

$$r\hat{\mu} + r\hat{\tau}_i + \sum_{j=1}^{b} n_{ij}\hat{\beta}_j = T_i$$

$$k\hat{\mu} + \sum_{i=1}^{t} n_{ij}\hat{\tau}_i + k\hat{\beta}_j = B_j,$$

where n_{ij} takes the value 1 if treatment i occurs in block j and is otherwise zero, S is the sum of all observed responses, T_i is the sum of the responses for experimental units administered treatment i, and B_j is the sum of responses for experimental units in block j. As with randomized complete block designs, the $\{\tau_i\}$ in (1) are not estimable (*see* ESTIMABILITY); the symmetry requirement for balanced incomplete block designs, however, results in simplified normal equations. Indeed, the desire for simplicity of analysis in the precomputer era was one argument in favor of balanced incomplete block designs vis-á-vis alternatives lacking in balance. Today, it is known [10] that a balanced incomplete block design is optimal under a broad range of optimality criteria*.

For any contrast in the $\{\tau_i\}$, say

$$L = \sum_{i=1}^{t} c_i \tau_i, \tag{2}$$

with $\sum c_i = 0$, its least-squares estimator is given by

$$T = \sum_{i=1}^{t} c_i Q_i / (\lambda t), \tag{3}$$

where

$$Q_i = kT_i - \sum_{j=1}^{b} n_{ij} B_j. \qquad (4)$$

If I_i is the set of block indices for blocks containing treatment i, then (4) may be rewritten as

$$Q_i = \sum_{j \in I_i} (ky_{ij} - B_j). \qquad (5)$$

The quantity in parentheses in (5) is a contrast of the responses within the jth block. Consequently, Q_i is obtained from intrablock* information, and T is an intrablock estimator. Being a least-squares estimator, T is unbiased* for L; its variance is

$$\text{var}(T) = \{k/(\lambda t)\}\left(\sum_{i=1}^{t} c_i^2\right)\sigma^2.$$

For example, the variance of the estimate of a simple difference $(\tau_i - \tau_j)$, $i \neq j$, is $2k\sigma^2/(\lambda t)$, irrespective of i and j.

It is easy to verify that

$$\sum_{i=1}^{t} Q_i = 0.$$

Hence the least-squares estimate in (3) of the ith treatment effect (relative to the average treatment contribution), i.e., of

$$\tau_i - \left(\sum_{j=1}^{t} \tau_j/t\right),$$

reduces simply to $Q_i/(\lambda t)$ and has variance $k(t-1)\sigma^2/(\lambda t^2)$.

Discussions of balanced incomplete block designs appear in the five texts cited as general references for randomized complete block designs: Box et al. [2], Cochran and Cox [4], Cox [5], John [8], and Kempthorne [9]. Where tables of these designs are presented, an "estimation efficiency factor" E is frequently included. Yates [14] formulated this factor by comparing the variance of a simple treatment difference estimate, i.e., of $\tau_i - \tau_j$, $i \neq j$, in a randomized complete block design to the variance of an estimate of such a difference in a balanced incomplete block design when the two designs have an equal number of treatments and an equal number of observations.

The ratio of these variances is

$$(2\sigma_{\text{RCB}}^2/r)/(2k\sigma_{\text{BIB}}^2/(\lambda t)), \qquad (6)$$

where σ_{RCB}^2 and σ_{BIB}^2 are the intrablock variances that would obtain in the randomized complete block and balanced incomplete block designs, respectively. The efficiency factor E is the value of (6) if $\sigma_{\text{RCB}}^2 = \sigma_{\text{BIB}}^2$, i.e.,

$$E = \lambda t/(kr) = (1 - k^{-1})/(1 - t^{-1}). \qquad (7)$$

Its reciprocal represents the proportionate increase in variance of a simple treatment difference *if no reduction in intrablock variability is achieved* when a randomized complete block design experiment is replaced by a balanced incomplete block design experiment. Since it is intuitively clear that in almost all cases $\sigma_{\text{BIB}}^2 \leqslant \sigma_{\text{RCB}}^2$ because of the smaller block size, E represents a lower bound to the efficiency of balanced incomplete block designs compared to randomized complete block designs. Of course, for cases where one cannot employ blocks of size t, this comparison is meaningless. Nevertheless, one can still use (7) to compare alternative balanced incomplete block designs.

As with point estimation*, the partition of observational variability into block components, treatment differences, and residual variability* is made more complex than for randomized complete block designs by the lack of orthogonality of blocks and treatments. Formulas for the constituents of the analysis of variance* are presented in Table 1, where the notation

$$\bar{y} = S/rt = \sum_{i=1}^{t} \sum_{j \in I_i} y_{ij}/rt$$

is employed.

If, in addition to the earlier error assumptions, the observational errors are assumed to be normally distributed, then the hypothesis of no differences among treatments, i.e.,

$$H_0 : \tau_i = \tau \qquad \text{(all } i\text{)}$$

can be tested by the ratio

$$F = \frac{\text{(treatment mean square)}}{\text{(error mean square)}}.$$

Table 1 Analysis of Variance for a Balanced Incomplete Block Design

Source	Degrees of Freedom	Sum of Squares	Mean Square
Blocks (unadjusted)	$b-1$	$k^{-1}\sum_{j=1}^{b}(B_j - k\bar{y})^2$	$[k(b-1)]\sum_{j=1}^{b}(B_j - k\bar{y})^2$
Treatments (adjusted for blocks)	$t-1$	$(\lambda tk)^{-1}\sum_{i=1}^{t} Q_i^2$	$[\lambda tk(t-1)]^{-1}\sum_{i=1}^{t} Q_i^2$
Residual or error	$rt-t-b+1$	Residual SS obtained by subtraction	(Residual SS)$/(rt-t-b+1)$
Corrected total	$rt-1$	$\sum_i \sum_{j\in I_i}(y_{ij}-\bar{y})^{-2}$	

Under H_0, this statistic has an F-distribution* with $(t-1)$ and $(rt-t-b+1)$ degrees of freedom. Confidence limits* or tests of hypotheses for a contrast* in (2) are based on the result that

$$(T-L)/\Big[\text{error mean square} \cdot (k/(\lambda t)) \cdot \sum c_i^2\Big]^{\frac{1}{2}}$$

has a Student's t-distribution* with $(rt-t-b+1)$ degrees of freedom. As with randomized complete blocks, inferences about more than one contrast raise the issue of multiple comparisons*.

The discussion of an analysis of a balanced incomplete block design, to this point, has been predicated on the assumption that block contributions are fixed unknowns. As mentioned earlier, there are diverse research studies for which it is reasonable to assume that the block contributions are normal random variables, independent of each other and of the observational errors, and that each has mean zero and variance σ_B^2. All other aspects of (1) remain unchanged. Yates [15] formulated this analysis for balanced incomplete block designs with $b > t$, and labeled it an analysis with recovery of interblock information*. One consequence of the new assumption regarding block contributions is that the variance of a difference between two observations from different blocks is $2(\sigma^2 + \sigma_B^2)$, whereas the variance of a difference between two observations in the same block is smaller, equaling $2\sigma^2$. A second and most important consequence of the randomness of block contributions is that

$$E(B_j) = k\mu + \sum_{i=1}^{t} n_{ij}\tau_i. \tag{8}$$

The block totals $\{B_j\}$, which are independent of the intrablock estimates (3) obtained earlier, can be used to obtain an unbiased estimate T^* of L in (2) independent of T:

$$T^* = \sum_{i=1}^{t} c_i \sum_{j=1}^{b} n_{ij}B_j/(r-\lambda). \tag{9}$$

It is these estimates that represent the recovered interblock information. The variance of T^* is

$$\text{var}(T^*) = \left(\sum_{i=1}^{t} c_i^2\right)(k\sigma^2 + k^2\sigma_B^2)/(r-\lambda).$$

Were σ^2 and σ_B^2 known, one would combine T and T^* by weighting them proportionally to the reciprocals of these variances to obtain

$$\tilde{T} = (w_T T + w_{T^*}T^*)/(w_T + w_{T^*}) \tag{10}$$

for $w_T = \lambda t/(k\sigma^2)$ and $w_{T^*} = (r-\lambda)/(k\sigma^2 + k^2\sigma_B^2)$. Whereas (10) was originally proposed by Yates [15] simply as a weighted estimate, Rao [12] later showed that (10) is the maximum likelihood* estimate of L for known σ^2 and σ_B^2.

It is informative to consider the behavior of (10) under two extremes. If σ_B^2 is very large compared to σ^2, i.e., the variability among block contributions is great compared to observational error variability, then $\tilde{T}$ differs little from T. In this case, there is little interblock information to recover. At the other extreme, if σ_B^2 is essentially equal to 0, then $\tilde{T}$ reduces to a contrast in the unadjusted means of the observations from units administered treatment i, $i = 1, \ldots, t$, i.e.,

$$\sum_{i=1}^{t} c_i T_i / r.$$

This is the estimator of L that would be obtained in a completely randomized design* or one-way layout, in which blocking is entirely ignored.

In reality, σ^2 and σ_B^2 are unknown. Standard practice has been to estimate these parameters from the data (see John [8, p. 238] or Kempthorne [9, p. 536], and to replace the parameters in (10) by estimates, without worrying about this substitution. Brown and Cohen [3] have shown that certain weighted combinations of T and T^*, with weights estimated from the data, always have smaller variance than T alone for any balanced incomplete block design, unless $b = 3$.

Two final remarks are in order. First, as with randomized complete block designs, nonparametric (distribution-free*) methods of analysis are available for balanced incomplete block designs (see Durbin [6] or Noether [11, p. 54]. Second, numerous cases exist where no balanced incomplete block design exists that satisfies the research project's constraints on the number of blocks. In that event, it is necessary to sacrifice the symmetry of balanced incomplete block designs in order to reduce the number of blocks. The partially balanced incomplete block* designs of Bose and Nair [1] represent the next level of compromise (see Cochran and Cox [4], Cox [5], and John [8], and Kempthorne [9]).

References

[1] Bose, R. C., and Nair, K. R. (1939). *Sankhyā*, **4**, 337–372.

[2] Box, G. E. P., Hunter, W. G., and Hunter, J. S. (1978). *Statistics for Experimenters*. Wiley, New York.

[3] Brown, L. D. and Cohen, A. (1974). *Ann. Statist.*, **2**, 963–976.

[4] Cochran, W. G. and Cox, G. M. (1957). *Experimental Designs*, 2nd ed. Wiley, New York.

[5] Cox, D. R. (1958). *Planning of Experiments*. Wiley, New York.

[6] Durbin, J. (1951). *Br. J. Statist. Psychol.*, **4**, 85–90.

[7] Fisher, R. A. and Yates, F. (1953). *Statistical Tables for Biological, Agricultural and Medical Research*, 4th. ed. Oliver & Boyd, Edinburgh.

[8] John, P. W. M. (1971). *Statistical Design and Analysis of Experiments*, Macmillan, New York.

[9] Kempthorne, O. (1952). *The Design and Analysis of Experiments*. Wiley, New York.

[10] Kiefer, J. (1959), *J. R. Statist. Soc. B*, **21**, 272–319.

[11] Noether, G. E. (1967). *Elements of Nonparametric Statistics*, Wiley, New York.

[12] Rao, C. R. (1947). *J. Amer. Statist. Ass.*, **42**, 541–561.

[13] Scheffé, H. (1959). *The Analysis of Variance*, Wiley, New York.

[14] Yates, F. (1936). *Ann. Eugen. (Lond.)*, **7**, 121–140.

[15] Yates, F. (1940). *Ann. Eugen. (Lond.)*, **10**, 317–325.

[ANALYSIS OF VARIANCE (ANOVA)
BLOCKS, RANDOMIZED COMPLETE
DESIGN OF EXPERIMENTS
FIXED-EFFECT MODELS
PARTIALLY BALANCED DESIGNS
RANDOM-EFFECT MODELS]

BARRY H. MARGOLIN

BLOCKS, RANDOMIZED COMPLETE

The concept of *blocking** in statistically designed experiments originated during the intensive agricultural field trial research conducted at Rothamsted Experimental Station during Sir R. A. Fisher's* tenure as Chief Statistician. It was during this period that Fisher [4, p. 269] perceived the "peculiarity of agricultural field experiments"—"that the area of ground chosen for the experimental plots may be assumed to be markedly heterogeneous, in that its fertility varies in a systematic, and often a complicated manner from point to point."

Fisher first concluded that complete randomization* of the allocation of treatments to experimental units, here the plots of land, would overcome the potential bias in assessing treatment differences due to systematic variation* in fertility. He then realized that far more was achievable; it was possible to insulate the estimates of treatment differences from much of the effect of soil heterogeneity, and thereby to increase the sensitivity of the experiment, by first grouping the experimental plots so that plots within a common group or *block* were nearly homogeneous, and then randomly applying all treatments to different plots within each block. This blocking permitted Fisher to partition the variability inherent in the plot yields after treatment into components attributable to (1) treatment differences, (2) within-block variability, and (3) between-block variability. With this conception, the *randomized complete block design* was born.

Agricultural research is not unique in having potentially heterogeneous experimental units that invite blocking; today, randomized complete block designs may involve blocking experimental units on a diversity of physical, chemical, genetic, socioeconomic, psychological, or temporal characteristics. Reports of only those studies that have used human subjects or litters of mice as blocks would fill this encyclopedia. Discussion of this topic may be found in most books on the design and analysis of experiments (*see* DESIGN OF EXPERIMENTS), including Box et al. [1], Cochran and Cox [2], Cox [3], John [6], and Kempthorne [7].

The term *block* is used with great generality to refer to any group of experimental units that share a set of characteristics thought to possibly affect the response to be observed after treatment. A randomized complete block design is a design in which each of t treatments under study is applied to a distinct experimental unit once in each of b blocks. For such a design, the number of experimental units per block is constant, say k, and $k = t$; moreover, the number of replications per treatment is constant, say r, and $r = b$. Allocation of the t treatments to the individual experimental units within each block is determined by randomization. The reader should note that certain authors, e.g., Cochran and Cox [2, p. 105], adopt a broader definition of a randomized complete design permitting certain of the treatments to occur more than once within each block.

In the following example, the blocking characteristic is temporal. Margolin et al. [8] reported on what was conceivably the first published on-line experiment upon a computer system servicing its normal user population. Two algorithms to manage computer free storage, labeled Old and New, represented the treatments of interest. Blocking on both weeks and days of the week was employed because it was thought that the user workload might vary substantially over time. For present purposes the first blocking characteristic, weeks, will be ignored. The experiment ran on two Mondays, two Tuesdays, two Wednesdays, and two Thursdays, so that $k = 2 = t$, and $b = 4 = r$. At 5:00 A.M. each morning of a designated test day, a computer operator powered up a computer system utilizing either the New or Old free storage algorithm, and allowed the system to run for 24 hours. The data in Table 1 represent the average supervisor time in microseconds needed to obtain required free storage for each day of the experiment.

Table 1 Free Storage Experiment Average Timings (microseconds)

	Blocks			
Treatment	Monday	Tuesday	Wednesday	Thursday
New	37	50	45	45
Old	325	346	313	426

Without employing formal statistical inference, it is clear that the New algorithm substantially reduced the average supervisor time needed to obtain free storage.

Although the blocking characteristic clearly must be specified before experimentation, it need not represent a proven source of heterogeneity for the response of interest. In this respect, blocking, at times, may be likened to an insurance policy taken to guard against a possible eventuality, here heterogeneity of experimental units; such was the case with the blocking in the free storage experiment. As with all insurance, there is a cost. To better appreciate this cost, it is necessary to understand the model implicitly assumed in the analysis of a randomized complete block design. To achieve this understanding, one must first arbitrarily number the treatments and blocks 1 to t and 1 to b, respectively.

The statistical model underlying the common parametric analysis of a randomized complete block design assumes that the response y_{ij} observed when the ith treatment is applied to a randomly chosen experimental unit in the jth block is simply a sum of an overall mean μ, a treatment contribution τ_i, a block contribution β_j, and an observational error ϵ_{ij}:

$$y_{ij} = \mu + \tau_i + \beta_j + \epsilon_{ij}. \tag{1}$$

Block contributions may be considered as fixed or random quantities; this distinction in no way alters inferences concerning the treatment contributions in (1). The observational errors are assumed to be independent, each with mean zero and variance σ^2; normality of the errors is sometimes assumed as well.

Treatment and block contributions enter (1) *additively*. This implies that the difference between two observations in any one block, say y_{ij} and y_{hj}, is unaltered by the block contribution. Therefore, up to observational errors, $(y_{ij} - y_{hj})$ measures solely the difference between the corresponding treatment contributions. This is clear mathematically from (1), for the model implies that

$$y_{ij} - y_{hj} = (\tau_i - \tau_h) + (\epsilon_{ij} - \epsilon_{hj}). \tag{2}$$

The model in (1) does not uniquely specify the treatment contributions, since a constant c may be added to each τ_i and subtracted from μ to produce a model of the form in (1) but with different parameter values:

$$\begin{aligned} y_{ij} &= (\mu - c) + (\tau_i + c) + \beta_j + \epsilon_{ij} \\ &= \mu^* + \tau_i^* + \beta_j + \epsilon_{ij}. \end{aligned} \tag{3}$$

This lack of uniqueness or identifiability implies that the treatment contributions are not estimable (*see* ESTIMABILITY). Any linear combination of the $\{\tau_i\}$ of the form

$$L = \sum_{i=1}^{t} c_i \tau_i \tag{4}$$

for which $\sum c_i = 0$, however, is estimable; such a linear combination is labeled a contrast*. The least-squares estimator* of L is given by

$$T = \sum_{i=1}^{t} c_i \bar{y}_{i\cdot}, \tag{5}$$

where $\bar{y}_{i\cdot} = \sum_{j=1}^{b} y_{ij}/b$.

Particular contrasts of the $\{\tau_i\}$ that are frequently of interest to researchers are:

(1) Simple differences: $\tau_i - \tau_j$, $i \neq j$;

(2) The ith treatment effect (relative to the average treatment contribution): $\tau_i - \bar{\tau}$, for $\bar{\tau} = \sum_{i=1}^{t} \tau_i / t$.

The estimator in (5) has mean and variance given by

$$E(T) = \sum_{i=1}^{t} c_i \tau_i \qquad \text{(unbiased)}$$

$$\operatorname{var}(T) = \left(\sum_{i=1}^{t} c_i^2 / b \right) \sigma^2,$$

and is normally distributed if the observational errors are assumed normally distributed. The estimator T may be written equivalently in the form

$$T = b^{-1} \sum_{j=1}^{b} \left(\sum_{i=1}^{t} c_i y_{ij} \right). \tag{6}$$

The quantity in parentheses is a contrast of the responses within the jth block, and is unchanged by the addition of a common constant to each observation in that block. Consequently, under (1), the behavior of T is unaffected by the block contributions β_j, so T is insulated from heterogeneity of the experimental units to the extent that (1) holds. This should improve the detection of treatment differences.

The partition of observational variability into the three components Fisher envisioned —those attributable to (a) block differences, (b) treatment differences, and (c) within-block or residual variability—is formalized in an analysis-of-variance table. Table 2 presents a prototype of such an analysis for a randomized complete block design, where $\bar{y}_{\cdot j}$ and $\bar{y}_{\cdot\cdot}$ denote the average in the jth block and the overall average, respectively.

Under the normal error* assumption, the partitioned components are independent. The hypothesis of no differences among treatments, i.e.,

$$H_0 : \tau_i = \tau \qquad \text{(all } i\text{)},$$

can then be tested against an all-inclusive alternative hypothesis by the ratio

$$F = \frac{\text{(treatment mean square)}}{\text{(error mean square)}} .$$

Under H_0, this statistic has an F-distribution* with $(t - 1)$ and $(b - 1)(t - 1)$ degrees of freedom. In addition, the error mean square is an unbiased* estimator of σ^2, and can be used together with (5) to set confidence limits on or test hypotheses for contrasts* in (4). The main result employed in either computation is that

$$(T - L)/\left(\text{error mean square} \cdot \sum c_i^2/b\right)^{1/2}$$

has a Student's t-distribution* on $(b - 1)(t - 1)$ degrees of freedom. If inferences are to be made concerning more than one contrast, the issue of *multiple comparisons** must be faced.

A researcher may wish to determine whether his or her blocking has isolated an important component of variability; this can be done only after the fact. He can form an F-test* with $(b - 1)$ and $(b - 1)(t - 1)$ degrees of freedom based on the ratio

$$R = \frac{\text{(block mean square)}}{\text{(error mean square)}} \tag{7}$$

to assess whether the blocking, together with any other extraneous contributions that may be confounded with blocks, represented a source of substantial variability.

Recall that the insurance afforded by blocking comes at a cost; this cost is the reduction in the degrees of freedom available for estimation of σ^2, which makes the detection of treatment differences more difficult. Again, after the fact, this cost may be compared with the intended gain of reduced variability for contrasts achieved through the isolation of the block component of variability. Various measures have been proposed (e.g., Cochran and Cox [2, p. 112]) to assess the efficiency of a randomized complete block design relative to a comparable completely randomized design, i.e., a design

Table 2 Analysis of Variance for a Randomized Complete Block Design

Source	Degrees of Freedom	Sum of Squares	Mean Square
Blocks	$b - 1$	$t\sum_{j=1}^{b}(\bar{y}_{\cdot j} - \bar{y}_{\cdot\cdot})^2$	$\frac{t}{b-1}\sum_{j=1}^{b}(\bar{y}_{\cdot j} - \bar{y}_{\cdot\cdot})^2$
Treatments	$t - 1$	$b\sum_{i=1}^{t}(\bar{y}_{i\cdot} - \bar{y}_{\cdot\cdot})^2$	$\frac{b}{t-1}\sum_{i=1}^{t}(\bar{y}_{i\cdot} - \bar{y}_{\cdot\cdot})^2$
Residual or error	$(t - 1)(b - 1)$	$\sum_i\sum_j(y_{ij} - \bar{y}_{i\cdot} - \bar{y}_{\cdot j} + \bar{y}_{\cdot\cdot})^2$	$\frac{1}{(t-1)(b-1)}\sum_i\sum_j(y_{ij} - \bar{y}_{i\cdot} - \bar{y}_{\cdot j} + \bar{y}_{\cdot\cdot})^2$
Corrected total	$bt - 1$	$\sum_i\sum_j(y_{ij} - \bar{y}_{\cdot\cdot})^2$	

with the same number of experimental units but no blocking. As far as is known, all such measures are monotonically increasing functions of (7).

The computation of the analysis of variance* in Table 2 is frequently available in packages of statistical computer programs, although it may not be identified as the analysis of a randomized complete block design. This is so because model (1) is also standardly adopted for the more common two-way layout with one observation per cell*. The analysis-of-variance table is the same in both cases, as are normal probability plots* to assess the normality assumption; and checks for outliers* and lack of additivity. Box et al. [1, Sec. 6.5] discuss diagnostic checking of the basic model in (1). The major distinction between the two designs is that in a two-way layout, there are two sets of treatments of roughly equal interest, both of whose allocation to the experimental units is determined by randomization; whereas in a randomized complete block design, the blocks represent some intrinsic property of the experimental units, and randomization is employed to allocate the single set of treatments to experimental units so that each treatment occurs once per block. Thus since the blocks themselves are in no sense randomized, the test in (7) is not a valid test for block effects, but rather, only for the combined effects of blocks as labeled, plus any other properties of the experimental units confounded with blocks.

A last comment concerns situations where the response of interest is unlikely to satisfy the normality assumption, even after transformation. In the extreme, for example, the observed responses for an entire block may only be a ranking of the t treatments within that block. In these situations, one may turn to nonparametric (distribution-free*) methods to analyze a randomized complete block design [5, Chap. 7, Sec. 1].

References

[1] Box, G. E. P., Hunter, W. G., and Hunter, J. S. (1978). *Statistics for Experimenters*. Wiley, New York.

[2] Cochran, W. G. and Cox, G. M. (1957). *Experimental Designs*, 2nd ed. Wiley, New York.

[3] Cox, D. R. (1958). *Planning of Experiments*. Wiley, New York.

[4] Fisher, R. A. (1938). *Statistical Methods for Research Workers*. Oliver & Boyd, Edinburgh.

[5] Hollander, M. and Wolfe, D. A. (1973). *Nonparametric Statistical Methods*. Wiley, New York.

[6] John, P. W. M. (1971). *Statistical Design and Analysis of Experiments*. Macmillan, New York.

[7] Kempthorne, O. (1952). *The Design and Analysis of Experiments*. Wiley, New York.

[8] Margolin, B. H., Parmelee, R. P., and Schatzoff, M. (1971). *IBM Syst. J.*, **10**, 283–304.

(ANALYSIS OF VARIANCE
BLOCKS, BALANCED INCOMPLETE
DESIGN AND ANALYSIS OF EXPERIMENTS
DISTRIBUTION-FREE METHODS
MULTIPLE COMPARISONS)

BARRY H. MARGOLIN

BLUMEN'S BIVARIATE SIGN TEST *See* BENNETT'S BIVARIATE SIGN TEST

BMDP *See* STATISTICAL SOFTWARE

BODE'S LAW

An example of a data-suggested model or a post-data model construction. Polish-German astronomer J. D. Titius (1729–1796) discovered in 1766 that the mean distances of planets from the sun, in order of increasing distance, are approximately in ratios

$$d_n = 4 + 3(2^n) \qquad (n = -\infty, 0, 1, 2, \ldots). \tag{1}$$

For the first eight planets this formula implies mean distances proportional to 4, 7, 10, 16, 28, 52, 100, and 196. The seven planets known up to 1800 had mean distances of 3.9, 7.2, 10 (the Earth), 15.2, 52, 95, and 192 and fit the theoretical sequence remarkably well. The exception was the missing planet located 28 units from the sun. Remarkably, J. E. Bode (1747–1826) and five other German astronomers discovered a small planet,

Ceres, on January 1, 1801, in the vicinity of 28 units from the sun. Since then several other small planets (the "minor planets") have been located in this area, which are conjectured to be fragments of a single large planet. Consequently, the "law" given in (1) is now referred to as Bode's law (or the Titius–Bode law). (According to Good [3], the actual discovery was made by Piazzi and Bode merely publicized it.) There is a heated controversy as to the extent to which Bode's discovery adds significantly to the believability of the "law". Moreover, it would seem that it is necessary to assess the degree of believability it had prior to 1801 and that the real test should be based on future observations. A detailed analysis of controversy surrounding Bode's law, taking into account that the data were "instigated", has been given by Leamer [5, pp. 300–305]. This work and refs. 1–4 and 6 present fascinating reading and touch on the outstanding points of difficulty in the contemporary theory and methods of data analysis.

References

[1] Efron, B. (1971). *J. Amer. Statist. Ass.*, **66**, 552–568. (Has comments by Good, Bross, Stuart, Danby, Blyth, and Pratt.)

[2] Good, I. J. (1969). *J. Amer. Statist. Ass.*, **64**, 23–66. (With discussion.)

[3] Good, I. J. (1972). Letter to the Editor, *Amer. Statist.*, **26**, 48–49.

[4] Leamer, E. E. (1974). *J. Amer. Statist. Ass.*, **69**, 122–131.

[5] Leamer, E. E. (1978). *Specification Searches*. Wiley, New York, Sect. 9.5.

[6] Polanyi, M. (1964). *Personal Knowledge*. Harper & Row, New York.

(DEGREES OF BELIEF
METASTATISTICS
POSTDATA MODEL
STATISTICS IN ASTRONOMY)

BOLTZMANN, LUDWIG EDWARD

Born: February 20, 1844, in Vienna, Austria.

Died: September 5, 1906, in Duino, near Trieste (now Italy).

Contributed to: physics (especially statistical mechanics).

Boltmann's statistical contributions are associated with his efforts, following on from J. C. Maxwell*, to explain the thermodynamics of gases on the basis of kinetic theory: i.e., to view gases as particles undergoing movement at different velocities and collisions according to the principles of mechanics. Maxwell, considering an equilibrium situation, had arrived at the probability density function $f(v) = 4\alpha^{-3}\pi^{-1/2}v^2 \exp(-v^2/\alpha^2)$, $v \geqslant 0$, for the (root-mean-square) velocity V of a particle in three dimensions, with $V = (X^2 + Y^2 + Z^2)^{1/2}$, where X, Y, Z are independent, identically distributed (i.i.d.) zero-mean normal random variables, corresponding to velocity components along the coordinate axes. The density of kinetic energy, U (of a randomly chosen particle), since it is proportional to V^2, is in this situation of the form const. $u^{1/2} \exp(-hu)$, $u \geqslant 0$, which is essentially the density of a chi-square* distribution with 3 degrees of freedom (*see* MAXWELL DISTRIBUTION). Boltzmann [1] in 1881 actually wrote down the density of the equilibrium kinetic energy distribution in the case of movement in n dimensions, apparently unaware of the work of earlier authors (Abbe*, Bienaymé*, and Helmert*), who had obtained expressions for the chi-square density of the sum of squares of n i.i.d. $N(0, 1)$ random variables.

His major contribution in the kinetic theory of gases, however, is in connection not with the equilibrium state itself, but with approach to equilibrium, and, consequently, with the relation between the evolution of thermodynamic entropy and the probability distribution of velocities. If $f(x, y, z; t)$ is the joint density of velocity components X, Y, Z at time t, he asserted that

$$E(t) = -\int\int\int f(x, y, z; t) \times \log f(x, y, z; t)\, dx\, dy\, dz$$

is nondecreasing to the value given by $f(x,$

$y, z; t) =$ const. $\exp[-h(x^2 + y^2 + z^2)]$: i.e., the equilibrium-state velocity distribution. This is essentially Boltzmann's H-theorem, with $E(t)$ corresponding to modern probabilistic usage of entropy* in connection with information*. Boltzmann's derivation was not rigorous and the fact that in classical mechanics collisions between particles are reversible, anticipating recurrence of any overall configuration, seemed inconsistent with the overall irreversibility predicted by increasing entropy. As a result of the ensuing controversy, the celebrated paper [2], which introduced the Ehrenfest urn model* as an illustrative example, suggested a probabilistic formulation in Markov chain* terms. The appropriate Markov chain $\{\mathbf{X}_n\}$, $n \geqslant 0$, describes the evolution of the relative-frequency table of velocities assumed by N particles at a specific time, where the states of $\mathbf{X}_n$ are vectors describing all such possible frequency tables, it being assumed that each particle must have one of r velocity values (such subdivision into quanta, leading to combinatorial treatment, is characteristic of Boltzmann's work). The chain $\{\mathbf{X}_n\}$ is such that each state will recur with probability 1, but a suitable scalar function $\phi(\mathcal{E}\mathbf{X}_n)$ ("entropy") is nondecreasing with n. The probability distribution, $\mathcal{E}\mathbf{X}_n$, reflects the density $f(v,t)$ of root-mean-square velocity at time t. The limiting-stationary distribution of $N\mathbf{X}_n$ is the multinomial with general term

$$\frac{N!}{x_1! \cdots x_r!} r^{-N},$$

now identified with "Maxwell–Boltzmann statistics" in physics, and manifests the ultimate (equilibrium) tendency of any one particle to have any of the possible velocities equiprobably. (See FERMI-DIRAC STATISTICS.)

We owe to Boltzmann also the word "ergodic"*, although its meaning has evolved considerably up to the present.

Additional information on Boltzmann's contributions to statistics can be found in refs. 3–6.

References

[1] Boltzmann, L. (1909). *Wissenschaftliche Abhandlungen*, F. Hasenöhrl, ed., 3 vols. Leipzig. (Boltzmann's collected technical papers; the expression appears in Vol. 2, p. 576.)

[2] Ehrenfest, P. and T. (1907). *Phys. Zeit.*, **8**, 311–314.

[3] Gnedenko, B. V. and Sheynin, O. B. (1978). In *Matematika XIX veka* [*Mathematics of the 19th Century*]. Izd. Nauka, Moscow, pp. 184–240. (Pages 229–232 contain a picture of Boltzmann and an assessment of his statistical contributions.)

[4] Kac, M. (1959). *Probability and Related Topics in Physical Sciences*. Interscience, London.

[5] Moran, P. A. P. (1961). *Proc. Camb. Philos. Soc.*, **57**, 833–842. (A modern exploration of the theorem in a Markovian setting.)

[6] Sheynin, O. B. (1971). *Biometrika*, **58**, 234–236. (Historical manifestations of the chi-square distribution.)

(CHI-SQUARE DISTRIBUTION
ENTROPY
ERGODIC THEOREMS
EXPECTED VALUE
FERMI-DIRAC STATISTICS
INFORMATION THEORY
MARKOV PROCESSES
MAXWELL, JAMES CLERK
MULTINOMIAL DISTRIBUTION)

E. SENETA

BONFERRONI INEQUALITIES AND INTERVALS

BONFERRONI INEQUALITIES

If $E_1, E_2, \ldots, E_n$ are n random events, the probability that *exactly* r of them occur is

$$P_{[r]} = \sum_{j=r}^{n} (-1)^{j-r} \binom{j}{j-r} S_j, \qquad (1)$$

where

$$S_j = \sum_{1 \leqslant \alpha_1 < \cdots < \alpha_j \leqslant n} \sum \Pr\left[\bigcap_{h=1}^{j} E_{\alpha_h}\right]$$

and $S_0 = 1$. The probability that *at least* r of them occur is

$$P_r = \sum_{i=r}^{n} P_{[i]} = \sum_{j=r}^{n} (-1)^{j-r} \binom{j-1}{j-r} S_j. \qquad (2)$$

(*See* INCLUSION-EXCLUSION METHOD, BOOLE

INEQUALITY, and, for an example of application, MATCHING PROBLEMS.)

Bonferroni [5, 6] showed that

$$\left.\begin{aligned} &0 \leqslant P_{[0]} \leqslant 1 \\ &1 - S_1 \leqslant P_{[0]} \leqslant 1 - S_1 + S_2, \\ &\text{and generally} \\ &1 - S_1 + S_2 - \cdots - S_{2i-1} \\ &\quad \leqslant P_{[0]} \leqslant 1 - S_1 + S_2 - \cdots + S_{2i} \end{aligned}\right\} \tag{3}$$

for $i = 1, 2, \ldots, n/2$.

Galambos [12, 13] gives an overview of methods of proving these inequalities, and also some improved bounds.

The inequality

$$1 - S_1 \leqslant P_{[0]}$$

is equivalent to

$$1 - \sum_{i=1}^{n} \Pr[E_i] \leqslant \Pr\left[\bigcap_{i=1}^{n} \bar{E}_i\right].$$

Replacing the complement $\bar{E}_i$ by A_i (and so E_i by $\bar{A}_i$), we have

$$1 - \sum_{i=1}^{n} \Pr[\bar{A}_i] \leqslant \Pr\left[\bigcap_{i=1}^{n} A_i\right]. \tag{4}$$

This result has been the most frequently used of Bonferroni's inequalities and is often referred to as "*the* Bonferroni inequality."

The set of inequalities (3) generalize to

$$\left.\begin{aligned} &S_r - \binom{r+1}{1} S_{r+1} \leqslant P_{[r]} \leqslant S_r \\ &S_r - \binom{r+1}{1} S_{r+1} + \binom{r+2}{2} S_{r+2} \\ &\quad - \binom{r+3}{3} S_{r+3} \\ &\quad \leqslant P_{[r]} \leqslant S_r - \binom{r+1}{1} S_{r+1} \\ &\quad + \binom{r+2}{2} S_{r+2}, \end{aligned}\right\} \tag{5}$$

and so on.

Similar sets of inequalities can be obtained for P_r. Specifically,

$$\left.\begin{aligned} &S_r - \binom{r}{1} S_{r+1} \leqslant P_r \leqslant S_r \\ &S_r - \binom{r}{1} S_{r+1} + \binom{r+1}{2} S_{r+2} \\ &\quad - \binom{r+2}{3} S_{r+3} \\ &\quad \leqslant P_r \leqslant S_r - \binom{r}{1} S_{r+1} \\ &\quad + \binom{r+1}{2} S_{r+2}, \end{aligned}\right\} \tag{6}$$

and so on.

Proofs of these inequalities are given by Fréchet [11] and Feller [10].

Meyer [16] has extended Bonferroni's inequalities to classes of events $\{E_{ij}, i = 1, \ldots, N_j\}$, $j = 1, 2, \ldots, k$. For nonnegative integers $r_j, 0 \leqslant r_j \leqslant N_j$, he has obtained Bonferroni inequalities for $P_{[r_1, r_2, \ldots, r_k]}$ and $P_{r_1, r_2, \ldots, r_k}$, where $P_{[r_1, r_2, \ldots, r_k]}$ denotes the probability that exactly r_1 of the E_{i1}'s, exactly r_2 of the E_{i2}'s, ..., and exactly r_k of the E_{ik}'s will occur; $P_{r_1, r_2, \ldots, r_k}$ is defined analogously with "at least" replacing "exactly."

From a statistical point of view, it is the Bonferroni inequality stated in (4) that is of paramount importance because of its use in simultaneous inference to maintain the family confidence level, at least in a bounded sense. *See* JOINT CONFIDENCE STATEMENTS. An extensive treatment of simultaneous inference is given in Miller [17] with updated references in ref. 18.

BONFERRONI INTERVALS

Let $\boldsymbol{\theta} = (\theta_1, \theta_2, \ldots, \theta_k)'$ be a $(k \times 1)$ vector of parameters—e.g., multinomial* proportions, elements in an expected value vector or a variance–covariance matrix* or partial regression coefficients in a general linear model*. If separate two-sided confidence intervals are constructed for each of the k parameters, each with confidence coefficient $100(1 - \alpha)\%$, and if A_i denotes the event that the interval for θ_i includes the actual value of θ_i, then, from the Bonferroni inequality (4), it follows that the probability ($\Pr[\cap_{i=1}^{k} A_i]$) that every interval includes the value of the parameter it estimates is at least $(1 - k\alpha)$. Formally,

$$\Pr\left[\bigcap_{i=1}^{k} A_i\right] \geqslant 1 - k\alpha. \qquad (7)$$

Thus the "family confidence coefficient"* is at least $100(1 - k\alpha)\%$, whatever be the dependence among the statistics used in constructing the confidence intervals. If the confidence level for each separate interval is increased to $100(1 - \alpha k^{-1})\%$, then the family confidence coefficient is at least $100(1 - \alpha)\%$. The resulting confidence intervals are called *Bonferroni intervals*.

In obtaining Bonferroni intervals, it is not necessary that all the separate confidence coefficients $[100(1 - \alpha_i)\%, i = 1, \dots, k]$ be equal, only that $\sum_{i=1}^{k} \alpha_i = \alpha$. Thus, if a few of the parameters warrant greater interest than the others, then the confidence coefficients for these parameters could be larger. Regardless of the allocation of the α_i's, the conservative Bonferroni intervals provide a viable alternative for achieving a family confidence coefficient of at least $(1 - \alpha)$.

To illustrate the preceding remarks, consider the normal error, multiple regression model (*see* REGRESSION):

$$Y_i = \beta_0 + \beta_1 X_{i1} + \beta_2 X_{i2} + \cdots + \beta_{p-1} X_{i,\,p-1} + \epsilon_i,$$

$i = 1, \dots, n$, where the ϵ_i's are mutually independent $N(0, \sigma^2)$.

A $100(1 - \alpha)\%$ confidence interval for β_i, $i = 1, \dots, p - 1$, is

$$b_i \pm s_i t_{n-p,\,1-\alpha/2},$$

where $s_i^2 =$ (residual mean square) $\times$ ($(i + 1, i + 1)$th diagonal element of $(\mathbf{X'X})^{-1}$), and $t_{n-p,\,1-\alpha}$ is the corresponding t-percentile*.

The corresponding set of Bonferroni intervals with family (joint) confidence coefficient at least $100(1 - \alpha)\%$ is

$$\left(b_i \pm s_i t_{n-p,\,1-\alpha/(2(p-1))}\right) \quad (i = 1, \dots, p - 1). \qquad (8)$$

These may be compared with the ellipsoidal confidence region*

$$(\mathbf{Ab} - \mathbf{A}\boldsymbol{\beta})'\left(\mathbf{A}(\mathbf{X'X})^{-1}\mathbf{A}'\right)^{-1}(\mathbf{Ab} - \mathbf{A}\boldsymbol{\beta}) < (p - 1)(\text{residual mean square}) \times F_{p-1,\,n-p,\,1-\alpha}, \qquad (9)$$

where $\mathbf{A}$ is a $((p - 1) \times p)$ matrix whose first column is the 0 vector and whose last $(p - 1)$ columns constitute the identity matrix of dimension $(p - 1)$, and $F_{p-1,\,n-p,\,1-\alpha}$ is the corresponding F-percentile*.

Comparison between the rectangular confidence region formed by intersection of the intervals (8) and the ellipsoidal confidence region (9) depends somewhat on the correlations among the b_i's (see Fig. 1).

Formulas will now be presented for constructing Bonferroni intervals for k linear combinations of the parameters, k mean responses, and the prediction of k new observations.

If the k linear functions of the β's are $\mathbf{C}'\boldsymbol{\beta}$, where $\mathbf{C}'$ is a full (k) rank $(k \times p)$ matrix, the Bonferroni intervals with family confidence coefficient at least $100(1 - \alpha)\%$ are

$$\mathbf{c}_i'\mathbf{b} \pm (\text{residual mean square})^{1/2} \times \left[\mathbf{c}_i'(\mathbf{X'X})^{-1}\mathbf{c}_i\right]^{1/2} t_{n-p,\,1-\alpha/(2k)} \qquad (10)$$

$(i = 1, \dots, k)$, where $\mathbf{c}_i'$ is the ith row of $\mathbf{C}'$.

To estimate $E(Y_i)$ at k different $\mathbf{x}$ vectors, $\mathbf{x}_1, \mathbf{x}_2, \dots, \mathbf{x}_k$, use the following Bonferroni intervals:

$$\mathbf{x}_i'\mathbf{b} \pm (\text{residual mean square})^{1/2} \times \left[\mathbf{x}_i'(\mathbf{X'X})^{-1}\mathbf{x}_i\right]^{1/2} t_{n-p,\,1-\alpha/(2k)}, \qquad (11)$$

$i = 1, 2, \dots, k$. Bonferroni prediction intervals for the Y_i at k different values of $\mathbf{x}$ are given by

$$\mathbf{x}_i'\mathbf{b} \pm (\text{residual mean square})^{1/2} \times \left[1 + \mathbf{x}_i'(\mathbf{X'X})^{-1}\mathbf{x}_i\right]^{1/2} t_{n-p,\,1-\alpha/(2k)}. \qquad (12)$$

In all cases, the intervals are of standard form with the exception of the t-percentile.

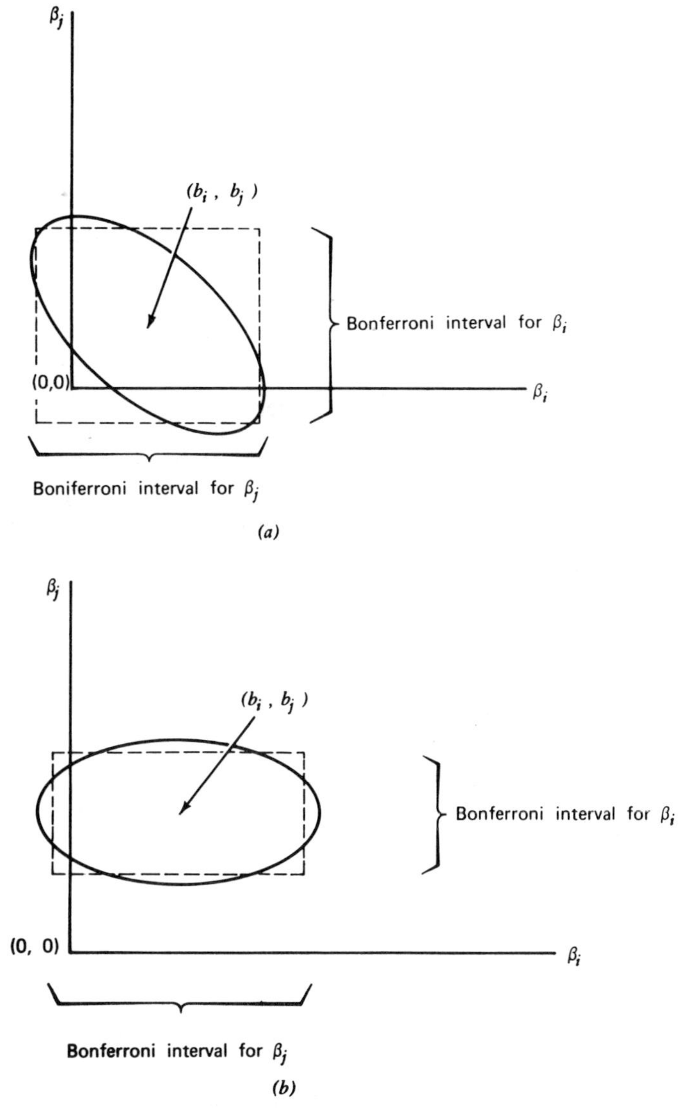

Figure 1 Rectangular Bonferroni confidence region and the elliptical region of the F-statistic in the presence of: (*a*) high correlation between b_i and b_j; (*b*) no correlation between b_i and b_j.

Neter and Wasserman [21] utilize the foregoing formulas in a detailed, numerical example for two independent variables. Seber [25] empirically compares Bonferroni intervals with maximum modulus t-intervals* and intervals generated by Scheffé's S-method*. In general, the Scheffé intervals are widest, and the maximum modulus intervals fare only slightly better than Bonferroni intervals.

Dunn and Clark [9] give detailed numerical examples showing how to construct Bonferroni intervals for model parameters and their contrasts when dealing with multifactor, randomized block* and Latin square* designs. Neter and Wasserman [21] also present several numerical examples for these cases. They point out that the superiority of one procedure over the other hinges on the number of comparisons to be investigated. Again, Miller's works [17, 18] are excellent references, as is the paper by O'Neill and Wetherill [22].

Because of flexibility, Bonferroni's inequality has received wide usage in almost all aspects of multivariate statistics. For example, suppose that a random sample of size n is drawn from a p-variate normal population*, where $\boldsymbol{\mu}' = (\mu_1, \mu_2, \ldots, \mu_p)'$ is the $(p \times 1)$ vector of unknown means. If the $(p \times p)$ positive definite covariance matrix $\boldsymbol{\Sigma}$ is known completely, then $100(1-\alpha)\%$ simultaneous confidence intervals* for k distinct linear combinations $\mathbf{a}_i'\boldsymbol{\mu}$ of $\boldsymbol{\mu}$ are given by

$$\mathbf{a}_i'\overline{\mathbf{x}} \pm \sqrt{\mathbf{a}_i'\boldsymbol{\Sigma}\mathbf{a}_i/n}\, z_{1-\alpha/(2k)} \quad (i = 1, 2, \ldots, k), \tag{13}$$

where $\overline{\mathbf{x}}$ denotes the $(p \times 1)$ vector of sample means. If intervals are desired only for the p individual means (μ_i), as would be the case when the $H_0 : \boldsymbol{\mu} = \boldsymbol{\mu}_0$ is rejected, then these are obtained from (13) by successively letting $\mathbf{a}_i$ be the $(p \times 1)$ vector whose entries are 0's except for a 1 in the ith position. Thus $100(1-\alpha)\%$ simultaneous confidence intervals for $\boldsymbol{\mu}$ are of the form

$$\bar{x}_i \pm \left(\sigma_i/\sqrt{n}\right) z_{1-\alpha/(2p)} \qquad (i = 1, 2, \ldots, p), \tag{14}$$

whatever be the covariance in Σ. Alt and Spruill [1] have shown that, for $p \geqslant 2$ and confidence levels above 50%, the Bonferroni intervals in (14) are shorter than the corresponding Scheffé intervals, which are obtained by replacing $z_{1-\alpha/(2p)}$ by $\sqrt{\chi^2_{p,1-\alpha}}$.

In the more frequently occurring situation when the covariance matrix is completely unknown and the only information available is the elements $\mathbf{X}_1, \mathbf{X}_2, \ldots, \mathbf{X}_n$ of a random sample from the population, one can construct Bonferroni intervals for the k linear combinations of the form

$$\mathbf{a}_i'\overline{\mathbf{x}} \pm \sqrt{\mathbf{a}_i'\mathbf{S}\mathbf{a}_i/n}\, t_{n-1,1-\alpha/(2k)} \quad (i = 1, 2, \ldots, k), \tag{15}$$

where $\mathbf{S} = (n-1)^{-1}\sum_{i=1}^{n}(\mathbf{X}_i - \overline{\mathbf{X}})(\mathbf{X}_i - \overline{\mathbf{X}})'$.

Bonferroni intervals for the p means are

$$\bar{x}_i \pm \left(s_i/\sqrt{n}\right) t_{n-1,1-\alpha/(2p)} \quad (i = 1, 2, \ldots, p), \tag{16}$$

where s_i^2 is simply the sample variance for the ith variable. To obtain the Scheffé intervals* (in this context sometimes called the Roy–Bose intervals) corresponding to (15) and (16), just replace the t-percentile by $\sqrt{p(n-1)F_{p,n-p,1-\alpha}/(n-p)}$. Determination of which procedure yields shorter intervals is easily accomplished by comparing the t-percentile with the $\sqrt{F\text{-}percentile}$ multiplied by $\sqrt{p(n-1)/(n-p)}$. In general, when intervals are obtained only for the p-means, the Bonferroni intervals are shorter. However, when k linear combinations are to be estimated and k is much larger than p, then the Scheffé intervals may be shorter [17, 19].

Bonferroni's inequality has also been used when there is more than one p-variate normal population [19].

Occasionally, simultaneous confidence intervals are needed for the variances of a p-variate normal distribution. Bens and Jensen [4] and Jensen and Jones [14] show that using the usual univariate intervals, each with confidence coefficient $1-(\alpha/p)$, yields satisfactory results.

Bonferroni's inequality plays an important role in the construction of simultaneous confidence intervals for multinomial* proportions. Let $\pi_1, \pi_2, \ldots, \pi_p$ be the multinomial cell probabilities and let $n_1 n_2, \ldots, n_p$ denote the corresponding observed cell frequencies for a sample of size N from this multinomial population. Then asymptotic $(N \to \infty)$ confidence intervals for the π_i are of the form

$$\hat{\pi}_i \pm \sqrt{\hat{\pi}_i(1-\hat{\pi}_i)/N}\, z_{1-\alpha/(2p)} \quad (i = 1, \ldots, p). \tag{17}$$

Alt and Spruill [1] demonstrate the superiority of the Bonferroni intervals over the Scheffé intervals in this case. Bonferroni intervals for $\pi_i - \pi_j$ are given in Miller [17]. He also treats the case when there are r multinomial populations. Pairwise median tests controlling type I error* by using Bonferroni inequalities were studied by Ryan [23]. See also Ryan [24] for historical remarks.

The preceding examples illustrate the wide use of Bonferroni's inequality, but the list is by no means exhaustive. For example, the conservative Bonferroni approach has also been used to obtain prediction intervals* [2, 7]. Its use in reliability problems is demonstrated in Kwerel [15]. Most of the applications require percentage points of Student's t-distribution which are not commonly available. Bailey [3] provides tables for $\alpha = 0.05$ and 0.01 and for a range of values of ν, the number of degrees of freedom. When the error allocation is unequal, the tables of Dayton and Schafer [8] may prove useful. The charts of Moses [20] should also prove helpful.

For multinormal distributions*, Bonferroni intervals have shown themselves to be strong competitors to intervals generated using other specialized techniques, but slightly shorter intervals are obtained by using Šidák's inequality [26] when it is applicable. If **Y** is distributed as a p-variate normal with zero means and arbitrary covariance matrix, then Šidák's inequality states that

$$\left.\begin{aligned} P(|Y_1| \leqslant c_1, |Y_2| \leqslant c_2, \dots, |Y_p| \leqslant c_p) \\ \geqslant P(|Y_1| \leqslant c_1) \cdot P(|Y_2| \leqslant c_2) \cdots \\ P(|Y_p| \leqslant c_p), \end{aligned}\right\} \tag{18}$$

for any positive numbers $c_1, c_2, \dots, c_p$. Simultaneous confidence intervals for the individual means of any p-variate normal distribution with known covariance matrix Σ are obtained by putting $Y_i = \sqrt{n}\,(\overline{X}_i - \mu_i)/\sigma_i$. The intervals obtained are those presented in (14) but with $z_{1-\alpha/(2p)}$ replaced by z_*, where z_* is such that $P(Z \leqslant z_*) = [1 + (1-\alpha)^{1/p}]/2$. This form of Šidák's inequality also applies to the intervals for multinomial proportions presented in (17). Since $(1-\alpha_1)\cdots(1-\alpha_p) > 1 - \sum_{i=1}^{p}\alpha_i$, for $p > 1$, the Šidák intervals are slightly shorter than the Bonferroni intervals. For example, if $p = 2$ and $\alpha_1 = \alpha_2 = 0.05$, then $(1-\alpha_1)(1-\alpha_2) = 0.9025$ while $1 - \sum_{i=1}^{2}\alpha_i = 0.90$. Šidák also provided a more general inequality which can be used to construct confidence intervals for t-statistics. *See* ŠIDÁK'S INEQUALITY.

In summary, when the number of intervals (k) is small, the Bonferroni intervals are competitive with intervals obtained using other methods. The Bonferroni intervals use a confidence level of $(1 - \alpha_i)$ per interval, resulting in a family confidence level of at least $1 - \sum_{i=1}^{k}\alpha_i$. When an equal error allocation of α/k per interval is used, the family confidence level is at least $1 - k(\alpha/k) = 1 - \alpha$. When k is large, however, the Bonferroni intervals are unnecessarily long.

References

[1] Alt, F. and Spruill, C. (1977). *Commun. Statist. A*, **6**, 1503–1510.

[2] Angers, C. and McLaughlin, G. (1979). *Technometrics*, **21**, 383–385.

[3] Bailey, B. (1977). *J. Amer. Statist. Ass.*, **72**, 469–478.

[4] Bens, G. and Jensen, D. (1967). Percentage Points of the Bonferroni Chi-Square Statistics. *Tech. Rep. No 3*, Dept. of Statistics, Virginia Polytechnic Institute, Blacksburg, Va.

[5] Bonferroni, C. E. (1936). *Pubbl. Ist Sup. Sci. Econ. Commun. Firenze*, **8**, 1–62.

[6] Bonferroni, C. E. (1936). Il Calcolo delle assicurazioni su gruppi di teste. In *Studii in onore del prof. S. O. Carboni*. Roma.

[7] Chew, V. (1968). *Technometrics*, **10**, 323–331.

[8] Dayton, C. and Schafer, W. (1973). *J. Amer. Statist. Ass.* **68**, 78–83.

[9] Dunn, O. and Clark, V. (1974). *Applied Statistics: Analysis of Variance and Regression*. Wiley, New York. (This introductory text provides a detailed treatment of the use of Bonferroni's inequality in analysis of variance.)

[10] Feller, W. (1968). *An Introduction to Probability Theory and its Applications*, 3rd ed., Vol. 1. Wiley, New York. (Chapter IV of this classic text covers combinations of events and states the generalized inequalities of Bonferroni.)

[11] Fréchet, M. (1940). *Les Probabilitiés associées à un système d'événements compatibles et dépendants*. (Actualités scientifiques et industrielles, no. 859.) Hermann & Cie, Paris. [This monograph (in French) gives a unified treatment of the inequalities of Bonferroni and others, such as Boole's and Gumbel's.]

[12] Galambos, J. (1975). *J. Lond. Math. Soc.*, **9**(2), 561–564.

[13] Galambos, J. (1977). *Ann. Prob.*, **5**, 577–581.

[14] Jensen, D. and Jones, M. (1969). *J. Amer. Statist. Ass.*, **64**, 324–332.

[15] Kwerel, S. (1975). *J. Amer. Statist. Ass.*, **70**, 472–479.

[16] Meyer, R. M. (1969). *Ann. Math. Statist.*, **40**, 692–693.

[17] Miller, R. (1966). *Simultaneous Statistical Inference*. McGraw-Hill, New York. (This frequently referenced book provides a lucid summary of the theory, methods, and applications of simultaneous inference through 1966. Although numerical examples are not given, specific formulas for using Bonferroni's inequality are presented for numerous scenarios.)

[18] Miller, R. (1977). *J. Amer. Statist. Ass.*, **72**, 779–788. (Over 250 references are given.)

[19] Morrison, D. (1976). *Multivariate Statistical Methods*, 2nd ed. McGraw-Hill, New York. (Although this book develops the multivariate linear hypothesis through the union-intersection principle of Roy and uses the corresponding simultaneous intervals, emphasis is also given to the Bonferroni intervals. Numerical examples are given.)

[20] Moses, L. (1976). Charts for Finding Upper Percentage Points of Student's t in the Range .01 to .00001. *Tech. Rep. No. 24 (5 R01 GM21215-02)*, Stanford University, Stanford, Calif.

[21] Neter, J. and Wasserman, W. (1974). *Applied Linear Statistical Models*. Richard D. Irwin, Homewood, Ill. (This introductory text contains many numerical examples and provides a good treatment of the importance and use of Bonferroni's inequality.)

[22] O'Neill, R. and Wetherill, G. (1971). *J. R. Statist. Soc. B*, **33**, 218–250.

[23] Ryan, T. A. (1959). *Psychol. Bull.*, **56**, 26–47.

[24] Ryan, T. A. (1980). *Amer. Statist.*, **34**, 122–123.

[25] Seber, G. (1977). *Linear Regression Analysis*. Wiley, New York. (Chapter 5 of this theoretical treatment of regression compares the use of Bonferroni t-intervals, maximum modulus t-intervals, and Scheffé's S-method in constructing simultaneous confidence intervals. Numerical examples are not given.)

[26] Šidák, Z. (1967). *J. Amer. Statist. Ass.*, **62**, 626–633.

(JOINT CONFIDENCE INTERVALS
MULTIPLE COMPARISONS
SCHEFFÈ'S S-METHOD
WORKING–HOTELLING
CONFIDENCE BAND)

F. B. ALT

BONFERRONI t-STATISTIC

A commonly used tool for simultaneous statistical inference when the parent distribution is normal. The technique is based on the following specialization of the Bonferroni inequalities*.

Let $Y_1, \dots, Y_k$ be normally distributed (not necessarily independent) random variables with means $\mu_1, \dots, \mu_k$ and variances $\sigma_1^2, \dots, \sigma_k^2$, respectively. Let $s_1^2, \dots, s_k^2$ be independent sample estimators of $\sigma_1^2, \dots, \sigma_k^2$, respectively, with $\nu_i s_i^2/\sigma_i^2$ distributed as χ^2 variables* with $\nu_i = n_i - 1$ degrees of freedom ($i = 1, \dots, k$). It is assumed that Y_i is independent of s_i^2, $i = 1, \dots, k$, so that $T_i = (Y_i - \mu_i)/s_i$, $i = 1, \dots, k$ are t variables* with $\nu_i = n_i - 1$ degrees of freedom. (s_i^2 may depend on $Y_{i'}$, $i' \neq i$.)

If $t_{r_i, \alpha/(2k)}$, $i = 1, \dots, k$, are the *upper* $\alpha/(2k)$ percentile points (or equivalently $\pm t_{r_i, \alpha/(2k)}$ are two-tailed α/k percentile points) of the t-distribution with r_i degrees of freedom, $i = 1, \dots, k$, then

$$\Pr\left(\bigcap_{i=1}^{k}\left[\left|\frac{Y_i - \mu_i}{s_i}\right| \leqslant t_{r_i, \alpha/(2k)}\right)\right) \geqslant 1 - \left(k \times \frac{\alpha}{k}\right)$$
$$= 1 - \alpha$$

since $\Pr\{|Y_i - \mu_i|/s_i| \leqslant t_{r_i, \alpha/(2k)}$ is by definition equal to $1 - \alpha/k$. In other words, with probability greater than or equal to $1 - \alpha$, $\mu_i \in Y_i \pm t_{r_i, \alpha/2k} s_i$ simultaneously for all $i = 1, \dots, k$. This is valid if the significance levels* for the component intervals are not all the same (α/k), provided only that $\alpha_1 + \alpha_2 + \cdots + \alpha_k = \alpha$.

Tables of $t_{\nu, \alpha/(2k)}$ for $\alpha = 0.05$, $k = 1(1)10$, 15, 20, 50 and $\nu = 5$, 10, 15, 20, 24, 30, 40, 60, 120, $+\infty$ are given by Dunn [1]. More extensive tables are given in Dunn [2], Dunn and Massey [3], and Miller [4].

References

[1] Dunn, O. J. (1959). *J. Amer. Statist. Ass.*, **54**, 613–621.

[2] Dunn, O. J. (1961). *J. Amer. Statist. Ass.*, **56**, 52–64.

[3] Dunn, O. J. and Massey, F. J. (1965). *J. Amer. Statist. Ass.*, **60**, 573–583.

[4] Miller, P. G. (1966). *Simultaneous Statistical Inference*. McGraw-Hill, New York, pp. 67–70.

(BONFERRONI INEQUALITIES
JOINT CONFIDENCE INTERVALS
STUDENTIZED MAXIMUM
MODULUS)

BOOLE'S INEQUALITY

If $A_1, A_2, \ldots, A_n$ belong to an algebra of events*, then

$$P(A_1 \cup A_2 \cup \cdots \cup A_n) \leqslant P(A_1) + P(A_2) + \cdots + P(A_n).$$

Boole's inequality is widely used in combinatorial applied probability.

(BONFERRONI INEQUALITIES
WARING'S INEQUALITY)

BOOTSTRAPPING

Many statistical procedures are optimal only for specific values of certain parameters: e.g., kurtosis*, skewness*. When these are not known, as is often the case, it is sometimes claimed that one can estimate their values from the data, and then use those estimated values *as if* they were the true ones, constructing the appropriate optimal procedures on that basis. This operation is known as "bootstrapping."

The operation can be used in a variety of situations. A few examples are:

1. When applying goodness-of-fit tests*, with only the form of the hypothesized population distribution (e.g., normal) known but not specific parameter values (e.g., mean, standard deviation).
2. When using ridge regression* techniques, where the value of the multiplier (k) depends on some population parameters that have to be estimated.
3. Many adaptive* procedures are of the nature of bootstrapped procedures.

It is very desirable that properties of the bootstrapped procedures be known at least to such an extent that their likely deviations from those of optimal procedures can be assessed and allowed for.

Bibliography

Efron, B. (1979). *SIAM Rev.*, **21**, 460–480.

Fuch, C. (1978). *Technometrics*, **20**, 291–299.

LaMotte, L. R. (1978). *Technometrics*, **20**, 281–290. (Applications to Bayesian procedures—regards the estimators of the parameters as "prior information.")

[ADAPTIVE METHODS
RIDGE REGRESSION
STEIN (SHRINKAGE) ESTIMATORS]

BOREL ZERO-ONE LAW *See* ZERO-ONE LAW

BOREL–CANTELLI LEMMA

Let $A_1, A_2, \ldots$ be an infinite sequence of events. The event $A^* = \{A_n$ occurs infinitely often$\}$ is defined by $\cap_{n=1}^{\infty} \cup_{j \geqslant m}^{\infty} A_j$ (or $\limsup_{n\to\infty} A_n$.The Borel–Cantelli lemma states that

1. If $\sum_{j=1}^{\infty} P(A_j) < \infty$, then $P(A^*) = 0$.
2. If $\sum_{j=1}^{\infty} P(A_j) = \infty$ and $\{A_n\}$ is a sequence of *independent* events, then $P(A^*) = 1$.

It is sometimes called the Borel–Cantelli zero–one law. It is a useful tool for establishing strong convergence for random sequences and is applied in the derivation of asymptotic distributions* of various test statistics.

Bibliography

Borel, E. (1909). *Rend. Circ. Math. Palermo*, **27**, pp. 247–271.

Cantelli, F. P. (1917). *Rend. Accad. Naz. Lincei*, **26**, 39–45.

Durbins, E. and Freedman, D. A. (1965). *Ann. Math. Statist.*, **36**, 800–807.

Freedman, D. (1973). *Ann. Prob.*, **1**, 910–925.

Loève, M. (1977). *Probability Theory*, 4th ed. Springer-Verlag, New York.

(CONVERGENCE OF DISTRIBUTIONS
LAWS OF LARGE NUMBERS)

BOREL'S STRONG LAW OF LARGE NUMBERS

A particular case of the strong law of large numbers* applicable to independent, identically distributed Bernoulli random variables*. Specifically, let $\{X_n\}$ be a sequence of independent, identically distributed Bernoulli random variables with $\Pr[X_i = 1] = p$ for all i (and $\Pr[X_i = 0] = 1 - p \equiv q$ for all i), then

$$\Pr\left[\lim_{n\to\infty} \frac{\sum_{i=1}^{n} X_i}{n} = p\right] = 1.$$

(CHEBYSHEV'S LAW OF LARGE NUMBERS
LAWS OF LARGE NUMBERS)

BOREL–TANNER DISTRIBUTION

A random variable X possesses a Borel–Tanner distribution with parameters r, α, where r is a positive integer and $0 < \alpha < 1$ if

$$\Pr(X = k) = \frac{r}{(k-r)!} k^{k-r-1} e^{-\alpha k} \alpha^{k-r} \qquad (k = r, r+1, \dots).$$

The expected value of this variable is $\mu_1' = E(X) = r/(1-\alpha)$ and variance $V(X) = \alpha r/(1-\alpha)^3$.

In queueing theory* the Borel–Tanner distribution arises as the distribution of the number of customers served in a queueing system with Poisson* input with parameter α and a constant service time, given that the length of the queue at the initial time is r.

Bibliography

Johnson, N. L. and Kotz, S. (1969). *Discrete Distributions*. Wiley, New York, pp. 253–255. (This volume discusses additional properties and contains references to original papers as well as information about tables of this distribution.)

(POISSON DISTRIBUTION
LAGRANGE DISTRIBUTIONS)

BORTKIEWICZ, LADISLAUS VON

German statistician and economist. Born in St. Petersburg, Russia, August 7, 1868, Bortkiewicz graduated from the University of St. Petersburg Faculty of Law in 1890, and subsequently studied under Lexis* in Göttingen, where he defended his doctoral thesis in 1893. Privatdozent in Strasbourg from 1895 to 1897, Bortkiewicz then returned to Russia until 1901, when he was appointed to a professorial position at the University of Berlin, where he taught statistics and economics until his death on July 15, 1931.

Best known for his modeling of rare-event phenomena by the Poisson distribution*, Bortkiewicz also made numerous other contributions to mathematical statistics, notably the statistical analysis of radioactivity, the theory of runs*, and the distributional properties of extreme values*. His work on population theory, actuarial science, and political economy was also noteworthy; for the latter, see Gumbel [2].

Bortkiewicz's monograph *Das Gesetz der kleinen Zahlen* (The Law of Small Numbers) [1] is unquestionably his best known work and was in large part responsible for the subsequent popularity of the Poisson distribution. Contrary to popular belief, Bortkiewicz meant by the expression "law of small numbers" not the Poisson distribution itself, but the tendency of data in binomial* sampling with small and variable success probability to appear as though the success probability were constant when the sample size was large. The Lexis ratio* was advocated as a means of detecting the presence of variable success probabilities.

A meticulous scholar, Bortkiewicz wrote in a difficult style that may have lessened his influence. In England, one of the few to appreciate his work was John Maynard Keynes*, who devoted a chapter of *A Treatise on Probability* [3] to describing the results of Lexis and Bortkiewicz.

Literature

Thor Andersson's lengthy obituary in the *Nordic Statistical Journal*, 1931 (**3**:9–26), includes an essentially complete bibliography. The biography by E. J. Gumbel in the *International Encyclopedia of Statistics* (1978, Free Press, New York) contains a useful list of secondary literature; see also the entry by O. B. Sheynin in the *Dictionary of Scientific Biography* (1970, Scribner's, New York). Bortkiewicz's contributions to dispersion theory* and the Lexis ratio are discussed within a historical perspective by C. C. Heyde and E. Seneta in *I. J. Bienaymé: Statistical Theory Anticipated* (1977, Springer-Verlag, New York), pp. 49–58. Other historical comments, including a description of Bortkiewicz's applications of the Poisson distribution and the controversies with Gini and Whittaker, may be found in Frank Haight's *Handbook of the Poisson Distribution* (1967, Wiley, New York), Chap. 9.

Bortkiewicz was a member of the International Statistical Institute and participated at its meetings; William Kruskal and Frederick Mosteller describe his criticism of A. N. Kiaer's concept of "representative sample"* at one such meeting in their "Representative Sampling IV: The History of the Concept in Statistics," *International Statistical Review*, 1980 (**48**: August).

References

[1] Bortkiewicz, L. von (1898). *Das Gesetz der kleinen Zahlen* (The Law of Small Numbers). Teubner, Leipzig.

[2] Gumbel, E. J. (1978). In *International Encyclopedia of Statistics*. Free Press, New York.

[3] Keynes, J. M. (1921). *A Treatise on Probability*. Macmillan, London.

(LAW OF SMALL NUMBERS
POISSON DISTRIBUTION
POISSON LIMIT)

SANDY L. ZABELL

BOSCOVICH, RUGGIERO GIUSEPPE

Born: May 18, 1711, in Ragusa (now Dubrovnik), Dalmatia (now in Yugoslavia).

Died: February 13, 1787, in Milan, Italy.

Contributed to: astronomy, geodesy, and physics.

Boscovich was a son of a Serb who settled in Ragusa (we use above the Italian version of his name). His early education (in his native city) was at a Jesuit school, and subsequently he attended the Collegium Romanum. Ordained a priest within the Society of Jesus in Rome in 1744, he propounded a very early version of the atomic theory of matter and was recognized by the foremost scientific bodies in Europe; he was made a Fellow of the Royal Society of London in 1760. After the suppression of the Jesuits in 1773, Boscovich, then in his sixties, went to Paris at the invitation of the King of France and spent 9 years as Director of Optics for the Marine before returning to Italy.

In 1750, Pope Benedict XIV commissioned Boscovich and his fellow Jesuit, C. Le Maire, to carry out several meridian measurements in Italy; part of the report of this journey, reprinted as ref. 1, contains his presently known contribution to statistical methodology. (Several other considerations on the effects of chance* are reported in ref. 3.) Boscovich considered the simple version $Z_i = \alpha + \beta w_i + \epsilon_i$, $i = 1, \ldots, n$ of the linear model*, where the data set is (Z_i, w_i),

$i = 1, \ldots, n$, the ϵ_i are residuals* due to errors in measurement, and α and β are to be determined from the conditions

$$\text{(a)} \qquad \sum_{i=1}^{n} |\epsilon_i| = \text{min.}$$

$$\text{(b)} \qquad \sum_{i=1}^{n} \epsilon_i = 0.$$

The second of the conditions became familiar in the eighteenth and nineteenth centuries. Noting, as was done by Boscovich, that it implies that $\bar{Z} = \alpha + \beta\bar{w}$, which yields by substitution in the model that $Y_i = Z_i - \bar{Z} = \beta(w_i - \bar{w}) + \epsilon_i, i = 1, \ldots, n$, where it is automatically satisfied, it can be ignored. Putting $x_i = w_i - \bar{w}$, Boscovich therefore proposes to determine β in this model according to criterion (a), i.e., by minimizing the sum of absolute deviations* (the l_1 criterion). Boscovich's method for achieving this was geometric; Laplace* gave an analytical solution, acknowledging Boscovich, and used it in his own astronomical writings as early as 1789. In the second supplement (1818) to the *Théorie Analytique des Probabilités*, he calls it the "method of situation" to distinguish it from the l_2 (least-squares*) procedures, which he calls the "most advantageous method." This robust* method of estimation (in principle, tantamount to the use of the sample median as a measure of location of a sample rather than the least-squares measure, the sample mean), whose efficient general application requires linear programming* methodology, therefore predates least squares in the linear model framework. After Laplace, it was taken up [3] by Gauss*, who indicated the form of analytic solution in the general case of the rank $r \geqslant 1$ linear model.

Additional information on Boscovich's life and work can be found in refs. 2 and 4.

References

[1] Bosković, R. J. (1757). In *Geodetski rad R. Boskovića*, N. Cubranić, ed., Zagreb, 1961. (Printed with Serbo-Croatian translation. See especially pp. 90–91.)

[2] Eisenhart, C. (1961). In *R. J. Boscovich: Studies of His Life and Work*, L. L. Whyte, ed. London, pp. 200–213.

[3] Sheynin, O. B. (1973). *Arch. History Exact Sci.*, **9**, 306–324. (The most comprehensive overview of Boscovich's statistical concepts.)

[4] Stigler, S. M. (1973). *Biometrika*, **60**, 439–445. (Section 3 discusses Laplace's treatment of the "method of situation.")

(ABSOLUTE DEVIATION
GENERAL LINEAR MODEL
LEAST SQUARES
LINEAR PROGRAMMING
REGRESSION
ROBUST ESTIMATION)

E. SENETA

BOSE–EINSTEIN STATISTICS *See* FERMI–DIRAC STATISTICS

BOWKER'S TEST FOR SYMMETRY

Consider first the cross classification of N experimental units on two associated scales. Each scale is divided into c categories so that the data can be arranged in a $c \times c$ contingency table*. For example, unaided distance vision can be scaled on a c-point scale from weak to strong and the two classifications correspond to the left and right eyes.

Next, let P_{ij} denote the probability that a randomly chosen experimental unit belongs to the ith category of the first scale and the jth category of the second scale. The null hypothesis* of symmetry is specified by $P_{ij} = P_{ji}$. This means that cells symmetrically located with respect to the main diagonal of the table are equally probable. Under the hypothesis of symmetry, the expected cell size is estimated by $(n_{ij} + n_{ji})/2$, where n_{ij} denotes the number observed in the (i, j) cell of the table. By substituting the estimated expected cell values into the usual chi-squared test* statistic, Bowker [2] derived the following statistic:

$$\chi^2 = \sum_{i<j}\sum \frac{(n_{ij} - n_{ji})^2}{n_{ij} + n_{ji}}.$$

Under the null hypothesis of symmetry, χ^2 has an approximate chi-squared distribution* with $c(c-1)/2$ degrees of freedom and rejects the null hypothesis for large values.

In the 2×2 table, the hypothesis of symmetry implies the equality (homogeneity) of the marginal distributions, and vice versa. Hence Bowker's test for symmetry will serve as a test for marginal homogeneity* in this case. Further, in the 2×2 case, the test for symmetry is identical to McNemar's test* [5] for significant change. Conditional on $n_{12} + n_{21} = n$, the test can be based on just n_{12} rather than Bowker's χ^2. Under the hypotheses of symmetry, marginal homogeneity, and no change, n_{12} has a binomial distribution with parameters n and 0.5. Hence an exact rather than approximate test can be carried out. For larger tables, exact tests are generally not available.

In the $c \times c$ case with $c > 2$, symmetry again implies marginal homogeneity but the reverse is no longer true; so, in general, Bowker's test is not appropriate for testing marginal homogeneity. Stuart [8] develops an appropriate test. For an extension of McNemar's test to the case $c > 2$ and a discussion of its relationship to the tests of Bowker and Stuart, see Hettmansperger and McKean [3].

A likelihood ratio test* rather than a chi-squared test can also be used to test for symmetry. The likelihood ratio statistic is given by

$$G^2 = 2\sum_{i \neq j}\sum n_{ij} \ln\left[\frac{2n_{ij}}{n_{ij} + n_{ji}}\right].$$

Under the null hypothesis of symmetry, G^2 has an approximate chi-squared distribution with $c(c-1)/2$ degrees of freedom and rejects the null hypothesis for large values. This test statistic is also derived by Kullback [4] as the minimum information discrimination* statistic.

Stuart [7] provides the following unaided distance vision data recorded on 7477 women, ages 30 to 39, employed in Royal Ordnance factories in Britain.

Right Eye Grade	Left-Eye Grade			
	Highest	Second	Third	Lowest
Highest	1520	266	124	66
Second	234	1512	432	78
Third	117	362	1772	205
Lowest	36	82	179	492

For example, one of the relationships implied by the hypothesis of symmetry is that the population proportion of individuals with highest left-eye grade and lowest right-eye grade is the same as the proportion with lowest left-eye grade and highest right-eye grade. To test for symmetry at the 0.05 significance level, either $\chi^2 = 19.1$ or $G^2 = 19.3$ is referred to 12.6, the chi-squared critical value corresponding to $c(c-1)/2 = 6$ degrees of freedom. Hence both Bowker's test and the likelihood ratio test reject the hypothesis of symmetry*. A closer look at the data shows that 71% of the women had balanced vision, and of the remaining women whose vision differed, 54% had a better right eye, reflecting the asymmetry present in the data. Note that Bowker's χ^2 and G^2 are numerically very close. Kullback [4], who approaches the problem of symmetry through information theory, shows that Bowker's χ^2 is always approximately equal to G^2. No information on the power of these tests is available at present.

It is possible to formulate the hypothesis of symmetry in terms of a log-linear model*. In this framework symmetry can be related to marginal homogeneity and quasi-symmetry* and appropriate tests can be developed. For a detailed discussion, see Biship et al. [1].

References

[1] Bishop, Y. M. M., Fienberg, S. E., and Holland, P. W. (1975). *Discrete Multivariate Analysis, Theory*

and Practice. MIT Press, Cambridge, Mass. (A good source for the log-linear model, the likelihood ratio test, and an analysis of the unaided distance vision data.)

[2] Bowker, A. H. (1948). *J. Amer. Statist. Ass.*, **43**, 572–574.

[3] Hettmansperger, T. P. and McKean, J. W. (1973). *Commun. Statist.*, **2**(6), 551–560.

[4] Kullback, S. (1959). *Information Theory and Statistics*. Wiley, New York.

[5] McNemar, Q. (1947). *Psychometrika*, **12**, 153–157.

[6] Marascuilo, L. A. and McSweeney, M. (1977). *Nonparametric and Distribution-Free Methods for the Social Sciences*. Brooks/Cole, Monterey, Calif. (An excellent reference for the various tests mentioned above.)

[7] Stuart, A. (1953). *Biometrika*, **40**, 105–110.

[8] Stuart, A. (1955). *Biometrika*, **42**, 421–426.

(BUTLER TEST
CATEGORICAL DATA
CONTINGENCY TABLES
LIKELIHOOD RATIO TEST
McNEMAR'S TEST
SYMMETRY, TESTS FOR)

THOMAS P. HETTMANSPERGER

BOX AND COX TRANSFORMATION

The usual assumptions when one analyzes data are the standard assumptions of the linear model (*see* GENERAL LINEAR MODEL), i.e., the additivity of effects, the constancy of variance, the normality of the observations, and the independence of observations. If these assumptions are not met by the data, Tukey [8] suggested two alternatives: either a new analysis must be devised to meet the assumptions, or the data must be transformed to meet the usual assumptions. If a satisfactory transformation is found, it is almost always easier to use than to develop a new method of analysis.

Tukey suggested a family of transformations with an unknown power parameter and Box and Cox [1] modified it to

$$y^{(\lambda)} = \begin{cases} (y-1)/\lambda & \text{for } \lambda \neq 0 \\ \log y, & \text{for } \lambda = 0, \end{cases}$$

where y is an original observation, $y^{(\lambda)}$ the "new" observation, and λ a real unknown parameter.

Box and Cox assumed that for some λ the n transformed observations

$$\mathbf{y}^{(\lambda)} = \begin{bmatrix} y_1^{(\lambda)} \\ y_2^{(\lambda)} \\ \vdots \\ y_n^{(\lambda)} \end{bmatrix}$$

are independent and normally distributed with constant variance σ^2 and mean vector

$$E\mathbf{y}^{(\lambda)} = A\boldsymbol{\theta},$$

where A is a known $n \times p$ matrix and $\boldsymbol{\theta}$ is a vector of parameters associated with the transformed observations.

Box and Cox [1] estimate the parameters by maximum likelihood* as follows. First, given λ,

$$\hat{\boldsymbol{\theta}}(\lambda) = (A'A)^{-1}A'\mathbf{y}^{(\lambda)}$$

and

$$\hat{\sigma}^2(\lambda) = \mathbf{y}'^{(\lambda)}\left[I - A(A'A)^{-1}A'\right]\mathbf{y}^{(\lambda)} \quad (1)$$

are the estimators of $\boldsymbol{\theta}$ and σ^2, respectively. Second, λ is estimated by maximizing the log likelihood function*

$$l(\lambda) = -n\log\hat{\sigma}^2(\lambda)/2 + \log J(\lambda : y),$$

where

$$J(\lambda : y) = \prod_{i=1}^{n} y_i^{\lambda-1}.$$

The maximum likelihood estimator of λ, say $\hat{\lambda}$, is then substituted into (1), which determines the estimators of the other parameters.

Their ideas are nicely illustrated with the analysis of the survival times* of animals that were exposed to three poisons.

Since the original Box–Cox article, there have been many related papers. For example, Draper and Cox [3] derive the precision of the maximum likelihood estimator of λ for a simple random sample, i.e., $Ey_i^{(\lambda)} = \mu$ for $i = 1, 2, \ldots, n$. They found that the approximate variance is

$$V(\hat{\lambda}) = \frac{2}{3n\Delta^2}\left(1 - \frac{1}{3}\gamma_1^2 + \frac{7}{18}\gamma_2\right)^{-1},$$

where

$$\gamma_1 = \mu_3/\sigma^3,$$
$$\gamma_2 = \mu_4/\sigma^4 - 3,$$
$$\Delta = \lambda\sigma/(1 + \lambda\mu),$$

σ^2 is the variance, and μ_i is the ith central moment of the original observations.

Hinkley [5] generalized the Box–Cox transformation to include power transformations to symmetric distributions and showed his "quick" estimate of λ to be consistent* and have a limiting normal distribution. He discusses the same exponential distribution* as Draper and Cox.

Literature

There are many references related to the Box and Cox transformation. Poirier [6] has extended the Box–Cox work to truncated normal distributions*. Some recent results related to large-sample behavior are given by Harnández and Johnson [4]. For information about other power transformations, the reader is referred to POWER TRANSFORMS and MULTIVARIATE POWER TRANSFORMATIONS.

The Box and Cox [1], Draper and Cox [3], Hinkley [5], and Poirier [6] articles are quite informative but highly technical.

At the textbook level, Chapter 10 of Box and Tiao [2] and Chapter VI of Zellner [9] are excellent introductions to the Box–Cox theory of transformations. For computer programs, see Tran Cong Liem [7] and references therein.

References

[1] Box, G. E. P. and Cox, D. R. (1964). *J. R. Statist. Soc. B*, **26**, 211–252.

[2] Box, G. E. P. and Tiao, G. C. (1973). *Bayesian Inference in Statistical Analysis*. Addison-Wesley, Reading, Mass.

[3] Draper, N. R. and Cox, D. R. (1969). *J. R. Statist. Soc. B*, **31**, 472–476.

[4] Harnández, F. and Johnson, R. A. (1979). *Tech. Rep. No. 545*, Dept. of Statistics, University of Wisconsin, Madison, Wis.

[5] Hinkley, D. V. (1975). *Biometrika*, **62**, 101–111.

[6] Poirier, D. J. (1978). *J. Amer. Statist. Ass.*, **73**, 284–287.

[7] Tran Cong Liem (1980). *Amer. Statist.*, **34**, 121.

[8] Tukey, J. W. (1957). *Ann. Math. Statist.*, **28**, 602–632.

[9] Zellner, A. (1971). *An Introduction to Bayesian Inference in Econometrics*. Wiley, New York.

(POWER TRANSFORMS
TRANSFORMATIONS TO
NORMALITY)

LYLE D. BROEMELING

BOX–JENKINS MODEL

Consider a time series* of "length" n denoted by $z_1, \ldots, z_n$. For example, the z_i $(i = 1, \ldots, n)$ might be the weekly wage bills for a large company during a calendar year, with $n = 52$ in that case. Then, if the series $\{z_i : i = 1, \ldots, n\}$ is taken as a realization of some underlying stochastic (time) process* $\{Z_i\}$, we are evidently interested in the statistical structure of this process.

One way of describing such a structure is to obtain a parametric model for the process, and the method of time-series analysis associated with and advocated by Box and Jenkins first postulates a plausible class of models for initial investigation. It then proceeds to tentatively choose or "identify" a promising member from this class, whose particular parameters are next efficiently estimated; and finally, the success of the resulting fit is assessed. The now precisely defined model (identified and estimated) is either accepted by this verification stage or the diagnostic checks carried out will find it wanting in certain respects and should then suggest a sensible modified identification—after which further estimation and checking takes place, and the cycle of identification, estimation, and verification is repeated until a satisfactory fit obtains.

It is important not to overparameterize* the model, since, although this might improve the goodness of fit* for the series history at hand, it is likely to result in the model portraying spurious features of the sampled data, which may detract from the usefulness of the achieved fit. For example, unnecessarily poor forecasts for future observations on the series are a typical outcome of ignoring the principle of parsimonious parameterization.

(Although it is frequently not realized, part of the verification should be to ensure that the fitted model does make sense in the context from which the data were drawn. This model interpretation is sometimes thought of as an extra fourth stage to the Box–Jenkins cycle.)

The univariate models, allowed for by Box and Jenkins, are a very general class of linear process* which may be taken as being driven by some completely random sequence of unknown "shocks," denoted by $\{A_i\}$, say. These A_i are uncorrelated and identically distributed zero-mean random variables, all with the same variance, σ_A^2 say, and $\{A_i\}$ is then referred to as a "white noise"* process. It is frequently convenient to assume that the shocks are in fact normally distributed*, but this assumption then needs to be justified in applications.

The process of interest, $\{Z_i\}$, is considered to be obtained by applying a linear filter* to the shocks $\{A_i\}$, according to

$$Z_i = A_i + \psi_1 A_{i-1} + \psi_2 A_{i-2} + \cdots \quad (1)$$

for suitable choices of $\psi_1, \psi_2, \ldots$. In this representation, each Z_i is taken as being formed from a weighted sum of the current and previous shocks, with psi weights $\psi_0 = 1$, $\psi_1, \psi_2, \ldots$.

The simplest case is when Z_i is itself completely random, giving for all i,

$$Z_i = A_i,$$

which is just white noise.

Next, we have the so-called moving average models* of general order $q \geqslant 0$, denoted by MA(q) and satisfying

$$Z_i = A_i + \theta_1 A_{i-1} + \cdots + \theta_q A_{i-q}, \quad (2)$$

which of course reduces to white noise in the special case with $q = 0$. One usually restricts the theta parameters $\theta_1, \ldots, \theta_q$ to values such that the polynomial $\theta_q(\zeta) \equiv 1 + \theta_1\zeta + \cdots + \theta_q\zeta^q$, in the complex variable ζ, has no zeros within the unit circle. This is the "invertibility" condition.

It is convenient to introduce the backshift operator B such that, for any function $f(\cdot)$ (e.g., z_i or A_i) and all integers i and j, $B^j f(i) \equiv f(i-j)$. Then, for instance, (2) can be rewritten as

$$Z_i = \theta_q(B) A_i, \quad (3)$$

where $\theta_q(B)$, a polynomial in B of degree q, is an operator obtained by writing B in place of ζ in $\theta_q(\zeta)$.

If θ_j is put at φ_1^j for $j = 1, \ldots, q$ and q is allowed to go to infinity, $\theta_q(B)$ becomes $(1 - \varphi_1 B)^{-1}$ and we can then rewrite (3) as

$$(1 - \varphi_1 B) Z_i = A_i \quad (4)$$

or, alternatively, as

$$Z_i = \varphi_1 Z_{i-1} + A_i. \quad (5)$$

Expressions (4) and (5) represent the first-order autoregressive model, or AR(1), in which each Z_i is "autoregressed" on its previous Z_{i-1} value. (*see* ARMA MODELS).

This type of model generalizes to AR($p + d$), with $p + d \geqslant 0$, the $(p + d)$th-order autoregressive model

$$Z_i = \varphi_1 Z_{i-1} + \cdots + \varphi_{p+d} Z_{i-p-d} + A_i \quad (6)$$

or, in backshift notation,

$$\varphi_{p+d}(B) Z_i = A_i,$$

$$\varphi_{p+d}(B) = 1 - \varphi_1 B - \cdots - \varphi_{p+d} B^{p+d}$$

and, again, none of the zeros of $\varphi_{p+d}(B)$ must lie within the unit circle. Once more, when $p + d = 0$, (6) reduces to white noise.

Box and Jenkins in fact distinguish between those zeros of $\varphi_{p+d}(B)$ that lie on the unit circle, say d of them, as opposed to those lying outside it, the remaining p zeros. They would then write $\varphi_{p+d}(B)$ in the associated factored form $\phi_p(B) S_d(B)$, where $\phi_p(B)$, the "stationary" autoregressive part, has no zeros on the unit circle. If $d = 0$, this condition ensures the "stationarity"* of the process.

A typical example of $S_d(B)$ is $(1 - B)^d$, where the operator $(1 - B)$ effects a (unit) differencing of the series under study. Thus $(1 - B)z_i = z_i - z_{i-1}$ and $(1 - B)^d$ results in d successive (unit) differences being taken. (It is very rare that a degree of differencing d needs to be greater than 1 or 2.)

Operators of the form $S_d(B)$ are termed "simplifying operators." They represent ways in which the raw series should be transformed initially before detailed statistical analysis is begun. As they yield linear transformations of the data, they do not lead to inefficient analysis in the ways that nonlinear transformations do.

However, in certain instances, notably for the purpose of stabilizing the variance of a series and obtaining more nearly Gaussian behavior, nonlinear transformations of the raw data might be made initially. These are usually of the Box–Cox* [6] form: i.e., $z_i \to z_i^{(\lambda,m)}$, where

$$z_i^{(\lambda,m)} = \begin{cases} (z_i + m)^\lambda & (\lambda \neq 0) \\ \ln(z_i + m) & (\lambda = 0) \end{cases}$$

and m and λ need to be chosen by the analyst. *See* BOX AND COX TRANSFORMATION. But it should be noted that, unless the resulting change in metric appears desirable from other considerations, such transformation tends to be controversial. (For instance, good forecasts might be obtained for the transformed series but the advantage lost on transforming back.) To avoid cumbrous notation, we will moreover suppress the transformation superscripts in what follows.

Note that the general stationary AR(p) model

$$Z_i = \phi_1 Z_{i-1} + \cdots + \phi_p Z_{i-p} + A_i \quad (7)$$

can also be written as an MA(∞), $Z_i = \phi_p^{-1}(B)A_i$, which is a special case of (1), with $\psi(B) = (1 + \psi_1 B + \psi_2 B^2 + \cdots) = \phi_p^{-1}(B)$. A further generalization, to (7) and (2), is the mixed (stationary) autoregressive moving average model of order (p, q), or ARMA (p, q),

$$Z_i = \phi_1 Z_{i-1} + \cdots + \phi_p Z_{i-p} + A_i + \theta_1 A_{i-1} + \cdots + \theta_q A_{i-q},$$

which, in operator notation, becomes

$$\phi_p(B)Z_i = \theta_q(B)A_i.$$

Introducing unit circle nonstationary zeros into the autoregressive part, we first get models of the form

$$\phi_p(B)(1 - B)^d Z_i = \theta_q(B)A_i, \quad (8)$$

which are termed autoregressive integrated moving average* models of order (p, d, q), or ARIMA (p, d, q); when $(1 - B)^d$ is replaced by more general $S_d(B)$, the models have been described as ARUMA (p, d, q)*.

Should B be replaced by B^T in (8), where T is some integer greater than unity, we get a purely seasonal model of period T. Such models are usually denoted by

$$\Phi_P(B^T)(1 - B^T)^D Z_i = \Theta_Q(B^T)A_i, \quad (9)$$

where $(1 - B^T)$ effects a seasonal differencing, according to $(1 - B^T)z_i = z_i - z_{i-T}$, and capital letters help to distinguish (9) from the earlier discussed nonseasonal models. Thus the purely seasonal first-order stationary autoregressive process of period 12, conveniently written as AR$_{12}$(1), would have the form $Z_i = \Phi_1 Z_{i-12} + A_i$.

Mixed nonseasonal seasonal models can occur. These may be expressed as

$$\phi_p(B)\Phi_P(B^T)(1 - B)^d(1 - B^T)^D Z_i = \theta_q(B)\Theta_Q(B)A_i \quad (10)$$

and indeed, models with more than one seasonal period are possible. For instance, hourly electricity demand over the year would be likely to depend, not only on B and B^{24} (daily), but on B^{168} (weekly) as well. Also, note that multiplicative factors such as

$$(1 - \phi_1 B)(1 - \Phi_1 B^T) = 1 - \phi_1 B - \Phi_1 B^T + \phi_1\Phi_1 B^{1+T}$$

may be generalized to, say,

$$1 - \phi_1 B - \Phi_1 B^T + \alpha B^{1+T}.$$

Finally, univariate models can be written with a deterministic trend $t(i)$ on the right-hand side, although this is frequently removed (at least in part) by the unit and seasonal differencing operators* which are often employed. For instance, if a series contains a linear trend ci, with slope c, simple differencing of the raw data will reduce this to just a constant c, since $(1 - B)ci = c[i - (i - 1)] = c$. The simplest "trend"* occurs when $E[Z_i]$ exists and is a nonzero constant μ. Then $t(i) = \mu$ and to achieve models such as (1), Z_i is replaced by $Z_i - \mu$. So when $\bar{z}$, the series mean, is significantly different from zero (as would be the case in our wages example), the z_i are considered to be replaced by the mean-corrected series, $\{\tilde{z}_i = z_i - \bar{z}\}$, which is a (linear) Box–Cox transform* of the original data.

The family of linear models of the types described above are commonly referred to as Box–Jenkins models. Although they were mostly originally due to earlier workers, Box and Jenkins deserve the credit for bringing together, developing and popularizing an extensive methodology (known as the Box–Jenkins approach), which has been highly successful as a means of analyzing time series met with in a very wide range of application areas. This success is founded on the fact that the various Box–Jenkins models can, among them, mimic the behaviors of diverse types of series—and do so adequately, usually without requiring very many parameters to be estimated in the final choice of model. The disadvantage, however, is that successful analysis generally requires considerable skill—although some quite promising automatic modeling computer packages are beginning to appear.

The formal objective of a Box–Jenkins analysis may be considered as discovering that parsimoniously parameterized filter* which satisfactorily reduces the original series to a residual white noise series $\{a_i\}$, with small variance. What is satisfactory will depend on the context from which the data were drawn, and on the purpose of the analysis, as well as on purely statistical criteria.

The main analytical tool for series identification is the sequence of sample serial correlations*, $\{r_1, \ldots, r_{n-1}\}$, where

$$r_k = \sum_{i=1}^{n-k} \tilde{z}_i \tilde{z}_{i-k} \Big/ \sum_{i=1}^{n} \tilde{z}_i^2,$$

although frequently only about the first quarter of them are computed. These $\{r_k : k = 1, \ldots, n-1\}$ are taken to mimic the theoretical autocorrelations $\{\rho_k\}$, defined as $\text{cov}[Z_i, Z_{i-k}]/\text{var}[Z_i]$, so the task of identification is, given the observed sample correlation pattern for the series, to try to match it with the known population values for some particular process.

For instance, a proper MA(1) model is characterized by $\rho_1 \neq 0$, $\rho_k = 0$ $(k > 1)$. So a set of serials with r_1 substantial, but later r_k negligible, would suggest that an MA(1) should be tentatively tried. What count as substantial or negligible serial "spikes" depend on the particular model being considered and the length of the observed series. Given these facts, significance tests are available.

Certain structures in the sampled correlations can suggest that a simplifying operator first be applied to the raw data. For example, a slow, roughly linear declining sequence of positive values, for the early r_k, is often taken as an indication that unit differencing is necessary.

Another useful tool for identification is the sequence of partial correlations—sampled $\{p_k\}$, theoretical $\{\pi_k\}$. Thus an AR(1) is characterized by $\pi_1 \neq 0$, $\pi_k = 0$ $(k > 0)$. So if p_1 is significant, but none of the later p_k are, an AR(1) model would be indicated.

A frequent purpose for analyzing time series is to obtain good forecasts*. Given a series $\{z_1, \ldots, z_n\}$ running up to time n = now, the aim then, typically, is to forecast z_{h+n} at h time intervals hence. If we assume that the generating process has the form (1), it can be shown that the optimal least-squares* forecast, ${}_hf_n$ say, is the expected value of Z_{h+n} conditional on the information available at time n. Now

$$Z_{h+n} = A_{h+n} + \psi_1 A_{h-1+n} + \cdots + \psi_{h-1} A_{1+n} + \psi_h A_n + \cdots .$$

So ${}_hf_n = \psi_h a_n + \cdots$ (since the expectations of future shocks are all zero, whereas those for past and present ones take the actual values that have already occurred) and the forecast error, ${}_h\epsilon_n = z_{h+n} - {}_hf_n$, is given by

$${}_h\epsilon_n = a_{h+n} + \psi_1 a_{h-1+n} + \cdots + \psi_{h-1} a_{1+n},$$

with variance ${}_hV_n = (1 + \psi_1^2 + \cdots + \psi_{h-1}^2) \cdot \sigma_A^2$, from which probability limits for the forecasts can be obtained, on replacing σ_A^2 by its estimate, the sample variance* of the estimated shock series, $\{\hat{a}_i : i = 1, \ldots, n\}$, which turns out to be just the series of residuals, $\{\hat{a}_i = z_i - \hat{z}_i\}$, where the $\hat{z}_i$ are the estimated values for the original series, as obtained from the fitted model.

Currently, there is much interest in extending the linear models discussed so far, to cater for at least part of the nonlinearity common in some applications areas. One extension gives the bilinear model, which is achieved by introducing additional product terms $Z_{i-u}A_{i-v}$ into the right of the linear model. Other popular generalizations involve substituting time-varying parameters for the constants in (10).

However, Box and Jenkins themselves (in conjunction with their co-workers) have developed certain major extensions to the univariate modeling described above. First, they considered building transfer function* models, which would perhaps improve the forecasts obtained for the series of interest $\{z_i\}$ by extracting relevant information contained in some appropriate leading indicator series $\{y_i\}$, say. This is done by relating the current z to current and previous y, according to a model

$$Z_i = \omega(B)B^{\delta}Y_i + E_i,$$

where $\omega(B)$ is a linear filter $(\omega_0 + \omega_1 B + \cdots)$, with $\omega_0 \neq 0$, and the B^{δ} factor indicates that there is a delay of δ units before a y-value can begin to affect the observed z. $\{E_i\}$ is a sequence of error terms, which is assumed to follow some ARMA(p,q) process. In general, parsimonious parameterization can be achieved by writing $\omega(B)$ as the quotient of two finite-length operators, $\alpha(B)$ and $\beta(B)$, say, so that the model fitted has the form

$$Z_i = \frac{\alpha(B)}{\beta(B)} Y_{i-\delta} + \frac{\theta_q(B)}{\phi_p(B)} A_i. \quad (11)$$

Box and Jenkins [7] provide a well-defined iterative model-building procedure for estimating these transfer function-noise processes, which is analogous to that for the univariate case. Identification of the relationship between the "input" y and "output" z series relies heavily on the cross-correlations between the two series, preferably after a procedure called "prewhitening"* has been effected. Here the filter needed to reduce the y_i to white noise is initially determined and then this prewhitening filter is applied to both the input and the output series before the cross-correlations are computed. (However, not all experts are agreed that this is the best way to "prewhiten.")

Equation (11) can be simply generalized to cater for several input series; again, a deterministic trend can be incorporated on the right, the Z_i and Y_i can be transformed initially, seasonal factors can be introduced into the various filters, and differencing can be employed.

The univariate models can also be generalized to mutlivariate ones with the basic form

$$\boldsymbol{\phi}(B)\mathbf{Z}_i = \boldsymbol{\theta}(B)\mathbf{A}_i, \quad (12)$$

where the matrix operators $\boldsymbol{\phi}(B)$ and $\boldsymbol{\theta}(B)$ have elements that are polynomials in B of general finite order, with the restrictions that those along the leading diagonals start with unity, whereas the rest start with powers of B, and the stationarity and (strict) invertibility conditions are, respectively, that all the zeros of the determinants $|\boldsymbol{\phi}(B)|$ and $|\boldsymbol{\theta}(B)|$ lie outside the unit circle.

A further extension is to consider multivariate transfer function models whose basic structure is

$$\mathbf{Z}_i = \boldsymbol{\Omega}(B) \otimes \mathbf{Y}_i + \mathbf{E}_i, \quad (13)$$

where the transfer function matrix, $\boldsymbol{\Omega}(B)$, has elements of the form $\omega(B)B^{\delta}$, $\otimes$ denotes the Kronecker product*, and $\mathbf{E}_i$ follows a multivariate model such as (12).

Again, for models (12) and (13), as previously, there is a three-stroke cycle of identification, estimation, and diagnostic checks for obtaining satisfactory fits; and, as well as cross-correlations, partial cross-correlations between individual pairs of residual series are also used. When building any of the models (11), (12), or (13), univariate stochastic modeling—of the sort extensively discussed above—is required for the individual series. So skillful univariate analysis is a prerequisite for all of the more advanced Box–Jenkins methodology.

Finally, in all the processes so far mentioned, it may be necessary to take account of "abnormal" events, such as strikes, changes in the law or freak weather conditions. Box–Jenkins "intervention" models allow such effects to be represented by dummy variables, typically introduced as a filtered pulse on the input side. For instance, to model a step change in level of magnitude Λ, occurring at time I, one needs to include a term $(1-B)^{-1}\Lambda\prod_i^{(I)}$ on the right, where

$$\prod_i^{(I)} = \begin{cases} 1 & \text{for } i = I \\ 0 & \text{otherwise.} \end{cases}$$

Interventions are thus treated just as particularly simple cases of leading indicator series. (Outliers* and missing observations* can also be dealt with by interpolating realistic values which are estimated from the remaining data.)

Literature

We complete this discussion of the Box–Jenkins model with a brief look at the literature. First, some readable articles: Anderson [2] provides a formula-free introduction, and a more mathematical treatment of the same material is found in Anderson [3]. Newbold [10] gives an excellent treatment at about the same level, and Anderson [4] states a later updated view.

As for books, Anderson [1] and Nelson [9] supply simple introductions, Box and Jenkins [7] is the best work for reference purposes, and Jenkins (in Anderson [5]) provides about the only reputable published account of the more advanced topics, although Granger and Newbold [8] is of related interest for multivariate* modeling, as is Newbold and Reed (in Anderson [5]).

References

[1] Anderson, O. D. (1975). *Time Series Analysis and Forecasting: The Box–Jenkins Approach*. Butterworths, London. Second Edition, 1982.

[2] Anderson, O. D. (1976). *Statist. News*, No. 32 (February), 14–20.

[3] Anderson, O. D. (1976). *Math. Sci.*, **1** (January), 27–41.

[4] Anderson, O. D. (1977). *The Statistician*, **25** (December), 285–303.

[5] Anderson, O. D., ed. (1979). *Forecasting*. North-Holland, Amsterdam.

[6] Box, G. E. P. and Cox, D. R. (1964). *J. R. Statist. Soc. B*, **26**, 211–252.

[7] Box, G. E. P. and Jenkins, G. M. (1970). *Time Series Analysis: Forecasting and Control*. Holden-Day, San Francisco.

[8] Granger, C. W. J. and Newbold, P. (1977). *Forecasting Economic Time Series*. Academic Press, New York.

[9] Nelson, C. R. (1973). *Applied Time Series Analysis for Managerial Forecasting*. Holden-Day, San Francisco.

[10] Newbold, P. (1975). *Operat. Res. Quart.*, **26**, 397–412.

(AUTOREGRESSIVE–INTEGRATED MOVING AVERAGE MODELS AUTOREGRESSIVE MOVING AVERAGE MODELS)

O. D. ANDERSON

BOX–MULLER TRANSFORMATION

If U_1 and U_2 are independent standard uniform* variables then

$$X_1 = (-2\ln U_1)^{1/2}\cos 2\pi U_2$$

$$X_2 = (-2\ln U_1)^{1/2}\sin 2\pi U_2$$

are independent unit normal* variables [1].

By means of this transformation, random normal deviates can be derived directly from a source of random uniform variates, such as a table of random numbers* or a generator program in a digital computer.

In ref. 1 it is pointed out that simple extensions of the method provide values of random variables distributed as χ^{2*}, F^*, and t^*. For a more recent discussion, see Golder and Settle [2].

References

[1] Box, G. E. P. and Muller, M. E. (1958). *Ann. Math. Statist.*, **29**, 610–611.

[2] Golder, E. R. and Settle, J. G. (1976). *Appl. Statist.*, **25**, 12–20.

(RANDOM-NUMBER GENERATION)

BRADFORD DISTRIBUTION

A special truncated (from above) Pearson Type VI distribution*, with density function

$$[\theta/\log(1+\theta)](1+\theta x)^{-1}$$
$$(0 < x < 1; \theta > -1).$$

It is related to the Yule* and Zipf* distributions. Bradford [1] used it in connection with distributions of frequency of citation of references.

Additional information on the Bradford distribution is provided by Johnson and Kotz [2].

References

[1] Bradford, S. C. (1948). *Documentation*. Crosby Lockwood, London.

[2] Johnson, N. L. and Kotz, S. (1970). *Continuous Univariate Distributions*, Vol. 2. Wiley, New York, p. 89.

(PARETO DISTRIBUTION
STATISTICS IN LIBRARY SCIENCE)

BRADLEY–TERRY MODEL

A model representing the results of experiments in which responses are pairwise rankings of treatments (so-called paired comparison experiments*). It assigns preference probabilities to each of the $\binom{t}{2}$ pairs among t treatments in terms of a set of $t-1$ parameters. Specifically, it postulates "treatment parameters" $\pi_1, \ldots, \pi_t$, $\pi_i \geqslant 0$, $i = 1, \ldots, t$, associated with treatments $T_1, \ldots, T_t$. These parameters represent relative selection probabilities for the treatments, subject to the constraints $\pi_i \geqslant 0$ $(i = 1, \ldots, t)$ and $\sum_{i=1}^t \pi_i = 1$. The probability that treatment T_i is preferred over treatment T_j in a single comparison is $\pi_i/(\pi_i + \pi_j)$ for all i and j.

Let a_{ij} denote the *number* of times treatment T_i is preferred over treatment T_j in n_{ij} comparisons of T_i and T_j. The likelihood function* is then given by

$$L(\boldsymbol{\pi}, \mathbf{A}) = \prod_{i=1}^{t} \pi_i^{a_i} \prod_{i<j} (\pi_i + \pi_j)^{-n_{ij}},$$

where $a_i = \sum_{j \neq i} a_{ij}$ and $\boldsymbol{\pi}$ and $\mathbf{A}$ are vectors of π's and a's, respectively. The maximum likelihood estimators* $\hat{\pi}_1, \ldots, \hat{\pi}_t$ of $\pi_1, \ldots, \pi_t$ are solutions of the equations

$$\hat{\pi}_i = \frac{a_i}{\sum_{j \neq i} n_{ij}/(\hat{\pi}_i + \hat{\pi}_j)} \qquad (i = 1, \ldots, t)$$

subject to $\sum_{i=1}^t \hat{\pi}_i = 1$.

For testing the null hypothesis* of equality of treatment selection probabilities,

$$H_0 : \pi_1 = \pi_2 = \cdots = \pi_t = 1/t$$

against the alternative,

$$H_1 : \pi_i \neq \pi_j \quad \text{for some } i, j, i \neq j, \quad i, j = 1, \ldots, t,$$

the likelihood ratio statistic* is

$$-2 \ln \lambda_1 = 2N \ln 2 - 2B,$$

where $N = \sum_{i<j} n_{ij}$ and $B_1 = \sum_{i<j} n_{ij} \ln(\hat{\pi}_i + \hat{\pi}_j) - \sum a_i \ln \pi_i$. For large n_{ij}, the statistic $-2\ln\lambda_1$ is distributed under H_0 approximately as a χ^2_{t-1} variable. Tables of values of B_1, together with exact significance levels, are given by Bradley and Terry [3] and Bradley [1]. Factorial experiments* based on this model are discussed by Littell and Boyett [5]. Extensions were also studied by van Baaren [6].

References

[1] Bradley, R. A. (1954). *Biometrika*, **41**, 502–537.

[2] Bradley, R. A. (1976). *Biometrics*, **32**, 213–232. (A comprehensive article covering various extensions as well.)

[3] Bradley, R. A. and Terry, M. E. (1952). *Biometrika*, **39**, 324–345.
[4] Dykstra, O. (1960). *Biometrics*, **16**, 176–188.
[5] Littell, R. C. and Boyett, J. M. (1977). *Biometrika*, **64**, 73–77.
[6] van Baaren, A. (1978). *Statist. Neerlandica*, **32**, 57–66.

(PAIRED COMPARISONS)

BRADSTREET INDEX *See* INDEX NUMBERS

BRANCH-AND-BOUND METHOD

The method of branch and bound was originally devised by Land and Doig [7] and independently by Little et al. [8] to solve integer programs*. It has since proved to be so powerful and general that its use has broadened enormously, and now includes a number of statistical applications. Since all but illustrative examples must be run on computers, this extension has also gone hand in hand with the development in computer science of tree-processing algorithms.

Branch and bound is useful for solving problems (usually combinatoric in nature) that can be manipulated into the following form:

1. One wishes to find the minimum of some function over a denumerable but usually large discrete set of possibilities. (Maximization problems can be solved by a trivial modification.)
2. The possibilities can be regarded as the terminal nodes of a tree.
3. The function that one wishes to minimize can be extended to the interior nodes* of a tree* (*see* GRAPH THEORY) in such a way that its value at any node cannot exceed that at any successor node and must equal or exceed that at any predecessor node.

Selection of a subset regression* provides a good example. Suppose that we have a problem with 11 predictors and seek the best subset of 5. We can set up a tree whose first node represents the regression on all 11 predictors. This has 11 successor nodes corresponding respectively to the deletion of predictors 1, 2, . . . , 11. Each of these nodes generates a number of successors corresponding to the deletion of a second variable, and so on. (It is obviously sensible, but is not necessary theoretically, to apply additional rules to ensure that each such subregression appears only once in the tree.) Continuing to define additional levels of the tree as corresponding to the deletion of further variables, we end up with a tree whose terminal nodes correspond to the 462 possible subsets of 5 predictors. The function we wish to minimize is the residual* sum of squares for that subset, and this extends at once to the interior nodes and satisfies requirement 3. The problem is then ripe for solution by branch and bound.

The solution of a problem by branch and bound involves traversing the tree (conceptually at least) from top to bottom. From the time the method first reaches a terminal node, it maintains a record of v, the function value at the best node located so far, and this is used to prune the tree. If any interior node has a value exceeding v, then so by property 3 must all its successors, so these need not be looked at. The secret of a successful application of branch and bound lies in pruning so successfully that only a tiny fraction of the branches need be followed down to their terminal nodes.

An example adapted from Narula and Wellington [10] is illustrated in Fig. 1. It shows a four-predictor regression problem; at each node the circled numbers show the variables included in the regression, while the superscript number is the sum of relative errors (SRE) for that subregression. By introducing variables in stepwise regression* order, we get successively the subsets 1, 12, 123, and 1234 with SREs 2.47, 2.45, 2.22, and 2.20, respectively.

Let us suppose that we wish to find the best subset of size 2. The bound v is initially 2.45 (from the set 12 used on the way to the root of the tree). Traversing the tree from the right, we may then prune at the 234 node

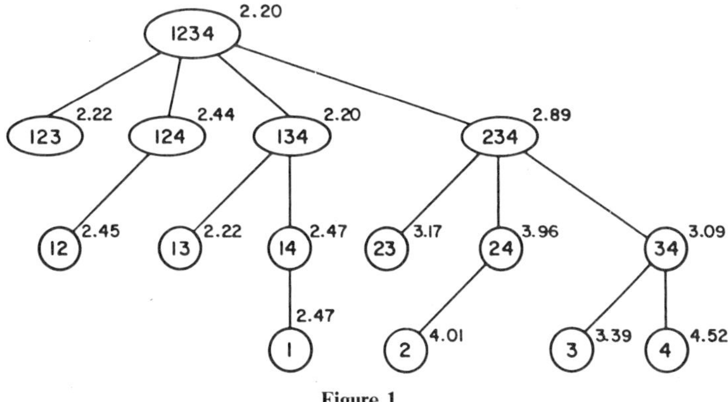

Figure 1

at once since its SRE of 2.89 exceeds v. The 134 node passes the pruning test and its successor nodes are investigated. Node 14 fails the test, but 13 has an SRE of 2.22, which becomes the new value of v. The 134 node has now been "fathomed" (i.e., analyzed completely). Next, the 124 node is investigated, but as its SRE of 2.44 exceeds the current value of $v = 2.22$, it is pruned. The 123 node may be ignored, as it has no successors corresponding to two-predictor regressions, so the problem is solved.

In this example we used the subset of size 2 found by stepwise regression* to provide what turned out to be an excellent starting bound v, and by putting the best predictor first and the worst last, were able to prune very successfully. If it were not convenient to compute this bound initially, it would be better to reorder the variables from worst to best. This ensures that the 12 node is the first of size 2 investigated by the algorithm, and again leads to effective pruning.

Another important principle is illustrated. If one wants not the best solution, but any one guaranteed to be within say 5% of optimum, this is given by any two-variable solution within 5% of the four-variable SRE or the best three-variable SRE. This possibility of finding not necessarily optimal but good solutions is often useful.

Much of the successful application of branch and bound is an art. First, it is necessary to manipulate the problem into a tree structure and extend the function's definition to the interior nodes, which are usually artificial constructs. Then it is extremely important to arrange the tree so that good terminal nodes are found quickly. Logically equivalent formulations that differ in one or both of these two aspects may lead to solution times differing by a factor of 1000 or more. Unfortunately, although the general principles involved are very simple, each application area of branch and bound has its own special structure and features, and a successful implementation depends on a recognition of these.

Further Reading

There is little published on branch and bound per se, but a welter of information on its applications, especially to integer programming problems: numbers of the latter may be found, for example, in every volume of *Operations Research**, which may be considered the natural home for mathematical writings on the topic. A useful exposition of the mathematical properties of branch and bound has been given by Mitten [9], and a very readable commentary on computer implementation of the method by Baker [2].

A discussion of the solution of subset regressions problems by branch and bound is given in the extremely important paper by Furnival and Wilson [4], and Narula and Wellington [10] extend the ideas to include use of the L_1 norm instead of least squares*. In the examples of the first of these papers, the number of operations in a p predictor regression was 88×1.4^p, a much more slowly varying function that the $O(2^p)$ required for investigating all subsets, and one that implies that as many as 30 predictors may be handled quite comfortably. Both these papers are quite technical and devote much attention to "labeling," which is computationally important because it defines the way in which the tree is traversed. The regression problem in which coefficients are constrained to be nonnegative is set out in a readable account by Armstrong and Frome [1]. Other short and easy-to-follow papers describing the application of branch-and-bound methods to statistical problems are on finding the nearest neighbors* in a set of points A to the points in set B [3], cluster analysis* [6], and determination of stochastic rankings* [11], and Hawkins [5] points out that the Furnival and Wilson code may also be used for subset discriminant analysis*.

References

[1] Armstrong, R. D. and Frome, E. L. (1976). *Technometrics*, **18**, 447–450.

[2] Baker, K. R. (1974). *Introduction to Sequencing and Scheduling*. Wiley, New York.

[3] Fukunaga, K. and Narendra, P. M. (1975). *IEEE Trans. Computers*, **C-24**, 750–753.

[4] Furnival, G. M. and Wilson, R. W. (1974). *Technometrics*, **13**, 403–408.

[5] Hawkins, D. M. (1976). *J. R. Statist. Soc. B*, **38**, 132–139.

[6] Koontz, W. L. G., Narendra, P. M. and Fukunaga, K. (1975). *IEEE Trans. Computers*, **C-24**, 908–914.

[7] Land, A. and Doig, A. (1960). *Econometrica*, **28**, 497–520.

[8] Little, J. D. C., Murty, K. G., Sweeney, D. W., and Karel, C. (1963). *Operat. Res.*, **11**, 972–989.

[9] Mitten, A. (1970). *Operat. Res.*, **18**, 24–34.

[10] Narula, S. C. and Wellington, J. F. (1979). *Technometrics*, **21**, 299–306.

[11] Singh, J. (1976). *Ann. Statist.*, **4**, 651–654.

(GRAPH THEORY
INTEGER PROGRAMMING
STEP-DOWN PROCEDURE
STEP-UP PROCEDURE
STEPWISE REGRESSION
TRAVELING-SALESMAN PROBLEM)

D. M. HAWKINS

BRANCHING PROCESSES

Branching processes originate as models for the development of a population, the members of which are termed *individuals* (or *particles*) and which are each capable of reproducing. The basic assumption, the *branching property*, states that different individuals reproduce *independently* of each other and, more generally, that their whole lines of descent are independent.

Most applications arise in biology and physics. Obvious examples in biology are the growth of a population of cells, say in a tumor, or the time evolution of a species within a closed area (in examples of this type, one often considers a single-sex model in order to maintain the branching property). The population modeled by the branching process could also be the individuals in some larger population having some specific characteristic, say carrying a mutant gene inherited by some of the offspring or having an infectious disease that may be transmitted to healthy individuals. A notable example in physics is the neutron chain reaction in nuclear reactors and bombs; and a number of special cascade phenomena, in particular in cosmic rays (electron–photon), have received considerable attention.

THE GALTON–WATSON PROCESS

The prototype of a branching process in discrete time is the Galton–Watson process $Z_0, Z_1, \ldots$ (historically connected also with the name of Bienayme)*. The individuals reproduce independently and each according to the *offspring distribution*, specified by the probabilities p_k of getting k children. That is, if $X_{n,i}$ represents the number of children of the ith individual of the nth generation, then

$$Z_{n+1} = \sum_{i=1}^{Z_n} X_{n,i}$$

and the $X_{n,i}$ are independent with $P(X_{n,i} = k) = p_k$. The realizations of the process are often depicted in a *family tree* (Fig. 1).

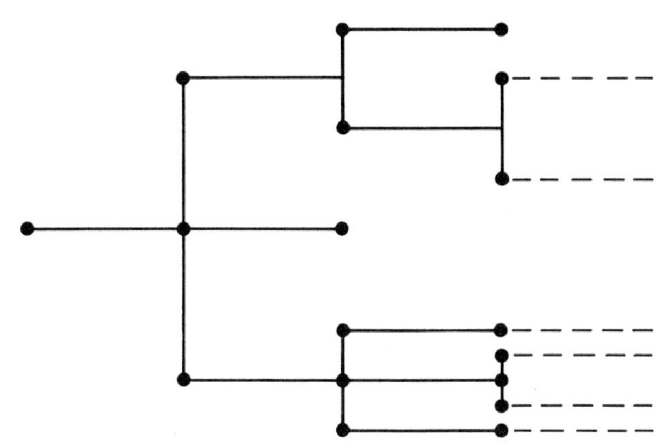

Figure 1 Family tree of a Galton–Watson path.

Because of the branching property, there is essentially no loss of generality in assuming that $Z_0 = 1$ since otherwise one just adds Z_0 independent copies. To avoid trivialities, assume also that $p_k < 1$ for all k.

The Galton–Watson process is a time-homogeneous Markov chain* with state space $\{0, 1, 2, \ldots\}$. State 0 is absorbing and all other states transient, and almost all realizations belong either to the set $E = \{Z_n = 0 \text{ eventually}\}$ of extinction or to $\{Z_n \to \infty\}$. Incidentally, this instability of the process (typical also of many generalizations) is a severe objection to the model in many applied situations since it predicts either infinite growth (irrespective of capacity of the environment) or ultimate extinction. For this reason, branching models are often appropriate only as approximations for the population development in its early stages.

Exact distribution results are available if the offspring distribution is modified geometric*, $p_k = cb^k$, $k > 0$, in which case the distribution of Z_n is again of the same type (with parameters depending on n). Simple recurrence formulas for the probability generating functions* (PGF) $f_n(s) = Es^{Z_n}$ follow immediately from the branching property (Fig. 2). In fact, letting $f(s) = f_1(s) = \sum_0^\infty s^k p_k$ be the offspring PGF, one gets $f_{n+1}(s) = f_n(f(s)) = \cdots = f^{o(n+1)}(s)$, the $(n + 1)$st functional iterate of f. The extinc-

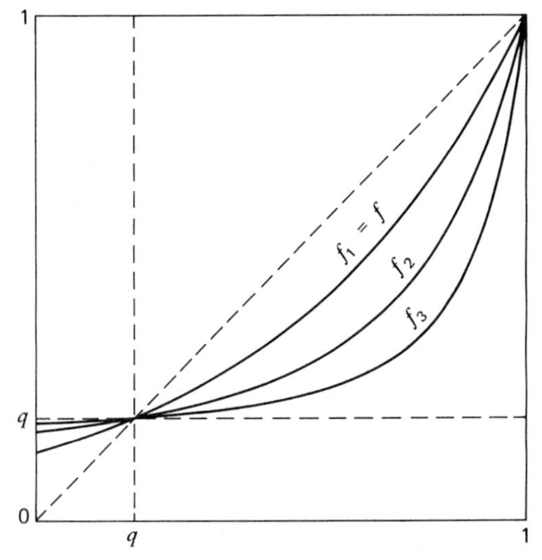

Figure 2 Iterated probability generating functions.

tion probability $q = P(Z_n = 0 \text{ eventually})$ can be found as the smallest root of the equation $s = f(s)$.

In a closer study of the limit theory, three main cases arise according to the value of the offspring mean $m = \sum_0^\infty kp_k$.

Case 1. The supercritical case $1 < m < \infty$. Here $q < 1$, $\{W_n\} = \{Z_n/m^n\}$ is a nonnegative martingale* with unit mean and hence a.s. convergent to a limit $W \in [0, \infty)$. The process indeed grows at rate m^n under weak regularity conditions. In fact, $\{0 < W < \infty\} = \{Z_n \to \infty\}$ a.s. if and only if

$$\sum_{k=2}^{\infty} k \log k \; p_k < \infty \qquad (1)$$

while $W = 0$ a.s. otherwise.

Case 2. The Critical Case $m = 1$. Here $q = 1$. More precisely (with a second moment assumption), $nP(Z_n \neq 0)$ has a limit and conditioned on $\{Z_n \neq 0\}$, Z_n/n has a limiting exponential distribution*.

Case 3. The Subcritical Case $0 < m < 1$. Here $q = 1$ and conditional on $\{Z_n \neq 0\}$, Z_n has a limiting distribution (the *Yaglom limit*). Furthermore, $P(Z_n \neq 0) \simeq m^n$ if and only if (1) holds.

Numerous generalizations and refinements of these results have been obtained. The flavor of the subcritical and critical case is analytic (expansions of generating functions), while in the supercritical case almost sure (a.s.) convergence and martingales also enter.

OTHER BRANCHING PROCESSES

The continuous-time counterpart of the Galton–Watson process* is the *Markov branching process** $\{Z_t\}_{t \geq 0}$, where each individual alive at time t reproduces in $[t, t+h]$ with probability $\beta h + o(h)$. A notable example is the linear birth-and-death process*, where $\beta = \lambda + \mu$ with λ the individual birth rate and μ the individual death rate. A Markov branching process has the property that any equidistant discrete skeleton $Z_0, Z_\delta, Z_{2\delta}, \ldots$ is a Galton–Watson process. (The skeletons of birth-and-death processes correspond to Galton–Watson processes with modified geometric offspring distribution.) As applied probability models, the Galton–Watson and Markov branching process are in general very crude approximations, so in most situations, one needs to incorporate features specific to the phenomenon under study. An obvious example is the need to study *age-dependent branching processes*. Examples of such are the *age-dependent birth-and-death process*, with individual birth and death rates $\lambda(a)$ and $\mu(a)$ depending on the age a, and the *Bellman–Harris process*, where reproduction takes place at the end of the life and the lifetime distribution is general (the exponential lifetime distribution corresponds to the Markov branching process). Both are special cases of the very general *Crump–Mode–Jagers–process*, where the distribution of the birth instants is a general point process, not necessarily independent of the life-length distribution. Also, one sometimes needs to introduce a spatial factor; e.g., to study the speed of spread of an epidemic. A simple way to do this is to superimpose on the branching a particle movement. *Branching diffusions* and *branching random walks* are examples of this and have received much attention.

An important way to generalize the simplest (one-type) models is by allowing for a finite number of types, each of which may produce offspring (of any type). This setting is very flexible from the application point of view: age-dependent models can be approximated by letting types be particles in various age groups, and spatial dependence can be treated by letting the types be particles in the elements of a partition of the region; and in epidemics, one can allow for several periods of the illness, such as incubation, latent, and infectious phases.

Passing to a continuum, we may consider a distribution of types and treat factors such as age and location within a unified setting. In some main cases (typical of which are age-dependent models and branching diffusions on bounded domains), the type distribution approaches a stable* one, the process grows from then on exponentially (the rate is the *Malthusian parameter*)*, and initial conditions enter through a factor, the *reproductive value*. But the type distribution may also (typically in branching random walks and diffusions on unbounded domains) drift off to infinity.

Also, branching processes with *immigration* have been considered. One of their main features is that in the subcritical case, a stationary distribution will typically exist. From the point of view of many applications, this is intuitively appealing, although the immigration will often be an artifice. Instead, it might be more appropriate to consider *controlled* branching processes, where the branching mechanism depends on the state, but the theory of such processes is so far comparatively incomplete.

Finally, we mention branching processes with *varying enviroment*, i.e., permitting time inhomogeneity, and with *random environment*, i.e., a varying environment determined as the outcome of a stochastic process*.

Bibliography

Athreya, K. B. and Ney, P. (1972). *Branching Processes*. Springer-Verlag, Berlin. (The standard monograph since its publication. Not all parts are still up to date.)

Harris, T. E. (1963). *The Theory of Branching Processes*. Springer-Verlag, Berlin. (The first standard reference.)

Heyde, C. C. and Seneta, E. (1977). *I. J. Bienaymé: Statistical Theory Anticipated*. Springer-Verlag, Berlin. (Gives some early history.)

Jagers, P. (1975). *Branching Processes with Biological Applications*. Wiley, New York. (Broad introduction.)

Joffe, A. and Ney, P. (eds.) (1978). *Branching Processes: Advances in Probability and Related Topics, Vol.* 5. Marcel Dekker, New York. (Contains a number of up-to-date surveys of special areas.)

Kendall, D. G. (1966). *J. Lond. Math. Soc.*, **41**, 385–406. (Historical.)

(BIRTH-AND-DEATH PROCESSES
EPIDEMIOLOGICAL STATISTICS
GALTON–WATSON PROCESS
GENETICS, STATISTICS IN
MARKOV PROCESSES)

S. ASMUSSEN

BROUILLON INDEX *See* DIVERSITY INDICES

BROWNIAN MOTION

Brownian motion with drift parameter μ and variance parameter $\sigma^2 > 0$ is the continuous-time, continuous-state stochastic process* $\{X(t);\ t \geqslant 0\}$ for which (1) $X(t) - X(0)$ is normally distributed with mean μt and variance $\sigma^2 t$; and (2) all nonoverlapping increments $X(t_1) - X(t_0), X(t_2) - X(t_1), \dots, X(t_n) - X(t_{n-1})$, where $0 \leqslant t_0 \leqslant t_1 \leqslant \cdots \leqslant t_n$, are independent random variables. It is common to assume that (3) $X(0) = 0$ unless explicitly stated otherwise (Brownian motion started at $X(0) = x$). *Standard Brownian motion* $\{B(t); t \geqslant 0\}$ refers to the case $\mu = 0$ and $\sigma^2 = 1$, and is related to the general Brownian motion by $X(t) = \mu t + \sigma B(t)$.

Brownian motion is important for a myriad of reasons, the primary ones being: (1) Numerous functionals of the process, such as distributions of first passage times, can be calculated explicitly. (2) Brownian motion is the analog in continuous time of sums of independent random variables, the statement being made precise by the *invariance principle**: If $S_n = \xi_1 + \cdots + \xi_n$ is a sum of independent identically distributed random variables $\{\xi_k\}$ having zero means and unit variances, then for n tending to infinity, the processes $B_n(t) = n^{-1/2} S_{[nt]}$ converge weakly in the appropriate function space to the standard Brownian motion. Thus functionals* of Brownian motion often serve as good approximations to analogous functionals for sums of independent random variables. (3) Brownian motion is the central example of a continuous-time, continuous-path strong Markov process* (a *diffusion process**). An arbitrary one-dimensional diffusion process can be constructed from Brownian motion by a homeomorphic transformation in the state space plus a random rescaling of time.

We provide a sampling of some of the known quantities concerning Brownian motion. Let $\phi(\xi) = (2\pi)^{-1/2} \exp(-\frac{1}{2}\xi^2)$ be

Table 1

Functional W	Distribution $\Pr[W \leqslant y]$
$B(t)$	$\Phi(y/\sqrt{t}), -\infty < y < \infty$
$\lvert B(t)\rvert$	$2\Phi(y/\sqrt{t}) - 1, y \geqslant 0$
$B(t) - B(s), 0 < s < t$	$\Phi\left(y/\sqrt{(t-s)}\right), -\infty < y < \infty$
$\sup\{B(s); 0 \leqslant s \leqslant t\}$	$2\Phi(y/\sqrt{t}) - 1, y \geqslant 0$
$B(t) - \inf\{B(s); 0 \leqslant s \leqslant t\}$	$2\Phi(y/\sqrt{t}) - 1, y \geqslant 0$
$\inf\{t > 0; B(t) = a\}, a > 0$	$2[1 - \Phi(a/\sqrt{y})], y > 0$
$t^{-1}m\{s \leqslant t; B(s) > 0\}$ (m = Lebesgue measure)	$(2/\pi)\arcsin\sqrt{y}, 0 \leqslant y \leqslant 1$
$t^{-1}\sup\{s \leqslant t; B(s) = 0\}$	$(2/\pi)\arcsin\sqrt{y}, 0 \leqslant y \leqslant 1$
$\sup\{\lvert B(s)\rvert; 0 \leqslant s \leqslant t\}$	$\sum_{k=-\infty}^{+\infty}(-1)^k\left[\Phi\left(\frac{(2k+1)y}{\sqrt{t}}\right) - \Phi\left(\frac{2(k-1)y}{\sqrt{t}}\right)\right]$

the standard normal density and $\Phi(z) = \int_{-\infty}^{z}\phi(\xi)\,d\xi$ the corresponding cumulative distribution function.

First consider standard Brownian motion $\{B(t)\}$ (Table 1).

Let $T_z = \inf\{t \geqslant 0;\ B(t) = z\}$ be the hitting time to a point z. Then the probability density function for T_z is

$$f(t) = \frac{z}{\sqrt{2\pi}}\, t^{-3/2}\exp\left(-\frac{z^2}{2t}\right),$$

sometimes called the *inverse Gaussian distribution**.

Also for $a > 0$ and $b < 0$,

$$\Pr[\text{Hit } a \text{ before } b] = \Pr[T_a < T_b] = |b|/(a + |b|),$$

$$E[\min\{T_a, T_b\}] = a|b|.$$

Turning to the general Brownian motion $\{X(t)\}$ with parameters μ and $\sigma^2 > 0$, we again let $T_z = \inf\{t > 0;\ X(t) = z\}$ be the hitting time to z. Of relevance in the *Wald approximation** in the *sequential probability ratio test** is the formula

$$\Pr[T_a < T_b] = \frac{1 - e^{-2\mu b/\sigma^2}}{e^{-2\mu a/\sigma^2} - e^{-2\mu b/\sigma^2}}.$$

When $\mu < 0$, then $M = \sup_{t \geqslant 0} X(t)$ is a finite random variable for which $\Pr[M > y] = e^{-2|\mu|y/\sigma^2}, y \geqslant 0$. When $\mu > 0$, then T_z is a finite random variable for $z > 0$ whose probability density function is

$$f(t) = \frac{z}{\sigma\sqrt{2\pi t^3}}\exp\left[-\frac{(z - \mu t)^2}{2\sigma^2 t}\right] \qquad (t \geqslant 0).$$

We complete this sampling of Brownian motion formulas with a quantity relevant to the one-sided cumulative sum control chart* method. Let $M(t) = \sup\{X(s);\ 0 \leqslant s \leqslant t\}$, $Y(t) = M(t) - X(t)$ and $\tau_a = \inf\{t \geqslant 0;\ Y(t) \geqslant a\}$. Then

$$E[\tau_a] = \frac{1}{\mu}\left[\frac{\sigma^2}{2\mu}(e^{2\mu a/\sigma^2} - 1) - a\right], \quad \mu \neq 0.$$

Some processes closely related to Brownian motion that appear frequently are the following.

REFLECTED BROWNIAN MOTION. $Y(t) = |B(t)|$. Then $E[Y(t)] = \sqrt{2t/\pi}$ and $\operatorname{var}\{Y(t)\} = (1 - 2/\pi)t$. Also arises as $Y(t) = B(t) - \inf\{B(s);\ 0 \leqslant s \leqslant t\}$.

ABSORBED BROWNIAN MOTION. $Y(t) = B(\min\{t, T_0\})$, where $\{B(t)\}$ is standard Brownian motion starting at $y > 0$ and $T_0 = \inf\{t \geqslant 0;\ B(t) = 0\}$ is the hitting time to zero.

GEOMETRIC BROWNIAN MOTION. $Y(t) = e^{X(t)}$. Arises in certain population models and in mathematical economics. $E[Y(t)] = e^{\alpha t}$, where $\alpha = \mu + \frac{1}{2}\sigma^2$ and $\operatorname{var}[Y(t)] = e^{2\alpha t}(e^{\sigma^2 t} - 1)$.

n-DIMENSIONAL BROWNIAN MOTION. $\mathbf{B}(t) = (B_1(t), \ldots, B_n(t))$, where $B_1(t), \ldots, B_n(t)$ are independent standard Brownian motions.

BESSEL PROCESS. $Y(t) = \|\mathbf{B}(t)\| = [B_1^2(t) + \cdots + B_n^2(t)]^{1/2}$, the radial distance of n-dimensional Brownian motion from the origin.

*Wiener measure** is the probability distribution on the Borel subsets of $C[0, \infty)$ that corresponds to standard Brownian motion. By considering Wiener measure one may assume that all trajectories $t \to B(t)$ are continuous. In contrast to this smoothness, however, with probability 1 the following statements hold: The paths are nowhere differentiable, and in fact,

$$\limsup_{h \downarrow 0} h^{-1/2}|B(t+h) - B(t)| = \infty$$

for all t. Consider a partition sequence $\pi_0 \subset \pi_1 \subset \cdots$ of $[0, t]$ where $\pi_k = \{0 = t_{k0} < t_{k1} < \cdots < t_{kn(k)} = t\}$ and for which $\max_{1 \leqslant j \leqslant n(k)} |t_{kj} - t_{kj-1}| \to 0$ as $k \to \infty$. Then

$$\lim_{k \to \infty} \sum_{j=1}^{n(k)} |B(t_{kj}) - B(t_{kj-1})|^2 = t, \quad (1)$$

while $\lim_{k \to \infty} \sum_{j=1}^{n(k)} |B(t_{kj}) - B(t_{kj-1})| = \infty$.

Let K be the class of absolutely continuous functions on $[0, 1]$ with $f(0) = 0$ and $\int_0^1 f'(t)^2\,dt \leqslant 1$. Strassen's law of the iterated logarithm* asserts that the family $\{Z_n\} \subset C[0, 1]$ defined by

$$Z_n(t) = (2n \log\log n)^{-1/2} \times B(nt)$$

for $0 \leqslant t \leqslant 1$ is relatively compact with limit set K. From this, the classical law of the iterated logarithm

$$\limsup_{t \to \infty} (2t \log\log t)^{-1/2} B(t) = 1,$$

$$\liminf_{t \to \infty} (2t \log\log t)^{-1/2} B(t) = -1,$$

follows.

Although the Brownian path is nowhere differentiable, a theory of stochastic integration* has been developed, defining such expressions as

$$\mathscr{I}(Y)(t) = \int_0^t Y(s)\,dB(s), \quad (2)$$

where $\{Y(s)\}$ is a stochastic process progressively measurable with respect to the Brownian motion and locally square integrable. There are two approaches to (2) in common use, both involving approximating sums of the form $\sum Y(s_j^*)[B(s_j) - B(s_{j-1})]$, where $0 = s_0 < s_1 < \cdots < s_n = t$ and $s_{j-1} \leqslant s_j^* \leqslant s_j$. In the Ito integral*, one takes $s_j^* = s_{j-1}$, the left endpoint. Then $\mathscr{I}(Y)(t)$ is a local martingale with respect to the Brownian motion. In the Stratonovich interpretation, denoted by $\mathscr{I}_S(Y)(t)$, one takes $s_j^* = (s_{j-1} + s_j)/2$, the midpoint of the interval. The Stratonovich calculus is more closely aligned with deterministic calculus and thus better suited for many models of physical systems perturbed by noise. The martingale property that the Ito calculus enjoys makes it the preferred approach in most theoretical work. It is important to recognize the distinction, as the two interpretations may lead to vastly differing conclusions, as exemplified by:

Ito	Stratonovich
$dB(t)^2 = dt$ [compare with (1)]	$dB(t)^2 = 0$
$\int_0^t 2B(s)\,dB(s) = B^2(t) - t$	$\int_0^t 2B(s)\,dB(s) = B^2(t)$

In the *semigroup** approach to Markov processes standard Brownian motion is characterized by the operator $T_t f(x) = \int p(t, x, y) \times f(y)\,dy$, on the space $BC(-\infty, +\infty)$ of bounded continuous real-valued functions f that corresponds to the *Gauss kernel**

$$p(t, x, y) = (2\pi t)^{-1/2} \exp\left[-(y - x)^2/(2t)\right].$$

Given f in $BC(-\infty, +\infty)$, the bounded solution of $\partial u/\partial t = \frac{1}{2}\partial^2 u/\partial x^2$, $t > 0$ and $u(0+, \cdot) = f$, is $u = u(t, x) = T_t f(x)$. The resolvent $R_\lambda f(x) = \int_0^\infty e^{-\lambda t} T_t f(x)\,dt$, $\lambda > 0$, has the kernel representation $R_\lambda f(x) = \int_{-\infty}^{+\infty} r(\lambda, x, y) f(y)\,dy$, where

$$r(\lambda, x, y) = \left(1/\sqrt{2\lambda}\right) \exp\left(-\sqrt{2\lambda}\,|y - x|\right).$$

R_λ maps $BC(-\infty, +\infty)$ one-to-one onto the set Δ of f in $BC(-\infty, +\infty)$ having two bounded continuous derivatives. Let R_λ^{-1} denote the operator on Δ that is inverse to R_λ. Then $\mathcal{Q} = \lambda - R_\lambda^{-1}$ together with the domain Δ is called the *infinitesimal generator**, and $\mathcal{Q}u(x) = \frac{1}{2}\partial^2 u/\partial x^2$ for u in Δ or, alternatively,

$$\mathcal{Q}u(x) = \lim_{\epsilon \downarrow 0} \epsilon^{-1}[T_\epsilon u(x) - u(x)],$$

where the convergence takes place boundedly and pointwise. Different notions of convergence, e.g., strong and uniform operator topologies, lead to different notions of infinitesimal generator.

The set $\Xi = \{t \geqslant 0;\ B(t) = 0\}$ of zeros of the Brownian path is a topological Cantor set, i.e., closed, uncountable, of topological dimension 0 and having no isolated points. Its Lebesgue measure is zero. However, its Hausdorff dimension* is $\frac{1}{2}$ and related to this fact is the possibility of assigning a nontrivial measure to the amount of time the process spends at 0, called the *local time** at 0 and denoted $l(0,t)$. For standard Brownian motion, the probability law of $l(0,t)$, is the same as that of $M(t) = \max_{0 \leqslant s \leqslant t} B(s)$.

Bibliography

Freedman, D. (1971). *Brownian Motion and Diffusion*. Holden-Day, San Francisco. (This is an advanced, difficult-to-read, but carefully written text. Contains some topics difficult to find elsewhere, but is suitable only for the mathematically advanced reader.)

Ito, K. and McKean, H. P., Jr. (1965). *Diffusion Processes and Their Sample Paths*. Springer-Verlag, Berlin. (The definitive book on Brownian motion and diffusion processes. A wealth of information, but very terse and mathematically advanced.)

Karlin, S. and Taylor, H. M. (1975). *A First Course in Stochastic Processes*, 2nd ed. Academic Press, New York. (Recommended starting point for an applied statistician having a modest mathematical background. The 40-page chapter on Brownian motion contains many derivations of functionals and examples of their use.)

Karlin, S. and Taylor, H. M. (1980). *A Second Course in Stochastic Processes*. Academic Press, New York. (Contains a complete discussion of Brownian motion as a diffusion process, stochastic calculus, and numerous examples of Brownian motion and related processes as they arise in applications. Intermediate level of presentation.)

The following three references contain recent results relating Brownian motion to Cusum control charts*:

Reynolds, M. R., (1975). *Ann. Statist.*, **3**, 382–400.
Reynolds, M. R., (1975). *Technometrics*, **17**, 65–72.
Taylor, H. M., (1975). *Ann. Prob.*, **3**, 234–246.

(DIFFUSION PROCESSES
HAUSDORFF DIMENSION
LOCAL TIME
MARKOV PROCESSES)

H. TAYLOR

BROWN–MOOD MEDIAN TEST

The Brown–Mood median test [2, 3, 16] is a distribution-free procedure (*see* DISTRIBUTION-FREE METHODS) for testing the null hypothesis that two mutually independent random samples measured on at least an ordinal scale are drawn from the same distribution. This procedure is also known as the two-sample median test and the Westenberg–Mood median test [19] and is a special case of Fisher's exact test (*see* FISHER'S EXACT TEXT). It is discussed in most books on nonparametric statistics, including Bradley [1], Conover [5], Gibbons [12], and Daniel [7].

Denote the two sets of sample observations by $X_1, X_2, \ldots, X_m$ and $Y_1, Y_2, \ldots, Y_n$ and let M denote the median of the pooled samples, i.e., the observation with rank $(N + 1)/2$ if N is odd and any number between the observations with rank $N/2$ and $(N+2)/2$ if N is even, where $N = m + n$. Let u and v be the number of observations in the X sample and Y sample, respectively, which are smaller than M. Then the data can be presented in a 2×2 table as follows:

	X	Y	Totals
$< M$	u	v	t
$\geqslant M$	$m-u$	$n-v$	$N-t$
Totals	m	n	N

Unless there are ties at the sample median, $t = [N/2]$, the largest integer not exceeding

$N/2$, i.e., $(N-1)/2$ or $N/2$ according as N is odd or even. (If there are ties at the sample median, they can be broken at random or in a manner that leads to a conservative test*. Alternatively, one might dichotomize at a value slightly different from M.)

Under the null hypothesis that the two samples are drawn from identical continuous populations, every set of t observations is equally likely to be smaller than M. Hence the null probability that exactly u of the X observations are smaller than M is found from the hypergeometric distribution* as

$$f_U(u) = \binom{m}{u}\binom{n}{t-u}\Big/\binom{m+n}{t} \quad \text{for } u = 0, 1, \ldots, t. \qquad (1)$$

A test based on the value of U is consistent against any alternative such that the medians M_X and M_Y of the X and Y populations, respectively, differ. Hence the null hypothesis is frequently written as $M_X = M_Y$ or as $\Pr[X < M] = \Pr[Y < M]$. The P-values* and rejection regions* for a test at level α are as follows:

Alternative	P-Value	Rejection Region
$M_X > M_Y$	$\Pr[U \leqslant u]$	$U \leqslant u_\alpha$
$M_X < M_Y$	$\Pr[U \geqslant u]$	$U \geqslant u'_\alpha$
$M_X \neq M_Y$		$U \leqslant c$ or $U \geqslant c'$

In this table, u_α and u'_α are, respectively, the largest and smallest integers such that $\Pr[U \leqslant u_\alpha] \leqslant \alpha$ and $\Pr[U \geqslant u'_\alpha] \leqslant \alpha$, and c and c' are any two integers such that $\Pr[U \leqslant c] + \Pr[U \geqslant c'] \leqslant \alpha$. The probabilities can be computed from (1) or from tables of the hypergeometric distribution (e.g., Lieberman and Owen [15]). A two-tailed P-value is not satisfactory because the distribution of U is not symmetric if $m \neq n$.

For large samples an approximate test can be based on the standard normal statistic

$$Z = \frac{U \pm 0.5 - mt/N}{\left[mnt(N-t)/N^3\right]^{1/2}},$$

where the $\pm$ term is a continuity correction* and is to be taken as + for a lower tail probability and − for an upper tail probability. Fisher and Yates [9, Table VIII] give the exact one-tailed critical values of Z for one-tailed tests at levels $\alpha = 0.025$ and $\alpha = 0.005$; the entries indicate that the normal approximation is quite accurate as long as the smaller of m and n is at least 12.

The asymptotic relative efficiency* (ARE) of this test relative to the two-sample Student's t-test* is $2/\pi = 0.637$ for normal populations with equal variances (see, e.g., Mood [17] and Chakravarti et al. [4]). The exact power* in small samples was investigated by Dixon [8], Leone et al. [14], and Gibbons [11] for various distributions and alternatives.

This test has a corresponding procedure for finding a confidence interval* estimate of $\theta = M_X - M_Y$ under the assumption that X and $Y - \theta$ are indentically distributed (the shift* assumption). The interval with confidence coefficient $1 - \alpha$ is

$$Y_{t-c'+1:n} - X_{c':m} \leqslant \theta \leqslant Y_{t-c:n} - X_{c+1:m},$$

where c and c' satisfy $\Pr[U \leqslant c] + \Pr[U \geqslant c'] = \alpha$.

The median test may be considered a special case of a general quantile test where M is replaced by a quantile of any order in the pooled sample*. The value of t is determined by the order of the quantile. These tests are useful in analyzing data from life-testing experiments because the experiment can be terminated as soon as t units have expired. The control median test [10, 13], the first-median test [10], and the Rosenbaum test [18] are variants of these two-sample quantile tests*. The median test can also be used as a one-sample test for trend over time by dichotomizing at the median point of time [6] (*see* TREND TESTS). A competing test which may be used for the same problem is the Mann–Whitney–Wilcoxon test* (*see also* DISTRIBUTION-FREE METHODS).

The median test can be extended to the case of three or more independent samples. The test statistic then follows a multivariate hypergeometric distribution* and a large-sample test can be based on a chi-square distribution*.

References

[1] Bradley, J. V. (1968). *Distribution-Free Statsitical Tests*. Prentice-Hall, New York.

[2] Brown, G. W. and Mood, A. M. (1948). *Amer. Statist.*, **2**(3), 22.

[3] Brown, G. W. and Mood, A. M. (1951). *Proc. 2nd Berkeley Symp. Math. Statist. Prob.* University of California Press, Berkeley, Calif., pp. 159–166.

[4] Chakravarti, I. M., Leone, F. C., and Alanen, J. D. (1962). *Ann. Math. Statist.*, **33**, 1375–1383.

[5] Conover, W. J. (1971). *Practical Nonparametric Statistics*. Wiley, New York.

[6] Cox, D. R. and Stuart, A. (1955). *Biometrika*, **42**, 80–95.

[7] Daniel, W. (1978). *Applied Nonparametric Statistics*. Houghton Mifflin, Boston.

[8] Dixon, W. J. (1954). *Ann. Math. Statist.*, **25**, 610–614.

[9] Fisher, R. A. and Yates, F. (1963). *Statistical Tables for Biological, Agricultural, and Medical Research*, 6th ed. Hafner, New York.

[10] Gastwirth, J. L. (1968). *J. Amer. Statist. Ass.*, **63**, 692–706.

[11] Gibbons, J. D. (1964). *J. R. Statist. Soc. B*, **26**, 293–304.

[12] Gibbons, J. D. (1971). *Nonparametric Statistical Inference*, McGraw-Hill, New York.

[13] Kimball, A. W., Burnett, W. T., Jr., and Doherty, D. G. (1957). *Radiat. Res.*, **7**, 1–12.

[14] Leone, F. C., Chakravarti, I. M., and Alanen, J. D. (1961). Exact Power of Some Quick Tests Based on Mood's and Massey's Statistics. *Bull. Int. Inst. Statist., 33rd Sess., Paris, No. 115.*

[15] Lieberman, G. J. and Owen, D. B. (1961). *Tables of the Hypergeometric Probability Distribution*. Stanford University Press, Stanford, Calif.

[16] Mood, A. M. (1950). *Introduction to the Theory of Statistics*. McGraw-Hill, New York.

[17] Mood, A. M. (1954). *Ann. Math. Statist.*, **25**, 514–521.

[18] Rosenbaum, S. (1954). *Ann. Math. Statist.*, **25**, 146–150.

[19] Westenberg, J. (1948). *Proc. Kon. Ned. Akad. Wet.*, **51**, 252–261.

(CONTINGENCY TABLES
FISHER'S EXACT TEXT
HYPERGEOMETRIC DISTRIBUTION
TWO-SAMPLE QUANTILE TESTS)

J. D. GIBBONS

BUFFON'S NEEDLE PROBLEM

A celebrated classical problem of geometric probability*. A plane is lined with parallel straight lines 1 unit apart and a thin needle N of length l is thrown "at random" on the plane. ["At random" here means that: (1) the center of the needle is uniformly distributed over the interval of unit length perpendicular to and between two neighboring straight lines, and (2) the probability that the angle ϕ between the needle and the straight lines is $\phi_1 \leqslant \phi \leqslant \phi_1 + \Delta\phi$ is proportional to $\Delta\phi$ for any ϕ_1, whatever be the position of the center.] What is the probability p that N intersects (at least) one of these lines? The simple answers are:

$$p = 2l/\pi \qquad \text{if } l \leqslant 1$$

$$p = 2\left[\cos^{-1}(1/l) + 1 - (1 - l^{-2})^{1/2}\right]/\pi \qquad \text{if } l \geqslant 1.$$

For a history of this problem, see Uspensky [4].

(Problems of this kind arise in some applications to ballistics which take the size of the shell into account.)

Additional information on the needle problem is provided by refs. 1–3.

References

[1] Buffon, L. (1777). *Essai d'arithmétique morale.*

[2] Santaló, L. A. (1976). Integral geometry and geometric probability In *Encyclopedia of Mathematics and Its Applications*, Vol. 1, G. C. Rota, ed. Addison-Wesley, Reading, Mass.

[3] Schuster, E. F. (1979). *Amer. Math. Monthly*, **81**, 26–29.

[4] Uspensky, J. V. (1937). *Introduction to Mathematical Probability*. McGraw-Hill, New York.

(GEOMETRIC PROBABILITY THEORY)

BULK SAMPLING

Bulk material is sampled by taking "increments" (small portions) of the material, blending these increments into a single composite sample, and then, if necessary, reducing this gross sample to a size suitable for laboratory testing. When dealing with bulk material there comes a point in the sampling procedure (either before or after the creation of the composite sample) when the material cannot be viewed as consisting of discrete constant preexistent and identifiable unique units that may be sampled as such. New units are created by means of some optional sampling device at the time of sampling. Another feature of bulk sampling is the physical mixing of the initial increments of the material. When the initial increments are blended into a single sample it is not possible to measure variation between the increments making up the composite. This is a disadvantage. In the case of bulk sampling, the object of the sampling is usually measurement of the mean quality of material sampled.

A detailed analysis of bulk sampling and formulation of the relevant theoretical models is given in Duncan [2]. For sampling standards, see ref. 1.

References

[1] American Society for Testing Materials, *Book of ASTM Standards*, 1958 Symposium on Bulk Sampling (ASTM Spec. Publ. No. 242). Philadelphia.

[2] Duncan, A. J. (1962). *Technometrics*, **4**, 319–343.

BULLETIN IN APPLIED STATISTICS (BIAS)

SCOPE OF THE JOURNAL

The scope of this journal is very wide and varied. Although it is a journal for descriptive statistics, it overlaps greatly with the *Journal of Applied Statistics** (*see* JOURNAL OF THE ROYAL STATISTICAL SOCIETY, SER. C) and *Statistical News and Notes*. This loose demarcation of the journal has provided greater interest and deeper stimulation to the applied statistician and others involved in practicing statistical methods in whatever field.

Applied statisticians in the course of their duties always come across the situation where the existing techniques either do not fit or have to be modified for their work. In other words, they have to engineer the existing statistical techniques to carry on their work. *BIAS* encourages such activities by way of publishing articles on statistical engineering and statistical case studies.

Table 1 should serve to indicate the nature and range of topics covered:

Table 1 Contents of *BIAS*, Vol. 6, No. 1, 1979

Title	Author
Experimental Design from the Time of Fisher to the 1970's	G. H. Freeman
Some Practical Aspects of Canonical Variate Analysis	N. A. Campbell
Age, Period and Cohort Effects—A Confounded Confusion	H. Goldstein
Cluster Analysis of Some Social Survey Data	B. Jones
Probability and Special Relativity	T. F. Cox
A Report on the 1978 IOS Conference on Time-Series Analysis	C. Chatfield
The Census of Production and Its Uses	B. Mitchell and E. Swires-Hennesy
FORUM: Treatment of Myopia	C. I. Phillips
Letter to the Editor	

EDITORIAL POLICY

BIAS, although a serious and informative journal, is also meant to be lively and entertaining, and is therefore aimed at all those who are using statistics in any sector of this modern world. The purpose of the *Bulletin* is to create a medium through which news, views, and ideas may be aired and shared and in this way acts as a link between people with a variety of statistical interest. It is endeavoring to produce *readable* articles about what is going on in statistics which will be of interest to practitioners of statistics. It is trying to reveal the reasons for the apparent failure of certain techniques in some areas, and it is also concerned with the problems of communication within the statistical profession and between statisticians and users of statistics.

The editorial policy is to encourage the submission of articles to the *Bulletin* regarding:

1. Statistical activities in schools, colleges, polytechnic institutes, universities, research units, industries, local and central government bodies, and so on.
2. General nontechnical articles of a statistical nature (semitechnical articles may be considered).
3. Research reports, short communications in the nature of notes, or brief accounts of work in progress (prior to more formal publication).
4. Statistical case studies.
5. Reviews of statistical techniques.
6. Book reviews.

The articles are refereed. Two issues of the journal are produced each year. Table 2 is a concise presentation of pertinent facts about *BIAS*.

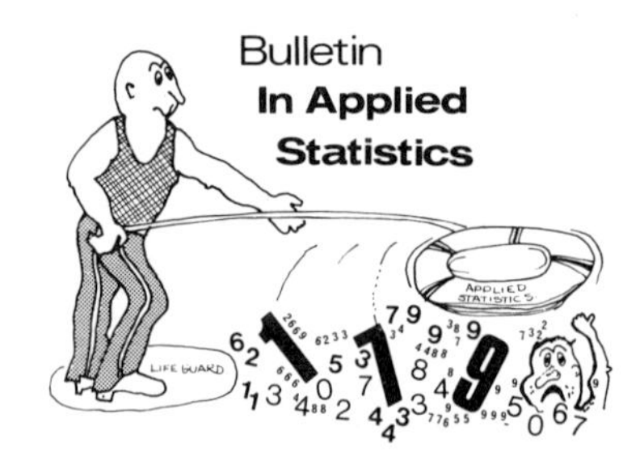

Figure 1 Picture on the cover page of *BIAS* portrays the editor. (Copyright © *BIAS*.)

Table 2 ***BIAS* in brief**

First issue	1974
Editor and Founder	G. K. Kanji, Department of Mathematics and Statistics, Sheffield City Polytechnic, Sheffield S1 1WB, United Kingdom
Publisher	Sheffield City Polytechnic
Number of issues per year/volume	Two
Number of pages per issue	100 (approx.)
Description of subscribers	schools; colleges; polytechnic and university teachers; industrial, social, business, and medical statisticians from over 21 countries
Description of contributors	From all spheres of the statistical world
Types of articles	Any aspects of applied statistics
Subscription rates/volume	Personal (UK = £3.00, foreign = £4.00) Library, etc. (UK = £4.00, foreign = £6.00)

Figure 1, from the cover page of *BIAS*, wordlessly portrays the editor's feelings and communicates his sentiments about applied statistics.

G. K. KANJI

BULLETIN ON MATHEMATICAL STATISTICS

A journal published by the Research Association of Statistical Sciences, Kyusha University, Fukuoka, Japan. Started in 1947, it appears somewhat irregularly; in recent years there has been one issue per year. All papers are in English and most (but not all) are authored by Japanese statisticians. The journal covers a wide spectrum of topics in probability theory and mathematical statistics at a rather high theoretical level. The most recent issues contain about 110 to 125 pages and 9 or 10 articles.

The founding and current editor is T. Kitagawa; there are 16 associate editors, including some well-known probabilists such as K. Ito and Y. Ishida.

BUREAU OF LABOR STATISTICS

Established by the U.S. Congress in 1884, and made a part of the newly created U.S. Department of Labor in 1913, the Bureau of Labor Statistics (BLS) is the nation's principal fact-finding agency in the field of labor statistics. The Bureau collects, processes, analyzes, and disseminates data relating to the labor force, including employment and unemployment, productivity and technological change, prices and living conditions, family expenditures, wages, industrial relations, industrial injuries and occupational illnesses, and economic growth. A principal criterion for the development of these data is their relevance to current social and economic needs.

Most of the Bureau's data are based on surveys conducted by its own field staff. In addition, some Bureau data are collected by the Bureau of the Census* on contract with BLS, or jointly by the Bureau of Labor Statistics and cooperating state agencies. Data collection methods include the use of both direct mail and personal interview surveys. Voluntary reporting and confidentiality of the reported data are hallmarks of the BLS data collection program.

In addition to producing statistical series, the Bureau conducts and publishes analytical studies based on these data. These analyses of economic, demographic, and social trends are designed to serve business, labor, the general public, Congress, and administrative and executive agencies of government.

MAJOR PROGRAMS

The Bureau's work is organized into seven subject-matter areas:

Current Employment Analysis

Current employment analysis statistics are based on the Current Population Survey (CPS), a monthly multistage probability sample of approximately 60,000 households, conducted on contract for the Bureau of Labor Statistics by the Bureau of the Census. In January 1980, the CPS sample was expanded to about 65,000 households. Future expansion is being considered as a result of recommendations of the National Commission on Employment and Unemployment Statistics. To produce these statistics, a variety of estimation techniques are employed: first- and second-stage ratio adjustment to population controls, composite estimation based on a rotation scheme of survey households, generalized variances*, estimation of nonsampling errors*, and seasonal adjustment*.

The CPS provides the most comprehensive national monthly data available on the employed, the unemployed, and persons not in the labor force. The survey is the source of unemployment rates, number and characteristics of discouraged workers, and labor force status of special groups such as persons of Hispanic origin, women, blacks, young workers, and poverty-area residents. The CPS also is the basis for long-run projections of the labor force, including work-life expectancy, and for socioeconomic studies of the American worker.

State and local area unemployment statistics, developed monthly in cooperation with the states, are based on a combination of unemployment insurance and Current Population Survey (CPS) data. The CPS is used as a bench mark for each state to ensure consistency with the national concept. The 10 largest states utilize the CPS directly on a monthly basis; the remaining 40 utilize on annual average from the CPS. State and local unemployment rates are an important element in the formulas used to distribute federal funds under the Comprehensive Employment and Training Act and economic development and countercyclical revenue-sharing statutes.

Wage and Salary Employment

Data on wage and salary employment are obtained largely by direct-mail surveys of samples of employers. The surveys are conducted in cooperation with state employment security agencies. Industry employment, earnings, and hours data are obtained from a sample of 165,000 establishments in all nonagricultural activities. Link-relative estimation is used between annual bench mark employment data.

The data are published for a large number of industries and for each state and many local areas. Labor turnover data are derived from a survey of 40,000 employers in manufacturing, mining, and communications industries. Data are published by industry for local areas and for 38 states.

The Bureau also collects and publishes data on the employment and wages of workers covered by unemployment insurance. Projections of industry and occupational employment for vocational guidance and educational planning are developed and the Bureau also collects data on occupational employment statistics.

Prices and Living Conditions

BLS price and living conditions statistics provide detailed measures of changes in consumer, industrial, and import and export prices.

The Consumer Price Index is a measure of the average change in prices of a constant market basket of goods and services, reflecting the buying habits of urban Americans. The Bureau publishes indexes for two groups of the civilian noninstitutional population: (1) the Consumer Price Index for All Urban Consumers (CPI-U), which covers 80% of the population, and (2) the Consumer Price Index for Urban Wage Earners and Clerical Workers (CPI-W), which covers about half of the CPI-U population group.

Prices for the CPI are collected from probability samples* of retail establishments in 85 urban areas across the country. The sampling frame* of retail establishments is obtained through a multistage probability sample* of households, referred to as the Point of Purchase Survey (POPS). The results of the POPS survey is used to systematically update the sample of establishments on an ongoing basis. All taxes directly associated with the purchase and use of items are included in the price. The CPI is widely used to adjust wages, pensions, social security benefits, food stamp allocation, and other payments for changes in prices.

Consumer Expenditure Surveys provide information on spending patterns of families grouped by size, income, and residence. The surveys are conducted by the Bureau of the Census for the Bureau of Labor Statistics. These survey data are used for selection of the CPI market basket and for composite estimation of CPI weights. As of October 1979, the Bureau of the Census instituted a survey of consumer expenditures on a continuing basis. The BLS will begin analyzing and publishing data from the survey in 1982.

The Bureau also publishes hypothetical annual budgets for an urban four-person family and a retired couple at three levels of living in different areas of the United States.

The Producer Price Indexes* measure average changes in prices received by producers of commodities in all stages of processing. The stage-of-processing structure organizes products by degree of fabrication.

An expansion of the Producer Price Index program, begun in 1978, is scheduled for completion in 1984. The expansion will provide coverage for all mining and manufacturing industries of the Standard Industrial Classification when complete, and will lead to the reporting of 165,000 price quotations by 35,000 producers each month. The revised methodology employs a two-stage probability sample* of producers and products. Producer Price Indexes frequently are used as escalators in long-term sales and purchase contracts.

U.S. import and export price indexes are released quarterly. The import price indexes currently account for over one-fourth of the value of all U.S. imports, while the export product indexes currently account for about 60% of the value of all U.S. exports.

Probability samples of importers and exporters and items to be priced are selected from import and export declarations filed with the Customs Bureau. Price data are collected for about 10,000 products from over 3000 companies. Most prices are collected by mail questionnaire and represent net transaction prices. Both indexes use a modified Laspeyres formula*.

Wages and Industrial Relations

Wages and industrial relations statistics measure workers' earnings and employers' expenditures for employee compensation and provide information on labor–management agreements and strikes. The Bureau's major statistical series in this area include occupational wages in large metropolitan areas and in selected industries. The Bureau conducts a national survey of selected professional, administrative, technical, and clerical (PATC) occupations in private employment. The Bureau uses the PATC survey in discharging its responsibility for measuring wage comparability for federal pay adjustment and other federal compensation reform legislation. The Bureau's Employment Cost Index measures rates of change in wage rates in the private sector, and has been expanded to include the cost of fringe benefits. Negotiated wage and benefit changes in major collective bargaining units are also studied, as are salary levels and trends for selected groups of government workers. Employee benefit statistics measure the incidence of a variety of provisions in employee benefits plans furnishing life and health insurance, retirement pensions, and salary continuation.

Industrial relations statistics include data on collective bargaining, contract provisions, a public file of agreements, and a directory of national unions and employee associations and membership statistics. Analyses of labor–management disputes include statistics on work stoppages—number of strikes, number of workers involved, and labor time lost.

Data are collected largely from samples of establishments, unions, and so on, stratified by industry and size class. In the Employment Cost Index, a two-stage sample with controlled selection is used to obtain occupational coverage. Statistics on wages and industrial relations are used for wage and salary administration, minimum wage policy guidance, union contract negotiations, conciliation and arbitration, and analysis of productivity changes.

Productivity and Technology

BLS productivity and technology statistics have two major aims: to measure productivity growth in the economy, in its major industrial sectors, and in individual industries; and to investigate the nature of technological changes within industries and across industry lines and their effect on employment. The Bureau produces indexes of labor productivity and compensation per hour, unit labor cost, and related measures. BLS also produces relevant international comparisons of productivity growth and related economic statistics, as well as studies of changing labor and materials requirements in major types of construction. Construction projects are sampled with a probability proportional to a value of construction size measure.

Occupational Safety and Health

Occupational safety and health statistics are obtained through a nationwide probability sample of employers, who provide data on job-related injuries and illnesses. BLS is responsible for providing employers with record-keeping forms and interpretations of regulations the Bureau has developed to maintain the system, for the collection and analysis of statistics based on the record-keeping system, and for related supplementary statistics from other sources. The occupational safety and health survey required by the Occupational Safety and Health Act of 1970 is the only mandatory survey carried out by BLS.

Statistics derived from the survey include estimates of injuries and illnesses, work fatalities, number of lost workdays, and other significant data by industry and size of establishment. A supplementary data system gathers information from workers's compensation records to identify specific hazards and dangerous work situations.

Economic Growth

BLS economic growth studies provide a framework for analyzing the implications of growth for employment. An integrated system of economic models is used to make medium- to long-range projections of demand, output, and employment under different assumptions about federal government policies and other factors. An input–output model estimates the effects of these economic conditions and policies on job requirements by industry.

DEVELOPING STATISTICAL METHODOLOGY

As a major collector of survey data, BLS conducts applied developmental research directed toward a variety of problems common to sample survey* activities. Examples of research topics receiving attention at BLS are (1) estimation from samples that are incomplete due to nonresponse*, (2) production of measures of collection process errors, (3) uses of improved variance estimates* in "optimal" sample allocation schemes, and (4) special estimation problems such as composite, link-relative, and model-based estimation.

COMPUTER PROCESSING OF DATA

The Bureau of Labor Statistics acquired its first small electronic computing machine in 1952. Since then, computer technology has moved the compilation of large-scale surveys from a clerical to a high-speed, computer-intensive operation. Along the way, the Bureau has made pioneering efforts in the field of computer systems, particularly in reducing the need for tailor-made computer programs.

The Bureau's Table Producing Language (TPL) coupled with Print Control Language (PCL) is used throughout the nation and in some foreign countries. TPL tabulates and automatically displays the results in any practical format. The TPL computer system does away with the need to write individual computer programs each time a new tabulation is needed. PCL permits users who have no training in either computing or printing technology to create electronically photocomposed tables with the style and font options typically found in typesetting. Other computer tools to aid in reducing the tailor-made approach to computer programming include: Statistical Charting System (SCS) and an extension of the electronic photocomposing capabilities to text preparation, editing, and word processing. (A system for automatic assembly of tables, text, and charts into full pages and entire publications as well as an electronic mail system are in process.)

These and other general tools for economic analysis have been linked to a data base of over 100,000 BLS time series, organized for easy access. The tools and data based together are called LABSTAT (for LABor STATistics), comprising the Bureau's "third-generation" computer environment. The combination offers analysts in Washington and in BLS Regional Offices access to published statistics through on-line computer terminals.

DISSEMINATION OF DATA

Results of the Bureau's research and data collection are disseminated to the public through a series of publications both in Washington, D.C., and eight BLS regional offices.

News releases provide the first quick summary of the Bureau's research and statistics, including the major economic indicators. More comprehensive reports and analyses of current data are published in periodicals, available by paid subscription. These include the *Montly Labor Review*, *Employment and Earnings*, *Current Wage Developments*, the *CPI Detailed Report*, *Producer Prices and Price Indexes*, and the *Occupational Outlook Quarterly*. Many of the Bureau's major survey and research studies are published as BLS bulletins and sold by the Government Printing Office and the BLS regional offices. Reference handbooks include the *Handbook of Labor Statistics*, *BLS Handbook of Methods*, and the *Occupational Outlook Handbook*. In addition, selected publications are available in microfiche and hard copy from the National Technical Information Service, Springfield, VA 22151. As they become available, data tapes, microfiche, and other publications are listed in *Just Published*, a monthly publication, and in a semiannual BLS catalog.

Literature

For history, see E. Clague, *The Bureau of Labor Statistics* (Praeger Library of U.S. Government, Departments and Agencies, 1968), and U.S. Department of Labor, *The Anvil and the Plow* (1964).

For a handy reference to BLS programs, showing their principal characteristics, see *Major Programs, Bureau of Labor Statistics* (U.S. Department of Labor, Bureau of Labor Statistics).

For more details on statistical methods within each subject area, see *BLS Handbook of Methods* (Bull. 1910, U.S. Department of Labor, Bureau of Labor Statistics, 1976).

For the most comprehensive presentation on the Current Population Survey, see *The Current Population Survey: Design and Methodology* (U.S. Department of Commerce, Bureau of the Census, 1978). For a general popular presentation, see *How the Government Measures Unemployment* (BLS Rep. 505, U.S. Department of Labor, Bureau of Labor Statistics, 1977).

For price index information, see *The Consumer Price Index: Concepts and Content over the Years* (BLS Rep. 517, U.S. Department of Labor, Bureau of Labor Statistics, 1977); *Escalation and the CPI: Information for Users* (U.S. Department of Labor, Bureau of Labor Statistics, 1978); J. F. Early, "Improving the Measurement of Producer Price Change" (*Monthly Labor Rev*., April 1978, pp. 7–15).

For an expanded discussion of the Employment Cost Index see V. J. Sheifer, "Employment Cost Index: A Measure of Change in the Price of Labor" (*Monthly Labor Rev*., July 1975, pp. 3–12).

For information on obtaining data tapes, write to Office of the Commissioner, U.S. Department of Labor, Bureau of Labor Statistics, Washington, D.C. 20212. Also see S.

D. Swatek, *BLS Data Bank Files and Statistical Routines* (BLS Rep. 620, U.S. Department of Labor, Bureau of Labor Statistics, 1981).

For further details on BLS use of computers, see R. C. Mendelssohn, *Data Processing at BLS* (BLS Rep. 471, U.S. Department of Labor, Bureau of Labor Statistics, 1976).

(BUREAU OF THE CENSUS
FEDERAL STATISTICS
INDEX NUMBERS
LABOR STATISTICS)

JANET L. NORWOOD

BUREAU OF THE CENSUS, U.S.

How many people live in the United States and where do they live? How old are they, what are their races and ethnic backgrounds, and what is their marital status? What kind of education do they have, what do they do for a living, and how much do they earn? And how many rent, how many own their home?

The answers to these and many other questions are available from a single source: the U.S. Bureau of the Census.

The Census Bureau is the oldest and most important fact-finding agency of the federal government. It produces detailed statistical profiles of the nation, which include information on population, housing, agriculture, manufacturing, service industries, wholesale trade, retail trade, foreign trade, mining, transportation, construction, and the revenues and expenditures of state and local governments.

The Census Bureau gets many of its facts through surveys*. In some, the information is collected by mail, or is partially obtained from the administrative records of other government agencies; in others, it is collected by direct interview. Some surveys are taken only once; others are taken every month or annually. Some require only a few hundred copies of a questionnaire; others require millions. Some are inexpensive; others are taken at great cost. The results of some can be reported in a leaflet or a press release a few days after the data are collected, whereas others require detailed publication.

Throughout the planning and implementation of its censuses and surveys the Bureau relies heavily on statistical methodology to help in fulfilling its mission. Probability sampling* is used extensively in the censuses and is the primary tool used for producing data between censuses. Through this methodology accurate data can be produced on a timely basis at minimum cost and require responses from only a small proportion of the public.

In the planning phase for censuses and surveys, controlled experiments are used, testing alternative methodologies to identify cost-effective methods for improving the quality of the data. During the operational stages, statistical quality controls* are used to assure that the procedures are applied accurately and that an acceptable level of quality is realized. An essential part of a census* or survey is an evaluation program to estimate the accuracy of the figures produced. Estimates of sampling error* are made and are published together with the estimates. Special studies are conducted to estimate the completeness of the count, the accuracy of the responses obtained, and other sources of error. They are published so that consumers can be informed about the strengths and weaknesses of the data, and thus make more intelligent decisions in their uses of the data.

The U.S. Constitution provides for the taking of a national census, and the specific authorizing legislation under which the Bureau operates is Title 13 of the *United States Code*. This same law requires census taking in the major areas mentioned above, and it also enables the Secretary of Commerce (who delegates Title 13 authority to the Director of the Census Bureau) to "make surveys deemed necessary to furnish annual and other interim current data on the subjects covered by the censuses."

The oldest and largest undertaking of the Bureau is the census of population and housing. Taken every 10 years since 1790, its goal is to collect, tabulate, and make available basic demographic and economic information aggregated from every home in the country. The most recent decennial census—the twentieth—took place in the spring of 1980. In conducting this enumeration, representatives of the Census Bureau contacted more than 86 million households and, within these households, gathered information on approximately 226.5 million people located in some 39,000 governmental units.

Results from this census will provide detailed, timely demographic and socioeconomic information for the nation as a whole as well as by state, county, city, census tract, neighborhood, and city block. These results will be used for reapportioning the seats in the U.S. House of Representatives (as has been the case since the first census in 1790), for redistricting state legislatures and local governmental bodies, and as a basis for the allocation of federal and state funds. Data will also be used widely for planning and many other activities conducted by government at all levels, business, the academic community, and many other organizations and the general public.

Many federal, state, and local programs depend on population counts. On the federal level alone, it is estimated that at least $50 billion is allocated each year to state and local governments, based partly on population data.

In 1980 the census was taken largely by mail. Every household received one of two census forms shortly before census day, April 1. Individuals were asked to fill it out and either mail it back in a preaddressed, postage-paid envelope or hold the form for a census taker to pick up.

The "short" form contained only a few basic demographic and housing questions and went to five out of every six households in metropolitan areas. The remaining households received a longer form that asked additional questions about demographic and socioeconomic characteristics as well as some additional housing questions.

Processing and tabulating the more than 86 million questionnaires were conducted between early June and late December 1980 in order to report the count for each state to the President by January 1, 1981, as required by law. These are the counts used to reapportion the seats in the House.

The 1980 questionnaires reflect the interests and priorities in American society that have changed since the last census. An example is the social and economic status of women. Since the male in a growing number of households no longer is automatically considered to be the head of the household, the 1980 form contained a more objective question to identify the household relationships of those listed. Also, in response to the growing needs and interests of minorities, the 1980 census form collected more data on racial and ethnic segments of the population. Finally, the census gathered supplementary information on transportation patterns in commuting to work to help better measure energy consumption.

Response to all censuses is required by law, but the law also requires that personal information furnished on report forms be kept strictly confidential. The law is very precise, requiring that information obtained from an individual or firm be used for statistical purposes only.

If another federal, state, or local government agency needs data not already provided by the Census Bureau, but does not have the facilities for collecting or tabulating them, it may contract with the Bureau to do this work. The other agency can then concentrate on analyzing the resulting data according to its own needs. Some examples at the federal level are the Annual Housing Survey conducted by the Bureau for the Department of Housing and Urban Development, the National Crime Survey for the Department of Justice, the Health Interview Survey for the Department of Health and Human Resources, and the Consumer Expenditure Survey and employment data for the Department of Labor.

The largest and most enduring of the Bureau's ongoing surveys is the Current Population Survey, started in 1942. Each month, enumerators visit some 76,000 households scattered throughout the country, selected by sampling methods* so that, taken together, they represent a cross section of all the people. When the questionnaires reach Washington, the data are transferred to magnetic tape and tabulated on electronic computers.

This survey enables the Bureau to report between censuses on the characteristics of the population. Changes in income, education, marital status, size of families, number of children, and migration are reported annually. Data on employment and unemployment are collected and tabulated each month and transmitted to the Department of Labor, where they are analyzed and released. Other subjects appear at various intervals or on a one-time basis.

The Bureau also takes special censuses of localities that request and pay for them. The city or town making such a request is usually a growing one that wishes to increase its entitlement to wider civic powers or a greater share of state tax receipts. More than 1700 of these censuses were taken between the 1970 and 1980 national censuses.

Two centuries of census taking and 79 years as a permanent agency have brought a wealth of experience to the Bureau of the Census, and wide recognition as the nation's major factfinder.

The Bureau collects data throughout the country, from year to year, and from one generation to the next; consequently, its statistics for different areas or time periods are useful for comparative study.

The Bureau has developed an extensive program for consulting with users of its statistics, primarily through advisory committees, and conferences and workshops in which the latest methods of handling census materials are studied to assure that the statistics are widely useful.

At the beginning of 1980 the Bureau had 2871 permanent, full-time employees at its headquarters in Suitland, Maryland, just outside Washington, D.C. There were 580 permanent, full-time employees at the Bureau's record-keeping and regional processing facility in Jeffersonville, Indiana; and employment in the Bureau's 12 permanent regional offices totaled 507. Employment at two regional processing centers set up to assist in the taking of the 1980 census totaled, respectively: 819 in Laguna Niguel, California; and 596 in New Orleans, Louisiana. More than 470,000 field workers were hired on a temporary basis to take the 1980 census out of 409 temporary district offices.

The Bureau's staff includes demographers, geographers, statisticians, mathematicians, economists, computer specialists, and members of other professional fields. Since 1946, the Bureau has provided training for hundreds of persons from statistical organizations in other parts of the world. It consults regularly with statistical agencies in other countries to take mutual advantage of the latest techniques being developed in the United States and abroad.

The Bureau also prepares population estimates and projections, conducts research on methods to improve the compilation of statistics from existing records, and maintains an extensive evaluation program so that statistical publications* can be accompanied by explanations of the quality and reliability of the data.

Census data, readily available to the public, can be obtained from local government offices, libraries, district offices of the U.S. Department of Commerce, the Government Printing Office, summary tape processing centers, and, of course, the Bureau itself. (The information comes in the form of printed reports, computer tapes, microfilm and microfiche, and by special tabulations.)

Congress has established a system of Government Depository Libraries to make federal government publications available for public reference. Libraries designated within this system may elect to receive from the Superintendent of Documents, U.S. Government Printing Office, those types of government publications which they deem appropriate.

To supplement this system the Bureau of the Census provides copies of its reports to an additional group of libraries, designated as *Census Depository Libraries*. These libraries may also elect to receive the types of

Census Bureau reports they deem most useful to the public they serve, together with certain basic census volumes, such as the *Statistical Abstract**, which are sent to all Census Depository Libraries. There are 107 of these libraries nationwide. A directory listing all depository libraries is available from the Subscriber Services Section (Publications), Bureau of the Census, Washington, D.C. 20233.

All census publications may be ordered through the Government Printing Office or by writing Bureau of the Census, Subscriber Services Section, Room 1121, Building #4, Washington, D.C. 20233. The telephone number is (301) 763-7472. (Make checks payable to "Superintendent of Documents.")

Bibliography of Selected Census Publications

Bureau of the Census Catalog

Bureau of the Census Guide to Programs and Publications: Subjects and Areas

Congressional District Data Book

The County and City Data Book

Current Population Reports—Consumer Income—Money Income and Poverty Status of Families and Persons in the United States

Current Population Reports—Population Characteristics—Demographic, Social, and Economic Profile of States

Current Population Reports—Population Characteristics—Household and Family Characteristics

Current Population Reports—Population Characteristics—Marital Status and Living Arrangements

Current Population Reports—Population Estimates and Projections—Estimates of the Population of States, by Age, Race, and Sex

Current Population Reports—Population Estimates and Projections—Projections of the Population of the United States: 1977 to 2050

Current Population Reports—The Social and Economic Status of the Black Population in the United States: An Historical View, 1790–1978

Directory of Federal Statistics for Local Areas—A Guide to Sources

Historical Statistics of the U.S. (Colonial Times to 1972)

Illustrative Projections of State Populations by Age, Race, and Sex: 1975 to 2000 (Current Population Reports—Population Estimates and Projections)

PC (1)—A, B, C, and D: 1980 Population Census Reports—Number of Inhabitants, General Population Characteristics, General Social and Economic Characteristics, and Detailed Population Characteristics

The Statistical Abstract of the United States

(BUREAU OF LABOR STATISTICS
CENSUS
FEDERAL STATISTICS
OFFICE OF POPULATION CENSUSES
AND SURVEYS)

VINCENT P. BARABBA

BURN-IN

A selection procedure, applicable when the survival distribution function* (SDF) has a decreasing hazard rate*. Items are put on test until a predetermined number fail, or for a predetermined time. The first of these corresponds to censored sampling (or censored data*), the second to truncated sampling*, but the purpose here is different. Rather than using the data to make inferences about the SDF, we are simply hoping to be left with items having greater survival probability than the original product.

Bibliography

Barlow, R. E., Madansky, A., Proschan, F., and Scheuer, E. M. (1968). *Technometrics*, **10**, 51–62. (A useful general presentation of estimation procedures based on data from burn-in processes.)

Marcus, R. and Blumenthal, S. (1974). *Technometrics*, **16**, 229–234. (Discusses a generalization of burn-in called sequential screening*.)

(CENSORED DATA
CENSORING
SEQUENTIAL SCREENING
TRUNCATED SAMPLING
TRUNCATION)

BURR DISTRIBUTIONS

The Burr system of distributions was constructed in 1941 by Irving W. Burr. Since the corresponding density functions have a wide variety of shapes, this system is useful for approximating histograms*, particularly when a simple mathematical structure for the fitted cumulative distribution function (CDF) is required. Other applications include simulation*, quantal response, approximation of distributions*, and development of nonnormal control charts*. A number of standard theoretical distributions are limiting forms of Burr distributions.

The original motivation for developing the Burr system was to provide a method for fitting CDFs to frequency data. This approach avoids the analytical difficulties often encountered when a fitted density (such as a Pearson curve*) is integrated to obtain probabilities or percentiles*. Burr [1] chose to work with the CDF $F(x)$ satisfying the differential equation

$$\frac{dy}{dx} = y(1-y)\,g(x,y) \qquad [y = F(x)],$$

an analog of the differential equation that generates the Pearson system (*see* PEARSON SYSTEM; SYSTEMS OF FREQUENCY CURVES). The function $g,(x, y)$ must be positive for $0 \leqslant y \leqslant 1$ and x in the support of $F(x)$. Different choices of $g(x, y)$ generate various solutions $F(x)$; e.g., when $g(x, y) = g(x)$,

$$F(x) = \left\{\exp\left[-\int_{-\infty}^{x} g(u)\,du\right] + 1\right\}^{-1}.$$

The solutions $F(x)$ of Burr's differential equation can be classified by their functional forms, each of which gives rise to a family of CDFs within the Burr system. Burr [1] listed 12 such families:

(I)	x	$(0 < x < 1)$
(II)	$(1+e^{-x})^{-k}$	$(-\infty < x < \infty)$
(III)	$(1+x^{-c})^{-k}$	$(0 < x)$
(IV)	$\left[1+\left(\frac{c-x}{x}\right)^{1/c}\right]^{-k}$	$(0 < x < c)$
(V)	$(1+ce^{-\tan x})^{-k}$	$(-\pi/2 < x < \pi/2)$
(VI)	$(1+ce^{-r\sinh x})^{-k}$	$(-\infty < x < \infty)$
(VII)	$2^{-k}(1+\tanh x)^{k}$	$(-\infty < x < \infty)$
(VIII)	$\left(\frac{2}{\pi}\tan^{-1}e^{x}\right)^{k}$	$(-\infty < x < \infty)$
(IX)	$1-\dfrac{2}{c[(1+e^{x})^{k}-1]+2}$	$(-\infty < x < \infty)$
(X)	$(1+e^{-x^2})^{k}$	$(0 < x)$
(XI)	$\left(x-\frac{1}{2\pi}\sin 2\pi x\right)^{k}$	$(0 < x < 1)$
(XII)	$1-(1+x^{c})^{-k}$	$(0 < x)$.

(In each case, $c > 0$; $k > 0$.) The Roman numeral designation for the 12 types was first used by Johnson and Kotz [8]. The parameters c, k, and r are positive; when fitting data, a parameter (ξ) for location* and a parameter (λ) for scale* are introduced by replacing x with $(x-\xi)/\lambda$.

The Burr I family consists of the uniform distributions* whose densities have a rectangular shape. The Burr III, IV, V, IX, and XII families yield a variety of density shapes and involve four parameters, the most that can be efficiently estimated with the method of moments*. Among these families, Types III and XII are the simplest functionally and thus the most attractive for statistical modeling. Only the Type XII family was studied in detail by Burr [1] in his original work.

Burr XII density functions are of the form

$$f(x) = \frac{ckx^{c-1}}{(1+x^{c})^{k+1}} \qquad (x > 0),$$

which is unimodal* (with a mode* at $x = [(c-1)/(ck+1)]^{1/c}$) if $c > 1$ and L-shaped if $c \leqslant 1$. Two typical Burr XII densities are illustrated in Fig. 1. The rth moment about the origin* of a Burr XII distribution is

$$\mu_r' = kB(r/c+1, k-r/c) \qquad (r < ck).$$

The skewness* ($\sqrt{\beta_1}$) and kurtosis* (β_2) are given by

$$\sqrt{\beta_1} = \frac{\Gamma^2(k)\lambda_3 - 3\Gamma(k)\lambda_2\lambda_1 + 2\lambda_1^3}{\left[\Gamma(k)\lambda_2 - \lambda_1^2\right]^{3/2}}$$

and

$$\beta_2 = \frac{\Gamma^3(k)\lambda_4 - 4\Gamma^2(k)\lambda_3\lambda_1 + 6\Gamma(k)\lambda_2\lambda_1^2 - 3\lambda_1^4}{\left[\Gamma(k)\lambda_2 - \lambda_1^2\right]^2},$$

where $\lambda_j = \Gamma(j/c+1)\Gamma(k-j/c)(j = 1,2,3,4)$ and $ck > 4$. Here skewness is defined as $\sqrt{\beta_1} = \mu_3/\mu_2^{3/2}$, and kurtosis is defined as $\beta_2 = \mu_4/\mu_2^2$; as usual, μ_2, μ_3, and μ_4 denote moments about the mean μ_1.)

Rodriguez [9] constructed a moment ratio* diagram for the coverage area in the ($\sqrt{\beta_1}$, β_2) plane corresponding to the Burr XII family. As shown in Fig. 2, the boundaries of the Burr XII area can be identified with limiting forms of Burr XII distributions. The northern boundary, labeled "Wei-

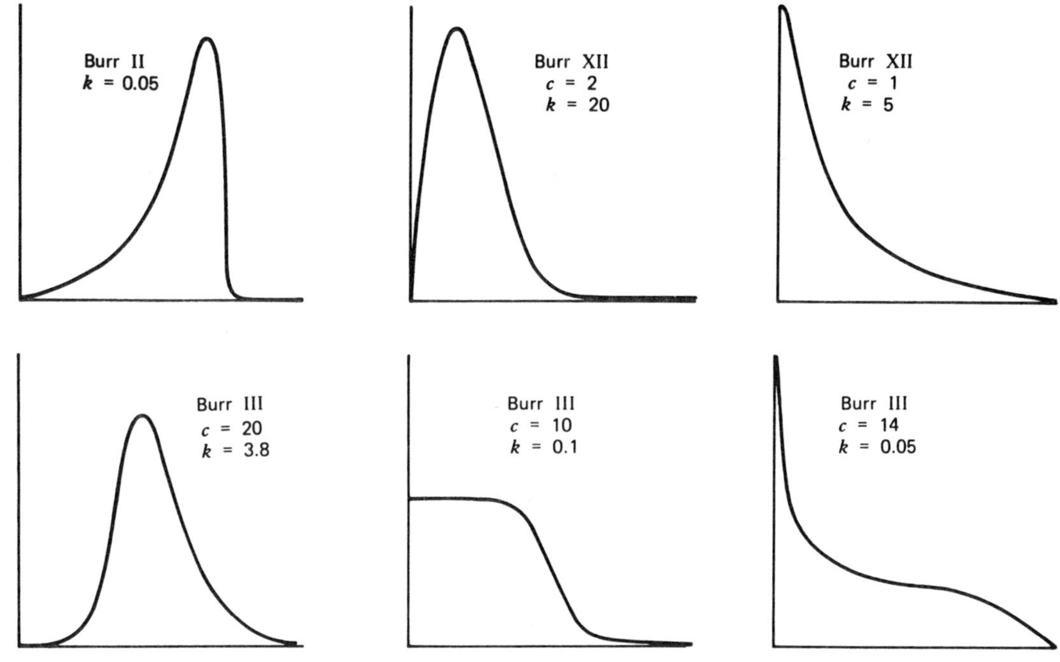

Figure 1 Some typical Burr density shapes

bull," is generated by the $(\sqrt{\beta_1}, \beta_2)$ points of the Weibull distributions*; if X has a Burr XII distribution, then for fixed k,

$$\lim_{k\to\infty} \Pr\left[X \leqslant (1/k)^{1/c}x\right] = 1 - \exp(-x^c).$$

The southwest boundary is part of a curve labeled "generalized logistic," which corresponds to the generalized logistic distributions*,

$$1 - (1 + e^x)^{-k}.$$

This family represents the limiting forms of Burr XII distributions as $c \to \infty$ with k fixed. The southeast boundary is generated by the Burr XII distributions with $k = 1$.

The curve labeled "generalized logistic" subdivides the Burr XII coverage area. In the single occupancy region north of this division, each $(\sqrt{\beta_1}, \beta_2)$ point corresponds to a unique (c, k) combination. In the double occupancy region south of this division, each $(\sqrt{\beta_1}, \beta_2)$ point corresponds to exactly two (c, k) combinations.

The Burr XII area occupies portions of the moment ratio diagram also covered by various well known families of distributions represented by points or curves in Fig. 2. These include the normal*, log-normal*, gamma*, logistic*, and extreme value Type I* distributions. Furthermore, the Burr XII region covers areas corresponding to the Type I, Type IV, and Type VI families of the Pearson system*, as well as the S_U and S_B families of the translation (Johnson) system*. *See* JOHNSON SYSTEM OF DISTRIBUTIONS.

Although the Burr XII family covers a large portion of the moment ratio diagram, a much greater area is covered by the Burr III family (see Fig. 2). These two families are related, in the sense that if X has a Burr XII distribution with parameters c and k, then X^{-1} has a Burr III distribution with parameters c and k. Burr III density functions have the form

$$g(x) = \frac{ckx^{ck-1}}{(1 + x^c)^{k+1}} \qquad (x > 0),$$

which is unimodal (with a mode at $x = [(ck - 1)/(c + 1)]^{1/c}$) if $ck > 1$, and twisted L-shaped if $ck \leqslant 1$. Figure 1 provides examples of these shapes. To obtain the rth moment about the origin, skewness,

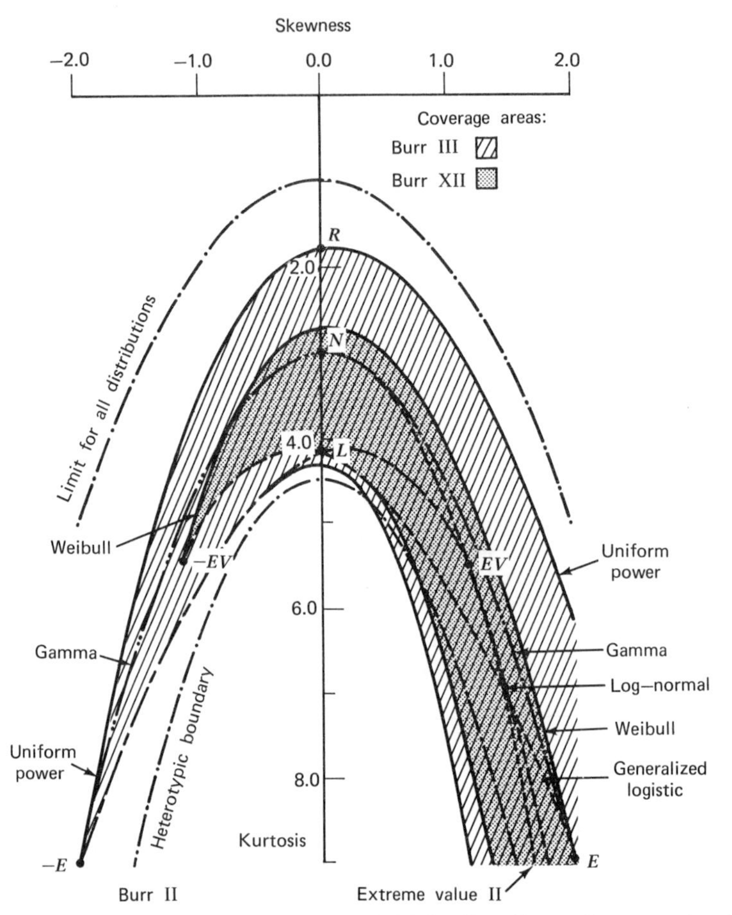

Figure 2 Moment ratio diagram for Burr II, III, and XII families. Oiunts: R, rectangular; N, normal; L, logistic; E, exponential; EV, extreme value 1. The symbol - denotes reflection.

and kurtosis of a Burr III distribution with parameters c and k, replace c with $-c$ in the epxressions for μ_r', $\sqrt{\beta_1}$, and β_2 given above for the Bur III family.

The northern boundary, labeled "uniform power," of the Burr III moment ratio coverage area in Fig. 2 is generated by the distributions of positive powers of a uniformly distributed random variable; these are limiting forms of Burr III distributions as $k \to 0$ with c fixed. The southwest boundary, labeled "Burr II," corresponds to the Burr II distributions with $k \leqslant 1$, whereas the southeast boundary is an envelope of $(\sqrt{\beta_1}, \beta_2)$ points for the Burr III family.

The Burr III area, like the Burr XII area, is subdivided into a northern single occupancy region and a southern double occupancy region. The division between these regions corresponds to the $(\sqrt{\beta_1}, \beta_2)$ points for two families of distributions: the Burr II distributions with $k > 1$, which are limiting forms as $c \to \infty$ of Burr III distributions, and the extreme value Type II* distributions, which are limiting forms of Burr III distributions as $k \to \infty$ with c fixed (see Rodriguez [10]).

Various methods are available for fitting* Burr distributions to frequency data. Burr [1] developed a cumulative moment technique for estimating the Burr XII parameters c and k from the observed skewness and kurtosis. This approach requires the use of a table (see Burr [4]), which can be difficult to interpolate due to the high nonlinearity of $\sqrt{\beta_1}$ and β_2 as functions of c and k. Other methods that can be used with both Burr XII and Burr III distributions include maximum likelihood estimation* and the method of percentiles*.

In addition to fitting frequency data, Burr distributions are useful for dealing with a number of statistical problems in which a class of distributions with functional simplicity and a variety of density shapes is required. For example, since the Burr XII and Burr III distributions can be inverted in closed form, they can be applied in simulation work, quantal response, and approximation of theoretical distributions whose moments are known, but whose functional forms cannot be expressed directly. See Drane et al. [6] for a discussion of Burr XII distributions as response* functions in analyzing quantal response experiments. This family was also used by Burr [2] to examine the effect of nonnormality on constants used in computing sample averages and ranges for plotting control charts.

It should also be noted that the sample medians* and ranges* for Burr III and Burr XII distributions have convenient distributional properties. For instance, if the sample size is $n = 2m + 1$, then the density function of the median is

$$\frac{(2m+1)!}{(m!)^2}(1+x^{-c})^{-mk}\left[1-(1+x^{-c})^{-k}\right]^m \times \frac{ckx^{ck-1}}{(1+x^c)^{k+1}} \qquad (x>0)$$

for Burr III distributions, and

$$\frac{(2m+1)!}{(m!)^2}\left[1-(1+x^c)^{-k}\right](1+x^c)^{-mk} \times \frac{ckx^{c-1}}{(1+x^c)^{k+1}} \qquad (x>0)$$

for Burr XII distributions. Related results were obtained by Rodriguez [10] for the Burr III family and by Burr [3] and Burr and Cislak [5] for the Burr XII family.

A multivariate* Burr XII distribution was constructed by Takahasi [12]; the joint density function for variables $X_1, X_2, \ldots, X_m$ with this distribution is

$$f(x_1, \ldots, x_m) = [\Gamma(k+m)/\Gamma(k)] \times \left(1 + \sum_{j=1}^{m} \alpha_j x_j^{c_j}\right)^{-(k+m)} \prod_{j=1}^{m} \left(\alpha_j c_j x_j^{c_j - 1}\right)$$

for $x_j > 0$, $c_j > 0$ ($j = 1, \ldots, m$), and $k > 0$. Here $\alpha_j^{1/c_j} X_j$ has a univariate Burr XII distribution with $c = c_j$, and any subset of the X_j's has a joint density of the foregoing form with an appropriate change of parameters. The regression function* of X_1 on $X_2, \ldots, X_m$ is

$$E[X_1 \mid X_2, \ldots, X_m] = \left[\alpha_1^{-1}\left(1 + \sum_{j=2}^{m} \alpha_j X_j^{c_j}\right)\right]^{1/c_1} (k+m-1) \times B\left(1 + c_1^{-1}, k+m-1-c_1^{-1}\right)$$

and

$$\operatorname{cov}(X_1, X_2) = \alpha_1^{-1/c_1}\alpha_2^{-1/c_2}\Gamma(1+c_1^{-1})\Gamma(1+c_2^{-1}) \times \left[\frac{\Gamma(k - c_1^{-1} - c_2^{-1})}{\Gamma(k)} - \left(\frac{\Gamma(k - c_1^{-1})}{\Gamma(k)}\right)\left(\frac{\Gamma(k - c_2^{-1})}{\Gamma(k)}\right)\right].$$

Durling [7] discussed a bivariate Burr XII distribution whose joint cumulative distribution function is

$$F(x_1, x_2) = 1 - (1 + x_1^{c_1})^{-k} - (1 + x_2^{c_2})^{-k} + (1 + x_1^{c_1} + x_2^{c_2} + r x_1^{c_1} x_2^{c_2})^{-k},$$

provided that $x_1 \geqslant 0$, $x_2 \geqslant 0$, and $0 \leqslant r \leqslant k+1$; otherwise, $F(x_1, x_2) = 0$. If $r = 0$, $F(x_1, x_2)$ reduces to the bivariate forms of the distribution introduced by Takahasi [12]. If $r = 1$, $F(x_1, x_2)$ can be written as the product of two independent Burr XII distributions.

The approach followed by Takahasi [12] can be used to derive a multivariate Burr III distribution whose joint density function is

$$g(x_1, \ldots, x_m) = [\Gamma(k+m)/\Gamma(k)] \times \left(1 + \sum_{j=1}^{m} \alpha_j x_j^{-c_j}\right)^{-(k+m)} \times \prod_{j=1}^{m} \left(\alpha_j c_j x_j^{-c_j - 1}\right)$$

for $x_j > 0$, $c_j > 0$ ($j = 1, \ldots, m$), and $k > 0$. See Rodriguez [11] for further details concerning this distribution and other multivariate Burr III distributions. *See* SYSTEMS OF FREQUENCY SURFACES.

References

[1] Burr, I. W. (1942). *Ann. Math. Statist.*, **13**, 215–232.

[2] Burr, I. W. (1967). *Ind. Quality Control*, **23**, 563–569.

[3] Burr, I. W. (1968). *J. Amer. Statist. Ass.*, **63**, 636–643.

[4] Burr, I. W. (1973). *Commun. Statist.*, **2**, 1–21.

[5] Burr, I. W. and Cislak, P. J. (1968). *J. Amer. Statist. Ass.*, **63**, 627–635.

[6] Drane, J. W., Owen, D. B., and Seibert, G. B., Jr. (1978). *Statist. Hefte*, **19**, 204–210.

[7] Durling, F. C. (1975). In *Statistical Distributions in Scientific Work*, Vol. 1, G. P. Patil, S. Kotz, and J. K. Ord, eds. D. Reidel, Dordrecht, Holland, 329–335.

[8] Johnson, N. L. and Kotz, S. (1970). *Continuous Univariate Distributions*, Vol. 1. Houghton Mifflin, Boston.

[9] Rodriguez, R. N. (1977). *Biometrika*, **64**, 129–134.

[10] Rodriguez, R. N. (1979). *The Moment Ratio Geography of Burr Type III Distributions. Forthcoming research publication*, General Motors Research Laboratories, Warren, Mich.

[11] Rodriguez, R. N. (1979). *Multivariate Burr III Distributions*. Forthcoming research publication, General Motors Research Laboratories, Warren, Mich.

[12] Takahasi, K. (1965). *Ann. Inst. Statist. Math. Tokyo*, **17**, 257–260.

(MOMENT-RATIO DIAGRAMS
FREQUENCY CURVES, SYSTEMS OF
FREQUENCY SURFACES,
SYSTEMS OF)

ROBERT N. RODRIGUEZ

BUTLER–SMIRNOV TEST

Let $X_1, \ldots, X_n$ be n independent, identically distributed (i.i.d.) real random variables with common cumulative distribution function* (CDF) F. The Butler–Smirnov test is designed to decide whether the unknown CDF is symmetric about a known (given) median*. The sample can be shifted so that this median becomes zero. The comparison of paired observations by taking differences is a typical application. The following example of such data is taken from Ury and Forrester [10]. For 10 rats, the tensile strength of taped closed wounds is compared with sutured wounds, obtaining the differences (taped minus sutured)

$$-63 \quad 64 \quad 70 \quad 96 \quad 202 \quad 207 \quad -213 \quad 245 \quad 397 \quad 442,$$

already ordered by their absolute value. Details of this example can also be found in Lehmann [6, Chap. 3, Ex. 3], where the use of the Wilcoxon signed-rank statistic* is demonstrated.

The Butler–Smirnov test is a test of Kolmogorov–Smirnov* type and has good power against all alternatives (omnibus test). It should be used for alternatives which are larger or different from the unimodal* shift alternatives. See the section "Power" for details.

Historical Remarks

Smirnov [9] described the test and obtained its distribution. Independently, Butler rediscovered this test in 1969 [2]. A detailed discussion of the Butler–Smirnov test is given in Chatterjee and Sen [3]. Power bounds for various alternatives have been published by Koul and Staudte [5].

HYPOTHESIS *H* AND TEST STATISTICS

Denote by $\mathfrak{F}$ the set of all continuous CDFs. Then $H := \{F \in \mathfrak{F} : F(x) = 1 - F(-x) \text{ for all } x \in R\}$ is the null hypothesis of symmetry (about zero). For the example above, this means the hypothesis of no difference in the tensile strength of tape-closed and sutured wounds.

Let $F_n(x)$ and $G_n(x)$ be the empirical distribution functions* of the samples $X_1, \ldots, X_n$ and $-X_1, \ldots, -X_n$. Define $B_n^+ := \sup_x\{G_n(x) - F_n(x)\}$, $B_n^- := \inf_x \cdot \{G_n(x) - F_n(x)\}$ and $B_n(x) = \max\{B_n^+, -B_n^-\}$. We denote the observed values of these statistics by b_n^+, b_n^-, and b_n. (For brevity, we use the same symbol for the statistic and the associated test.) Under H, these statistics are distribution-free*. Therefore, we can write $P_n(l,k) := \Pr_F(B_n^+ \geqslant l/n$ or $B_n^- \leqslant k/n)$ for all $F \in H$. Analogously, define the one- and two-sided probabilities by $P_n^+(l) = P_n(l, -n-1)$, $P_n^-(k) = P_n(n+1, k)$ $[= P_n^+(-k)]$, and $P_n(l) = P_n(l,l)$.

TEST PROCEDURE

One-sided tests*: Let $k^+(\alpha)$ be the smallest integer k such that $P_n^+(k) \leqslant \alpha$. For testing H against alternatives* contained in $A^+ := \{F \in \mathfrak{F}\backslash H : F(x) \leqslant 1 - F(-x) \forall x \in \mathbf{R}\}$, reject H if $nb_n^+ \geqslant k^+(\alpha)$. Then the level* of this test is not higher than α. Because B_n^+ is a discrete random variable with only $n+1$ different values, it might be preferable to base the decision on the significance* probability $P_n^+(b_n^+)$. For testing H versus alternatives contained in $\{F \in \mathfrak{F}\backslash H : F(x) \geqslant 1 - F(x) \forall x \in \mathbf{R}\}$, reject H if $-nb_n^- \geqslant k^+(\alpha)$. Again, the level is not higher than α.

Two-sided test*: Let $k(\alpha)$ be the smallest integer such that $P_n(k) \leqslant \alpha$. For testing H against alternatives contained in $\mathfrak{F}\backslash H$, reject H if $nb_n \geqslant k(\alpha)$. Then the level of this test is not higher than α. For asymmetric alternatives $P_n(b_n^+, b_n^-)$ is the significance probability.

OTHER FORMULATIONS OF THE TEST STATISTICS

$G_n(x) = 1 - F_n(-x-)$ yields $B_n^+ = \sup_x\{1 - F_n(x) - F_n(-x-)\}$ and analogous forms of B_n^- and B_n [3]. For practical computations of B_n^+ and B_n^-, the following description may be useful. Order the observations by their absolute value and define the random variables:

$$T_i = \begin{cases} 1 & \text{if the } i \text{ smallest observation is positive} \\ -1 & \text{if the } i \text{ smallest observation is negative.} \end{cases}$$

Then $B_n^+ = \max_{1 \leqslant j \leqslant n+1} \sum_{i=j}^{n+1} T_i$ and $B_n^- = \min_{1 \leqslant j \leqslant n+1} \sum_{i=j}^{n+1} T_i$, where $T_{n+1} = 0$. In the example, we obtain

j	T_j	$\sum_{i=j}^{n+1} T_i$
1	−1	6
2	1	7
3	1	6
4	1	5
5	1	4
6	1	3
7	−1	2
8	1	3
9	1	2
10	1	1
11	0	0

Hence $B_{10}^+ = 7$ and $B_{10}^- = 0$.

EXACT DISTRIBUTION UNDER H

For all integers l with $0 \leqslant l \leqslant n+1$,

$$\begin{aligned} P_n^+(l) &= P_n^-(-l) \\ &= B_{n,0.5}((n-1)/2) \\ &\quad + B_{n,0.5}((n-l-1)/2) \\ &= \begin{cases} P_{n+1}^+(l) & \text{if } n-l \text{ is even} \\ P_{n-1}^+(l) & \text{if } n-l \text{ is odd,} \end{cases} \end{aligned} \tag{1}$$

where $B_{n,0.5}(x) = 2^{-n}\sum_{i=0}^{[x]}\binom{n}{i}$, the binomial* CDF with probability $p = 0.5$ of success ($[x] = \max\{i \leqslant x, i$ an integer$\}$). For all integers l with $1 \leqslant l \leqslant n+1$,

$$P_n(l) = 2\sum_{j=0}^{u}(-1)^j P_n^+((2j+1)l) \tag{2}$$

$$= \begin{cases} P_{n+1}(l) & \text{if } n-l \text{ is even} \\ P_{n-1}(l) & \text{if } n-l \text{ is odd,} \end{cases} \tag{3}$$

where $u = [(n/l - 1)/2]$. Equations (1) and (2) can be both obtained from the asymmetric case

$$P_n(l,k) = 1 - 2^{-n} \sum_{j=-\infty}^{\infty} \sum_{i=k'}^{l'} \left(\binom{n}{i+js} - \binom{n}{i+js-k} \right), \qquad (4)$$

where l and k are integers, $-n \leqslant k \leqslant 0 \leqslant l \leqslant n$, $s = l - k$, $k' = [(n+k)/2] + 1$, and $l' = [(n + l - 1)/2]$.

Equation (1) shows that no special tables for P_n^+ are needed, because tables for the binomial CDF are widely available. P_n is given in Table 1 below for small and even n [use (3) for odd n].

For $n > 10$, see the next section.

Example 1. From (1) we obtain $P_{10}^+(7) = 2B_{10,0.5}(1) \approx 0.0215$. Therefore, the significance* probability for testing H against the alternative that the tape-closed wounds are stronger is ≈ 0.0215. This is slightly more significant* than the significance probability 0.0244 of the Wilcoxon signed-rank test [6].

ASYMPTOTICS AND APPROXIMATIONS UNDER H

The well-known asymptotic behavior of the binomial CDF yields in (4) the limit

$$\lim_{n\to\infty} P_n(xn^{1/2}, yn^{1/2}) = 2\Phi(y) + 2 \sum_{j=1}^{\infty} (-1)^j \times (\Phi(y - jz) - \Phi(-y - jz))$$

for all $y \leqslant 0 \leqslant x$, $z = x - y$ (Φ = standard normal* CDF). This specializes to

$$\lim_{n\to\infty} P_n(xn^{1/2}) = 4\textstyle\sum_{j=1}^{\infty}(-1)^{j+1} \times \Phi(-(2j-1)x)$$

and

$$\lim_{n\to\infty} P_n^+(xn^{1/2}) = 2\Phi(-x). \qquad (5)$$

For $n > 40$, (5) can be used as a satisfactory approximation. For $10 \leqslant n \leqslant 40$ we suggest the approximation in Molenaar [7, (5.5)]:

$$P_n^+(l) \sim \Phi(f(l^-) - f(-l^-)) + \Phi(f(l^+) - f(-l^+)), \qquad (6)$$

where $f(l) = ((6n - l)^2/(12n) - 2n + 1/6)^{1/2}$ and

$$l^{\pm} = \begin{cases} l \pm 1 & \text{if } n - l \text{ is even} \\ l & \text{if } n - l \text{ is odd.} \end{cases}$$

The two-sided case can be approximated by $P_n(l,k) \sim P_n^+(l) + P_n^+(-k)$ for $P_n(l,k) \leqslant 0.1$ with an absolute error less than 0.0005, if $n \geqslant 10$. For $l = -k$ this yields $P_n(l) \sim 2P_n^+(l)$.

Example 2. Approximating $P_{10}^+(7)$ with (5) yields $2\Phi(-7/\sqrt{10}) \approx 0.0272$. The more adequate approximation (6) gives $\approx 2\Phi(-2.30) \approx 0.0214$.

CONFIDENCE BAND

Let $F(x)$ be the common CDF of $X_1, \ldots, X_n$, and $G(x)$ the common CDF of $-X_1, \ldots, -X_n$. Denote the left inverse of G by $G^{-1}(u) = \inf\{x : G(x) \geqslant u\}$. Then $h(x) = G^{-1}(F(x))$ equals x under H. Define the critical value $k(\alpha)$ as in the section "Test

Table 1 $P_n(l)$, **Last Digit Rounded**

Z / n'	4	5	6	7	8	9	10
4	0.125						
6	0.250	0.063	0.031				
8	0.359	0.141	0.078	0.016	0.008		
10	0.453	0.219	0.131	0.043	0.023	0.004	0.002

Procedure." Under H, the function h lies with probability at least $1-\alpha$ in the confidence band [1] $-X_{n+1-i+k(\alpha):n} \leqslant h(x) \leqslant -X_{n-i-k(\alpha):n} \forall x \in [X_{i:n}, X_{i+1:n}),$

$$\left(i=0,\ldots,n;\ X_{j:n}=\begin{cases}-\infty & \text{if } j\leqslant 0\\ +\infty & \text{if } j>n\end{cases}\right).$$

POWER*

The Butler–Smirnov statistics are consistent* against all alternatives. In ref. 5 the power of B_n^+ is investigated with respect to the following (as well as some other) alternatives:

1. $\Omega(\Delta) = \{F \in A^+ : \sup_x\{1 - F(x) - F(-x)\} \geqslant \Delta\} \subset A^+ = \Omega(0)$.
2. $\Omega_u(\Delta) = \{F_\theta \in \Omega(\Delta) : F_\theta(x) = F(x-\theta)$, where $\theta \geqslant 0$ and F has a symmetric unimodal* density$\}$ (shift alternative).
3. $\Omega_L(\Delta) = \{F_\gamma \in \Omega(\Delta) : F_\gamma = F^\gamma(x)$, where $\gamma \geqslant 1$ and F is symmetric about zero$\}$ (Lehmann alternative*).

Let (Δ_n) be a real nonnegative sequence such that $\Delta_n n^{1/2} \to c < \infty$ as $n \to \infty$. Denote for any parametric family of alternatives $A(\Delta)$ and fixed α the asymptotic minimum power of B_n^+ by

$$\beta_c(A) = \lim_{n\to\infty} \inf\{\Pr_F(B_n^+ \geqslant k(\alpha)) : F \in A(\Delta_n)\}.$$

Then $\beta_c(\Omega_u)$ and $\beta_c(\Omega_L)$ are both bounded below by $\Phi(c - k(\alpha))$, which equals the asymptotic minimum power of the sign test* under shift alternatives. With respect to $\Omega(\Delta)$, B_n^+ is preferable to a large class of linear rank tests*, including the Wilcoxon signed-rank test* (under certain conditions, see ref. 5). (But one should keep in mind that these three families of alternatives favor B_n^+.) The Bahadur efficiency* of B_n is not smaller than the Bahadur efficiency of the sign test for testing H versus $\mathfrak{F}\backslash H$. It is equal for testing H against $\cup_{\Delta\neq 0}\Omega_u(\Delta)$ [3].

EXTENSIONS

1. If the independent random variables X_i have different continuous CDFs F_i $(i = 1, \ldots, n)$, which are symmetric about zero, then all the results in "Exact Distribution under H" and "Asymptotics and Approximations under H" and the Bahadur efficiency remains unchanged [3].
2. Let $\Psi(x, y)$ be a positive weight function. The weighted Butler–Smirnov statistic $T_n^+ := \sup\{(F_n(x) - G_n(x)) / \Psi(F_n(x), G_n(x)) : x \in \mathbf{R}$ such that $u/n \leqslant F_n(x) \leqslant v/n\}$ (u, v integers, $0 \leqslant u \leqslant v \leqslant n$) has been investigated in ref. 1, with special asymptotic results for the "equal-weight" statistic, where

$$\Psi^2(s,t) = \begin{cases}1 - |1 - s - t| & \text{if } 0 < s+t < 2\\ 1 & \text{otherwise.}\end{cases}$$

A recursion for the exact distribution of T_n^+ can be found in ref. 8.

References

[1] Aaberge, R., Doksum, K. A., and Fenstad, G. (1977). *Biometrika*, **64**, 473–487. (Includes the weighted statistic, confidence bands, and the case of unknown median.)

[2] Butler, C. C. (1969). *Ann. Math. Statist.*, **40**, 2209–2210. (Short statement of the exact and asymptotic results.)

[3] Chatterjee, S. K. and Sen, P. K. (1973). *Ann. Inst. Statist. Math. Tokyo*, **25**, 287–300. (Detailed proof of the exact, asymptotic, and Bahadur-efficiency results. Basic reference. Printing errors in Tables 1 and 2.)

[4] Doksum, K. and Thompson, R. (1971). *Ann. Math. Statist.*, **42**, 12–34. (Power bounds for monotone rank tests.)

[5] Koul, H. L. and Staudte, R. G. (1976). *Ann. Statist.*, **4**, 924–935.

[6] Lehmann, E. L. (1975). *Nonparametrics*. Holden-Day, San Francisco. (Excellent general introduction to rank statistics. Butler–Smirnov test is not explicitly stated.)

[7] Molenaar, W. (1970). *Math. Centre Tracts*, **31**, Mathematisch Centrum, Amsterdam. (Approximations to the binomial CDF.)

[8] Niederhausen, H. (1979). Sheffer Polynomials for Computing Exact Kolmogorov–Smirnov and

Rényi Type Distributions. *Tech. Rep. 6*, Dept. of Mathematics, MIT, Cambridge, Mass. (Proof of the general two-sided distribution. Recursion for the equal-weight distribution.)

[9] Smirnov, N. V. (1947). *Akad. Nauk. SSR, C. R. (Dokl.) Acad. Sci. URSS*, **56**, 11–14. (Original paper about the test, but hard to obtain. Also contains bounds for the distributions.)

[10] Ury, H. K., and Forrester, J. C. (1970). *Amer. Statist.*, **24**, 25–26.

(KOLMOGOROV–SMIRNOV (TYPE) TESTS
RANK TESTS
SIGN TEST
WILCOXON SIGNED RANK TEST)

H. NIEDERHAUSEN

BUYS–BALLOT TABLE *See* TIME SERIES

C

CALCUTTA STATISTICAL ASSOCIATION BULLETIN

The *Calcutta Statistical Association Bulletin* is a journal of statistics published by the Calcutta Statistical Association. The offices of the association are located in the Department of Statistics, Calcutta University, Calcutta 700019, India. The present editor is S. K. Chatterjee. The *Bulletin* is published in quarterly numbers (March, June, September, and December) and four numbers comprise a volume. Lately, combined numbers have often been issued to make up for dislocations in publication schedule. The average number of pages per volume (based on the last 10 volumes) is 162. The subscription rates (likely to be revised in the near future) are at present: current: Rs 30.00 or $8.00 per volume; previous 5 years: Rs 90.00 or $12.00 per volume; earlier years: Rs 112.50 or $15.00 per volume.

The Calcutta Statistical Association, as a registered society, was formed in 1945 by a group of persons interested in the development, application, and propagation of statistics. Among the specific purposes of the Association enunciated in its memorandum at that time were (1) to promote a study of the science of statistics, both pure and applied, and all matters and subjects allied thereto; and (2) to provide for research and instruction for the advancement of the study and dissemination of knowledge of statistics and all matters and subjects allied thereto. To realize these objectives, it was imperative for the Association to have a regularly published organ of its own, and accordingly, an editorial board comprising S. N. Roy, P. K. Bose, B. N. Ghosh, H. K. Nandi, M. N. Ghosh, and P. K. Banerjee was formed with H. K. Nandi as editor. The first number of the *Bulletin* came out in August 1947. Professor Nandi served as editor until 1977, guided the journal ably through many vicissitudes, and in spite of some dislocation in publication schedules that cropped up in later years, 26 volumes were published through 1978.

At first, the *Bulletin* sought to play a twofold role. At that time, many official authorities, technical personnel, and scientists, in general, did not have clear notions about the scope and applicability of statistics. This led the *Bulletin* to devote a considerable portion of its printed fare to the task of educating nonstatisticians. As such, many of the articles in the earlier volumes were application-oriented critical studies. With the passage of time, as statistics began to be accepted more

Table 1

Country	Percent
India	32.4
Other Asian countries	6.3
United States	27.9
Other North and South American Countries	3.4
United Kingdom	3.2
Germany	6.9
Other European countries	13.6
African countries	2.4
Australia and New Zealand	3.9

widely, the first need was gradually reduced, and with the opening of courses in statistics in many universities and institutions, the number of workers engaged in statistical research also grew. These workers were in need of an easily accessible organ for publishing their work and the *Bulletin* was ready to play its part. In serving the growing population of researchers in India, the *Bulletin* retained sufficient breadth to accommodate articles related to all branches of statistics. The only requirement was that the standard of the contributions should be of a high order. Over the years the *Bulletin* built up a fairly large clientele in different parts of the world. Table 1, a countrywise percentage distribution of number of copies sent to subscribers, members, and parties having an exchange arrangement, relates to Volume 25 (1976), and gives a view of the current circulation pattern of the *Bulletin*.

A subject-matter breakdown of the research articles published in successive quinquennia of the first 25 volumes of the *Bulletin* is revealing (Table 2). This classification excludes articles on general topics, reports, answers to queries, book reviews, and so on. The group "other" includes quality control, psychometry, econometrics, demography, operations research, and similar topics.

Table 3 presents the Table of Contents of Vol. 21, Nos. 83–84, September and December 1972.

Traditionally, the editorial policy has been to entertain all contributions pertaining to statistical theory and its applications. In the case of disciplines related to statistics, the policy has been to accommodate a paper unless the level of specialization makes its appeal too limited.

S. K. CHATTERJEE

CALIBRATION

Calibration is the process of assigning values to the response of an instrument or the property of an artifact relative to reference standards or measuring processes. This may involve determining the corrections to the scale (as with direct-reading instruments), determining the response curve* of an instrument or artifact as a function of changes in a second variable (as with platinum resis-

Table 2

	Number of Articles Published in Volumes:				
Subject	1–5	6–10	11–15	16–20	21–25
Probability	3	3	—	4	9
Statistical inference	23	39	26	28	26
Design and analysis of experiments	17	11	30	24	18
Sample surveys	12	7	9	7	10
Other	6	15	11	10	7
Applied studies	35	7	1	1	—
Total	96	82	77	74	70

Table 3

Article	Pages
A sampling scheme to realise a pre-assigned set of inclusion probabilities of first two orders—Parimal Mukhopadhyay	87–122
On the convex set in the extensions of unilateral and bilateral test statistics—Umesh D. Naik	123–131
Exact properties of sequential tests of some composite hypotheses—Nanak Chand	133–142
A minimax procedure for selecting the population with the largest (smallest) scale parameter—J. B. Ofosu	143–154
A two-stage attribute sampling plan—S. B. Bagchi	155–170
Miscellaneous Notes	
A Hájek–Rényi type inequality for generalized U-statistics—Pranab Kumar Sen	171–179
Asymptotic properties of linear functions of order statistics for m-dependent variables—Malay Ghosh	181–192
A note on restricted Kronecker product method of constructing statistical designs—Basudeb Adhikary	193–196
On identical blocks in a balanced incomplete block design—G. A. Patwardhan	197–200
Optimization of storage capacity with shortage cost included—B. Chandra and A. Ghoshal	201–205

tance thermometers), or assigning values to reference objects (as with standards of mass, voltage, etc.) In all cases the objective is to bring the local measurement process into line with that of an accepted reference. This is often a legal requirement; e.g., weights and measures laws throughout the world call for some form of calibration of the instruments used in commerce, and a number of federal agencies require their contractors to maintain calibration systems.

The object of calibration is to bring about "good" measurement, and it must be noted that calibration is just a means to that end. Unless the instrument or artifact is used correctly, "good" measurements will not abound regardless of how carefully the calibration has been done. One must not, therefore, lose sight of this when determining the contribution of the errors of calibration on practical measurement. This will bring into focus the appropriateness of the method of calibration—the possible errors in calibrating with pure substances when only contaminated samples* are to be measured comes to mind.

CALIBRATION OF REFERENCE STANDARDS

Perhaps the simplest form of calibration is that involving reference standards such as standards used in mass, voltage, resistance, length, etc. In these cases the standard is maintained by one or more reference items at a central laboratory, and calibration consists of comparing an "unknown" group with them using a comparator under carefully prescribed conditions. When the items for test are identical with those of the reference group, one is concerned with carrying out a pattern of difference measurements among the two groups which have the property that one has a check on stability of the group average, the level of random variation, and the dependence on environmental or other extraneous factors.

Example 1. In mass calibration a typical set of standards starting at the kilogram level would have weights for 1000, 500, 300, 200, 100, 50, 30, 20, and 10 grams and usually three lower decades. In high-precision work,

Table 1

Observation[a]	Nominal Weight 50	30	20	10	(5 + 3 + 2)	Check Standard 10
y_1	+	−	−	+	−	
y_2	+	−	−		+	−
y_3	+	−	−	−		+
y_4	+	−	−			
y_5	+		−	−	−	−
y_6		+	−	+	−	−
y_7		+	−	−	+	−
y_8		+	−	−	−	+
y_9			+	−	−	
y_{10}			+	−		−
y_{11}			+		−	−

[a] An observation consists of measuring the difference between the items marked + and those marked −, the others being omitted.

one can measure only differences between nominally equal groups, and because of the possible dependence of the variability on the load, each decade is done separately. If a check standard is added to each decade, Table 1 can be used to calibrate the set. Using the value of the (50 + 30 + 20) from the previous series enables one to solve for the unknown. One has 6 degrees of freedom in the standard deviation*, which when coupled with the value for the check standard gives the ingredients needed for a proper error analysis (*see* ERROR ANALYSIS). A list of designs useful in weight calibration is given in ref. 1.

Example 2. Voltage and resistance standards are maintained by a group of n standards and calibration consists of measuring differences between and among them and a set of k unknowns. The total of $\binom{n+k}{2}$ possible pairings is usually unnecessarily large, so that one would do either the set of nk differences between elements of the two groups or some fraction thereof. For $n = k = 6$ the design involving 24 intercomparisons are shown by the asterisks in Table 2. (Because only differences are measured, it is necessary to use the sum of the assigned value for the standards as a restraint to solve the system.) The differences between members of the reference group and the standard deviation from the deviations constitute controls on the measurement process. If the intercomparisons are done in a short time period, the standard deviation from the deviations may seriously underestimate that found when a check group is run periodically throughout the year.

Example 3. The determination of corrections to an instrument scale such as is done wih liquid-in-glass thermometers and direct-reading instruments usually involves setting up a sequence of conditions corresponding to a subset of the scale markings—temperature baths are an example. In this case, one usually wishes to calibrate several

Table 2

Group of Standards	Test Items 7	8	9	10	11	12
1	*	*		*		*
2	*	*	*		*	
3		*	*	*		*
4	*		*	*	*	
5		*		*	*	*
6	*		*		*	*

instruments simultaneously. Because one expects time-dependent changes in the conditions, a trend elimination design is to be recommended. If S constitutes the standard and $A, B, C, D, \ldots, N$ the unknowns, then the sequence

$$SABCD \cdots NSSN \cdots DCBAS$$

provides a trend-free calibration of A, B, $C \ldots$ relative to S.

What uncertainties should be assigned to these calibrations? In the case of mass, the uncertainty of the value assigned by the calibrating laboratory becomes a bound to the systematic error in uses involving items similar to the standard. In the electrical example, if the calibration is done apart from the point of use, one has no check on biases introduced by the operating conditions nor any knowledge of whether the units just calibrated survived the trip back home. In the scale correction example, if the conditions of calibration and of use are the same, then the uncertainty in the calibration becomes a possible systematic error for measurements made by the instrument.

CALIBRATION CURVES

Perhaps the most common calibration problem is that of determining a curve relating the response of an instrument or artifact in units of y (e.g., voltage or resistance) to the amount of another variable x (e.g., amount of radiation or temperature). (*See* CURVE FITTING.) Because one uses the curve to predict x from a reading of y (just the reverse of the way the curve is usually fitted) the value for x for an observed y is given by solving $y = f(x)$ for x, a result that is a nonlinear function of the parameters of the curve. Eisenhart [3] has discussed the case of inverting the usual least-squares regression* (see also refs. 1 and 4), and Scheffé [5] has provided a technique for interval estimates*. Platinum resistance thermometers and turbine flow meters are two examples of this form of calibration where a continuous curve relates the two variables. In other cases, such as the calibration of the volume of a cylinder or the response of a load cell, discontinuities in the curves may be present.

Because the deviations from the fitted curve may not be random errors but represent actual departures from the assumed curve, care must be exercised in the error analysis. If independent determinations of the curve reveal the same pattern of residuals, then the random error* in the coefficients or of a point on the curve should be determined directly from the variability revealed by such a sequence of repetitions of the whole curve-fitting* process. When the variation found by this direct method differs from that indicated by the variance estimates from the usual curve-fitting procedures, the observed deviations become the source for systematic error* in the use of the instrument.

The conditions of calibration may differ significantly from those of the use of the instrument so that the error of the measured value may not be predictable from the variation in the results obtained during calibration. When the phenomena being measured can be repeated so that an independent measurement can be made, or if a second instrument can be used in parallel with the first, then one has a proper set of repetitions for determining random error. The bias* introduced by the differences between conditions of calibration and conditions of use can be determined by varying the conditions of calibration so as to cover those expected in practice and measuring the departures from results under nominal conditions. These differences constitute a bound to possible systematic error in the use of the instrument when used without correction for the change in conditions.

The calibration operation is often done at nominally equal increments of Δx in the variable to be measured; i.e., the temperature, concentration, or volume are present at even intervals and the output resistance, voltage, or height is measured. There is always the concern that the order of running the points will introduce a bias if a time-dependent effect (e.g., due to temperature changes) is present. If the curve itself is assumed to be linear so that the expected results take the form α, $\alpha + \Delta$, $\alpha + 2\Delta$, $\alpha + 3\Delta, \ldots$ for increments $0, 1, 2, 3, \ldots$ in the quantity to be measured (corresponding to values at $x_0, x_0 + \Delta x$, $x_0 + 2\Delta x, \ldots$) and a time-dependent linear drift of 0, Δt, $2\Delta t$, $3\Delta t, \ldots$ exists, then the value for Δ will be biased. The order of running should in such cases avoid purely increasing or decreasing sequences. For four measurements the order $\alpha + 2\Delta$, α, $\alpha + 3\Delta$, $\alpha + \Delta$ can be seen to be orthgonal* to the linear time effect $-3\Delta t$, $-\Delta t$, Δt, $3\Delta t$. For five measurements the order $\alpha + \Delta$, $\alpha + 4\Delta$, $\alpha + 2\Delta$, α, $\alpha + 3\Delta$ is orthogonal to the time effect $-2\Delta t$, $-\Delta t$, 0, Δt, $2\Delta t$. These designs give efficient estimates of the time effect in addition to giving values for Δ that are free of the time effects.

INSTRUMENT CALIBRATION

When an instrument has only short-term stability, it is a common practice to use two (or more) standard reference samples or artifacts to provide a calibration along with each measurement. Thus two items with known values x_0 and x_1 are measured by the instrument to give values y_0 and y_1 (in units of resistance, voltage, or whatever the output of the instrument). The value of x_i for a measured response of y_i is then given by

$$x_i = x_0 + \frac{x_1 - x_0}{y_1 - y_0}(y_i - y_0)$$

for an instrument assumed to have a linear response curve. The uncertainty in the value, x_i, will involve systematic errors due to uncertainty in x_0 and $(x_1 - x_0)$, possible nonlinearity in the instrument's response, and any "mismatch" between the standardizing items and the test items. The random error affecting x_i may involve a between-occasion component so that a stable reference item or substance should be measured periodically and the variance of such a sequence should be compared with that obtained from the variation in the y's.

USE OF MEASUREMENT CONTROLS

The purpose of calibration is to ensure that an instrument will unambiguously measure the sought-after quantity, and provide results that are consistent with others making similar measurements. Calibration is a means to an end, and what constitutes proper calibration depends on the end use of the measurements. Once one has determined the uncertainty of a result from a calibrated instrument relative to that use, the efficacy of any calibration scheme can be judged.

To determine the uncertainty involves answering the questions one would expect if the measurement were the subject of legal or scientific inquiry. Such questions would include the following: Has the instrument changed since it was calibrated? Do the conditions of use involve a component of variation not present in the calibration? What checks are made of assumptions of linearity, independence of background variables, etc.? Is there a sequence of repetitions available which have variation in conditions comparable to that encountered in an independent verification of the value?

The need for "controls" measured under the full diversity of conditions under which the instrument is to be used cannot be overemphasized. Repetitions by a second instrument, the use of alternative standard reference materials or items, and any other techniques that lead to independent measurements to assure the adequacy of the error analysis should be a continuing effort if one's measurements are to "stand up in court."

The repeated use of some form of control —a reference item of phenomenon— provides the information needed to decide when recalibration is needed. "Accuracy" is not inherent in an instrument (only measurements have that property), so that it is necessary to determine the quality of measurements by some "sampling" of the measurements themselves. Measurement control systems based on fixed calibration intervals leave in doubt the time of "failure" of an instrument when it is found to be outside allowable tolerances upon recalibration. When the calibration is carried out by a higher-echelon laboratory, there is a possibility that its conditions of calibration are more restrictive than those of the user, so that unduly small uncertainties may be attributed to one's measurements.

References

[1] Acton, Forman, S. (1959). *Analysis of Straight-Line Data*. Wiley, New York. (Pages 43–52 discuss confidence and tolerance limits arising in the use of calibration curves.)

[2] Cameron, J. M., Croarkin, M. C., and Raybold, R. C. (1977). Designs for the Calibration of Standards of Mass. *Natl. Bur. Stand. (U.S.) Tech. Note 952* (Washington, D.C.).

[3] Eisenhart, C. (1939). *Ann. Math. Statist.*, **10**, 162.

[4] Mandel, J. (1964). *The Statistical Analysis of Experimental Data*. Wiley, New York. (Chapter 12 gives an excellent discussion of the statistical procedures involved in linear calibration curves.)

[5] Scheffé, H. (1973). *Ann. Statist.* **1**, 1–37.

(CHEMISTRY, STATISTICS IN
CURVE FITTING
ERROR ANALYSIS)

J. M. CAMERON

CALLBACKS

A term used in survey sampling*, referring to efforts made to obtain responses from individuals who do not provide an immediate response. The term applies directly to situations wherein there is a direct approach to each individual in a sample: e.g., by visiting their residence or calling them on the telephone. It is applied by extension to situations where there may be no actual callback but where there are repeated attempts at measurement, whatever the means employed.

When data have been obtained from callbacks, it is desirable to pay special attention to differences between callbacks and original responses. This may give a clue that may assist in the very difficult task of trying to assess responses that would have been obtained from individuals for whom no response (even on callback) has been obtained.

Bibliography

El-Badry, M. A. (1956). *J. Amer. Statist. Ass.*, **51**, 209–227. (See also various papers in the 1976 and 1977 ASA *Proc. Social Statist. Sect.*)

(SURVEY SAMPLING
SURVEYS, HOUSEHOLD)

$C(\alpha)$-TESTS *See* OPTIMAL $C(\alpha)$-TESTS

CAMP–MEIDELL INEQUALITY

An improvement on Chebyshev's inequality*, valid when the distribution is unimodal and the mode* equals the expected value. It states that more than $(1 - \frac{4}{9}k^{-2})$ of the distribution must fall within the closed range of $\mu \pm k\sigma$; (formally $\Pr[|X - \mu| > k\sigma] \leqslant \frac{4}{9}k^{-2}$). For the interval $\mu \pm 2\sigma$ the Chebyshev inequality yields 0.25 of the cases to fall outside this interval while the Camp-Meidell inequality yields 0.111; for $\mu \pm 3\sigma$, the numbers are 0.111 and 0.049, respectively. Many nonnormal distributions do come close enough to meet these conditions for the Camp–Meidell inequality to be applied with confidence.

The original forms of what is known in the literature as the "Camp–Meidell inequality" [1, 2, 4] are

$$\Pr[|X - \mu_0| > \lambda\tau] \leqslant \begin{cases} 1 - \lambda/\sqrt{3} & \text{for } \lambda < 2/\sqrt{3} = 1.1547 \\ 4/(9\lambda^2) & \text{for } \lambda \geqslant 2/\sqrt{3} \end{cases} \tag{1}$$

or

$$\Pr[|X - \mu| > \lambda\sigma] \leqslant \frac{4}{9}\,\frac{1 + s^2}{(\lambda - s)^2} \qquad \text{if } \lambda > s. \tag{2}$$

(It can be traced to Gauss [3] and is discussed in Narumi [5].) Here X is a random variable with a unimodal density function with the mode at μ_0, $\mu = EX$, $\sigma^2 = E(X - \mu)^2$, $\tau^2 = \sigma^2 + (\mu - \mu_0)^2$, and $s = |(\mu - \mu_0)/\sigma|$.

For a symmetric unimodal distribution $s = 0$ and (2) takes the form quoted above. (For refined applications of this inequality it is necessary to know the mode of the distribution.)

References

[1] Camp, B. H. (1922). *Amer. Math. Soc. Bull.*, **28**, 427–432.

[2] Camp, B. H. (1923). *Biometrika*, **15**, 421–423.

[3] Gauss, C. F. (1821). *Theoria Combinations Observationum*. Göttingen.

[4] Meidell, B. (1921). *Skand. Aktuarietidskr.*, **4**, 230–238.

[5] Narumi, S. (1923). *Biometrika*, **15**, 245–253.

[6] Savage, I. R. (1961). *J. Res. Natl. Bur. Stand.*, **65B**(3), 211–222.

(CHEBYSHEV'S INEQUALITY
GAUSS INEQUALITIES)

CANADIAN JOURNAL OF STATISTICS, THE (LA REVUE CANADIENNE DE STATISTIQUE)

The Canadian Journal of Statistics publishes original work in either English or French in the theory and applications of statistics, including applications in the social, physical, biological, and engineering sciences. All papers are refereed. The *Journal* also publishes reviews of books in the statistical sciences written by Canadian authors or published in Canada, as well as abstracts of papers presented at the annual meetings of the society. The circulation of the *Journal* for Volume 8 (1980) exceeded 850 copies, which were sent to over 40 different countries.

Sudhish G. Ghurye of the Department of Statistics and Applied Probability at The University of Alberta, is the editor of *The Canadian Journal of Statistics*, and all research papers intended for publication should be sent to him. George P. H. Styan of the Department of Mathematics at McGill University in Montreal is the managing editor. All new subscriptions, as well as all orders for back issues, should be sent to him.

The book reviews editor is Roger R. Davidson of the Department of Mathematics at the University of Victoria. The senior associate editor is James V. Zidek of the Department of Mathematics at The University of British Columbia; the other associate editors are: David R. Brillinger, Sandy R. Brubacher, John S. Chipman, Robert A. Cléroux, Marcel G. Dagenais, Donald A. Dawson, Charles W. Dunnett, Andrey Feuerverger, Donald A. S. Fraser, Jane F. Gentleman, Agnes M. Herzberg, John D. Kalbfleisch, Albert W. Marshall, John R. McGregor, Don L. McLeish, Marc Moore, André Plante, Ronald Pyke, J. N. K. Rao, Carl-Erik Särndal, A. R. Sen, M. P. Singh, Michael A. Stephens, Paul Switzer, H. Jean Thiébaux, Constance van Eeden, William G. Warren, and G. A. Whitmore.

The *Journal* began with Volume 1 in 1973 under the editorship of N. C. Giri (Université de Montréal), and continued with Volumes 2–5 (1974–1977) edited by A. M. Mathai (McGill University) and Volumes 6–8 (1978–1980) edited by Donald A. S. Fraser (University of Toronto). The present editorial management started with Vol. 8, No. 2, 1980. The *Journal* is published twice a year by the Statistical Society of Canada*, which was formed in early 1978 by the merger of the Statistical Science Association of Canada and the Canadian Statistical Society. The *Journal* is supported in part by the Natural Sciences and Engineering Research Council Canada.

Starting with Volume 7 (1979), the *Journal* has been printed by letterpress by Waverly Press, Inc., in Baltimore, Maryland. The average number of pages per volume is 256, with 24 research papers; the page size is 17 cm × 25 cm. [Through Volume 6 (1978) the *Journal* was printed by photooffset from typescript; the number of pages then averaged 296 with 22 research papers per volume.]

Over the past 2 to 3 years the average

Table 1

S. D. Walter	Some generalizations of the committee problem
B. G. Ong and L. L. Campbell	Estimation of frequencies of sinusoids in the presence of noise
David R. Brillinger	Analyzing point processes subjected to random deletions
Janice L. DuBien and William D. Warde	A mathematical comparison of the members of an infinite family of agglomerative clustering algorithms
M. Haseeb Rizvi and R. W. Shorrock	A note on matrix-convexity
James A. Koziol and Amanda F. Nemec	On a Cramér–von Mises type statistic for testing bivariate independence
N. Giri	Locally minimax tests for multiple correlations
G. P. Patil and M. V. Ratnaparkhi	On additive and multiplicative damage models and the characterizations of linear and logarithmic exponential families
Harold V. Henderson and S. R. Searle	Vec and vech operators for matrices, with some uses in Jacobians and multivariate statistics
Samaradasa Weerahandi and James V. Zidek	A characterization of the general mean
R. D. Gupta and D. G. Kabe	Goodness-of-fit tests of hypothesized canonical variables
Norman R. Draper and Agnes M. Herzberg	An investigation of first-order and second-order designs for extrapolation outside a hypersphere
Norman R. Draper and Agnes M. Herzberg	Designs to guard against outliers in the presence or absence of model bias
E. Richard Shillington	Testing lack of fit in regression without replication
Marc Moore and Maurice Legault	Sur l'estimation du risque encouru en reconstruisant une forme aléatoire
Luc Devroye	Recursive estimation of the mode of a multivariate density
Allan Donner	The design of a clinical trial with several patient categories
J. Ranganathan and B. K. Kale	Tests of hypotheses for reliability functions in two-parameter exponential models
T. A. Ramasubban	A comparison of alternative rate estimators in some frequently used growth models
Kam-Wah Tsui	Multiparameter estimation of discrete exponential distributions
Kam-Wah Tsui	Estimation of Poisson means under weighted squared error loss
Theodor D. Sterling	Statistical problems in multi-variate (etiological) surveys
André Plante	On the validation of fiducial techniques
André Plante	Structured probability statements

Abstracts of papers presented in Saskatoon, Saskatchewan, May 28–30, 1979
Book reviews

waiting time from submission of an accepted research paper until acceptance has been $3\frac{1}{2}$ months when no revisions are required and 9 months when revisions have been made; the average waiting time from acceptance until publication is 8 months.

Starting with 1980, all subscribers to *The Canadian Journal of Statistics* also receive, free of extra cost, the newsletter *Forthcoming Events–Activités prévues*, which lists conferences and symposia in the statistical and mathematical sciences. A recent issue lists over 100 events for the period February 1981 through August 1983.

For 1981 the annual dues of the Statistical Society of Canada and the annual subscription rate for libraries \$18 and \$40 , respectively. All back issues (Vols. 1–8, 1973–1980) are available, as are a cumulative index for these volumes and the 1979–1980 membership directory of the Statistical Society of Canada.

The table of contents for Volume 7 is presented in Table 1.

G. P. H. Styan

CANONICAL ANALYSIS

The purpose of a canonical analysis is to characterize the independent statistical relationships that exist between two (and possibly more) sets of random variables. The process involves the calculation of canonical variables, which are appropriately chosen linear functions of the variables in the separate sets, and associated canonical correlations, which measure the extent of the (linear) relationships between the sets. In a sense that will be made precise later, the canonical variables and correlations isolate and summarize, in the most complete and concise way possible, exactly how the sets are associated.

Canonical analysis is one of several major contributions to multivariate statistics made by Harold Hotelling*. His basic, elegant paper on the topic appeared in 1936 [8]. In this paper, Hotelling describes an example where one set of variables consists of mental tests and the other of physical measurements on a group of people.

> The questions then arise of determining the number and nature of the independent relations of mind and body shown by these data to exist, and of extracting from the multiplicity of correlations in the system suitable characterizations of these independent relations.

The data for a two-set canonical analysis consist of vectors of observations made on all of the variables. The largest or first canonical correlation is the maximum correlation attainable between a linear function of the first set and another of the second set of variables. These two linear combinations, which give rise to the maximum correlation, form the first pair of canonical variables. By convention, they are usually standardized* to have unit variances. The second canonical correlation and pair of variables are defined similarly subject to the constraint that the canonical variables within sets must be uncorrelated. Assuming that there are no singularities within either set, the number of canonical variables that can be defined in this way equals the number of variables in the smaller of the two sets. All of the associated canonical correlations will necessarily be nonnegative. If one set has more variables than the other, then additional canonical variables can be defined for the larger set which have the property that they are uncorrelated among themselves and with all the other canonical variables. In total, there are as many canonical variables as original variables in the two sets.

Hotelling [8] also mentions the problem of dealing with more than two sets simultaneously, but he does not suggest how to define canonical variables in this case. However, several proposals have been made in subsequent years. Each one is a generalization of the two-set canonical analysis procedure. One technique, for instance, is to define the canonical variables, one from each set, so as to maximize the sum of their pairwise correlations [7]. Additional canonical variables can be defined subject to the constraint that they be uncorrelated within sets.

Since canonical analysis is usually considered to be one of the "major methods" of multivariate analysis*, it is perhaps surprising that this technique does not in fact play a larger role in data analysis. One reason may be the difficulty one encounters in attempting to interpret a pair of canonical variables, both of which are likely to be complicated linear combinations of the original variables. They may possess nice mathematical properties but nevertheless be of no practical interest. Another reason may be the fact that very little *useful* distribution theory exists for purposes of making statistical inferences. For example, Fisher's z-transformation*, which allows one to routinely make tests and construct confidence intervals* about a single correlation coefficient*, fails for canonical correlations because the transformation is not "variance stabilizing"* for them [17]. No doubt there are other inhibiting factors, too, such as how to perform a canonical analysis effectively when there are missing values in the data. *See* MISSING DATA.

Perhaps the most appropriate role for canonical analysis is as an exploratory tool for suggesting unanticipated hypotheses about relations between sets and possible reductions in dimensionality by restricting attention to the most highly correlated canonical pairs. The practical utility of the analysis can often be enhanced by replacing the optimal canonical variables with related but simpler and more meaningful ones.

The details of the two-set canonical analysis problem are presented in the next section, including basic algebra, computational methods, mathematical properties, practical suggestions, and related problems and topics. The third section outlines some approaches to the canonical analysis of three or more sets. General references and summary remarks follow in the concluding section.

TWO SETS OF VARIABLES

Basic Theory and Computations

The variables in a two-set canonical analysis can be denoted by the vectors $\mathbf{x}_1(p_1 \times 1)$ and $\mathbf{x}_2(p_2 \times 1)$ for the two sets or in combined form as $\mathbf{x}' = (\mathbf{x}_1', \mathbf{x}_2')$, where $\mathbf{x}$ contains all $p = p_1 + p_2$ variables. The data consist of n observations on the p variables and can be represented in matrix form as

$$\mathbf{X} = \begin{pmatrix} \mathbf{X}_1 \\ \mathbf{X}_2 \end{pmatrix} = \begin{pmatrix} \mathbf{x}_{11} & \cdots & \mathbf{x}_{1n} \\ \mathbf{x}_{21} & \cdots & \mathbf{x}_{2n} \end{pmatrix}$$

with dimensions $(p \times n)$ for $\mathbf{X}$, $(p_1 \times n)$ for $\mathbf{X}_1$, and $(p_2 \times n)$ for $\mathbf{X}_2$. The ith observation consists of the vectors $\mathbf{x}_{1i}(p_1 \times 1)$ and $\mathbf{x}_{2i}(p_2 \times 1)$. For convenience, assume for now that $p_1 \leqslant p_2$, $n > p$, and the rank of $\mathbf{X}$ is p.

The sample mean vectors for the two sets are

$$\bar{\mathbf{x}}_i = (1/n)\mathbf{X}_i\mathbf{1} \qquad (i = 1, 2),$$

where $\mathbf{1}$ is a column vector of 1's. Also,

$$\overline{\mathbf{X}}_i = (\bar{\mathbf{x}}_i \cdots \bar{\mathbf{x}}_i) \qquad (i = 1, 2)$$

are $(p_i \times n)$ matrices whose identical columns are the sample mean vectors. The $(p \times p)$ sample covariance matrix $\mathbf{S}$ has the form

$$\mathbf{S} = \begin{pmatrix} \mathbf{S}_{11} & \mathbf{S}_{12} \\ \mathbf{S}_{21} & \mathbf{S}_{22} \end{pmatrix},$$

where

$$\mathbf{S}_{ij} = \frac{1}{n-1}(\mathbf{X}_i - \overline{\mathbf{X}}_i)(\mathbf{X}_j - \overline{\mathbf{X}}_j)' \qquad (i, j = 1, 2) \tag{1}$$

with dimensions $(p_i \times p_j)$. The corresponding correlation matrix obtained from $\mathbf{S}$ is denoted as

$$\mathbf{R} = \begin{pmatrix} \mathbf{R}_{11} & \mathbf{R}_{12} \\ \mathbf{R}_{21} & \mathbf{R}_{22} \end{pmatrix}.$$

The (sample) canonical variables can be expressed as

$$z_{1(i)} = \mathbf{a}'_{1(i)}\mathbf{x}_1 \qquad \text{and}$$
$$z_{2(j)} = \mathbf{a}'_{2(j)}\mathbf{x}_2$$

and their n observed values as

$$\mathbf{z}'_{1(i)} = \mathbf{a}'_{1(i)}\mathbf{X}_1 \quad \text{and}$$
$$\mathbf{z}'_{2(j)} = \mathbf{a}'_{2(j)}\mathbf{X}_2,$$

where $\mathbf{a}_{1(i)}$ and $\mathbf{a}_{2(j)}$ are coefficient vectors determined from the data, for $i = 1, \ldots, p_1$ and $j = 1, \ldots, p_2$. The vectors have the following properties:

$$\mathbf{a}'_{1(i)}\mathbf{S}_{11}\mathbf{a}_{1(j)} = \delta_{ij}, \tag{2}$$

$$\mathbf{a}'_{2(i)}\mathbf{S}_{22}\mathbf{a}_{2(j)} = \delta_{ij}, \tag{3}$$

and

$$\mathbf{a}'_{1(i)}\mathbf{S}_{12}\mathbf{a}_{2(j)} = \delta_{ij}r_{(i)},$$

where δ_{ij} is Kronecker's delta*. The ith pair of (sample) canonical variables consists of $(z_{1(i)}, z_{2(i)})$ and the associated ith (sample) canonical correlation is $r_{(i)}$, $i = 1, \ldots, p_1$, with $1 > r_{(1)} \geqslant \cdots r_{(p_1)} \geqslant 0$.

Two methods of computation will be mentioned for finding the canonical variables and correlations. The first method is based on decompositions of $\mathbf{S}$ (or $\mathbf{R}$). The calculations begin with Choleski decompositions* of $\mathbf{S}_{11}$ and $\mathbf{S}_{22}$:

$$\mathbf{S}_{ii} = \mathbf{T}_i\mathbf{T}'_i \qquad (i = 1, 2),$$

where each $\mathbf{T}_i(p_i \times p_i)$ is a lower triangular matrix of full rank. These are used to form

$$\mathbf{R}^*_{12} = \mathbf{T}_1^{-1}\mathbf{S}_{12}\mathbf{T}_2^{-1\prime}.$$

Then the singular value decomposition* is applied to decompose $\mathbf{R}^*_{12}$ into

$$\mathbf{R}^*_{12} = \mathbf{Q}_1(\mathbf{D} : \mathbf{O})\mathbf{Q}'_2,$$

where $\mathbf{Q}_i(p_i \times p_i)$ is an orthogonal matrix, $i = 1, 2$; $\mathbf{D}(p_1 \times p_1)$ is a diagonal matrix with entries $r_{(1)}, \ldots, r_{(p_1)}$, the canonical correlations; and $\mathbf{O}(p_1 \times (p_2 - p_1))$ is a matrix of zeros. The canonical coefficient vectors, $\mathbf{a}_{1(i)}$ and $\mathbf{a}_{2(j)}$, correspond to the ith column of $\mathbf{T}_1^{-1\prime}\mathbf{Q}_1$ and the jth column of $\mathbf{T}_2^{-1\prime}\mathbf{Q}_2$, respectively.

Example 1. To illustrate the calculations involved in this method, consider the case where

$$\mathbf{S} = \mathbf{R} = \left[\begin{array}{cc:cc} 1 & 1/2 & 1/2 & 0 \\ 1/2 & 1 & 0 & 0 \\ \hdashline 1/2 & 0 & 1 & 1/2 \\ 0 & 0 & 1/2 & 1 \end{array}\right]$$

Then, for $i = 1, 2$,

$$\mathbf{T}_i = \frac{1}{2}\begin{pmatrix} 2 & 0 \\ 1 & \sqrt{3} \end{pmatrix}, \qquad \mathbf{T}_i^{-1} = \frac{1}{\sqrt{3}}\begin{pmatrix} \sqrt{3} & 0 \\ -1 & 2 \end{pmatrix},$$

$$\mathbf{R}^*_{12} = \frac{1}{6}\begin{pmatrix} 3 & -\sqrt{3} \\ -\sqrt{3} & 1 \end{pmatrix},$$

$$\mathbf{Q}_i = \frac{1}{2}\begin{pmatrix} -\sqrt{3} & 1 \\ 1 & \sqrt{3} \end{pmatrix},$$

$$\mathbf{D} = \begin{pmatrix} 2/3 & 0 \\ 0 & 0 \end{pmatrix},$$

$$\mathbf{T}_i^{-1\prime}\mathbf{Q}_i = \frac{1}{\sqrt{3}}\begin{pmatrix} -2 & 0 \\ 1 & \sqrt{3} \end{pmatrix}.$$

Thus $r_{(1)} = \frac{2}{3}$ and the first pair of canonical variables is obtained using the coefficient vectors $\mathbf{a}'_{1(1)} = \mathbf{a}'_{2(1)} = 1/\sqrt{3}\,(-2 \quad 1)$. The second pair of canonical variables, which corresponds to the second of the original variables in each of the sets, has canonical correlation $r_{(2)} = 0$. The power of the canonical correlation method is evident even in this simple case, since $r_{(1)}$ substantially exceeds the largest—and the only—nonzero correlation in $\mathbf{R}_{12}$.

A convenient way of describing the calculations involved is that one first transforms the variables so that they are uncorrelated and have unit variances within sets:

$$\mathbf{Y}_i = \mathbf{T}_i^{-1}\mathbf{X}_i \qquad (i = 1, 2).$$

(The canonical variables and correlations are not affected by such preliminary transformations as long as they are nonsingular.) Then the observations on the canonical variables are obtained via orthogonal transformations* of $\mathbf{Y}_1$ and $\mathbf{Y}_2$:

$$\mathbf{Z}_i = \mathbf{Q}'_i\mathbf{Y}_i \qquad (i = 1, 2)$$

where the jth row of $\mathbf{Z}_i$ is $\mathbf{z}'_{i(j)}$.

The correlation matrix of $\mathbf{Y}' = (\mathbf{Y}_1' \quad \mathbf{Y}_2')$, denoted by $\mathbf{R}^*$, has the structure

$$\mathbf{R}^* = \begin{pmatrix} \mathbf{I}_{p_1} & \mathbf{R}_{12}^* \\ \mathbf{R}_{21}^* & \mathbf{I}_{p_2} \end{pmatrix}, \tag{4}$$

where $\mathbf{I}_{p_i}$ is an identity matrix of order p_i. The elements of $\mathbf{R}_{12}^*$ are denoted as $\mathbf{r}_{ij}^*$ and

$$\sum_{i=1}^{p_1} r_{(i)}^2 = \sum_{j=1}^{p_1} \sum_{k=1}^{p_2} r_{jk}^{*2}. \tag{5}$$

Equation (5) provides one specific sense in which the canonical variables account for all linear relations between the two sets.

The number of positive canonical correlations equals the rank of $\mathbf{R}_{12}^*$ (or $\mathbf{S}_{12}$ or $\mathbf{R}_{12}$). For instance, in Example 1, it is easy to verify that the rank of these matrices is equal to 1.

The second method of computation [2, Chap. 5] starts by making orthogonal decompositions* of the data:

$$(\mathbf{X}_i - \overline{\mathbf{X}}_i) = \sqrt{n-1}\,\mathbf{T}_i\mathbf{H}_i' \qquad (i = 1, 2).$$

Here $\mathbf{T}_i$ is the same lower triangle matrix previously defined, and $\mathbf{H}_i(n \times p_i)$ is a matrix with orthonormal columns. Then $\mathbf{H}_1$ and $\mathbf{H}_2$ are used to form

$$\mathbf{R}_{12}^* = \mathbf{H}_1'\mathbf{H}_2,$$

from which point the calculations proceed as for the first method. The main advantage of operating directly on $\mathbf{X}$ is that round-off errors that can accrue in forming $\mathbf{S}$ or $\mathbf{R}$ are avoided.

An equivalent, but computationally inconvenient formulation, often found in textbooks, can be stated in terms of the eigenvalue–eigenvector* equation

$$\mathbf{S}_{11}^{-1}\mathbf{S}_{12}\mathbf{S}_{22}^{-1}\mathbf{S}_{21}\mathbf{a}_1 = \mathbf{r}^2\mathbf{a}_1. \tag{6}$$

The ith largest eigenvalue is $r_{(i)}^2$ and the corresponding eigenvector can be taken as $\mathbf{a}_{1(i)}$, subject to (2). Moreover, for $r_{(i)} > 0$, the corresponding $\mathbf{a}_{2(i)}$ is simply

$$\mathbf{a}_{2(i)} = (r_{(i)})^{-1}\mathbf{S}_{22}^{-1}\mathbf{S}_{21}\mathbf{a}_{1(i)}, \tag{7}$$

and the remaining $\mathbf{a}_{2(j)}$ are arbitrary except for the constraint (3). If $\mathbf{S}_{ii}$ is singular, then (6) and (7) are still valid if a generalized inverse*, $\mathbf{S}_{ii}^{-}$, is substituted in place of $\mathbf{S}_{ii}^{-1}$ [10].

Frequently, the canonical analysis is performed on $\mathbf{D}_{11}^{-1/2}\mathbf{X}_1$ and $\mathbf{D}_{22}^{-1/2}\mathbf{X}_2$ instead of $\mathbf{X}_1$ and $\mathbf{X}_2$ or on $\mathbf{R}$ instead of $\mathbf{S}$. ($\mathbf{D}_{ii}$ is a diagonal matrix with diagonal terms equal to those of $\mathbf{S}_{ii}$.) The canonical variables and correlations are unaffected by this rescaling because of the invariance property mentioned earlier. However, the canonical coefficient vectors for the unscaled data need to be multiplied by $\mathbf{D}_{11}^{1/2}$ for the first set and $\mathbf{D}_{22}^{1/2}$ for the second to obtain the correct coefficients in terms of the scaled data.

RELATED PROBLEMS AND TOPICS

Many other criteria besides maximizing correlation also lead to canonical variables. The least-squares* estimate of $\mathbf{a}_2$ in the regression equation*

$$\mathbf{a}_{1(1)}'(\mathbf{X}_1 - \overline{\mathbf{X}}_1) = \mathbf{a}_2'(\mathbf{X}_2 - \overline{\mathbf{X}}_2) + \mathbf{e}',$$

where $\mathbf{e}$ is a vector of residuals, is obtained when $\mathbf{a}_2 = r_{(1)}\mathbf{a}_{2(1)}$. For other types of regression formulations involving canonical variables, see Brillinger [1, Chap. 10]. Another criterion is to choose a unit-variance linear combination $\mathbf{a}_2'\mathbf{X}_2$ which together with $\mathbf{X}_1$ has minimum generalized variance (the determinant of their covariance matrix). The minimum is achieved when $\mathbf{a}_2 = \mathbf{a}_{2(1)}$ [11].

The canonical correlation problem can be expressed completely in geometrical terms by treating $\mathbf{X}$ as n points in a p-dimensional Euclidean space with origin at their mean and having subspaces of dimensions p_1 and p_2 containing $\mathbf{X}_1$ and $\mathbf{X}_2$. The mutual orientation of the two subspaces is characterized by p_1 critical or canonical angles whose cosines are the canonical correlations. For related discussions, see Hotelling [8], Roy [19], Dempster [5, Sec. 5.6], and Davis and Kahan [3] plus their references.

The analysis of association in a two-way contingency table* with p_1 rows and p_2 columns, $p_1 \leqslant p_2$, can be viewed as a canonical analysis problem. The data in this case consist of n_{rc} observations in the rth row and cth column of the table with $n_{r.}$ in the rth row, $n_{.c}$ in the cth column, and n overall. An observation in the rth row and cth column can be represented using indicator vectors $\mathbf{x}_{1i}(p_1 \times 1)$ and $\mathbf{x}_{2i}(p_2 \times 1)$ all of whose elements are zero except for the rth element of $\mathbf{x}_{1i}$ and the cth element of $\mathbf{x}_{2i}$, both of which equal 1.

The covariance matrix of $(\mathbf{x}'_{1i}, \mathbf{x}'_{2i})$ from (1) (with n in place of $n - 1$) has the form

$$\mathbf{S}_{11} = n^{-1}\mathbf{D}_r - n^{-2}\mathbf{n}_r\mathbf{n}'_r,$$

$$\mathbf{S}_{22} = n^{-1}\mathbf{D}_c - n^{-2}\mathbf{n}_c\mathbf{n}'_c,$$

and

$$\mathbf{S}_{12} = n^{-1}\mathbf{N} - n^{-2}\mathbf{n}_r\mathbf{n}'_c,$$

where $\mathbf{D}_r = \operatorname{diag}(n_{1.}, \ldots, n_{p_1.})$, $\mathbf{D}_c = \operatorname{diag}(n_{.1}, \ldots, n_{.p_2})$, $\mathbf{n}_r = \mathbf{D}_r\mathbf{1}$, $\mathbf{n}_c = \mathbf{D}_c\mathbf{1}$, and $\mathbf{N} = ((n_{rc}))$. The ranks of $\mathbf{S}_{11}$ and $\mathbf{S}_{22}$ are $(p_1 - 1)$ and $(p_2 - 1)$. Convenient choices for the generalized inverses* to use in (6) and (7) are $\mathbf{S}_{11}^- = n\mathbf{D}_r^{-1}$ and $\mathbf{S}_{22}^- = n\mathbf{D}_c^{-1}$. There are $(p_1 - 1)$ canonical correlations, and they satisfy

$$\sum_{i=1}^{p_1-1} \mathbf{r}^2_{(i)} = \sum_{r=1}^{p_1} \sum_{c=1}^{p_2} \left(n^2_{rc}/n_{r.}n_{.c}\right) - 1$$
$$= n^{-1}\chi^2,$$

where χ^2 is the usual chi-squared statistic* for testing independence in a two-way table.

The canonical coefficient vectors $\mathbf{a}_{1(i)}$, $\mathbf{a}_{2(i)}$ are sometimes used to assign scores to the rows and columns of the table to aid interpretation*. This procedure is most useful when there is only one large canonical correlation and, therefore, only one scoring system to consider.

Additional discussion of the canonical analysis of a contingency table may be found in Kshirsagar [13, Chap. 9] and his references. For an example utilitizing canonical scores, see Klatzky and Hodge [12], who use them to assign weights to the categories of occupational mobility tables.

Certain aspects of the standard multigroup discriminant analysis* are also special cases of canonical analysis. Suppose that there are p variables, denoted as $\mathbf{u}$, and g groups with $n_i(p \times 1)$ vector observations in the ith group. Let $\mathbf{u}_{ij}$, $\bar{\mathbf{u}}_i$, and $\bar{\mathbf{u}}$ be the jth observation in the ith group, the ith group mean, and the overall mean, repsectively. The between-group (B), within-group (W), and total (T) dispersion of the data can be summarized in terms of the usual sums of products and cross-products matrices:

$$\mathbf{B} = \sum n_i(\bar{\mathbf{u}}_i - \bar{\mathbf{u}})(\bar{\mathbf{u}}_i - \bar{\mathbf{u}})',$$

$$\mathbf{W} = \sum\sum(\mathbf{u}_{ij} - \bar{\mathbf{u}}_i)(\mathbf{u}_{ij} - \bar{\mathbf{u}}_i)',$$

and

$$\mathbf{T} = \sum\sum(\mathbf{u}_{ij} - \bar{\mathbf{u}})(\mathbf{u}_{ij} - \bar{\mathbf{u}})' = \mathbf{W} + \mathbf{B}.$$

Differences among the group mean vectors can be completely and conveniently represented in terms of "discriminant variables," which are appropriately chosen linear functions of $\mathbf{u}$. The first discriminant variable, $\mathbf{a}'_{(1)}\mathbf{u}$, has the largest ratio of between- to within-group sums of squares (or, equivalently, the largest value of F, the ratio of between- to within-group mean squares) among all possible linear combinations. The second discriminant variable has the next largest ratio subject to being uncorrelated with the first one, etc. It is customary to normalize the $\mathbf{a}_{(i)}$ so that the variance of $\mathbf{a}'_{(i)}\mathbf{u}$ equals 1. Thus $\mathbf{a}'_{(i)}\mathbf{W}\mathbf{a}_{(j)} = \delta_{ij}(n - g)$. Generally, there are $\min(p, g - 1)$ discriminant variables that can be constructed in this way. Often, p is quite large relative to g, so that, in fact, a large reduction of dimensionality is achieved by representing the data in terms of their discriminant variables. Classification of an "unknown" observation $\mathbf{u}_0$ into the closest group can be made by measuring its squared Euclidean distance to each group mean in the discriminant variable space and assigning it to the group corresponding to the smallest distance. This procedure is equivalent to making the assignment on the basis of the minimum value of the generalized squared distances, $(\mathbf{u}_0 - \bar{\mathbf{u}}_i)'\mathbf{W}^{-1}(\mathbf{u}_0 - \bar{\mathbf{u}}_i)$, $i = 1, \ldots, g$, in the original p-dimensional space.

To make the connection with canonical analysis, relabel the observations with a single subscript i and denote them as $\mathbf{x}_{1i}$, $i = 1, \ldots, n$. Then match each with a $(g \times 1)$ indicator vector, $\mathbf{x}_{2i}$, as in the contingency table discussion, which identifies the group membership of $\mathbf{x}_{1i}$. The squared canonical correlations, $r^2_{(i)}$, between these two sets are the eigenvalues of

$$\mathbf{S}_{11}^{-1}\mathbf{S}_{12}\mathbf{S}_{22}^{-}\mathbf{S}_{21} = \mathbf{T}^{-1}\mathbf{B}$$

and will be $\min(p, g-1)$ in number. Since $\mathbf{T}^{-1}\mathbf{B}$ and $\mathbf{W}^{-1}\mathbf{B}$ have the same eigenvectors, it follows that $\mathbf{a}_{1(i)} \propto \mathbf{a}_{(i)}$, $i = 1, \ldots, \min(p, g-1)$. In view of this, it is not surprising that discriminant variables are sometimes referred to as canonical variables. Moreover, the eigenvalues of $\mathbf{W}^{-1}\mathbf{B}$ are $r^2_{(i)}/(1 - r^2_{(i)})$, so that there is also a simple link between the canonical correlations and the F-values mentioned earlier.

Associated with each discriminant variable is a "most significant" contrast* vector $\mathbf{c}_i(1'\mathbf{c}_i = 0,\ \mathbf{c}_i'\mathbf{c}_i = 1)$. For instance, $\mathbf{c}_i' = (2^{-1/2}, -2^{-1/2}, 0 \cdots 0)$ would indicate that the first and second groups are separated from each other and also from the remaining ones. The corresponding F-value measures the degree of the separation. Specifically,

$$\mathbf{c}_i' \propto \mathbf{a}_{1(i)}'\left[n_1(\bar{\mathbf{u}}_1 - \bar{\mathbf{u}}) \cdots n_g(\bar{\mathbf{u}}_g - \bar{\mathbf{u}})\right] \propto \mathbf{a}_{2(i)}'\mathbf{D}_g,$$

where $\mathbf{D}_g = \text{diag}(n_1 \cdots n_g)$. Thus apart from a constant, the elements of $\mathbf{c}_i$ are the sums of the canonical scores, as prescribed by $\mathbf{a}_{2(i)}$, for the individual groups.

Related discussions of discriminant analysis may be found in Kshirsagar [13, Chap. 9] and Gnanadesikan [6, Chap. 4]. *See also* DISCRIMINANT ANALYSIS.

A different form of canonical analysis [14; 16, Chap. X] can be defined for bivariate normal* random variables, x_1 and x_2, with correlation ρ. Assume that both x_1 and x_2 are standardized [i.e., $E(x_1) = E(x_2) = 0$ and $E(x_1^2) = E(x_2^2) = 1$]. Then one can define a sequence of canonical variables $(z_{1(i)}, z_{2(i)})$, $i = 1, 2, \ldots$, in the usual way, except that now $z_{1(i)}$ and $z_{2(i)}$ are standardized but otherwise arbitrary functions of x_1 and x_2. The canonical variables turn out to be

$$(z_{1(i)}, z_{2(i)}) = (\psi_i(x_1), \psi_i(x_2)) \qquad i = 1, 2, \ldots,$$

where ψ_i is the ith standardized Hermite polynomial (*see* CHEYBSHEV–HERMITE POLYNOMIALS), and the associated canonical correlation is

$$\rho_{(i)} = |\rho|^i.$$

In particular, $(\psi_1(x_1), \psi_1(x_2)) = (x_1, x_2)$, so that $|\rho|$ is the maximum canonical correlation.

A connection between these results concerning the bivariate normal variables (x_1, x_2) and the previous discussion of contingency tables can be made using the Mehler identity or tetrachoric series*:

$$f(x_1, x_2, \rho)/\{f(x_1)f(x_2)\} = 1 + \sum_{i=1}^{\infty} \rho^i \psi_i(x_1)\psi_i(x_2),$$

where f denotes a probability density function. A similar equation can be written, in matrix form, for a contingency table:

$$n\mathbf{D}_r^{-1}\mathbf{N}\mathbf{D}_c^{-1} = 11' + \sum_{i=1}^{p_1-1} r_{(i)}\tilde{\mathbf{a}}_{1(i)}\tilde{\mathbf{a}}_{2(i)}',$$

where $\tilde{\mathbf{a}}_{1(i)} = (\mathbf{I} - n^{-1}\mathbf{1n}_r')\mathbf{a}_{1(i)}$ and $\tilde{\mathbf{a}}_{2(i)} = (\mathbf{I} - n^{-1}\mathbf{1n}_c')\mathbf{a}_{2(i)}$. In other words, $\tilde{\mathbf{a}}_{1(i)}$ and $\tilde{\mathbf{a}}_{2(i)}$ are standardized canonical scores for the rows and columns of the table. More details may be found in Lancaster [14].

If $\mathbf{x}_1(p_1 \times 1)$ and $\mathbf{x}_2(p_2 \times 1)$, $p_1 \leqslant p_2$, have a joint multivariate normal* distribution, then canonical variables $(z_{1(i)}, z_{2(i)})$ with canonical correlation $\rho_{(i)}$ can be defined as in the bivariate case as general standardized functions of $\mathbf{x}_1$ and $\mathbf{x}_2$. However, if $\rho_{(k)} > \rho^2_{(1)}$ and $k \leqslant p_1$, then the first k pairs of canonical variables will be the usual linear ones as defined by Hotelling [15].

USE AND INTEPRETATION

It is virtually certain that additional analyses, beyond simply calculating the canonical variables and correlations, will be needed to fully appreciate the relations between the two sets. Relatively informal procedures are likely to be the most useful ones. These would include (1) modifying the canonical variables by tinkering with the coefficients to enhance their interpretability; (2) deleting variables and observations, especially ones that are unusual, to assess their impact; (3) jackknifing* [4]; (4) applying cross-validation techniques [21]; (5) comparing the results of robust* and standard analyses; and (6) plotting the canonical variables in a variety of ways. Some of these ideas are illustrated in the next example.

Example 2. The first set of variables consists of incidence rates for five types of cancers—esophagus, stomach, small intestine, colon, and rectum—for white males in 41 states during the period 1950–1967. The second set contains five variables that may be associated with the cancer rates: percent population in urban areas and per capita consumption of cigarettes, spirits, wine, and beer based on 1960 figures. The largest pairwise correlation between the sets is 0.78 and involves the last variable in each of them.

The first two canonical correlations are $r_{(1)} = 0.86$ and $r_{(2)} = 0.64$. A scatter plot of the first pair of canonical variables is shown in Fig. 1. Standard two-letter postal abbreviations are used to identify the states. The plot not only confirms the high degree of linear association suggested by $r_{(1)}$ but also reveals interesting geographical patterns such as the concentration of southern states in the lower left portion of the plot.

The observations on the second pair of canonical variables are plotted in Fig. 2. Evidently, the unusual position of Alaska (AK) explains most of $r_{(2)}$, since the rest of the scatter exhibits very little correlation.

Omitting Alaska and repeating the analysis produces revised values of $r_{(1)} = 0.87$ and $r_{(2)} = 0.74$. The coefficients for the first pair of canonical variables, after standardizing the original variables, are 0.63, 0.04, −0.08,

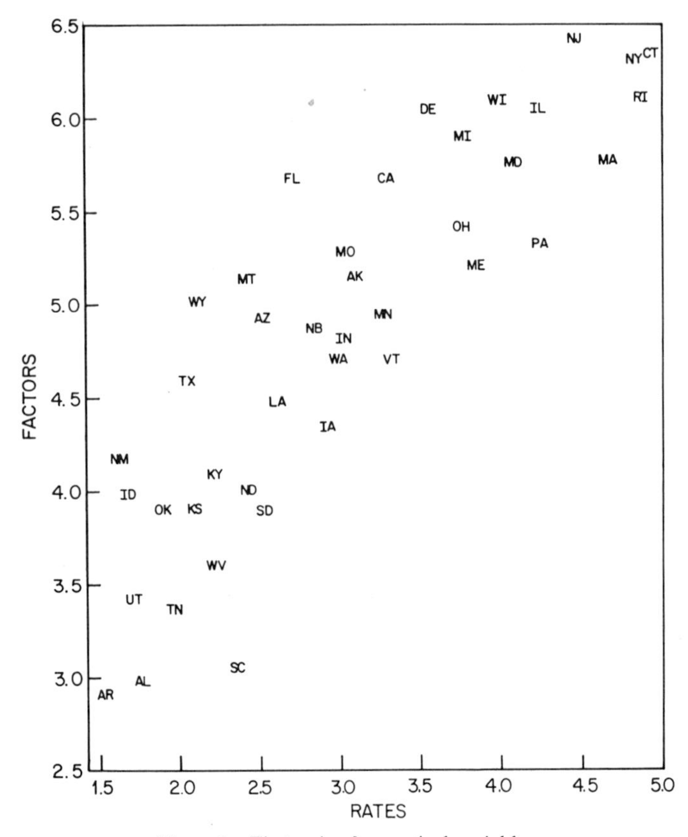

Figure 1 First pair of canonical variables.

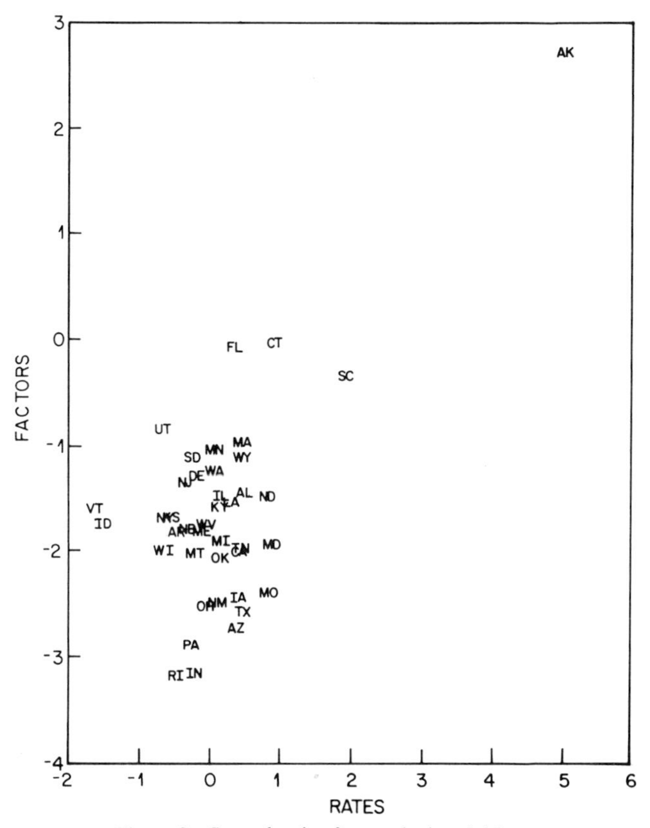

Figure 2 Second pair of canonical variables.

−0.64, and 1.03 for the first set and 0.05, 0.00, 0.48, 0.13, and 0.51 for the second set. The revised first pair of canonical variables results in basically the same geographical patterns as in Fig. 1, but the second pair of canonical variables is entirely different and no longer dominated by a single state (Fig. 3). Note that Alaska is shown in the plot even though it did not influence the analysis. The effect of Alaska on the first analysis was to diminish several of the original pairwise correlations and also $r_{(2)}$.

To gain insight into the relative importance of the variables, the analysis was repeated using various combinations of them. For instance, any pair of the cancer rates, together with the five variables from the second set, results in $r_{(1)} \geqslant 0.84$, except the pair involving stomach and small intestine, for which $r_{(1)} = 0.65$. This suggests that the first canonical variable for the cancer rates could assign coefficients of zero to these two rates with little loss.

This example illustrates, first, the value of plotting the observed canonical variables to increase understanding of them, as in Fig. 1, and to reveal idiosyncracies in the data, as in Fig. 2. It also shows how repeated analyses involving subsets of the variables and observations can help to improve the canonical representation of the data.

In particular, Fig. 2 shows how a small fraction of bad data in the form of outliers* can cause gross distortions. This potential for disaster stems from the basic sensitivity of the standard product-moment correlation coefficient to outliers, which can either inflate or deflate the coefficient severely. Outliers can be handled by culling them from the data, as illustrated in Example 2, or by employing robust versions of **R** or **S**, which are specifically designed to protect against them [6, Chap. 5 and 6].

The canonical coefficient vectors play a similar role to vectors of regression coefficients but are more difficult to assess statisti-

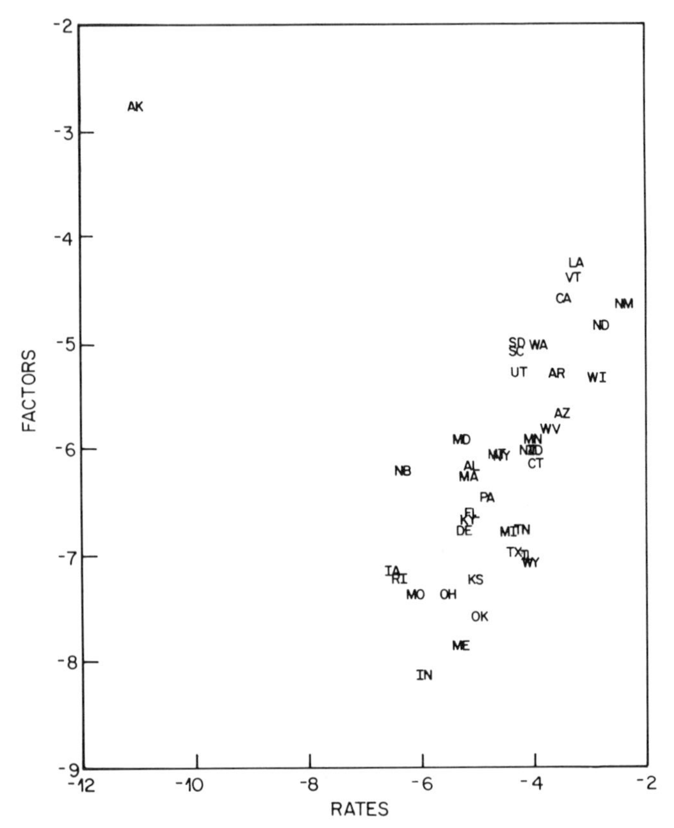

Figure 3 Revised second pair of canonical variables.

cally. If the analysis is done on standardized variables or $\mathbf{R}$, then it may help to look at the (squared) multiple correlations of each variable with the others in the same set or equivalently at the diagonal elements of $\mathbf{R}_{11}^{-1}$ and $\mathbf{R}_{22}^{-1}$. Large values would suggest possible multicollinearity* problems and instabilities in the corresponding coefficients.

Example 2 (Continued). For the cancer data, the diagonal elements of $\mathbf{R}_{11}^{-1}$ are roughly 5, 3, 2, 16, and 17, with the latter two corresponding to squared multiple correlations in excess of 0.9. This suggests that the standardized coefficients for these variables are not as well pinned down as the others for this set. (The correlation between the colon and rectal cancer rates is 0.93, which explains much of the problem.)

Another approach which is often suggested is to interpret a canonical variable in terms of its correlations with the variables in the set it represents. For example, $\mathbf{R}_{11}\mathbf{a}_{1((i)}$ is the vector of such correlations for the ith canonical variable from the first set (for an analysis based on $\mathbf{R}$). If $\mathbf{a}_{1(i)}$ is also an eigenvector* of $\mathbf{R}_{11}$, then the correlation vector will be proportional to the coefficient vector, and they both will suggest the same variables as the most important ones. Otherwise, their patterns can be quite different.

Example 1 (Continued). In this case, $\mathbf{a}'_{1(1)} \approx (-1.2 \quad 0.6)$ and $\mathbf{a}'_{1(1)}\mathbf{R}_{11} \approx (0.9 \quad 0.0)$. Thus $z_{1(1)}$ is highly correlated with the first variable and not at all with the second variable in the first set (which happens also to be $z_{1(2)}$). However, to conclude that the second variable contributes nothing to $z_{1(1)}$ would be clearly inappropriate.

Even after applying these or other aids for thinking about the canonical variables, it still may be difficult to understand what (if anything) they mean and whether they are capturing a relationship that is of practical interest. To illustrate, suppose that the first set contains test scores for job applicants and the second one consists of measures of success on the job. Then one would like to know what aspects of success are represented by the canonical variables from the second set. If the first of these canonical variables reflects the ability to make a good first impression, it may actually be of less interest than other aspects of success which are not so easily predicted.

Acceptable procedures for drawing formal statistical inferences about canonical correlations are rather limited and not as useful as one would like. The usual setting assumes that the data are a random sample* from a multivariate normal distribution with covariance matrix $\sum$. In particular, the sample canonical correlations, $r_{(i)}$, obtained from $\mathbf{S}$ are used as estimates for the population canonical correlations, $\rho_{(i)}$, derived from $\sum$. The statistic $\Lambda = \prod_{i=k+1}^{p_1}(1 - r_{(i)}^2)$ is often used to test the hypothesis that $\rho_{(i)} = 0$, $i = k+1, \ldots, p_1$. The test can be carried out using the fact that $-\{n - (p+1)/2\} \log_e \Lambda$ is distributed approximately as a χ^2 random variable* with $(p_1 - k)(p_2 - k)$ degrees of freedom, if the hypothesis is true. An exact test can also be made [20]. For more details and related tests, see Kshirsagar [13, Chaps. 7 and 8] and Williams [22].

Approximate expressions for the moments of $r_{(i)}$, $i = 1, \ldots, k$, for normal data were derived by Lawley [17], assuming that only $\rho_{(1)}, \ldots, \rho_{(k)}$ are positive and that they are neither too small nor too close together. His results imply that $E(r_{(1)}) > \rho_{(1)}$, and intuitive arguments, such as those presented by Dempster [4], suggest that this is true more generally. Moreover, their mean and variance can be expressed as $E[r_{(i)}] = \rho_{(i)} + O(n^{-1})$ and $\operatorname{var}(r_{(i)}) = n^{-1}(1 - \rho_{(i)}^2)^2 + O(n^{-2})$. The terms of order n^{-1} for the mean and n^{-2} for the variance, which may be necessary to consider explicitly unless n is large, involve $\rho_{(i)}$ and $(\rho_{(j)}^2 - \rho_{(i)}^2)$ for all $j \neq i$.

THREE OR MORE SETS OF VARIABLES

Several methods are available for extending the two-set canonical analysis method to treat three or more sets simultaneously and symmetrically. They operate by seeking out canonical variables, one from each set, which optimize some function of the correlation matrix of the selected variables.

Suppose, for instance, that the data consist of n observations on three sets of variables: $\mathbf{X}_1(p_1 \times n)$, $\mathbf{X}_2(p_2 \times n)$, and $\mathbf{X}_3(p_3 \times n)$, with $p_1 \leqslant p_2 \leqslant p_3$. Then (first-stage) canonical variables, with observed values $z'_{i(1)} = \mathbf{a}'_{i(1)}\mathbf{X}_i$, $i = 1, 2, 3$, are chosen which yield the "optimal" value of their (3×3) sample correlation matrix $\mathbf{R}_{(1)}$. If constraints such as (2) and (3) are adopted to assure that successive canonical variables are uncorrelated within sets, then p_1 stages of canonical variables can be defined, each with a canonical correlation matrix, $\mathbf{R}_{(i)}$, $i = 1, \ldots, p_1$.

Three possible optimality criteria for defining the canonical variables are to maximize the sum of the correlations (SUMCOR), the sum of the squares of the correlations (SSQCOR), or the largest eigenvalue (MAXVAR) of $\mathbf{R}_{(i)}$. Each is equivalent to Hotelling's procedure, if there are only two sets.

A simple model for the first-stage canonical variables is helpful for motivating the SUMCOR and MAXVAR criteria:

$$\mathbf{Z}_{(1)} = \mathbf{l}_{(1)}\mathbf{f}'_{(1)} + \mathbf{E}_{(1)}, \tag{8}$$

where $\mathbf{Z}'_{(1)} = (\mathbf{z}_{1(1)}\mathbf{z}_{2(1)}\mathbf{z}_{3(1)})$, $\mathbf{z}_{i(1)}$ is a standardized but otherwise arbitrary (for the moment) linear combination of $\mathbf{X}_i$, $\mathbf{l}_{(1)}$ is a nonnull (3×1) vector, $\mathbf{f}_{(1)}$ is a standardized $(n \times 1)$ vector, and $\mathbf{E}_{(1)}(3 \times n)$ is a matrix of residuals. If $\mathbf{l}_{(1)} \propto 1$, the unit vector, then the least-squares fit of (8) with respect to $\mathbf{Z}_{(1)}$ and $\mathbf{f}_{(1)}$ is attained when $\mathbf{Z}_{(1)}$ contains the observations on the first-stage SUMCOR variables and $\mathbf{f}_{(1)} = (\mathbf{1}'\mathbf{R}_{(1)}\mathbf{1})^{-1/2}\mathbf{1}'\mathbf{Z}_{(1)}$. If $\mathbf{l}_{(1)}$ is also allowed to vary, then the least-squares fit of (8) is provided by the first-stage MAXVAR variables with

$$\mathbf{l}_{(1)} = \{e_{1(1)}\}^{1/2}\mathbf{e}_{1(1)},$$

and

$$\mathbf{f}_1' = \{e_{1(1)}\}^{-1/2}\mathbf{e}_{1(1)}'\mathbf{Z}_{(1)},$$

where $e_{1(1)}$ is the largest eigenvalue and $\mathbf{e}_{1(1)}$ the corresponding eigenvector of $\mathbf{R}_{(1)}$.

Thus the SUMCOR method produces canonical variables with the best-fitting common factor, $\mathbf{f}_{(1)}$, assuming the factor contributes with the same weight to each of them. The MAXVAR method can be described similarly, except that the "loading" vector, $\mathbf{l}_{(1)}$, allows for the factor to contribute with varying weights in this case.

The SSQCOR criterion is equivalent to choosing the first-stage canonical variables to maximize the sum of the squares of the eigenvalues of $\mathbf{R}_{(1)}$. A three-term model such as (8), where the terms are fit sequentially, can be used to motivate it further.

The computations for the MAXVAR method are straightforward. They involve $\mathbf{R}^*$, which is formed as in (4) but for three sets. For instance, the first-stage variables are defined by subvectors of the first eigenvector of $\mathbf{R}^*$, which need to be normalized individually, and the corresponding eigenvalue gives the largest eigenvalue of the canonical $\mathbf{R}_{(1)}$. The SUMCOR and SSQCOR procedures require iterative computations which work by repeatedly fixing two of the three variables in turn and maximizing with respect to the third until all three have settled down.

More details on all aspects of these procedures, additional criteria for selecting the canonical variables, may be found in Kettenring [11] and Gnanadesikan [6]. Although the foregoing descriptions are in terms of three sets of variables, the extension to more sets is straightforward. As for the two-set case, close scrutiny of the results and additional computations, in the spirit of those suggested in the section on use and interpretation of two sets of variables, will be needed in most cases.

Example 3. Suppose that there are five standardized variables in each of three sets and that every pair of variables has the same correlation, $\frac{1}{2}$, between them. Then for all three methods

$$\mathbf{R}_{(1)} = \begin{bmatrix} 1 & \frac{5}{6} & \frac{5}{6} \\ & 1 & \frac{5}{6} \\ & & 1 \end{bmatrix}$$

and $\mathbf{a}_{1(1)} = \mathbf{a}_{2(1)} = \mathbf{a}_{3(1)} \propto 1$. For the other stages, $\mathbf{R}_{(i)} = \mathbf{I}$, $i = 2, \ldots, 5$. Thus the relations among the three sets in this simple case are entirely summarized by the first-stage canonical variables—the ultimate situation in terms of reduction of dimensionality. Moreover, these canonical variables and correlations are the same ones that would be found if the sets were looked at two at a time.

A slightly more complex situation is provided by the next example, which illustrates how the three criteria can lead to completely different first-stage canonical variables.

Example 4. In this case there are three variables in each three sets with

$\mathbf{R} =$

1.00	0.00	0.00	0.40	0.00	0.00	0.40	0.00	0.00
	1.00	0.00	0.00	0.49	0.00	0.00	0.49	0.00
		1.00	0.00	0.00	0.83	0.00	0.00	0.10
			1.00	0.00	0.00	0.40	0.00	0.00
				1.00	0.00	0.00	−0.49	0.00
					1.00	0.00	0.00	0.10
						1.00	0.00	0.00
							1.00	0.00
								1.00

The first-stage SUMCOR, SSQCOR, and MAXVAR variables are the first, second, and third variables, respectively, in the three sets. Among all possible linear combinations of the variables within sets, these are the ones that maximize the different criteria.

CONCLUSION

Canonical analysis is one of the classical methods in the arena of mutlivariate analysis*. The mathematical aspects of canonical analysis arise in many contexts, including geometry, vector space theory, and numerical analysis. Numerous methodological applications and spin-offs crop up in statistical areas such as contingency-table* analysis and discriminant analysis* (see "Two Sets of Variables"), analysis of variance* [9], factor analysis* [18], and time-series* analysis [1, Chap. 10].

The value of canonical analysis for actual data purposes is less firmly established. To achieve its potential, better methods are needed for selecting "canonical variables" which have practical as well as theoretical interest and for making statistical inferences about them.

Further Reading

Suggestions for further general reading on this topic include Dempster [5, Chap. 9], Gnanadesikan [6, Chap. 3], Hotelling [8], Kshirsagar [13, Chap.7], and Williams [22].

References

[1] Brillinger, D. R. (1975). *Time Series: Data Analysis and Theory*. Holt, Rinehart and Winston, New York. (This book provides an advanced theoretical treatment of canonical analysis in a time-series context; see Chaps. 8 and 10.)

[2] Chambers, J. M. (1977). *Computational Methods for Data Analysis*. Wiley, New York. (Chapter 5 shows how orthogonal and singular value decompositions can be used to compute canonical variables and correlations in a numerically sound manner.)

[3] Davis, C. and Kahan, W. M. (1970). *SIAM J. Numer. Anal.*, **7**, 1–46.

[4] Dempster, A. P. (1966). In *Multivariate Analysis*, P. R. Krishnaiah, ed. Academic Press, New York, pp. 315–334.

[5] Dempster, A. P. (1969). *Elements of Continuous Multivariate Analysis*. Addison-Wesley, Reading, Mass. (Chapter 9 provides a geometrically oriented discussion at an advanced level but with important practical points and examples.)

[6] Gnanadesikan, R. (1977). *Methods for Statistical Data Analysis of Multivariate Observations*. Wiley, New York. (This book treats many topics relevant to canonical analysis, such as plotting, outliers, and robustness. Methods for multiset analysis are developed in Chap. 3. The level is advanced but the emphasis is on practical tools.)

[7] Horst, P. (1961). *Psychometrika*, **26**, 129–149.

[8] Hotelling, H. (1936). *Biometrika*, **28**, 321–377. (This paper provides the original in-depth treatment of canonical analysis.)

[9] James, A. T. and Wilkinson, G. N. (1971). *Biometrika*, **58**, 279–294.

[10] Kettenring, J. R. (1969). Canonical Analysis of Several Sets of Variables. Ph.D. dissertation, University of North Carolina.

[11] Kettenring, J. R. (1971). *Biometrika*, **58**, 433–451.

[12] Klatzky, S. R. and Hodge, R. W. (1971). *J. Amer. Statist. Ass.*, **66**, 16–22.

[13] Kshirsagar, A. M. (1972). *Multivariate Analysis*. Marcel Dekker, New York. (Chapter 7 gives an algebraically oriented treatment which includes some distribution theory. Chapter 8 discusses standard tests and Chap. 9 deals with related topics. The level is advanced.)

[14] Lancaster, H. O. (1957). *Biometrika*, **44**, 289–292.

[15] Lancaster, H. O. (1966). *Biometrika*, **53**, 585–588.

[16] Lancaster, H. O. (1969). *The Chi-Squared Distribution*. Wiley, New York.

[17] Lawley, D. N. (1959). *Biometrika*, **46**, 59–66.

[18] Rao, C. R. (1955). *Psychometrika*, **20**, 93–111.

[19] Roy, S. N. (1947). *Sankhyā*, **8**, 177–191.

[20] Schatzoff, M. (1966). *Biometrika*, **53**, 347–358.

[21] Thorndike, R. M. and Weiss, D. J. (1973). *Educ. Psychol. Meas.*, **33**, 123–134.

[22] Williams, E. J. (1967). *J. R. Statist. Soc. B*, **29**, 199–242.

(CONTINGENCY TABLES
DISCRIMINANT ANALYSIS
FACTOR ANALYSIS
GENERALIZED CANONICAL VARIABLES
GEOMETRY IN STATISTICS
MULTIVARIATE ANALYSIS)

J. R. KETTENRING

CANONICAL CORRELATION *See* CANONICAL ANALYSIS

CANONICAL SCORES *See* CANONICAL ANALYSIS

CANONICAL VARIABLES *See* CANONICAL ANALYSIS; GENERALIZED CANONICAL VARIABLES

CANTELLI'S INEQUALITY

A probability inequality of the Chebyshev* type but one-sided. For any random variable

X,

$$\Pr[X - \mu \leqslant \lambda] \leqslant \frac{\sigma^2}{\sigma^2 + \lambda^2} \qquad \text{if } \lambda < 0$$

and

$$\Pr[X - \mu \leqslant \lambda] \geqslant 1 - \frac{\sigma^2}{\sigma^2 + \lambda^2} \qquad \text{if } \lambda \geqslant 0,$$

where $\mu = E(X)$ and $\sigma^2 = E(X - \mu)^2$.

This inequality is applicable when one is interested in one-sided alternatives—to detect large positive deviations from the mean; for example, in the construction of distribution-free one-sided confidence intervals* or one-sided test regions*. (See Savage [5] for more details.)

The inequality is also discussed in Cantelli's own works [1, 2] and in refs. 3 and 4.

References

[1] Cantelli, F. P. (1910). *Bol. Ass. Attuari Ital.*, 1–23.

[2] Cantelli, F. P. (1928). *Att. Cong. Int. Mat.*, **6**, 47–59.

[3] Cramèr, H. (1946). *Mathematical Methods of Statistics*. Princeton University Press, Princeton, N. J.

[4] Pearson, K. (1919). *Biometrika*, **12**, 284–296.

[5] Savage, I. R. *J. Res. Natl. Bur. Stand.*, **65B**(3), 211–222.

(CHEBYSHEV'S INEQUALITY)

CAPON TEST

The Capon test is a procedure used to test the hypothesis of equality of the scale parameters (or variances, if they exist) of two continuous populations. Let $X_1, X_2, \ldots, X_m$, and $Y_1, Y_2, \ldots, Y_n$, denote two independent random samples from two populations which differ only in scale. Furthermore, let $Z_{Ni} = 1$ if the ith smallest of the $N = m + n$ observations is an X, and $Z_{Ni} = 0$, otherwise. Then the Capon test statistic, denoted by C_N, is

$$C_N = m^{-1} \sum_{i=1}^{N} E(\xi_i^2) Z_{Ni}, \tag{1}$$

where ξ_i is the ith smallest observation in a sample of size N from the standard normal distribution, $N(0, 1)$. By Capon test, or C_N-test, is meant the test based on the statistic C_N. The statistic C_N in a *linear rank statistic** and is a special case of a wide class of *optimum* statistics considered by Capon [1].

Suppose that X and Y have cumulative distribution functions* (CDF's) F_1 and F_2, respectively, which differ only in scale, i.e., $F_1(x) = F[(x - \mu)/\theta_1]$ and $F_2(x) = F[(x - \mu)/\theta_2]$, where F is a specified CDF. The statistic in (1) was shown by Capon to be a locally most powerful rank test* statistic for testing $H_0: F_1 = F_2$ against either the alternative $H_1: \theta_1 > \theta_2$ or $H_2: \theta_1 < \theta_2$, when F is the standard normal* CDF, $\Phi(x) = \int_{-\infty}^{x} (\sqrt{2\pi})^{-1} \exp(-t^2/2)\, dt$. The test is to reject H_0 in favor of H_1 if $C_N > c_1$ or in favor of H_2 if $C_N < c_2$, where c_1 and c_2 are determined once the significance level α is stated, i.e., $\Pr[C_N > c_1 \mid H_0] = \alpha = \Pr[C_N < c_2 \mid H_0]$. Since C_N is a discrete random variable, α will be a *nominal* significance level*. The C_N-test can also be used to test H_0 against the two-sided alternative $H_3: \theta_1 \neq \theta_2$ and H_0 is rejected for small or large values of C_N.

The statistic C_N is asymptotically normal, both under the hypothesis H_0 and the alternatives. The mean and variance of C_N, under the hypothesis $H_0: F_1 = F_2$, are $E(C_N) = 1$ and

$$\operatorname{var}(C_N) = \frac{n}{mN(N-1)} \times \left[\sum_{i=1}^{N} \left\{ E(\xi_i^2) \right\}^2 - N \right]. \tag{2}$$

The C_N-test is asymptotically efficient (optimum), i.e., asymptotically it has the same power as the test based on the likelihood ratio test*. Thus the asymptotic relative efficiency (ARE)*, in the Pitman sense, of the Capon test relative to the classical F-test*, when the underlying CDF is normal, is 1. Because of the lack of robustness* of the F-test, it is not of interest to compare the C_N-test with the F-test for distributions other than normal. However, the ARE of the Mood test* relative to the C_N-test is 0.760, 0.896, 0.900, and 1.670, for normal, logistic*, double exponential*, and Cauchy* alternatives, respectively. The ARE of the

Table 1

Ordered Values	X or Y	Rank	$E(\xi_1^2)$	Ordered Values	X or Y	Rank	$E(\xi_1^2)$
2.1	Y	1	3.314	11.8	Y	9	0.130
4.2	X	2	1.736	12.5	X	10	0.218
5.6	X	3	1.039	12.6	Y	11	0.378
6.3	X	4	0.633	12.7	Y	12	0.633
7.3	X	5	0.378	14.2	Y	13	1.039
9.0	X	6	0.218	14.8	X	14	1.736
10.6	X	7	0.130	15.7	Y	15	3.314
10.9	Y	8	0.102				

Ansari–Bradley* test relative to the C_N-test is 0.608, 0.750, 0.774, and 1.783 for the same alternatives.

The Capon test is asymptotically equivalent (i.e., ARE = 1) to a test of Klotz [3], which is based on $K_N = m^{-1}\sum_{i=1}^{N}[\Phi^{-1}\{i/(N+1)\}]^2$. The Klotz test* is only asymptotically a locally most powerful rank test, however.

There are numerous other tests that have been proposed, e.g., the Sukhatme test* and the Siegel–Tukey test*. A survey of such tests is contained in Duran [2]. [This paper discusses $c(c \geqslant 2)$-sample scale tests and compares them with several parametric procedures. Suggestions on which procedures to use in various situations and 90 references on related topics are provided.]

The constants $E(\xi_i^2)$ needed to compute C_N can be found in Tietjen et al. [4] for $N \leqslant 50$. There appears to be no null distribution of C_N available in the existing literature needed to carry out a test of H_0. However, for sufficiently large values of N the normal approximation* may be used, i.e., the standardized C_N,

$$Z = \left[C_N - E(C_N \mid H_0)\right] / \sqrt{\mathrm{var}(C_N \mid H_0)}\,,$$

has a $N(0,1)$ distribution, approximately. Thus given the data $X_1, X_2, \ldots, X_m$ and $Y_1, Y_2, \ldots, Y_n$, Z can be computed and compared with the appropriate percentile point from the $N(0,1)$ distribution to determine whether H_0 is to be rejected or not. This is illustrated in the following example.

Example. Consider the observations 14.8, 7.3, 5.6, 6.3, 9.0, 4.2, 10.6, 12.5 from population 1 and 12.7, 14.2, 12.6, 2.1, 15.7, 11.8, 10.9 from population 2. We have $m = 8$ and $n = 7$. Suppose that we wish to test $H_0: \theta_1 = \theta_2$ against $H_3: \theta_1 \neq \theta_2$. The results are summarized in Table 1.

Summing the $E(\xi_i^2)$ for the X-values, we have $C_N = 0.761$. Furthermore, $\sum_{i=1}^{15} \cdot \{E(\xi_i^2)\}^2 = 31.378$ and by (2), $[\mathrm{var}(C_N \mid H_0)]^{1/2} = (0.0682)^{1/2} = 0.2612$. Thus the standardized value of C_N is

$$Z = (0.761 - 1)/0.2612 = -0.915$$

Since the 97.5th percentile point from $\Phi(x)$ is 1.96, the hypothesis $H_0: \theta_1 = \theta_2$ is not rejected at the 5% significance level.

References

[1] Capon, J. (1961). *Ann. Math. Statist.*, **32**, 88–100.

[2] Duran, B. S. (1976). *Commun. Statist. A*, **5**, 1287–1312.

[3] Klotz, J. (1962). *Ann. Math. Statist.*, **33**, 498–512.

[4] Tietjen, G. L., Kahaner, D. K., and Beckman, R. J. (1977). *Select. Tables Math. Statist.*, **5**, 1–73.

(BARTON–DAVID TEST
KLOTZ TEST
LINEAR RANK STATISTICS
MOOD TEST
SIEGEL–TUKEY TEST)

BENJAMIN S. DURAN

CAPTURE–RECAPTURE METHODS

The idea of obtaining information about a population by marking or labeling* some of its members can be traced back several centuries. However, two noteworthy applications to ecology were made by Petersen in 1896, using tagged plaice, and Lincoln [19], who used band returns to estimate the size of the North American waterfowl population. Since the 1940s the capture–recapture method has been widely used for estimating population numbers and related parameters such as survival and immigration rates. Extensive reviews of the methods are given by Cormack [11, 14] and Seber [40, 41], and the technique has been recently considered in relation to estimating the size of a human population from several incomplete lists of the population (see Wittes [42] for epidemiological applications and El-Khorazaty et al. [17] for a general review). A historical overview of the subject is given by Otis et al [24].

In discussing the methodology it is convenient to classify populations as either open or closed. A closed population is one that remains unchanged during the experiment so that the population size remains constant and the effects of process, such as migration, mortality, birth, and recruitment, are negligible; otherwise, the population is said to be open. The notation and material in this article come from refs. 40 and 41.

CLOSED POPULATION: SINGLE RECAPTURE

The simplest mark-recapture experiment, which we shall call the Petersen experiment, for estimating the size N of a closed population consists of catching, marking, and releasing a sample (sample 1) of n_1 animals. After allowing the marked and unmarked to mix, a second representative sample is taken from the population. Equating the proportion of marked in the second sample with the population proportion, n_1/N, of marked leads to an estimate of N. Mathematically, if n_2 is the size of sample 2 and m_2 are found to be marked in this sample, then $m_2/n_2 \approx n_1/N$, leading to the estimate

$$\hat{N} = n_1 n_2 / m_2, \qquad (1)$$

the so-called Petersen estimate or Lincoln index (although the term index is misleading.) If n_1 and n_2 are regarded as constants and sample 2 is a simple random sample (i.e., a random sample without replacement*), m_2 has a hypergeometric distribution* with probability function

$$f(m_2 \mid n_1, n_2) = \binom{n_1}{m_2}\binom{N - n_1}{n_2 - m_2} \Big/ \binom{N}{n_2}. \qquad (2)$$

It transpires that $\hat{N}$, suitably rounded to an integer, is the maximum likelihood* estimate of N for this distribution. However, a modification

$$N^* = \frac{(n_1 + 1)(n_2 + 1)}{(m_2 + 1)} - 1 \qquad (3)$$

is unbiased* when $n_1 + n_2 \geqslant N$, and approximately unbiased otherwise. An estimate of the variance of N^* with similar properties [37] is

$$V^* = \frac{(n_1 + 1)(n_2 + 1)(n_1 - m_2)(n_2 - m_2)}{(m_2 + 1)^2 (m_2 + 2)}.$$

Large-sample confidence intervals* based on N^* or $E[m_2](= n_1 n_2 / N)$ (e.g., Chapman [8, 9]) can be constructed. Charts are available for choosing n_1 and n_2 to achieve a prescribed accuracy of $\hat{N}$ (see ref. 40, Chap. 3, or the original paper of Robson and Regier [31]).

In using a Petersen experiment a number of basic assumptions must be satisfied. These can be expressed various ways, but from a practical viewpoint they amount to the following: (a) the population is closed, so that N is constant; (b) all animals have the same probability of being caught in sample 1; (c) marking does not affect the catchability of an animal; (d) sample 2 is a simple random sample, i.e., each of the $\binom{N}{n_2}$ possible samples has an equal chance of being chosen; (e) animals do not lose their marks between samples; and (f) all marks (or tags) are reported on recovery in sample 2. We next discuss briefly some of the main problems associated with departures from these assumptions; further details are given in ref.

40, Chap. 3, and in the original papers mentioned below.

If assumption (d) is likely to be false, N will be overestimated. However, double marking can be used to provide a correction factor for m_2: those which have lost one tag can be used to estimate the number losing both (see also Robson and Regier [32]). Assumption (f), which is appropriate when the tags are reported by hunters, commercial fishermen, etc., can be tested if part of sample 2 has a 100% tag reporting rate—say by experienced observers (see Paulik [25]). Assumption (d) will depend on (b) and (c) as any variation in the catchability of the animals, whether natural or induced by the handling and marking, will lead to a nonrandom sample 2. In fact, departures from (b) and (c) are difficult to differentiate, as a variable catchability means that the more catchable animals are caught in sample 1 and, for a similar method of taking sample 2, the marked will tend to have a higher probability of capture in sample 2: this leads to the underestimation* of N. It transpires that in just one case can the effect of variable catchability be eliminated: when entirely different methods are used for taking each sample: the catchability in sample 2 is then independent of mark status [37].

Use of the hypergeometric distribution (2) emphasizes the fact that it is basically the activity of the experimenter that brings about random sampling. However, another approach in which randomness* is related to the activity of the animals considers the N animals in the population as N independent multinomial* trials each with the same probability of belonging to a given capture–recapture category. In the Petersen experiment there are four categories: caught in sample 1 only, caught in sample 2 only, caught in both samples, and caught in neither sample, with respective probabilities P_1, P_2, P_{12}, and Q $(= 1 - P_1 - P_2 - P_{12} = 1 - P)$. Thus if a_1 $(= n_1 - m_2)$, a_2 $(= n_2 - m_2)$, a_{12} $(= m_2)$, and $N - a_1 - a_2 - a_{12}$ $(= N - r$, where r is the total number of different animals caught in the experiment) are the numbers caught in the four categories, then the joint probability function of the random variables a_1, a_2, and a_{12} is

$$f(a_1,a_2,a_{12}) = \frac{N!}{a_1!a_2!a_{12}!(N-r)!} \times P_1^{a_1}P_2^{a_2}P_{12}^{a_{12}}Q^{N-r}. \quad (4)$$

If p_i $(= 1 - q_i)$ is the probability of capture in sample i, and samples are independent so that capture in sample 1 does not affect capture in sample 2, then $P_1 = p_1q_2$, $P_2 = q_1p_2$, $P_{12} = p_1p_2$, and $Q = q_1q_2$. Now the triple (a_{12}, a_2, a_2) determines the triple $(a_{12}, a_1 + a_{12}, a_2 + a_{12})$ and vice versa, so that substituting for the P's gives us

$$\begin{aligned} f(m_2,n_1,n_2) &= f(a_1,a_2,a_{12}) \\ &= \frac{N!}{(n_1-m_2)!(n_2-m_2)!m_2!(N-r)!} \\ &\quad \times p_1^{n_1}q_1^{N-n_1}p_2^{n_2}q_2^{N-n_2} \\ &= f(m_2 \mid n_1,n_2)\prod_{i=1}^{2}\left\{\binom{N}{n_i}p_i^{n_i}q_i^{N-n_i}\right\} \\ &= f(m_2 \mid n_1,n_2)f(n_1,n_2), \end{aligned} \quad (5)$$

where $f(m_2 \mid n_1, n_2)$ is given by (1). Thus (5) follows from (2) by simply treating the sample sizes n_i as random variables. If the sample sizes are fixed, (2) is the appropriate model. On the other hand, if the experimenter controls the sampling procedure by the effort he puts in, i.e., fixed p_1 and p_2, then (5) is the appropriate model. However, under (5), $\hat{N}$ of (1) is still the maximum likelihood estimate with essentially the same asymptotic variance*. The fact that sample sizes can be treated as fixed or random with little change in the theory is a common feature of capture–recapture models. It stems from the fact that for large samples n_i and $E[n_i]$ are essentially interchangeable.

In practice, samples are frequently neither completely fixed nor completely random, as other factors, such as finance, weather, etc., can affect the length of an experiment once it has started. For this reason there has been some interest in sequential* type experiments in which sample 2 is continued until a prescribed number of marked, or unmarked, are caught in sample 2. Sometimes, sample 2 is taken with replacement, e.g., when the animals are merely observed and not actually captured. In this case $f(m_2 | n_1, n_2)$ is binomial* with parameters n_2 and $p = n_1/N$ [2]. Further details of these modifications are given in refs. 40 and 41.

CLOSED POPULATION: MULTIPLE RECAPTURES

A natural extension of the Petersen experiment is the so-called Schnabel experiment (see Schnabel [34]) or multiple recapture census in which s $(s > 2)$ consecutive samples are taken from the population. If n_i animals are caught in sample i, and m_i are the number found to be marked on a previous sampling occasion, then the u_i $(= n_i - m_i)$ unmarked are given a mark and the whole sample returned to the population. If individual numbered marks or tags are used, then animals have to be tagged only once, the first time they are caught. Depending on whether the n_i are regarded as fixed or random, both the hypergeometric and multinomial models readily generalize to this case [15]. For example, (4) now becomes

$$f(\{a_w\}) = \frac{N!}{\prod_w a_w!(N-r)!} Q^{N-r} \prod_w P_w^{a_w} \tag{6}$$

$$= \frac{N!}{\prod_w a_w!(N-r)!} \prod_{i=1}^{s} p_i^{n_i} q_i^{N-n_i}, \tag{7}$$

where w is a label describing membership of one of the $2^s - 1$ capture categories. The maximum likelihood estimate, $\hat{N}$, is close to the unique root, greater than r, of the $(s-1)$th-degree polynomial,

$$\left(1 - \frac{r}{N}\right) = \prod_{i=1}^{s} \left(1 - \frac{n_i}{N}\right).$$

This equation is simply an expression of the fact that $Q = q_1 q_2 \cdots q_s : r$ is binomial with parameters N and P $(= 1 - Q)$. For further details, see refs. 40 and 41.

The corresponding hypergeometric model $f(\{a_w\} | \{n_i\})$ leads to the same maximum likelihood estimate $\hat{N}$ and asymptotic variance. In this case r is sufficient for N, so that $f(\{a_w\} | r)$ contains the same information about N. Using some general theorems about a class of factorial series* distributions, of which $f(\{a_w\} | r)$ is a member, Berg [3] obtains a ratio-type estimate $\tilde{N}$ which is minimum-variance-unbiased if $N \leqslant \sum n_i$. He provides a useful recurrence relation for overcoming some of the computational difficulties in finding $\tilde{N}$ and its variance estimate. When $s = 2$, $\tilde{N}$ reduces to N^* of (3).

Another method of estimating N from a Schnabel census or a set of lists is to use the theory of incomplete contingency tables*; incomplete as the number of animals in the category "not caught at all" $(= N - r)$ is unobservable. By fitting a log-linear* model to the numbers in the observable categories, N can be estimated (see Bishop et al. [4, Chap. 6] and Cormack [14] for readable discussions of this approach). Finally, a number of useful regression* models are available for estimating N [26, 27; 40, Chap. 6].

We note that the assumptions underlying the Petersen method must apply to all the samples in a Schnabel census so that any departures from these assumptions can seriously affect the validity of $\hat{N}$. Since variation in catchability seems to be a fact of life, a general package of models has been developed by Otis et al. [24], which allows this variation to be modeled in various ways. They present six models: M_t [variation with trapping occasion or time, given by (7)], M_b (variation by behavioral responses, e.g., trap shyness or addiction), M_h (variation by individual response or heterogeneity, e.g., size selectivity in electro-fishing), and various combinations M_{tb}, M_{bh}, M_{th}, and M_{tbh}. The authors also include the model M_0 in which there is no variation [i.e., (7) with $p_i = p$], and a generalized removal model in which removal corresponds to tagging.

If p_{ij} is the probability that the ith animal $(i = 1, 2, \ldots, N)$ is caught in the jth sample $(j = 1, 2, \ldots, s)$, and we can assume that the animals are independent of one another as far as catching is concerned, then the likelihood* function is $\prod_i \prod_j p_{ij}^{x_{ij}}(1 - p_{ij})^{1-x_{ij}}$, where $x_{ij} = 1$ if the ith animal is caught in the jth sample and $x_{ij} = 0$ otherwise. The various models can now be described mathematically by specifying p_{ij}, e.g., $M_t(p_{ij} = p_j)$, $M_h(p_{ij} = p_i$, where the p_i are a random sample from some distribution), $M_{bh}(p_{ij} = p_i$ for any first capture and $p_{ij} = c_i$ for any recapture), etc. A standard algorithm is used to calculate the maximum likelihood estimates for the models M_0, M_t, M_b, and M_{bh}, the last model being regarded as the most realistic and the most useful. The method used for analyzing M_{bh} can also be applied to so-called "removal" experiments in which animals are removed from the population on capture. Suitable estimation procedures are not yet available for M_{bt}, M_{th}, and M_{tbh}, but a generalized jackknife* estimator due to Burnham (see Burnham and Overton [7]) is available for M_h. This method utilizes the data on frequency of capture and may replace various other ad hoc frequency of capture models which have been shown to be unreliable for estimating the number $(N - r)$ with zero frequency (see Cormack [14] and Seber [41]). Otis et al. [24] discuss the robustness* of these estimates and the usual confidence intervals* *estimate* ± 1.96 *standard deviations**. They give a procedure for selecting a model and discuss the important aspects of design relating to number of traps, trap spacing, and number of samples s. Included is a useful nonlinear regression* method of converting animal numbers to densities using recapture data from selected subgrids. Their method seems to be superior to the various ad hoc procedures of determining the effective trapping area from home range estimates based on a variety of home-range models (e.g., circles or ellipses, capture distributions) or the use of assessment lines (see ref. 41). The authors also discuss various tests of closure, including one of their own, and conclude from the general insensitivity of these tests that the best evidence for closure is biological rather than statistical. The experiment should be designed to approximate closure; otherwise, more general methods such as those described below should be used.

OPEN POPULATION: MULTIPLE RECAPTURES

When the population is open we are now faced with the estimation of additional parameters, and the notation, which we now give, becomes considerably more complex. Let

M_i = size of the marked population just prior to sample i

U_i = size of the unmarked population just prior to sample i

$N_i = M_i + U_i$

m_i = number of marked caught in sample i

u_i = number of unmarked caught in sample i

$n_i = m_i + u_i$

ϕ_i = probability that a marked animal survives from the release of sample i to the capture of sample $i+1$ and is in the population at the time of sample $i+1$, given that it is alive and in the population immediately after the ith release

p_i = probability that a marked animal is caught in sample i

ν_i = probability that an animal (marked or unmarked) caught in sample i is returned to the population

B_i = number of *new* animals joining the population between samples i and $i+1$ which are still alive and in the population at sample $i+1$

R_i = number of marked released from sample i

r_i = number of R_i subsequently recaptured

z_i = number of different animals caught before sample i which are not caught in sample i but are caught subsequently

χ_i = probability that a member of R_i is not caught again

$$= 1 - \phi_i p_{i+1} - \phi_i q_{i+1} \phi_{i+1} p_{i+2} - \cdots - \phi_i q_{i+1} \cdots \phi_{s-2} q_{s-1} \phi_{s-1} p_s$$

Cormack [14] notes that, in contrast to the historical development of models for the closed population, the first models for open populations were essentially multinomial*, i.e., the special cases of Darroch [16], Seber [35], and Cormack [10], leading to the general models of Jolly [18] and Seber [36]. Since Jolly and Seber developed their theory from different, but equivalent viewpoints, and arrived at the same estimates, their method of analysis is sometimes called the Jolly–Seber or J–S method (see ref. 40, Chap. 5, for details). There are several practical features of their method which should be noted. In the first instance it is assumed that all emigration is premanent, so that animals enter and leave the population only once. This is a strong assumption and is absolutely essential for a valid application of the J–S method. Because of the assumption we can lump births and immigrants together as "new animals" ($= B_i$), and regard deaths and emigrants collectively as "losses" with associated probabilities $1 - \phi_i$. Also, there is an allowance in the model for the possibility of deaths on handling, with associated probabilities $1 - \nu_i$. We note that inherent in the foregoing notation is the assumption that all marked animals have the same survival and capture probabilities ϕ_i and p_i.

A general multinomial-type model can be constructed by first considering the subsequent capture history of the u_i unmarked animals in each sample. This leads to the conditional probability* function $f(\{a_w\} | \{u_i\})$, a product of multinomial distributions (one for each u_i). Assuming that p_i also applies to unmarked animals, and treating the U_i as unknown constants, the joint distribution of the u_i is

$$f(\{u_i\}) = \prod_{i=1}^{s} \binom{U_i}{u_i} p_i^{u_i} q_i^{U_i - u_i}.$$

Thus

$$f(\{a_w\}) = f(\{a_w\} | \{u_i\}) f(\{u_i\}) = L_1 \times L_2,$$

say where L_1 is proportional to

$$\prod_{i=1}^{s-1} \left\{ \chi_i^{R_i - r_i} (\phi_i q_{i+1})^{z_{i+1}} (\phi_i p_{i+1})^{m_{i+1}} \right\} \times \prod_{i=1}^{s} \nu_i^{R_i} (1 - \nu_i)^{n_i - R_i}.$$

Thus the marked population, represented by L_1, supplies information on the $\{\phi_i, p_i, \nu_i\}$ via the statistics $\{R_i, r_i, z_i, m_i, n_i\}$, while the $\{u_i\}$ supply information on the parameters $\{U_i\}$. The maximum likelihood estimates of U_i, ϕ_i, and p_i, obtained by maximizing the logarithm of $f(\{a_w\})$, can also be obtained intuitively as follows.

Immediately after the ith sample there are two groups of marked animals, the $M_i - m_i$ not caught in the ith sample of which z_i are subsequently caught, and the R_i just released, of which r_i are subsequently caught. Since the chances of recapture are assumed to be the same for both groups, we would expect $z_i/(M_i - m_i) \approx r_i/R_i$, which leads to the estimate $\hat{M}_i = (R_i z_i / r_i) + m_i$. Analogous to (1) we have $\hat{N}_i = \hat{M}_i n_i / m_i$, leading to $\hat{U}_i = \hat{N}_i - \hat{M}_i$. Also, we have the intuitive relationships $\hat{\phi}_i = \hat{M}_{i+1}/(\hat{M}_i - m_i + R_i)$, $\hat{p}_i = m_i/\hat{M}_i = n_i/\hat{N}_i$, $\hat{\nu}_i = R_i/n_i$, and assuming that the ϕ_i also apply to the unmarked, $\hat{B}_i = \hat{N}_{i+1} - \hat{\phi}_i(\hat{N}_i - n_i + R_i)$. It should be noted that $\hat{M}_i$ and $\hat{N}_i$ are *not* maximum likelihood estimates but are simply used as intermediate steps in the calculation of the maximum likelihood estimates $\hat{\phi}_i, \hat{p}_i$, and $\hat{U}_i$. Large-sample variances, small-sample modifications similar to (3), and details of the computations are given in ref. 40; a useful computer package is also available [1]. The foregoing estimates and their variances have been studied by a number of people using simulation* and Manly [22] gives a promising jackknife technique for improving the reliability of estimates and their confidence intervals* (see ref. 41, Chap. 13).

An alternative approach that leads to a generalized hypergeometric*-type model is to regard the sizes of the various groups which have the same capture and survival characteristics as constants and assume that simple random samples are taken from these groups. The triples $\{m_i, r_i, z_i\}$ are sufficient statistics* for $\{M_i, U_i\}$ and, not surprisingly, we obtain the same estimates for M_i and N_i as before. However, this approach forms the basis of a number of very general models proposed by Robson [30] and Pollock [28]. Cormack [12, 13] utilized similar ideas to develop intuitive estimates for certain "nonstandard" situations. Robson showed that the foregoing model could be modified to allow animals with different capture histories to have different survival rates. Pollock took this a step further and allowed the catchability to also vary with capture history.

A number of special cases and variations of the Schnabel census have been considered by various authors, particularly Manly (see refs. 20, 21, and 23), and these are described in ref. 41. One of the most useful models is the so-called multisample single-recapture census, developed independently by Robson [29] and Seber [35]. Here the samples n_i are completely removed from the population (100% losses on capture), as in commercial fishing and hunting, and releases of similar animals are made independently by scientists after each sample. By an appropriate reparametrization* this model was applied to bird banding independently by Seber [38] and Robson and Youngs [33] (see also Seber [39], and Youngs and Robson [43] for a fisheries application). This particular model, together with extensive modifications due to Brownie and Robson [5] to allow for some degree of age-dependent survival and band recovery rates, are given by Brownie et al. [6].

References

[1] Arnason, A. N. and Baniuk, L. (1978). *POPAN-2, A Data Maintenance and Analysis System for Recapture Data, (Release 3)*. Charles Babbage Research Centre, St. Pierre, Manitoba, Canada.

[2] Bailey, N. T. J. (1951). *Biometrika*, **38**, 293–306.

[3] Berg, S. (1976). *Scand. J. Statist.*, **3**, 86–88.

[4] Bishop, Y. M. M., Fienberg, S. E., and Holland, P. W. (1975). *Discrete Multivariate Analysis: Theory and Practice*. MIT Press, Cambridge, Mass.

[5] Brownie, C. and Robson, D. S. (1976). *Biometrics*, **32**, 305–323.

[6] Brownie, C., Anderson, D. R., Burnham, K. P., and Robson, D. S. (1977). Statistical Inference from Band Recovery Data: A Handbook. *U.S. Fish Wildl. Serv. Resour. Publ. No. 131*. (Their important monograph describes in detail a comprehensive range of models together with appropriate goodness-of-fit tests. There is also a good practical discussion on matters relating to the design of banding experiments.)

[7] Burnham, K. P. and Overton, W. S. (1978). *Biometrika*, **65**, 625–633.

[8] Chapman, D. G. (1948). *Int. Pacific Salmon Fish. Commun. Bull.*, **2**, 69–85.

[9] Chapman, D. G. (1951). *Univ. Calif. Publ. Statist.*, **1**, 131–160.

[10] Cormack, R. M. (1964). *Biometrika*, **51**, 429–438.

[11] Cormack, R. M. (1968). *Oceanogr. Mar. Biol. Annu. Rev.*, **6**, 455–506.

[12] Cormack, R. M. (1972). *Biometrics*, **28**, 337–343.

[13] Cormack, R. M. (1973). In *The Mathematical Theory of the Dynamics of Biological Populations*, M. S. Bartlett and R. W. Hiorns, eds. Academic Press, New York, pp. 225–234.

[14] Cormack, R. M. (1979). In *Sampling Biological Populations*, R. M. Cormack, G. P. Patil, and D. S. Robson, eds. Satellite Program in Statistical Ecology. International Co-operative Publishing House, Fairland, MD. (To appear.)

[15] Darroch, J. N. (1958). *Biometrika*, **45**, 343–359.

[16] Darroch, J. N. (1959). *Biometrika*, **46**, 336–351.

[17] El-Khorazaty, M. N., Imrey, P. B., Koch, G. G., and Wells, H. B. (1977). *Int. Statist. Rev.*, **45**, 129–157.

[18] Jolly, G. M. (1965). *Biometrika*, **52**, 225–247.

[19] Lincoln, F. C. (1930). *U.S. Dept. Agric. Circ. No. 118*, 1–4.

[20] Manly, B. F. J. (1974). *Biom. Zeit.*, **16**, 185–190.

[21] Manly, B. F. J. (1975). *Biom. Zeit.*, **17**, 431–435.

[22] Manly, B. F. J. (1977). *Acta Theriol.*, **22**, 215–223.

[23] Manly, B. F. J. (1977). *Biom. J.* (formerly Biom. Zeit.), **19**, 687–692.

[24] Otis, D. L., Burnham, K. P., White, G. C., and Anderson, D. R. (1978). Statistical Inference for Capture Data from Closed Populations. *Wildl. Monogr. No. 62*.

[25] Paulik, G. J. (1961). *J. Fish. Res. Board Canada*, **18**, 817–829.

[26] Paulik, G. J. (1963). *Biometrics*, **19**, 28–57.

[27] Paulik, G. J. (1963). In North Atlantic Fish Marking Symposium, *ICNAF, Spec. Publ. No. 4*, pp. 230–237.

[28] Pollock, K. H. (1975). *Biometrika*, **62**, 577–583.

[29] Robson, D. S. (1963). In North Atlantic Fish Marking Symposium, *ICNAF Spec. Publ. No. 4*, 330–335.

[30] Robson, D. S. (1969). In *New Developments in Survey Sampling*, N. L. Johnson and H. Smith, Jr., eds. Wiley-Interscience, New York, pp. 120–140.

[31] Robson, D. S. and Regier, H. A. (1964). *Trans. Amer. Fish. Soc.*, **93**, 215–226.

[32] Robson, D. S. and Regier, H. A. (1966). *Trans. Amer. Fish. Soc.*, **95**, 56–59.

[33] Robson, D. S. and Youngs, W. D. (1971). Statistical Analysis of Reported Tag-Recaptures in the

Harvest from an Exploited Population. *Biom. Unit Rep. BU-369-M*, Cornell University, Ithaca, N. Y.

[34] Schnabel, Z. E. (1938). *Amer. Math. Monthly*, **45**, 348–352.

[35] Seber, G. A. F. (1962). *Biometrika*, **49**, 339–349.

[36] Seber, G. A. F. (1965). *Biometrika*, **52**, 249–259.

[37] Seber, G. A. F. (1970). *Biometrics*, **26**, 13–22.

[38] Seber, G. A. F. (1970). *Biometrika*, **57**, 313–318.

[39] Seber, G. A. F. 1972). *J. Wildl. Manag.*, **36**, 405–413.

[40] Seber, G. A. F. (1973). *The Estimation of Animal Abundance and Related Parameters*. Charles Griffin, London. (Contains an up-to-date treatment of capture–recapture models with emphasis on applications, together with an extensive bibliography.)

[41] Seber, G. A. F. (1980). *The Estimation of Animal Abundance and Related Parameters*, Vol 2. Charles Griffin, London, (in press). (The new material covers recent developments from about 1972–1979.)

[42] Wittes, J. T. (1974). *J. Amer. Statist. Ass.*, **69**, 93–97.

[43] Youngs, W. D. and Robson, D. S. (1975). *J. Fish. Res. Board Canada*, **32**, 2365–2371.

(FISHERIES, STATISTICS IN
HYPERGEOMETRIC DISTRIBUTION
MULTINOMIAL DISTRIBUTION
SAMPLE-SIZE ESTIMATION)

G. A. F. SEBER

CARLEMAN'S CRITERION

A sufficient condition for the uniqueness of determination of a distribution $F_x(\cdot)$ by its moments (*see* MOMENT PROBLEM*). Carleman [1] proved that the distribution is unique if the series $\sum_{n=1}^{\infty}(\mu'_{2n})^{-1/(2n)}$ diverges, where μ'_n is the nth crude moment*. This result was extended to the multidimensional case by Cramér and Wold [2]. The condition is clearly satisfied for all distributions with finite range of variation.

References

[1] Carleman, T. (1925). *Les Fonctions quasi-analytiques*. Collection Borel. Gauthier-Villars, Paris.

[2] Cramér, H. and Wold, H. (1936). *J. Lond. Math. Soc.*, **11**, 290–294.

[3] Hausdorff, F. (1921). *Math. Zeit.*, **9**, 74–109.

(MOMENT PROBLEM)

CARNAP'S THEORY OF PROBABILITY

See DEGREE OF CONFIRMATION

CARRIERS

An alternative term for "independent" variables in regression analysis* to emphasize the fact that the latter may not be independent in any reasonable sense.

(REGRESSION ANALYSIS)

CARRY-OVER EFFECT

In repeated measurement designs* there is the possibility of several types of interference*. One type is connected with the preceding treatment of treatments. This type of interference is called the carry-over effect. Allowing sufficient time between treatments may reduce these effects.

(CHANGEOVER DESIGNS
DESIGN AND ANALYSIS OF
EXPERIMENTS)

CATASTROPHE THEORY

During the 1970s, articles appeared in the press and in scientific journals extolling the virtues of "catastrophe theory" (CT), and describing, sometimes overzealously, its impact on applications as diverse as phase transitions in heat, national war policies, laser physics, the splitting of biological cells, the crash of stock markets, and the psychological dynamics of self-pity. It was variously described as the greatest discovery in mathematics since Newton and Leibnitz "invented" the calculus and, in an apparent backlash of opinion, as an emperor without any clothes. Although these claims may both

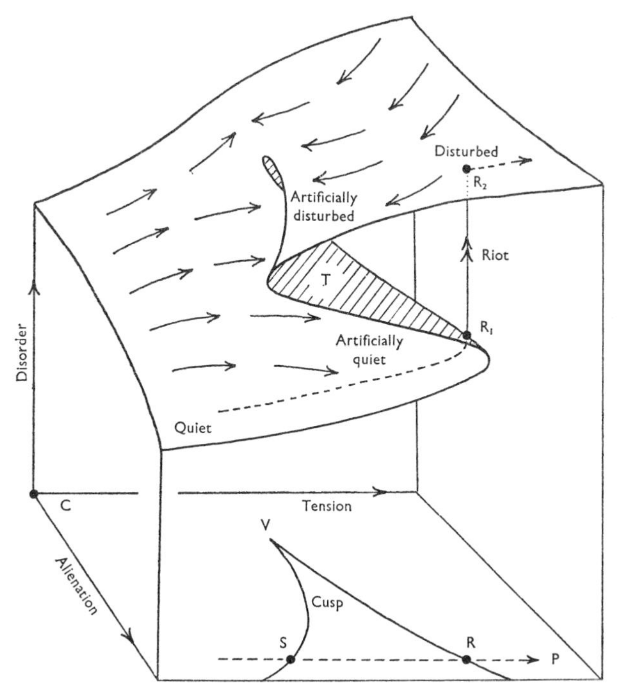

Figure 1 Theoretical graph of disorder as a function of tension and alienation. Reproduced from [14] with permission of the authors.

be exaggerated, the attendant publicity made many more people aware of this new field than would ever have come to hear of it otherwise.

CT is the study of discontinuous effects caused by continuous changes, and was developed largely by the French topologist René Thom [7, 8]. The following summary is based on Zeeman [11]. Let E^k be a control space of dimension k and E^n a behavior space of n responses* $(x_1, \ldots, x_n)$. In Fig. 1, $k = 2$ and $n = 1$, with control variables "alienation" a and "tension" b, and with "disorder" as the single response x in a model for describing prison disturbances [14].

Let f be a real-valued potential or energy function representing a dynamical system, where $f: E^k \times E^n \to (-\infty, \infty)$. In applications such as Zeeman and Isnard's war model [13, pp. 302 et seq.], $-f$ may be a likelihood* function; assume that the system seeks to minimize f locally, so that the behavior state flows toward minimum energy or most likely response values, for every combination of control variables. Let S_f be the surface of points in $E^k \times E^n$ such that $\partial f/\partial x_i = 0$; $i = 1, \ldots, n$. Let $g: S_f \downarrow E^k$ be the projection of S_f on E^k. Singularities occur in S_f where $\partial^2 f/\partial x_i^2 = 0$; $i = 1, \ldots, n$; the projection of the singularity set in S_f on E^k is the bifurcation set or catastrophe map of f on E^k.

Example. Let $k = 2$ and $n = 1$, with control variables a (alienation) and b (tension) in E^2 and a single response variable x (disorder) in E^1. If

$$f(a, b; x) = \tfrac{1}{4}x^4 - \tfrac{1}{2}ax^2 + bx$$

canonically, the behavior surface S_f is given by

$$\partial f/\partial x \equiv x^3 - ax + b = 0. \qquad (1)$$

This is the surface in Fig. 1 with a fold; singularities on S_f occur where

$$\partial^2 f/\partial x^2 \equiv 3x^2 - a = 0; \qquad (2)$$

the bifurcation set in E^2, on eliminating x from (1) and (2), is the cusp $4a^3 = 27b^2$.

If alienation is fixed and tension increases along the path P in Fig. 1 to the point R on the cusp, x increases smoothly from a quiet state on the behavior surface until at R_1 on the lower-fold curve it jumps catastrophically to R_2, and a riot ensues. The cusp catastrophe has a hysteresis effect; if tension is high and decreases, disturbance changes suddenly, perhaps by a truce, from a point on the higher-fold curve to the lower part of the sheet; however, the change occurs, not at R, but at S on the other branch of the cusp. The shaded portion T of S_f is inaccessible, representing least likely behavior, and corresponds to maxima of f, which is bimodal at (a, b)-points inside the cusp.

In Fig. 1, b and a are normal and splitting factors, respectively; the behavior variable increases with tension, while as alienation increases through the cusp point V, a small change in tension up or down may lead to divergence of behavior onto the upper or the lower part of the surface. In other models, the a and b axes in E^2 rotate through 45° to represent conflicting factors.

Thom and others [1, 4, 7] proved that for all suitably smooth functions, and when $k = 2$, 3, 4, or 5, any singularity of g is equivalent in a certain sense to one of 2, 5, 7, or 11 elementary catastrophes, respectively; and that g has a stability property under small changes in f. If $k \geqslant 6$, the number of classifications is infinite; most applications have been in elementary catastrophe theory (ECT), where $k \leqslant 5$ and Thom's theorem can be used to best effect.

Thus, in the example, whatever the function f might be (provided only that f is smooth enough), the form of the catastrophe map is a fold or a cusp. If $n = 1$ and $k \leqslant 4$, the forms are cuspoids, including the butterfly ($k = 4$), which involves a trimodal function f, and next to the cusp is the most useful for applications. If $n = 2$ and $k \leqslant 4$, the forms are umbilics. (A rigorous discussion of CT requires an understanding of several mathematical concepts [1, 4, 7, 13].)

Several books and expository papers on CT and ECT are listed in the bibliography [1, 4, 7, 11–13]. Poston and Stewart [4] give the most comprehensive account, with detailed discussions of many applications; in physics, to fluid mechanics, the stability of ships and oil rigs, shockwaves, caustics in optics, and the buckling of elastic beams; in biology, to cell differentiation and ecological frontiers (see also ref. 7); in sociology, to the effects of alcohol on driving ability. Zeeman [11–14] has done more research on applications of ECT to the inexact sciences, including the collapse of a stock market, the voting behavior of committees with actions a and b by supporters and opponents as conflicting factors, the retreat–aggression behavior of a dog as a cusp catastrophe with rage and fear as conflicting factors, and fasting–gorging behavior in the disease anorexia nervosa as a butterfly catastrophe with hunger level, abnormality, loss of self-control, and reassurance as control variables. These models are qualitative, in that one or more variables cannot be measured quantitatively. (It is this class of applications of ECT to the social and behavioral sciences which has aroused the greatest criticism [6, 10, 15].)

A statistical approach to fitting quantitative CT models is needed if data sets that arise from an underlying CT dynamic are to be properly analyzed. Linear model techniques based on smooth regression surfaces (*see* GENERAL LINEAR MODEL AND HYPOTHESIS) would then be inadequate and misleading [1, p. 385; 5]. The model for Fig. 1 allows for external events or incidents inside the prison to be incorporated as stochastic noise*, and Lewis [3] has investigated regression procedures for identifying a cusp or butterfly catastrophe to describe psychological crisis changes. Another ECT model, augmented by statistical mechanics*, predicts changes in a protein-denaturant system as the constraints are changed, with illustrations from two data sets [2]. A unified statistical approach to ECT has been developed by Loren Cobb, however (*see* [0] and STATISTICAL CATASTROPHE THEORY).

Are qualitative ECT models to be dismissed because of a "lack of precision in the ideas that they use" [7, p. 6]? Thom believes that, because of progress in topology and differential analysis, we can define forms, determine whether two functions have the same form or not, and hence present qualitative results rigorously. Although he sees only a slight hope for quantitative modeling in the social sciences, he claims that CT has a scientific status because of its internal mathematical consistency [9, pp. 384–388]. This justifies modeling which leads to qualitative predictions or which reduces the arbitrariness of the description of behavior. The last, he says, is the proper definition of scientific explanation. He sees the role of the statistician to be rather like that of the diviner in primitive societies, and suggests elsewhere [8, pp. 197–198] that CT with its morphological approach may be helpful, since "a complete statistical theory will necessarily call for intermediate situations between a strict determinism, and a completely incoherent behavior." Thom's approach (see also Golubitsky [1]) should give food for thought to statisticians, who customarily require numerical data as one prerequisite for accepting mathematical models.

References

[0] Cobb, L. (1981). In *1980 Proc. of the Section on Survey Research Methods*, American Statistical Association. (This includes several references to Cobb's research in statistical catastrophe theory.)

[1] Golubitsky, M. (1978). *SIAM Rev.*, **20**, 352–387. (A concise but advanced mathematical summary of ECT, with applications in physics.)

[2] Kozak, J. J. and Benham, C. J. (1976). *J. Theor. Biol.*, **63**, 125–149. (A cusp catastrophe in biology.)

[3] Lewis, M. (1977). *Science*, **196** (June 17), 1270.

[4] Poston, T. and Stewart, I. (1978). *Catastrophe Theory and Its Applications*. Pitman, London. (An excellent, well-written, and comprehensive book on CT; the mathematical exposition is lucid, at the advanced calculus level. Examples are discussed in detail. Complete bibliography of CT up to 1977.)

[5] Stewart, I. (1975). *New Sci.*, **68**, 447–454.

[6] Sussman, H. J. and Zahler, R. S. (1978). *Synthèse*, **37**, 117–216. (An extensive criticism of ECT which pulls no punches, and is readable by nonmathematicians.)

[7] Thom, R. (1975). *Structural Stability and Morphogenesis*, C. H. Waddington, trans. W. A. Benjamin, Reading, Mass. (Classical first treatise on CT; mathematical ideas are discussed and applied mainly to biology, with some physics and philosophy. Thought-provoking.)

[8] Thom, R. (1977). *SIAM Rev.*, **19**, 189–201. (A philosophical discussion of CT, including Thom's assessment of whether statistics fits in.)

[9] Thom, R. and Zeeman, E. C. (1975). *Lect. Notes Math.*, **468**, 366–389. (The discussion is nonmathematical, but presupposes some understanding of CT.)

[10] Zahler, R. S. and Sussman, H. (1977). *Nature (Lond.)*, **269**, 759–763. (Criticisms of CT along the lines of ref. 6.)

[11] Zeeman, E. C. (1973). In *Manifolds*. University of Tokyo Press, Tokyo, pp. 11–23. (A brief summary of the mathematical framework and of some applications of CT.)

[12] Zeeman, E. C. (1976). *Sci. Amer.*, April, 65–83. (An elementary, lucid, nonmathematical introduction to ECT, with several applications.)

[13] Zeeman, E. C. (1977). *Catastrophe Theory: Selected Papers* (1972–1977). Addison-Wesley, Reading, Mass. (A collection of the author's contributions to CT. The applications can be interpreted by those who are not mathematically minded, but the book is not a systematic development of the subject.)

[14] Zeeman, E. C., Hall, C. S., Harrison, P. J., Marriage, G. M., and Shapland, P. H. (1976). *Brit. J. Math. Statist. Psychol.*, **29**, 66–80. (A full discussion and analysis of the model underlying Fig. 1.)

[15] Zeeman, E. C., Bellairs, R., Goodwin, B., Mackley, M. R., Stewart, I., Berry, M., Guckenheimer, J., and Woodcock, A. E. R. (1977). *Nature (Lond.)*, **270**, 381–384. (These reply to criticisms in ref. 10.)

(MATHEMATICAL PROGRAMMING
MODELING
OPTIMIZATION
STATISTICAL CATASTROPHE THEORY)

CAMPBELL B. READ

CATCH CURVE

The frequency distribution of the catch curve by age from a stationary animal or fish population. The term originated in Edser's note [2] and is now well known in fisheries work. An analytic treatment of these curves and a study of statistical problems associated with them is presented in Chapman and Robson [1].

References

[1] Chapman, D. G. and Robson, D. S. (1960). *Biometrics*, **16**, 354–368.

[2] Edser, T. (1908). *J. R. Statist. Soc. A*, **71**, 686–690.

[3] Jackson, C. H. N. (1939). *J. Animal Ecol.*, **8**, 238–246.

(CHAPMAN ESTIMATOR
FISHERIES, STATISTICS IN
JACKSON ESTIMATOR)

CATEGORICAL DATA

Categorical data consist of counts* rather than measurements*. Categorical variables may have categories that are naturally ordered (ordinal* variables) or have no natural order (nominal* variables). For example, the variable "height," with categories "small," "medium," and "tall" is an ordinal variable, as is an attitudinal variable with categories "agree," "neutral," and "disagree,". On the other hand, variables such as "sex" and "hair color," which have no natural category order, are examples of nominal variables. A fuller classification of variable types is provided by Stevens [33]. Examples of some types of categorical data are given in the following section, together with some possible models of interest.

Most recent work on categorical data is concerned with the situation where people or things are classified simultaneously by two or more attributes*. For example, we might categorize people simultaneously by sex, political opinion, and social class. The results of such a *cross-classification* can be conveniently arranged as a table of counts known as a contingency table*. The pattern of association between the classificatory variables may be measured by computing some measure of association* or by the fitting of a log-linear model*. The special problem of the cross-classification of two *dichotomous** (two-category) variables is treated in detail in the section "2 × 2 Table," and the general problem is discussed briefly in "Multiway Tables." A fuller discussion is to be found in the article CONTINGENCY TABLES. The present article concludes with brief remarks concerning the relevant computer programs.

THREE EXAMPLES OF CATEGORICAL DATA

The examples in this section emphasize the relevance of the scale of measurement to the type of analysis.

The first example is concerned with gene heredity. Peas may be yellow or green, round or wrinkled, short or tall. Within each of these classifications the dominant gene is believed to be the first-named category, and a theory suggests that these characteristics are mutually independent with, for each characteristic, the ratio of dominant to recessive being 3 : 1. Suppose that we have a random sample of 64 peas whose characteristics are given in Table 1. The expected frequencies according to the genetic theory are included in the table and we would wish to enquire whether the observed data are consistent with the theory. Denoting an observed frequency by O_i and the corresponding expected frequency by E_i, the usual

Table 1. Nominal Data: Pea Types

	Yellow				Green			
	Round		Wrinkled		Round		Wrinkled	
	Short	Long	Short	Long	Short	Long	Short	Long
Observed count	36	12	8	4	2	1	1	0
Expected count	27	9	9	3	9	3	3	1

goodness-of-fit statistic* is that proposed by Pearson* [28]:

$$x^2 = \sum_i (O_i - E_i)^2 / E_i$$

The distribution of $X^2$is closely approximated by a χ^2 distribution provided that the individual expected frequencies are not too small (*see* CHI-SQUARE TESTS). Recent work [21] suggests that not too small means at least of size 3.

In the present case the expected frequency of 1 in the final category rules out the use of the χ^2 approximation. Common practice would be to combine this category with some adjacent category, but because the data are nominal and not ordinal, there is no category that can be said to be adjacent, and arbitrary clumping of categories seems unsatisfactory.

Lawal and Upton [22] have shown that a reasonable alternative approximation to the distribution of X^2 is obtained by fitting a log-normal distribution* having the same mean and variance. Writing n for the number of categories, N for the total sample size, and R for the sum of the reciprocals of the expected frequencies, it is necessary to calculate first $\theta = 2\ln(n-1)$ and $\psi = \ln\{n^2 - 1 + (R - n^2 - 2n + 2)/N\}$, and then $\mu = \theta - \frac{1}{2}\psi$ and $\sigma^2 = \psi - \theta$. The upper α percentage point of the distribution of X^2 is then estimated as $\exp\{\mu + \sigma\Phi^{-1}(\alpha)\}$, where $\Phi^{-1}(\cdot)$ is the upper α percentage point* of a unit normal* random variable. For the pea data $N = 64$, $n = 8$, and $R = 2.37$, giving $\theta = 3.89$, $\psi = 4.12$, and hence $\mu = 1.83$ and $\sigma^2 = 0.23$. Consequently, the upper 1% point of the distribution of X^2 is estimated as 19.13 compared to the χ^2 approximation of 18.48. Lawal [21] gives tables of the distribution function of X^2 which permit the calculation of the more accurate value 19.35. Since the observed value of X^2 is 13.56, there is no cause to doubt the theory in the present case.

The second example is taken from Haberman [17, p. 85], who presents data concerning the political views of subjects interviewed during the U.S. 1975 General Social Survey [24]. The data are given in Table 2.

Haberman suggests testing the symmetry model* that categories 1 and 7 should be equally likely, as should categories 2 and 6 and categories 3 and 5. The expected frequencies for this model are given in the table, and the value of X^2 is 7.08.

The problems experienced with the previous set of data do not arise here because the expected frequencies are very large. Even with small expected frequencies the innate ordering of the categories would make the amalgamation of adjacent categories both simple and logical. With the present data there are 3 degrees of freedom for the approximating χ^2, and the model seems acceptable. ($\chi^2_{0.95;\,3} = 7.81$).

The final example concerns ordinal data of a cyclic nature, relating to the distribution of suicides by day of the week, which are reproduced in Table 3. The data were originally presented by Durkheim [6]. Haberman [17] presents one possible analysis. An alternative is to consider the possibility that the data follow a von Mises (circular normal) distribution*. This distribution has two parameters and leads to the expected frequencies shown in the table. The value of X^2 is

Table 2 Ordinal Data: Political Views

	Extremely Liberal	Moderate					Extremely Conservative
	(1)	2	3	4	5	6	(7)
Observed count	46	179	196	559	232	150	35
Expected count	40.5	164.5	214	559	214	164.5	40.5

Table 3 Cyclic Data: Suicides

	Day of Week						
	M	T	W	Th	F	S	S
Observed count	1001	1035	982	1033	905	737	894
Expected count	955	1041	1042	968	883	823	870

21.0 and the approximating χ^2 has 4 degrees of freedom: we conclude that this explanation is inappropriate. ($\chi^2_{0.975;4} = 11.14$.)

Haberman [17, Chap. 1] gives some further examples of analyses of data sets of this type using log-linear models*. Plackett [30] discusses the general analysis of Poisson* data. Gani [9] discusses the analysis of biological data arising from a Markov process*. McCullagh [23] discusses the various type of regression models* that may be used for ordinal data.

2 × 2 TABLE

In this section we consider the case where there are two dichotomous classificatory variables providing four category combinations. A simple example is given in Table 4.

Before considering this table in detail it is helpful to set out some notation for the general case, as shown in Table 5. Each entry in the body of a table is said to refer to a *cell* of that table.

Interest centers around whether the two classificatory variables are *independent* of one another, in which case the conditional proportion of being in column 1, given that an experimental unit belongs to a known row, is the same for both rows, i.e.,

$$p_{11}/p_{1\cdot} = p_{21}/p_{2\cdot}.$$

Simple algebraic manipulation reveals this to be equivalent to the statement that $p_{11}/p_{\cdot 1} = p_{12}/p_{\cdot 2}$ and also to the statement

$$\alpha = p_{11}p_{22}/p_{12}p_{21} = 1.$$

The parameter α was first introduced by Yule [39] and is variously termed the cross-product ratio or odds ratio*, the latter term arising from the rearrangement $\alpha = (p_{11}/p_{12})/(p_{21}/p_{22})$. If $\alpha > 1$, there is said to be *positive association* between the variables. If $\alpha < 1$, there is *negative association*: the terms positive and negative may be taken to correspond to the sign of $\ln(\alpha)$ (*see* ASSOCIATION, MEASURES OF). A consistent* estimate of α is provided by $\{(f_{11}+\frac{1}{2})\cdot(f_{22}+\frac{1}{2})\}/\{(f_{12}+\frac{1}{2})(f_{22}+\frac{1}{2})\}$, where the inclusion of the $\frac{1}{2}$ with each cell frequency avoids problems with zero cell frequencies.

If $\alpha = 1$ (independence), the expected frequency in cell (i, j) is given by

$$e_{ij} = f_{i\cdot}\, f_{\cdot j}/f_{\cdot\cdot}.$$

Provided that all the expected frequencies are large, either the X^2 or the likelihood

Table 4 2 × 2 Contingency Table: Butter vs. Margarine

		Toast spread with: Butter	Margarine	
Individual states that toast is spread with:	Butter	7	3	10
	Margarine	3	7	10
		10	10	20

Table 5 Notation for the 2 × 2 Contingency Table

Observed Frequencies			Population Probabilities		
f_{11}	f_{12}	$f_{1\cdot}$	p_{11}	p_{12}	p_1
f_{21}	f_{22}	$f_{2\cdot}$	p_{21}	p_{22}	p_2
$f_{\cdot 1}$	$f_{\cdot 2}$	$f_{\cdot\cdot}$	$p_{\cdot 1}$	$p_{\cdot 2}$	1

ratio* Y^2 goodness-of-fit statistics, defined in the present case by

$$X^2 = \sum_i \sum_j (f_{ij} - e_{ij})^2 / e_{ij}$$

$$= f_{\cdot\cdot}(f_{11}f_{22} - f_{12}f_{21})^2 / f_{1\cdot}\, f_{2\cdot}\, f_{\cdot 1} f_{\cdot 2}$$

$$Y^2 = 2\sum_i \sum_j f_{ij} \ln(f_{ij}/e_{ij}),$$

may be assumed to have a χ_1^2 distribution (chi-squared distribution with 1 degree of freedom), and hence provide a simple test of independence.

With small expected frequencies the distributions of X^2 and Y^2, which are discrete, diverge from χ_1^2, and the method of generation of the sample data becomes especially relevant. There are at least four possible situations that might have resulted in the figures given in Table 4.

1. Ten pieces of toast were spread with butter and 10 with margarine. These pieces were presented to the individual for his assessment and *he was told* that there were 10 of each.
2. As in 1, but the individual was *not* told how many there were of each type.
3. As in 2, but the 20 pieces of toast were chosen at random* from a pile containing both types of spread.
4. Pieces of toast chosen at random from the pile were presented to the individual one after another until he had stated that 10 had been spread with butter.

The differences in the situations lie in the numbers fixed by the schemes. In situation 1 each of $f_{\cdot 1}$, $f_{\cdot 2}$, $f_{1\cdot}$, and $f_{2\cdot}$ are fixed. In situation 2 only $f_{\cdot 1}$ and $f_{\cdot 2}$ are fixed. In situation 3 only $f_{\cdot\cdot}$ is fixed, and in situation 4 only f_{11} is fixed. A particularly clear discussion of the conceptual differences in these sampling schemes* has been given by Pearson [27].

The underlying distributions for these four situations are, respectively, the hypergeometric*, binomial*, multinomial*, and negative binomial*. The hypergeometric case has had most publicity, although genuine examples in which both sets of marginal totals are fixed are very rare. The individual probability of a particular configuration is

$$\frac{f_{1\cdot}!\, f_{2\cdot}!\, f_{\cdot 1}!\, f_{\cdot 2}!}{f_{\cdot\cdot}!\, f_{11}!\, f_{12}!\, f_{21}!\, f_{22}!}$$

Fisher's exact test* [8] consists of calculating the sum of the probability of thc observed configuration and all more extreme configurations. If this total probability is small, then the observed configuration is extreme and the hypothesis of independence is rejected. In the butter/margarine example this tail probability is (0.078 + 0.011 + 0. 001 + 0.000) = 0.090 and the hypothesis is accepted.

With larger numbers exact calculation is awkward and it is simpler to use the quite accurate χ_1^2 approximation due to Yates [38]. The Yates's correction* to X^2 involves calculating $X_y^2 = f_{\cdot\cdot}\{|f_{11}f_{22} - f_{12}f_{21}| - \frac{1}{2}f_{\cdot\cdot}\}^2 / f_{1\cdot}\, f_{2\cdot}\, f_{\cdot 1} f_{\cdot 2}$ (*see* CONTINUITY CORRECTION). In the present case $X_y^2 = 1.8$, corresponding to a one-sided χ_1^2 tail probability of 0.090, which agrees with the exact value.

Tocher [34] showed that this "hypergeometric" test is the uniformly most powerful (UMP) test in all situations, *provided* that it is augmented by randomization*. To understand the randomization idea, suppose that we wish to test the hypothesis of independence for the butter/margarine data at the 5% level. The figures given earlier show that the configuration $(7,3\,|\,3,7)$ has a tail probability of 9%, while that of $(8,2\,|\,2,8)$ has a tail probability of 1.2%. No configuration corresponds exactly to 5%.

Suppose that the data have arisen from situations 2 or 3. The UMP test requires us to reject the independence hypothesis (H_0) if we observe the configurations $(10,0\,|\,0,10)$, $(9,1\,|\,1,9)$ or $(8,2\,|\,2,8)$ and to accept H_0 if we observe $(5,5\,|\,5,5)$ or $(6,4\,|\,4,6)$. However, with $(7,3\,|\,3,7)$ we are required to accept H_0 on a proportion $(9-5)/(9-1.2) = 0.51$ of occasions on which this configuration occurs, and otherwise to reject H_0. Next we consider alternative approaches.

Suppose that just one set of marginal totals is fixed by the sampling method. Several authors [3, 13, 30] have demonstrated that the correction factor of $-\frac{1}{2}f_{..}$ in X_m^2 is inappropriate in this case. Recent work [4] suggests that a preferable correction is $-\frac{1}{2}h$, where h is the highest common factor of $f_{.1}$ and $f_{.2}$. The empirical study of Upton [36] suggests that while Cook's correction is vastly preferable to that of Yates, it rarely improves on the following scaled version of X^2:

$$X_s^2 = (f_{..} - 1)(f_{11}f_{22} - f_{21}f_{12})^2/(f_{1.}\, f_{2.}\, f_{.1} f_{.2}).$$

In the social sciences it is more usual for no marginal totals to be fixed. Regarding the overall total, $f_{..}$, as fixed, each f_{ij} is now an observation from a multinomial distribution*, and Yates' correction* $(-\frac{1}{2}f_{..})$ should be replaced by $-\frac{1}{2}$. See Pirie and Hamdan [29].

In the present example the effects of these modifications are to give the values $X_{\text{Cook}}^2 = 2.45, X_s^2 = 3.04$, and $X_{\text{P-H}}^2 = 3.12$ compared to the uncorrected $X^2 = 3.20$ and the Yates corrected $X_y^2 = 1.80$.

The fourth (negative binomial)* situation is discussed by Kudô and Tanimi [20].

MULTIWAY TABLES

The traditional method of analyzing multiply cross-classified data has been to collapse the multidimensional contingency table over all but two of the variables, and to follow this with the computation of some measure of association* between these variables. For an N-dimensional table this results in $N(N-1)/2$ separate analyses. The result is a very crude and possible fallacious picture of the linkages* between the variables [32, 37].

A preferable analysis is based on log-linear models*, and details are given in the main entry on CONTINGENCY TABLES. The idea is to find as simple a model as possible that describes the pattern of cell counts in the multiway table. The description is based on linear combinations of the logarithms of odds* or ratios of odds. Recall that the parameter α was an odds ratio (p_{11} to p_{12} divided by p_{21} to p_{22}). The class of log-linear models is extensive and enables one to identify associations not just between a pair of variables but between three or more variables. Introductory accounts are given by Fienberg [7], Haberman [17], and Upton [35], with a more advanced data analytic treatment being provided by Bishop et al. [1]. The theory is developed in detail in the books by Haberman [16], Plackett [30] and Goodman [10].

The particular case of a multiway table arising from a repeated measurement design* is discussed by Halperin and Nehrke [18]. The case of a table arising from stratified random sampling* is discussed by Imrey et al. [19].

COMPUTER PROGRAMS

The most familiar software package for the handling of data is probably SPSS [26]. The most relevant command within this package is the CROSSTABS command, which constructs a multidimensional contingency table from a standard coded data file. The STATISTICS option of the CROSSTABS command produces the values of a huge number of measures of association without comment on their usefulness. Very pertinent comments on measures of association in general are given by Goodman and Kruskal [12].

The SPSS package expects precoded data. Another package, GURU which is described by Riedel and Dodson [31], is designed to allow greater flexibility in handling responses to open-ended techniques.

Several packages have been designed specifically for the analysis of contingency tables. Of these, the simplest is probably ECTA [11]. More flexible, but therefore more complex, are CTAB [15], GLIM* [25], and MULTIQUAL [2]. There are doubtless others, and choice will be dictated largely by availability. ECTA is based upon the iterative scaling routine suggested by Deming and Stephen [5]. This routine is so simple that it can easily be incorporated into tailor-made programs for one-off data analyses, or recourse can be made to the FORTRAN version of the algorithm provided by Haberman [14]. See STATISTICAL SOFTWARE.

References

[1] Bishop, Y. M. M., Fienberg, S. E., and Holland, P. W. (1975). *Discrete Multivariate Analysis: Theory and Practice*. MIT Press, Cambridge, Mass.

[2] Bock, R. D. and Yates, G. (1973). *MULTIQUAL: Log-Linear Analysis of Nominal or Ordinal Qualitative Data by the Method of Maximum Likelihood*. International Education Services, Chicago.

[3] Camilli, G. and Hopkins, K. D. (1978). *Psychol. Bull.* **80**, 163 – 167.

[4] Cook, I. T. (1980). On the continuity correction for bivariate discrete distributions. In preparation.

[5] Deming, W. E. and Stephen, F. F. (1940). *Ann. Math. Statist.*, **11**, 427–444.

[6] Durkheim, E. (1951 [1897]). *Suicide*, J. A. Spaulding and G. Simpson, trans. Free Press, Glencoe, Ill.

[7] Fienberg, S. E. (1980). *The Analysis of Cross-Classified Categorical Data*, 2nd ed. MIT Press, Cambridge, Mass.

[8] Fisher, R. A. (1970 [1925]). *Statistical Methods for Research Workers*, 14th ed. Hafner, New York.

[9] Gani, J. (1978). *Commun. Statist.*, **A7**, 905–915.

[10] Goodman, L. A. (1979). *Analysing Qualitative/Categorical Data: Log-Linear Models and Latent-Structure Analysis*, Abt Associates, Cambridge, Mass.

[11] Goodman, L. A. and Fay, R. E. (1975). ECTA Program—Description for Users. Dept. of Statistics, University of Chicago, Chicago.

[12] Goodman, L. A. and Kruskal, W. E. (1979). *Measures of Association for Cross Classifications*. Springer-Verlag, New York.

[13] Grizzle, J. E. (1967). *Amer. Statist.*, **21**(4), 28–33.

[14] Haberman, S. J. (1972). *Appl. Statist.*, **21**, 218–225.

[15] Haberman, S. J. (1973). *CTAB: Analysis of Multidimensional Contingency Tables by Log-Linear Models: User's Guide*. International Educational Services, Chicago.

[16] Haberman, S. J. (1974). *The Analysis of Frequency Data*. University of Chicago Press, Chicago.

[17] Haberman, S. J. (1978). *Analysis of Qualitative Data*, Vol. 1: *Introductory Topics*. Academic Press, New York.

[18] Halperin, S. and Nehrke, M. F. (1976). *Exper. Aging Research*, **2**, 105–118.

[19] Imrey, P. B., Sobel, E., and Francis, K. E. (1979). *Commun. Statist. A*, **8**(7), 653–670.

[20] Kudô, A. and Tanimi, T. (1978). *Commun. Statist. A*, **7**(10), 977–986.

[21] Lawal, H. B. (1980). *Appl. Statist.* **29**, 292–298.

[22] Lawal, H. B. and Upton, G. J. G. (1980). *Biometrika*, **67**, 447–453.

[23] McCullagh, P. (1980). *J. R. Statist. Soc. B*, **42**, 109–127.

[24] National Opinion Research Center (1975). *Codebook for the Spring 1975 General Social Survey*. National Opinion Research Center, University of Chicago, Chicago.

[25] Nelder, S. A. (1975). *Appl. Statist.*, **24**, 259–261.

[26] Nie, N. H., Hull, C. H., Jenkins, J. G., Steinbrenner, K., and Bent, D. H. (1975). *SPSS: Statistical Package for the Social Sciences*, 2nd ed. McGraw-Hill, New York.

[27] Pearson, E. S. (1947). *Biometrika*, **34**, 139 – 167.

[28] Pearson, K. (1900). *Philos. Mag.*, 5th Ser., **50**, 157–175.

[29] Pirie, W. R. and Hamdan, M. A. (1972). *Biometrics*, **28**, 693 – 701.

[30] Plackett, R. L. (1974). *The Analysis of Categorical Data*. Charles Griffin, London.

[31] Riedel, J. A. and Dodson, J. D. (1977). *Educ. Psychol. Meas.*, **37**, 779–782.

[32] Simpson, E. H. (1951). *J. R. Statist. Soc. B*, **13**, 238 – 241.

[33] Stevens, S. S. (1968). *Science*, **161**, 849–856.

[34] Tocher, K. D. (1950). *Biometrika*, **37**, 130–144.

[35] Upton, G. J. G. (1978). *The Analysis of Cross-Tabulated Data*. Wiley, New York.

[36] Upton, G. J. G. (1981). *J. R. Statist. Soc. A*, **144**.

[37] Whittemore, A. S. (1978). *J. R. Statist. Soc. B*, **40**, 328–340.

[38] Yates, F. (1934). *J. R. Statist. Soc. Suppl.*, **1**, 217–223.

[39] Yule, G. U. (1900). *Philos. Trans. Roy. Soc. Lond. A*, **194**, 257–319.

(ASSOCIATION, MEASURES OF
CHI-SQUARE TESTS
CONTINGENCY TABLES
DISTRIBUTION-FREE METHODS
FISHER'S EXACT TEST)

G. J. G. UPTON

CAUCHY, AUGUSTIN-LOUIS

Born: August 21, 1789, in Paris, France.

Died: May 22, 1857, in Sceaux, France.

Contributed to: mathematics, mathematical physics, celestial mechanics, probability, mathematical statistics.

The contributions of this great mathematician to mathematical statistics occur in connection with the problem of estimation* (using modern notation) of the $(r \times 1)$ vector $\boldsymbol{\beta} = \{\beta_i\}$ from a vector $\boldsymbol{Y}$ of n observations in the classical linear model*: $\boldsymbol{Y} = X\boldsymbol{\beta} + \boldsymbol{\epsilon}$, where $\boldsymbol{\epsilon} = \{\epsilon_i\}$ is an error vector, and $X = \{x_{ij}\}$, a known fixed $(n \times r)$ matrix, $n \geqslant r$, of full column rank r. The nineteenth-century mathematicians (Gauss*, Laplace*, Bienaymé*, and Cauchy) regarded this problem as one of finding an $(r \times n)$ matrix $K = \{k_{ij}\}$ (or as they would put it, a system of "multipliers") such that

$$KX = I, \tag{1}$$

so that consequently $\boldsymbol{\beta}$ is estimated by (the linear estimate) $\bar{\boldsymbol{\beta}} = \{\bar{\beta}_i\}$, where $\bar{\boldsymbol{\beta}} = K\boldsymbol{Y} = \boldsymbol{\beta} + K\boldsymbol{\epsilon}$, K being chosen under the constraint (1), in some optimal way. The least-squares* choice of K— that minimizing $\boldsymbol{\epsilon}^T\boldsymbol{\epsilon}$ and leading to $K = (X^TX)^{-1}X^T$—had been justified by Gauss on probabilistic grounds, in 1809 as the most probable (maximum likelihood*) estimator if $\boldsymbol{\epsilon}$ is $N(\mathbf{0}, \sigma^2 I)$, and in 1821 as the minimum variance* (best linear unbiased) estimator if $E\boldsymbol{\epsilon} = \mathbf{0}, \operatorname{var} \boldsymbol{\epsilon} = \sigma^2 I$. Intermediate to these justifications of Gauss were approaches of Laplace in 1812, which assume that independent, identically distributed (i.i.d.) ϵ_i are described by a symmetric density confined to a finite interval, $[-g, g]$. Laplace shows (heuristically, and at least for $r = 1$) that the standardized random variable

$$(\bar{\beta}_i - \beta_i)\Big/\left(\sigma^2 \sum_{h=1}^{n} k_{ih}^2\right)^{1/2} \equiv \sum_{h=1}^{n} k_{ih}\epsilon_h \Big/\left(\sigma^2 \sum_{h=1}^{n} k_{ih}^2\right)^{1/2} \to N(0,1)$$

as $n \to \infty$, so that, for large n, a symmetric, fixed-probability-level interval about the origin for the error $\bar{\beta}_i - \beta_i$ is of the form $\pm z_0(\sigma^2\sum_{h=1}^n k_{ih}^2)^{1/2}$, z_0 constant, whose length is therefore minimized by choosing K to minimize under constraint (1), the quantity $\sigma^2\sum_{h=1}^n k_{ih}^2$. Since this last is var $\bar{\beta}_i$, simultaneous minimization for all $i = 1, \ldots, r$ leads again to the least-squares choice of K. (*Compare* CENTRAL LIMIT THEOREM.)

Cauchy had started to write on the choice of K in 1853, one of his intentions being to show that there are situations where the least-squares choice is not optimal. Bienaymé resented the apparent criticism of Laplace, and there ensued a heated controversy between them in the meetings of the Academy of Sciences, and on the pages of its journal, the *Comptes Rendus*, for that year [4]. Within it, Cauchy shows, using heuristic and involved reasoning, that if the i.i.d. ϵ_i, $i = 1, \ldots, n$, have characteristic function* $\exp(-c|\theta|^\alpha)$, $c, \alpha > 0$, then the "most probable" estimate $\bar{\boldsymbol{\beta}}$, in a certain sense, comes about from choosing K to satisfy (1) and to simultaneously minimize $\sum_h |k_{ih}|^\alpha$ for each i. This marks the first time in the history of probability that the (symmetric) stable laws* appear; Cauchy's "most probable" estimate $\bar{\boldsymbol{\beta}}$ has been rediscovered recently in econometric* contexts, as that which minimizes the scale parameter of the distribution of each of the $(\bar{\beta}_i - \beta_i) \equiv \sum_{h=1}^n k_{ih}\epsilon_h$ when the common distribution of the ϵ_i is such a symmetric stable law.

Toward the close of the controversy, Cauchy considers the same asymptotic problem as Laplace (see above), i.e., the distribution of a linear function of residuals $\sum_{h=1}^n \lambda_h\epsilon_h$ as $n \to \infty$ under Laplace's distributional assumptions on the ϵ_i, but allowing the λ_i's to depend on n, using characteristic functions, inversion formulas*, and careful estimates of the various integrals that occur. All these ideas were rather innovative for their time, but, restricted to a small amount of space, Cauchy's presentation was a very concentrated sketch; and since it occurred in the context of his dispute with Bienaymé, and within the framework of least squares, it is not surprising that this rigorous work of Cauchy on the central limit theorem was not understood, and was passed over at the time. In 1892, I. V. Sleshinsky [7] took up Cauchy's sketch, filled in the proofs, adding new steps as necessary, to produce by characteristic function methods, a first rigorously proved, if somewhat restricted, version of the central limit theorem.

A final contribution to mathematical statistics by Cauchy, alluding also to the nonoptimality of least squares, arises from noting that (for fixed n) if the ϵ_i again satisfy Laplace's distributional assumptions, then the maximum attainable error of estimate in $\bar{\beta}_i - \beta_i$ is $g\sum_{h=1}^{n}|k_{ih}|$, which is therefore minimized for each i by choosing K to satisfy (1), and to minimize $\sum_{h=1}^{n}|k_{ih}|$ for each i. Cauchy gives a correct solution for the appropriate K in the case $r = 1$, and proposes, without proof, a correct solution in general [6]. Note that this K is the same as that needed to resolve the case $\alpha = 1$ in the above-described stable-laws setting.

Cauchy made certain other contributions to the theory of the classical linear model from an *interpolational* (error-theoretic) rather than *probabilistic* (as hitherto described) point of view. An early contribution along these lines is an attempt, dating from 1814, although not published until 1824 and 1831 [1, 2], to extend by geometric reasoning to the case $r = 2$ Laplace's treatment of the case $r = 1$, with a view to obtaining the estimate of $\boldsymbol{\beta}$ from the criterion $\sum_{i=1}^{n}|\epsilon_i| = \min.$, which procedure has been revived in modern times since the estimate is robust* although nonlinear. In 1835, Cauchy formulated and ingeniously resolved a then-new aspect of the interpolation view of the classical linear model, of how many β_i, $i = 1, 2, \ldots,$ to fit in succession until an "adequate" fit to the data obtains, proceeding in such a way that the estimates for $\beta_1, \ldots, \beta_r$ are unaffected in going to that for β_{r+1}. This idea was revived within the context of the controversy of 1853; and is thence directly linked [4], via Bienaymé, with the idea of successive orthogonalization of the columns of the design matrix X, and its connection with least-squares estimates, in the work of Chebyshev, at least in a polynomial interpolation* setting.

Although the term "Cauchy distribution"* arises from the 1853 controversy, in which it occupies a prominent place, it is known to have occurred in earlier writings of Poisson* [8]. The equation $f(x) + f(y) = f(x + y)$, variants of which are encountered in the theory of Markov processes*, and of the stable laws, was shown by Cauchy to have general solution, under prior assumption of continuity of f, $f(x) = cx$; and is often called Cauchy's functional equation (*see* FUNCTIONAL EQUATIONS), although it was known earlier also (to d'Alembert, for example).

References

[1] Cauchy, A. *Oeuvres complètes d'Augustin Cauchy*. Gauthier-Villars, Paris. (Collected works; volumes in several series, with various dates of publication.)

[2] Freudenthal, H. (1971). In *Dictionary of Scientific Biography*, Vol. 3, C. C. Gillispie, ed. Scribner's, New York, pp. 131–148. (Cauchy's life and scientific work surveyed in their entirety; contains a brief sketch of his "error theory.")

[3] Gnedenko, B. V. and Sheynin, O. B. (1978). In *Matematika XIX veka* [*Mathematics of the 19th Century*]. Nauka, Moscow, pp. 184–240. (Pages 205–207 of this Russian-language account sketch some of Cauchy's contributions.)

[4] Heyde, C. C. and Seneta, E. (1977). *I. J. Bienaymé: Statistical Theory Anticipated*. Springer-Verlag, New York. (Chapter 4 contains an extensive modern analysis of Cauchy's work and its consequences in reference to the classical linear model and the central limit theorem, with particular emphasis on 1853.)

[5] Seal, H. L. (1967). *Biometrika*, **54**, 1–24. (A well-known account endowing the historical development of least-squares theory with somewhat fictitious continuity in regard to the roles of Bienaymé and Cauchy.)

[6] Seneta, E. (1976). *Ann. Soc. Sci. Brux.*, **90**, 229–235.

[7] Sleshinsky [Sleschinsky, Sleszyński], I. V. (1892). *Zap. mat. otd. novoross. obshch. estestvoispyt.* (*Odessa*), **14**, 201–264 (in Russian).

[8] Stigler, S. M. (1974). *Biometrika*, **61**, 375–380.

[9] Valson, C. A. (1868). *La Vie et les travaux de Baron Cauchy*. Gauthier-Villars, Paris. (Well-known hagiography.)

(CAUCHY DISTRIBUTION
GENERAL LINEAR MODEL

INTERPOLATION
LIMIT THEOREMS, CENTRAL
LINEAR ESTIMATORS
LINEAR LEAST SQUARES
MARKOV PROCESSES
NONLINEAR ESTIMATION
ROBUSTNESS
SYMMETRIC STABLE LAWS)

E. SENETA

CAUCHY DISTRIBUTION

The two-parameter Cauchy distribution is given by the density

$$f(x) = \left[\pi\lambda \left(1 + \left(\frac{x-\theta}{\lambda} \right)^2 \right) \right]^{-1}$$

or by the CDF

$$F(x) = \frac{1}{2} + \frac{1}{\pi} \tan^{-1}\left(\frac{x-\theta}{\lambda} \right) \qquad (\lambda > 0).$$

It is considered as a possible model whenever one needs a density function with heavier tails than the normal distribution allows. The distribution does not possess (finite) moments or cumulants*. However, the order statistics* $X_{(r:n)}$ possess finite expected values $\mu_{(r:n)} = E(X_{(r:n)})$ except for $r = 1$ and $r = n$. It is a unimodal, symmetric distribution, stable* and infinitely divisible*. The characteristic function is given by $\phi(t) = \exp(it\theta - |t|\lambda)$. There are numerous characterizations* of this distribution. The basic one, based on the infinite divisibility* of this distribution, is that a random variable X has the Cauchy distribution iff the mean of the sample $X_1, \ldots, X_n$ from this distribution has a Cauchy distribution. Another characterization is based on the property of conditional expectations: A random variable X has the standard Cauchy distribution (with the CDF $F(x) = \frac{1}{2} + (1/\pi)$ arc tan x, and the PDF $f(x) = [\pi(1 + x^2)]^{-1}$ iff for every $y < z$ there exists the conditional expectation $E[X \mid y < X \leqslant z]$ given by

$$\log\left(\frac{1+z^2}{1+y^2} \right) \times \left[\arctan\left(\frac{z-y}{1+zy} \right) \right]^{-1}.$$

Yet another, based on the property that if X is Cauchy and $X = \tan Z$, then Z is uniformly distributed on $(-\frac{1}{2}\pi, \frac{1}{2}\pi)$, is as follows. If for some real a such that a is not $\tan K$, where K is a rational multiple of π, $(1 + aX)/(a - X)$ has the same distribution as X, then X is a "standard" Cauchy random variable. Moreover, identical distribution of X and $(X - X^{-1})/2$ characterizes the Cauchy distribution among absolutely continuous distributions [1].

Additional characterizations are given in Johnson and Kotz [3, p. 162], and Kagan et al. [4, p. 488].

The density f is complete* with respect to θ. A confidence interval* for θ (when λ is known) is given by

$$\Pr\left[\hat{\theta} - a_{1-(\alpha/2)} \frac{\lambda}{\sqrt{n}} \leqslant \theta \leqslant \hat{\theta} - a_{\alpha/2} \frac{\lambda}{\sqrt{n}} \right] = 1 - \alpha,$$

where $\hat{\theta}$ is the maximum likelihood estimator* of θ and 100γ percentage points a_γ for selected values of γ are given in Haas et al. [2] for $n = 5(5)30(10)50(25)100$.

References

[1] Arnold, B. (1979). *Aust. J. Statist.*, **21**(2), 166–169.

[2] Haas, G., Bain, L., and Antle, C. (1970). *Biometrika*, **57**, 403–408.

[3] Johnson, N. L. and Kotz, S. (1970). *Continuous Distributions*, Vol. 1. Wiley, New York. (Contains extensive bibliography up to 1969.)

[4] Kagan, A. M., Linnik, Yu. V., and Rao, C. R. (1973). *Characterization Problems of Mathematical Statistics*. Wiley, New York.

[5] Kotlarski, I. I. (1972). *Sankhyā*, **34**, 461–466.

CAUCHY–SCHWARZ INEQUALITY

The Cauchy–Schwarz inequality is useful in various combinatorial and probabilistic calculations.

For real sequences, let $\mathbf{a} = (a_1, a_2, \ldots, a_n)$ and $\mathbf{b} = (b_1, \ldots, b_n)$ be two sequences of real numbers. Then

$$\left(\sum_{i=1}^{n} a_i^2 \right) \left(\sum_{i=1}^{n} b_i^2 \right) \geqslant \left(\sum_{i=1}^{n} a_i b_i \right)^2.$$

Equality holds if and only if the sequences are linearly dependent, i.e., there exists a real constant c such that

$$a_i = cb_i \qquad \text{for } i = 1, \ldots, n.$$

For real-valued square integrable functions, let $g(\mathbf{x})$ and $h(\mathbf{x})$ be square integrable with respect to a given measure P over a set S in R_n. Then

$$\left(\int_S gh\, dP\right)^2 \leqslant \int_S g^2\, dP \cdot \int_S h^2\, dP.$$

For random variables, let (X, Y) be a bivariate random variable. Then

$$E\left[|XY|^2\right] \leqslant E\left[|X|^2\right]E\left[|Y|^2\right]$$

provided that the expectation on the left-hand side exists.

(HÖLDER'S INEQUALITY)

CAUCHY SEQUENCE *See* EXCHANGEABILITY

CAUSATION

That correlation* is not causation is perhaps the first thing that must be said. If, in sampling a population, we consistently find a larger proportion of sufferers from lung cancer among smokers than we do among nonsmokers, we are not entitled to conclude that smoking predisposes that population toward lung cancer. It may be that persons whose genetic constitution predisposes them to smoke also predisposes them to lung cancer or even that incipient lung cancer causes people to smoke. It also may be that there is a subpopulation of the original one sampled within which smoking does cause lung cancer, while for the majority of the original population, smoking has no effect on the incidence of lung cancer. Even if, in response to the suggestion of common genetic predisposition, it is pointed out that the genetic composition of the population must have remained largely unchanged over the past 50 years, while during that 50 years both the incidence of lung cancer and the incidence of smoking have increased enormously, there have been many other changes over the past 50 years and included among these may be an increased incidence of some pollution factor that is necessary to the development of lung cancer among those genetically predisposed to it.

It is not only true that correlation does not imply causation; it is also true that causation does not necessarily imply correlation as the latter is usually measured. If x and y are connected by the relationship

$$y = \sin \pi x$$

and x varies uniformly over its range from -1 to $+1$, y will then vary between -1 and 1 and the correlation between x and y will be 0. More complicated examples show that this can happen even when x and y, separately, are normally distributed.

We refer to articles in general encyclopedias for discussion of the manifold senses that the word "cause" has had in the history of philosophy and theology. We are here concerned with the way in which the word "cause" is used in scientific contexts. In scientific work we attempt to represent reality by means of models, usually of a mathematical kind. In these models some variables may be thought of as having values that can be chosen arbitrarily. Then the values of other variables appearing in the model can be derived, in a deterministic or in a statistical sense, from the values of these "input variables."If the model is judged correct, the input variables may be said to "cause" the variables whose values are derived.

For the causation concept to be fully applicable, the model in question must have some degree of generality of application. This is why the concept of causation has its clearest application in the physical sciences, where models representing the behavior of atoms, molecules, electrons, etc., are applicable over a very wide range of conditions of space and time. In the social sciences models tend to have a much more restricted domain of application and here, if circumstances change so that a given model is no longer applicable, we may be in doubt whether to say that an input variable in the model ceased to be the cause of a given output variable, or whether some aspect of the surrounding circumstances that made the model applicable should have been regarded as the cause. If, for example, it were discovered that a particular constituent of tobacco was responsible for precipitating lung cancer, and if it were possible to develop a form of tobacco from which this constituent was absent so that those who smoked the purified tobacco did not incur the risk of lung cancer, the proposition "smoking causes lung cancer" would cease to be true. It would be a mere dispute over terminology to argue whether one should say that smoking never has caused lung cancer, but only smoking tobacco of the old kind, or whether to express the facts in the way already indicated.

The fact that a clear concept of causation requires widely applicable models of a class of phenomena implies that by far the surest, if not the only way to establish correlation is by a series of experiments in which situations to which the model is supposed to apply are set up, and the relevant input variables are deliberately modified by the experimenter. If over a long series of repetitions under widely different circumstances changes in the input variables are found to be associated with changes in the values or in the distribution of the output variables, then causal relations can be said to be established. In other cases we can speak of causal relations only insofar as we consider we have evidence which in our view amounts to the same thing as such a series of repeated experiments. The many difficulties involved in experiments with human beings make it especially difficult to establish causal relations in this area. The need for experimental manipulation of some kind is what underlies experiments, for example, in which some dogs are made to smoke while others are not and the incidence of lung cancer in these animals is observed. Clearly, in relating the results to humans assumptions of similarity of physiology must be made.

In econometrics*, for example, a demonstration that a particular set of equations fits well to a given set of economic time series* does not by itself constitute evidence of causal connection. We need, in addition, assurance that the equations of the model represent, at least approximately, credible aspects of economic behavior, our judgment on these being based on our everyday experiences of interactions with our fellow human beings.

Attempts have been made to define concepts of causation without reference to the experimental modification of input variables. Perhaps the most sophisticated of these is associated with the name of Wiener and applied to econometrics* by Granger. According to Granger [2], y is causing x if we are better able to predict the value of x using the value of y than we are using all the information we have at our disposal other than y. Feedback occurs if x causes y and y causes x. For the applicability of Granger's definition, however, assumptions concerning stationarity and other properties of the stochastic series* involved are necessary and the validity of these may be difficult to establish. Path analysis* has also been thought at times to constitute a method for establishing causation. Any path analysis must, however, be based on a recursive model of some kind which in principle needs justification by experimental or equivalent means.

In the experimental verification of a supposed causal relation, in the common case when the perfectly uniform experimental units are not available, it is necessary to randomize the allocation of units between the various values or states of the input variable. It is sometimes thought that the main purpose of randomization* is to provide an objective probability distribution for a test of significance*. Its crucial function is in fact, to assure, with high probability, that differences in the output variables associated with changes in the input variables really are due to these changes and not to other factors. *See* DESIGN OF EXPERIMENTS. The results must be subjected to a test of significance to check that the differences between the output variables corresponding to different values of input variables are not such as could reasonably arise from the variability of experimental units. And it must be remembered that a single test of significance cannot by itself be considered to establish the reality of a phenomenon. For, as R. A. Fisher* says [1], "the 'one change in a million' will undoubtedly occur with no less and no more than its appropriate frequency, however, surprised we may be that it should occur to *us*. In relation to the test of significance we may say that a phenomenon is experimentally demonstrable when we know how to conduct an experiment that will rarely fail to give us a statistically significant result."

Literature

The Grammar of Science (Meridian Books, New York, 1957) by K. Pearson* is a classic, by a giant of statistics. More recent general discussion of causality in science is to be found in H. Feigl and M. Brodbeck, eds. *Readings in the Philosophy of Science* (Appleton-Century-Crofts, New York, 1953, Secs. V, VII, and VIII). General surveys of the notion of causality are to be found, among others, in *Encyclopaedia Britannica* (1972 edition; not in the 15th edition) and (in French) in the *Encyclopedia Universalis*, Paris, 1969. Causality in econometrics and social science, in relation to statistical issues, is discussed in C. W. J. Granger and P. Newbold, *Forecasting Economic Time Series* [2], in H. Simon, *Causal Ordering and Identifiability*, and in Wm. C. Hood and T. C. Koopmans, eds., *Studies in Econometric Method* (Cowles Comm. Monogr. No. 14, Wiley, New York, 1953).

Scientific models in common use in the seventeenth and eighteenth centuries were typically mechanistically deterministic. By contrast, scientific models in the past 100 years have typically been stochastic*. The transition from deterministic to stochastic models has given rise to much discussion in which the deterministic "causal disposition" is contrasted with the stochastic "chance propensity" (*see* CHANCE). For a very clear discussion of the usage of "causal" in this context, see the article by Mario Bunge in Harper and Hooker, eds., *Foundations of Probability Theory, Statistical Inference, and Statistical Theories of Science*, Vol. 3, (D. Reidel, Dordrecht, Holland, 1976, pp. 17–33).

References

[1] Fisher, R. A. (1951). *The Design of Experiments* (6th edn.) Oliver and Boyd, Edinburgh, U. K. (See the discussion on significance tests, pp. 13–15.)

[2] Granger, C. W. J. and Newbold, P. (1977). *Forecasting Economic Time Series*, Academic, New York.

(CHANCE
CORRELATION
DESIGN AND ANALYSIS OF EXPERIMENTS
RANDOMIZATION)

G. A. BARNARD

c CHARTS

A term used in quality control for a chart in which the number *c* of defective (or "nonconforming") items in each of a sequence of samples is plotted against the serial number of the sample. They are also called "number-defective charts." If the samples are not all of the same size, this should be clearly indicated.

It is often assumed that a Poisson distribution* is appropriate for *c*, and control limits based on this assumption are inserted on the chart, which is then a special kind of control chart*.

(QUALITY CONTROL)

CENSORED DATA

DEFINITIONS AND TYPES OF CENSORING

A data sample is said to be censored when, either by accident or design, the value of the random variable under investigation is unobserved for some of the items in the sample. A *censored* observation is distinct from a *missing* observation (*see* MISSING DATA) in that the order of the censored observation relative to some of the uncensored observations is known and conveys information regarding the distribution being sampled. Censored data arise most often when the random variable of interest is life* or survival* time, and our discussion generally assumes this context.

In medical studies of the effect of various treatment methods on the survival of a group of patients or laboratory animals, the eventual time to death of some of the subjects may not be observable for many reasons, e.g.: they move away, they are eaten by the other subjects, or the allocated time for the study elapses prior their death. The text by Gross and Clark [6] is a useful reference for the treatment of censored data as they arise in biomedical survival studies.

In life testing conducted in the physical and engineering sciences, censoring is frequently used intentionally to shorten the elapsed time to test completion. The book by Nelson [18] gives a thorough summary of the treatment of censored data as they arise in engineering applications.

Censoring is said to be *on the right* when the item or subject is removed from test or observation prior to failure or death. Censoring on the left means that the death or failure occurred prior to some designated "left-censoring" time. Left censoring might be employed to burn off "duds," i.e., early failing items atypical of the population targeted for study. Left censoring occurs in other than life-testing contexts when the lower limit of resolution of a measuring device prevents the measurement of a fraction of the smallest ordered values in a sample. *See also* CENSORING, LEFT; CENSORING, RIGHT.

In *type I* censoring on the right, testing is suspended when a preestablished lifetime is reached. If this time is the same for all units, the sample is *single-censored*. If a different censoring time is used for each unit, the sample is *multiple*-censored.

In type I censoring *on the left*, life is observed only if it exceeds a fixed threshold time. It is characteristic of type I censoring that the number of failures is a random variable.

In *type II* censoring (or truncation*) the number of failures is fixed and the test time is random. With type II censoring on the right, testing ceases at the occurrence of the rth-ordered failure time in a sample of n items ($r < n$). Correspondingly, for type II censoring on the left the first observed lifetime is that of the qth-ordered failure ($q < n$). *See also* TRUNCATION.

In *progressive censoring* a fixed number of surviving units is removed from test upon the occurrence of each successive failure. This is conceptually equivalent to the practice known as sudden-death testing, wherein the test apparatus simultaneously accommodates several test specimens and it is administratively convenient to remove them all when the first one fails.

Unplanned multiple censoring occurs when removal from test is necessitated by the occurrence of an alternative mode of failure that competes with the mode of interest to end the life of the specimen. In the fatigue testing of rolling bearings the cracking of an improperly heat-treated ring is considered an extraneous failure mode. In the analysis of competing mode data it is assumed that each item has a "latent" life at which it would fail by each of the different possible causes if that cause were the only one acting.

The resultant life of the item is the minimum of these latent lives. It is usual, although not always reasonable, to assume that the latent lives are independently distributed random variables. Birnbaum [4] has written a useful expository report on competing risk* analysis. Moeschberger and David [16] give a generalized methodology for maximum likelihood estimation of the parameters of a primary life distribution in the presence of a number of dependent or independent competing risks. *See also* COMPETING RISKS.

ESTIMATION

A nonparametric* technique for estimating the cumulative distribution function* (CDF) in the presence of multiple censoring was developed by Kaplan and Meier [8]. It is called a product limit estimate* and is applicable when the latent times to failure and censoring are independent. The Kaplan–Meier* (K-M) method is a specialization of the actuarial method [6] for the analysis of *interval* data*.

The calculations of the K-M method are illustrated in Table 1. Column 1 of Table 1 gives the life order of each of the six items in the sample. Column 2 contains the lives at failure (F) or suspension (S). Column 3 contains the "reverse ranks," i.e., the life-order numbers of column 1 listed in reverse order. Column 4 contains the recursively calculated estimates of the reliability or survival probability at those times at which failures occurred. At the ith failure the reliability R_i is estimated as

$$\hat{R}_i = [(r_i - 1)/r_i] \cdot \hat{R}_{i-1},$$

where $\hat{R}_0 \equiv 1$.

The CDF F_i is estimated as the complement of the reliability function*. The variance of $\hat{F}_i$ may also be estimated recursively (see Gross and Clark [6]).

To investigate appropriate parametric forms for the CDF, one could plot $\hat{F}_i$ against t_i on different types of probability paper. A disadvantage of the K-M method is that if the last point in the sample is a failure, the associated estimate $\hat{F}_i = 1.0$ will not be plottable on most probability papers. *See also* KAPLAN–MEIER ESTIMATOR.

The hazard* plotting technique due to Nelson [17] was developed primarily for graphical estimation* but can, like the K-M method, be used just to estimate the CDF at the failure times. These calculations are also illustrated in Table 1. In column 6 the "hazard" h_i corresponding to each failure time is computed as the reciprocal of the associated reverse rank. Column 7 contains the cumulative hazard H_i computed as the sum of the hazards for the failure at time t_i and all previous failure times. The CDF at the failure times is estimated from

$$\hat{F}_i = 1 - \exp(-H_i).$$

A method similar to the K-M method is termed the Herd–Johnson method by Nelson [18]. It differs from the K-M method in using a slightly different form for the recursive calculation of the reliability estimate:

$$\hat{R}_i = r_i/(r_i + 1) \cdot \hat{R}_{i-1}.$$

The H-J estimates of the CDF are shown for comparison with the other two methods in Table 1.

Parametric estimation methods for censored data include estimates constructed as

Table 1 Comparison of Nonparametric Estimates of CDF from Multicensored Samples

			Kaplan–Meier		Hazard Plotting			Herd–Johnson	
Life Order Number	Time t_i	Reverse Rank r_i	$\hat{R}_i$	$\hat{F}_i = 1 - \hat{R}_i$	$h_i = 1/r_i$	$h_i = \sum H_i$	$\hat{F}_i = 1 - e^{-h_i}$	$\hat{R}_i$	$\hat{F}_i = 1 - \hat{R}_i$
1	31.7 F	6	0.833	0.167	0.167	0.167	0.154	0.857	0.143
2	65.0 S	5	—	—	—	—	—	—	—
3	75.0 S	4	—	—	—	—	—	—	—
4	88.3 F	3	0.555	0.445	0.333	0.500	0.393	0.643	0.357
5	105.8 S	2	—	—	—	—	—	—	—
6	130.0 F	1	0.	1.00	1.000	1.500	0.777	0.322	0.678

linear combinations of selected sample percentiles. These estimates are, of course, applicable only if the censoring pattern leaves the relevant percentiles unambiguously defined. They are therefore inapplicable to multicensored data. Best linear unbiased estimates (BLUE)* are a special case of these estimates. (see Lloyd in ref. 21). The coefficients of BLUE estimators for type II right-, and in some cases also left-censored samples of various sizes have been developed for a number of distributions, including the exponential*, normal*, log-normal*, and extreme-value* distributions (see refs. 2, 15, and 18).

The method of maximum likelihood* is probably the most widely used technique for parametric estimation with censored samples. It is applicable to all types of censoring, asymptotically efficient, and generally as good or better than other methods, even for small samples. As noted, maximum likelihood is sufficiently general to account for cases wherein the censoring and failure times are stochastically dependent.

For a censored sample having a possibly vector-valued parameter θ, PDF $f(t;\theta)$, and CDF $F(t;\theta)$, an item failing at life t_i contributes a multiplicative term $f(t_i;\theta)$ to the likelihood function*. Assuming independence of failure and censoring times, an item censored at time y_i contributes a term $1 - F(y_i;\theta)$. The likelihood function for a sample of size n having a total of r failures is of the form

$$L = c \cdot \prod_{i}^{r} f(t_i;\theta) \cdot \prod_{i}^{n-r} [1 - F(y_i;\theta)],$$

where c is a constant which, in the case of random censoring*, will depend on the parameters of the censoring distribution.

The ML estimate $\hat{\theta}$ is the value of θ for which L (or $\log L$) is a maximum, irrespective of the kind of censoring. The distribution of the estimator and its large sample variance do, however, depend on the type and amount of censoring. For type II right-censored samples from the single-parameter exponential distribution, the ML estimator was found by Epstein and Sobel [5] to be

$$\hat{\theta} = \left[\sum_{i=1}^{r} t_i + (n-r)t_r \right] / r.$$

This estimator is a linear combination of the ordered sample values and in fact is identical to the BLUE estimator. It is generally true for the exponential distribution that irrespective of how the censoring is performed, the ML estimate is calculated as the total time on test divided by the number of failures. Estimation of the parameters of the gamma*, two- and three-parameter Weibull*, normal and log-normal distributions under various types of censoring is considered in the literature (see the *Bibliography*). In general, the likelihood equations must be solved numerically to determine the ML estimates.

INFERENCE FROM CENSORED SAMPLES

For percentile and BLUE estimates, their approximate normality and estimated variance based on the covariance of pairs of order statistics are used to set approximate confidence intervals* for the estimated parameters. The asymptotic joint normality* and limiting covariance matrix* of the ML estimates is often used for setting confidence limits based on the ML estimates.

The limiting covariance matrix is the inverse of the Fisher information matrix* defined for a two-parameter distribution as

$$F = \times \begin{bmatrix} E\{-\partial^2 \log L/\partial\mu^2\} & E\{-\partial^2 \log L/\partial\mu\,\partial\sigma\} \\ E\{-\partial^2 \log L/\partial\sigma\,\partial\mu\} & E\{-\partial^2 \log L/\partial\sigma^2\} \end{bmatrix},$$

where the expectations are taken with respect to the true values of the parameters μ and σ.

To approximate the variance, one may:

1. Perform the expectations in the information matrix considering the effect of the censoring, evaluate them at the ML estimated values $\hat{\mu}$ and $\hat{\sigma}$, and invert the matrix. This yields the *maximum likeli-*

hood estimates of var($\hat{\mu}$), var($\hat{\sigma}$), and cov($\hat{\mu}, \hat{\sigma}$). Halperin [7] discusses the calculation of expectations under type I and type II censoring.

2. Estimate the information matrix as the sample value of each corresponding term, e.g.,

$$E\{-\partial^2 \log L/\partial \mu^2\} \simeq -\partial^2 \log L/\partial \mu^2$$

The inverse of this matrix yields the *local estimate* of the covariance matrix. Method I is preferred if the expectations are tractable.

For some distributions under some types of censoring, exact small-sample inference is possible. The exponential distribution under single type II censoring on the right is a notable case. For this case $2r\hat{\theta}/\theta$ follows a χ^2 distribution* with $2r$ degrees of freedom, allowing an exact $100(1-\alpha)\%$ confidence interval* to be set on θ as

$$2r\hat{\theta}/\chi^2_{1-\alpha/2}(2r) < \theta < 2r\hat{\theta}/\chi^2_{\alpha/2}(2r).$$

This interval depends on r only and not on sample size n. This permits trading specimens for test time to achieve equal precision. With n large the time to obtain the first r failures is short, but $n - r$ items will be "wasted." For n small ($\geqslant r$) the expected test time will be longer but fewer specimens will be needed. Under type I censoring the number of failures r is a random variable and the distribution of $\hat{\theta}$ derived by Bartholomew [3] is quite complicated.

Exact inference is possible for type II singly censored samples from the two-parameter Weibull distribution* having shape parameter β, scale parameter η, and pth percentile $x_p = \eta\,[\ln\{1/(1-p)\}]^{1/\beta}$. It is made possible by the fact that the functions

$$v(r,n) = \hat{\beta}/\beta$$

and

$$u(r,n,p) = \hat{\beta}\ln(\hat{x}_p/x_p)$$

are pivotal* (i.e., are distributed independently of the population parameters) and can be determined for r, n, and p by Monte Carlo sampling*. Tables are in refs. 2, 10, 23 and 24. Exact procedures have been developed for paired* [9, 20, 22] and multiple comparisons* [11], analysis of one-way classifications [13], and regression* [14] based on type II singly right-censored Weibull samples.

McCool [12] has also developed exact confidence limits for the Weibull parameters for the two-mode competing risk problem when the latent times are independently Weibull-distributed with a common shape parameter.

Techniques for exact inference from type II censored samples were developed for the normal distribution by Schmee and Nelson [21], and for the logistic* distribution by Antle et al. [1] and Schafer and Sheffield [19].

References

[1] Antle, C. E., Klimko, L., and Harkness, W. (1970). Confidence intervals for the parameters of the logistic distribution. *Biometrika*, **57**, 397–402.

[2] Bain, L. J. (1978). Statistical Analysis of Reliability and Life Testing Models: Theory and Methods. Marcel Dekker, New York.

[3] Bartholomew, D. J. (1963). The sampling distribution of an estimate arising in life testing. *Technometrics*, **5**, 361–374.

[4] Birnbaum, Z. W. (1979). On the Mathematics of Competing Risks. *DHEW Publ. No. (PHS)79–1351*. For sale by the Superintendent of Documents, U.S. Government Printing Office, Washington, DC 20402

[5] Epstein, B. and Sobel, M. (1953). Life testing. *J. Amer. Statist. Ass.*, **48**, 486–502.

[6] Gross, A. J. and Clark, V. A. (1975). *Survival Distributions: Reliability Applications in the Biomedical Sciences* Wiley, New York.

[7] Halperin, M. (1952). Maximum likelihood estimation in truncated samples. *Ann. Math. Statist.*, **23**, 226–238.

[8] Kaplan, E. L. and Meier, P. (1958). Nonparametric estimation from incomplete observations. *J. Amer. Statist. Ass.*, **53**, 457–481.

[9] McCool, J. I. (1970). Inference on Weibull percentiles and shape parameter from maximum likelihood estimates. *IEEE Trans. Rel.*, **R-19**, 2–19.

[10] McCool, J. I. (1974). Inferential Techniques for Weibull Populations. *Wright-Patterson AFB, Interim Tech. Rep. ARL TR 74-0180, AD-A009645.*

[11] McCool, J. I. (1975). Multiple comparisons for Weibull parameters. *IEEE Trans. Rel.* **R-24**, 186–192.

[12] McCool, J. I. (1976). Estimation of Weibull parameters with data censored by a competing Weibull failure mode. *IEEE Trans. Rel.* **R-25**, 25–31.

[13] McCool, J. I. (1979). Analysis of single classification experiments based on censored samples from the two-parameter Weibull distribution. *J. Statist. Plan. Infer.*, **3**, 39–68.

[14] McCool, J. I. (1980). Confidence limits for Weibull regression with censored data. *IEEE Trans. Rel.*, **R-29**, 145–150.

[15] Mann, N. R., Schafer, R. E., and Singpurwalla, N. D. (1974). *Methods for Statistical Analysis of Reliability and Life Data*. Wiley, New York.

[16] Moeschberger, M. L. and David, H. A. (1971). Life tests under competing causes of failure and the theory of competing risks. *Biometrics*, **27**, 909–933.

[17] Nelson, W. (1972). Theory and application of hazard plotting for censored failure data. *Technometrics*, **14**, 945–966.

[18] Nelson, W. (1981). *Life Data Analysis*. Wiley–Interscience, New York.

[19] Schafer, R. E. and Sheffield, T. S. (1973). Inferences on the parameters of the logistic distribution. *Biometrics*, **29**, 449–455.

[20] Schafer, R. E. and Sheffield, T. S. (1976). On procedures for comparing two Weibull populations. *Technometrics*, **18**, 231–235.

[21] Schmee, J. and Nelson, W. B. (1976). Confidence Limits for Parameters of (Log) Normal Life Distributions from Small Singly Censored Samples by Maximum Likelihood. *General Electric Co., Res. Dev. TIS Rep. 76CRD218*.

[22] Thoman, D. R. and Bain, L. J. (1969). Two sample tests in the Weibull distribution. *Technometrics*, **11**, 805–815.

[23] Thoman, D. R., Bain, L. J., and Antle, C. E. (1969). Inferences on the parameters of the Weibull distribution. *Technometrics*, **11**, 445–460.

[24] Thoman, D. R., Bain, L. J., and Antle, C. E. (1970). Maximum likelihood estimation, exact confidence intervals for reliability and tolerance limits in the Weibull distribution. *Technometrics*, **12**, 363–371.

Bibliography

The bibliography below, in addition to the references just given, is comprehensive but far from exhaustive. Numerous papers dealing with censored data are scattered in engineering and statistical periodicals, the overwhelming majority of them written in the last two decades.

The titles of papers and monographs cited below describe adequately in most cases the content of the particular contribution to the literature.

Bain, L. J. (1972). Inferences based on censored sampling from the Weibull or extreme-value distribution. *Technometrics*, **14**, 693–702.

Berkson, J. and Elveback, L. (1960). Competing exponential risks, with particular reference to the study of smoking and lung cancer. *J. Amer. Statist. Ass.*, **55**, 415–428.

Berman, S. M. (1963). Note on extreme values, competing risks and semi-Markov processes. *Ann. Math. Statist.*, **34**, 1104–1106.

Billman, B. R., Antle, C. E., and Bain, L. J. (1972). Statistical inference from censored Weibull samples. *Technometrics*, **14**, 831–840.

Chiang, C. L. (1970). Competing risks and conditional probabilities. *Biometrics*, **26**, 767–776.

Cohen, A. C. (1950). Estimating the mean and variance of normal populations from singly truncated and doubly truncated samples. *Ann. Math. Statist.*, **21**, 557–569.

Cohen, A. C. (1959). Simplified estimators for the normal distribution when samples are singly censored or truncated. *Technometrics*, **1**, 217–237.

Cohen, A. C. (1961). Tables for maximum likelihood estimates: singly truncated and singly censored samples. *Technometrics*, **3**, 535–541.

Cohen, A. C. (1963). Progressively censored samples in life testing. *Technometrics*, **5**, 327–339.

Cohen, A. C. (1965). Maximum likelihood estimation in the Weibull distribution based on complete and on censored samples. *Technometrics*, **7**, 579–588.

Cohen, A. C. (1966). Query 18: "Life testing and early failure." *Technometrics*, **8**, 539–545.

Cohen, A. C. (1975). Multi-censored sampling in the three-parameter Weibull distribution. *Technometrics*, **17**, 347 – 351.

Cox, D. R. (1959). The analysis of exponentially distributed life-times with two types of failure. *J. R. Statist. Soc. B*, **21**, 411–421.

David, H. A. (1970). *Order Statistics*, Wiley, New York.

David, H. A. and Moeschberger, M. L. (1979). *The Theory of Competing Risks*. Griffin's Statist. Monogr. No. 39. Methuen, London.

Dubey, S. D. (1967). Some percentile estimators for Weibull parameters. *Technometrics*, **9**, 119–129.

Englehardt, M. E. (1975). Simple linear estimation of the parameters of the logistic distribution from a complete or censored sample. *J. Amer. Statist. Ass.*, **70**, 899–902.

Englehardt, M. E. and Bain, L. J. (1973). Some complete and censored sampling results for the Weibull or extreme-value distribution. *Technometrics*, **15**, 541–549.

Glasser, M. (1965). Regression analysis with censored data. *Biometrics*, **21**, 300–307.

Gupta, A. K. (1952). Estimation of the mean and standard deviation of a normal population from a censored sample. *Biometrika*, **39**, 260–273.

Harter, H. L. and Moore, A. H. (1965). Maximum-likelihood estimation of the parameters of gamma and Weibull populations from complete and from censored samples. *Technometrics*, **7**, 639–643.

Harter, H. L. and Moore, A. H. (1966). Iterative maximum-likelihood estimation of the parameters of normal populations from singly and doubly censored samples. *Biometrika*, **53**, 205–213.

Harter, H. L. and Moore, A. H. (1967). A note on estimation from a type I extreme-value distribution. *Technometrics*, **9**, 325–331.

Harter, H. L. and Moore, A. H. (1967). Asymptotic variances and covariances of maximum-likelihood estimators from censored samples, of parameters of Weibull and gamma populations. *Ann. Math. Statist.*, **38**(2), 557–571.

Harter, H. L. and Moore, A. H. (1967). Maximum-likelihood estimation, from censored samples, of the parameters of a logistic distribution, *Amer. Statist. Ass.*, **62**, 675–684.

Harter, H. L. and Moore, A. H. (1968). Maximum-likelihood estimation, from doubly censored samples, of the parameters of the first asymptotic distribution of extreme values. *J. Amer. Statist. Ass.*, **63**, 889–901.

Herd, G. R. (1960). Estimation of reliability from incomplete data. *Proc. 6th Nat. Symp. Rel. Quality Control*, pp. 202–217.

Herman, R. J. and Patell, R. K. N. (1971). Maximum-likelihood estimation for multi-risk model. *Technometrics*, **13**, 385–396.

Hoel, David G. (1972). A representation of mortality data by competing risks. *Biometrics*, **28**, 475–488.

Johns, M. V. and Lieberman, G. J. (1966). An exact asymptotically efficient confidence bound for reliability in the case of the Weibull distribution. *Technometrics*, **8**, 135–175.

Johnson, L. G. (1964). *The Statistical Treatment of Fatigue Experiments*. Elsevier, New York.

Kimball, A. W. (1957), Disease incidence in populations subject to multiple causes of death. *Bull. Inst. Statist.*, **36**, 193–204.

Kimball, A. W. (1969), Models for the estimation of competing risks from grouped data. *Biometrics*, **25**, 329–337.

Lieblein, J. (1954). A New Method of Analyzing Extreme-Value Data. *Nat. Advis. Comm. Aeronaut. Tech. Note 3053*.

McCool, J. I. (1966). Inference from the third failure in a sample of size thirty from a Weibull distribution. *Ind. Quality Control*, **23**, 109–114.

McCool, J. I. (1970). Evaluating Weibull endurance data by the method of maximum likelihood. *ASLE Trans.*, **13**, 189–202.

McCool, J. I. (1970). Inference on Weibull percentiles from sudden death tests using maximum likelihood. *IEEE Trans. Rel.*, **R-19**, 177–179.

McCool, J. I. (1974). Analysis of sudden death tests of bearings endurance, *ASLE Trans.*, **17**, 8–13.

McCool, J. I. (1975). Inferential Techniques for Weibull Populations II. *Wright-Patterson AFB Final Rep. ARL TR 75-0233, AD-AO23937, No. 215.*

McCool, J. I. (1977). Analysis of variance for Weibull populations. In *The Theory and Applications of Reliability*, C. Tsokos and I. Shimi, eds. Academic Press, New York.

Mann, N. R. (1967). Tables for obtaining the best linear invariant estimates of parameters of the Weibull distribution. *Technometrics*, **9**, 629–645.

Mann, N. R. (1968). Exact three-order statistic confidence bounds on reliable life for a Weibull model with progressive censoring. *J. Amer. Statist. Ass.*, **64**, 306–315.

Mann, N. R. (1968). Point and interval estimation procedures for the two-parameter Weibull and extreme-value distributions. *Technometrics*, **10**, 231–256.

Mann, N. R. (1969). Cramèr–Rao efficiencies of best linear invariant estimators of parameters of the extreme-value distribution under type II censoring from above. *SIAM J. Appl. Math.*, **17**, 1150–1162.

Mann, N. R. (1969). Optimum estimators for linear functions of location and scale parameters. *Ann. Math. Statist.*, **40**, 2149–2155.

Mann, N. R. (1971). Best linear invariant estimation for Weibull parameters under progressive censoring. *Technometrics*, **13**, 521–533.

Mann, N. R. and Fertig, K. W. (1973). Tables for obtaining confidence bounds and tolerance bounds based on best linear invariant estimates of parameters of the extreme-value distribution. *Technometrics*, **15**, 87–101.

Moeschberger, M. L. (1964). Life tests under dependent competing causes of failure. *Technometrics*, **16**, 39–47.

Nelson, W. (1970). Hazard plotting methods for analysis of life data with different failure modes. *J. Quality Tech.*, **2**, 126–149.

Nelson, W. and Hahn, G. T. (1972). Linear estimation of a regression relationship from censored data, Part I: Simple methods and their application. *Technometrics*, **14**, 247–269.

Nelson, W. and Schmee, J. (1979). Inference for (log) normal life distributions from small singly censored samples and BLUE's. *Technometrics*, **21**, 43–54.

Sampford, M. R. (1952). The estimation of response-time distributions, II: Multi-stimulus distributions. *Biometrics*, **8**, 307–369.

Sarhan, A. E. and Greenberg, B. G. (1956). Estimation of location and scale parameters by order statistics from singly and doubly censored samples, Part I. *Ann. Math. Statist.*, **27**, 427–451.

Sarhan, A. E. and Greenberg, B. G. (1957). Tables for best linear estimates by order statistics of the parameters of singly exponential distributions from singly and doubly censored samples. *J. Amer. Statist. Ass.*, **52**, 58–87.

Sarhan, A. E. and Greenberg, B. G. (1958). Estimation of location and scale parameters by order statistics from singly and doubly censored samples, Part II. *Ann. Math. Statist.*, **29**, 79–105.

Sarhan, A. E. and Greenberg, B. G., eds. (1962). *Contributions to Order Statistics*, Wiley, New York.

Thoman, D. R. and Wilson, W. M. (1972). Linear order statistic estimation for the two-parameter Weibull and extreme-value distributions from type II progressively censored samples. *Technometrics*, **14**, 679–691.

Wingo, D. R. (1973). Solution of the three-parameter Weibull equations by constrained modified quasilinearization (progressively censored samples). *IEEE Trans. Rel.*, **R-22**(2), 96–102.

Zelen, M. (1969). Factorial experiments in life testing. *Technometrics*, **11**, 269–288.

(COMPETING RISKS
EXPONENTIAL DISTRIBUTION
TESTS FOR CENSORING
TRIMMING
TRUNCATION
WEIBULL DISTRIBUTION
WINSORIZATION)

JOHN I. MCCOOL

CENSORING

When certain values among a set X_1, $X_2, \ldots, X_n$ are not used (or available), the selection of values for omission being determined by their relative magnitudes, the data are said to be *censored*. It is important to distinguish censoring, as just defined, from *truncation**, in which omission is determined by absolute magnitude (e.g., omission of all values greater than some constant value, or outside a specified interval).

Specific types of censoring are conveniently described in terms of the *order statistics** corresponding to $X_1, X_2, \ldots, X_n$. Denoting these by $X'_1, X'_2, \ldots, X'_n$ with $X'_1 \leqslant X'_2 \leqslant \cdots \leqslant X'_n$, the most common types of censoring are:

1. From above: omission of the r greatest values, $X'_{n-r+1}, \ldots, X'_n$.
2. From below: omission of the r least values, $X'_1, \ldots, X'_r$.
3. Symmetrical: omission of equal numbers (r) of values at each extreme—$X'_1, \ldots, X'_r$ and $X'_{n-r+1}, \ldots, X'_n$.

Other types of censoring—e.g., omission of some order statistics in the body of the distribution—are much less common. It may be noted, however, that the use of statistics such as median* and interquartile distance* may be regarded as use of heavily censored sample values.

(CENSORED DATA
TESTS FOR CENSORING)

CENSORING, INDIRECT

If sample values are censored according to values of a variable X, but a related variable Y is observed, the Y is said to be *indirectly* censored. The effect on Y depends, of course, on the relationship between X and Y. If this relationship is both sufficiently close and sufficiently well-known, it may be possible to use sets of observed values of Y to test whether there has been censoring with respect to X.

CENSORING, LEFT

Censoring from below, by omission of the first r order statistics.

(CENSORING)

CENSORING, PROGRESSIVE *See* PROGRESSIVE CENSORING

CENSORING, RANDOM

Censoring* in which the order statistics to be omitted are chosen randomly. In certain competing risk* models, values in excess of the value of a certain random variable (called the "censoring variable") are not observed (and so omitted). Although this is called "random censoring," it might better be called "random truncation."

CENSORING, RIGHT

Censoring from above, by omission of the last r order statistics.

(CENSORING)

CENSORING, SYMMETRICAL

Censoring by omission of the first r order statistics (the r least values) and the last r order statistics (the r greatest values).

(CENSORING)

CENSUS

The terms means "a count," usually of people but sometimes of dwellings or of consumer durables (e.g., automobiles, refrigerators, telephones). In agriculture it may refer to a count of cattle or of areas under particular types of cultivation. The term should therefore be qualified by a noun/adjective referring to the objects counted. This entry refers to a *population census*, but most of what follows applies to any kind of census.

A population census is taken in most developed countries at regular intervals, usually of 5 or more years, and this is the primary source of information about the population of a country. Although the population census considered as a field operation has much in common with other kinds of social survey*, it is nevertheless sharply distinguished by its traditional background, legal sanctions, coverage, and by the whole scale of the operation and the resources normally devoted to it, which permit a far greater content and depth of analysis than can normally be encompassed in other types of field study.

DEFINITION

The official U.N. definition (1967) is as follows: A census of population may be defined as "the total process of collecting, compiling, evaluating, analyzing and publishing demographic, economic and social data pertaining, at a specified time, to all persons in a country of in a well-delimited part of a country." Certain essential features follow from this definition. An official census is sponsored by the government of the area or of some larger region within which the area falls. The area covered is precisely defined. The principle of universality applies; i.e., the enumeration should either include every member of the community to which the census relates without omission or duplication or, if sampling is used, it must give every member of a stratum equal likelihood of inclusion. The concept is one of individual rather than group enumeration. (The objection to group enumeration is that it tends to result in underenumeration of the total population, but there are also statistical objections, e.g., the restriction of cross-tabulation of individual characteristics.) Relating the census to a point of time implies simultaneity of enumeration of all persons included in the census (or at least a close approximation to simultaneity). Population growth and change cannot be measured by a census taken at a single point of time; so there is an implication also that there should be a series of censuses of regular periodicity.

LEGAL BASIS

There is usually a legal basis for the census in order to render public participation enforceable. The legal sanctions that can be applied distinguish a census from other voluntary types of social survey which are more likely to suffer from incompleteness and bias. The penalties that may be imposed upon those who fail to complete a census schedule are not usually heavy nor are they actually applied except in a minority of exemplary cases, but the fact that the sanctions exist and that there is a history of their existence is sufficient to ensure, in most developed countries, that noncooperation is minimal.

PUBLICITY

Completeness and accuracy are, however, distinguishing characteristics of the census that depend less upon legal sanctions than upon public acceptance of the need for the collection of the information in the interest, ultimately, of public welfare. For this reason it is necessary to ensure the retention of public confidence by careful prior explanation of the information needs and the subsequent demonstration that all the information that has been gathered is needed and has actually been used in the public interest.

CONFIDENTIALITY AND PRIVACY

Much of the information required in a modern census involves questions bearing upon intrafamily and extrafamily relationships and upon educational attainment, earning capacity, and employment status. Such questions tend to be regarded as bordering upon an invasion of privacy. To retain public confidence there has to be a demonstrably valid assurance that the personal records will not be accessible to third parties and that the processing of the records into statistics will not permit the identification of individual persons or households. Public sensitivity on this issue has been more acute since the advent of computers and, in the early days, some notable instances of malpractice. Since then the electronic security arrangements for preventing unauthorized access to computer files have been rigorously tightened. But public ignorance about computers remains, and so does their nervousness about confidentiality; moreover, there undoubtedly *have* been breaches either due to failures in the tightness of the code arrangements for restricting entry into files or to outright carelessness. One unfortunate side effect of this nervousness is public opposition to the use of the census as a population sampling frame* for other surveys. In countries without population registers*, the census provides the only total listing of the population and it is an ideal frame for the selection of unbiased samples* of special population strata which could not otherwise be identified. There is a need for continued public education in these matters, and prior to a population census there has to be a vigorous and well-prepared publicity program.

SCOPE AND COVERAGE

As economic development progresses, society becomes more complex and government information needs widen in their scope and detail. However, there is a practical restriction on the extent of census enquiries. Merely to ask an additional question in the census schedule does not ensure a correct answer. Any progressive extension of the schedule is likely to reach a stage at which indifference, if not resentment, will introduce inaccuracy. This is a very important consideration where the head of the household is required to complete the schedule (e.g., the U.K. "self-enumeration" system, but even where visiting interviewers are employed (e.g., the U. S. "canvas" system), steps still have to be taken to reduce the burden of questions to be directed to any one household. If the number of aspects on which population statistics are sought (additional to the basic details of age, sex, marital condition, size of households and dwelling, etc.) are too numerous to be covered at one census without excessive complexity in the schedule, it is better to cover them some at a time by a set of supplemental questions at successive censuses, especially if these censuses by virtue of their simplicity can be held more frequently than otherwise. Alternatively, if it is considered essential that these supplementary topics be cross-classifiable with all others, then, at the one census, a system of interlocking samples* can be used, each sample covering a different selection of questions so that no schedule covers the full range.

The final decision as to scope is made by the government, which authorizes and pays for the census (the proposed U.K. census of 1976 was canceled altogether at the last moment because of financial retrenchment by the government of the day). Clearly, priority has to be given to topics that are of direct relevance to current problems of government. At the same time, since in a democracy, schedule fillers are also voters, a government studiously avoids any question which is known to be regarded as offensive by a substantial proportion of the electorate. For this reason, few censuses cover such topics as personal income, religion, or affinity to minority groups (ethnic or otherwise). *See*, however, BUREAU OF CENSUS. The census authority usually consults widely among all bodies which may have an interest in the use of the census as a vehicle for obtaining population information, before making proposals to government.

PILOTING

Prior to the actual census it is usual to conduct a number of small-scale sample surveys* (small in scale because no hard information is required) in order not only to test the design of the census schedule layout but also to ascertain public reaction to questions that have not been asked in any previous census and on which therefore there is no documented national experience. (A proposed ethnic minority question for the 1981 U.K. census was withdrawn, because a pilot test produced an adverse reaction from members of ethnic minorities.)

FIELDWORK

A census is a head count ("... and David numbered the people ...") and an organization of field workers (enumerators) has to be provided to carry out this basic operation of identifying buildings, deciding which are, in part or whole, used as dwellings (i.e., for living and sleeping), and making contact with the persons who occupy each such dwelling in order to direct census questions to them. Although the initial contact may be through the postal services, the main contact is a physical one; a vast army of enumerators must approach each dwelling on foot.

MAPPING

This means that, as part of census preparation, there is a large-scale and very careful mapping operation in which the entire country is divided into parts (enumeration districts) to be covered by the individual enumerators. The delineation of these enumeration districts must satisfy the following criteria:

1. In the interests of speed of making contact and of the simultaneity of the whole census operation, no district should be larger than can, reasonably, be covered on foot by an enumerator in 1 or 2 days.
2. The boundary of the district must be clearly recognizable on the ground. Each enumerator will be given a map of his or her district and there must be no ambiguity as to where this district ends and another begins.
3. Each district must be completely contiguous with others. There must be no gaps between defined districts leading to underenumeration, and no overlapping which would heighten the risk of double enumeration.
4. Because much of the census information is required for local as well as central government purposes, the districts must be so designed that they can be aggre-

gated to exactly complete local administrative areas. It is a data-processing advantage if the maps make it possible to apply a grid reference to each dwelling, although this is not a district shape criterion. For planning purposes statistics are often required for small areas that do not appear to conform to local administrative area boundaries but are capable of being fitted into a grid coordinate system. If the coordinates are recorded on the maps, they can be transferred to the census schedule for each dwelling, the occupants of which can then be allocated (as part of the computer processing) to any required combination of grid squares.

TRAINING OF ENUMERATORS

The level of skill required of an enumerator varies according to whether a self-enumeration or a canvas system is used. In the self-enumeration system, the enumerator must be of average intelligence, literate, legible, neat, and meticulous without being labored and slow, sensitive to the difficulties of those of lesser or greater educational attainment than himself (experienced census workers maintain that university professors find more options within census questions than do road laborers), articulate in explanation, and at least to some degree dedicated to the success of the census operation. In the canvas system the enumerator will need to know a little about the skill of a social survey interviewer; e.g., how to avoid leading questions, how to avoid giving the impression of invading privacy, etc. The enumerators work part-time for a period at most of a few weeks. They are persons already in full-time employment who have the required qualities, are prepared to work in their spare time, and can take a full day off from their full-time employment to concentrate on specific tasks. They comprise primarily local government officials, school teachers, clerks in industry or commerce, insurance agents, and the like.

It is usual to arrange for the organization to be hierarchical with the country divided into regions with a full-time census officer responsible in each region for the recruitment, training, and direction of the enumerators of all the districts in the region. The amount of training is usually limited by cost consideration to a few hours of lectures and demonstration by the supervising census officer. In particular, possible difficulties of interpretation in the application of the questions are reviewed.

Each enumerator will normally be issued with an enumeration book containing a map of his or her district, and a brief review of the questions, their purpose, and probable difficulties of interpretation. Each enumerator will also have a book in which to record a listing of the households seriatim as he or she identifies them with certain details (1) to facilitate reference back at a later date and (2) to facilitate the provision of a rapid but manually produced preliminary estimate of the total population enumerated, perhaps by age and sex.

TIMING

Population changes are measured by comparison of the census enumeration time with that of another time. It is essential, therefore, that the census show the picture of population size and structure at a fixed point and not a picture of "shifting sand." A time reference must be defined and all answers to census questions must be related to that time reference. It is usual to select a time at which population movement is minimal and at which most people are at home within their de jure households. In U.K. censuses, the time reference is always midnight on a Sunday in April. Midnight on Sunday finds most people, even commercial travelers, in their own homes. April is chosen because this is before the beginning of the "early holiday" season. It is the earliest month in which warmer weather and lighter evenings make the conditions suitable for the footwork of the enumerators.

INITIAL REVIEW OF SCHEDULES

When the census schedules have been completed, they are checked visually for incompleteness or obvious errors so that, if necessary, callbacks* for corrections can be made. A quick review by the enumerator at the point of collection (or immediately on completion, in the canvas system) will avoid many callbacks. A second and more thorough visual check is made by the census officer, who will also see that all districts, and all known dwellings within the district, have been covered. At this stage, again manually, even if machine-assisted, a rapid preliminary summation of numbers may be made.

CLASSIFICATION AND CODING

Many of the census questions require verbal answers, and these must be converted to a numerical form for handling by computer. At an early stage in the census preparatory period, many classifications of verbal data will have been made to achieve this conversion. If the classification is of short range (e.g., tenure of dwelling: owner occupied, rented from private landlord, etc.), it can be reproduced on the census schedule and is then self-coding. Larger classifications (e.g., that of occupation, which may contain several hundred rubrics) cannot be so reproduced and the rubric numbers must be written on the schedules. This is usually done at census headquarters by a cadre of well-trained and experienced coders before the schedules are consigned to conversion to the appropriate computer input.

EDITING

Although errors will have been reduced by visual inspection of the schedules, there will be some errors which are not immediately visible but which would be visible on close inspection. These are limited to errors of inconsistency (mistatements which are not inconsistent with other information would not be detectable however close the inspection.) The computer can be programmed to provide this close inspection of the input presented to it. What is required is for the statistician to list the possible inconsistencies (e.g., baby born to a man, clergyman aged 12, etc.) or the very unlikely circumstances (e.g., age 110+, family of 20+ children). The computer is programmed to look for these, to print out a query message together with details to identify the schedule, and to mark the tape or disk record to indicate that it has been queried. The schedules relative to the queries are then inspected and decisions for correction are made on the basis of other information on the schedule. These corrections are then fed into the computer, which writes in the new information and removes the query marks. *See also* EDITING STATISTICAL DATA.

TABULATIONS

Although reference to tabulations* has been left until late in this article, drafting of the scheme of tabulation is actually one of the earliest stages in the preparation for the census. There are three reasons for this. First, it enables estimates to be made of the computer resources required. This helps the preparation of estimates of cost and enables recruitment of staff to begin in good time. Second, it helps to sharpen the definition of the questions to be asked on the census schedule. It is not until the statistician begins to consider how he will tabulate the information that he gets a clear idea of what to ask for. Third, it enables decisions to be made of the eventual scale and format of publication.

PUBLICATION AND DATA-BANK STORAGE

Before the introduction of computers, the publication of large volumes of tables was regarded as the best way of storing the census results permanently and in an accessible form. Many tables were printed for record purposes, although they might never be used and were, in any case, not specific to any enquiry.

The computer records are rapidly accessible at will and extracts can be made in a format specific to the information need, specific, that is, to both the range of statistics and the locality to which they relate. It is a common practice for the census authority to store a wide range of tabulations for the smallest convenient unit of area—either the 1- or 10-kilometer grid square (if grid references have been recorded) or the enumeration district. For planning purposes there is a great demand for statistics for small areas that do not conform to administrative boundaries, and this kind of data bank enables the census authority to satisfy the demand very quickly and at low cost. Given this facility the census authority can restrict its publication program to those tabulations that are of universal and immediate interest.

CHECKS ON ACCURACY

Because the census provides the benchwork to control the accuracy of intercensal estimates of population, it is important that the census authority should attempt to calibrate* the accuracy of the census itself. It is therefore common practice to carry out a reenumeration of a small sample of the population very quickly after the main census. This sample is on an area basis in order to check total coverage (underenumeration or duplicate enumeration.) In order to assess the accuracy of the answers to questions in the main census, the postenumeration test is carried out on an interview basis with the questions being led into in a manner *different* from that in the main census. The responses are then compared with those of the same respondents in the main census.

Bibliography

Benjamin, B. (1970). *The Population Census*. Social Science Council Review, Heinemann Educational Books, London.

Benjamin, B. (1968). *Demographic Analysis*. George Allen and Unwin, London.

United Nations (1967). Principles and recommendations for the 1970 population censuses. *Statistical Papers Series M44*. UN Statistical Office, New York.

(BUREAU OF THE CENSUS, U.S.
SURVEY SAMPLING)

B. BENJAMIN

CENTRAL LIMIT THEOREMS *See* LIMIT THEOREMS, CENTRAL

CENTRAL MOMENTS

The hth central moment of a random variable is the expected value* of the hth power of its differences from its mean. Symbolically, $\mu_h = E[(X - E[X])^h]$. If h is a positive integer,

$$\mu_h = \sum_{j=0}^{h-2} (-1)^j \binom{h}{j} \mu'_{h-j} \mu'^{j}_{1} + (-1)^{h-1}(h-1)\,\mu'^{h}_{1},$$

where $\mu'_r = E[X^r]$ is the rth *crude moment**.

(CUMULANTS
FACTORIAL MOMENTS
MOMENT GENERATING FUNCTION)

CESARO AVERAGING *See* ERGODIC THEOREMS

CHAIN INDEX NUMBERS *See* INDEX NUMBERS

CHAIN SAMPLING

"A system of sampling inspection in which the criteria for accepting and rejecting the lot depends on the results of the inspection of the immediately preceding lots." (This is the definition suggested by the Standards Committee of ASQC in 1978.

The original chain sampling plan (ChSP-1) was introduced by H. F. Dodge [1]. It is applied in situations in which relatively small sample sizes are necessary because the cost or destructiveness of testing is very high, but other factors make a large sample desirable. The inspected product comprises a series of successive lots, produced by a continuous process, and the lots are expected to be of essentially the same quality. This sampling plan allows significant reduction in sample size under the conditions of a continuing succession of lots from a stable and trusted supplier.

The plan is implemented as follows:

1. From each lot select a sample of n units and test each unit for conformity to the specified requirement.
2. Accept the lot if the observed number of defectives d is zero in a sample of n units and reject if $d > 1$.
3. Accept the lot if $d = 1$ *and* if no defectives are found in the *immediately* preceding i samples of size n.

OC curves* for these plans and tables for determining the associated AOQL (Average Outgoing Quality Limit*) were developed by Soundararajan [3].

References

[1] Dodge, H. F. (1955). *Ind. Quality Control*, **12**(4), 10–13.

[2] Dodge, H. F. and Stephens, K. S. (1966). *Ind. Quality Control*, **23**(2), 61–67.

[3] Soundararajan. V. (1978). *J. Quality Control*, **10**, 56–60.

(ACCEPTANCE SAMPLING
AVERAGE OUTGOING QUALITY LIMIT
OC CURVES
QUALITY CONTROL)

CHANCE (I)

Ancient and medieval, as well as recent, writers have seen an element of chaos in the universe. This view, it seems, has also much appealed to the popular mind in all ages.

Of the ancient philosophers, Heraclitus and the Stoics held principally to determinism and necessity [14, p. 180]. Chance played a legitimate role in Aristotle's and Plutarch's views, both as one of several possible causes of events [14, p. 180] and as the necessary antecedent of free will [13, pp. 12, 16]. Epicurus also saw in chance one of several possible forms of causation* [4, p. 172]. Indeed, in Epicurus' prophetic revision of Democritus' atomic doctrine, chance, as "permissive cause" allowing for "atomic swerve," is placed in essentially the position that it occupies today in statistical mechanics [14, p. 180].

Chance is in evidence as well in ancient popular mythology. The fickle goddess Tyche shared Mount Olympus with the Greeks' other theomorphisms of the major elements of the human condition. She was followed by the goddess Fortuna, who flourished, especially in the waning days of Rome, as the embodiment of cynicism and doubt [13, pp. 12–13].

Fortuna held her own after the fall of Rome, at least as a "force," if not as deity [13, p. 34]. Certainly, the ordinary folk of the Frankish and Holy Roman empires had no less reason to doubt that anything much fairer, more predictable, than chance held sway over their lives. In the prevalent view of the time, Fortuna was imagined as spinning her wheel of fortune, which propelled people at her whim upward to high estate or downward [13, Chap. V].

Although the Church was antipodal to that despairing view, many of its fathers found it not inappropriate to make their accommodations with the idea of chance; St. Augustine, Thomas Aquinas [13, p. 16], and Boethius [13, p. 18], e.g., because they subscribed to the notion of free will and its Aristotelean link to chance.

The Church itself, although not during the Reformation [2, p. 2], maintained its endorsement, if not of chance at least of free will, up to and beyond the Renaissance, as is illustrated by Bramhall's stand against Thomas Hobbes [2]. And if the Church sees free will as implicitly assumed in its teachings, John Locke see it as the source of all liberty: "For the mind . . . is [able] to consider the objects of [desire], examine them on all sides, and weigh them with others. . . . This seems to me the source of all liberty; in this seems to consist that which is . . . called free will" [12, p. 267].

Nonetheless, it was largely a climate of determinism, not a climate of free will or chance, that was to nurture the spirit of inquiry of the Age of Reason: "Chance produceth nothing; . . . all events and actions have their necessary causes" [2, a]. And so chance was largly robbed of its legitimacy, and found itself a sort of makeshift palliative for human ignorance. The universe, while seen as orderly and structured, is still thought well beyond our grasp; although the uncomprehended part is not believed to be governed by chance, it is conceded to appear to be, and becomes the natural arena for probabilistic modeling [15, pp. 64–70], an arena, in economists' terms, for risk* as opposed to uncertainty* [10].

Laplace* apparently saw such a universe —intrinsically orderly, in part hidden, in effect probabilistic: "All events . . . are a result of [the great laws of nature]" [11, p. vi]; "A thing cannot occur without a cause which produces it" [11, p. vi]; " . . . the [vast] intelligence from which (the human mind) will always remain infinitely removed" [11, p. vii]; "The principal means for ascertaining truth . . . are based on probabilities" [11, p. v].

Einstein, too, held to an ordered, inaccessible universe: " . . . the sublimity and marvelous order which reveal themselves . . . in nature . . . " [6, p. 38]; "I, at any rate, am convinced that He is not playing at dice" [1, p. 91]; "Even the great initial success of the quantum theory does not make me believe in the fundamental dice-game . . . " [1, p. 149]; " . . . the grandeur of reason incarnate in existence . . . is inaccessible to man" [6, p. 49]. And he too, nevertheless, resorted to stochastic modeling [5].

Recent times find us driven once more to take refuge in chance. The great majority of physicists see in chance and randomness the correct conceptual foundation of their discipline. And Einstein's unwavering deterministic faith has few adherents: "I still do not believe that the statistical method of quantum mechanics is the last word, but for the time being I am alone in my opinion" [1, p. 125]. Perhaps the strongest recent advocate of chance has been the American pragmatist Charles Peirce*. Indeed, "absolute chance" is the "warp of a total philosophical system Peirce calls Tychism" [14, p. 202]. Particularly with regard to cosmology, Peirce regards chance as synonymous with its manifestations—spontaneity and diversity—and finds it the most plausible primordial feature of the universe: "Chance is first, Law is second . . . " [14, p. 177]; or, in M. R. Cohen's interpretation, " . . . such original or underived individuality and diversity is precisely what Peirce means by chance; and from this point of view chance is prior to law . . . " [14, p. xii]. Much of the universe's evolution, in addition to its origins, is tied by Peirce to chance, essentially now in Darwinian terms: "This Darwinian principle is plainly capable of great generalization . . . " [14, p. 163]; and also: " . . . pure spontaneity as a character of the universe, acting . . . everywhere, producing infinitesimal departures from law continually . . . " [14, p. 196]. Indeed, natural law is itself seen as subject to this process: "Now the only possible way of accounting for the laws of nature . . . is to suppose them results of evolution . . . . It makes an element of indeterminacy, spontaneity, or absolute chance in nature . . . " [14, pp. 162–163]. A recent existentialist view is that of Karl Jaspers: "The determined conditions of my situation occur to me in the course of time as chance events" [8, p. 216].

There has never been a serious challenge raised to the idea, since its inception in the seventeenth century, of at least describing much of the physical and societal world through probabilistic modeling. The more radical step of not just *describing*, but in fact *directing* parts of our lives through chance, had already been initiated in ancient times. Be it by Caesar's dice at the Rubicon, or other means, we gladly trust to chance for our next move, in an essentially religious act of faith, when reason fails us: "My head's begun to ache. I believe it would be best to cast lots. Rely on God's will in everything. Whichever is drawn shall be my husband . . . " [7, p. 154]. This resort is taken most often, in fact, when evenhandedness and fairplay are at issue, be it at kickoff time in a football game or in assigning treatments* to plots in a variety trial*.

Chance moves also are brought into play when we attempt to confound each other, when it is a matter not so much of mapping a good or fair next move as it is of foiling a malevolent opponent. Examples here include the random elements of military codes, such as the Enigma code of World War II [9], as well as the "bluffing" randomization components of optimal strategies for stylized versions of poker [16, p. 204].

Finally, in a somewhat different vein, the convexity guaranteed by admitting randomized decisions* no doubt has been a unifying and simplifying factor in the development of statistical decision theory*.

References

[1] Born, M. (1971). *The Born–Einstein Letters*, Walker, New York.

[2] Bramhall, Bishop of Derry and Thomas Hobbes of Malmes (1656). *The Questions Concerning Liberty, Necessity and Chance*. Andrew Crook, London.

[3] Cohen, M. R. (1923). Introductory Chapter of *Chance, Love and Logic*, by Charles S. Peirce. Harcourt Brace, New York.

[4] DeWitt, N. W. (1954). *Epicurus and His Philosophy*, University of Minnesota Press, Minneapolis, Minn.

[5] Einstein, A. (1926). *Investigations on the Theory of the Brownian Movement*. E. P. Dutton, New York.

[6] Einstein, A. (1954). *Ideas and Opinions*, Crown, New York.

[7] Gogol, N. (1927). *The Government Inspector and Other Plays*. Alfred A. Knopf, New York.

[8] Jaspers, K. (1973). *Philosophie*, 4th ed., Vol. 2. Springer-Verlag, New York.

[9] Kahn, D. (1974). *The Code Breakers*. Weidenfeld and Nicolson, London.

[10] Knight, F. H. (1921). *Risk, Uncertainty and Profit*. Houghton Mifflin, New York.

[11] Laplace, P. S. de (1847). *Oeuvres de Laplace*, 2nd ed., Vol 7. Imprimerie Royale, Paris.

[12] Locke, J. (1963). *Works*, Vol. 1. Scientia Verlag Aalen, Darmstadt.

[13] Patch, H. R. (1967). *The Goddess Fortuna in Medieval Literature*. Octagon Books, New York.

[14] Peirce, C. S. (1923). *Chance, Love and Logic*, Harcourt Brace, New York.

[15] Poincaré, H. (1914). *Science and Method*, Thomas Nelson, London.

[16] von Neumann, J. and Morgenstern, O. (1947). *Theory of Games and Economic Behavior*. Princeton University Press, Princeton, N.J.

(AXIOMS OF PROBABILITY
CHANCE (II)
DECISION THEORY
FOUNDATIONS OF PROBABILITY)

H. T. DAVID
WALTER MORRIS

CHANCE (II)

The term "chance" is used in a variety of ways, but chiefly as a kind of probability, and this is the use discussed here. This kind of probabilty, sometimes called "physical" or "statistical" (as opposed to "inductive") probability, has long been familiar and is a concept central to statistical science. *See* FOUNDATIONS OF PROBABILITY. It has nonetheless not proved easy to make sense of, and its still current rival theories are all open to objection. The problem is that chance is credited with several characteristics that are not readily combined, and no one theory accounts well for all of them.

The concept of chance applies to events* (trials*) with a number of possible consequences (outcomes), no one of which is certain to ensue: e.g., heads or tails as outcomes of tossing a coin; the decay or otherwise of a radium atom in a specific stretch of time. There is a distribution of chances over these possible outcomes, and this is a probability distribution* in the sense of being a measure satisfying the axioms of the standard mathematical calculus of probabilities. *See* AXIOMS OF PROBABILITY. The chance distribution is supposed to be a property of the trial and to have the following characteristics. The chance of an outcome is the fair betting quotient for a bet that the outcome will occur. It is also the limit toward which the proportion or relative frequency* of occurrence of that outcome would tend if sufficiently similar trials were repeated endlessly. The property is supposed to be empirical: the fair betting quotients and relative frequencies involved cannot be deduced a priori. It is also objective: these quotients and frequencies exist whether we know it or not, and their actual values are independent of what we believe them to be. And although the content and value of our beliefs about the trial and its properties are no doubt relative to the evidence we have about it, the chance distribution is not.

No theory of chance makes equally good sense of all these aspects of it, and which theory one adopts will depend on which one takes to be its central and which its peripheral and dispensable aspects. In what follows the historically more important theories are sketched and what seem to the author the most promising current developments are outlined.

In Laplace's classical theory [3], chance is not an objective feature of the real world, which Laplace* took to be deterministic, i.e., such that only the actual outcome of a trial was ever really possible. Chance on this theory is merely a measure of our ignorance of that outcome. A number of outcomes would be consistent with what we do know about a trial: e.g., a number of trajectories of a coin tossed in a certain way. Of these, some proper fraction will result in the coin landing heads, and this fraction, on the classical theory, is the chance of that outcome.

Since proper fractions trivially satisfy the probability calculus, the classical theory explains at once why chances are probabilities. Indeed, historically it was under this classical conception that the probability calculus was originally developed, with application especially to gambling on games of chance*. In contexts of drawing cards and throwing what are supposed to be unloaded dice, the classical idea of chances as fractions of a finite number of possibilities is very natural. The laws of large numbers* in particular have a natural interpretation in these contexts and under this theory, since repeating a trial simply multiplies the possibilities; and by these laws the theory readily explains why the chance of an outcome is also its limiting frequency in endlessly repeated trials. It also seems intuitively right to match one's betting odds to the proportion of ways one thinks a bet can be won rather than lost, and to that extent the theory explains why a classical chance is a fair betting quotient.

The classical theory fails, however, to explain the *objective* and nonrelational aspects of chance. Indeed, as already remarked, it denies them, since it presupposes determinism and makes chances relative to the evidence we have about a trial. Moreover, once given the evidence, the empirical character of chance is obscure: it is a fraction, not of actual events that could be counted empirically, but of *abstract* possibilities which have somehow to be counted a priori. Before possibilities can be counted, they must be identified, and this the theory attempts to do by means of a principle of indifference*. Each possibility is equally probable, since any two outcomes that can each occur in only one way have by definition the same chance. The principle of indifference (or "insufficient reason") therefore attempts to identify these equally probable possibilities as those that we are, in Laplace's words, "equally undecided about in regard to their existence." The eventual supersession of the classical theory by frequency theories of chance resulted largely from the seemingly insuperable difficulties of giving this principle a usable, intelligible and nontrivial content, objective and free of paradox.

Another theory, more influential as an account of other kinds of probability, is the logical theory that treats probability as a quasi-logical relation of partial entailment between evidence and some hypothesis to which the evidence relates. The object of the theory, advanced by Keynes [2] and later developed by Carnap [1] and others, is to account in terms of this relation for how logically inconclusive evidence may still support (or "confirm") a hypothesis to a greater or lesser extent. Where the relation is taken to have a measure satisfying the probability calculus, it forms a basis for probabilistic confirmation theories, and the probabilities involved are called "inductive" probabilities.

These are prima facie quite a different kind of probability from chance. The two are alike in being objective, but there the similarity ends. Chance is empirical and nonrelational, whereas inductive probability is a priori and relative to evidence. Indeed, hypotheses about chances will be among these which have inductive probabilities relative to inconclusive evidence about chance trials. Carnap thought as a result that there were two distinct concepts of probability: inductive, to which the logical theory applied; and statistical, to which the frequency theory applied. These concepts must of course be connected: in particular, an event's inductive probability, relative to the evidence that its chance is p, must obviously be p; and theories of chance and of inductive probability must between them explain why this is so.

It is pertinent to note that the concept of inductive probability has proved so problematic that many thinkers have wished to dispense with it. Some have abandoned the idea of objective probabilities in this context, and resorted instead to Bayesian principles for changing subjective probabilities*, i.e., degrees of belief*, in acquiring new evidence about a hypothesis. But others have tried to make chance do the work of inductive probability. The Neyman–Pearson theory* is a notable instance, and the role of chance in this theory should be briefly illustrated here. Prima facie a measurement of a length, for example, makes various possible values of the actual length more or less probable, depending on the accuracy of the measuring device. These probabilities measure the reliability of inferences drawn from the measurement: e.g., that, with 98% probability, the actual length is betwen 25 and 26 centimeters. But these probabilities are clearly inductive. They cannot be chances, because the measurement cannot be regarded as a trial with different actual values of the length as possible outcomes. The one and only actual length was there before the measurement was made, and is in no sense an outcome of it (setting aside the disputed case of quantum measurement). So if inductive probability is taboo, a different way must be found of assessing the safety of inferences drawn from the measurement, these still being that the actual length lies between limits containing the measured value. Now the result of the measurement, the measured value, can certainly be regarded as an outcome of a chance trial; given the actual length, there may well be definite chances of an interval, centered on the measured value, including or excluding it. So sense can be made of the chance of a measurement leading one, by a rule of inference, to say that the length is between certain limits when it is not. This is the chance of the *rule* leading one into the error of accepting a false hypothesis about the length; and there may also be a definite chance of the rule leading one into the error of rejecting a true hypothesis. The same account can be given of measuring any quantity, including in particular the measurement of chance itself by observing, e.g., the frequency of heads in many tosses of a coin. So in terms of the chances of these *two types of error**, and of the relative importance of avoiding them, the Neyman–Pearson theory is enabled to evaluate the reliability of a number of important rules of statistical inference* without appealing to inductive probabilities and without retreating into subjectivism. Chance therefore has an important, if controversial, role in the foundations of statistical inference.

In many respects the most obvious account of chance is that given by the frequency theory* of Venn [5] and his successors, which superseded the classical theory and which has remained the dominant theory until quite recently. On it the chance of a trial of some kind F having an outcome of kind G is the relative frequency f with which G outcomes occur in the class of all F trials. Thus the chance of heads being the outcome of tossing a coin is taken to be the relative frequency with which heads is the outcome of all tosses of that coin and of coins just like it. Chances are thus identified with the content of statistical laws to the effect that $100f$ percent of F trials have G outcomes.

For some F, the set of F events may be infinite (e.g., all possible trials to see whether any radium atom decays in any period of a year), and proportions in infinite sets are strictly undefined. Here the chance of a *G* outcome is identified with the limiting proportion of *G* outcomes in a sequence of successively more inclusive finite sets of *F* trials. Not all such proportions or limits are chances, however; we should not speak of chance if a *G* event were the outcome of every second *F* event. So arbitrary selections from the set of all *F* events are also required to have the same limiting proportion of *G* outcomes—a proviso known for obvious reasons as the principle of the impossibility of gambling systems! *See* GAMES OF CHANCE.

A single trial of course belongs to many sets, with generally different proportions of their members having *G* outcomes. To get an individual trial's chance of being so followed, these sets are intersected until further intersections cease to give different proportions. Only then are the statistics taken to yield chance as a property of an individual trial, a statistical law applicable to the so-called "single case." Suppose, for example, that only smoking and gender affect the proportion of people getting cancer, and that I am a nonsmoking male. My chance of cancer is equated with the eventually cancerous proportion of nonsmoking males, regardless of their other attributes.

If the world is in fact deterministic, this process of repeated intersection will not yield nontrivial chances unless some limit is put on the sets to be intersected. Otherwise, any apparent chance other than 1 or 0 will be destroyed by intersection with the sets corresponding to different values of whatever "hidden variables" in fact determine the actual outcome of the trial (e.g., the imperceptible differences of initial conditions that make the difference between a tossed coin landing heads rather than tails). In practice, however, intersections are usually limited by the taxonomy of some theory within which the supposed statistical law is explicitly or implicitly formulated. (The completeness of the taxonomy usually remains an open question, and so therefore do the theory's implications for determinism.)

The frequency theory explains many of the characteristics of chance listed earlier. Like the classical theory, it explains chance being a kind of probability, because relative frequencies automatically satisfy the probability calculus. It obviously explains why a chance is a limiting frequency on repeated trials, since it makes this the definition of chance. This property, moreover, has the virtues of being plainly empirical, objective, and not relative to the evidence we have about the trial; and we have seen above how a frequency sense can be made of chance as a property of a single trial.

The theory's Achilles' heel is its inability to explain why the chance of an outcome is the fair quotient for betting that it will occur. The reason the frequency theory fails to explain this aspect of chance is that it makes the chance of a *G* event depend logically on the existence and fate of all other *F* trials in the whole universe: past, present, and future; and this seems quite irrelevant to the prospects of winning a bet on the outcome of one particular trial.

The usual response to this objection is to abandon *actual* frequencies (and their limits) and to redefine chance instead as the limiting frequency of *G* outcomes in an endlessly long run of sufficiently similar *hypothetical* trials. This, however, is to abandon the frequency theory in all but name, since "sufficiently similar" means "similar enough to have the same chance." But then the laws of large numbers* will suffice to get as high a chance as anyone may demand for the individual's chance being as close as one likes to such a hypothetical limiting frequency; and these laws are theorems of the probability calculus, however probability is then interpreted. Chance's connection with merely hypothetical long-run frequencies neither depends on nor supports a frequency interpretation of what chance itself is.

Once the shift is made from actual to merely hypothetical frequencies, chance is being regarded as a disposition or "propensity". The theory of chance as a propensity was explicitly distinguished from frequency theories by Popper in 1957 [4], since when several propensity theories of chance have been developed. The analogy is with nonstatistical dispositions such as fragility: what gives an object this property is that it would break if dropped, but it has the property whether it is dropped or not. Analogously, a coin has its chance of landing heads on a toss whether it is tossed again or not, even though the chance is defined by what the limiting frequency of heads would be in an infinite class of similar tosses.

This type of propensity theory, however, is still based on frequency and, although it shares the virtues of frequency theory, it also fails to explain why chances are fair betting quotients. Why should a proportion of cancers in other people, actual or hypothetical, be the measure of *my* prospects of getting the disease? The other type of propensity theory defines chance instead in terms of the prospects of an actual *F* trial having a *G* outcome, letting the laws of large numbers take care of chance's connection with the limiting frequencies in hypothetical sets of similar trials.

This alternative approach starts from the subjective or personalist concept of degree of belief. My degree of belief in a trial's having a *G* outcome measures how much I expect it to. To say there is a definite prospect that it will is to say that the trial is such as to make some definite degree of belief* in the outcome objectively right; and this degree of belief is what the chance is taken to be.

In saying this we do not need to credit people with actual degrees of belief admitting of indefinitely precise measurement, any more than ordinary objects actually have indefinitely precise shapes, temperatures, masses, etc. Any quantitative state, whether physical or psychological, is no doubt best represented by an interval of values rather than by a single value. But we can still say that chances make some degrees of belief objectively right, meaning that the intervals of values representing the strengths of people's actual belief states should include these values.

Chances defined in this way are probabilities, because degrees of belief have been shown by subjectivists to have a measure satisfying the probabilty calculus. Specifically, so-called coherent betting quotients (CBQs) satisfy the calculus. *See* COHERENCE.

The next problem is to say what makes a degree of belief objectively right in these contexts, and to do that this theory also invokes statistical laws. What such a law says in this theory is that all trials similar to this one in a certain respect (e.g., in being *F*) have the same prospect, i.e., chance of having a *G* outcome. These chances being supposed to be objective, and the trials independent, the laws of large numbers apply to them. It can then be shown that in repeated bets on such trials (under the restrictions prescribed to make CBQs measure degrees of belief), a gambler can know he or she will eventually break even only at a CBQ equal to the chance, and this in the circumstances is the best result he or she could know of. The law therefore gives this CBQ a peculiar virtue in these hypothetical compulsory bets, and consequently gives an objective rightness to the degree of belief which the CBQ measures.

The advantage this propensity theory has over its frequency-based rivals is that it can explain why chances are objectively fair betting quotients. The objection to it is that, when it says that certain trials have an empirical objective, nonrelational property of chance, it fails to say what this property is. In other words, it fails to say what in the real world, if not frequencies, makes statements of chance objectively true; and unlike its rivals cannot even offer hypothetical frequencies in lieu of actual ones.

It does not, however, follow that nothing objective makes probability statements true and therefore that subjective theories of probability must be adopted *faute de mieux*. Truth conditions can be supplied for any kind of chance statement at least; i.e., by statistical laws. Thus the laws of radioactivity show the nuclear structure of radium to be what makes true the statement of its half-life, i.e., makes objectively appropriate a degree 0.5 in the belief that a radium atom will decay in that time. There is again an analogy with nonstatistical dispositions such as fragility: what makes true the statement that a glass is fragile (when it is not being dropped and its fragility is not directly observable) is something about its microstructure. The microstructure cannot, of course, be deduced from the meaning of "fragile," nor need any one microstructure be common to all kinds of fragile objects; yet statements about the fragility of objects can be objectively true, and it is the microstructure of those objects that makes them so. So it is with statements about chances.

Probability, then, need not be denied objectivity just because it corresponds rather to degrees of belief in other things than to full belief in some one thing called "probability.". We might, on the contrary, say that a full belief is true just in case a degree close to 1 in that belief is objectively appropriate. So far from objective truth being beyond the reach of probability statements, truth can plausibly be regarded as an extreme case of objective probability.

So much may briefly be said in defense of objective probability, and of a belief-based propensity theory's declining to say with what other objective property of trials chances are to be identified. Objective chance may be distinguished from merely inductive probabilities by the fact that chances, like other physical attributes of events, are credited with having causes and effects. Thus it is held that smoking causes cancer, although the connection between the two is only statistical, not deterministic. What this means is that smoking causes an increase in the chance that the smoker will contract cancer. Conversely, an atomic explosion is caused by changing the chance of a mass of fissile material absorbing its own fission products. No merely subjective or relational account of chance does justice to the way chances are thus embedded in the network of causes and effects which determine the history of the world and both enable and limit our actions within it.

Indeed, the clearer it becomes that the fundamental laws of physics are irreducibly statistical (*see* STATISTICAL PHYSICS), the greater the pressure to involve chance itself in the analysis of causation. The relation of cause to effect can no longer be credibly restricted to deterministic situations; rather, these must be regarded as extreme cases of probabilistic causation. Therefore, attempts have recently been made to develop theories of physical causation based on the concept of chance; these involve distinguishing causal factors by how the chances of events depend on their presence or absence.

These developments make it unlikely that the concept of chance can be analyzed in terms of causation, since causation is itself being taken to depend on chance. They do seem to the author to strengthen the appeal of a belief-based propensity theory of chance, since a central function of the cause–effect relation is to give reason for expecting an effect once its cause is observed. If being a cause involves raising the chance of an effect, where this is understood as raising the objectively right degree of belief in the effect's occurrence, this aspect of causation will be readily accounted for.

Serious problems remain to be overcome in the theory of chance, not least in the problematic area of quantum theory. But the progress made in recent decades in devising an objective theory, free of certain difficulties inherent in the frequency theory, which have driven many philosophers and statisticians to deny the existence of objective probabilities, bodes well for the rehabilitation of the concept of chance.

Literature

The following are classic sources or provide useful expositions and further references for the theories of chance discussed. For the classical theory, see P. S. de Laplace, *A Philosophical Essay on Probabilities* (Dover, New York, 1951), and W. Kneale, *Probability and Induction* (Clarendon Press, Oxford, 1949). For the logical relation theory, see J. M. Keynes, *A Treatise on Probability* (Macmillan, London, 1921). For the distinction between statistical and inductive probability, see R. Carnap, *Logical Foundations of*

Probability (2nd ed., University of Chicago Press, Chicago, 1962). For more general discussion of kinds of probability, see B. Russell, *Human Knowledge* (Humanities Press, New York, 1948). For the frequency theory, see J. Venn, *The Logic of Chance* (3rd ed., Chelsea, New York, 1962), and W. C. Salmon, *The Foundations of Scientific Inference* (University of Pittsburgh Press, Pittsburgh, Pa., 1967). For frequency-based propensity theory, see K. R. Popper, "The propensity interpretation of the calculus of probability . . . " and discussion in S. Körner, ed., *Observation and Interpretation* (Dover, New York, 1962). I. Hacking, *Logic of Statistical Inference* (Cambridge University Press, Cambridge, 1965), and I. Levi, *Gambling with Truth* (Alfred A. Knopf, New York, 1967). For belief-based propensity theory, see D. H. Mellor, *The Matter of Chance* (Cambridge University Press, Cambridge, 1971). For subjective theory, see F. P. Ramsey, *Foundations* (Routledge & Kegan Paul, London, 1978), and L. J. Savage, *The Foundations of Statistics* (Wiley, New York, 1954). For the link between chance and causation, see W. C. Salmon, "Theoretical explanation" and discussion in S. Körner, ed., *Explanation* (Blackwell, Oxford, 1975).

References

[1] Carnap, R. (1962). *Logical Foundations of Probability*, 2nd ed. University of Chicago Press, Chicago.

[2] Keynes, J. M. (1921). *A Treatise on Probability*. Macmillan, London.

[3] Laplace, P. S. de (1951). *A Philosophical Essay on Probabilities*. Dover, New York.

[4] Popper, K. R. (1962). In *Observation and Interpretation*, S. Körner, ed. Dover, New York.

[5] Venn, J. (1962). *The Logic of Chance*, 3rd ed. Chelsea, New York.

Acknowledgment

This article was written during the author's tenure of a Radcliffe Fellowship and of a British Academy Overseas Visiting Fellowship, for which he is indebted to the Radcliffe Trust and the British Academy.

(AXIOMS OF PROBABILITY
BAYESIAN INFERENCE
CHANCE (I)
FOUNDATIONS OF PROBABILITY
FREQUENCY INTERPRETATION OF PROBABILITY
STATISTICAL INFERENCE)

D. H. MELLOR

CHANGEOVER DESIGNS

In most experimental designs (*see* DESIGN AND ANALYSIS OF EXPERIMENTS) each subject is allocated a single treatment or combination of treatments and receives no other during the course of the experiment. Suppose, for example, that a new drug is to be tested in a clinical trial*. Patients are divided into two groups, one receiving the drug and the other a control. The contrast of drug vs. control is confounded* with differences between the two groups of patients that persist throughout the trial. If these differences are large, estimates of the effects of the drug are imprecise.

In another type of design the total duration of the experiment is divided into two or more periods and the treatment of each subject changed from one period to the next. A design of this type is called a changeover design. (The term crossover design is also used.) In the clinical example one group of subjects receives the new drug only in the second period; the other group receives the new drug in the first period but then reverts to control status in the second (Design 1).

More complicated changeover designs are also available. We denote by t the number of treatments and by p the number of periods; the total number of subjects is n. In all designs subjects are represented by rows and

Design 1 Two-Period Design for Two Treatments

	Group of Subjects	
Period	i	ii
I	1	2
II	2	1

periods by columns. Each treatment is allocated to an equal number of subjects in each period. Most but not all designs are binary in the sense that no treatment is allocated to the same subject in more than one period.

Changeover designs are most effective when the following conditions are met: (a) there are large differences between subjects, (b) a subject responds fully to a treatment soon after application starts, and (c) ceases to respond soon after the treatment is discontinued. We also require that (d) the experimental environment is stable so that treatment effects* are the same in each period. Under these conditions changeover designs test and estimate treatment responses more efficiently than do continuous designs. When the conditions are not met, interpretation of results may be complicated and some or all of the gain in accuracy is lost.

Before introducing changeover designs into a new field of application it is advisable to check their efficiency; Finney [8] gives an example of the calculations.

DESIGNS FOR MORE THAN TWO TREATMENTS

Design 1 is a 2×2 Latin square*. Larger Latin squares can also be used, but the total time available for a trial often imposes a constraint on the number of periods.

Use of designs with $t > p$ results in loss of information on some treatment comparisons. The four treatments of Design 2, for example, are two preparations A and B, each tested at a high level and a low level. The design gives sensitive tests of average differences in responses between the two preparations and between the two dose levels, but the difference in slope is confounded with subjects.

In Design 3 the loss of information is spread evenly over all treatment contrasts.

Design 2 Two-Period Design for a Four-Point Assay

A_1	A_2	B_1	B_2
B_2	B_1	A_2	A_1

Design 3 Youden Square* Design

1	2	3	4	5	6	7
2	3	4	5	6	7	1
4	5	6	7	1	2	3

This design is a balanced incomplete block design (*see* BLOCKS, BALANCED INCOMPLETE) arranged in rows and columns. Designs of this type with $n = t$ are available for only a small number of combinations of t and p. Cyclic* incomplete block designs [14] can also be arranged in rows and columns and used as changeover designs. Amounts of information confounded with subject differences vary slightly from one treatment comparison to another. In all these designs a conventional row-and-column* analysis is appropriate, provided that conditions (b), (c), and (d) are met.

ESTIMATION OF RESIDUAL EFFECTS

Suitably chosen, a changeover design can still be useful when residual effects* of treatments are carried over from one period to the next. The residual effects must, however, be small. They are usually estimated by incorporating additional constants into the row-and-column analysis [4]. For example, the total response of subject 4 in the third period of Design 3 is taken to be the sum of the direct effect of treatment 7 and the residual of treatment 5 carried over from the second period.

Direct effects are adjusted in the analysis to allow for differences between residual effects. Unnecessary adjustment results in loss of accuracy, but failure to adjust for real residual effects can be even more harmful; treatment differences are underestimated and errors overestimated, so that the chance of detecting treatment differences is reduced in two ways.

BALANCED CHANGEOVER DESIGNS

Special designs are required to ensure accuracy in the adjustment for residual effects. Design 5, for example, is five times as effi-

Design 4 Latin Square Design for Four Treatments

1	2	3	4
2	3	4	1
3	4	1	2
4	1	2	3

Design 5 Williams' Balanced Design for Four Treatments

1	2	3	4
2	4	1	3
3	1	4	2
4	3	2	1

cient as Design 4 in the estimation of adjusted direct effects. Efficiency depends on the values of λ_{ij}, the number of subjects receiving treatment i in some period followed by treatment j in the next. The narrower the range of different values of λ_{ij}, the more efficient the designs. Design 5 is efficient because all λ_{ij} with $i \neq j$ are equal in value. By contrast, Design 4 is inefficient because the λ_{ij} vary widely (from 0 to 3).

A Latin square with equal λ_{ij} is said to be column-complete [7, Sec. 2.3]. This description can also be applied to changeover designs. For example, Designs 5 and 6 are column-complete.

Column completeness is a combinatorial property. Statisticians are more interested in variance properties such as balance or efficiency. A changeover design is balanced if (1) all normalized contrasts between adjusted direct effects have equal variance, and (2) all normalized contrasts between residual effects have equal variance. A column-complete changeover design is balanced when $t = p$ but must satisfy additional conditions when $t > p$ [16].

In general, two-period changeover designs do not provide useful estimates of residual effects, but Grizzle [9] has pointed out that residual effects can be estimated when differences between periods can be ignored.

Balanced designs exist for any p larger than 2 and any t such that $t \geqslant p$ and a complete set of $t - 1$ orthogonal $t \times t$ Latin squares exist [4, 16]. The number of subjects is $t(t - 1)$.

Williams [18] introduced more economical balanced designs such as Design 5 for any even t and $n = p = t$. Column-complete Latin squares and hence balanced changeover designs with $n = t$ exist for some odd t (e.g., $t = 9$; see Hedayat and Afsarinejad [11]) but not for $t = 3$, 5, or 7 [7]. *See* LATIN SQUARE DESIGNS. Williams [18] also provided a series of balanced designs for any odd t and $n = 2t$, $p = t$.

Design 6 is a balanced changeover design with $t > p$. This design is column-complete and has the balance properties of a balanced incomplete block design with blocks given by the columns; the incomplete block design obtained by deleting the fourth period is also balanced.

Design 6 Balanced Design

1	2	3	4	5	6	7	1	2	3	4	5	6	7
2	3	4	5	6	7	1	7	1	2	3	4	5	6
4	5	6	7	1	2	3	5	6	7	1	2	3	4
7	1	2	3	4	5	6	2	3	4	5	6	7	1

NEARLY BALANCED CHANGEOVER DESIGNS

Many more designs are available if the balance condition is relaxed. Davis and Hall [6] give details of cyclically generated designs with n equal to t or $2t$ and $t > p$. *See* CYCLIC DESIGNS. These designs are not balanced but are chosen so that all normalized contrasts between adjusted direct effects have approximately equal variance; similarly normalized contrasts between residual effects have approximately equal variance.

Values of λ_{ij} again provide guidance on the efficiency of a design. They should vary as little as possible over different pairs of treatments. For example, all λ_{ij} in Design 7 are either 0 or 1.

The Davis and Hall [6] designs require a slightly more complicated analysis than balanced designs, but this is a small price to pay for their economy and availability.

Design 7 Nearly Balanced Design

1	2	3	4	5	6	1	2	3	4	5	6
4	5	6	1	2	3	6	1	2	3	4	5
5	6	1	2	3	4	2	3	4	5	6	1

For any combination of t, p, and n such that n/t is an integer, designs exist that are in some sense optimal, e.g., with minimum average variance among direct effects adjusted for residual effects. There is some interest in identifying optimal designs. Hedayat and Afsarinejad [11] report recent work in this field.

Design 8 Orthogonal* Design for Two Treatments

1	2	1	2
2	2	1	1
2	1	2	1
1	1	2	2

OTHER CHANGEOVER DESIGNS

Design 8 is the simplest example of a class of designs with the valuable property that direct effects and residual effects are orthogonal [1]. These designs differ from conventional changeover designs in that a treatment can be repeated in consecutive periods. Orthogonality is achieved by arranging that the λ_{ij} are equal for all pairs of treatments (i, j), including those with $i = j$.

Design 9 is a double changeover design based on Design 8. It provides orthogonal estimates of the direct effects of A, B, and AB and the residual effects of A and B. The residual effects interaction AB is aliased* with the direct-effects interaction.

Design 9 2 × 2 Factorial* Design

(1)	*a*	*b*	*ab*	(1)	*a*	*b*	*ab*
a	*ab*	(1)	*b*	*b*	(1)	*ab*	*a*
ab	*b*	*a*	(1)	*ab*	*b*	*a*	(1)
b	(1)	*ab*	*a*	*a*	*ab*	(1)	*b*

Replacement of the symbols (1), a, b, and ab in Design 9 by 1, 2, 3, and 4 gives Berenblut's [2] design for four equally spaced doses or amounts. Direct effects and the linear component of residual effects are orthogonal. When quadratic and cubic residual effects can be ignored, this design provides more accurate estimates of direct effects than does Design 5.

Literature

Changeover designs have been used primarily in clinical trials* [12], bioassay* [8], and animal nutrition experiments. Hedayat and Afsarinejad [10] give many references for these and other applications. General accounts are given in several textbooks including those by Cochran and Cox [3], Cox [5], John [15], and John and Quenouille [13]. Tables are available of balanced and partially balanced designs [17] and nearly balanced designs* [6]. Many of the cyclic block designs cataloged by John et al. [14] are suitable for changeover trials when there are no residual effects.

References

[1] Berenblut, I. I. (1964). *Biometrics*, **20**, 707–712.

[2] Berenblut, I. I. (1967). *J. R. Statist. Soc. B*, **29**, 370–373.

[3] Cochran, W. G. and Cox, G. M. (1966). *Experimental Designs*, 2nd ed. Wiley, New York, Chap. 4.

[4] Cochran, W. G., Autrey, K. M. and Cannon, C. Y. (1941). *J. Dairy Sci.*, **24**, 937–951.

[5] Cox, D. R. (1958). *Planning of Experiments*, Wiley, New York, Chap. 13.

[6] Davis, A. W. and Hall, W. B. (1969). *Biometrika*, **56**, 283–293.

[7] Dénes, J. and Keedwell, A. D. (1974). *Latin Squares and Their Applications*. English Universities Press, London/Academic, New York.

[8] Finney, D. J. (1978). *Statistical Methods in Biological Assay*, 3rd ed. Charles Griffin, London, Chap. 10.

[9] Grizzle, J. (1965). *Biometrics*, **21**, 467–480.

[10] Hedayat, A. and Afsarinejad, K. (1975). In *A Survey of Statistical Design and Linear Models*, J. N. Srivastava, ed. North-Holland, Amsterdam.

[11] Hedayat, A. and Afsarinejad, K. (1978). *Ann. Statist.*, **6**, 619–628.

[12] Hill, A. B. (1966). *Principles of Medical Statistics*, 8th ed. Oxford University Press, Oxford.

[13] John, J. A. and Quenouille, M. H. (1977). *Experiments: Design and Analysis*. Charles Griffin, London, Chap. 11.

[14] John, J. A., Wolock, F. W., and David, H. A. (1972). Cyclic Designs. *Natl. Bur. Stand. (U.S.) Appl. Math. Ser. 62* (Washington, D.C.).

[15] John, P. W. M. (1971). *Statistical Design and Analysis of Experiments*. Macmillan, New York, Chap. 6.

[16] Patterson, H. D. (1952). *Biometrika*, **39**, 32–48.

[17] Patterson, H. D. and Lucas, H. L. (1962). Change-Over Designs *N.C. Agric. Exper. Stn. Tech. Bull. No. 147.*

[18] Williams, E. J. (1949). *Aust. J. Sci. Res.*, **2**, 149–168.

(BIOASSAY
CHI-SQUARE TESTS
CLINICAL TRIALS
CYCLIC DESIGNS
LATIN SQUARE DESIGNS
REPEATED MEASUREMENT DESIGNS
ROW AND COLUMN DESIGNS)

H. D. PATTERSON

CHANGE-POINT MODEL

A model in which there is a change in the mathematical form at one or more values of certain variables. A common form of change-point model arises in representation of time series* as a regression* on time of form

$$E[Y \mid t] = f(t),$$

where t represents time, and $f(t)$ has a specific form, depending on the interval $t_i < t \leqslant t_{i+1}$ wherein t lies. The interval boundaries t_i are the *change points*.

As a simple example, let Y be normally distributed with expected value η and variance σ^2, where

$$\eta = \eta_i \qquad \text{for } t_i < t < t_{i+1}.$$

This might be an appropriate model for results of monitoring a continuous production process subject to sudden shifts in quality. Problems of statistical inference* associated with detection of such sudden changes have been discussed by Bartholomew [1].

Reference

[1] Bartholomew, D. J. (1959). *Biometrika*, **46**, 36–48, 328–335.

CHAPMAN–KOLMOGOROV EQUATION

Let $P_{ij}(t)$ be the transition probabilities* of a discrete-state continuous-time Markov process*. The Chapman–Kolmogorov equation states that for any s in $(0, t)$, and any i and j (including $i = j$),

$$P_{ij}(t) = \sum_{k=0}^{\infty} P_{ik}(s) P_{kj}(t - s). \qquad (1)$$

This is a basic property of Markov processes with stationary transition probabilities.

The discrete-time version of the Chapman–Kolmogorov equation is obtained if it is supposed that t and s are integers. If we denote the matrix $\{P_{ij}(t)\}$ by $\mathbf{P}^{(t)}$, then (1) can be written $\mathbf{P}^{(t)} = \mathbf{P}^{(s)}\mathbf{P}^{(t-s)}$.

Equation (1) is used in deriving the differential equations in birth or birth-and-death processes*. Another important application is in the derivation of the higher-order stationary transition probabilities for discrete-time Markov chains.

Bibliography

Chapman, S. (1928). *Philos. Mag.*, 7th Ser. **5**, 630–636.

Chung, K. L. (1967). *Markov Chains with Stationary Transition Probabilities*, 2nd ed. Springer-Verlag, New York.

Kolmogorov, A. N. (1931). *Math. Ann.*, **104**, 415–458 (in German).

Kolmogorov, A. N. (1937). *Bull. Mosc. State Univ.*, **1**(3), 1–16 (in Russian).

(BIRTH-AND-DEATH PROCESSES
MARKOV PROCESSES)

CHARACTERISTIC FUNCTIONS

Let X be a (real-valued) random variable and let F_X be the distribution function of X

given by the relation $F_X(x) = \Pr[X \leqslant x]$, $x \in \mathbb{R}$. Then we note that F_X is a nondecreasing, right-continuous function on $\mathbb{R}$ satisfying the relations $F_X(-\infty) = 0$ and $F_X(+\infty) = 1$. The characteristic function ϕ_X of the random variable X (or of the distribution function F_X) is a complex-valued function defined on $\mathbb{R}$ by the formula

$$\phi_X(t) = E(e^{itX}) = \int_{-\infty}^{\infty} e^{itx}\, dF_X(x)$$
$$= \int_{-\infty}^{\infty} \cos tx\, dF_X(x)$$
$$+ i\int_{-\infty}^{\infty} \sin tx\, dF_X(x)$$
$$t \in \mathbb{R}, \qquad \text{where } i = \sqrt{-1}\,.$$

SOME ELEMENTARY PROPERTIES

1. The characteristic function ϕ of a distribution function F is uniformly continuous on $\mathbb{R}$ and moreover satisfies the relations (a) $\phi(0) = 1$; (b) $|\phi(t)| \leqslant 1$ for all $t \in \mathbb{R}$ and (c) $\phi(-t) = \overline{\phi(t)}$ for all $t \in \mathbb{R}$, where $\overline{\phi(t)}$ denotes the complex conjugate of $\phi(t)$.
2. (Relationship between the characteristic function of a distribution function and its moments). Let F be a distribution function and let ϕ be the characteristic function of F. Suppose that F has finite moments μ_k' up to order n, then ϕ has continuous derivatives up to order n, and the relation $\phi^{(k)}(0) = i^k \mu_k'$ holds for $1 \leqslant k \leqslant n$. Moreover, in this case, ϕ admits the expansion

$$\phi(t) = 1 + \sum_{k=1}^{n} \mu_k' \frac{(it)^k}{k!} + o(t^n)$$
$$\text{as } t \to 0.$$

Conversely, suppose that the characteristic function ϕ of a distribution function F admits an expansion of the form

$$\phi(t) = 1 + \sum_{k=1}^{n} \alpha_k \frac{(it)^k}{k!} + o(t^n)$$
$$\text{as } t \to 0.$$

Then F has finite moments μ_k' up to order n, if n is even, but up to order $n-1$, if n is odd. Moreover, in this case $\alpha_k = \mu_k'$ for $k \geqslant 1$.

For the proofs of properties 1 and 2, we refer to Laha and Rohatgi [5, Chap. 3, Sec. 2] and Lukács [7, Chap. 2, Secs. 1 and 3].

BASIC THEOREMS ON CHARACTERISTIC FUNCTIONS

Theorem 1 (Inversion Theorem). Let F be a distribution with characteristic function ϕ. Then the relation

$$F(a+h) - F(a) = \lim_{T\to\infty} \frac{1}{2\pi} \int_{-T}^{T} \frac{1 - e^{-ith}}{it} \times e^{-ita}\phi(t)\, dt$$

holds for $a \in \mathbb{R}$ and $h > 0$, whenever the points a and $a + h$ are continuity points of F.

The following important corollaries are immediate consequences of Theorem 1.

Corollary 1 (Uniqueness Theorem). Let F_1 and F_2 be two distribution functions with characteristic functions ϕ_1 and ϕ_2, respectively. Suppose that $\phi_1 = \phi_2$. Then $F_1 = F_2$.

In other words, a distribution function F is determined uniquely by its characteristic function ϕ.

Remark 1. Suppose that $\phi_1(t) = \phi_2(t)$ for all t belonging to some finite interval $[-a, a] \subset \mathbb{R}(a > 0)$. Then F_1 and F_2 are not necessarily identical.

Corollary 2 (Fourier Inversion Theorem). Suppose that the characteristic function ϕ is absolutely integrable on $\mathbb{R}$, that is, $\int_{-\infty}^{\infty}|\phi(t)|\, dt < \infty$. Then the corresponding distribution function F is absolutely continuous on $\mathbb{R}$. Moreover, in this case the probability density function $f = F'$ of the distribution function F is bounded and is uniformly continuous on $\mathbb{R}$ and is given by the formula

$$f(x) = \frac{1}{2\pi} \int_{-\infty}^{\infty} e^{-itx}\phi(t)\, dt \quad (x \in \mathbb{R}).$$

Remark 2. There exist absolutely continuous distribution functions whose characteristic functions are not absolutely integrable.

For the proof of Theorem 1 and its corollaries, we refer to Laha and Rohatgi [5, Chap. 3, Sec. 3] and Lukács [7, Chap. 3, Secs. 1 and 2].

Next, we introduce the concept of convolution of two distribution functions.

Let F_1 and F_2 be two distribution functions. Then the convolution of F_1 and F_2 is the distribution function F defined by the formula

$$F(x) = \int_{-\infty}^{\infty} F_1(x-y)\,dF_2(y) \quad (x \in \mathbb{R}),$$

where the integral on the right-hand side is an improper Riemann–Stieltjes integral. In this case, we write $F = F_1 * F_2$.

Theorem 2 (Convolution Theorem). Let F, F_1, and F_2 be three distribution functions with characteristic functions ϕ, ϕ_1, and ϕ_2, respectively. Then $F = F_1 * F_2$ if and only if $\phi = \phi_1\phi_2$.

Remark 3. The operation of convolution $*$ is commutative and associative.

Remark 4. Let X_1 and X_2 be two independent random variables with distribution functions F_1 and F_2 and characteristic functions ϕ_1 and ϕ_2, respectively. Then the random variable $X = X_1 + X_2$ has the distribution function $F = F_1 * F_2$ and characteristic function $\phi = \phi_1\phi_2$.

For the proof of Theorem 2 and the remarks following it, we refer to Laha and Rohatgi [5, Chap. 3, Sec. 4] and also Lukács [7, Chap. 3, Sec.3].

Next, we introduce the concepts of weak and complete convergence* of a sequence of distribution functions. *See* CONVERGENCE OF SEQUENCES OF RANDOM VARIABLES.

Let $\{G_n\}$ be a sequence of uniformly bounded, nondecreasing, right-continuous functions defined on $\mathbb{R}$. We say that $\{G_n\}$ converges weakly to a bounded, nondecreasing, right-continuous function G on $\mathbb{R}$ if $\lim_{n\to\infty} G_n(x) = G(x)$ at all continuity points x of G. In this case, we write $G_n \xrightarrow{w} G$, as $n \to \infty$. Moreover, the sequence $\{G_n\}$ is said to converge completely to G on $\mathbb{R}$, if (a) $G_n \xrightarrow{w} G$; (b) $G_n(\mp\infty) \to G(\mp\infty)$, as $n \to \infty$. In this case, we write $G_n \xrightarrow{c} G$, as $n \to \infty$.

Remark 5. The weak limit of the sequence $\{G_n\}$, if it exists, is unique.

The following theorem, which is due to P. Lévy, gives an important necessary and sufficient condition for the complete convergence of a sequence of distribution functions to a distribution function.

Theorem 3 (Lévy Continuity Theorem). Let $\{F_n\}$ be a sequence of distribution functions and let $\{\phi_n\}$ be the sequence of corresponding characteristic functions. Then the sequence $\{F_n\}$ converges completly to a distribution function F if and only if the sequence $\{\phi_n\}$ converges (pointwise) to a function ϕ on $\mathbb{R}$ (as $n \to \infty$), where ϕ is continuous at the point $t = 0$. Moreover, in this case the limit function ϕ is the characteristic function of the limit distribution function F.

An alternate version is the following.

Theorem 3′. Let $\{F_n\}$ be a sequence of distribution functions and let $\{\phi_n\}$ be the sequence of corresponding characteristic functions. Then the sequence $\{F_n\}$ converges completely to a distribution function F if and only if the sequence $\{\phi_n\}$ converges to a function ϕ (as $n \to \infty$) uniformly in every finite t-interval $[-T, T]$, $T > 0$. In this case, the limit function ϕ is the characteristic function of the limit distribution function F.

For the proofs of Theorems 3 and 3′ and the bibliography, we refer to Laha and Rohatgi [5, Chap. 3, Secs. 1 and 5] and Lukács [7, Chap. 3, Secs. 4–6].

Remark 6. Theorem 3 and its alternative version 3′ play a fundamental role in the study of the limit distributions* of sums of independent random variables. (See Gnedenko and Kolmogorov [2].)

For various applications of Theorems 1, 2, and 3 to probability theory and mathematical statistics, we refer the reader to Laha and Rohatgi [5, Chap. 3, Sec. 7]. For their applications to characterization problems in mathematical statistics, we refer to Kagan et al. [3] and also Laha and Lukács [4].

SOME CRITERIA FOR CHARACTERISTIC FUNCTIONS

We next give an important necessary and sufficient condition for a complex-valued function to be the characteristic function of a distribution function. For this purpose, we first introduce the concept of a positive-definite function on $\mathbb{R}$ which is due to S. Bochner.

Let ϕ be a complex-valued function defined on $\mathbb{R}$. Then ϕ is said to be positive-definite on $\mathbb{R}$ if for every positive integer $N \geqslant 1$, for every real $t_1, t_2, \ldots, t_N$ and for every complex $\lambda_1, \lambda_2, \ldots, \lambda_N$, the sum $\sum_{j=1}^{N} \sum_{k=1}^{N} \lambda_j \bar{\lambda}_k \phi(t_j - t_k) \geqslant 0$.

Remark 7. Let ϕ be positive-definite on $\mathbb{R}$. Then (a) $\phi(0) \geqslant 0$; (b) $\phi(-T) = \overline{\phi(t)}$; $|\phi(t)| \leqslant \phi(0)$.

A positive-definite function ϕ on $\mathbb{R}$ ($\phi \not\equiv 0$) is said to be normalized if $\phi(0) = 1$.

Theorem 4 (Bochner). Let ϕ be a complex-valued function defined on $\mathbb{R}$. Then ϕ is a continuous normalized positive-definite function on $\mathbb{R}$ if and only if ϕ is the characteristic function of a distribution function F.

Remark 8. Let $\{\phi_n : n \geqslant 1\}$ be a sequence of characteristic functions and let $\{\alpha_n : n \geqslant 1\}$ be a sequence of real numbers such that $\alpha_n \geqslant 0$ for $n \geqslant 1$ and $\sum_{n=1}^{\infty} \alpha_n = 1$. Then it follows at once from Theorem 4 that $\sum_{n=1}^{\infty} \alpha_n \phi_n$ is also a characteristic function. In particular, if ϕ is a characteristic function, then $e^{\alpha(\phi - 1)}$is also a characteristic function for any $\alpha > 0$.

Next we give a useful sufficient condition, due to G. Pólya, for a real-valued function to be a characteristic function.

Theorem 5 (Pólya). Let ϕ be a real-valued continuous function defined on $\mathbb{R}$ satisfying the following conditions:

(a) $\phi(0) = 1$.
(b) $\phi(-t) = \phi(t)$ for all $t \in \mathbb{R}$.
(c) ϕ is convex on $(0, \infty)$.
(d) $\lim_{|t| \to \infty} \phi(t) = 0$.

Then ϕ is the characteristic function of an absolutely continuous distribution function F.

Remark 9. Some examples of real-valued functions that satisfy the conditions of Theorem 5 are as follows:

$$\text{(a)} \quad \phi(t) = e^{-|t|}; \qquad \text{(b)} \quad \phi(t) = \frac{1}{1 + |t|}$$

$$\text{(c)} \quad \phi(t) = \begin{cases} 1 - |t| & \text{for } 0 \leqslant |t| \leqslant \frac{1}{2} \\ \dfrac{1}{4|t|} & \text{for } |t| > \frac{1}{2} \end{cases}$$

$$\text{(d)} \quad \phi(t) = \begin{cases} 1 - |t| & \text{for } |t| \leqslant 1 \\ 0 & \text{for } |t| > 1 \end{cases}$$

Here we note that the characteristic functions in (b) and (c) are not absolutely integrable, but still they are characteristic functions of absolutely continuous distribution functions (see Remark 2). For the proofs of Theorems 4 and 5, we refer to Laha and Rohatgi [5, Chap. 3, Sec. 6) and Lukács [7, Chap. 4, Secs. 2 and 3].

INFINITELY DIVISIBLE DISTRIBUTIONS

Note that in view of Theorem 2 (the convolution theorem), the product of two characteristic functions is also a characteristic function. We consider the problem of decomposing (or factorizing) a characteristic function as a product of two or more characteristic functions. A characteristic function ϕ is said to be decomposable if it can be written as the product $\phi = \phi_1 \phi_2$, where ϕ_1 and ϕ_2 are characteristic functions of some nondegenerate distributions. In this case ϕ_1 and ϕ_2 are called the factors of ϕ. A characteristic function that admits only trivial decomposition is said to be indecomposable. For example, the characteristic function of a purely discrete distribution function with only two discontinuity points is indecomposable.

A distribution function F is said to be infinitely divisible (i.d.) if for every positive integer $n \geqslant 1$, there exists a distribution function F_n such that $F = F_n * F_n * \cdots * F_n$ (n times). Equivalently, a characteristic function ϕ is said to be infinitely divisible (i.d.) if for every positive integer $n \geqslant 1$, there exists a characteristic function ϕ_n such that $\phi = \phi_n^n$.

The normal*, Poisson*, gamma*, and Cauchy* distributions are well-known examples of i.d. distribution functions. The binomial distribution is not i.d.

Some elementary properties of i.d. characteristic functions are as follows:

1. An i.d. characteristic function has no real zeros.
2. The product of a finite number of i.d. characteristic functions is also an i.d. characteristic function.
3. A characteristic function that is the limit of a sequence of i.d. characteristic functions is i.d.

For the proofs of these properties, we refer to Laha and Rohatgi [5, Chap. 4, Sec. 1] and Lukács [7, Chap. 5, Secs. 2 and 3].

Next we give a basic result on the representation of i.d. characteristic functions which is due to P. Lévy and A. Ya. Khintchine.

Theorem 6 (Lévy–Khintchine Representation). A complex-valued function ϕ defined on $\mathbb{R}$ is an infinitely divisible characteristic function if and only if its logarithm $\ln \phi$ admits the representation

$$\ln \phi(t) = i\alpha t + \int_{-\infty}^{\infty} \left(e^{itx} - 1 - \frac{itx}{1+x^2} \right) \times \frac{1+x^2}{x^2}\, dG(x), \qquad t \in \mathbb{R},$$

where $\alpha \in \mathbb{R}$ and G is a bounded, nondecreasing right-continuous function on $\mathbb{R}$ such that $G(-\infty) = 0$ and $G(+\infty) < \infty$. Here the value of the integrand at $x = 0$ is defined by continuity as

$$\left(e^{itx} - 1 - \frac{itx}{1+x^2} \right) \frac{1+x^2}{x^2} \Bigg]_{x=0} = -\frac{t^2}{2}.$$

Moreover, α and G are determined uniquely by ϕ.

For the proof and bibliography, we refer to Laha and Rohatgi [5, Chap. 4, Sec. 1] and also Lukács [7, Chap. 5, Sec. 5]. *See also* INFINITE DIVISIBILITY for further details.

GENERAL DECOMPOSITION THEOREMS

We now give some general theorems concerning the decomposition of characteristic functions which are due to A. Ya. Khintchine.

Theorem 7 (Khintchine). Let ϕ be an arbitrary characteristic function. Then ϕ admits a decomposition $\phi = \theta\psi$, where ψ is a characteristic function without any indecomposable factors, while θ can be written as the convergent product of at most a countable sequence of indecomposable characteristic functions.

Theorem 8 (Khintchine). A characteristic function that has no indecomposable factors is i.d.

Remark 10. The decomposition of the characteristic function ϕ in Theorem 7 is not unique. The converse statement of Theorem 8 is not necessarily true. (There exist i.d. characteristic functions that can be written as the product of indecomposable factors.)

For the proofs and bibliography, we refer to Laha and Rohatgi [5, Chap. 4, Sec. 3] and Lukács [7, Chap. 6, Sec. 2].

ANALYTIC CHARACTERISTIC FUNCTIONS

In the following t and v denote real variables and $z = t + iv \, (i = \sqrt{-1})$ a complex variable.

A characteristic function ϕ is said to be an analytic characteristic function if there exists a complex-valued function $\theta(z)$ of the complex variable z which is holomorphic (or regular) in a circle $|z| < \rho$ $(\rho > 0)$ and a positive real number $\delta > 0$ such that $\phi(t) = \theta(t)$ for $|t| < \delta$. In other words, an analytic characteristic function is a characteristic function that coincides with a holomorphic function in some neighborhood of zero.

As some well-known examples of distributions with analytic characteristic functions, we mention the binomial*, Poisson*, normal*, and gamma* distributions. On the other hand, the Cauchy* distribution is an example of a distribution whose characteristic function is not analytic. We first give an important result on analytic characteristic functions which is due to D. A. Raikov.

Theorem 9 (Raikov). Let ϕ be an analytic characteristic function and let F be the corresponding distribution function. Then ϕ can be continued analytically in the complex z-plane as a function that is holomorphic in a horizontal strip $-\alpha < \operatorname{Im} z < \beta$ $(\alpha > 0, \beta > 0)$ containing the real axis. Moreover, ϕ admits the Fourier integral representation

$$\phi(z) = \int_{-\infty}^{\infty} e^{izx}\, dF(x) \qquad (-\alpha < \operatorname{Im} z < \beta)$$

in this strip.

This strip is either the whole complex z-plane, in which case ϕ is an entire characteristic function, or it has one or two horizontal boundary lines. In the latter case, the purely imaginary points on the boundary of this strip are the singularities of the function $\phi(z)$ nearest to the real axis.

Remark 11. Let ϕ be an analytic characteristic function which is holomorphic in the strip $-\alpha < \operatorname{Im} z < \beta$; then the inequality

$$\sup_{t \in \mathbb{R}} |\phi(t + iv)| \leqslant \phi(iv)$$

holds for $v(-\alpha < v < \beta)$.

The following results deal with the relationship between an analytic characteristic function and the corresponding distribution function.

Theorem 10. Let ϕ be a characteristic function and let F be the corresponding distribution function. Then ϕ is analytic if and only if

(a) F has finite moments μ_k' of all orders $k \geqslant 1$.

(b) There exists a positive real number $\gamma > 0$ such that the inequality $|\mu_k'| \leqslant k!\gamma^k$ holds for all $k \geqslant 1$.

Remark 12. We note that the conditions (a) and (b) of Theorem 10 are a set of necessary and sufficient conditions for the existence of the moment generating function of F. Hence it follows that a distribution function has an analytic characteristic function if and only if its moment generating function * exists.

Theorem 11. Let ϕ be a characteristic function and let F be the corresponding distribution function. Then ϕ is analytic if and only if there exists a positive real number $R > 0$ such that the relation

$$\begin{cases} 1 - F(x) = 0(e^{-rx}) & \text{as } x \to \infty \\ F(-x) = 0(e^{-rx}) & \text{as } x \to \infty \end{cases}$$

hold for any $r (0 < r < R)$. In this case, ϕ is holomorphic at least in the strip $|\operatorname{Im} z| < R$.

In particular, if $R = +\infty$, then the foregoing relations hold for all $r > 0$ and $\phi(z)$ is an entire characteristic function.

Remark 13. Let F be a distribution function which is concentrated in a finite interval $[a, b]$, that is, $F(x) = 0$ for all $x \leqslant a$ and $F(x) = 1$ for all $x \geqslant b$. Then clearly the condition of Theorem 11 is trivially satisfied for all $r > 0$, so that the corresponding characteristic function is entire.

Next we give two results on entire characteristic functions.

Theorem 12. Let ϕ be an entire characteristic function. Then either $\phi(z) \equiv 1$, or the order of ϕ must be $\geqslant 1$.

Theorem 13 (Marcinkiewicz). Let $P_n(t)$ be a polynomial in t of degree $n > 2$ and with complex coefficients. Then the function $\phi(t) = \exp\{P_n(t)\}$ cannot be a characteristic function.

Remark 14. As an immediate consequence of Theorem 13, we note that if a characteristic function ϕ is of the form $\phi(t) = \exp\{P(t)\}$, where P is a polynomial in t with complex coefficients, then P must be a polynomial of degree $\leqslant 2$. Consequently, ϕ must be the characteristic function of a normal (or possibly degenerate) distribution, which yields a characterization* of a normal distribution.

For the proofs and bibliography, we refer to Laha and Rohatgi [5, Chap. 4, Sec. 2] and Lukács [7, Chap. 7, Secs. 1–3].

Next we give some useful results on the decomposition of analytic characteristic functions. The following theorem is due to D. A. Raikov.

Theorem 14 (Raikov). Let ϕ be an analytic characteristic function that is holomorphic in the strip $-\alpha < \operatorname{Im} z < \beta(\alpha > 0, \beta > 0)$. Suppose that ϕ admits the decomposition $\phi = \phi_1\phi_2$. Then each of ϕ_1 and ϕ_2 is an analytic characteristic function that is holomorphic at least in the strip $-\alpha < \operatorname{Im} z < \beta$.

In particular, let ϕ be an entire characteristic function that admits a decomposition $\phi = \phi_1\phi_2$. Then each of ϕ_1 and ϕ_2 is an entire characteristic function whose order cannot exceed the order of ϕ.

The following theorem deals with the decomposition of a normal distribution which is due to H. Cramér.

Theorem 15 (Cramér). Let ϕ be the characteristic function of a normal distribution and let $\phi = \phi_1\phi_2$ be a decomposition of ϕ. Then each of ϕ_1 and ϕ_2 is the characteristic function of a normal distribution.

The next result is due to D. A. Raikov and deals with the decomposition of a Poisson distribution.

Theorem 16 (Raikov). Let ϕ be the characteristic function of a Poisson distribution and let $\phi = \phi_1\phi_2$ be a decomposition of ϕ. Then each of ϕ_1 and ϕ_2 is the characteristic function of a Poisson distribution.

For the proofs and bibliography, we refer to Laha and Rohatgi [5, Chap. 4, Sec. 3] and Lukács [7, Chap. 8, Secs. 1 and 2].

SOME GENERALIZATIONS

The characteristic functions of distribution functions on $\mathbb{R}_n$ (multidimensional distribution functions), the representation and properties of infinitely divisible distributions on $\mathbb{R}_n$ and also the theory of arithmetic of distribution functions on $\mathbb{R}_n$ are studied systematically by Cuppens [1] and also by Ostrovski (see ref. 6).

The characteristic functions (Fourier transforms) of probability measures in the general framework of abstract spaces (particularly Hilbert spaces* and locally compact topological groups); the representation and properties of infinitely divisible probability measures on such spaces and also some problems associated with decomposition of characteristic functions on these spaces are systematically studied by Parthasarathy (see ref. 8). Some results on characteristic functions in the framework of abstract spaces are also discussed by Laha and Rohatgi [5, Chaps. 3 and 7].

References

[1] Cuppens, R. (1975). *Decomposition of Multivariate Probability*. Academic Press, New York.

[2] Gnedenko, B. V. and Kolmogorov, A. N.(1954). *Limit Distributions for Sums of Independent Random Variables*. (English translation). Addison-Wesley, Reading, Mass.

[3] Kagan, A. M., Linnik, Yu. V., and Rao, C. R. (1972). *Characterization Problems in Mathematical Statistics* (in Russian). Moscow. English translation: Wiley, New York, 1973.

[4] Laha, R. G. and Lukács, E. (1964). *Applications of Characteristic Functions*. Charles Griffin, London.

[5] Laha, R. G. and Rohatgi, V. K. (1979). *Probability Theory*. Wiley, New York.

[6] Linnik, Yu. V. and Ostrovski, I. V. (1972). *Decomposition of Random Variables and Vectors* (in Russian). Moscow. English translation: *Amer. Math. Soc. Transl.*, **48** (1977).

[7] Lukács, E. (1970). *Characteristic Functions*, 2nd ed. Charles Griffin, London/Hafner, New York.

[8] Parthasarathy, K. R. (1967). *Probability Measures on Metric Spaces*. Academic Press, New York.

This work was partly supported by the National Science Foundation through Grant NSF-MCS 78-01338.

(CHARACTERIZATIONS OF DISTRIBUTIONS
INFINITE DIVISIBILITY)

R. G. LAHA

CHARACTERIZATIONS OF DISTRIBUTIONS

In most cases of statistical evaluation, decisions based on a set of observations on a random quantity X (which may represent a vector as well) depend on the assumption that the distribution function $F(x)$ of X is known. Knowing $F(x)$ may mean to the applied scientist just a subjective choice that may be supported by the data by using some empirical method (probability paper* or goodness-of-fit* tests). Another method of "determining" $F(x)$ is an approximation result from probability theory. However, the only method of finding $F(x)$ exactly is a characterization theorem. A theorem is a characterization of the distribution function $F(x)$ if it concludes that a set of conditions is satisfied by $F(x)$ and only by $F(x)$. Here we do not consider parameters; hence, $F(x)$ may signify a family of distributions.

In order to see the danger of a purely empirical evaluation of data, and thus to emphasize the need for the application of characterization theorems in model building, let us quote an example from Galambos [8, p. 90].

Let X be the weekly return on a proposed investment. Let $X_1, X_2, \ldots, X_{50}$ be the actual returns in a year (accounting for 2 weeks of vacation), where the X_j are assumed to be independent. A business manager would go ahead with the investment if $P(\max(X_1, X_2, \ldots, X_{50}) > 2.6) \geqslant 0.3$ but he would reject it if this same probability were smaller than 0.25. He hires a statistical consultant for a decision. Assume that the nature of the investment is such that $1 + 0.1X > 0$ and the following two properties hold: (1) if X_1 and X_2 are two independent observations on X, then except perhaps for location and scale, the distribution of $\log\{(1 + 0.1X_1)(1 + 0.1X_2)\}$ is the same as that of $\log(1 + 0.1X)$ and (2) $U = 10\log(1 + 0.1X)$ has zero expectation and unit variance. We shall see in the next section that these two properties imply that U is a standard normal* variate and thus the distribution of X is that of $10(Y^{0.1} - 1)$, where $U = \log Y$ is standard normal. Now, assume that the consultant does not inquire about the circumstances that lead to the properties above, but rather he looks at the 50 observations and takes the following course of action. He assumes that X is standard normal and he carries out a test. His test will almost certainly accept his hypothesis since $10(Y^{0.1} - 1)$ is approximately standard normal. Now since $P(\max(X_1, X_2, \ldots, X_{50}) > 2.6) \sim 0.22$ under the assumption of normality, the advice is to reject the planned investment. However, this advice is against the interest of the business manager because the actual value of the probability above is about 0.4 (for the computations see the quoted book). It is evident that the only appropriate action of the consultant would have been to inquire about the circumstances that affect X, and thus a model should have been built in which the distribution of X is determined by a characterization theorem rather than by an empirical procedure.

Not all characterization theorems are as significant as this example indicates. However, more and more problems require more accurate decisions than those obtainable by classical statistical methods (floods, reliability*, effects of chemicals and food additives, fatigue failure of metals, and others). For such problems, the theory of characterizations has an important role to play. Another important consequence of characterization theorems is that these results help us in better understanding the structures and implications of the choice of a distribution for a special problem.

Some comments are made in the present article which refer to the history of the theory of characterizations. In particular, it can be seen from those remarks that the theory is quite new. In fact, the first book on the subject that is not dominated by the normal distribution* is the one by J. Galambos and S. Kotz, published in 1978 [9]. Readers are referred to this book (pp. 1–5) for a detailed account of the history of the theory of characterizations of distributions as well as for a comprehensive bibliography.

CHARACTERIZATIONS THROUGH LIMIT THEOREMS

Limit theorems* are usually applied as approximations. It is much less known that they can also be applied to finding the exact distribution of certain random phenomena. Let us first see an example.

Every physical measurement is subject to random fluctuations due to the inaccuracy of the instrument by which the measuring is done. Hence if the exact value is d, then $D = d + X$ is measured, where X is a random variable with distribution function $F(x)$. It is evident that one can assume that both $E(X)$ and $V(X) > 0$ are finite. We first set up a model for X.

For simplicity of description, let us assume that d is the length of item A whose end points are marked as a and b. Let c be an inner point of A, and the lengths from a to c and from c to b are denoted by d_1 and d_2, respectively. Now if we measure A by first measuring the part from a to c and then that from c to b, then the obtained values D_1 and D_2 yield for the length of A the value $D_1 + D_2 = (d_1 + X_1) + (d_2 + X_2)$. By the nature of the problem, X_1 and X_2 are two independent copies of X, and $X_1 + X_2$ should also have the same distribution as X except perhaps for its location and scale parameters. This therefore leads to the following mathematical questions. Let X_1 and X_2 be independent and identically distributed random variables with common distribution function $F(x)$ and with finite expectation E and variance $V > 0$. Assume that the distribution function of $X_1 + X_2$ is $F(A + Bx)$ with some constants A and $B > 0$. Is $F(x)$ uniquely determined?

As was mentioned earlier, an appeal to the central limit theorem shows that $F(x)$ is unique and in fact it is normal. Consider n independent copies $X_1, X_2, \ldots, X_n$ of X. The assumption on the distribution of $X_1 + X_2$ can be rephrased as saying that, for any pair $i \neq j$, the distribution of $(X_i + X_j - A)/B$ is $F(x)$. But then $(X_1 + X_2 - A)/B$ and $(X_3 + X_4 - A)/B$ can also be considered as independent copies of X and thus

$$\left(\frac{X_1 + X_2 - A}{B} + \frac{X_3 + X_4 - A}{B} - A\right)\Big/ B$$
$$= (X_1 + X_2 + X_3 + X_4 - A_2)/B_2$$

is also distributed like X. By repeating this argument we get that there are constants A_n and $B_n > 0$ such that the distribution of

$$(X_1 + X_2 + \cdots + X_{2^n} - A_n)/B_n$$

is $F(x)$ itself. Hence it has a limiting distribution which is evidently $F(x)$. On the other hand, by the central limit theorem

$$(X_1 + X_2 + \cdots + X_m - mE)/(mV)^{1/2}$$

is asymptotically normally distributed. Since, apart from the normalizing constants, we obtained the limiting distribution of the same random variable once as $F(x)$ and once as the normal distribution $\phi(x)$, an elementary lemma from probability theory (see, e.g., Galambos [8, p.61]) implies that $F(x) = \phi(a + bx)$ with some constants a and $b > 0$, which was to be proved.

The basic idea of the preceding proof can be extended to a general method of obtaining characterization theorems. Since it is based on a limit theorem, we call this the method of limit laws. The nature of this method is very simple. From the assumptions of a model, we generate an infinite sequence T_n of random variables such that, with some numbers A_n and $B_n > 0$, $(T_n -$

$A_n)/B_n$ has the same distribution $F(x)$ for all n. Next we prove (or apply) a limit theorem for the distribution of $(T_m - c_m)/d_m$, where c_m and $d_m > 0$ are two other sequences of constants and n runs through a subsequence of $m = 1, 2, \ldots$. If this limiting distribution is $G(x)$, then the lemma quoted at the end of the preceding paragraph yields that $F(x) = G(a + bx)$ with some a and $b > 0$.

The actual application of the method is not always as simple as it sounds in this short description. The proper generation of the sequence T_n may require nontrivial arguments, whereas in other cases the proof of the required limit theorem can be very involved.

There are a number of examples in the literature where the method of limit laws is applied in the proof of a characterization theorem. One is the rarefaction model* of Rényi [23]. Assume that telephone calls arrive at a large company at random times $0 < \tau_1 < \tau_2 < \cdots$, where the differences $\tau_j - \tau_{j-1}$, $j \geqslant 1$ $(\tau_0 = 0)$, are independent, identically distributed (i.i.d) with common distribution function $F(x)$ (a so-called renewal process*). At the company, each call is recorded with probability $0 < p < 1$ and ignored with probability $1 - p$. The decision on each call's being recorded is done independently of each other and of the process. It can then be proved that the remaining points also form a renewal process. Now if this new renewal process is such that its first point, apart from a new unit of scale, is distributed as τ_1, then $F(x)$ is exponential*, i.e., the process is Poisson*. For proving this result, the method of limit laws is applied as follows. Repeat the described rarefaction procedure several times and let T_n be the first point in the nth rarefaction. With a change of scale B_n, its distribution is always $F(x)$. On the other hand, it can be proved that, whatever be the initial renewal process, repeated rarefactions lead to the Poisson process in the limit. Rényi's work induced much research in this area. See Galambos and Kotz [9, pp. 95–100] for references. On those pages it is also shown that the rarefaction model is strongly related to the so-called damage models*, which have also been investigated extensively. In a damage model, one investigates whether the distribution of a random variable X can be reconstructed through another random variable Y, which is the observable part of X. For example, in the case of the telephone example, if X is the number of calls in the time interval $(0, t)$, then only a fraction Y of them reaches the management. What is known to the management is that if $X = n$ calls arrived, then the recorded value Y is a binomial variable* with parameters n and p. The problem is to find the distribution of X. The first work on this line is due to Rao and Rubin [22]. In addition to the quoted pages of Galambos and Kotz, see Chap. 6 in the Calgary Proceedings [5], which is devoted entirely to damage models. *See also* DAMAGE MODELS.

Other characterization theorems that use the method of limit laws are due to Arnold [3] and independently to Gupta [11] (for the exponential* distribution); to Arnold and Meeden [4] (for the uniform* distribution) and to Galambos [8, p. 189], in which the foundations are laid for the first rigorous proof that the strength of sheets of metals has a Weibull* distribution.

NORMAL DISTRIBUTION

The early theory of statistics was dominated by investigations in which the population distribution was assumed to be normal. This evidently had a dominating effect on the theory of characterizations as well. Unfortunately, this effect was not a positive one. Since the major role of a characterization theorem is in model building, if one assumes *a priori* that the population is normal, the importance of a characterization theorem is authomatically neglected. For example, the classical t-test* is based on the property that the sample mean and the sample standard deviation are independent in the case of a normal population. The discovery by Geary [10], Lukács [17], and Kawata and Sakamoto [14] that the independence property described above is a characteristic one for the normal distribution did not serve as a signal warning of the danger of routinely assuming normality for the population. On the contrary, statisticians played down characterization theorems as too mathematical to be applicable. Therefore, the characterizations of the normal distribution developed as a branch of mathematics and they had very little influence on applied sciences. Strangely enough, the application of characterization theorems in goodness-of-fit tests contributed to the acceptance of these theorems as "applicable." However, as Csörgö et al. [6] rightly point out, the *property* contained in a characterization but not the characterization itself is applied in a goodness-of-fit test. Tests are affected by characterization theorems only if the distributional assumption is a part of the model rather than that of the null hypothesis.

The characterization of the normal distribution by the independence of the sample mean and the sample standard deviation induced much research. Directions of investigation were to characterize normality by independence of two linear statistics in n independent variables (the Darmois–Skitovich theorem*) and by the independence of the sample mean and a general class of quadratic statistics (several results on this line were obtained by R. G. Laha). The reader is referred to the monograph by Lukács and Laha [20, particularly Chaps. 5 and 6], where all early results are formulated and some representative proofs are given. Extensions to the case where independence is replaced by constant regression are also mentioned by Lukács and Laha*. Several results have been extended to the multivariate normal distribution* as well. References to these can be found in the monographs by Kagan et al. [13], Mathai and Pederzoli [21], and the Calgary Proceedings [5; see, in particular, the contributions by C. G. Khatri and B. Gyires].

The literature on characterizations of the normal distribution is huge, and thus no attempt can be made to give a representative account because of space limitations. For an introductory textbook-style reading, see the book by Mathai and Pederzoli [21]; the research-oriented reader may want to consult the book by Kagan et al. [13], as well as the Calgary Proceedings [5].

EXPONENTIAL DISTRIBUTION

The no-aging character of the exponential distribution makes it very widely applicable in accident and warranty insurance*, while the mathematical simplicity of many of its characteristics led to its acceptance in engineering applications as an alternative to the normal distribution. This latter role, however, is being taken over by a variety of distributions which have shape parameters (such as Weibull, log-normal*, logistic*, and others).

The fact that "age" has no effect on a positive random variable X can be expressed in several ways. In terms of distributions it means that the conditional probability* $P(X - t > s \mid X > t)$ does not depend on t, which is known as the lack of memory property*, while by means of expectations, the equation

$$E(X - t \mid X > t) = E(X), \qquad |E(X)| < \infty$$

is a proper translation of the no-aging character. A third possibility of expressing this same fact is to assume that the distribution function $F(x)$ of X has a constant hazard rate* or mortality rate. Since these three properties express the same physical assumption, one can expect that they lead to the same distributional property of X. In particular, if one characterizes the distribution of X, then so does the other (and the same distribution should be obtained). This is indeed the case: each of the three properties above characterizes the exponential distribution $F(x) = 1 - e^{-\lambda x}, x > 0$. The fact that such a unifying approach is possible to the theory of characterizing the exponential distribution was only very recently recognized by Galambos and Kotz [9], who then carried further this unifying theory by including some characterization in terms of order statistics*. One, of course, cannot expect to reduce the vast literature to the lack of memory (which is equivalent to the Cauchy functional equation (*see* FUNCTIONAL EQUATIONS) contained in several elementary texts on calculus). Since the book by Galambos and Kotz [9] gives a detailed account of the literature on exponentiality, we mention only a few of the trends in its theory of characterizations.

Order statistics* play a central role for the exponential distribution. Let $X_{r:n}$ denote the rth order statistic in an independent sample of size n, when observations are arranged in an increasing order. Then each of the following properties characterizes the exponential distribution: (a) the distribution of $nX_{1:n}$ does not depend on n; (b) $nE(X_{1:n}) = a$ (finite) for all n; and (c) the differences $X_{j+1:n} - X_{j:n}, j \geqslant 0$, where $X_{0:n} = 0$, are independent for a fixed $n \geqslant 2$. For (a) and (b), no assumption is needed on $F(x)$ except that it is not degenerate at $x = 0$. For (c), the assumption of continuity is sufficient (see Chap. 3 in Galambos and Kotz [9] for more general statements and for references).

The characterization stated at (a) can be proved by the method of limit laws discussed in the first section. If we modify the assumption of (a) as $(X_{1:n} - a_n)/b_n$ has the same distribution for all n, where a_n and $b_n > 0$ are suitable constants, then, again by the method of limit laws, the so called extreme-value distributions* are characterized. See Chap. 2 in Galambos [8] for these distributions. (Interestingly, the extreme-value distributions were the first continuous and nonnormal distributions that entered the literature but, in spite of their practical importance, they became accepted by applied statisticians only several decades after their discovery.)

If the condition of (a) is modified to the simpler assumption that $X_{1:n}$ has a specific distribution, then it is a simple result that the population distribution is uniquely determined. However, the assumption that $X_{j+1:n} - X_{j:n}, 2 \leqslant j < n$, has a specific distribution (say, exponential) alone does not lead to a characterization theorem. With a somewhat involved mathematical argument under an analytical assumption, Rossberg [24] obtained a characterization of exponential populations based on the exponentiality of $X_{j+1:n} - X_{j:n}$. More recently, Ahsanullah, in a series of papers (see, e.g., ref. 2) discussed characterizations of the exponential distribution among those distributions that have a monotonic hazard rate. Although these results are somewhat limited, they are very practical characterizations since failure distributions of equipment, for quite complicated structures, are of monotonic hazard rate (see Galambos [8, p. 196], and HAZARD RATES.)

TRANSFORMATION TO EXPONENTIALITY

Several characterizations of other distributions can be stated in an equivalent form by means of the exponential distribution. This is due to the fact that if a random variable X has a continuous distribution function $F(x)$, the random variable

$$Y = -\log[1 - F(X)]$$

is a unit exponential variate. Since this transformation is monotonic, it preserves inequalities. Thus an order statistic of a sample on X is transformed into the appropriate Y-order statistic.

As an example for this transformation, we mention the following characterization theorem. The lack of memory characterization of the exponential distribution is equivalent to the following statement: The distribution of X is uniform*, if and only if for all $0 \leqslant u, v \leqslant 1, P(X \leqslant uv \mid X \leqslant v) = P(X \leqslant u)$ and if X is nondegenerate.

GAMMA DISTRIBUTIONS

Gamma distributions* have two important properties which helped them achieve a prominent role in applied statistics. One is the fact that the sample variance of normal observations has a gamma distribution. The other important property is that the general family of gamma distributions has a shape parameter which makes it possible to fit a large variety of data to one of the gamma distributions. In this latter aspect, they recently found competitors in other popular families such as the log-normal, the logistic, the Weibull, and other distributions.

Another close relation of the gamma distributions to the normal is expressed in a characterization theorem of Lukács and Laha [20, p. 106] in which the regression of the sample mean on a special quadratic statistic is assumed to be constant. In this same monograph (pp. 73–74), it is shown by a simple application of characteristic functions* that if $S(X_1, X_2, \ldots, X_n)$ is a scale-invariant statistic, then $X_1 + \cdots + X_n$ and $S(X_1, X_2, \ldots, X_n)$ are independent if $X_1, \ldots, X_n$ is an independent sample from a gamma population. It is, however, not known in general under which additional assumptions on $S = S(X_1, X_2, \ldots, X_n)$ the independence of S and the sample mean would imply that the population is gamma. Some special cases are known, out of which we mention only two. The case $S = (\sum a_{ij} X_i X_j)/(\sum X_j)^2$ with suitable assumptions on the matrix $A = (a_{ij})$ is treated by Lukács and Laha [20, p. 106]. Another interesting case is a result of Lukács [18]. Let $n = 2$ and let $S = S(X_1, X_2) = X_1/(X_1 + X_2)$. Now if $X_j > 0$ and is nondegenerate and if S and $X_1 + X_2$ are independent, then the population is gamma. The result still holds even if X_1 and X_2 are not assumed to be identically distributed, in which case X_1 and X_2 may have different parameters, but both are gamma variates under the foregoing independence assumption. A vector version of results of this nature is due to Khatri and Rao [15], who we quote in a special case. If $n \geqslant 3$, then $X_1 + X_2 + \cdots + X_n$ is independent of the vector $(X_2/X_1, X_3/X_2, \ldots, X_n/X_{n-1})$ if, and only if, the population is gamma. The general form of their result is reproduced in the book by Kagan et al. [13, Sec. 6.2], where further results and references are also given. See also the Calgary Proceedings [5], in particular the contributions by L. Bondesson (p. 185) and A. B. Godambe and G. P. Patil (p. 339).

Further Reading

Space limitations do not permit an introduction to all areas of characterizations. The interested reader can, however, find good collections of material on several other topics not mentioned here.

Several characterizations of the Poisson process are given by Galambos and Kotz [9]. Discrete distributions* are discussed in Galambos [7] and in several other contributions in the Calgary Proceedings [5]. So-called stability theorems*, in which an assumption is modified "slightly" and one investigates the extent of the effect of this change on a characterization theorem, are surveyed by Lukács [19]. Among the multivariate cases, we mentioned the normal distribution. Characterizations for other multivariate distributions are not well developed. The only exceptions are the multivariate extreme-value distributions* (See Chap. 5 in Galambos [8]) and some multivariate exponential families* (see Chap. 5 in Galambos and Kotz [9]).

In addition to the above-mentioned books by Lukács and Laha [20], Kagan et al. [13], Mathai and Pederzoli [21], Galambos [8], and Galambos and Kotz [9], the reader can find a large variety of results in the Calgary Proceedings [5]. Furthermore, a detailed survey of the literature is given by Kotz [16] as a supplement to Kagan et al. [13]. See also the four-volume set by Johnson and Kotz [12], where descriptions of distributions often contain characterization theorems.

One of the basic tools of characterizations is the solution of functional equations*. The book by Aczél [1] is a useful reference for such results.

References

[1] Aczél, J. (1966). *Lectures on Functional Equations and Their Applications*. Academic Press, New York.

[2] Ahsanullah, M. (1978). *J. Appl. Prob.*, **15**, 650–653.

[3] Arnold, B. C. (1971). Two Characterizations of the Exponential Distribution Using Order Statistics. *Tech. Report*, Iowa State University, Ames, Iowa.

[4] Arnold, B. C. and Meeden, G. (1976). *Aust. J. Statist.*, **18**, 173–175.

[5] Calgary Proceedings (1975). Published as Vol. 3 in the series *Statistical Distributions in Scientific Work*, G. P. Patil, S. Kotz, and J. K. Ord, eds. D. Reidel, Dordrecht, Holland.

[6] Csörgő, M., Seshadri, V., and Yalovsky, M., (1975). In *Statistical Distributions in Scientific Work*, Vol. 2, G. P. Patil, S. Kotz, and J. K. Ord, eds. D. Reidel, Dordrecht, Holland, pp. 79–90.

[7] Galambos, J., (1975). In *Statistical Distributions in Scientific Work*, Vol. 3, G. P. Patil, S. Kotz, and J. K. Ord, eds. D. Reidel, Dordrecht, Holland, pp. 89–101.

[8] Galambos, J., (1978). *The Asymptotic Theory of Extreme Order Statistics.* Wiley, New York.

[9] Galambos, J., and Kotz, S. (1978). *Characterizations of Probability Distributions*. Lec. Notes Math., **675**, Springer-Verlag, Heidelberg.

[10] Geary, R. C. (1936). *J. R. Statist. Soc. B*, **3**, 178–184.

[11] Gupta, R. C. (1973). *Sankhyā B*, **35**, 365–366.

[12] Johnson, N. L. and Kotz, S., (1969–1972). *Distributions in Statistics*, Vols. 1–4. Wiley, New York.

[13] Kagan, A. M., Linnik, Y. V., and Rao, C. R. (1973). *Characterization Problems in Mathematical Statistics*. Wiley, New York.

[14] Kawata, T. and Sakamoto, H. (1949). *J. Math. Soc. Japan*, **1**, 111–115.

[15] Khatri, C. G. and Rao, C. R. (1968). *Sankhyā A*, **30**, 157–166.

[16] Kotz, S. (1974). *Rev. Inst. Int. Statist.*, **42**, 39–65.

[17] Lukács, E. (1942). *Ann. Math. Statist.*, **13**, 91–93.

[18] Lukács, E. (1955). *Ann. Math. Statist.*, **26**, 319–324.

[19] Lukács, E. (1977). *Adv. Appl. Prob.*, **9**, 336–361.

[20] Lukács, E. and Laha, R. G. (1964). *Applications of Characteristic Functions*. Charles Griffin, London.

[21] Mathai, A. M. and Pederzoli, G. (1977). *Characterizations of the Normal Probability Law*. Wiley, New York.

[22] Rao, C. R. and Rubin, H. (1964). *Sankhyā A*, **26**, 294–298.

[23] Rényi, A. (1956). A Characterization of the Poisson Process. (Original in Hungarian. Translated into English in *Selected Papers of Alfréd Rényi*, Vol. 1. Akadémiai Kiadó, Budapest, 1976.)

[24] Rossberg, H. J. (1972). *Math. Operationsforschung. Statist.*, **3**, 207–216.

Acknowledgment

This work was supported by the Air Force Office of Scientific Research under Grant AFOSR-78-3504.

(CHARACTERISTIC FUNCTIONS
EXPONENTIAL DISTRIBUTION
FUNCTIONAL EQUATIONS
GAMMA DISTRIBUTION
LIMIT THEOREMS
NORMAL DISTRIBUTION)

JANOS GALAMBOS

CHARLIER'S SERIES *See* APPROXIMATIONS TO DISTRIBUTIONS

CHATTERJEE TEST *See* BENNETT'S BIVARIATE SIGN TEST

CHAUVENET'S CRITERION

Introduced by W. Chauvenet [1, p. 564], who regarded it as an approximation to the use of Peirce's criterion*, for the case of rejection of a single "doubtful observation" among a sample of n observations, assuming normal variation*. The idea is that an outlying value should be rejected if it differs from the mean by an amount exceeding, in absolute value, $\theta_n \times$ (standard deviation), where θ_n is chosen so that the expected number of such values is $\frac{1}{2}$. This means that θ_n satisfies the equation

$$2\left[1 - \Phi(\theta_n)\right] = \tfrac{1}{2}n^{-1}$$

so that

$$\theta_n = \Phi^{-1}\left(1 - \tfrac{1}{4}n^{-1}\right)$$

where $\Phi(\cdot)$ is the normal probability integral*, and $\Phi^{-1}(\cdot)$ its inverse.

In a random sample of size n from the normal population with the assumed mean and standard deviation, there is a probability equal to $[1 - \{1 - 1/(2n)\}^n]$ of rejecting an observation if Chauvenet's criterion is used. For $n = 5$, this probability is 0.41; for $n = 10$, it is 0.40; as n increases, it tends to $1 - e^{-1/2} = 0.39$. By present-day standards these would generally be regarded as rather high values.

Although this technique might be applied sequentially, with appropriate reduction in the value of n at each step, Chauvenet notes that some modifications are needed to make it approximate to the results of using Peirce's criterion. Chauvenet also points out that in contrast to Peirce's criterion, his criterion does not allow for effects of estimation of unknown parameters (e.g., population standard deviation). For recent investigations, see Bol'shev and Ubaidullaeva [2].

References

[1] Chauvenet, W. (1864). *A Manual of Spherical and Practical Astronomy*, Vol. 2, Lippincott, Philadelphia.

[2] Bol'shev, L. N. and Ubaidullaeva, M. (1974). *Theor. Prob. Appl.*, **19**, 683–686.

(OUTLIERS
PEIRCE'S CRITERION
THOMPSON'S CRITERION)

CHEBYSHEV (or TCHÉBICHEF), PAFNUTY LVOVICH

Born: May 26 (n.s.), 1821, in Okatovo (Kaluga region), Russia.

Died: December 8 (n.s.), 1894, in St. Petersburg, Russia.

Contributed to: number theory and analysis, the theory of mechanisms, approximation theory, probability theory.

Chebyshev's contributions to probability theory form a relatively small portion of his contributions to mathematics and practical mechanics, although in regard to mathematical statistics, this is supplemented by his interpolational work in the context of linear least squares*. He was a leading exponent of the Russian tradition of treating the probability calculus as an integral part of mathematical training, and through the "Petersburg Mathematical School," of which he was the central figure, his overall influence on mathematics within the Russian Empire was enormous. His distinguished disciples within this framework included A. A. Markov* and A. M. Liapunov*, who, *inter alia*, extended his own remarkable probabilistic work.

In his early mathematical education at Moscow University, where he enrolled in 1837, Chebyshev was strongly influenced by N. D. Brashman. Under this influence he produced his first two contributions to probability in 1846, his master's thesis and the article that is ref. 8, both of which seek to give an elementary but rigorous analytical discussion of some of the then-principal aspects of probability theory. Reference 8 in particular is notable in that it contains an analytical deduction of the weak law of large numbers* (WLLN) of Poisson*: that if X is the number of successes in n independent trials, where p_i, $i \geqslant 1$, is the probability of success in the ith, then

$$\Pr[\,|(X/n) - \bar{p}(n)| < \epsilon] \to 1$$

as $n \to \infty$ for any $\epsilon > 0$, where $\bar{p}(n) = \sum_{i=1}^{n} p_i/n$. This publication, however, passed unnoticed, and the law, with its "floating mean" $\bar{p}(n)$, remained an object of controversy among the French mathematicians who dominated probability theory at the time, for years to come [5; Sec. 3.3]. The proof is also interesting insofar as it proceeds by obtaining upper bounds for the lower- and upper-tail probabilities of the number of successes. (The modern approach, via the Bienaymé–Chebyshev inequality* of which we shall speak shortly, was then unknown.) Indeed, both early works already display one of the features peculiar to Chebyshev's work: the estimation for finite n of the *deviation* from its limit of a quantity that approaches the limit as $n \to \infty$, which was evidently communicated to both Markov* and Liapunov*.

In 1847, Chebyshev began to teach at St. Petersburg University, eventually becoming full professor in 1860, in which year he took over the course in probability theory (on the retirement of V. Ya. Buniakovsky), which reawakened his interest in the subject area. He had been promoted to the highest academic rank of the St. Petersburg Academy of Sciences in the previous year. Subsequently, he wrote only two papers [11, 12] in probability theory, which, nevertheless, had great influence.

In ref. 11 he obtains the Bienaymé–Chebyshev inequality for the arithmetic mean* of independently but not necessarily identically distributed random variables, each of which has only a finite number of sample points, and uses it to deduce the corresponding WLLN, with subsequent application to both the cases of Poisson (which he had treated in ref. 8) and of J. Bernoulli*. I. J. Bienaymé* had arrived at both the inequality and a WLLN by the simple reasoning still used in 1853 for general probability distributions, and may have arranged the juxtaposition of a reprinting of his own article next to a French printing of Chebyshev's in Liouville's journal. Indeed, the year 1858 [9] seemed to mark the beginning of a mutual correspondence and admiration between the two men, leading to the eventual election of each to a membership in the other's Academy of Science. Even though in 1874 Chebyshev gave Bienaymé credit in print for arriving at the inequality via "the method of moments"* whose discovery he ascribed to Bienaymé, and this view was later reiterated by Markov, it is a valid point that it was more clearly stated and proved by Chebyshev. In any case, through the subsequent writings of the strong Russian probability school, Chebyshev's paper has undeniably had the greater publicity, to the extent that the inequality has often borne Chebyshev's name alone (*see* CHEBYSHEV'S INEQUALITY).

In ref. 12, Chebyshev uses "the method of moments" for the first time as a tool in the proof of the central limit theorem* for not necessarily identically distributed summands $X_i, i \geqslant 1$ (under the implicit assumption of independence, to which, in the manner of the times, he never alluded explicitly). His assumptions and proof are incomplete, and have been the subject of much discussion in the Soviet historical literature. In 1898, Markov used the same method to overcome the inadequacies, and shortly after this the very general version, proved with the aid of characteristic functions, was obtained by Liapunov. Chebyshev's attempt at *rigorous* proof and Markov's follow-up have often been thought to be the first such, although in a more restricted setting, a rigorous proof had been largely given by A. L. Cauchy* in 1853, and completed by I. V. Sleshinsky [7], who recognized the gaps in Chebyshev's treatment, by characteristic function methods. Chebyshev's paper is also notable for his pointing out the possibility of refining the central limit theorem by means of an asymptotic expansion in what are now known as the Chebyshev–Hermite polynomials*.

In respect to mathematical statistics, Chebyshev's influence stems from the paper that is ref. 9, in which he is concerned with fitting a polynomial

$$y = \sum_{j=1}^{q} \beta_j x^{j-1}$$

to n pairs of observations (Y_i, x_i), $i = 1, \dots, n$, and does so by producing from the q powers $1, x, \dots, x^{q-1}$, a set of q polynomials $T_0(x) = 1, T_1(x), \dots, T_{q-1}(x)$ which are *orthogonal* with respect to the points $x_1, \dots, x_n$ in that

$$\sum_{i=1}^{n} T_s(x_i) T_t(x_i) = 0 \qquad (s \neq t)$$

(assuming equal weights). These orthogonal polynomials, at least in the case where x_i, $i = 1, \dots, n$, are at equidistant intervals, have come to bear his name. He is aware that the coefficients produced by his procedure are those arising out of a linear least-squares fit*, but is more concerned with the connection of the problem with his continued-fractions theory approach, another central theme of his *oeuvre*. The statistical significance of this work is finally made clear, through the catalytic effect of Bienaymé, in Chebyshev's paper given here as ref. 10, in which he recognizes the ease of modifying the expression for residual sum of squares* with increasing q by progressive orthogonalization, in the process of deciding where to stop the "expansion." In the setting of the general linear model* $\mathbf{Y} = \mathbf{X}\boldsymbol{\beta} + \boldsymbol{\epsilon}$, in which the $(n \times r)$ design matrix $\mathbf{X} = \{x_{ij}\}$ has the special form $x_{ij} = x_i^{j-1}$ in the polynomial context, the interpolational problem of how many β_i, $i = 1, 2, \dots,$ to fit in succession until an "adequate" fit to the data obtains, proceeding in such a way that the estimates $\beta_1, \dots, \beta_r$ are unaffected in going to that for β_{r+1}, had been revived by Cauchy in 1853 in the course of his controversy with Bienaymé. (However, Cauchy's own, earlier, solution does not have the numerous interpolational and statistical advantages of orthogonalization.)

Chebyshev had a deep belief in the mutual benefits of the interaction of theory and practice; a famous quotation [2] reads in part: "And if theory gains much when new applications or new developments of old methods occur, the gain is still greater when new methods are discovered; and here science finds a reliable guide in practice." When he retired from teaching in St. Petersburg University in 1882, he continued to maintain close contact with his disciples and young scientists; and he died greatly loved and esteemed among his colleagues. However, to the English-speaking world the significance of his probabilistic and statistical work, at least, was not immediately apparent [1], although in other areas he had an international reputation.

References

[1] Anonymous (1895). *Nature (Lond.)*, **52**, 345. (An English-language obituary.)

[2] Chebyshev, P. L. (1944–1951). *Polnoe Sobranie Sochineniy*, 5 vols. Izd. AN SSSR, Moscow. (Russian-language collected works with commentaries; the quotation is in Vol. 5, p. 150.)

[3] Chebyshev, P. L. (1955). *Izbrannie Trudy*. Izd. AN SSSR, Moscow. (Russian-language selected works, with commentaries.)

[4] Gnedenko, B. V. and Sheynin, O. B. (1978). In *Matematika XIX Veka*. [*Mathematics of the 19th Century*]. Nauka, Moscow, pp. 184–240. (Pages 216ff. contain, with a portrait, a more balanced Russian-language view than usual of the question of Chebyshev's priority and significance in probability theory.)

[5] Heyde, C. C. and Seneta, E. (1977). *I. J. Bienaymé: Statistical Theory Anticipated*. Springer-Verlag, New York. (Contains a detailed account of the interaction between Bienaymé, Cauchy, and Chebyshev.)

[6] Maistrov, L. E. (1974). *Probability Theory: A Historical Sketch*. Academic Press, New York. (Translated and edited from the Russian-language work of 1967 by S. Kotz. Strong bias to Russian contributions.)

[7] Sleshinsky (Sleschinsky, Sleszyński), I. V. (1892). *Zap. Mat. Otd. Novoross. Obshch. Estestvoispyt. (Odessa)*, **14**, 201–264 (in Russian).

[8] Tchébichef, P. L. (1846). *Crelle's J. Reine angew. Math.*, **33**, 259–267. (Also in refs. 2, 3, and 13.)

[9] Tchébichef, P. L. (1858). *Liouville's J. Math. Pures Appl.*, **3**, 289–323. (Translation by Bienaymé of a Russian article of 1855 with prefatory footnote by the translator. Also in refs. 2, 3, and 13.)

[10] Tchébichef, P. L. (1859). *Mem. Acad. Sci. St. Pétersbourg*, (8) **1**(15), 1–24. (Also in refs. 2 and 3.)

[11] Tchébichef, P. L. (1867). *Liouville's J. Math. Pures. Appl.*, (2) **12**, 177–184. (Published simultaneously in Russian in *Mat. Sb.*, (2) **2**, 1–9. Also contained in refs. 2, 3, and 13.)

[12] Tchébichef, P. L. (1890-1891). *Acta Math.*, **14**, 305–315. [Originally published in Supplement to *Zap. Imp. Akad. Nauk (S.P. – B.)*, **55**(6) (1887). Also contained in refs. 2, 3, and 13.]

[13] Tchébichef, P. L. (n.d.). *Oeuvres*, 2 vols. A. Markov and N. Sonin, eds. Chelsea, New York.

[14] Youshkevitch, A. P. (1971). In *Dictionary of Scientific Biography*, Vol 3, C. C. Gillispie, ed. Scribner's, New York, pp. 222–232. (Chebyshev's life and scientific work surveyed in their entirety; contains a valuable list of secondary materials.)

(BIENAYMÉ, IRENÉE-JULES
BIENAYMÉ–CHEBYSHEV INEQUALITY
CHEBYSHEV–HERMITE POLYNOMIALS
LIAPUNOV'S THEOREM
LIMIT THEOREMS, CENTRAL
LINEAR LEAST SQUARES
POLYNOMIAL REGRESSION
WEAK LAW OF LARGE NUMBERS)

E. SENETA

CHEBYSHEV–HERMITE POLYNOMIALS

The polynomial of mth degree is defined by the formula

$$H_m(x) = (-1)^m e^{x^2/2} \frac{d^m}{dx^m} e^{-x^2/2}$$

$$(m = 0, 1, 2, \dots)$$

or equivalently by

$$H_m(x) = m! \sum_{k=0}^{m/2} \frac{(-1)^k x^{m-2k}}{k!(m-2k)!2^k},$$

where $[\frac{1}{2}m]$ is the integral part of $\frac{1}{2}m$. Thus $H_m(x)$ is the coefficient of $t^m/m!$ in the expansion of $\exp(tx - \frac{1}{2}t^2)$. If $\phi(x) = (2\pi)^{-1/2}e^{-x^2/2}$, the standard normal density*, then

$$(-1)^m H_m(x)\phi(x) = \frac{d^m}{dx^m}\phi(x).$$

The recurrence relation

$$H_m(x) = xH_{m-1}(x) - (m-1)H_{m-2}(x) \quad (m = 2, 3, \dots)$$

holds. The first five Chebyshev-Hermite polynomials are

$$H_0(x) = 1; \qquad H_1(x) = x;$$
$$H_2(x) = x^2 - 1; \qquad H_3(x) = x^3 - 3x;$$
$$H_4(x) = x^4 - 6x^2 + 3;$$
$$H_5(x) = x^5 - 10x^3 + 15x.$$

Draper and Tierney [2] give expressions for $H_m(x)$ for $0 \leqslant m \leqslant 27$; see also Kendall and Stuart [4, pp. 167–168.] Fisher and Cornish [3] have tabulated values of $H_m(x_p)$ for $1 \leqslant r \leqslant 7$ and $0.0005 \leqslant p \leqslant 0.5$, where x_p is the quantile of the standard normal distribution* having a probability p in the right tail. One of the main uses of Chebyshev-Hermite polynomials is expanding cumulative distribution* and probability density functions* of random variables and normalized sums of absolutely continuous random variables in Edgeworth* and Gram-Charlier series* (see Kendall and Stuart [4, Chap. 6], Patel and Read [5, Chap. 6], and Cramér [1].

References

[1] Cramér, H. (1970). *Random Variables and Probability Distributions*, 3rd ed., Cambridge University Press, London.

[2] Draper, N. R. and Tierney, D. E. (1973). *Comm. Stat.*, **1**, 495–524.

[3] Fisher, R. A. and Cornish, E. A. (1960). *Technometrics*, **2**, 209–225.

[4] Kendall, M. G. and Stuart, A. (1977). *The Advanced Theory of Statistics*, 4th ed. Vol. 1, Macmillan, New York.

[5] Patel, J. K. and Read, C. B. (1982). *Handbook of the Normal Distribution*, Dekker, New York.

(CORNISH-FISHER/EDGEWORTH EXPANSION
GRAM-CHARLIER SERIES
NORMAL DISTRIBUTION)

CHEBYSHEV'S INEQUALITY

(Also known as the Bienaymé–Chebyshev inequality—see Heyde and Seneta [1].

For any distribution of a random variable X with finite expected value ξ, and standard deviation σ, the probability that the variable differs from the expected value by no more than $t\sigma$ is at least $1 - t^{-2}$. Formally,

$$\Pr[\,|X - \xi| < t\sigma\,] \geqslant 1 - t^{-2}.$$

This is a special case (for $r = 2$) of the inequality

$$\Pr\left[|X - \xi| < t\nu_r^{1/r}\right] \geqslant 1 - t^{-r},$$

where ν_r is the rth central absolute moment* of the distribution of X.

Reference

[1] Heyde, C. C. and Seneta, E. (1977). *I. J. Bienaymé: Statistical Theory Anticipated*. Springer-Verlag, New York.

(BERNSTEIN INEQUALITY
BIENAYMÉ, IRENÉE-JULES
BIRNBAUM–RAYMOND–ZUCKERMAN INEQUALITY
CAMP–MEIDELL INEQUALITY
CHEBYSHEV, PAFNUTY LVOVICH
GAUSS INEQUALITY)

CHEBYSHEV'S LAW OF LARGE NUMBERS

A particular case of the weak law of large numbers* applicable to a sequence of independent random variables with a uniformly bounded variance. Specifically, let $\{X_n\}$ be a sequence of independent random variables with $E(X_i) = m_i$ and $\text{var}(X_i) = \sigma_i^2$, $i = 1, \ldots, n$, and let $\sigma_i^2 \leqslant c < \infty$; then for any $\epsilon > 0$,

$$\lim_{n\to\infty} P\left(\left| \overline{X}_n - \frac{1}{n} \sum_{i=1}^{n} m_i \right| < \epsilon \right) = 1,$$

$$\text{where } \overline{X}_n = \frac{\sum_1^n X_i}{n}.$$

(LAWS OF LARGE NUMBERS
POISSON'S LAW OF LARGE NUMBERS)

CHEMISTRY, STATISTICAL METHODS IN

The use of statistics in chemistry generally involves the estimation* of unknown parameters. Hypothesis testing* is most often used by a chemist as an aid in combining data (*see* COMBINATION OF DATA) and checking calibrations*. Since the major emphasis is on parameter estimation, much interest centers around the concepts of the precision* and accuracy* of the estimators [4]. Precision is defined by the variability in the estimator. In many cases the precision and accuracy are expressed in relative terms; i.e., the error is expressed as the ratio of the error in the estimator (either precision or accuracy) to the value of the estimate.

In chemistry almost all measurements are made with reference to a "known" standard or standards. This process is called calibration and is usually accomplished through the use of calibration curves. A calibration curve is formed by measuring with an instrument or chemical procedure a "known" attribute of several physical standards. Then a line or curve is drawn, either by "eye" (*see* EYE ESTIMATE) or by least squares*, which relates the instrument measurements (usually the dependent variable) to the "known" values of the standards (usually the independent variable) (*see* REGRESSION ANALYSIS). Using the calibration curve, subsequent instrument measurements on one axis are projected onto the other axis, thus giving a "corrected" or "calibrated" reading. To illustrate, suppose k standards (having known concentrations of a chemical) are measured on a certain instrument, the measurements being expressed as voltages. Let $X_1, \ldots, X_k$ denote the concentrations and $Y_1, \ldots, Y_k$ the corresponding voltages. A calibration curve $Y = \hat{\alpha} + \hat{\beta}x$ is fitted to the k pairs of data. A measurement Y_{k+1} is then taken on an unknown solution and the concentration estimated as $\hat{X}_{k+1} = (Y_{k+1} - \hat{\alpha})/\hat{\beta}$. The actual uncertainty in $\hat{X}_{k+1}$ includes the uncertainty in fitting the calibration curve (i.e., the variability in $\hat{\alpha}$ and $\hat{\beta}$). We call this actual uncertainty the unconditional variance* of X_{k+1}. However, in chemistry, as elsewhere, it is a frequent practice to use a single calibration curve repeatedly. One can conceive of taking a large number of readings $Y_{k+1}, Y_{k+2}, \ldots$ on a given unknown and transforming these to $\hat{X}_{k+1}, X_{k+2}, \ldots$ through the use of a single calibration curve $Y = \hat{\alpha} + \hat{\beta}x$. When the calibration line is considered fixed, the variance of the X_i's is the conditional variance* of X_i given $\hat{\alpha}$ and $\hat{\beta}$, but this *apparent* variance is too small. Furthermore, the X_i's generated from a single calibration curve are correlated.

The problems encountered in using a calibration curve repeatedly also occur in using estimates based upon National Bureau of Standards* "constants" (speed of light, purity of a chemical standard, etc.). These "constants" are not constants at all, but random variables which are estimates of physical parameters. As such, each of these "constants" has an uncertainty that can be expressed as the standard deviation of the estimator (i.e., the "constant"). Some authors like to refer to the increase in variability in the estimate due to the variability in the "constants" as systematic error*, but the term "systematic error" is often equated with *bias*, which is a constant rather than a random variable; hence this choice of words is a poor one. The terms "unconditional variance" and "conditional variance" are more easily understood and better convey the ideas involved.

There is yet another area in which variances are larger than they appear to be. A chemical measurement may well be a result of a multiple-stage process. "Replication"* of such a measurement frequently duplicates only a part of this process; hence the variability among such measurements is only a fraction of that which would be found among true replicates. As a consequence, error terms in the analysis of variance are small, and the hypothesis is rejected more often than it should be.

Many times in chemistry, an estimate of the error is needed for a *function* of random variables. In most cases the estimate is obtained through a process called "propagation of error"* (see, e.g., Mandel [6, p. 72]). Propagation of error generally refers to an estimate of the variance derived from a Taylor series* approximation to the function of random variables. It is given as follows. Let $f(X_1, X_2, \ldots, X_n)$ be a function of the random variables $X_1, X_2, \ldots, X_n$ which have variances $\sigma_1^2, \sigma_2^2, \ldots, \sigma_n^2$. Let the variance between X_i and X_j be σ_{ij}. Then the variance of f is given approximately by

$$\operatorname{var}(f) = \sum_i \left(\frac{\partial f}{\partial x_i} \right)^2 \sigma_i^2 + \sum_{i \neq j}\sum \left(\frac{\partial f}{\partial x_i} \right)\left(\frac{\partial f}{\partial x_j} \right)\sigma_{ij}.$$

It is not quite clear how well this approximation estimates the variance of $f(X_1, X_2, \ldots, X_n)$. In addition to being an approximation, it is almost always true that the derivatives $(\partial f/\partial x_i)$ are functions of the random variables X_i and that σ_1^2 and σ_{ij} are unknown. In practice, estimates are used in place of σ_i^2 and σ_{ij}, while the observations x_j are used in place of the random variables X_j. All these things have some influence on the quality of the estimator of the variance.

It is not uncommon for a chemist to produce data that are *multiples* of the actual measurements. For reasons of convenience, the chemist frequently measures only a part of the sample. For example, a chemist may divide a certain solution into four equal parts and refer to each part as a "25% aliquot." He or she then measures the amount of a given substance in the 25% aliquot and multiplies by 4 to estimate the amount of substance in the entire solution. Consider, for example, the estimation of the number of disintegrations (λ) per unit of time in a solution containing a radioactive substance. In a 25% aliquot there are approximately $\lambda/4$ disintegrations. If the process has a Poisson distribution* with parameter λ, then λ represents both the mean and the variance of the original solution and $\lambda/4$ is the mean and variance of the 25% aliquot. If we let x be the observed count in the aliquot and let $y = 4x$ be the estimated count in the entire solution, we have $\sigma_y^2 = 4^2\sigma_x^2 = 16(\lambda/4) = 4\lambda$, so that the variance of the total count, as estimated from a 25% aliquot, is four times the variance of the count estimated from the entire solution. This implies that the largest possible aliquots should be taken. When combining estimates from samples using different aliquot sizes, a weighted mean* should be used for which the weights are inverse variances.

Other problems in parameter estimation come from the chemist who is measuring very small quantities of some material and is working near "background" or near the "detection limit" of the instrument. To be more specific, suppose that the chemist is counting the number of disintegrations from a radioactive substance for a given time period. There is, in this case, a background of cosmic radiation. It is usually assumed that the background is a Poisson process* with parameter λ, the average number of counts per unit of time. The $(1 - \alpha)100$th percentile L of the background distribution is frequently used as a detection limit. A sample whose count falls below this point is indistinguishable from background, and we say of such a sample that "nothing was detected." (The point $L_c = L - \lambda$ has been referred to by Currie [3] as the critical level, and by Altschuler and Pasternak [1] as the minimum significant measured activity.) Another type of limit is the (hypothetical) count, which (if it were present in a sample) would be detected "most of the time." In other words, given a Poisson process with parameter η, $(1 - \epsilon)100\%$ of the distribution is above the point L. The count then represents the *capability* of the instrument; it is the count the instrument will "detect" $(1 - \epsilon)$ 100% of the time. For each sample a decision is made, using the point L, about whether something was detected or not; the parameter η is not used for decision making. Given a minimum detection limit L, it is a frequent practice not to report measurements below L. Typically, the chemist wishes to estimate the mean and standard deviation of the sample given $(n - k)$ actual measurements and k observations which are known only to be "below the detection limit." Badly biased results are obtained by setting the k observations equal to either zero or L. Ignoring the k observations also leads to bias. If normality* can be assumed, a reasonable practice is to plot the ordered $(n - k)$ points on normal probability paper*, using as ordinates $(k + 1)/(n + 1)$, $(k + 2)/(n + 1), \ldots, n/(n + 1)$. A straight line is then fitted to the points by eye (*see* EYE ESTIMATE), the 50th percentile serving as an estimate of the mean, and the distance between the 84th and 50th percentiles serving as an estimate of the standard deviation. Fitting the measured values to a cumulative normal by nonlinear least squares* is a nearly equivalent but preferred procedure. For samples of size 20 or less a better procedure is to use the best linear estimates based on the normal order statistics*. Sarhan and Greenberg [7] have given the coefficients for the linear combinations of the $(n - k)$ measurements that yield these estimates.

As an attempt to detect gross errors in the chemical laboratory, it is routine practice to make either duplicate or triplicate analyses. If the duplicates "agree" reasonably well with each other, they serve as a sort of "check" on each other. In some laboratories a third measurement is made only if the first two are not in sufficient agreement, whereas in others a third measurement is always made. We deal first with the latter case.

A widespread but dangerous practice is to select the "best" (closest) two-out-of-three measurements and average them as an estimate of the mean. In terms of mean squared error*, Lieblein [5] has shown that (barring an outlier*) the following estimates (in the order given) are all to be preferred to averaging the best two out of three: (1) the mean of the three, (2) the midrange* of the three, (3), the median* of the three, (4) the average of the first two (which are true duplicates), and (5) the average of the lowest pair (which is biased). The only case in which the average of the best two out of three can be recommended is when one of the observations is clearly an outlier*.

In the case of routine duplicates, the mean is estimated by their average unless they are discrepant. In that case a third observation (which we call the referee's value) is taken. Assuming that the third observation is not itself an outlier, we define the following seven estimates of the mean: R_1 = the referee's value, R_2 = the median of the three observations, R_3 = the average of the referee's value and the one closer to it, R_4 = the average of the closest pair, R_5 = the mean of the three observations, R_6 = the mean of the lowest pair, and R_7 = the midrange of the three. R_2 corresponds to trimming* the sample, R_4 to rejecting the outlier, R_5 to ignoring the outlier. Tietjen and Beckman [8] used a Monte Carlo study to evaluate the rules in terms of mean squared error. They concluded that (1) R_5 and R_7 are heavily biased and should not be used; (2) if it is *known* that outliers occur only on the high side, R_6 is the best rule; and (3) if no prior knowledge is available, it is uniformly better to report the mean of the closest pair than to report the median or the referee's values. The situation just described is also applicable to the selling of ore, say, in which the seller makes one estimate and the buyer another. If the estimates are in good agreement, they compromise. Otherwise, they hire a referee to adjudicate the dispute.

To summarize, the statistical analysis of chemical data can be much more complicated than it appears on the surface. Unless he or she asks the right questions, the statistician may not realize that he or she is dealing with correlated data*, multiples of data, functions of data, or averages of the "best two out of three." Behind nearly every measurement there is a hidden calibration curve* whose contribution to the error is not apparent and may go unnoticed unless the statistician mentions it.

For a general (somewhat outdated) textbook in statistics with applications in chemistry, see Bennett and Franklin [2].

References

[1] Altschuler, B. and Pasternak, B. (1963). *Health Phys.*, **9**, 293–298.

[2] Bennett, C. A. and Franklin, N. L. (1954). *Statistical Analysis in Chemistry and the Chemical Industry*. Wiley, New York.

[3] Currie, L. A. (1968). *Anal. Chem.*, **40**, 586–592.

[4] Ku, H. H., ed. (1969). Precision Measurement and Calibration. *Natl. Bur. Stand. Spec. Publ. 300*, Vol. 1.

[5] Lieblein, J. (1952). *J. Res. Natl. Bur. Stand.*, **48**, 255–268.

[6] Mandel, J. (1964). *The Statistical Analysis of Experimental Data.* Wiley, New York.

[7] Sarhan, A. and Greenberg, B. (1962). *Contributions to Order Statistics.* Wiley, New York, Sec. 10C.

[8] Tietjen, G. and Beckman, R. (1974). *Technometrics*, **16**, 53–56.

(CALIBRATION
ERROR ANALYSIS
STANDARDS
TRIMMING)

R. J. BECKMAN
G. L. TIETJEN

CHERNOFF EFFICIENCY *See* BAHADUR EFFICIENCY

CHERNOFF FACES

It is often important to represent data graphically* (*see* GRAPHICAL REPRESENTATION OF DATA) in order to facilitate the observer's ability to comprehend the essence of the data and to aid him or her in selecting proper methods of analysis. This is relatively difficult to do if the data are multivariate of dimension greater than 2. The method of faces was developed to cope with this problem [3].

The method applies a computer program which leads to the drawing of a cartoon of a face. The cartoon is determined by 18 parameters, such as the length of the nose, the curvature of the mouth, the size of the eyes, etc. Varying the 18 parameters changes the appearance of the face. Thus each face represents a point in 18-dimensional space. Given a sample of points in 10-dimensional space, these points may be mapped into 18 dimensions (e.g., by adjoining 8 constants to the 10 components of the points) and the corresponding cartoons represent the original data points.

In Fig. 1 some of the set of 88 data points are represented by cartoons based on six measurements on nummulited specimens from the Eocene Yellow Limestone Formation of northwestern Jamaica [9]. The components of each data point consist of the inner diameter of the embryonic chamber, total number of whorls, number of chambers in the first whorl, etc. It was desired to do a cluster analysis on the data and it is apparent from the figure that there are three major clusters and that the size of the eyes will play an important role in separating the three clusters.

Graphical methods may have several possible functions. A typical one is that of communicating a good deal of information that has been studied and analyzed. Here simplicity, familiarity, and lack of distortion are important attributes of such methods. Faces were developed for a different purpose. The object was to take multivariate data*, among whose components complex regularities and relations exist, and to enable the observer to discern these relations and regularities.

The cartoons of faces seem to be very effective for this latter purpose. People grow up studying and reacting to faces. Small and barely measurable differences are easily detected and evoke emotional reactions from a long catalog buried in the memory. Relatively large differences go unnoticed in cir-

Figure 1 Faces for fossil specimens.

cumstances where they are not important. This suggests that the human mind subconsciously operates as a high-speed computer filtering out insignificant phenomena and focusing on the potentially important ones. Particularly valuable is this flexibility in disregarding noninformative data and searching for useful information. (It is this flexibility that is lacking in standard computer programs.)

The ability to relate faces to emotional reactions seems to carry a mnemonic advantage. Experience with caricatures and cartoons indicates that they seem to have an advantage over more realistic faces. It is conjectured that these cartoons may tend to more nearly resemble the memory one has of a face than does a real face and that distortions of faces that resemble our memory of faces are more effective than real faces (or other bizarre distortions.)

In the early experience with faces, it was pointed out in many applications that the key insights obtained derived from the prominence of certain features (such as the size of eyes or shape of face), which *fortunately* coincided with the important components of the data vectors. A natural question that frequently arose was whether a rearrangement of the components of the vector would create a substantial effect in the ability to discern relationships. Chernoff and Rizvi [4] investigated the effect of random permutations* of the data vector on the ability of subjects to separate two groups of corresponding faces. It was discovered that the error rate in carrying out this task might be affected by about 25%. That is, for a task for which the average number of errors would be four, the number might be reduced or increased by one.

Another question that is often raised concerns the number of dimensions that can be comfortably handled using faces. Experience does not seem to indicate any difficulty in coping with relatively high dimensions. The author has on occasion used two faces side by side to deal with more than 18 dimensions. However, there is a difficulty in comprehension when the number of data points becomes large. It is difficult to deal with more than 50 or 100 faces at a time.

In recent years faces have been used in a variety of applications, including the study of moon rocks, electrocardiograms, and business data. Some of these applications discussed at a symposium on graphical representations concern Soviet foreign policy in Africa [8] and the representation of psychiatric diagnostic data [5].

Many applications have used faces as a communicator of information rather than as a tool for analysis. For this purpose the use of faces has some severe limitations. Information communicated this way may be easily distorted, and the receiver must be trained to interpret the data properly. Jacob [5] applied faces in an innovative way to reduce the tendency for distortion and the need for training. Basically, he did a prestudy to see how psychiatrists would diagnose a large class of faces. Then he represented patients with a specified diagnosis by the face most likely to be diagnosed that way by psychiatrists. In this fashion he avoided such dangers as that of representing a depressed patient by a smiling face.

When faces are used to communicate information, the face must act as a generalized model of the information being represented which the receiver learns to interpret after some training. Some models are easier to adapt to than others. As an extension of the face idea, Goode (see Marshall [6]) developed a cartoon figure of a football player to present the strengths and weaknesses of a team. Here a strong left arm indicates success in passing and a strong right arm indicates a good defense against passing. Bruckner and Montoya used an oil truck figure to represent companies involved in offshore oil leasing [2]. There is a trade-off between the excellent ability of people to react to faces and changes in faces, on the one hand, and the relation between a model and its natural analogs on the other hand.

The original faces were designed so that the results appeared on the output of a plotter driven by a computer. Turner and Tidmore [7] have designed a program where a coarser version of the faces appear in the output of a line printer. Various modifications of the original faces have been created. For example, Bruckner and Mills [1] describe the use of a modified version where the nose is a triangle and circular ears are appended to the face.

References

[1] Bruckner, L. A. and Mills, C. F. (1979). The Interactive Use of Computer Drawn Faces to Study Multidimensional Data. Informal Rep. *Los Alamos Sci. Lab. LA-7752-M*. pp. 1–27.

[2] Bruckner, L. A. and Montoya, A. F. (1979). The Use of an Oil Truck Figure to Represent Companies Involved in Offshore Oil Leasing. Informal Rep. *Los Alamos Sci. Lab. LA-7653-MS*, pp. 1–13.

[3] Chernoff, H. (1973). *J. Amer. Statist. Ass.*, **68**, 361–368.

[4] Chernoff, H. and Rizvi, M. H. (1975). *J. Amer. Statist. Ass.*, **70**, 548–554.

[5] Jacob, R. J. R. (1978). In *Graphical Representation of Multivariate Data*, P. C. C. Wang, ed. Academic Press, New York, pp. 143–168.

[6] Marshall, J. (1974). *Sports Illustrated*, Jan. 14, pp. 42–49.

[7] Turner, D. W. and Tidmore, F. E. (1977). *Proc. Statist. Sect. Amer. Statist. Ass. Annu. Meet.*, Chicago.

[8] Wang, P. C. C. and Lake, G. E. (1978). In *Graphical Representation of Multivariate Data*, P. C. C. Wang, ed. Academic Press, New York, pp. 13–58.

[9] Wright, R. M. and Switzer, P. (1971). *Ass. Math. Geol.*, **3**, 297–311.

(ANDREWS FUNCTION PLOT
BIPLOTS
GRAPHICAL REPRESENTATION OF DATA)

HERMAN CHERNOFF

CHERNOFF–SAVAGE THEOREM

This theorem establishes the asymptotic normality* of linear functions of the rank statistics under mild conditions. This result appeared in "Asymptotic normality and efficiency of certain non-parametric test statistics" *Ann. Math. Statist.*, **29**, 972–994 (1958).

(ASYMPTOTIC NORMALITY
RANK ORDER STATISTICS)

CHERNOFF THEOREM

Let Y be a real-valued random variable, let F be its *left-continuous* distribution function [i.e., $F(y) = \Pr[Y < y](-\infty < y < \infty)$], and let $\phi(t)$ be the moment generating function* of F [i.e. $\phi(t) = E(e^{tY})$]. [This function satisfies $0 < \phi(t) \leqslant \infty$ for each t and $\phi(0) = 1$.] Let $\rho = \inf\{\phi(t) : t \geqslant 0\}$, $0 \leqslant \rho \leqslant 1$. Let $Y_1, Y_2, \ldots$ be a sequence of independent replicates of Y and for each $n = 1, 2, \ldots$ let $P_n = \Pr[Y_1 + \cdots + Y_n \geqslant 0]$. Chernoff's theorem [2] states that

$$n^{-1} \log P_n \to \log \rho \qquad \text{as } n \to \infty.$$

M. Wichura (see, e.g., Bahadur [1]) has shown by counterexample that the theorem becomes false if "$\geqslant$" is replaced by "$>$" in the definition of P_n.

Chernoff's theorem is a basic tool in large deviation theory* and in investigations of asymptotic (large-sample) properties of statistical tests of hypotheses and efficiency of estimators.

References

[1] Bahadur, R. R. (1971). *Some Limit Theorems in Statistics*, Reg. Ser. Appl. Math. No. 4, SIAM, Philadelphia.

[2] Chernoff, H. (1952). *Ann. Math. Statist.*, **23**, 493–507.

(LARGE DEVIATIONS LIMIT THEOREMS)

CHI DISTRIBUTION

The distribution of the positive square root of a variable having a chi-square distribution*. Since the PDF of the latter distribution is

$$\frac{x^{(\nu/2)-1} \exp\left(-\tfrac{1}{2}x\right)}{2^{\nu/2}\Gamma\left(\tfrac{1}{2}\nu\right)} \qquad (x > 0)$$

the PDF of the chi distribution is

$$2x \frac{(x^2)^{(\nu/2)-1} \exp\left(-\tfrac{1}{2}x^2\right)}{2^{\nu/2}\Gamma\left(\tfrac{1}{2}\nu\right)}$$

$$= \frac{x^{\nu-1} \exp\left(-\tfrac{1}{2}x^2\right)}{2^{\nu/2}\Gamma\left(\tfrac{1}{2}\nu\right)} \qquad (x > 0).$$

The rth moment about zero is

$$E[X^r] = E\left[(X^2)^{r/2}\right] = 2^r\Gamma\left(\frac{\nu + r}{2}\right) / \Gamma\left(\frac{\nu}{2}\right).$$

(CHI-SQUARE DISTRIBUTION)

CHI-SQUARE DISTRIBUTION

The χ^2 distribution with ν degrees of freedom is the distribution of the sum of squares of ν independent unit normal variables [variables having the distribution $N(0, 1)$]. It is, in fact, a gamma distribution* with shape parameter $\frac{1}{2}\nu$ and scale parameter 2. Its properties can therefore be derived directly from those of the gamma distribution. A few are given here, for purposes of convenience.

Probability density function:
$\{2^{\nu/2}\Gamma(\frac{1}{2}\nu)\}^{-1} x^{(\nu/2)-1} e^{-x/2}$ $(0 < x)$

rth moment about zero:
$\nu(\nu + 2) \cdots (\nu + 2r - 2) = 2^r(\frac{1}{2}\nu)^{[r]}$

Expected value: ν; *variance*: 2ν

rth cumulant about zero ($r \geqslant 2$):
$2^{r-1}(r - 1)!$

Mode: $\nu - 2$; *median*: $\nu - \frac{2}{3}$ (approximately)

Approximations to the χ^2 distribution include Fisher's χ^2 approximation* and the Wilson–Hilferty formula*.

If ν is even, the probability that the χ^2 exceeds x is equal to the probability that a Poisson* variable with expected value $\frac{1}{2}x$ does not exceed $\frac{1}{2}\nu - 1$.

The distribution of the sum of two independent χ^2 variables with ν_1 and ν_2 degrees of freedom, respectively, is a χ^2 distribution with $(\nu_1 + \nu_2)$ degrees of freedom. The ratio of either of the variables to its sum has a beta distribution*; the ratio of the two variables (multiplied by ν_2/ν_1) has an F-distribution*.

The sum of squares of ν independent normal variables $N(\omega_j; 1)$ $(j = 1, \ldots,)$ has a *noncentral* χ^2 distribution* with ν degrees of freedom and noncentrality parameter

$\sum_{j=1}^{\nu} \omega_j^2$. (More details and a discussion of tables are given in Johnson and Kotz [5].)

K. Pearson [9] introduced the test statistic,

$$X^2 = \sum (\text{observed} - \text{expected})^2/(\text{expected})$$
$$\equiv \sum_{i=1}^{n+1} (M_i - Np_i)^2/(Np_i), \qquad (1)$$

with summation over the cells of an empirical frequency table to test the departure from the null hypothesis, H_0 say, (*see* CHI-SQUARE TESTS). The probability model is that there are N independent experiments at each of which there is a positive probability p_i that the ith event occurs, $\sum p_i = 1$; this is the common multinomial distribution*; the number of times the ith event is observed is M_i. The variance–covariance matrix*, $\mathbf{V}$, of the indicator variables is given by

$$v_{ii} = p_i(1 - p_i), \qquad v_{ij} = -p_i p_j \qquad (2)$$

and the expectation of each is p_i. The $(n + 1)$ indicator variables* are linearly dependent so that only the first n of them are considered; there are n degrees of freedom. $\mathbf{V}$ is thus of size n, and positive-definite. The expectation of the cell number is Np_i and the covariance matrix is $N\mathbf{V}$. The random variables can be chosen as $Y_i = N^{-1/2} \times (M_i - Np_i)$, $i = 1, 2, \ldots, n$. An asymptotic distribution can be obtained by the use of the multivariate central limit theorem*. Any linear form in the indicator variables in a single experiment obeys the condition for the central limit theorem. However, each M_i is the sum of indicator variables of the ith event at the N experiments, and so any arbitrary linear form $\mathbf{b}^T\mathbf{Y}$ is asymptotically normal; but this is sufficient for the asymptotic distribution of $\mathbf{Y}$ to be jointly normal; moreover, its covariance matrix is $\mathbf{V}$, so there is a density,

$$f(Y) = \text{constant} \times \exp\left[-\tfrac{1}{2}\mathbf{Y}^T\mathbf{W}\mathbf{Y}\right], \quad \text{say}$$
$$= \text{constant} \times \exp\left[-\tfrac{1}{2}\mathbf{Y}^T\mathbf{V}^{-1}\mathbf{Y}\right], \qquad (3)$$

which is proved by considering the moment generating function* $E(\exp \mathbf{y}^T\mathbf{t})$. Further, $\mathbf{Y}^T\mathbf{V}^{-1}\mathbf{Y}$ is then distributed as χ^2 with n degrees of freedom. This is usually known as *Pearson's lemma*. (Pearson's lemma can be derived by calculation of the generating function or by a linear transformation yielding $\mathbf{Y}^T\mathbf{V}^{-1}\mathbf{Y} \to \sum Y^{*2}$.) $\mathbf{Y}^T\mathbf{V}^{-1}\mathbf{Y}$ can now be identified with the expressions in (1). Let $\mathbf{D} = \operatorname{diag}(p_1^{1/2}, \ldots, p_n^{1/2})$ and $\mathbf{p}$ be the unit vector with elements, $p_j^{1/2}(1 - p_{n+1})^{-1/2}$. Then $\mathbf{p}\mathbf{p}^T$ is idempotent.

$$\mathbf{V} = (p_i\delta_{ij} - p_ip_j)$$
$$= \mathbf{D}\left(\mathbf{1} - (1 - p_{n+1})\mathbf{p}\mathbf{p}^T\right)\mathbf{D},$$

so its inverse is

$$\mathbf{D}^{-1}\left(\mathbf{1} + p_{n+1}^{-1}(1 - p_{n+1})\mathbf{p}\mathbf{p}^T\right)\mathbf{D}^{-1}.$$

There follows

$$\mathbf{Y}^T\mathbf{V}^{-1}\mathbf{Y} = \mathbf{Y}^T \operatorname{diag}\left(p_1^{-1}, \ldots, p_n^{-1}\right)\mathbf{Y}$$
$$+ p_{n+1}^{-1}\mathbf{Y}^T(\mathbf{1})\mathbf{Y}$$
$$= \sum_1^{n+1} Y_i^2/p_i, \qquad (4)$$

the expression given in (1), for $\mathbf{Y}^T(\mathbf{1})\mathbf{Y} = (\sum_1^n Y_i)^2 = Y_{n+1}^2$ by the linear dependence. In the preceding expressions, $\mathbf{1}$ is a square matrix, with every element equal to 1.

[The use of Fréchet's definition of the joint normal distribution as one in which every linear form is normal avoids the use of the false assumption of Pearson [9] that marginal normality is sufficient for joint normality of the form (3).]

The multinomial can be represented as the joint distribution of $(n + 1)$ independently distributed Poisson* variables with parameters $\lambda p_1, \ldots, \lambda p_{n+1}$, conditional on their sum being N. Each standardized Poisson variable, $Z_i = (M_i - \lambda p_i)(\lambda p_i)^{-1/2}$, is asymptotically standard normal and the set is mutually independent, so that asymptotic joint normality* is assured. The joint asymptotic distribution is then obtained by supposing that $\sum p_i^{1/2}Z_i = 0$ or alternatively by estimating $\hat{\lambda} = N$. An orthogonal transformation*, $\mathbf{X} = \mathbf{H}^T\mathbf{Z}$, $h_{i1} = p_i^{1/2}$, then yields n standardized mutually independent normal variables as required and $\mathbf{Z}^T\mathbf{Z} = \sum_2^{n+1} X_i^2$. The theory of estimation can be avoided by noting that the conditional distribution does not contain the parameter λ, so the orthogonal transformation, $\mathbf{X} = \mathbf{H}^T\mathbf{Z}$, can be made and λ chosen so that $X_1 = 0$. This is the proof of H. E. Soper, cited by Pearson [11], later to be popularized and extensively used by R. A. Fisher [2]. A proof by factorization of the multinomial has been given by Lancaster [6], which is equivalent to choosing **H** as a Helmert matrix* [4]. The asymptotic distribution can also be verified by a calculation of the moments of X^2 of (1). These proofs are all detailed in Lancaster [7]. Let $x^{(0)} = 1$ and suppose that $\{x^{(i)}\}$ is a basis or complete orthonormal set of functions with respect to a distribution function. Then if N independent observations are made, and if S is the summation operator,

$$X_i = N^{-1/2} S x^{(i)}, \tag{5}$$

is asymptotically normal and $\{X_i\}$ is a mutually independent set. $X^2 = \sum_1^n X_i^2$ is then distributed asymptotically as χ^2 with n degrees of freedom. The Pearson χ^2 is based on the generalized Haar functions*, but other systems are possible. Neyman [8] introduced the standardized Legendre polynomials* on the unit interval; the Hermite system may be used on the normal distribution (*see* CHEBYSHEV–HERMITE POLYNOMIALS), the trigonometric functions on the unit interval, and so on. The standardized sums from these distributions can be readily generated from ungrouped data or approximated from grouped data. Such test functions seem especially applicable in the analysis of bivariate data.

The parameters may be unknown and have to be estimated from the data. In general, it may be said that the justification of the use of the χ^2-test is easy if the statistics estimated are sufficient, but it may be more troublesome in the general case. A beginning was made by Pearson [11], who showed that homogeneous linear restrictions on the random variables would lead to fewer "variables" and other linear restrictions would introduce constants into the X^2-sum. Fisher in 1922 [2] and 1925 [3] used the ideas of hyperspace to obtain the distribution and the correct number of degrees of freedom. A more formal proof was given by Cramér [1].

Pearson [10] defined

$$\phi^2 = \chi^2/N \tag{6}$$

as an absolute estimate of the deviation of the observed from the theoretical distribution; it can also be interpreted as the sum of squares of the Fourier coefficients* of the orthonormal functions used in the test.

An $m \times n$ array of nonnegative integers $\{a_{ij}\}$ is called a contingency table*. It is usually supposed that it is generated by some probabilistic process of which there are three principal models: unrestricted sampling from a parent population; comparative trial, by sampling from parallel multinomial distributions; and by a permutation. The joint distribution of $\{a_{ij}\}$ conditional on fixed marginal totals is given in each case by the same combinatorial formula. The conditional expectation of an entry, a_{ij}, is $a_{i\cdot}\, a_{\cdot j}/N, N = \sum\sum a_{ij}$.

$$\begin{aligned} X^2 &= \sum_{i=1}^{m} \sum_{j=1}^{n} (a_{ij} - Ea_{ij})^2 / Ea_{ij} \\ &\equiv \sum_{i=1}^{m} \sum_{j=1}^{n} a_{ij}^2 / Ea_{ij} - N \end{aligned} \tag{7}$$

is then the test function for mutual independence. The test is much used in practical statistics, especially as a preliminary test of homogeneity* between populations (*see* CHI-SQUARE TESTS for more details). Many theories based on information-theoretical* ideas finally obtain a test function with the asymptotic χ^2 distribution.

The results of n independent experiments can often be combined by calculating

$$\sum Z_i = -2 \sum \log_e P_i, \tag{8}$$

where P_i is the probability (*see* P-VALUE) assigned to an individual experiment. Z_i is distributed as χ^2 with 2 degrees of freedom and $\sum Z_i$ with $2n$ degrees of freedom [3].

An extensive bibliography, especially of the classical literature, is available in Lancaster [7].

References

[1] Cramér, H. (1946). *Mathematical Methods of Statistics*. Princeton University Press, Princeton, N.J.

[2] Fisher, R. A. (1922). *J. R. Statist. Soc.*, **85**, 87–94.

[3] Fisher, R. A. (1925). *Statistical Methods for Research Workers*. Oliver & Boyd, Edinburgh. (14th ed., 1970.)

[4] Irwin, J. O. (1949). *Biometrika*, **36**, 130–134.

[5] Johnson, N. L. and Kotz, S. (1970). *Continuous Univariate Distributions*, Vol. 1. Wiley, New York, Chap. 17.

[6] Lancaster, H. O. (1949). *Biometrika*, **36**, 117–129.

[7] Lancaster, H. O. (1969). *The Chi-Squared Distribution*. Wiley, New York.

[8] Neyman, J. (1937). *Skand. Aktuarietidskr.*, **20**, 149–199.

[9] Pearson, K. (1900). *Philos. Mag.*, Ser. 5, **50**, 157–175.

[10] Pearson, K. (1904). Mathematical Contributions to the Theory of Evolution: XIII. On the Theory of Contingency and its Relation to Association and Normal Correlation. *Draper's Company Res. Mem. Biom. Ser. 1.*

[11] Pearson, K. (1916). *Biometrika*, **11**, 145–158.

(APPROXIMATIONS TO
DISTRIBUTIONS
CHI-SQUARE TESTS
CONTINGENCY TABLES
GAMMA DISTRIBUTION
GOODNESS OF FIT)

H. O. LANCASTER

CHI-SQUARE TESTS

Chi-square tests are a widely used statistical method for an extremely broad range of applications. However, almost all of these share one common feature. It is the availability of sufficient sample size for them to have approximately a chi-square distribution* under a specific null hypothesis through multivariate normal central limit theory* for the underlying statistics on which they are based (*see* CHI-SQUARE DISTRIBUTION). When this research design requirement is satisfied, chi-square tests can be formulated for data from many different types of sampling or observational/measurement processes. Moreover, such methods can be directed at either univariate or multivariate data involving all types of measurement scales. For all of these different situations, the various types of chi-square test statistics that are used in practice can be viewed as being equivalent to quadratic forms* of the type

$$Q = Q(\mathbf{y}) = \mathbf{y}'\mathbf{V}^{-1}\mathbf{y}, \tag{1}$$

where $\mathbf{y}$ is a $(u \times 1)$ vector of random variables which approximately has the multivariate normal distribution* $N(\mathbf{0}_u, \mathbf{V})$, where $\mathbf{0}_u$ is a $(u \times 1)$ vector of 0's and $\mathbf{V}$ is a $(u \times u)$ positive-definite covariance matrix; so Q has the chi-squared distribution* with u degrees of freedom (d.f.) because it can be expressed as the sum of the squares of u independent normal* $N(0, 1)$ random variables.

One important type of chi-square test is the goodness-of-fit* statistic $Q_p = \sum_{j=1}^{r} (n_j - m_j)^2 / m_j$ due to Karl Pearson [36]; this statistic is used as an index of discrepancy between a set of observed counts $n_1, n_2, \dots, n_r$ in r mutually exclusive categories and corresponding expected counts $m_1, m_2, \dots, m_r$ for a specific hypothesis H_0 under investigation. For situations where the observed and expected counts are at least moderately large, Pearson suggested rejection of H_0 at significance level α if Q_p exceeded the critical value $\chi^2_{1-\alpha}(r-1)$. R. A. Fisher [14] provided an adjustment in the degrees of freedom for Q_p to d.f.$=(r-t-1)$ if the values $m_1, m_2, \dots, m_r$ were based upon asymptotically efficient estimates for t unknown, independent parameters; see, e.g., Cochran [9] for further discussion, and also CHI-SQUARE DISTRIBUTION.

A well-known example of the Pearson statistic is the chi-square test for the hypothesis of no association (or independence) for two-dimensional *contingency tables**. Its specific form for this application is mentioned in the context of a random partition of a finite population subsequent to (24) and (30) and in the context of independent samples having a common distribution (or homogeneity) in (51). The nature of the Pearson statistic for general hypotheses pertaining to a sample from a single discretely distributed population is given in (43) together with some other asymptotically equivalent statistics.

The subsequent sections of this entry describe some of the statistical settings wherein chi-square tests are typically used. More specifically, in the next section, attention is directed to cases involving random variables which are known to be normally distributed regardless of sample size. Although this assumption is often not realistic, the methods based upon it are of reference interest for more general extensions. The other sections deal with the manner in which chi-square tests are formulated with respect to asymptotic considerations for several types of frequently used (or hypothetically presumed) sampling processes. These include finite population random sampling without replacement*, infinite population stratified simple random (or distributional/likelihood) sampling*, and complex probability random sampling*. Attention is directed at the common logic which chi-square tests from many diverse settings share. Illustrative applications for these methods are given in the accompanying entry for CHI-SQUARE TESTS: NUMERICAL EXAMPLES, as well as those for CATEGORICAL DATA and CONTINGENCY TABLES. Bibliographies of related references have been prepared by Killion and Zahn [21] and Singer [40].

Finally, the use of the term "chi-square test" in this entry pertains primarily to situations in which the corresponding test statistic has an approximate (or true) chi-square distribution*. However, for some applications where this condition usually does not hold because of insufficient sample size or some other reason, the reader should note that some other term may be preferable, since evaluation of statistical significance must be undertaken either by exact methods* or some other type of approximation. See Fienberg [13] for a review of the properties of goodness-of-fit statistics such as Q_p for small-sample situations.

NORMAL PROBABILITY MODEL SAMPLING SITUATIONS

This section is concerned with the formulation of chi-square statistics for two types of situations in which the data can be presumed to have normal distributions. First, the standard test for a specific variance for independent random variables with unknown mean is discussed. Following that is a discussion of general linear hypotheses* concerning the mean vector for a multivariate normal distribution* with known covariance matrix. Since it is relatively theoretical, it may not be of interest to applications-oriented readers. Its primary purpose is to describe three basic principles pertaining to the underlying framework for essentially all chi-square tests. These are:

1. The direct formulation of quadratic form chi-square statistics in the sense of (1).
2. The indirect formulation of goodness-of-fit chi-square statistics via the demonstration of their identity to direct quadratic-form counterparts.
3. The partition of chi-square statistics into two or more components directed at corresponding alternatives of interest.

These principles are then used for broadening the coverage of various types of chi-square tests and also for linking them to a unified theme.

Tests for Specified Variance for Simple Random Samples with Replacement from a Univariate Normal Population

Let $y_1, y_2, \ldots, y_n$ be observed independent, identically distributed (i.i.d.) random variables from a population with the univariate normal distribution $N(\mu, v)$, where μ is its unknown mean and v is its unknown variance. For this framework, a chi-square test for the hypothesis $H_0: v = v_0$, where v_0 is a specified value, can be undertaken with

$$Q = Q(\mathbf{y} \mid H_0 : v = v_0) = \sum_{l=1}^{n} (y_l - \bar{y})^2 / v_0, \tag{2}$$

which has the chi-square distribution with d.f. $= (n-1)$. If $v > v_0$ is the only alternative of practical interest, then H_0 is rejected if $Q \geqslant \chi^2_{1-\alpha}(n-1)$, where α is the significance level*; whereas if $v < v_0$ is the only alternative of practical interest, then H_0 is rejected if $Q \leqslant \chi^2_\alpha(n-1)$; if both directions are of interest, then α is split into components α_1 and α_2 such that $\alpha_1 + \alpha_2 = \alpha$ according to their relative importance and H_0 is rejected if either $Q \geqslant \chi^2_{1-\alpha_1}(n-1)$ or $Q \leqslant \chi^2_{\alpha_2}(n-1)$. Finally, the specified variance test statistic (2) can be expressed in the general form (1) as

$$Q = Q(\mathbf{Cy}) = \mathbf{y}'\mathbf{C}'\left[v_0\mathbf{CC}'\right]^{-1}\mathbf{Cy}, \quad (3)$$

where $\mathbf{C}$ is any $[(n-1) \times n]$ matrix that is a basis of contrast* space, e.g., $\mathbf{C} = [\mathbf{I}_u, \mathbf{1}_u]$, where $u = n - 1$, $\mathbf{I}_u$ is the uth-order identity matrix and $\mathbf{1}_u$ is a $u \times 1$ vector of 1's [see the Appendix for a lemma implying equality of (2) and (3)].

Some applications for which this type of test concerning a specified variance is often useful are as follows:

1. Quality control* studies concerned with whether measures of product performance are in accordance with variability standards (see Natrella [33, Chap. 4]).
2. Educational testing studies concerned with whether a particular study population has the same extent of variability as a standard reference population (*see* EDUCATIONAL STATISTICS).
3. Reliability* studies to determine whether newly trained observers obtain data in accordance with variability standards.

Tests for Linear Hypotheses Involving Population Means for Multivariate Random Variables with Known Covariance Structure

Let $y_1, y_2, \ldots, y_u$ be observed, possibly correlated random variables from a sampling or observation/measurement process which can be assumed to follow the multivariate normal distribution $N(\boldsymbol{\mu}, \mathbf{V})$, where $\boldsymbol{\mu}$ is an unknown vector of population means and $\mathbf{V}$ is a known $(u \times u)$ positive-definite covariance matrix. For this framework, a chi-square test for the general linear hypothesis*

$$H_0 : \mathbf{W}\boldsymbol{\mu} - \boldsymbol{\eta} = \mathbf{0}_w, \quad (4)$$

where $\mathbf{W}$ is a prespecified $(w \times u)$ matrix of known coefficients with full rank $w \leqslant u$ and $\boldsymbol{\eta}$ is a prespecified $(w \times 1)$ vector of known coefficients, can be undertaken with

$$\begin{aligned} Q &= Q(\mathbf{Wy} - \boldsymbol{\eta}) \\ &= (\mathbf{Wy} - \boldsymbol{\eta})'(\mathbf{WVW}')^{-1}(\mathbf{Wy} - \boldsymbol{\eta}). \quad (5) \end{aligned}$$

Since $(\mathbf{WVW}')$ can be written as the product $\mathbf{TT}'$ of a $(w \times w)$ lower-triangular matrix $\mathbf{T}$ and its transpose $\mathbf{T}'$, the statistic Q can be expressed as

$$\begin{aligned} Q &= \left[\mathbf{T}^{-1}(\mathbf{Wy} - \boldsymbol{\eta})\right]'\mathbf{T}'(\mathbf{WVW}')^{-1}\mathbf{T} \\ &\quad \cdot \left[\mathbf{T}^{-1}(\mathbf{Wy} - \boldsymbol{\eta})\right] = \mathbf{g}'\mathbf{g} \quad (6) \end{aligned}$$

with $\mathbf{g} = \mathbf{T}^{-1}(\mathbf{Wy} - \boldsymbol{\eta})$. Under H_0, the random vector $\mathbf{g}$ has the multivariate normal distribution $N(\mathbf{0}_w, \mathbf{I}_w)$ by construction; so Q has the chi-square distribution with d.f. $= w$ because it can be written as the sum of the squares of w independent $N(0, 1)$ random variables.

For most applications, $\boldsymbol{\eta} = 0$, and $\mathbf{W}$ is a matrix of contrasts, i.e, $\mathbf{W}\mathbf{1}_u = \mathbf{0}_w$. In this case, H_0 in (4) means that the variation among the elements of $\boldsymbol{\mu}$ can be characterized by the linear regression model*

$$\boldsymbol{\mu} = \mathbf{X}\boldsymbol{\beta}, \quad (7)$$

where $\mathbf{X}$ is any $(u \times t)$ matrix with full rank $t = (u - w)$ such that $\mathbf{W}$ and $\mathbf{X}'$ are orthogonal and $\boldsymbol{\beta}$ is a $(t \times 1)$ vector of unknown coefficients. The specification (7) is often called the freedom equation or *model* formulation of H_0, and the specification (4) is called the constraint formulation. Efficient*, linear unbiased* estimates $\mathbf{b}$ for the parameters $\boldsymbol{\beta}$ can be obtained by applying weighted least squares*. These estimates, which minimize the quadratic function,

$$\begin{aligned} Q &= Q\left[\mathbf{y}, \mathbf{b} \mid H_0 : \boldsymbol{\mu} = \mathbf{X}\boldsymbol{\beta}\right] \\ &= (\mathbf{y} - \mathbf{Xb})'\mathbf{V}^{-1}(\mathbf{y} - \mathbf{Xb}), \quad (8) \end{aligned}$$

have the form

$$\mathbf{b} = \mathbf{b}(\mathbf{y}) = (\mathbf{X}'\mathbf{V}^{-1}\mathbf{X})^{-1}\mathbf{X}'\mathbf{V}^{-1}\mathbf{y}. \quad (9)$$

Also, $\mathbf{b}$ is the maximum likelihood estimator* for $\boldsymbol{\beta}$. Finally, the minimized value of (8), which may be expressed as

$$Q = \mathbf{y}'\left[\mathbf{V}^{-1} - \mathbf{V}^{-1}\mathbf{X}(\mathbf{X}'\mathbf{V}^{-1}\mathbf{X})^{-1}\mathbf{X}'\mathbf{V}^{-1}\right]\mathbf{y}, \quad (10)$$

is identical to the chi-square test statistic in (5) with d.f. $= w = (u - t)$ for the case $\boldsymbol{\eta} = \mathbf{0}$ since

$$\left[\mathbf{V}^{-1} - \mathbf{V}^{-1}\mathbf{X}(\mathbf{X}'\mathbf{V}^{-1}\mathbf{X})^{-1}\mathbf{X}'\mathbf{V}^{-1}\right] = \mathbf{W}'(\mathbf{W}\mathbf{V}\mathbf{W}')^{-1}\mathbf{W}, \quad (11)$$

as a consequence, for example, of the matrix lemma given in the Appendix. Thus Q in (10) represents a goodness-of-fit* statistic for the model (7), since it is equal to the test statistic (5) for the corresponding constraints (4).

If the model (7) adequately characterizes the variation among the elements of $\boldsymbol{\mu}$, attention can be directed at linear hypotheses

$$H_0 : \mathbf{C}\boldsymbol{\beta} = \mathbf{0}_c, \quad (12)$$

where $\mathbf{C}$ is a prespecified $(c \times t)$ matrix of full rank $c \leqslant t$. Since $\mathbf{b}$ has the multivariate normal distribution $N(\boldsymbol{\beta}, (\mathbf{X}'\mathbf{V}^{-1}\mathbf{X})^{-1})$, a chi-square test statistic for the hypothesis (12) is

$$Q = Q(\mathbf{Cb}) = \mathbf{b}'\mathbf{C}'\left[\mathbf{C}(\mathbf{X}'\mathbf{V}^{-1}\mathbf{X})^{-1}\mathbf{C}'\right]^{-1}\mathbf{Cb}, \quad (13)$$

which has the chi-square distribution with d.f. $= c$ under H_0. However, the hypothesis (12) is equivalent to the parameter vector model $\boldsymbol{\beta} = \mathbf{Z}\boldsymbol{\gamma}$, where $\mathbf{Z}$ is any $[t \times (t - c)]$ matrix such that $\mathbf{C}$ and $\mathbf{Z}'$ are orthogonal, which together with (7) implies the population mean vector model

$$\boldsymbol{\mu} = \mathbf{X}\boldsymbol{\beta} = \mathbf{XZ}\boldsymbol{\gamma}. \quad (14)$$

The statistic Q in (13) can be interpreted as the difference between the goodness-of-fit statistic of type (8) for the model (14) and that for the model (7). In other words, if $\mathbf{W}_C$ is a $[(w + c) \times u]$ full rank matrix which is orthogonal to $(\mathbf{XZ})'$, then

$$\begin{aligned} Q(\mathbf{Cb}) &= Q(\mathbf{W}_C\mathbf{y}) - Q(\mathbf{Wy}) \\ &= \mathbf{y}'\mathbf{W}_C'(\mathbf{W}_C\mathbf{V}\mathbf{W}_C')^{-1}\mathbf{W}_C\mathbf{y} \\ &\quad - \mathbf{y}'\mathbf{W}'(\mathbf{WVW}')^{-1}\mathbf{Wy}; \quad (15) \end{aligned}$$

so $Q(\mathbf{Cb})$ corresponds to a test statistic for the additional constraints implied by the model (14) given that those implied by the model (7) hold. For this reason, the statistics (8) and (13) represent a *partition of the chi-square test statistic* $Q(W_C\mathbf{y})$ for the goodness of fit of the model (14). Finally, such partitions can be extended to involve additional components through the specification of appropriate hierarchical sequences of either constraints $\mathbf{W}$ or models $\mathbf{X}$.

Some applications where the types of chi-square test statistics for linear hypotheses described here are sometimes useful are as follows:

1. Multifactor experiments* that involve processes for which the covariance structure of the data can be presumed known on the basis of measurement process standards or the availability of valid historical estimates from previous investigations, with an important special case being $\mathbf{V} = v_0\mathbf{I}_u$, where v_0 is known,
2. Multifactor studies for which covariance structure is known on the basis of the performance of some standard reference population.

RANDOMIZATION MODEL SITUATIONS

In this section, chi-square tests based upon finite population randomization model concepts are discussed in general terms which encompass nominal, ordinal, and interval measurement scales. These include such well-known methods as the Kruskal–Wallis one-way rank analysis-of-variance statistic*, the Friedman two-way rank analysis-of-variance statistic*, the Spearman rank correlation test statistic*, the Pearson chi-square statistic* for contingency tables, the Cochran Q-statistic* for equality of matched proportions, and the Mantel–Haenszel statistic* for sets of 2×2 contingency tables.

Tests of Randomness or No Association between a Response Variable and a Subpopulation Partition Structure for a Finite Population

Let $y_1, y_2, \ldots, y_n$ denote the observed values of some response variable for the study units (or subjects) in some finite population of size n. Let $i = 1, 2, \ldots, s$ index a set of subpopulations for which the relationship to the response variable is to be investigated; and let $n_1, n_2, \ldots, n_s$ denote the corresponding numbers of subjects who belong to them, and so $\sum_{i=1}^{s} n_i = n$. Let $\mathbf{U}$ be a $(n \times s)$ matrix of subpopulation indicator random variables

$$U_{li} = \begin{cases} 1 & \text{if subject } l \text{ belongs to} \\ & i\text{th subpopulation} \\ 0 & \text{otherwise} \end{cases} \qquad (16)$$

where $l = 1, 2, \ldots, n$. For this type of framework, a basic question is whether or not the response variable is distributed at random with respect to the subpopulations. A more specific formulation is the hypothesis (17) of no association:

$\boldsymbol{H_0}$: There is no relationship between the subpopulations and the response variable in the sense that the observed partition of the response values $y_1, y_2, \ldots, y_n$ into the subpopulations can be regarded as equivalent to a successive set of simple random samples* of $n_1, n_2, \ldots, n_s$ subjects. (17)

Since the hypothesis H_0 implies that the $(n!/\prod_{i=1}^{s} n_i!)$ possible allocations of the $y_1, y_2, \ldots, y_n$ to the subpopulations are equally likely, it follows from the theory of simple random sampling (see, e.g., Cochran [10]) that the sample means

$$\bar{y}_i = \frac{1}{n_i} \sum_{l=1}^{n} U_{li} y_l \qquad (18)$$

have the same expected value

$$\mu_i = \mu_i(H_0) = \frac{1}{n_i} \sum_{l=1}^{n} y_l E(U_{li} \mid H_0)$$

$$= \frac{1}{n} \sum_{l=1}^{n} y_l = \bar{y}, \qquad (19)$$

which is the overall population mean of the y's, and the covariance structure

$$v_{ii'} = v_{ii'}(H_0)$$

$$= \frac{1}{n_i n_{i'}} \sum_{l=1}^{n} \sum_{l'=1}^{n} y_l y_{l'} \operatorname{cov}(U_{li}, U_{l'i'} \mid H_0)$$

$$= \frac{n\delta_{ii'} - n_i}{n_i(n-1)n} \left[\sum_{l=1}^{n} (y_l - \bar{y})^2 \right]$$

$$= \frac{v}{n-1} \left(\frac{n\delta_{ii'}}{n_i} - 1 \right) \qquad (20)$$

where $\delta_{ii'} = 1$ if $i = i'$ and $\delta_{ii'} = 0$ otherwise, and v denotes the finite population variance. If the sample sizes $n_1, n_2, \ldots, n_s$ for the respective subpopulations are sufficiently large, randomization central limit theory* (as discussed in Hájek and Šidák [19] or Puri and Sen [38]) implies that the vector $\bar{\mathbf{y}}$ of means $\bar{y}_1, \bar{y}_2, \ldots, \bar{y}_s$ has approximately the multivariate normal distribution $N(\boldsymbol{\mu}, \mathbf{V})$, with $\boldsymbol{\mu}$ and $\mathbf{V}$ having the structure given in (19) and (20). Thus a chi-square test for H_0 can be undertaken with

$$Q = Q(\mathbf{C}\bar{\mathbf{y}}) = \bar{\mathbf{y}}'\mathbf{C}'[\mathbf{C}\mathbf{V}\mathbf{C}']^{-1}\mathbf{C}\bar{\mathbf{y}}$$

$$= \frac{n-1}{nv} \bar{\mathbf{y}}'\mathbf{C}'[\mathbf{C}\mathbf{D}_{\mathbf{n}}^{-1}\mathbf{C}']^{-1}\mathbf{C}\bar{\mathbf{y}}, \qquad (21)$$

where $\mathbf{C} = [\mathbf{I}_{s-1}, -\mathbf{1}_{s-1}]$ is a contrast matrix basis and $\mathbf{D}_{\mathbf{n}}$ is a diagonal matrix with $n_1, n_2, \ldots, n_s$ on the diagonal. Since Q in (21) has the same structure for large-sample situations as Q in (5) has for normal populations, its distribution is approximately chi-square with d.f. $= (s - 1)$.

The matrix expression (21) can be verified to be identical to

$$Q = \frac{n-1}{nv} \sum_{i=1}^{s} n_i(\bar{y}_i - \bar{y})^2, \qquad (22)$$

by the use of the matrix lemma in the Appendix. Thus Q can be interpreted as a one-way analysis-of-variance* test statistic. In this regard, two special cases for the y's are of particular interest. If $y_l = l$ for $l = 1, 2, \ldots, n$ so that the y's are ranks (without ties), then

$$Q = \frac{12}{n(n+1)} \sum_{i=1}^{s} n_i(\bar{y}_i - \bar{y})^2 \qquad (23)$$

is the Kruskal and Wallis [26] rank analysis-of-variance* statistic. Alternatively, if the y's correspond to a binary attribute such that $y_l = 1$ for $l = 1, 2, \dots, m$ and $y_l = 0$ for $l = m + 1, \dots, n$, then

$$Q = \frac{n(n-1)}{m(n-m)} \sum_{i=1}^{s} \frac{1}{n_i}(n_{i1} - m_{i1})^2$$
$$= \left(\frac{n-1}{n}\right) \sum_{i=1}^{s} \sum_{k=0}^{1} \frac{(n_{ik} - m_{ik})^2}{m_{ik}}, \quad (24)$$

where n_{i1} is the number of study units with the attribute in the ith subpopulation, $m_{i1} = n_i(m/n)$ is its expected value under H_0, and $n_{i0} = (n_i - n_{i1})$ and $m_{i0} = (n_i - m_{i1})$ are the corresponding numbers without the attribute. In this setting, $Q_p = [nQ/(n-1)]$ is the Pearson chi-square statistic for the hypothesis of no association in the $(s \times 2)$ contingency table for the subpopulations vs. the binary attribute.

The statistic Q in (21) can be directed at specific types of comparisons by use of the linear models framework in (4) to (6). For this purpose, let $\mathbf{A}$ denote a full rank $(a \times s)$ matrix of prespecified linear combinations for $a < s$ contrasts among the s subpopulations, i.e., $\mathbf{A1}_s = \mathbf{0}_a$. Let

$$\mathbf{F} = \mathbf{AD_n\bar{y}} = [n_1\mathbf{a}_1, \dots, n_s\mathbf{a}_s]\bar{\mathbf{y}}$$
$$= \sum_{i=1}^{s} n_i\bar{y}_i\mathbf{a}_i \quad (25)$$

denote a set of summary measures for the association of the response with the scores $\mathbf{A}$. Under H_0, $\mathbf{F}$ has approximately the multivariate normal distribution $N(\boldsymbol{\mu}_\mathbf{F}, \mathbf{V}_\mathbf{F})$, where

$$\boldsymbol{\mu}_\mathbf{F} = \boldsymbol{\mu}_\mathbf{F}(H_0) = \bar{y}(\mathbf{An}),$$
$$\mathbf{V}_\mathbf{F} = \mathbf{V}_\mathbf{F}(H_0)$$
$$= \frac{v}{n-1}\mathbf{A}[n\mathbf{D_n} - \mathbf{nn}']\mathbf{A}' \quad (26)$$

if the sample sizes $n_1, n_2, \dots, n_s$ are sufficiently large for the application of randomization model central limit theory. Thus a chi-square test of H_0 with respect to $\mathbf{F}$ is

$$Q = Q(\mathbf{F} - \boldsymbol{\mu}_\mathbf{F})$$
$$= (\mathbf{F} - \boldsymbol{\mu}_\mathbf{F})'\mathbf{V}_\mathbf{F}^{-1}(\mathbf{F} - \boldsymbol{\mu}_\mathbf{F}), \quad (27)$$

which has approximately the chi-square distribution with d.f. $= a$. Moreover, by letting $\mathbf{S} = [n\mathbf{D_n} - \mathbf{nn}']$, it can be verified that Q in (27) can be expressed as

$$Q = (n-1)R_{y,\mathbf{A}}^2$$
$$= \frac{n-1}{n^2 v}\left\{\bar{\mathbf{y}}'\mathbf{SA}'[\mathbf{ASA}']^{-1}\mathbf{AS}\bar{\mathbf{y}}\right\}, \quad (28)$$

where $R_{y,\mathbf{A}}^2$ is the squared multiple correlation coefficient* for the multiple linear regression* of the observed responses y_l on the scores $\mathbf{a}_i$ for the corresponding subpopulations. Hence if $a = 1$, then Q in (28) represents a test statistic for H_0 that is based on the first-order Pearson product-moment correlation coefficient* of the response and a single factor. Similarly, if both the y's and the single score vector are ranks, then Q represents a test statistic that is based on the Spearman rank correlation* coefficient.

In the form (28), the statistic Q is often applicable to more general situations where the overall population size n is large but the separate subpopulation sizes $n_1, n_2, \dots, n_s$ are relatively small. In fact, one special case of considerable interest is where $n_1 = n_2 = \cdots n_s = 1$. However, the use of an approximate chi-square distribution for Q in (28) under these conditions requires that the score matrix $\mathbf{A}$ satisfy certain limit process assumptions which are discussed in Hájek and Šidák [19] and Puri and Sen [38] in order to justify central limit theory for $\mathbf{F}$.

Finally, either the overall Q in (21) or the multiple correlation Q in (28) can be partitioned into components like those in (8) and (13). Here it would seem appropriate to let $\mathbf{W}$ correspond to the higher-order interaction* effects and/or trend* effects that might be anticipated a priori to be null and to let $\mathbf{X}$ correspond to the lower-order effects of interest. However, these types of statistics should be applied carefully because they are *not* specifically tests of the effects, but rather of H_0 in (17), since it is the underlying framework for the randomization distribution* of $\bar{\mathbf{y}}$. In other words, the partition of chi-square in this randomization model setting permits power* to be focused in the direction of those alternatives to H_0 which are of the most interest, but it does not provide valid tests for the null counterparts of such alternatives. For these more narrow hypotheses, other methods such as those discussed in the section "Likelihood Model Sampling Situations," are required.

Some applications for which chi-square test statistics with respect to a randomization framework are useful are:

1. Experimental design data for which only randomization is assumed, as opposed to some underlying probability distribution structure (*see* CHI-SQUARE TESTS: NUMERICAL EXAMPLES, and Koch et al. [25]).
2. Observational and/or historical data from restricted populations such as all reported motor vehicle accidents in a specific area during a specific year, all arrests for a specific type of criminal offense in a specific area during a specific year,.etc. Here randomization itself represents a hypothesis of interest which permits the statistical evaluation of apparent associations for the experience of a *fixed* population without any underlying probability distribution (or superpopulation sampling) assumptions (see Koch et al. [25] for specific examples).

Further discussion of the chi-square tests described in this section and their application is given in Landis et al. [28].

Tests of Randomness for a Multivariate Set of Response Variables vs. a Subpopulation Partition Structure

For the same framework, considered in the preceding discussion, let $\mathbf{y}_1, \mathbf{y}_2, \ldots, \mathbf{y}_n$ denote $(d \times 1)$ vectors of d response variables. Then a chi-square test statistic for the multivariate version of H_0 in (17) is

$$Q = \left(\frac{n-1}{n}\right) \sum_{i=1}^{s} n_i (\bar{\mathbf{y}}_i - \bar{\mathbf{y}})' \mathbf{V}^{-1} (\bar{\mathbf{y}}_i - \bar{\mathbf{y}}), \tag{29}$$

where the $\bar{\mathbf{y}}_i$'s, and $\bar{\mathbf{y}}$ and $\mathbf{V}$ have analogous definitions to their univariate counterparts in (18), (19), and (20). If the sample sizes $n_1, n_2, \ldots, n_s$ are sufficiently large, Q has approximately the chi-square distribution with d.f. $= d(s-1)$.

If the d response variables all involve ranks, then Q is the multivariate Kruskal–Wallis statistic (see, e.g., Puri and Sen [38]). In addition, the difference between Q in (29) for all d responses and an analogous Q for some subset of c responses which are covariables with no anticipated association with subpopulations represents a multivariate rank analysis of covariance* statistic for the $(d-c)$ responses after adjustment for the c covariables. This type of test statistic has an approximate chi-square distribution with d.f. $= (d-c)(s-1)$ under H_0, given that H_0 can be initially presumed for the c covariables. Other properties of this type of statistic are discussed in Quade [39].

If the d response variables are binary indicators (*see* BINARY DATA) for d of the $(d+1)$ possible outcomes of a categorical variable and if $m_1, m_2, \ldots, m_d$ denote the respective numbers of study units in the overall population with these outcomes and m_0 denotes the remainder, then Q in (29) can be verified to be identical to

$$Q = \left(\frac{n-1}{n}\right) \sum_{i=1}^{s} \sum_{k=0}^{d} \frac{(n_{ik} - m_{ik})^2}{m_{ik}}, \tag{30}$$

where n_{ik} is the number of study units in the ith subpopulation with the outcome for the kth indicator, $m_{ik} = n_i(m_k/n)$ is its expected value under H_0, and n_{i0} and m_{i0} are the corresponding numbers for the study units without any of the d attributes. In this setting, $Q_p = [nQ/(n-1)]$ is the Pearson chi-square statistic for the hypothesis of no association in the $[s \times (d+1)]$ contingency table for the subpopulations vs. the categorical variable.

The multivariate methods described here are of interest for the same types of applications as their univariate counterparts in the section "Test of Randomness or No Association . . . ."

Tests of No Partial Association for a Stratified Finite Population

Let $h = 1, 2, \ldots, q$ index a set of strata for some prespecified partition of a finite population. Let $y_{h1}, y_{h2}, \ldots, y_{hn_h}$ denote the observed values of some response variable for the study units in the hth stratum. Let $i = 1, 2, \ldots, s$ index a set of subpopulations for which the relationship to the response variable in a within-stratum sense is to be investigated, and let $n_{hi} \geqslant 1$ denote the number of study units in the ith subpopulation of the hth stratum. One hypothesis of interest for this type of situation is *no partial association*, for which a specific formulation is

H_0: For each of the strata $h = 1, 2, \ldots, q$, there is no relationship between the subpopulations and the response variable in the sense of (17). (31)

Since the hypothesis (31) implies the hypothesis (17) for each stratum, chi-square tests for it may be undertaken by various types of combination of the frameworks discussed in the preceding two subsections for the separate strata.

One type of chi-square test for H_0 is the *total partial association* statistic

$$Q_T = \sum_{h=1}^{q} \left(\frac{n_h - 1}{n_h v_h} \right) \sum_{i=1}^{s} n_{hi} (\bar{y}_{hi} - \bar{y}_h)^2, \tag{32}$$

where the $\bar{y}_{hi}$, $\bar{y}_h$, and v_h are the hth-stratum counterparts of (18), (19), and (20). If all the sample sizes n_{hi} are sufficiently large, then Q_T has approximately the chi-square distribution with d.f. $= q(s-1)$.

For situations where H_0 in (31) is of interest, a somewhat different approach may be more useful. It is based on the across-strata summary measures

$$G_i = \sum_{h=1}^{q} n_{hi} \bar{y}_{hi}. \tag{33}$$

If the combined strata sample sizes $n_{+i} = \sum_{h=1}^{q} n_{hi}$ are sufficiently large, then the vector $\mathbf{G}$ of summary measures $G_1, G_2, \ldots, G_s$ has approximately the multivariate normal distribution $N(\boldsymbol{\mu}_\mathbf{G}, \mathbf{V}_\mathbf{G})$ with

$$\boldsymbol{\mu}_\mathbf{G} = \sum_{h=1}^{q} \mathbf{n}_h \bar{y}_h,$$
$$\mathbf{V}_\mathbf{G} = \sum_{h=1}^{q} \frac{v_h}{n_h - 1} \left[n_h \mathbf{D}_{\mathbf{n}_h} - \mathbf{n}_h \mathbf{n}_h' \right], \tag{34}$$

where $\mathbf{n}_h' = (n_{h1}, n_{h2}, \ldots, n_{hs})$. Thus an *average partial association* chi-square test for H_0 is

$$Q_G = (\mathbf{G} - \boldsymbol{\mu}_\mathbf{G})' \mathbf{C}' [\mathbf{C} \mathbf{V}_\mathbf{G} \mathbf{C}']^{-1} \mathbf{C} (\mathbf{G} - \boldsymbol{\mu}_\mathbf{G}), \tag{35}$$

where $\mathbf{C} = [\mathbf{I}_{s-1}, -\mathbf{1}_{s-1}]$ is a contrast matrix basis. Under H_0, Q_G has approximately the chi-square distribution with d.f. $= (s-1)$. The statistic Q_G has two principal advantages relative to Q_T. Its sample-size requirements are less stringent (only the n_{+i} are required to be large rather than the n_{hi}). It is more powerful with respect to alternatives for which the pattern of subpopulation differences is similar in the respective strata.

Several special cases of Q_G are well known chi-square statistics:

1. If all the $n_{hi} = 1$ and the y_{hl} are within-stratum ranks, then Q_G is the Friedman [15] two-way rank analysis-of-variance statistic; its specific form is

$$Q_G = (s-1) \frac{\sum_{i=1}^{s} (G_i - q(s+1)/2)^2}{\sum_{h=1}^{q} \sum_{i=1}^{s} (y_{hi} - (s+1)/2)^2}.$$

2. If all the $n_{hi} = 1$ and the y_{hl} are all either 0 or 1, then Q_G is the Cochran [8] statistic for matched proportions; its specific form is

$$Q_G = (s-1) \frac{s \sum_{i=1}^{s} G_i^2 - (\sum_{i=1}^{s} G_i)^2}{s \sum_{i=1}^{s} G_i - \sum_{h=1}^{q} (\sum_{i=1}^{s} y_{hi})^2}.$$

Also, if $s = 2$, then

$$Q_G = (G_1 - G_2)^2 \Big/ \left\{ G_1 + G_2 - \sum_{h=1}^{q} 2 y_{h1} y_{h2} \right\}$$

is the McNemar statistic* for pair-matched samples.

3. If $s = 2$ and the y_{hl} are all either 0 or 1, then Q_G is the Mantel and Haenszel [31] statistic; its specific form is

$$Q_G = \frac{\left\{ \sum_{h=1}^{q} \left(n_{h11} - \dfrac{n_{h1} n_{h+1}}{n_h} \right) \right\}^2}{\sum_{h=1}^{q} \dfrac{n_{h1} n_{h2} n_{h+1} n_{h+2}}{n_h^2 (n_h - 1)}},$$

where n_{hi1} denotes the number of 1's in the ith subpopulation of the hth stratum, $n_{h+1} = (n_{h11} + n_{h21})$ and $n_{h+2} = (n_h - n_{h+1})$.

Other types of partial association statistics, including extensions to multivariate responses, can be formulated by applying strategies analogous to those indicated in earlier sections. The types of applications where partial association methods are of interest are the stratified population counterparts of those described in the section "Test of Randomness or No Association . . . ." Further discussion of these topics and/or illustrative examples are given in Birch [4], Koch et al. [25], Landis et al. [28], and Mantel [30]. Documentation for related computer programs is given in Landis et al. [29] for contingency table summary data and in Amara and Koch [1] for subject-wise raw data arrays.

LIKELIHOOD MODEL SAMPLING SITUATIONS

This section is concerned with chi-square tests for situations in which the general probability distribution structure for the data or the likelihood function* can be presumed known. Primary emphasis is given to discrete (or categorical) data for which several types of asymptotically equivalent test statistics are discussed. These include Wald statistics*, Neyman modified chi-square statistics*, Pearson chi-square statistics, and likelihood ratio statistics*.

Contingency Table Tests for Functions of Cell Probabilities for Simple Random Samples from a Discretely Distributed Population

Let $\boldsymbol{\pi}' = (\pi_1, \pi_2, \ldots, \pi_r)$ denote the vector of nonzero probabilities for the population distribution of a possibly multivariate, discrete (or categorical) response variable. Let $\mathbf{n}' = (n_1, n_2, \ldots, n_r)$ denote the vector of frequencies for the sample distribution of a set of n subjects for whom this response variable is observed. If this sample can be viewed as coming from a framework or process equivalent to simple random sampling with replacement* (or from an infinite superpopulation), then $\mathbf{n}$ has the multinomial distribution*

$$\phi(\mathbf{n} \mid \boldsymbol{\pi}) = n! \prod_{j=1}^{r} (\pi_j^{n_j} / n_j!), \qquad (36)$$

where $\sum_{j=1}^{r} \pi_j = 1$. For most applications hypotheses concerning $\boldsymbol{\pi}$ can be expressed generally in the *constraint form*

$$H_0 : \mathbf{F}(\boldsymbol{\pi}) = \mathbf{0}_u, \qquad (37)$$

where $\mathbf{F}(\cdot)$ is a prespecified set of u functions of $\boldsymbol{\pi}$ which are of interest. Also, the functions $\mathbf{F}(\cdot)$ are required to have continuous partial derivatives through order 1 in an open region containing $\boldsymbol{\pi}$ and such that the asymptotic covariance matrix

$$\mathbf{V_F}(\boldsymbol{\pi}) = \frac{1}{n} \left[\left. \frac{\mathbf{dF(y)}}{\mathbf{dy}} \right|_{\mathbf{y}=\boldsymbol{\pi}} \right] \cdot \left[\mathbf{D}_{\boldsymbol{\pi}} - \boldsymbol{\pi}\boldsymbol{\pi}' \right] \left[\left. \frac{\mathbf{dF(y)}}{\mathbf{dy}} \right|_{\mathbf{y}=\boldsymbol{\pi}} \right]' \qquad (38)$$

of the sample estimator $\mathbf{F} = \mathbf{F}(\mathbf{p})$ of $\mathbf{F}(\boldsymbol{\pi})$ is nonsingular where $\mathbf{p} = (\mathbf{n}/n)$. Here $\mathbf{F}$ represents the *unrestricted maximum likelihood estimator* for the functions $\mathbf{F}(\boldsymbol{\pi})$ with respect to the model (36) without regard to the hypothesis (37) since $\mathbf{p}$ is the corresponding unrestricted maximum likelihood estimator of $\boldsymbol{\pi}$.

If the sample size n is sufficiently large, the vector $\mathbf{F}$ approximately has the multivariate normal distribution $N(\mathbf{0}_u, \mathbf{V_F}(\boldsymbol{\pi}))$ under H_0. Also, $\mathbf{V_F} = \mathbf{V_F}(\mathbf{p})$, which is obtained by replacing $\boldsymbol{\pi}$ by $\mathbf{p}$ in (38), is a consistent estimator for $\mathbf{V_F}(\boldsymbol{\pi})$. Thus given a sufficiently large sample for $\mathbf{V_F}$ to be almost certainly nonsingular, a chi-square test statistic for H_0 is

$$Q_W = Q_W(\mathbf{F}) = \mathbf{F}'\mathbf{V}_{\mathbf{F}}^{-1}\mathbf{F}, \qquad (39)$$

which approximately has the chi-square distribution with d.f. $= u$, as a consequence of large-sample theory for functions of random variables*. Test statistics such as Q_W, are often called Wald [42] statistics, and their application to problems involving the analysis of discrete or categorical data is discussed in Grizzle et al. [17] and Koch et al. [24]. Details concerning their general theoretical properties are given in Stroud [41].

For situations where all $n_j > 0$, the Wald statistic Q_W is of additional interest because it has been shown by Bhapkar [3] to be *identical* to the Neyman [35] modified chi-square statistic

$$Q_N = Q_N(\tilde{\boldsymbol{\pi}}) = \sum_{j=1}^{r} (n_j - n\tilde{\pi}_j)^2/n_j$$
$$= \sum_{j=1}^{r} (n_j - \tilde{m}_j)^2/n_j = Q_w, \qquad (40)$$

where the $\tilde{m}_j = n\tilde{\pi}_j$ are estimates of the outcome expected values $n\pi_j$, which are determined to minimize Q_N under the condition that $\tilde{\boldsymbol{\pi}}$ satisfies the linearized version of the hypothesis (37); i.e., $\tilde{\boldsymbol{\pi}}$ satisfies the linear constraints

$$\mathbf{F}(\mathbf{p}) - \left[\left. \frac{\mathbf{dF(y)}}{\mathbf{dy}} \right|_{\mathbf{y}=\mathbf{p}} \right] (\mathbf{p} - \tilde{\boldsymbol{\pi}}) = \mathbf{0}_u, \qquad (41)$$

where $\mathbf{p}$ is regarded as a vector of constants. The estimates $\tilde{\pi}_j$ that are obtained by this process are usually called *linearized minimum modified (or Neyman) chi-square estimates*. Furthermore, they were shown by Neyman [35] to belong to the class of *best asymptotic normal* (or BAN) *estimates* for $\boldsymbol{\pi}$; i.e., they are *asymptotically unbiased* and *asymptotically efficient* (*see* ESTIMATION, POINT).

Another type of well-known BAN estimate for $\boldsymbol{\pi}$ is the *restricted maximum likelihood estimate* $\hat{\boldsymbol{\pi}}$ with respect to the hypothesis (37). These estimates are determined to maximize the likelihood (36) under the condition that $\hat{\boldsymbol{\pi}}$ satisfies the constraints (37). Similarly, a related type of well-known chi-square test for H_0 is the Wilks log-likelihood ratio criterion*

$$Q_L = Q_L(\hat{\boldsymbol{\pi}}) = -2\log_e \left[\phi(\mathbf{n}\,|\,\hat{\boldsymbol{\pi}})/\phi(\mathbf{n}\,|\,\mathbf{p}) \right]$$
$$= 2n \sum_{j=1}^{r} p_j \left[\log_e (p_j/\hat{\pi}_j) \right]$$
$$= 2n \sum_{j=1}^{r} p_j \left[\log_e p_j - \log_e p_j - \frac{\hat{\pi}_j - p_j}{p_j} + \frac{(\hat{\pi}_j - p_j)^2}{2p_j^2} \right] + o(1)$$
$$= \left[\sum_{j=1}^{r} (n_j - n\hat{\pi}_j)^2/n_j \right] + o(1)$$
$$= Q_N(\hat{\boldsymbol{\pi}}) + o(1), \qquad (42)$$

where o(1) denotes terms that are approximately 0 for large n under H_0. Thus Q_L is asymptotically equivalent to the Neyman chi-square statistic for which $\boldsymbol{\pi}$ is estimated by $\hat{\boldsymbol{\pi}}$ instead of $\tilde{\boldsymbol{\pi}}$. Since both $\hat{\boldsymbol{\pi}}$ and $\tilde{\boldsymbol{\pi}}$ are BAN estimates of $\boldsymbol{\pi}$, it follows that $Q_N(\tilde{\boldsymbol{\pi}})$ and $Q_N(\hat{\boldsymbol{\pi}})$ are asymptotically equivalent to each other, which implies that Q_W and Q_L are asymptotically equivalent. A third type of test statistic for H_0 in (37) is the Pearson chi-square statistic

$$Q_p = Q_p(\hat{\boldsymbol{\pi}}) = \sum_{j=1}^{r} (n_j - n\hat{\pi}_j)^2/n\hat{\pi}_j$$
$$= \sum_{j=1}^{r} (n_j - \hat{m}_j)^2/\hat{m}_j \qquad (43)$$

with respect to the maximum likelihood estimates $\hat{\boldsymbol{\pi}}$. Both this statistic and its counterpart $Q_P(\tilde{\boldsymbol{\pi}})$ are also asymptotically equivalent to Q_L. In general, all test statistics involving any BAN estimate of $\boldsymbol{\pi}$ in conjunction with the criteria Q_L, Q_p, and Q_W ($= Q_N$) are all asymptotically equivalent. For some applications, there exist other test statistics, such as those based on the *minimum discrimination information criterion** discussed by Gokhale and Kullback [16], which are also asymptotically equivalent to these chi-square test statistics. (See Bishop et al. [5], Neyman [35], and the entry for CONTINGENCY TABLES for further details.)

Test statistics of the type Q_L, Q_p, and Q_W may also be used to test hypotheses

$$H_0 : \mathbf{F}(\boldsymbol{\pi}) = \mathbf{X}\boldsymbol{\beta}, \qquad (44)$$

where $\mathbf{X}$ is a prespecified $(u \times t)$ matrix of known coefficients with full rank $t \leqslant u$ and $\boldsymbol{\beta}$ is a $(t \times 1)$ vector of unknown parameters. Here the basic issue is that the hypotheses (44) can be expressed in the constraint form (37) as

$$H_0 : \mathbf{G}(\boldsymbol{\pi}) = \mathbf{W}\mathbf{F}(\boldsymbol{\pi}) = \mathbf{W}\mathbf{X}\boldsymbol{\beta} = \mathbf{0}, \qquad (45)$$

where $\mathbf{W}$ is any $(w \times u)$ matrix with $w = (u - t)$ such that $\mathbf{W}$ and $\mathbf{X}'$ are orthogonal. Thus if the sample size n is sufficiently large, then the chi-square statistics of the type Q_L, Q_p, and Q_W with respect to the hypothesis (44) all have approximate chi-square distributions with d.f. $= w$.

As indicated in the section "Tests for Linear Hypotheses Involving Population Means . . . ," test statistics for the hypothesi. (45) may be interpreted as goodness-of-fit statistics for the linear model (44). Moreover, if such a model is considered adequate, linear hypotheses such as (12) become of interest for its parameters. Wald statistics Q_V for these types of hypotheses may be constructed by using (15). Alternatively, analogous Q_L and Q_p statistics can be obtained by the same type of subtraction operation for their goodness-of-fit counterparts for the reduced model* such as (14) implied by the hypothesis vs. the original model (44). Thus the chi-square statistics Q_L, Q_p, and Q_W can be partitioned into a set of components that pertain to a sequence of constraints $\mathbf{W}$ or models $\mathbf{X}$ for the functions $\mathbf{F}(\boldsymbol{\pi})$.

When linear model hypotheses such as (44) are under consideration, estimates for the parameter vector $\boldsymbol{\beta}$ are also of interest. *Linearized minimum modified chi-square BAN estimates* of $\boldsymbol{\beta}$ may be obtained by applying the weighted least-squares computations (8) and (9) to $\mathbf{y} = \mathbf{F}(\mathbf{p})$ with $\mathbf{V} = \mathbf{V}_\mathbf{F}$; and their asymptotic covariance matrix can be estimated consistently with $\mathbf{V}_\mathbf{b} = (\mathbf{X}'\mathbf{V}_\mathbf{F}^{-1}\mathbf{X})^{-1}$. *Maximum likelihood estimates* $\hat{\boldsymbol{\beta}}$ for $\boldsymbol{\beta}$ may be obtained by either maximizing (36) as a function of $\boldsymbol{\beta}$ (after expressing $\boldsymbol{\pi}$ as functions of $\boldsymbol{\beta}$) or by transforming maximum likelihood estimates $\hat{\boldsymbol{\pi}}$ which maximize (36) subject to the constraints (45) to $\hat{\boldsymbol{\beta}}$ via the weighted least-squares equation (9) with $\mathbf{y} = \mathbf{F}(\hat{\boldsymbol{\pi}})$ and $\mathbf{V} = \mathbf{V}_\mathbf{F}(\hat{\boldsymbol{\pi}})$; their asymptotic covariance matrix can be estimated consistently with $\mathbf{V}_{\hat{\beta}} = \{\mathbf{X}'[\mathbf{V}_\mathbf{F}(\hat{\boldsymbol{\pi}})]^{-1}\mathbf{X}\}^{-1}$.

An important special case of (37) is the log-linear model*

$$H_0 : \mathbf{W} \log \boldsymbol{\pi} = \mathbf{0}, \qquad (46)$$

where $\boldsymbol{\pi}$ corresponds to a multiway cross-classification of response variables and $\mathbf{W}$ is a basis for certain higher-order interactions among them. For this situation, H_0 can be also expressed as

$$H_0 : \boldsymbol{\pi} = \exp(\mathbf{X}\boldsymbol{\beta})/\mathbf{1}'_r[\exp(\mathbf{X}\boldsymbol{\beta})], \qquad (47)$$

where $\mathbf{X}$ is any basis of the vector space orthogonal to $[\mathbf{1}_r, \mathbf{W}]$. Extensive discussion of maximum likelihood methods for this situation is given in Bishop et al. [5] and documentation for a related computer program is given in Brown [7]; analogous minimum modified chi-square methods are discussed in Landis et al. [27] and Landis et al. [28].

Finally, for some applications (stochastic process models for biological phenomena, the occurrence of vital events), the hypotheses of interest are expressed in the *freedom equation form*

$$\boldsymbol{\pi} = \mathbf{g}(\boldsymbol{\theta}), \qquad (48)$$

where $\mathbf{g}(\cdot)$ is a prespecified set of r functions of t nonredundant unknown parameters that satisfy the underlying constraint on $\boldsymbol{\pi}$ that $\mathbf{1}'_r\boldsymbol{\pi} = \mathbf{1}'_r[\mathbf{g}(\boldsymbol{\theta})] = 1$. If the hypothesis (48) can be expressed explicitly in the form (37), then chi-square tests for it may be constructed by using the Q_L, Q_P, or Q_W for the corresponding constraints. However, if identification of the constraints (37) is not feasible, then chi-square tests for the hypothesis (48) can be undertaken by first obtaining a BAN estimate $\hat{\boldsymbol{\theta}}$ and then calculating Q_L, Q_P, or Q_N with respect to the corresponding estimates $\hat{\boldsymbol{\pi}} = \mathbf{g}(\hat{\boldsymbol{\theta}})$. In this regard, $\hat{\boldsymbol{\theta}}$ can be obtained by either maximizing the likelihood (36) as a function of $\boldsymbol{\theta}$ or minimizing the Neyman criterion (40) or the Pearson criterion (43). (See Plackett [37] for more details.)

Some applications of chi-square tests described here are as follows:

1. Multivariate analysis of the association among a set of response variables; see Bishop et al. [5], Gokhale and Kullback [16], and CATEGORICAL DATA and CONTINGENCY TABLES.
2. Multivariate analysis of the differences among first-order marginal distributions of response variables for repeated measurements experiments*; see Koch et al. [24], and CHI-SQUARE TESTS: NUMERICAL EXAMPLES.
3. Evaluation of the goodness of fit of certain probability distributions, such as the Poisson*, negative binomial*, or normal* to data for observed phenomena; see Dahiya and Gurland [12], Gurland et al. [18], Moore [32], and Plackett [37].
4. Evaluation of the goodness of fit of certain models for stochastic processes; see Cox [11] and Bishop et al. [5].
5. Analysis of measures of survival experience for life-table* data; see Johnson and Koch [20] and Koch et al. [25].

Contingency-Table Tests for Functions of Cell Probabilities for Stratified Simple Random Samples from a Set of Discrete Distributions

Let $i = 1, 2, \ldots, s$ index a set of s subpopulations. Let $\boldsymbol{\pi}_i' = (\pi_{i1}, \pi_{i2}, \ldots, \pi_{ir})$ denote the vector of probabilities for the distribution of some possibly multivariate discrete response variable within the ith subpopulation. Let $\mathbf{n}_i' = (n_{i1}, n_{i2}, \ldots, n_{ir})$ denote the vector of frequencies for the sample distribution of a set of n_i subjects from the ith subpopulation for whom this response variable is observed. If the combined sample for all subpopulations can be viewed as coming from a framework or process equivalent to stratified simple random sampling with replacement, then $\mathbf{n}' = (\mathbf{n}_1', \mathbf{n}_2', \ldots, \mathbf{n}_s')$ has the multinomial distribution*

$$\phi(\mathbf{n} \mid \boldsymbol{\pi}) = \prod_{i=1}^{s} n_i! \prod_{j=1}^{r} (\pi_{ij}^{n_{ij}} / n_{ij}!), \qquad (49)$$

where $\boldsymbol{\pi}' = (\boldsymbol{\pi}_1', \boldsymbol{\pi}_2', \ldots, \boldsymbol{\pi}_s')$ and $\boldsymbol{\pi}_i' \mathbf{1}_r = 1$ for $i = 1, 2, \ldots, s$.

For these situations, hypotheses concerning $\boldsymbol{\pi}$ can be expressed either in terms of constraints such as (37) or freedom equations such as (48). Similarly, corresponding chi-square tests can be undertaken in terms of the stratified sample counterparts of Q_L, Q_P, and Q_W (or Q_N). Here one particular hypothesis of interest is the hypothesis

$$H_0 : \pi_{ij} - \pi_{sj} = 0 \qquad \begin{pmatrix} i = 1, 2, \ldots, s-1 \\ j = 1, 2, \ldots, r-1 \end{pmatrix} \qquad (50)$$

of homogeneity* (or no differences among the subpopulations) for the distribution of the response variable. Since the maximum likelihood estimates for the π_{ij} under H_0 are the combined sample proportions $p_{+j} = n_{+j}/n$, where $n_{+}j = \sum_{i=1}^{s} n_{ij}$ and $n = \sum_{i=1}^{s} n_i$, the corresponding Pearson chi-square statistic analogous to (43) for H_0 has the classical form

$$Q_P = \sum_{i=1}^{s} \sum_{j=1}^{r} \frac{(n_{ij} - m_{ij})^2}{m_{ij}}, \qquad (51)$$

where $m_{ij} = n_i n_{+j}/n$ is the estimated expected value for n_{ij}. If the sample sizes $n_1, n_2, \ldots, n_s$ are sufficiently large (e.g., all $m_{ij} \geqslant 5$), then Q_{P} approximately has the chi-square distribution with d.f. $= (r-1)(s-1)$ under H_0. See CONTINGENCY TABLES for further discussion.

Chi-Square Tests for Other Types of Likelihood Sampling Situations

For many general situations, maximum likelihood methods can be used to determine estimates $\hat{\boldsymbol{\theta}}$ for some $(t \times 1)$ parameter vector $\boldsymbol{\theta}$ which pertains to the behavior of a set of observed random variables. If the sample size is sufficiently large that $\hat{\boldsymbol{\theta}}$ has an approximate normal distribution (with the necessary assumptions being presumed), then chi-square test statistics for linear hypotheses involving $\boldsymbol{\theta}$ can be constructed by using log-likelihood ratio statistics. Alternatively, if a consistent estimate of variance $\mathbf{V}_{\hat{\theta}}$ is also available, then Wald statistics analogous to (39) can be used. Both types of chi-square statistics can be partitioned into components in a sense analogous to (15). An important class of applications for this framework are generalized linear models* as discussed by Nelder and Wedderburn [34]. For this class, which includes the normal, multinomial, Poisson, and gamma probability distributions, general methods are provided for maximum likelihood estimation of the parameters of certain types of linear regression models and for log-likelihood ratio chi-square tests for linear hypotheses concerning them. Related computer program documentation is given in Baker and Nelder [2]. Finally, the proportional hazards linear model* for survival data, as reviewed in Breslow [6], is another type of likelihood model sampling situation for which chi-square tests can be constructed.

CHI-SQUARE TESTS FOR COMPLEX PROBABILITY RANDOM SAMPLING SITUATIONS

In the health and social sciences, data are often obtained by means of complex probability samples involving possibly multistage selection of clusters of study units with not necessarily equal probabilities. Also, estimates for various population characteristics are often formulated in terms of compound ratio statistics (e.g., ratios of linear functions of other ratios) or related multiplicative adjustment procedures. Since the sample size for these investigations is usually sufficiently large that such ratio statistics approximately have a multivariate normal distribution, chi-square tests for linear hypotheses concerning them can be undertaken via Wald statistics* with respect to any consistent estimate of their asymptotic covariance matrix. In addition, the weighted least-squares method described in the section "Tests for Linear Hypotheses Involving Populations Means . . . " can be used to obtain estimates of parameters for the corresponding linear models. Further discussion of these methods for chi-square tests for complex probability samples is given in Koch et al. [23]; and illustrative examples are given in Koch et al. [25] and CHI-SQUARE TESTS: NUMERICAL EXAMPLES. A review of some of the properties of the log-likelihood ratio criterion statistic (42) and the Pearson criterion statistic (43) for such situations is given in Fienberg [13].

APPENDIX: A LEMMA CONCERNING THE EQUALITY OF TWO MATRICES

The following lemma is a useful tool for showing the equality of the defining matrices for alternative expressions for quadratic forms.

Lemma. *Let* $\mathbf{A}_1$ *and* $\mathbf{A}_2$ *be two* $(u \times t)$ *matrices satisfying the conditions shown below, which are sufficient to imply that* $\mathbf{A}_1 = \mathbf{A}_2$.

(a) Rank $\mathbf{A}_1$ = Rank $\mathbf{A}_2 = u_0 \leqslant t \leqslant u$.

(b) *There exists a* $(t \times (t - u_0))$ *matrix* $\mathbf{K}_R$ *of full rank* $(t - u_0)$ *such that* $\mathbf{A}_1\mathbf{K}_R = \mathbf{A}_2\mathbf{K}_R = \mathbf{0}$, *where* $\mathbf{0}$ *is a zero matrix.*

(c) *There exists a* $((u - u_0) \times u)$ *matrix* $\mathbf{K}_L$ *of full rank* $(u - u_0)$ *such that* $\mathbf{K}_L\mathbf{A}_1 = \mathbf{K}_L\mathbf{A}_2 = \mathbf{0}$.

(d) *There exists a* $(t \times u)$ *matrix* $\mathbf{H}$ *such that* $\mathbf{A}_1\mathbf{H}\mathbf{A}_1 = \mathbf{A}_1$ *and* $\mathbf{A}_2\mathbf{H}\mathbf{A}_2 = \mathbf{A}_2$.

A proof of this lemma is given in Koch [22].

The situations in which application of this lemma arises are those where one type of expression for a quadratic form is convenient for theoretical purposes while another is convenient for computational purposes. Some examples of its use in this entry are as follows:

1. Equality of (2) and (3) with

$$\mathbf{A}_1 = (n\mathbf{I}_n - \mathbf{1}_n\mathbf{1}'_n)/n,$$

$$\mathbf{A}_2 = \mathbf{C}'[\mathbf{C}\mathbf{C}']^{-1}\mathbf{C},$$

$$\mathbf{K}_R = \mathbf{1}_n = \mathbf{K}'_L,$$

and $\mathbf{H} = \mathbf{I}_n$.

2. Equality of (5) and (10) when $\eta = \mathbf{0}$ with

$$\mathbf{A}_1 = \mathbf{W}'(\mathbf{WVW}')^{-1}\mathbf{W},$$

$$\mathbf{A}_2 = \left[\mathbf{V}^{-1} - \mathbf{V}^{-1}\mathbf{X}(\mathbf{X}'\mathbf{V}^{-1}\mathbf{X})^{-1}\mathbf{X}'\mathbf{V}^{-1}\right],$$

$$\mathbf{K}_R = \mathbf{X} = \mathbf{K}'_L,$$

and $\mathbf{H} = \mathbf{V}$.

3. Equality of (13) and (15) with

$$\mathbf{A}_1 = \mathbf{V}^{-1}\mathbf{X}(\mathbf{X}'\mathbf{V}^{-1}\mathbf{X})^{-1} \cdot \mathbf{C}'\left[\mathbf{C}(\mathbf{X}'\mathbf{V}^{-1}\mathbf{X})^{-1}\mathbf{C}'\right]^{-1} \cdot \mathbf{C}(\mathbf{X}'\mathbf{V}^{-1}\mathbf{X})^{-1}\mathbf{X}'\mathbf{V}^{-1},$$

$$\mathbf{A}_2 = \mathbf{W}'_C(\mathbf{W}_C\mathbf{V}\mathbf{W}'_C)^{-1}\mathbf{W}_C - \mathbf{W}'(\mathbf{WVW}')^{-1}\mathbf{W},$$

$$\mathbf{K}_R = \mathbf{XZ} = \mathbf{K}'_L,$$

and $\mathbf{H} = \mathbf{V}$.

4. Equality of (21) and (22) with $\mathbf{A}_1 = \mathbf{C}'[\mathbf{C}\mathbf{D}_{\mathbf{n}}^{-1}\mathbf{C}']^{-1}\mathbf{C}$, $\mathbf{A}_2 = [n\mathbf{D}_n - \mathbf{nn}']/n$, $\mathbf{K}_R = \mathbf{I}_s = \mathbf{K}'_L$, and $\mathbf{H} = \mathbf{D}_n^{-1}$.

References

[1] Amara, I. A. and Koch, G. G. (1980). *Proc. 5th Ann. SAS Users Group Int. Conf.*, pp. 134–144. (Provides documentation for an SAS* macro for computing several types of partial association test statistics with respect to subject-wise raw data arrays.)

[2] Baker, R. J. and Nelder, J. A. (1978). *The GLIM System Manual (Release 3).* The Numerical Algorithms Group/Royal Statistical Society, Oxford. (Provides documentation for a general computer program for maximum likelihood estimation of parameters of linear models pertaining to exponential family distributions and log-likelihood ratio chi-square tests for hypotheses concerning them.)

[3] Bhapkar, V. P. (1966). *J. Amer. Statist. Ass.*, **61**, 228–235. (Proves formally the algebraic identity of the Neyman modified chi-square statistic (40) and the Wald statistic (39) for testing the hypothesis (37) for linear functions **F** when all $n_j > 0$; the result is then extended to the general case for the Neyman statistic using linearized estimates.)

[4] Birch, M. W. (1965). *J. R. Statist. Soc. B*, **27**, 111–124. (A comprehensive theoretical discussion of the statistical properties of alternative test statistics for the hypothesis (31) of no partial association.)

[5] Bishop, Y. M. M., Fienberg, S. E., and Holland, P. W. (1975). *Discrete Multivariate Analysis: Theory and Practice.* MIT Press, Cambridge, Mass. (A systematic exposition and development of the log-linear model for Poisson and multinomial data, primarily using maximum likelihood estimation, likelihood ratio, and Pearson chi-square test criteria.)

[6] Breslow, N. E. (1975). *Int. Statist. Rev.*, **43**, 45–58. (Reviews statistical methodology for the analysis of survival data under the proportional hazards model with emphasis on maximum likelihood procedures.)

[7] Brown, M. (1977). In *BMDP Biomedical Computer Programs (P Series)*, W. J. Dixon and M. B. Brown, eds. University of California Press, Los Angeles, Chap. 11.3. (Provides documentation for a general computer program for maximum likelihood estimation of cell expected values with respect to a broad class of log-linear models for multiway contingency tables and corresponding chi-square goodness-of-fit statistics.)

[8] Cochran, W. G. (1950). *Biometrika*, **37**, 256–266. (A randomization argument is used to develop a conditional test criterion for the hypothesis (31) of no partial association for s stratification-matched, binary responses; it is shown to have a limiting $\chi^2(s-1)$ distribution. This test is a special case for the average partial association statistic (35), which is oriented toward the comparison of correlated proportions. If $s = 2$, this statistic is the same as McNemar's test for pair-matched samples.)

[9] Cochran, W. G. (1952). *Ann. Math. Statist.*, **23**, 315–345. (This expository discussion of the chi-square test of goodness of fit describes the historical development of the distribution theory of the test criterion and deals with some matters in the practical application of the test; e.g., minimum number per cell, subdivision of chi-square into appropriate components, etc.)

[10] Cochran, W. G. (1977). *Sampling Techniques.* Wiley, New York.

[11] Cox, D. R. (1970) *The Analysis of Binary Data.* Methuen, London. (A concise treatment of data from dichotomous response variables using exact test procedures and also asymptotic criteria based on logistic transforms.)

[12] Dahiya, R. C. and Gurland, J. (1973). *J. Amer. Statist. Ass.*, **68**, 707–712. (Discusses power properties and the number of class intervals to be used in applying a modified Pearson chi-square method to test for normality.)

[13] Fienberg, S. E. (1979). *J. R. Statist. Soc. B*, **41**, 54–64. (Reviews the properties of goodness-of-fit statistics such as (42) and (43) pertaining to categorical data situations where they do not have asymptotic chi-square distributions; e.g., small samples, large contingency tables with small cell counts, complex probability sample surveys, etc.)

[14] Fisher, R. A. (1924). *J. R. Statist. Soc.*, **87**, 442–450. (An adjustment is proposed for the degrees of freedom of the χ^2 criterion to test a partially specified hypothesis.)

[15] Friedman, M. (1937). *J. Amer. Statist. Ass.*, **32**, 675–701. (A rank analog of the two-way analysis-of-variance statistic is developed for the comparison of s treatments in a randomized blocks experiment using intra-blocks (i.e., strata) ranks; it is shown to have a limiting χ^2 $(s-1)$ distribution.)

[16] Gokhale, D. V. and Kullback, S. (1978). *The Information in Contingency Tables*. Marcel Dekker, New York. (Elaborates the minimum discrimination information procedure for estimation and testing of linear and log-linear models in contingency tables; these estimation and test procedures are similar to maximum likelihood estimation and likelihood ratio tests, respectively.)

[17] Grizzle, J. E., Starmer, C. F., and Koch, G. G. (1969). *Biometrics*, **25**, 489–504. (Provides a computationally convenient matrix framework for the application of Wald test statistics and weighted least-squares methods of estimation to linear and log-linear models in contingency tables; these procedures are illustrated for several examples.)

[18] Gurland, J., Lee, I. and Dahm, P. A. (1960). *Biometrics*, **16**, 382–398. (Describes methods analogous to minimum chi-square for fitting normal or logistic tolerance distributions to ordinal data.)

[19] Hájek, J. and Šidák, Z. (1967). *Theory of Rank Tests*. Academic Press, New York.

[20] Johnson, W. D. and Koch, G. G. (1978). *Int. Statist. Rev.*, **46**, 21–51. (Illustrates the application of Wald test statistics and weighted least-squares methods of estimation to the analysis of grouped survival data.)

[21] Killion, R. A. and Zahn, D. A. (1976). *Int. Statist. Rev.*, **44**, 71–112.

[22] Koch, G. G. (1969). *J. Amer. Statist. Ass.*, **64**, 969–970.

[23] Koch, G. G., Freeman, D. H., Jr., and Freeman, J. L. (1975). *Int. Statist. Rev.*, **43**, 59–78. (Describes the rationale for the application of Wald test statistics and weighted least-squares methods of estimation to data from complex probability sample surveys.)

[24] Koch, G. G., Landis, J. R., Freeman, J. L., Freeman, D. H., Jr., and Lehnen, R. (1977). *Biometrics*, **33**, 133–158. (Illustrates the application of Wald test statistics and weighted least-squares methods of estimation to repeated measurement experiments for which multivariate data are obtained from the same subject for the same conceptual response variable under two or more observational conditions.)

[25] Koch, G. G., Gillings, D. B. and Stokes, M. E. (1980). *Ann. Rev. Public Health*, **1**, 163–225. (An expository review paper dealing with several types of applications of chi-square tests.)

[26] Kruskal, W. H. and Wallis, W. A. (1952, 1953). *J. Amer. Statist. Ass.*, **47**, 583–621; **48**, 907–911. (A rank analog of the one-way analysis of variance statistic for the comparison of s groups in a completely randomized experiment is developed using the combined sample rankings; it is shown to have a limiting χ^2 $(s-1)$ distribution.)

[27] Landis, J. R., Stanish, W. M., Freeman, J. L. and Koch, G. G. (1976). *Computer Programs Biomed.*, **6**, 196–231. (Provides documentation for a general computer program for Wald test statistics with respect to functions of categorical data proportions and weighted least-squares estimation of parameters for corresponding linear models.)

[28] Landis, J. R., Heyman, E. R. and Koch, G. G. (1978). *Int. Statist. Rev.*, **46**, 237–254. (Reviews univariate and multivariate randomization test statistics for the hypothesis of no partial association (31) for sets of $(s \times r)$ contingency tables.)

[29] Landis, J. R., Cooper, M. M., Kennedy, T., and Koch, G. G. (1979). *Computer Programs Biomed.*, **9**, 223–246. (Provides documentation for a general computer program for several types of partial association test statistics for contingency tables, particularly those with ordinally scaled categories.)

[30] Mantel, N. (1963). *J. Amer. Statist. Ass.*, **58**, 690–700. (Describes a randomization method for constructing chi-square tests with one degree of freedom for the hypothesis of no partial association (31) for sets of $(s \times r)$ contingency tables with ordinally scaled categories.)

[31] Mantel, N. and Haenszel, W. (1959). *J. Nat. Cancer Inst.*, **22**, 719–748. (Describes a randomization chi-square statistic with one degree of freedom for a set of q 2×2 contingency tables via its application to retrospective epidemiological studies.)

[32] Moore, D. S. (1977). *J. Amer. Statist. Ass.*, **72**, 131–137. (Discusses certain generalizations of Wald statistics and their application to goodness-of-fit tests for certain probability distributions.)

[33] Natrella, M. G. (1963). *Experimental Statistics. Nat. Bur. Stand. Handb. 91*, U. S. Government Printing Office, Washington, D.C.

[34] Nelder, J. A. and Wedderburn, R. W. (1972). *J. R. Statist. Soc. A*, **135**, 370–384. (For the usual distributions in the exponential family with parameter $\boldsymbol{\theta}$, the paper discusses linear models $\boldsymbol{\mu} = \mathbf{X}\boldsymbol{\beta}$ with $\boldsymbol{\mu}$ a suitable parametric function $\boldsymbol{\mu} = \boldsymbol{\mu}(\boldsymbol{\theta})$ linking the parameter $\boldsymbol{\theta}$ of the distribution with $\boldsymbol{\mu}$ of the linear model. Maximum likelihood estimates are then obtained by iterative weighted least squares.)

[35] Neyman, J. (1949). *Proc. Berkeley Symp. Math. Statist. Prob.* University of California Press, Berkeley, Calif., pp. 239–273. (Develops the BAN property, i.e., the asymptotic normality and efficiency of the maximum likelihood estimators and

also of minimum chi-square and modified chi-square estimators; also establishes asymptotic equivalence in some sense of the likelihood-ratio, minimum chi-square, and modified chi-square test criteria for hypotheses in multinomial distributions.)

[36] Pearson, K. (1900). *Philos. Mag.*, Ser. 5, **50**, 157–175. (The statistic $\sum\{(n-m)^2/m\}$ is proposed for testing a completely specified hypothesis for the r-category multinomial distribution; and the asymptotic $\chi^2(r-1)$ distribution is derived under the hypothesis. The extension of this criterion to test partially specified hypotheses contains an error in the degrees of freedom of the aymptotic χ^2 distribution as discussed in this entry.)

[37] Plackett, R. L. (1974). *The Analysis of Categorical Data.* Charles Griffin, London. (A concise treatment of a broad range of theoretical and methodological topics for categorical data, involving both exact and asymptotic methods.)

[38] Puri, M. L. and Sen, P. K. (1971). *Non-parametric Methods in Multivariate Analysis.* Wiley, New York.

[39] Quade, D. (1967). *J. Amer. Statist. Ass.*, **62**, 1187–1200. (A rank analog of one-way analysis of covariance for the comparison of s groups in a completely randomized experiment is developed using least-squares residuals of combined sample response variable rankings on rankings for one or more concomitant variables.)

[40] Singer, B. (1979). *Brit. J. Math. Statist. Psychol.*, **32**, 1–60.

[41] Stroud, T. W. F. (1971). *Ann. Math. Statist.*, **42**, 1412–1424. (Discusses theoretical aspects of Wald statistics.)

[42] Wald, A. (1943). *Trans. Amer. Math. Soc.*, **54**, 426–482. (Develops the asymptotic distribution theory of the statistic of type (39) for testing hypotheses of type (37) concerning parameters $\boldsymbol{\theta}$ in distributions satisfying certain regularity assumptions, and establishes that the statistic (39) is asymptotically as efficient as the likelihood ratio statistic.)

Acknowledgements

This research was supported in part by the U.S. Bureau of the Census (JSA-79-16). The authors would like to thank John Darroch and P. K. Sen for helpful comments with respect to the preparation of this paper. They would also like to express their appreciation to Joyce Hill and Jo Ann DeGraffenreidt for their conscientious typing of the manuscript.

(CATEGORICAL DATA
CHI-SQUARE DISTRIBUTION
CHI-SQUARE TESTS: NUMERICAL EXAMPLES
CONTINGENCY TABLES
GENERALIZED LINEAR MODELS)

GARY G. KOCH
VASANT P. BHAPKAR

CHI-SQUARE TESTS: NUMERICAL EXAMPLES

In this entry, the range of application of chi-square tests is illustrated in terms of three examples. The first of these is based on large-sample estimates from a national health survey. The variation among these is investigated by linear models methods for multivariate normal* data as discussed in the second section of the CHI-SQUARE TESTS entry. The second example is from a randomized clinical trial concerned with an aspect of heart disease. Its analysis is undertaken with chi-square approximations to randomization nonparametric (distribution-free) rank tests* as discussed in the third section of CHI-SQUARE TESTS. The third example is based on a three-dimensional contingency table* from a clinical trial* involving dairy cows for which three types of analysis are discussed. One of these is randomization model methods for contingency tables as discussed in the third section of CHI-SQUARE TESTS. The other two are concerned with chi-square tests pertaining to functions of contingency table cell probabilities and corresponding statistical models as discussed in the fourth section of CHI-SQUARE TESTS. Specifically, the use of Wald statistics* and weighted least-squares* methods to test linear hypotheses* is illustrated, followed by the log-likelihood ratio* chi-square statistics and Pearson chi-square statistics to test log-linear hypotheses*.

EXAMPLE 1: TESTS FOR LINEAR HYPOTHESES FOR CROSS-CLASSIFIED DOMAIN ESTIMATES FROM A NATIONAL SAMPLE SURVEY

This example is based upon the Health and Nutrition Examination Survey (HANES) which was undertaken in the United States during 1971–1974. Specific attention is directed at estimates pertaining to a general well-being scale for a two-way (age × sex) cross-classification of domains (i.e., subpopulations that do not correspond to strata in the survey design). The underlying data for these estimates are the responses to 18 psychological questions on the general well-being questionnaire of HANES. This survey instrument was administered to each 24 to 74-year-old subject in the survey on the day of examination. From this information, the general well-being scale was constructed as a summary composite index. These quantities were then combined across the subjects in this national probability sample to produce estimates for the civilian, noninstitutional United States target population. Also, the method used to obtain them involved post-stratification to adjust for oversampling components of the HANES design with respect to preschool children, women of child-bearing age, elderly people, and low-income people. (See ref. 15 for further details.)

The vector **y** of general well-being estimates for the respective age × sex domains and the estimated covariance matrix **V** are shown in Table 1. This matrix was obtained by the method of balanced repeated replications* as described in McCarthy [12] and Kish and Frankel [5]. Since the sample size for HANES was relatively large (i.e., over 10,000), the vector **y** can be viewed as approximately having a multivariate normal distribution* with essentially known covariance matrix **V**. Thus as discussed in CHI-SQUARE TESTS, linear hypotheses involving **y** can be tested via Wald (or quadratic form) statistics having approximate chi-square distributions under them.

One class of questions of preliminary interest for the domain estimates is concerned with the identification of statistically important sources of variation among them; i.e., sources of variation for which equivalence to sampling variability is contradicted by significant* chi-square test statistics. For this purpose, attention is focused on linear hypotheses that can be formulated either directly in terms of constraints for the domain estimates or indirectly in terms of the goodness of fit* of corresponding linear models implying constraints. An example of a direct specification is hypothesis (1):

$\boldsymbol{H_{01}}$: There is no variation between the sex subdomains of each age domain in the sense that the pairwise differences between males and females for the five age domains are all equivalent to 0 except for sampling variability. (1)

Table 1 General Well-Being Scale Estimates and Estimated Covariance Matrix[a]

Domains		General Well-Being	Balanced Repeated Replication Estimated Covariance Matrix x 10^4 for General Well-Being Scale Estimates									
Sex	Age	Estimates										
Male	25–34	7.937	73.9	1.9	14.6	− 8.2	7.6	18.9	11.8	14.0	− 14.0	3.9
Male	35–44	7.925		117.2	18.3	2.9	8.3	− 12.3	− 62.9	− 8.8	− 23.2	3.4
Male	45–54	7.828			105.0	− 17.3	1.1	43.4	− 5.9	− 5.5	2.3	− 1.3
Male	55–64	7.737				133.5	14.0	15.8	21.2	21.1	6.6	24.0
Male	65–74	8.168					143.0	− 5.0	− 9.8	23.9	− 1.0	21.3
Female	25–34	7.250						111.0	10.1	17.7	− 1.8	− 8.2
Female	35–44	7.190				Symmetric			234.2	14.4	36.9	25.3
Female	45–54	7.360								106.0	15.7	22.6
Female	55–64	7.319									229.8	91.8
Female	65–74	7.552										192.1

[a] From the 1971–1974 Health and Nutrition Examination Survey of the U.S. population, cross-classified according to age and sex of examined person.

For this hypothesis, the constraint specification matrix is given in (2):

$$\mathbf{W}_1 = \begin{array}{c} \begin{array}{ccccc} \text{Male} & \text{Male} & \text{Male} & \text{Male} & \text{Male} \\ 25\text{–}34 & 35\text{–}44 & 45\text{–}54 & 55\text{–}64 & 65\text{–}74 \end{array} \\ \left[\begin{array}{ccccc} 1 & 0 & 0 & 0 & 0 \\ 0 & 1 & 0 & 0 & 0 \\ 0 & 0 & 1 & 0 & 0 \\ 0 & 0 & 0 & 1 & 0 \\ 0 & 0 & 0 & 0 & 1 \end{array}\right. \\ \begin{array}{ccccc} \text{Female} & \text{Female} & \text{Female} & \text{Female} & \text{Female} \\ 25\text{–}34 & 35\text{–}44 & 45\text{–}54 & 55\text{–}64 & 65\text{–}74 \end{array} \\ \left.\begin{array}{ccccc} -1 & 0 & 0 & 0 & 0 \\ 0 & -1 & 0 & 0 & 0 \\ 0 & 0 & -1 & 0 & 0 \\ 0 & 0 & 0 & -1 & 0 \\ 0 & 0 & 0 & 0 & -1 \end{array}\right], \end{array} \tag{2}$$

where the respective rows correspond to differences between males and females for the five age domains. Thus the Wald chi-square statistic with d.f. = 5 [see (5) of CHI-SQUARE TESTS] has the form

$$\begin{aligned} Q_1 &= \mathbf{y}'\mathbf{W}_1'[\mathbf{W}_1\mathbf{V}\mathbf{W}_1']^{-1}\mathbf{W}_1\mathbf{y} \\ &= \mathbf{g}'\mathbf{V}_g^{-1}\mathbf{g} = 68.76, \end{aligned} \tag{3}$$

where $\mathbf{g} = \mathbf{W}_1\mathbf{y} = (0.687, 0.735, 0.469, 0.418, 0.616)'$ is the vector of estimated differences between males and females and $\mathbf{V}_g = (\mathbf{W}_1\mathbf{V}\mathbf{W}_1')$ is its corresponding estimated covariance matrix. Since Q_1 in (3) is significant with $p < 0.01$, the hypothesis H_{01} is contradicted. Furthermore, the knowledge of the existence of such sex variation provides the rationale for its further assessment. For example, hypothesis (4) of no age × sex interaction:

H_{02}: There is no variation among the age domains for the differences between males and females in the sense that the corresponding second-order comparisons for any one of the age domains (e.g., 25–34) vs. each of the others are all equivalent to 0 except for sampling variability. (4)

may be of interest. For this hypothesis, the constraint specification matrix is $\mathbf{W}_2$ in (5):

$$\mathbf{W}_2 = \begin{array}{c} \begin{array}{ccccc} \text{Male} & \text{Male} & \text{Male} & \text{Male} & \text{Male} \\ 25\text{–}34 & 35\text{–}44 & 45\text{–}54 & 55\text{–}64 & 65\text{–}74 \end{array} \\ \left[\begin{array}{ccccc} 1 & -1 & 0 & 0 & 0 \\ 1 & 0 & -1 & 0 & 0 \\ 1 & 0 & 0 & -1 & 0 \\ 1 & 0 & 0 & 0 & -1 \end{array}\right. \\ \begin{array}{ccccc} \text{Female} & \text{Female} & \text{Female} & \text{Female} & \text{Female} \\ 25\text{–}34 & 35\text{–}44 & 45\text{–}54 & 55\text{–}64 & 65\text{–}74 \end{array} \\ \left.\begin{array}{ccccc} -1 & 1 & 0 & 0 & 0 \\ -1 & 0 & 1 & 0 & 0 \\ -1 & 0 & 0 & 1 & 0 \\ -1 & 0 & 0 & 0 & 1 \end{array}\right] \end{array} \tag{5}$$

Since the Wald chi-square statistic $Q_2 = 2.92$ with d.f. = 4 for $\mathbf{W}_2$ is nonsignificant with $p > 0.25$, the hypothesis H_{02} is judged to be compatible with the general well-being estimates. This condition implies that the variation among these estimates $\mathbf{y}$ can be represented in terms of a linear regression model $\mathbf{X}$ that is an orthocomplement to $\mathbf{W}_2$ (i.e., $\mathbf{W}_2\mathbf{X} = 0$). One such model is given in (6) together with descriptors for its parameters β:

$$\mathbf{X}\beta = \begin{bmatrix} 1 & 0 & 0 & 0 & 0 & 0 \\ 1 & 0 & 1 & 0 & 0 & 0 \\ 1 & 0 & 0 & 1 & 0 & 0 \\ 1 & 0 & 0 & 0 & 1 & 0 \\ 1 & 0 & 0 & 0 & 0 & 1 \\ 1 & 1 & 0 & 0 & 0 & 0 \\ 1 & 1 & 1 & 0 & 0 & 0 \\ 1 & 1 & 0 & 1 & 0 & 0 \\ 1 & 1 & 0 & 0 & 1 & 0 \\ 1 & 1 & 0 & 0 & 0 & 1 \end{bmatrix} \cdot \begin{bmatrix} \text{reference value for males, 25–34} \\ \text{increment for females} \\ \text{increment for age 35–44} \\ \text{increment for age 45–54} \\ \text{increment for age 55–64} \\ \text{increment for age 65–74} \end{bmatrix}. \tag{6}$$

This model can be interpreted as equivalent to the alternative specification (7) of H_{02} in (4).

H_{02}: There is no interaction between age and sex in the sense that the variation among the age × sex domains can be represented by a linear model with additive age and sex effects. (7)

For the specification (7) of H_{02}, the test statistic Q_2 is obtained as the minimized weighted residual goodness-of-fit statistic for the model (6) with d.f. = (10 − 6) = 4 [see (8) of CHI-SQUARE TESTS]. It has the form

$$Q_2 = (\mathbf{y} - \mathbf{Xb})'\mathbf{V}^{-1}(\mathbf{y} - \mathbf{Xb}) = 2.92, \quad (8)$$

where

$$b = (\mathbf{X'V}^{-1}\mathbf{X})^{-1}\mathbf{X'V}^{-1}\mathbf{y} = (7.914, -0.580, -0.039, -0.026, -0.102, 0.206)'$$

is the vector of weighted least-squares estimates for the model parameters. Thus the result in (8) is the same as its constraint formulation counterpart with respect to $\mathbf{W}_2$ in (5); see the identity (11) of CHI-SQUARE TESTS.

Since the variation among the general well-being estimates can be characterized by the model (6), its further analysis can be based on the estimated parameters **b**. For this framework, one hypothesis of interest is:

H_{03}: There is no variation among the age $\leqslant$ 64 subdomains of the sex domains with respect to the model **X** in the sense that the increment parameters for the 35–44, 45–54, and 55–64 age ranges are equivalent to 0 except for sampling variability. (9)

The constraint matrix for this hypothesis with respect to **b** is **C**, where **C** =

Male 25–34	Sex: Females	Age 35–44	Age 45–54	Age 55–64	Age 65–74
0	0	1	0	0	0
0	0	0	1	0	0
0	0	0	0	1	0

; (10)

and the corresponding chi-square test statistic with d.f. = 3 [see (13) of CHI-SQUARE TESTS] is

$$Q_3 = \mathbf{b'C'}\left[\mathbf{C}(\mathbf{X'V}^{-1}\mathbf{X})^{-1}\mathbf{C'}\right]^{-1}\mathbf{Cb} = 0.72. \quad (11)$$

Since Q_3 is nonsignificant with $p > 0.25$, the hypothesis H_{03} is concluded to be compatible with the general well-being estimates. Thus their variation can be represented by the simplified linear regression model $\mathbf{X}_R$ in (12) obtained by deleting columns 3 to 5 of **X** in (6) together with the corresponding parameters:

$$\mathbf{X}_R\boldsymbol{\beta}_R = \begin{bmatrix} 1 & 1 & 1 & 1 & 1 & 1 & 1 & 1 & 1 & 1 \\ 0 & 0 & 0 & 0 & 0 & 1 & 1 & 1 & 1 & 1 \\ 0 & 0 & 0 & 0 & 1 & 0 & 0 & 0 & 0 & 1 \end{bmatrix}' \cdot \begin{bmatrix} \text{reference values for males, 25–34} \\ \text{increment for females} \\ \text{increment for age 65–74} \end{bmatrix}. \quad (12)$$

For the model specification (12), the weighted residual goodness-of-fit statistic $Q_4 = 3.64$ with d.f. = 10 − 3 = 7 is nonsignificant. Also, it can be verified that $Q_4 = 3.64 = 2.92 + 0.72$ is the sum of the test statistics (8) and (11). Since the estimates are compatible with the model $\mathbf{X}_R$, its corresponding parameter vector $\mathbf{b}_R = (7.868, -0.560, 0.261)'$ can be used to interpret the variation among them. These quantities indicate that the general well-being estimates were 0.56 higher for males than females for all age domains, and 0.26 higher for 65 to 74-year-old persons for both sexes than the other age ranges, which were generally similar.

(A computer program for obtaining the results based on (1) to (12) is documented in Landis et al. [9]; see Koch et al. [8] for another application to health survey data.)

EXAMPLE 2: TESTS OF RANDOMNESS FOR A MULTIPERIOD CHANGEOVER STUDY

This example is concerned with the application of nonparametric (distribution-free) randomization chi-square tests to a multiperiod changeover* clinical trial* undertaken to compare an investigational drug (B) and an active control drug (A) for patients with ventricular arrhythmias and organic heart disease. Its research design involved the random (and double-blind) partition of 26 patients into two sequence groups (A:B and B:A) of 13 patients each. Of these, 20 patients completed the study protocol (9 in group A:B and 11 in group B:A) which involved seven visits over four evaluation periods as follows: two during a pretreatment period, two during the first treatment period, one at the end of a drug-free period following the first treatment period, and two during the second treatment period. Otherwise, the patients in group A:B received A during the first treatment period and B during the second, while those in group B:A received B during the first treatment period and A during the second. One measure of interest obtained from each patient at each visit during the study was peak heart rate for a bicycle exercise test, and its analysis is the focus of this example. For this purpose, the relevant information is the data array in Table 2 of average values (over visits) of peak heart rate with respect to each evaluation period for the 20 patients.

The questions of interest for the data in Table 2 are concerned with whether or not there is any association between the sequence groups and the peak heart rate data during the respective evaluation periods. If there is no difference of any type between the effects of the investigational drug (B) and the control drug (A) for each patient under study, then there should be no association between the sequence groups and the corresponding peak heart rate data; i.e., the observed values for each patient would be expected to be the same as they would be if

Table 2 Peak Heart Rate during Bicycle Exercise Test for the Four Periods of a Changeover Design Pertaining to the Treatment of Patients with Documented Ventricular Arrhythmias

		Peak Heart Rate during Bicycle Exercise Test			
Sequence Group[a]	Patient	Pretreatment Period	First Treatment Period	Drug-Free Period	Second Treatment Period
A:B	1	101.5	79.0	95.0	85.5
A:B	2	122.0	152.5	110.0	108.0
A:B	3	110.0	110.0	96.0	96.5
A:B	4	100.0	114.0	102.0	91.0
A:B	5	110.5	105.0	130.0	112.0
A:B	6	113.0	110.0	114.0	93.0
A:B	7	96.0	96.0	100.0	78.5
A:B	8	98.0	97.5	106.0	90.0
A:B	9	81.0	106.0	95.0	70.0
A:B	Mean	103.6	107.8	105.3	91.6
B:A	10	144.0	109.0	140.0	130.0
B:A	11	130.0	85.5	120.0	115.0
B:A	12	114.0	108.0	155.0	125.0
B:A	13	99.0	87.5	108.0	105.0
B:A	14	107.0	93.5	116.0	108.5
B:A	15	138.0	115.0	121.0	139.0
B:A	16	113.0	103.0	110.0	115.0
B:A	17	144.0	96.0	120.0	107.0
B:A	18	110.0	74.0	88.0	95.5
B:A	19	107.0	92.0	104.0	120.0
B:A	20	80.0	77.5	95.0	102.0
B:A	Mean	116.9	94.6	116.1	114.7

[a] A = control drug; B = investigational drug.

assignment had been to the other sequence group. Since this framework implies that the observed distributions for each sequence group is (by design) a simple random sample from the finite population corresponding to their pooled combination, the hypothesis of no association can be expressed as follows:

H_0: The two sequence groups are equivalent in the sense that their respective sets of peak heart rate data are compatible with equally likely realizations for the (20!/11!9!) possible random partitions of the 20 patients under study. (13)

[See (17) of CHI-SQUARE TESTS for a more general statement.] Since the patients in this changeover study were observed under different conditions during the four study periods (i.e., no treatment, drug A, or drug B), the hypothesis (13) has different interpretations according to the particular aspects of the data at which it is directed. Some important specific formulations of H_0 are addressed in Table 4, which also contains corresponding test statistics based on the across-subject ranks (with ties handled via midranks*) displayed in Table 3 for the combined set of 20 patients in both sequence groups. Such hypotheses include:

1. Equivalence of the two sequence groups at pretreatment and during the drug-free period (to verify the nonexistence of prior sequence group differences relative to the first treatment period and the

Table 3 Rank Matrix for Peak Heart Rate Data (across Patients in the Combined Sequence Groups) for Changeover Design Pertaining to Treatment of Ventricular Arrhythmias

		Across-Patient Ranks for Peak Heart Rate					
Sequence Group[a]	Patient In Group	Pretreatment Period	First Treatment Period	Drug-Free Period	Second Treatment Period	Difference between Treatment Periods	Sum of Treatment Periods
A:B	1	7	3	3	3	13	1
A:B	2	16	20	11.5	12	20	20
A:B	3	10.5	16.5	5	8	15	13
A:B	4	6	18	7	5	18	12
A:B	5	12	12	18	14	12	15
A:B	6	13.5	16.5	13	6	16	10.5
A:B	7	3	8.5	6	2	17	3
A:B	8	4	10	9	4	14	6
A:B	9	2	13	3	1	19	4
A:B	Mean	8.2	13.1	8.4	6.1	16	9.4
B:A	10	19.5	15	19	19	6	18
B:A	11	17	4	15.5	15.5	1	8
B:A	12	15	14	20	18	8	17
B:A	13	5	5	10	10	7	7
B:A	14	8.5	7	14	13	9	9
B:A	15	18	19	17	20	4	19
B:A	16	13.5	11	11.5	15.5	10	16
B:A	17	19.5	8.5	15.5	11	11	10.5
B:A	18	10.5	1	1	7	5	2
B:A	19	8.5	6	8	17	2	14
B:A	20	1	2	3	9	3	5
B:A	Mean	12.4	8.4	12.2	14.1	6	11.4

[a] A = active control drug; B = investigational drug.

Table 4 Test Statistics for Comparison of Changeover Design Sequence Groups with Respect to Peak Heart Rate

Sequence Groups Comparisons	Chi-Square Test Statistic	Degrees of Freedom	Chi-Square Approximate *P*-Value
Univariate tests[a]			
Pretreatment (PT)	2.43	1	0.119
First treatment period (T1)	3.06	1	0.080
Drug-free period (ND)	2.09	1	0.148
Second treatment period (T2)	9.01	1	0.003
Treatment period difference (T1 − T2)	14.14	1	0.001
Treatment period sum (T1 + T2)	0.58	1	0.447
Multivariate tests[b]			
(T1, PT)	8.80	2	0.012
(ND, PT)	2.61	2	0.271
(T1, T2)	14.29	2	0.001
(T1 + T2, PT, ND)	3.04	3	0.386
(PT, T1, ND, T2)	14.42	4	0.006
Covariance adjusted tests[c]			
(T1 adjusted for PT)	6.37	1	0.012
(ND adjusted for PT)	0.18	1	0.670
(T1 + T2 adjusted for PT, ND)	0.42	1	0.516
(T1, T2 adjusted for PT, ND)	11.81	2	0.003

[a] These tests are obtained by applying (22) of CHI-SQUARE TESTS to the corresponding column of Table 2; see expression (16) here for its specific application to (PT).
[b] These tests are obtained by applying (29) of CHI-SQUARE TESTS to the corresponding sets of columns of Table 2. See expression (14) here for its specific application to (PT, ND).
[c] These tests are obtained by subtracting the test statistic for the covariables alone from the overall joint test statistic for response variables and covariables; see Amara and Koch [1] for further discussion of computational aspects of rank analysis of covariance statistics.

nonexistence of carryover effects* relative to the second treatment period).

2. No difference between the effects of the investigational drug (B) and the control drug (A) during the first treatment period as expressed in terms of the corresponding equivalence of the two sequence groups.

3. No difference between the effects of drug B and drug A during both the first and second treatment periods as expressed in terms of the equivalence of the two sequence groups for the within-patient differences between the first and second treatment periods.

The test statistics for the hypotheses displayed in Table 4 are obtained by applying the one-way analysis-of-variance* randomization criterion given in (21) and (22) of CHI-SQUARE TESTS for a univariate response and in (29) for multivariate responses. Their usage presumes only the research design conditions of randomization and measurement with either no error or purely independent error. Otherwise, since the results in Table 4 are based on ranks, they represent large sample chi-square approximations for univariate and multivariate Wilcoxon rank sum tests*.

Further insight concerning the computation of the test statistics in Table 4 can be gained by considering the bivariate comparison of the sequence groups with respect to the pretreatment and drug-free periods as a specific illustration. For this hypothesis, the multivariate randomization chi-square statistic (29) of CHI-SQUARE TESTS with d.f. = 2 has the form shown in (14).

$$Q = \left(\frac{n-1}{n}\right)\sum_{i=1}^{2} n_i(\bar{\mathbf{y}}_i - \bar{\mathbf{y}})'\mathbf{V}^{-1}(\bar{\mathbf{y}}_i - \bar{\mathbf{y}})$$
$$= 2.61 \qquad (14)$$

where $n_1 = 9$, $n_2 = 11$, $n = n_1 + n_2 = 20$ and

$$\bar{\mathbf{y}}_1 = \begin{bmatrix} 8.22 \\ 8.39 \end{bmatrix}, \qquad \bar{\mathbf{y}}_2 = \begin{bmatrix} 12.36 \\ 12.23 \end{bmatrix},$$
$$\bar{\mathbf{y}} = \begin{bmatrix} 10.50 \\ 10.50 \end{bmatrix}, \qquad \mathbf{V} = \begin{bmatrix} 33.15 & 24.70 \\ 24.70 & 33.10 \end{bmatrix}. \qquad (15)$$

Here the $\bar{\mathbf{y}}_i = (\bar{y}_{i1}, \bar{y}_{i2})'$ are the mean rank vectors for the ith sequence group where $i = 1, 2$ (within which $\bar{y}_{i1}$ pertains to pretreatment and $\bar{y}_{i2}$ to drug-free), $\bar{\mathbf{y}} = (\bar{y}_{\cdot 1}, \bar{y}_{\cdot 2})' = (n_1\bar{\mathbf{y}}_1 + n_2\bar{\mathbf{y}}_2)/n$ is the corresponding finite population mean vector for all 20 patients, and $\mathbf{V}$ is the finite population covariance matrix [see (20) of CHI-SQUARE TESTS]. Similarly, the chi-square test statistic (with d.f. = 1) for the equivalence of the two sequence groups at pretreatment is computed via (22) of CHI-SQUARE TESTS as

$$Q = \left(\frac{n-1}{n}\right)\sum_{i=1}^{2} n_i(\bar{y}_{i1} - \bar{y}_{\cdot 1})^2/v_{11}$$
$$= \left(\frac{19}{20}\right)\left\{9(2.28)^2 + 11(1.86)^2\right\}/(33.15)$$
$$= 2.43. \qquad (16)$$

Finally, the chi-square test statistic (with d.f. = 1) for the equivalence of the two sequence groups during the drug-free period after covariance adjustment for their assumed pretreatment equivalence in the sense of Quade [14] is the difference (17) between the statistics (14) and (16):

$$Q = 2.61 - 2.43 = 0.18. \qquad (17)$$

All the test statistics (14) to (17) are nonsignificant with $p \geqslant 0.10$. Thus the pretreatment and drug-free data are interpreted to be compatible with the corresponding formulations of H_0, and this conclusion supports the validity of the changeover design* for the comparison of the investigational drug (B) and the active control drug (A). As indicated in Table 4, the statistical tests for the equivalence of these treatments indicate that drug (B) is associated with significantly lower peak heart rates for both the first treatment period (via the rank analysis of covariance statistic $Q(\text{d.f.} = 1) = 6.37$ with adjustment for pretreatment for which $p < 0.05$) and for the two treatment periods combined (via, e.g., the analysis of within-patient differences between the first and second treatment periods for which $Q(\text{d.f.} = 1) = 14.14$ with $p < 0.05$). Additional discussion of the application of rank methods to changeover designs is given in Koch [6]. A computer program for obtaining the results in Table 4 is documented in Amara and Koch [1].

EXAMPLE 3: CONTINGENCY TABLE TESTS FOR A CLINICAL TRIAL INVOLVING CATEGORICAL DATA

The data summarized in Table 5 are from a clinical trial concerned with the treatment of mastitis in dairy cows. They were obtained from a research design in which cows from a specific geographic area were assigned sequentially as they were identified (rather than randomly) to one of nine drug combinations of novobiocin and/or penicillin. One of these was a no-drug (control) treatment; three were separate doses of penicillin alone [100,000, 200,000, or 400,000 international units (IU)]; two were separate doses of novobiocin alone [400 or 600 milligrams (mg)]; and three were combination doses of penicillin (100,000, 200,000, or 400,000 IU) and novobiocin (400 mg). For each cow in the study, the infection status of each quarter of the udder was evaluated before and after treatment. Here this information has been summarized in terms of the pretreatment number of infected quarters and the post-treatment dichotomy for absence of infection in all quarters vs. its presence in any quarter. The post-treatment status of the cows is the outcome (or response) variable of interest for evaluation purposes. The pretreatment number of infected quarters is a

Table 5 Contingency Table from Clinical Trial Concerning Mastitis in Dairy Cows

Treatment		Pretreatment Number of Infected Quarters / Post-treatment Status Re Presence of Infection									
		0		1		2		3		4	
Novobiocin (mg)	Penicillin (IU)	No	Yes	No	Yes	No	Yes	No	Yes	No	Yes
None	None	2	6	2	4	0	3	0	4	0	5
None	100, 000	5	4	5	2	6	8	1	3	3	5
None	200, 000	5	6	4	3	0	3	1	5	3	4
None	400, 000	5	4	4	3	7	1	2	2	3	4
400	None	7	1	4	5	5	1	4	2	3	5
400	100, 000	6	1	2	3	1	3	1	2	4	4
400	200, 000	6	1	3	2	3	2	1	1	3	5
400	400, 000	5	3	3	2	1	2	3	0	4	3
600	None	5	1	2	5	1	0	4	0	0	2

covariable that potentially provides a framework for more accurate comparisons among the treatments through its inherent association with post-treatment status. (See Heald et al. [4] and Koch et al. [7] for more details.)

TESTS OF RANDOMNESS*

A straightforward analysis strategy for direct comparisons among the treatments are randomization model tests as discussed in the third section of CHI-SQUARE TESTS and illustrated in the third section of this entry for subject-wise data arrays. With this approach, one hypothesis of interest is:

H_{01}: The nine treatment groups are equivalent at pretreatment in the sense that their respective distributions for number of infected quarters are compatible with an exhaustive set of simple random samples (without replacement) from the corresponding fixed population distribution for all cows (at the beginning of the study). (18)

Under the hypothesis H_{01}, the pretreatment distributions for number of infected quarters have the multiple hypergeometric distribution*, where $\Pr[\{n_{hi+}\}] =$

$$\left\{\prod_{h=0}^{4} n_{h++}! \prod_{i=1}^{9} n_{+i+}! \Big/ \left(n! \prod_{h=0}^{4}\prod_{i=1}^{9} n_{hi+}!\right)\right\}, \tag{19}$$

where the $n_{hi+} = n_{hi1} + n_{hi2}$ are the pretreatment status $(h = 0, 1, 2, 3, 4) \times$ treatment $(i = 1, 2, \ldots, 9)$ marginal sums of observed frequencies n_{hij} with no $(j = 1)$ or at least one $(j = 2)$ infected quarters at post-treatment, the

$$n_{+i+} = \sum_{h=0}^{4} n_{hi+}$$
$$= (26, 42, 34, 35, 37, 27, 27, 26, 20)$$

are the sample sizes for the respective treatment groups, and the

$$n_{h++} = \sum_{i=1}^{9} n_{hi+} = (73, 58, 47, 36, 60)$$

are the frequencies for the pretreatment distribution of the $n = 274$ cows. From (19), it follows that the frequencies n_{hi+} have expected values and covariance structure as shown in (20).

$$E\{n_{hi+} \mid H_{01}\} = (n_{h++}/n)n_{+i+} = m_{hi*} \tag{20}$$

$$\begin{aligned}\operatorname{cov}\{n_{hi+}, n_{h'i'+} \mid H_{01}\} &= n_{h++}n_{+i+}(n\delta_{hh'} - n_{h'++}) \\ &\quad \times (n\delta_{ii'} - n_{+i'+})/\{n^2(n-1)\},\end{aligned}$$

where $\delta_{hh'} = 1$ if $h = h'$ and $\delta_{hh'} = 0$ if $h \neq h'$ and $\delta_{ii'} = 1$ if $i = i'$ and $\delta_{ii'} = 0$ if $i \neq i$[via arguments similar to those given for (19) and (20) of CHI-SQUARE TESTS]. Thus the randomization chi-square test statistic Q for H_{01} (with respect to all the $\{n_{hi+}\}$) can be computed via (30) of CHI-SQUARE TESTS as

$$Q = \left(\frac{n-1}{n}\right) Q_p$$

$$= \left(\frac{n-1}{n}\right) \sum_{h=0}^{4} \sum_{i=1}^{9} \frac{(n_{hi+} - m_{hi*})^2}{m_{hi*}}$$

$$= 21.80, \qquad (21)$$

where Q_p denotes the Pearson chi-square statistic for the 9×5 contingency table for treatment vs. pretreatment (as obtained by adding successive pairs of columns of Table 5). Since all of the expected frequencies m_{hi*} are $\geqslant 2.50$ and many are $\geqslant 5.00$, the chi-square distribution with d.f. $= 32$ provides a reasonable approximation for the distribution of Q in (21) under H_0. Accordingly, Q is interpreted to be nonsignificant with $p > 0.25$; and thus the nine treatment groups are concluded to have equivalent pretreatment distributions for numbers of infected quarters.

Another method that can be used to test H_{01} is the one-way analysis-of-variance randomization criterion given in (21) and (22) of CHI-SQUARE TESTS. This statistic is directed at comparisons among the treatment groups with respect to their pretreatment mean numbers of infected quarters

$$\bar{y}_i = \left\{ \sum_{h=0}^{4} h n_{hi+} / n_{+i+} \right\},$$

where $i = 1, 2, \ldots, 9$. These means are displayed in Table 6 together with their within-group estimated standard errors:

$$(v_{y,i})^{1/2} = \left\{ \sum_{h=0}^{4} n_{hi+} (h - \bar{y}_i)^2 / n_{+i+}^2 \right\}^{1/2}. \qquad (22)$$

The randomization chi-square statistic (with d.f. $= 8$) in reference to the mean scores $\bar{y}_i$ is computed via (23) as

$$Q = \left(\frac{n-1}{n v_y}\right) \sum_{i=1}^{9} n_{+i+} (\bar{y}_i - \bar{y})^2 = 2.38, \qquad (23)$$

where $\bar{y} = \{\sum_{h=0}^{4} h n_{h++} / n\} = 1.82$ is the finite population pretreatment mean number of infected quarters for all $n = 274$ cows and $v_y = \{\sum_{h=0}^{4} n_{h++} (h - \bar{y})^2 / n\} = 2.2540$ is the finite population variance. Since Q in (23) is nonsignificant with $p > 0.25$, it supports the conclusion that the pretreatment distributions for number of infected quarters are compatible with H_{01}; thus its interpretation is the same as that for the frequency distri-

Table 6 Pretreatment Mean Numbers of Infected Quarters and Post-Treatment Proportions with No Infected Quarters from Clinical Trial for Dairy Cows

Treatment Novobiocin (mg)	Treatment Penicillin (IU)	Number of Cows in Group	Pretreatment Number of Infected Quarters: Mean	Pretreatment Number of Infected Quarters: SE	Post-treatment Proportion with No Infected Quarters: Estimate	Post-treatment Proportion with No Infected Quarters: SE
None	None	26	1.69	0.30	0.154	0.071
None	100,000	42	1.88	0.21	0.476	0.077
None	200,000	34	1.74	0.27	0.382	0.083
None	400,000	35	1.80	0.25	0.600	0.083
400	None	37	1.92	0.24	0.622	0.080
400	100,000	27	2.00	0.31	0.519	0.096
400	200,000	27	1.96	0.30	0.593	0.095
400	400,000	26	1.85	0.32	0.615	0.095
600	None	20	1.45	0.30	0.600	0.110

bution statistic (21). This result is of additional interest because the mean score statistic in (23) has two advantages relative to (21). First, its sample-size requirements for chi-square approximations are somewhat less stringent since they are in reference to the means $\bar{y}_i$ (which are linear combinations of the n_{hi+}) rather than the n_{hi+} separately. It is also more powerful with respect to location shift alternatives as expressed in terms of expected differences among the $\bar{y}_i$, since its smaller degrees of freedom are specifically targeted at them.

In view of the pretreatment equivalence of the respective treatment groups, comparisons among them with respect to post-treatment status can be meaningfully undertaken either with or without adjustment for pretreatment status. In the latter case, the hypothesis of interest is:

H_{02}: The nine treatment groups are equivalent at post-treatment in the sense that their respective distributions for absence or presence of any infected quarters are compatible with an exhaustive set of simple random samples (without replacement). (24)

By the same rationale summarized in (19) to (21) for H_{01}, it follows that the randomization chi-square test statistic (with d.f. = 8) for H_{02} can be computed via (24) or (30) of CHI-SQUARE TESTS as

$$Q = \left(\frac{n-1}{n}\right) Q_p = \left(\frac{n-1}{n}\right) \sum_{i=1}^{9} \sum_{j=1}^{2} \frac{(n_{+ij} - m_{*ij})^2}{m_{*ij}} = 21.04, \tag{25}$$

where $n_{+ij} = \sum_{h=0}^{4} n_{hij}$ and $m_{*ij} = n_{+i+}(\sum_{i=1}^{9} n_{+ij}/n)$ are the observed and expected frequencies for the (9×2) contingency table for treatment vs. post-treatment (as obtained by adding columns 1, 3, 5, 7, 9 and columns 2, 4, 6, 8, 10 of Table 5) and Q_p is the corresponding Pearson chi-square statistic. Since Q in (25) is significant with $p < 0.01$, the hypothesis H_{02} is contradicted in the sense that some treatment groups have larger proportions $g_i = (n_{+i1}/n_{+i+})$ of cows with no infected quarters than the others. The nature of such variation can be seen in Table 6, where the g_i and their estimated standard errors $(v_{g,i})^{1/2} = \{g_i(1-g_i)/n_i\}^{1/2}$ are displayed. Examination of these quantities suggests that the eight active treatments have relatively similar proportions with no infected quarters and that their range of values 0.38–0.62 is substantially higher than the 0.15 value for the control group. These conclusions can be given more formal support through tests of corresponding hypotheses. In this regard, the equivalence of the eight active treatments can be tested by a chi-square statistic (with d.f. = 7) analogous to (25) for which the corresponding $Q = 6.79$ is nonsignificant with $p > 0.25$. The combined set of active treatments can be compared to the control treatment by a chi-square statistic (with d.f. = 1) which is directed at the linear combination

$$F = \left\{ \frac{1}{n - n_{+1+}} \sum_{i=2}^{9} n_{+i+}\, g_i \right\} - g_1 = 0.391.$$

This test statistic is computed via (27) of CHI-SQUARE TESTS as

$$Q = F^2/v_F = 14.31, \tag{26}$$

where v_F is the variance of F, and is significant with $p < 0.01$.

Since the post-treatment status of a cow has a natural association with the pretreatment number of infected quarters, treatment comparisons which are adjusted for pretreatment status are of some interest. In these situations, attention is directed at the hypothesis

H_{03}: For each of the pretreatment status categories 0, 1, 2, 3, 4, the nine treatment groups are equivalent at post-treatment in the sense of H_{02}. (27)

[See (31) of CHI-SQUARE TESTS for a more general statement.] An effective method for testing this hypothesis is the average partial association chi-square statistic (with d.f. = 8), which is computed by applying (35) of CHI-SQUARE TESTS to the set of five 9×2 tables given in columns 1 to 10 of Table 5. Since $Q_G = 22.47$ is significant with $p < 0.01$, the hypothesis H_{03} is contradicted in the sense that some treatment groups have larger proportions $p_{hi1} = (n_{hi1}/n_{hi+})$ of cows with no infected quarters than others in a similar manner for all pretreatment status categories $h = 0, 1, 2, 3, 4$. Thus this test statistic with adjustment for pretreatment status has a somewhat stronger but generally the same interpretation as that given in (25) for H_{02} without adjustment.

Finally, it can be verified that the average partial association statistic for comparing the eight active treatments is nonsignificant with $p > 0.25$ (which supports their equivalence); and that for comparing the combined set of active treatments with the control treatment is significant with $p < 0.01$. The latter test statistic can be readily computed as the Mantel–Haenszel* chi-square (d.f. = 1) statistic

$$Q_G = \frac{\left\{\sum_{h=0}^{4}\left(n_{h11} - \frac{n_{h1+}n_{h+1}}{n_{h++}}\right)\right\}^2}{\sum_{h=0}^{4}\frac{n_{h1+}(n_{h++} - n_{h1+})n_{h+1}n_{h+2}}{n_{h++}^2(n_{h++} - 1)}} = 15.21 \qquad (28)$$

for the set of five 2×2 tables corresponding to the control treatment vs. the pooled active treatments.

In summary, randomization model tests represent a useful strategy for investigating a broad class of hypotheses. However, they do not provide a framework for describing the variation that is detected when hypotheses such as H_{02} and H_{03} are contradicted. For this purpose, methods involving statistical models like those discussed in the following two sections are needed.

Wald Chi-Square Tests* for Linear Hypotheses Pertaining to Analysis of Covariance

One approach for formulating statistical models for contingency tables like Table 5 is through constraints for some set of functions **F** of the frequencies $\{n_{hij}\}$ that are implied by hypotheses with which the data are compatible. Here the pretreatment means $\bar{y}_i$ and the post-treatment proportions g_i with no infection are functions of interest for summarizing the information for the respective treatment groups $i = 1, 2, \ldots, 9$. These functions may be computed from the $\{n_{hij}\}$ via the linear transformation

$$\begin{bmatrix} g_i \\ \bar{y}_i \end{bmatrix} = \frac{1}{n_{+i+}}\begin{bmatrix} 1 & 0 & 1 & 0 & 1 & 0 & 1 & 0 & 1 & 0 \\ 0 & 0 & 1 & 1 & 2 & 2 & 3 & 3 & 4 & 4 \end{bmatrix}\mathbf{n}_i = (\mathbf{A}\mathbf{n}_i / n_{+i+}), \qquad (29)$$

where $\mathbf{n}_i' = (n_{0i1}, n_{0i2}, n_{1i1}, n_{1i2}, n_{2i1}, n_{2i2}, n_{3i1}, n_{3i2}, n_{4i1}, n_{4i2})$. If the joint pretreatment and post-treatment data for the cows in each group $i = 1, 2, \ldots, 9$ are considered representative of some broader population for which they can be viewed as a stratified simple random sample* (with replacement), then the $\{n_{hij}\}$ can be assumed to have the product multinomial distribution*

$$\phi(\mathbf{n} \mid \boldsymbol{\pi}) = \prod_{i=1}^{9} n_{+i+}! \prod_{h=0}^{4}\prod_{j=1}^{2}(\pi_{hij}^{n_{hij}}/n_{hij}!), \qquad (30)$$

where $\sum_{h=0}^{4}\sum_{j=1}^{2}\pi_{hij} = 1$ for $i = 1, 2, \ldots, 9$. For this framework, the π_{hij} denote the respective probabilities that a randomly observed cow with the ith treatment had the hth pretreatment status and the jth post-treatment status; also, $\mathbf{n} = (\mathbf{n}_1', \mathbf{n}_2', \ldots, \mathbf{n}_9')'$ denotes the concatenated vector of all frequencies $\{n_{hij}\}$ (ordered according to j in h in i) and $\boldsymbol{\pi}$ denotes a similarly arranged concatenated vector of all the $\{\pi_{hij}\}$. From the structure of (30), it follows that the function vector

$$\mathbf{F} = (g_1, \bar{y}_1, g_2, \bar{y}_2, \ldots, g_9, \bar{y}_9)' \qquad (31)$$

has expected value vector $\boldsymbol{\mu}_{\mathbf{F}}(\boldsymbol{\pi})$ and covariance matrix* $\mathbf{V}_{\mathbf{F}}(\boldsymbol{\pi})$, where $\boldsymbol{\mu}_{\mathbf{F}}(\boldsymbol{\pi})$ is a concatenated vector with successive segments $\boldsymbol{\mu}_i(\boldsymbol{\pi}_i) = \mathbf{A}\boldsymbol{\pi}_i$ for $i = 1, 2, \ldots, 9$ and $\mathbf{V}_{\mathbf{F}}(\boldsymbol{\pi})$ is a block diagonal matrix with successive blocks

$$\mathbf{V}_{\mathbf{F},i}(\boldsymbol{\pi}_i) = \mathbf{A}\left[\mathbf{D}_{\boldsymbol{\pi}_i} - \boldsymbol{\pi}_i\boldsymbol{\pi}_i'\right]\mathbf{A}'/n_{+i+}, \qquad (32)$$

for which $\mathbf{D}_{\boldsymbol{\pi}_i}$ is a diagonal matrix with the elements of $\boldsymbol{\pi}_i$ on the main diagonal.

As noted in the preceding section, the nine treatment groups were equivalent at pretreatment in the sense of H_{01}. Thus the function vector **F** is compatible with the constraints that pairwise differences in the expected values of the $\bar{y}_i$ are all 0. However, these constraints imply that the variation among the elements of $\boldsymbol{\mu}_F(\boldsymbol{\pi})$ can be represented by the linear regression model*

$$\boldsymbol{\mu}_i(\boldsymbol{\pi}_i) = \mathbf{A}\boldsymbol{\pi}_i = \begin{bmatrix} \beta_i \\ \beta_{10} \end{bmatrix} = \begin{bmatrix} E\{g_i\} \\ E\{\bar{y}_i\} \end{bmatrix}, \quad (33)$$

where $\beta_1, \beta_2, \ldots, \beta_9$ are adjusted mean value parameters for the post-treatment status of the respective treatment groups and β_{10} is the common mean parameter for their equivalent pretreatment status. If **X** denotes the specification matrix for the model (33), then its Wald goodness-of-fit statistic Q with d.f. = 8 may be computed via (8) and (39) of CHI-SQUARE TESTS as

$$Q = (\mathbf{F} - \mathbf{X}\mathbf{b})'\mathbf{V}_F^{-1}(\mathbf{F} - \mathbf{X}\mathbf{b}) = 2.60, \quad (34)$$

where $\mathbf{V}_F = \mathbf{V}_F(\mathbf{f})$ is the consistent estimate for the covariance matrix of **F** obtained by replacing the $\boldsymbol{\pi}_i$ in $\mathbf{V}_F(\boldsymbol{\pi})$ by the corresponding sample estimates $\mathbf{f}_i = (\mathbf{n}_i / n_{+i+})$ and

$$\mathbf{b} = \left(\mathbf{X}'\mathbf{V}_F^{-1}\mathbf{X}\right)^{-1}\mathbf{X}'\mathbf{V}_F^{-1}\mathbf{F}$$
$$= (0.144, 0.481, 0.379, 0.600, 0.629,$$
$$0.532, 0.609, 0.615, 0.586, 1.819) \quad (35)$$

is the vector of weighted-least-squares estimates for the model parameters $\boldsymbol{\beta}$ [see (9) of CHI-SQUARE TESTS]. Since Q in (34) is nonsignificant with $p > 0.25$, the model X is considered to provide an adequate characterization of the variation among the elements of $\boldsymbol{\mu}_F(\boldsymbol{\pi})$. Thus further analysis can be based on its parameters $\boldsymbol{\beta}$. In this regard, one hypothesis of interest is the equivalence of the nine treatment groups at post-treatment given the model **X** in (33). The constraint matrix for this hypothesis with respect to $\boldsymbol{\beta}$ is $\mathbf{C}_1 = [\mathbf{1}_8, -\mathbf{I}_8, \mathbf{0}_8]$ where $\mathbf{1}_8$ is an (8×1) vector of 1's, $\mathbf{I}_8$ is the (8×8) identity matrix, and $\mathbf{0}_8$ is an (8×1) vector of 0's. The corresponding test statistic with d.f. = 8 [see (13) of CHI-SQUARE TESTS] is

$$Q = \mathbf{b}'\mathbf{C}_1'\left[\mathbf{C}_1\left(\mathbf{X}'\mathbf{V}_F^{-1}\mathbf{X}\right)^{-1}\mathbf{C}_1'\right]^{-1}\mathbf{C}_1\mathbf{b} = 38.13, \quad (36)$$

which is significant at $p < 0.01$. Another hypothesis of interest is equivalence of the eight active treatment groups at post-treatment, for which the constraint matrix is $\mathbf{C}_2 = [\mathbf{0}_7, \mathbf{1}_7, -\mathbf{I}_7, \mathbf{0}_7]$, where $\mathbf{0}_7$ is a (7×1) vector of 0's, $\mathbf{1}_7$ is a (7×1) vector of **1**'s and $\mathbf{I}_7$ is the (7×7) identity matrix. The corresponding test statistic with d.f. = 7 is $Q = 7.61$, which is nonsignificant with $p > 0.25$. Thus the post-treatment proportions of cows with no infected quarters for the eight active treatments are generally similar to each other and significantly larger than that for the control treatment.

Logistic Regression Analysis

Another analysis strategy for the data in Table 5 is the formulation of a statistical model for post-treatment status which encompasses the joint effects of treatment and pretreatment status. In this case, attention is directed at the variation among the conditional post-treatment distributions $\mathbf{n}'_{hi} = (n_{hi1}, n_{hi2})$ for subpopulations corresponding to the pretreatment status × treatment cross-classification. These quantities are assumed to have the product binomial distribution*, given by $\phi(\mathbf{n} \mid \boldsymbol{\theta}) =$

$$\prod_{h=0}^{4} \prod_{i=1}^{9} \left\{ n_{hi+}!\, \theta_{hi}^{n_{hi1}} (1 - \theta_{hi})^{n_{hi2}} / n_{hi1}!\, n_{hi2}! \right\}, \quad (37)$$

where the θ_{hi} denote the respective probabilities that a randomly observed cow with the hth pretreatment status and the ith treatment has no infected quarters at post-treatment; **n** denotes the concatenated vector of all frequencies defined previously in reference to (30), and

$$\boldsymbol{\theta} = (\theta_{01}, \theta_{11}, \theta_{21}, \theta_{31}, \theta_{41}, \theta_{02}, \ldots,$$
$$\theta_{42}, \ldots, \theta_{09}, \ldots, \theta_{49})' \quad (38)$$

is a similarly arranged vector of the $\{\theta_{hi}\}$. One model of interest for the θ_{hi} is the parallel-line logistic, which can be expressed as

$$\theta_{hi} = \{1 + \exp(\tau_i + \xi h)\}^{-1}, \quad (39)$$

where the τ_i are intercept* parameters for the respective treatment groups corresponding to the post-treatment status of cows with no infected quarters at pretreatment and ξ is the common slope parameter for the extent to which the logit transformations* $\log_e\{\theta_{hi}/(1-\theta_{hi})\}$ decrease as the number of infected quarters at pretreatment increases. As can be seen from the mathematical structure of (39), an important property of the linear logistic model* is that all values of the τ_i and/or ξ in $(-\infty, \infty)$ yield values for the $\{\theta_{hi}\}$ in the (0, 1) interval.

If most of the frequencies n_{hij} were sufficiently large (i.e., $\geqslant 5$) to have approximately normal distributions, then either weighted least-squares methods or maximum likelihood methods could be equivalently used to obtain estimates for the parameters τ_i and ξ and test statistics for hypotheses of interest. However, for this example, many of the $\{n_{hij}\}$ are small, so maximum likelihood (ML) methods are preferable. These estimates can be expressed as the solution of the nonlinear equations derived from substituting the model expression (39) for θ_{hi} into the expression (37) for ϕ, differentiating $\log_e \phi$ with respect to the τ_i and ξ and equating the result to 0. After some simplifications, these equations can be shown to have the form

$$\sum_{h=0}^{4} n_{hi+}\hat{\theta}_{hi} = n_{+i+}\, g_i,$$

$$\sum_{h=0}^{4}\sum_{i=1}^{9} hn_{hi+}\hat{\theta}_{hi} = \sum_{h=0}^{4}\sum_{i=1}^{9} hn_{hi1} = \sum_{i=1}^{9} n_{+i+}\,\bar{y}_{*i1}, \quad (40)$$

where the $\hat{\theta}_{hi} = \{1 + \exp(\hat{\tau}_i + \hat{\xi}h)\}^{-1}$ are the model-predicted ML estimates of the θ_{hi} based on the ML parameter estimates $\hat{\tau}_i$ and $\hat{\xi}$, and g_i are the post-treatment proportions with no infection (ignoring pretreatment status), and the $\bar{y}_{*i1}$ are the average numbers of infected quarters at pretreatment for cows with no infected quarters at post-treatment. Thus the $\hat{\tau}_i$ and $\hat{\xi}$ are similar in spirit to the g_i and $\bar{y}_i$ considered in the preceding section, being functions of the g_i also and the analogs $\bar{y}_{*i1}$ to the $\bar{y}_i$.

In view of the nonlinear nature of (40), iterative procedures are required for the computation of the $\hat{\tau}_i$ and $\hat{\xi}$. For this purpose, one generally useful approach is the Newton–Raphson* method (or iterative weighted least squares) described in Nelder and Wedderburn [13]. This type of computing procedure yielded the ML estimates $\hat{\xi} = 0.25$ and $\hat{\tau}_i$ as follows:

$i = 1$	$i = 2$	$i = 3$	$i = 4$	$i = 5$
1.32	− 0.38	0.06	− 0.88	− 1.00

$i = 6$	$i = 7$	$i = 8$	$i = 9$
− 0.59	− 0.89	− 0.96	− 0.79

From the $\hat{\tau}_i$ and $\hat{\xi}$, ML estimates $\hat{\theta}_{hi}$ for the θ_{hi} can be obtained by substitution into (39). These quantities can then be used to calculate the log-likelihood ratio chi-square statistic Q_L with d.f. = 45 − 10 = 35 for the goodness of fit of the model (39) via the stratified sample extension of (42) of CHI-SQUARE TESTS:

$$Q_L = \sum_{h=0}^{4}\sum_{i=1}^{9}\sum_{j=1}^{2} 2n_{hij}\left[\log_e(n_{hij}/\hat{m}_{hij})\right] = 43.18, \quad (41)$$

where $\hat{m}_{hi1} = n_{hi+}\hat{\theta}_{hi}$ and $\hat{m}_{hi2} = n_{hi+} \times (1-\hat{\theta}_{hi})$ are ML estimates for the expected values of the n_{hij} under the model (39); or they can be used to calculate the Pearson chi-square goodness-of-fit statistic Q_P (which is asymptotically equivalent to Q_L) via the stratified sample extension of (43) of CHI-SQUARE TESTS.

$$Q_P = \sum_{h=0}^{4}\sum_{i=1}^{9}\sum_{j=1}^{2} (n_{hij} - \hat{m}_{hij})^2/\hat{m}_{hij} = 36.56. \quad (42)$$

Although both Q_L and Q_P are nonsignificant with $p > 0.10$, some caution needs to be given to their interpretation because the extent to which many of the frequencies n_{hij} in Table 5 (or the $\hat{m}_{hij}$) are small tends to contradict the strict validity of chi-square approximations to their distributions. In this regard, numerical studies (e.g., Larntz [11]) suggest that chi-square approximations are reasonable for goodness-of-fit assessment for a broad range of small sample situations, particularly in reference to Q_P for cases where most of the $\hat{m}_{hij} > 2$ and few are < 1 (*see* CATEGORICAL DATA). Thus the variation among the θ_{hi} is judged to be compatible with the model (39).

Since the model (39) provides an adequate characterization of the variation among the θ_{hi}, further analysis can be undertaken in terms of its parameters τ_i and ξ. Accordingly, one hypothesis of interest is:

$\boldsymbol{H_{04}}$: There is no variation among the treatment groups in the sense that

$$\tau_1 = \tau_2 = \cdots = \tau_9. \tag{43}$$

This hypothesis implies the model

$$\theta_{hi} = \{1 + \exp(\tau + \xi h)\}^{-1}, \tag{44}$$

where τ and ξ are the common intercept and slope* parameters for all treatment groups.

The parameters for the simplified model (44) can be estimated by applying maximum likelihood with respect to it in a manner analogous to that discussed previously for (39). The resulting ML estimates were $\hat{\tau} = -0.433$ and $\hat{\xi} = 0.221$. These quantities can be used to calculate the log-likelihood ratio chi-square statistic* $Q_L = 67.13$ with d.f. $= 43$ for the goodness of fit of the model (44) in a manner analogous to (41). More important, the log-likelihood ratio reduction statistic Q_L with d.f. $= 8$ for the hypothesis H_{04} in (43) can be obtained as the difference between the log-likelihood ratio statistics for the models (44) and (39); i.e. $Q_L = 67.13 - 43.18 = 23.95$. Alternatively, the equivalent Pearson chi-square reduction statistic is the analogous difference $Q_P = 55.71 - 36.56 = 19.15$. Since both of these results are significant with $p < 0.01$, the hypothesis H_{04} is contradicted in the sense that some of the treatment groups have larger treatment effect parameters τ_i than do the others. Another hypothesis of interest is equivalence of the eight active treatments in the sense that $\tau_2 = \tau_3 = \cdots = \tau_9$. The corresponding log-likelihood ratio reduction statistic (with d.f. $= 7$) is $Q_L = 50.45 - 43.18 = 7.27$ and its Pearson chi-square reduction counterpart is $Q_P = 42.66 - 36.56 = 6.10$, both of which are nonsignificant with $p > 0.25$. Thus for each pretreatment status category, the proportions of cows with no infected quarters for the eight active treatments are similar to each other and significantly larger than that for the control treatment.

Concluding Comments

Although the statistical methods illustrated in the preceding sections may appear to be competing strategies for obtaining equivalent results for essentially the same hypotheses, they involve fundamentally different underlying frameworks. The randomization procedures in the first section are based on hypergeometric models such as (19), which are applicable to finite populations and thereby do not require any assumptions concerning the extent to which the cows under study are representative of a larger population. On the other hand, the methods in the second and third sections are based on multinomial models such as (30) and (37), which presume that the data in Table 5 can be viewed as a random sample of a larger population. Each of the methods discussed here has specific advantages relative to the others for certain types of applications, with the choice among them being mostly a matter of the presumed sampling framework and analytical objectives. For further discussion of such issues, see Koch et al. [8].

Finally, the results in the first section were obtained by the partial association computer program PARCAT documented in Landis et al. [10]; the results in the second section were obtained by the weighted least-squares computer program GENCAT documented in Landis et al. [9]; and the results in the third section were obtained by maximum likelihood logistic regression* computer programs such as those documented by Baker and Nelder [2] and Harrell [3].

References

[1] Amara, I. A. and Koch, G. G. (1980). *Proc. 5th Annu. SAS Users Group Int. Conf.*, 134–144.

[2] Baker, R. J. and Nelder, J. A. (1978). *The GLIM System Manual (Release 3)*. The Numerical Algorithms Group/Royal Statistical Society, Oxford.

[3] Harrell, F. (1980). LOGIST. *SAS Supplemental Library Users' Guide*, 1980 ed.

[4] Heald, C. W., Jones, G. M., Nickerson, S., and Bibb, T. L. (1977). *Canad. Vet. J.*, **18**, 171.

[5] Kish, L. and Frankel, M. (1970). *J. Amer. Statist. Assoc.*, **65**, 1071–1094.

[6] Koch, G. G. (1972). *Biometrics*, **28**, 577–584.

[7] Koch, G. G., Grizzle, J. E., Semenya, K. and Sen, P. K. (1978). *J. Dairy Sci.*, **61**, 829–847.

[8] Koch, G. G., Gillings, D. B., and Stokes, M. E. (1980). *Ann. Rev. Public Health*, **1**, 163–225.

[9] Landis, J. R., Stanish, W. M., Freeman, J. L., and Koch, G. G. (1976). *Computer Programs Biomed.*, **6**, 196–231.

[10] Landis, J. R., Cooper, M. M., Kennedy, T., and Koch, G. G. (1979). *Computer Programs Biomed.*, **9**, 223–246.

[11] Larntz, K. (1978). *J. Amer. Statist. Ass.*, **73**, 253–263.

[12] McCarthy, P. (1969). *Int. Statist. Rev.*, **37**, 239–264.

[13] Nelder, J. A. and Wedderburn, R. W. M. (1972). *J. R. Statist. Soc. A*, **135**, 370–384.

[14] Quade, D. (1967). *J. Amer. Statist. Ass.*, **62**, 1187–1200.

[15] United States National Center for Health Statistics (1973). Plan and Operation of the Health and Nutrition Examination Survey. *Vital Health Statist. Ser. 1*, Nos. 10A, 10B, 14. DHEW Publ. No (HRA) 73–1310.

Acknowledgment

This research was supported in part by the U. S. Bureau of the Census through Joint Statistical Agreement JSA-79-16. The authors thank Dwight Brock for providing the HANES estimates in Example 1, William Shapiro for providing the data in Example 2, and William Vinson for providing the data in Example 3. They also express their appreciation to Suzanne Edwards for statistical computing assistance and to Jo Ann Degraffenreidt for typing the original manuscript.

(CHI-SQUARE TESTS)

GARY G. KOCH
MAURA E. STOKES

CHUNG PROCESSES

Chung processes form a special class of Markov processes* with continuous-time parameter, discrete-state space[1], and stationary transition probabilities*. As such, their theory should be subsumed by that in the arbitrary state space case. But for Markov processes with arbitrary state spaces, it is impossible to do much without assuming certain regularity conditions, which conditions turn out to be too strong in the discrete-space case. Thus the theory of Chung processes takes advantage of the simplicity of discrete-state spaces in order to better examine other key concepts in the absence of strong simplifying conditions.

We start the exposition by assuming right continuity and the existence of left-hand limits. This case coincides with the standard one, and all the major notions are easy to describe. Then we move on to arbitrary Chung processes, describe the novel features involved, and give a sample of their treacherous beauty.

MARKOV PROCESSES*

Throughout this article, E is a discrete space, $(\Omega, \mathcal{H}, P)$ is a probability space, and $X = (X_t)_{t \geqslant 0}$ is a stochastic process* defined on $(\Omega, \mathcal{H}, P)$ and taking values in E. To avoid trivialities we assume that E is the minimal state space: for every i in E there is $t \geqslant 0$ such that $P[X_t = i] > 0$. The value of X_t corresponding to the outcome ω in Ω is denoted by $X_t(\omega)$.

The process X is said to be a time-homogeneous Markov process if

$$P[X_{t+u} = j \mid X_s; s \leqslant t] = P_u(i, j) \quad \text{on } \{X_t = i\} \qquad (1)$$

for all states i, j and all times $t, u \geqslant 0$, for some family $(P_t)_{t \geqslant 0}$ of matrices on $E \times E$. Then (P_t) is called the transition function of X, and satisfies

$$P_t(i,k) \geqslant 0, \qquad \sum_j P_t(i,j) = 1,$$
$$\sum_j P_t(i,j) P_u(j,k) = P_{t+u}(i,k) \qquad (2)$$

for all states i, k and times t, u. The last

equation in (2) is called the Chapman–Kolmogorov equation*.

The property (1) implies that the future $\{X_{t+u}; u \geqslant 0\}$ after t and the past $\{X_s; s \leqslant t\}$ before t are conditionally independent given the state X_t at t. This is called the Markov property.

SIMPLE CHUNG PROCESSES

The Markov process X is said to be a simple Chung process provided that, for every outcome ω, the path $X(\omega): t \to X_t(\omega)$ is right-continuous and has left-hand limits in E at all times t. Throughout this section we assume that X is a simple Chung process.

Strong Markov Property

A random time* $T:\Omega \to [0, \infty]$ is said to be a stopping time* of X (also called Markov time or optional time) if the knowledge of the past history* $\mathcal{H}_u = \sigma\{X_s : s \leqslant u\}$ before u is sufficient to tell whether the event $\{T \leqslant u\}$ has occurred or not. The Markov property for fixed times t extends to such random times: the future $\{X_{T+u}; u \geqslant 0\}$ and the past $\mathcal{H}_T$ are conditionally independent given X_T. This is called the strong Markov property, and it follows from the Markov property, the right continuity of X, and the discreteness of the state space.

Structure of the Process

Fix an outcome ω. The assumed regularity of the path $X(\omega)$ and the discreteness of E imply that every discontinuity of $X(\omega)$ is of the jump type* and that there are at most finitely many jumps during any finite time interval. Thus the path $X(\omega)$ is a step function*. Let $S_0(\omega) = 0$, and let $X_n(\omega)$ be the time of the nth jump [if there is no nth jump, set $S_n(\omega) = \infty$] for $n = 1, 2, \dots$. Let $Y_n(\omega) = X_{S_n}(\omega)$ if $S_n(\omega) < \infty$; otherwise, set $Y_n(\omega) = Y_{n-1}(\omega)$. Then for every time t there is a unique n such that $S_n(\omega) \leqslant t < S_{n+1}(\omega)$, in which case $X_t(\omega) = Y_n(\omega)$.

It follows that the sequence (Y_n, S_n) carries the same information as (X_t). We now consider the law of the former. Let R_t be the length of the interval between t and the first jump after t; note that $R_0 = S_1$ and $R_{S_n} + S_n = S_{n+1}$. The event $\{X_t = i, R_t > u + v\}$ is the same as $\{X_t = i, R_t > u, X_{t+u} = i, R_{t+u} > v\}$. Thus by the Markov property, the conditional distribution of R_t given that $X_t = i$ is exponential* with some parameter $q(i)$. By the strong Markov property, we may replace t by S_n to arrive at the same conclusion. A similar reasoning including the next state to be visited yields

$$P\{Y_{n+1} = j, S_{n+1} > S_n + u \mid Y_m, S_m; m \leqslant n\} = e^{-q(i)u} Q(i,j) \tag{3}$$

on $\{Y_n = i\}$ for some numbers $q(i)$ and $Q(i,j)$ satisfying

$$0 \leqslant q(i) < \infty; \qquad Q(i,j) \geqslant 0; \qquad \sum_j Q(i,j) = 1. \tag{4}$$

In other words, (Y_n) is a Markov chain* with state space E and transition probability matrix Q, and on $\{S_n < \infty\}$, the difference $S_{n+1} - S_n$ is exponentially distributed with parameter $q(i)$ if $Y_n = i$. Hence the following is the structure of (X_t). Suppose that X starts at i; then X stays at i an exponentially distributed amount of time with mean $1/q(i)$ and then jumps to some new state j with probability $Q(i,j)$; it stays at j an exponentially distributed amount of time with mean $1/q(j)$ and then jumps to a new state k with probability $Q(j,k)$; and so on.

Continuity Properties of (P_t)

The right continuity of X implies that $t \to P_t(i,j)$ is right continuous and

$$\lim_{t \to 0} P_t(i,j) = I(i,j) \qquad (i, j \in E), \tag{5}$$

where I is the identity matrix*. This right continuity in fact implies infinite differentiability of $t \to P_t$.

Generator of X

Using (3) with $n = 1$ and a renewal argument at S_1 yields the following basic formula:

$$P_t(i,k) = e^{-q(i)t}\left[I(i,k) + q(i)\int_0^t e^{q(i)u} \times \sum_j Q(i,j)P_u(j,k)\,du\right] \tag{6}$$

Taking derivatives on both sides of (6) yields

$$\frac{d}{dt}P_t(i,k) = \sum_j G(i,j)P_t(j,k) \quad (i,k \in E), \tag{7}$$

where

$$G(i,j) = -q(i)I(i,j) + q(i)Q(i,j) \quad (i,j \in E). \tag{8}$$

In particular, (7) and (5) give

$$G(i,j) = \lim_{t\downarrow 0} \frac{P_t(i,j) - I(i,j)}{t}. \tag{9}$$

The matrix G is called the *generator* of X. Since (8) can be solved for the $q(i)$ and $Q(i,j)$ in terms of the $G(i,j)$, and since the numbers $q(i)$ and $Q(i,j)$ define the probability law of (Y_n, S_n), and therefore of (X_t), the matrix G describes the law of X. Similarly, the differential equations (7), called *Kolmogorov's backward differential equations**, can be solved for (P_t) and yield a unique solution with initial conditions $P_0(i,j) = I(i,j)$. Thus the derivatives $G(i,j)$ in (9) of the functions $P_t(i,j)$ at $t = 0$ define the whole function P_t. Hence the $G(i,j)$ may be regarded as the fundamental data that define everything else, thus justifying the term "generator" for G. Moreover, it can be shown that if $i \neq j$ and if the process X is recurrent*,

$$G(i,j) = \lim_{t\to\infty} \frac{\sum_{u<t} I_{\{X_{u-}=i,\,X_u=j\}}}{\int_0^t I_{\{X_u=i\}}\,du}, \tag{10}$$

i.e., $G(i,j)$ is the long-term ratio of the number of jumps from i to j to the amount of time spent in i. This fact may be used for estimating $G(i,j)(i \neq j)$ by statistical means, and then $G(i,i)$ can be computed by noting that $G(i,i) = -q(i) = -\sum_{j\neq i} G(i,j)$ in view of (7).

CHUNG PROCESSES

Let X be a Markov process with transition function (P_t) and the minimal state space E as in the introduction. We no longer assume that X is a simple Chung process, but without some assumption on something there is very little that can be said. Chung processes are obtained by basically assuming that the intuitively appealing property (5) holds.

A transition function* (P_t) is called *standard* if (5) holds. Then the paths $X(\omega)$ can be modified in such a way that the new paths have certain desirable regularity properties and the modified process is still Markov with the same transition function (P_t). The modification involved usually requires enlarging the minimal state space E by adding an extra point Δ as "the point at infinity," and the new paths take values in $\bar{E} = E \cup \{\Delta\}$; but in contrast to the states in E, the new point is such that $P\{X_t = \Delta\} = 0$ for all $t > 0$.

The process X is said to be a *Chung process* if its transition function is standard and its paths have been so modified that, for every t and ω, $s \to X_s(\omega)$ has at most one limiting value in E as s decreases to t, $X_t(\omega)$ is equal to that limiting value in E if it exists, and $X_t(\omega) = \Delta$ otherwise.

Every simple Chung process is a Chung process. Every Chung process with finitely many states is a simple Chung process. A Chung process may be right-continuous and have left-hand limits and still not be a simple Chung process (because limits are now being taken in $\bar{E}$).

Standard Transition Functions

For a standard transition function (P_t), $t \to P_t(i,j)$ is continuous and the limits (9) exist and satisfy

$$0 \leqslant q(i) = -G(i,i) \leqslant +\infty;$$

$$0 \leqslant G(i,j) < \infty \quad (i \neq j); \tag{11}$$

$$\sum_j G(i,j) \leqslant q(i). \tag{12}$$

It is possible to give examples of G where every equality or inequality above is satisfied. The matrix G is called the generator of X, even though the term "generator" is now misleading. Generally, the differential equations (7) do not hold; and even in cases where (7) does hold, it is possible that (7) has more than one solution (P_t). To gain insight into the latter possibility, consider the following process X. Suppose that $E = \{0, 1, 2, \dots\}$, $X_0 = 0$; X goes through the states $0, 1, 2, \dots$ in that order, spending an exponential amount of time in each state i with parameter $q(i) = i^2$. Let $S_1, S_2, \dots$ be the times of successive jumps. Then, for $T_1 = \lim S_n$, we have $E[T_1] = \sum_i 1/q(i) < \infty$, which means that $T_1 < \infty$ almost surely and $\lim_{t \to T_1} X_t = +\infty = \Delta$. Suppose that the state at T_1 is chosen independent of the previous history according to some distribution $\pi(j)$, $j \in E$, i.e., $P\{X_{T_1} = j\} = \pi(j)$. If the state at T_1 is j, then X goes through the states $j, j+1, \dots$ in that order, spending exponential amounts with parameters $q(j)$, $q(j+1), \dots$ as before. The jump times following T_1 will have a finite limit T_2 by the reasoning given for T_1. At T_2, X again selects a new state according to the same distribution π, and the whole behavior is repeated. This process X is a Chung process, and the limits $G(i, j)$ are given by

$$G(i,i) = -i^2, \qquad G(i,i+1) = i^2. \quad (13)$$

In this case, the transition function (P_t) satisfies (7) but is not the only solution of (7). Note that G does not depend on the distribution π. For different choices of π, we will have different transition functions (P_t^{π}), for each one of which (9) and (7) hold with G as given by (13).

The problem of obtaining all transition functions corresponding to a given G is still unresolved despite much effort spent since 1938 by Kolmogorov*, Doob, Feller, Chung, and others. This problem also points out the bankruptcy of the "semi-group theoretic" approach to Markov processes, which some authors hold out as a unifying treatment. In fact, such an approach is applicable to Chung processes only if the $q(i)$ in (11) are bounded, in which case the corresponding process is a special case of simple Chung processes.

Strong Markov Property

For a Chung process, the strong Markov property holds for stopping times T such that X_T is in E almost surely on $\{T < \infty\}$. Since it is possible to find stopping times such that $P\{X_T = \Delta\} > 0$ (even though $P\{X_t = \Delta\} = 0$ for every fixed time t), this is a weaker form of a strong Markov property.

Behavior at a State

A state i in E is said to be *stable* if $q(i) < \infty$ and *instantaneous* if $q(i) = \infty$, where $q(i)$ is as in (10). If X is a simple Chung process, then all states are stable. But generally, there are Chung processes with any number of instantaneous states, and there are Chung processes all of whose states are instantaneous.

To understand the behavior of the process at a fixed state i, we examine the random time set $K_i = \{t : X_t = i\}$. Let C_t be the Lebesgue measure of $K_i \cap [0, t]$ and let A_u be the first time the process $t \to C_t$ exceeds the level u. Then (A_u) is a strictly increasing right-continuous Lévy process* (i.e., A has stationary and independent increments); this follows from the strong Markov property of X. The process A has drift rate 1, and its Lévy measure is finite if i is stable and infinite if i is instantaneous. Now basic facts about Lévy processes yield, together with the fact that $K_i = \{t : A_u = t \text{ for some } u \geqslant 0\}$, the following description of the set K_i.

Suppose that i is stable. Then the set K_i is just as in the case of a simple Chung process: K_i is the union of a countable number of intervals; each component interval has the form $[\cdot)$; the lengths of the component intervals are independent and exponentially distributed with parameter $q(i)$; and the lengths of the intervals contiguous to K_i are independent of each other and of the lengths of the component intervals, and with the possible exception of the first one, all have the same distribution.

Suppose that i is instantaneous. Then the

structure of K_i is more complicated. For almost every ω, the time set $K_i(\omega)$ does not contain any open intervals; thus its interior is empty. However, $K_i(\omega)$ is uncountable, and in fact its Lebesgue measure is positive and may even be infinite. More important, for every $t \in K_i(\omega)$ there are t_n in $K_i(\omega)$ such that t_n decreases strictly to t; and if a sequence of times t_n in $K_i(\omega)$ decreases to t, then that time t belongs to $K_i(\omega)$.

Local Behavior

The description of the behavior given above at a fixed state can be used to deduce the behavior of a Chung process near a time t. Fix a typical ω and a time t, and suppose that $X_t(\omega) = i \in E$. If i is stable, then t must belong to one of the component intervals of $K_i(\omega)$ and each such interval has the form $[\cdot)$. Thus if i is stable, then there is an $\epsilon > 0$ such that $X_{t+u}(\omega) = i$ for all $u \leqslant \epsilon$.

If i is instantaneous, $K_i(\omega)$ does not contain any open intervals, and thus there is no such $\epsilon > 0$: for every $\epsilon > 0$ there is $0 < u < \epsilon$ such that $X_{t+u}(\omega) \neq i$; in other words, the process leaves i immediately. Yet $X_t(\omega) = i$ means that there are t_n decreasing to t such that $X_{t_n}(\omega) = i$ for all n. In particular, this implies that for i instantaneous the interval $(t, t+\epsilon)$ cannot contain infinitely many component intervals of a K_j for some stable j. A closer analysis shows that, when i is instantaneous, there is in fact no sequence of times t_n decreasing to t such that $X_{t_n}(\omega) = j$ for some other state j for all n; hence there must exist a sequence (t_n) decreasing to t such that $X_{t_n}(\omega) \to \Delta$. Thus such a t is a time of discontinuity of the second type.

Literature

The best reference for Chung processes is still [1]. For the case where X is right-continuous and has left-hand limits (in $\overline{E}$), one may consult [2] and [6]. For detailed analytical facts about $t \to P_t(i, j)$ the best reference is [5]. At this time, the theory seems far from satisfactory. An approach based on the general theory of processes and random sets [4] appear promising; for a small progress in that direction, see [3] and [6].

NOTE

1. On the topic of Markovian processes, we can construct the following 2×2 table:

Time Parameter	State Space	
	Discrete	Arbitrary
Discrete	DD	DA
Continuous	CD	CA

According to Kolmogorov*, Lévy, Feller, almost all of the French probabilists, and the author of this article, processes of the type CD and CA are called Markov processes, and those of type DD and DA are called Markov chains. According to Chung and Doob, however, processes of type DD and CD are Markov chains, and those of type DA and CA are Markov processes. This causes some confusion in the literature. The use of the term "Chung process" for type CD might lessen the confusion. See also MARKOV PROCESSES.

References

[1] Chung, K. L. (1967). *Markov Chains*, 2nd ed. Springer-Verlag, New York.

[2] Chung, K. L. (1970). *Lectures on Boundary Theory for Markov Chains*, Princeton University Press, Princeton, N.J.

[3] Çinlar, E. (1976). *Ann. Prob.*, **4**, 402–417.

[4] Dellacherie, C. (1972). *Capacités et processus stochastiques*. Springer-Verlag, Berlin.

[5] Kingman, J. F. C. (1972). *Regenerative Phenomena*. Wiley, New York.

[6] Meyer, P. A. (1974). *Lect. Notes Math.*, **381**, 242–261. (Séminaire de Probabilités VIII, Université de Strasbourg.)

(KOLMOGOROV'S DIFFERENTIAL EQUATIONS
MARKOV PROCESSES
SEMI-MARKOV PROCESSES)

E. ÇINLAR

CHUPROV (or TSCHUPROW), ALEXANDER ALEXANDROVICH

Born: February 18 (n.s.), 1874, in Mosal'sk, Russia.

Died: April 19, 1926, in Geneva, Switzerland.

Contributed to: mathematical statistics, demography, rural economics.

The formative years of Chuprov's education were heavily influenced by his father, A. I. Chuprov (1842–1908), for many years a professor of political economy and statistics at Moscow University. Alexander graduated from the physico-mathematical faculty of Moscow University in 1896 with a dissertation on probability theory as a basis for theoretical statistics, then traveled to Germany to study political economy. During a semester at the University of Berlin, he established a lasting friendship with L. Bortkiewicz*, who introduced him to W. Lexis* in Göttingen. The years 1897–1901 were spent at Strasbourg University, where among his teachers were Bortkiewicz and G. F. Knapp, who supervised his doctoral dissertation "Die Feldgemeinschaft, eine morphologische Untersuchung," published in 1902. In 1902, in order to gain a teaching position in Russia, Chuprov completed magisterial examinations at the University of Moscow, concentrating on theoretical economics and the application of mathematical methods.

He was offered a position in 1902 in the newly formed Economics Section of the St. Petersburg Polytechnic Institute, where he was in charge of the organization and teaching of statistics until 1916. In this area he developed a remarkably broad and modern-style pedagogical approach to the subject and a fine library, and later produced a number of notable disciples, including O. N. Anderson and N. S. Chetverikov (or Tschetwerikoff). In 1909, his work, *Ocherki po Teorii Statistiki* [2], for which he was awarded a doctor's degree by Moscow University, was published and ran into a second edition within a year. This work had enormous influence for a number of years in Russia due to its stress on the logical and mathematical approach to statistics. It includes principles of probability theory with strong emphasis on the frequency interpretation* of objective probability on the basis of the law of large numbers*; and an account of the stability theory of statistical series of trials (dispersion theory*) of Lexis and Bortkiewicz. A response in 1910 to the work by A. A. Markov* brought Chuprov into awareness of Markov's work, and marked the beginning of an intense correspondence between the two [8] that resulted in a focusing of Chuprov's researches on theory.

In May 1917, he went to Scandinavia for a vacation but never returned to Russia, because of the revolution. Most of Chuprov's work in mathematical statistics was published during his emigré period. The initial thrust of it is in the framework of dispersion theory. If X_{ij} is the jth observation $(j = 1, \ldots, n)$ in the ith set of observations $(i = 1, \ldots, m)$, then interest within this theory focused on the *dispersion coefficient* L (*see* DISPERSION THEORY). Chuprov showed, *inter alia*, that in the case of general i.i.d. variables X_{ij}, indeed $EL = 1$, but that this could still hold in the case of $N = mn$ variables that are not i.i.d. The notable conceptual advances here consist of an extension to general variables, and the investigation (under Markov's influence) of the moment structure of a sample statistic. Clearly in evidence also are ideas of analysis of variance*.

The work on dispersion theory led to an extensive investigation of expectations* of sample moments in terms of population moments (i.e., the "method of moments"*) under, ultimately, very general conditions embracing samples of correlated observations [10]. An obvious application of such results to sample survey theory*, in which Chuprov had always been interested, anticipated several results of J. Neyman*, especially the well-known formula for optimal allocation $\{n_h\}$ among t strata* of a sample of fixed size $n: n_h = nN_h S_h / \sum_{r=1}^{t} N_r S_r$, $h = 1, \ldots, t$, where N_h and S_h are the population size and standard deviation of stratum h [10, Chap. V, Sec. III]. His investigations in correlation theory are presented in ref. 11.

Work of this general nature led to his propagation of the modern "stochastic" view of statistical theory, based on the notion of empirical observations as manifestations of random variables following a probability distribution, and leading to a sampling distribution for sample statistics, with statistical inference based on conceptual repetitions of samples to accord with the law of large numbers. From the law of large numbers, he was also led to the notion of consistency* of an estimator; and in his writings a limiting Gaussian form of a sampling distribution is adduced from a consideration of moments. Indeed, his lasting contribution to mathematical statistics would seem to rest on an early recognition of such fundamental ideas and their lucid and extensive popularization on the continent. He is often credited, on the one hand, with unifying several streams of statistical thinking, i.e., the Russian probabilistic stream, the German stream typified by Lexis and Bortkiewicz, and that of the English biometric school* led by Karl Pearson*, although his technical contributions were soon overshadowed by those of R. A. Fisher*.

On the other hand, he is also often regarded as the main proponent of the "Continental direction" of statistics, which sought to develop statistical methods free of those extensive assumptions about underlying distributions of the kind later developed by the English school.

In the area of demography, Chuprov presented to the International Statistical Institute*, during his St. Petersburg period, the results of an extensive data-analytic consideration of the problem of decrease in the surplus of male births.

Chuprov was a high-minded ascetic. That part of his emigré period which he spent in Dresden, which began in mid-1920, was spent in solitary and intense productivity without steady income. He appears to have been apolitical, maintaining extensive professional contacts with his homeland and publishing in the Russian journal *Vestnik Statistiki**, while sporadically participating in Russian emigré–professional–academic organizations and publishing in their journals. Following his Dresden period, he took up an appointment with the Russian College in Prague in 1925. There soon followed a serious decline in his health, leading to his death.

He exerted a profound influence on the evolution of statistics, in Scandinavia in particular, and many of his papers were published in *Nordisk Statistisk Tidskrift (N.S.T.)*, founded in 1922. His work was known to most of the notable English statisticians of the time and was championed by L. Isserlis in particular.

The most comprehensive account of Chuprov's life and work is ref. 5; more readily accessible is the obituary [9]; and Vol. 5 of *N.S.T.* (1926) contains much relevant material, including the obituary [7]. (Several other obituaries are derivative of refs. 7 and 9.) Volume 18 of *Biometrika** (1926) has a portrait of Chuprov. Russian versions of many of his papers originally published in *N.S.T.* are collected in ref. 3; refs. 2 and 11 were reprinted in the USSR.

References

[1] Chetverikov, N. S., ed. (1968). *O teorii dispersii*. Statistika, Moscow. (Presents evolution of dispersion theory through a series of key papers, in Russian versions, including Lexis's, Bortkiewicz's, and Chuprov's.)

[2] Chuprov, A. A. (1910). *Ocherki po teorii statistiki* (1st ed., 1909.) St. Petersburg. (Reprinted by Gosstatizdat, Moscow, 1959.)

[3] Chuprov, A. A. (1960). *Voprosy statistiki*. Gosstatizdat, Moscow. (Introductory remarks, and translations into Russian, by B. I. Karpenko and N. S. Chetverikov.)

[4] Heyde, C. C. and Seneta, E. (1977). *I. J. Bienaymé: Statistical Theory Anticipated*. Springer-Verlag, New York. (Chapter 3 gives an account of dispersion theory from a modern standpoint.)

[5] Karpenko, B. I. (1957). *Uch. zap. statist.*, **3**, 282–317 (in Russian). (Written with the cooperation of N. S. Chetverikov. Contains the most complete listing of Chuprov's writings.)

[6] Kendall, M. G. and Doig, A. (1968). *Bibliography*

of Statistical Literature, Vol. 3: *Pre-1940 with Supplements*. Oliver & Boyd, Edinburgh. Page 281 lists most of Chuprov's papers published during his emigré period.)

[7] Kohn, S. (1926). *Nord. statist. tidskr.*, **5**, 171–194.

[8] Ondar, Kh. O., ed. (1977). *O teorii veroiatnostei i matematicheskoi statistike*, Nauka, Moscow. (Annotated correspondence between Chuprov and Markov, Nov. 1910–Feb. 1917.)

[9] Tschetwerikoff, N. S. (1926). *Metron*, **6**, 314–320.

[10] Tschuprow, A. A. (1923). *Metron*, **2**, 461–493, 646–680.

[11] Tschuprow, A. A. (1925). *Grundbegriffe und Grundprobleme der Korrelationstheorie*. Teubner, Leipzig-Berlin. (Published in Russian version: *Osnovnie problemi teorii korrelatsii*, M. and S. Sabashnikov, Moscow, 1926; and in English translation: *Principles of the Mathematical Theory of Correlation*, Hodge, London, 1939. Russian version reprinted by Gosstatizdat, 1960.)

[12] Vinogradova, N. M. (1957). *Uch. zap. statist.*, **3**, 318–324 (in Russian).

(ANALYSIS OF VARIANCE
CORRELATION
DISPERSION THEORY
LAW OF LARGE NUMBERS
METHOD OF MOMENTS
SAMPLE SURVEY THEORY)

E. SENETA

CIRCULAR COVERAGE FUNCTION *See* TARGET COVERAGE

CIRCULAR DISTRIBUTIONS *See* DIRECTIONAL DISTRIBUTIONS

CIRCULAR NORMAL DISTRIBUTION

The distribution of a random variable X with density function

$$[2\pi I_0(k)]^{-1}\exp[k\cos(y-\theta)]$$
$$(0 \leqslant y < 2\pi; 0 < \theta < 2\pi),$$

where $I_0(k) = \sum_{j=0}^{\infty}(j!)^{-2}(\frac{1}{2}k)^{2j}$ is a modified Bessel function* of first kind (and zero order). The distribution was derived by von Mises [3] to be such that the maximum likelihood estimator* $\hat{\theta}$ of θ, given values of n independent Y's—$Y_1, Y_2, \ldots, Y_n$—satisfies the equation

$$\sum_{j=1}^{n}\sin(Y_j - \hat{\theta}) = 0.$$

The parameters of the distribution are k and θ. As k increases, the distribution becomes more and more concentrated around the "pole" $Y = \theta$.

This distribution must be distinguished from bivariate normal distributions* with equal marginal standard deviations—also sometimes called "circular normal" (e.g., Dyer [1]).

Additional information on the circular normal distribution is provided by ref. 2.

References

[1] Dyer, D. D. (1974). *Operat. Res.*, **22**, 197–205.

[2] Johnson, N. L. and Kotz, S. (1970). *Continuous Univariate Distributions*, Vol. 2. Wiley, New York, Chap. 33.

[3] Mises, R. von (1918). *Phys. Zeit.*, **19**, 490–500.

(VON MISES DISTRIBUTION
"WRAPPED-UP" NORMAL
DISTRIBUTIONS)

CIRCULAR PROBABLE ERROR

When missiles are aimed at a target, the deviations X and Y (along two orthogonal directions) of the impact point from the target center are often assumed to be distributed according to a bivariate normal distribution*. The random variable $R = \sqrt{X^2 + Y^2}$ is called the radial error* and its distribution functions $F_R(\cdot)$ gives the probability of a missile falling within a circle with the target as a center and radius r. The median of this distribution is the radius of the circle such that the probability of a missile hitting inside it is 50%. It is called the circular probable error (C.E.P.). Tables of C.E.P. for selected values of the ratio of standard deviations of X and Y are given by Harter [2] and Lowe [3]. Additional information on circular probable error is provided by ref. 1.

References

[1] Groenwoud, C., Hoaglin, D. C., and Vitalis, J. A. (1967). *Bivariate Normal Offset Circle Probabilities*. Cornell Aeronautics Laboratory, Buffalo, N.Y.

[2] Harter, H. L. (1960). *J. Amer. Statist. Ass.*, **55**, 723–731.

[3] Lowe, J. R. (1960). *J. R. Statist. Soc. B*, **22**, 176–187.

(BIVARIATE NORMAL DISTRIBUTION
DIRECTIONAL DATA
TARGET COVERAGE)